T0297732

Manual of Percutaneous Coronary Interventions

A Step-by-Step Approach

Manual of Percutaneous Coronary Interventions

A Step-by-Step Approach

Emmanouil Brilakis, MD, PhD
Center for Complex Coronary Interventions,
Minneapolis Heart Institute,
Center for Coronary Artery Disease,
Minneapolis Heart Institute Foundation,
Minneapolis, MN, United States

Academic Press is an imprint of Elsevier
125 London Wall, London EC2Y 5AS, United Kingdom
525 B Street, Suite 1650, San Diego, CA 92101, United States
50 Hampshire Street, 5th Floor, Cambridge, MA 02139, United States
The Boulevard, Langford Lane, Kidlington, Oxford OX5 1GB, United Kingdom

Notices
Knowledge and best practice in this field are constantly changing. As new research and experience broaden our understanding, changes in research methods, professional practices, or medical treatment may become necessary.

Practitioners and researchers must always rely on their own experience and knowledge in evaluating and using any information, methods, compounds, or experiments described herein. In using such information or methods they should be mindful of their own safety and the safety of others, including parties for whom they have a professional responsibility.

To the fullest extent of the law, neither the Publisher nor the authors, contributors, or editors, assume any liability for any injury and/or damage to persons or property as a matter of products liability, negligence or otherwise, or from any use or operation of any methods, products, instructions, or ideas contained in the material herein.

British Library Cataloguing-in-Publication Data
A catalogue record for this book is available from the British Library

Library of Congress Cataloging-in-Publication Data
A catalog record for this book is available from the Library of Congress

ISBN: 978-0-12-819367-9

For Information on all Academic Press publications
visit our website at https://www.elsevier.com/books-and-journals

Publisher: Stacy Masucci
Acquisitions Editor: Katie Chan
Editorial Project Manager: Tracy I. Tufaga
Production Project Manager: Kiruthika Govindaraju
Cover Designer: Christian J. Bilbow

Typeset by MPS Limited, Chennai, India

Dedication

To Nicole, Stelios, and Thomas.
To my parents and my brother.

Contents

List of contributors

J. Dawn Abbott Warren Alpert Medical School at Brown University, Providence, RI, United States

Nidal Abi Rafeh St. George Hospital University Medical Center, Beirut, Lebanon

Mazen Abu Fadel Oklahoma Heart Hospital North Campus, Oklahoma City, OK, United States; University of Oklahoma Cardiovascular Institute, Oklahoma City, OK, United States

Pierfrancesco Agostoni HartCentrum, Ziekenhuis Netwerk Antwerpen (ZNA) Middelheim, Antwerp, Belgium

Sukru Akyuz University of Health Sciences, Dr. Siyami Ersek Thoracic and Cardiovascular Surgery Training and Research Hospital, Istanbul, Turkey

Khaldoon Alaswad Henry Ford Hospital, Detroit, MI, United States

Dimitrios Alexopoulos National and Kapodistrian University of Athens Medical School, Athens, Greece; Attikon University Hospital, Athens, Greece

Dominick J. Angiolillo University of Florida College of Medicine-Jacksonville, Jacksonville, FL, United States

Herbert D. Aronow Alpert Medical School of Brown University, Providence, RI, United States; Lifespan Cardiovascular Institute, Providence, RI, United States; Rhode Island and The Miriam Hospitals, Providence, RI, United States

Alexandre Avran Clinique Pasteur, Essey-lès-nancy, France

Lorenzo Azzalini Division of Cardiology, VCU Health Pauley Heart Center, Virginia Commonwealth University, Richmond, VA, United States

Avtandil M. Babunashvili Department of Cardiovascular Surgery, Center for Endosurgery and Lithotripsy, Moscow, Russian Federation

Jayant Bagai Vanderbilt University Medical Center, Nashville, TN, United States

Subhash Banerjee VA North Texas Health Care System and UT Southwestern Medical School, Dallas, TX, United States

Kenneth Baran Minneapolis Heart Institute, Minneapolis, MN, United States

Mir Babar Basir Henry Ford Health System, Detroit, MI, United States

Nicolas Boudou Clinique Saint Augustin, Bordeaux, France

Konstantinos Dean Boudoulas The Ohio State University, Columbus, OH, United States

Christos V. Bourantas Barts Heart Centre, Barts Health NHS Trust, London, United Kingdom; Institute of Cardiovascular Sciences, University College London, London, United Kingdom

Nenad Ž. Božinović University Clinical Center Nis, Niš, Serbia

Leszek Bryniarski Department of Cardiology and Cardiovascular Interventions, University Hospital, Institute of Cardiology, Jagiellonian University Medical College, Cracow, Poland

Alexander Bufe Heartcentre Niederrhein, Helios Clinic Krefeld, University Witten/Herdecke, Germany

M. Nicholas Burke Minneapolis Heart Institute and Minneapolis Heart Institute Foundation, Minneapolis MN, United States

Heinz Joachim Büttner Department of Cardiology and Angiology II University Heart Center Freiburg Bad Krozingen, Bad Krozingen, Germany

Pedro Pinto Cardoso Cardiology Division, Heart and Vessels Department, University Hospital, CHULN, Lisboa, Portugal; Faculty of Medicine, Universidade de Lisboa, Centro Cardiovascular da Universidade de Lisboa, Lisboa, Portugal

Mauro Carlino Interventional Cardiology Unit, Cardio-Thoracic-Vascular Department, IRCCS San Raffaele Scientific Institute, Milan, Italy

Jeff Chambers Metropolitan Heart and Vascular Institute, Mercy Hospital, Minneapolis, MN, United States

Konstantinos Charitakis University of Texas Health Science Center at Houston, Houston, TX, United States

Yiannis S. Chatzizisis Cardiovascular Division, University of Nebraska Medical Center, Omaha, NE, United States

Ivan J. Chavez Minneapolis Heart Institute at Abbott Northwestern Hospital and Minneapolis Heart Institute Foundation, Minneapolis, MN, United States

James W. Choi Baylor Scott & White Heart and Vascular Hospital, Dallas, TX, United States; Texas A&M College of Medicine, Bryan, TX, United States

Evald Høj Christiansen Department of Cardiology, Aarhus University Hospital, Aarhus, Denmark

Mauricio G. Cohen Cardiovascular Division, Department of Medicine, Elaine and Sydney Sussman Cardiac Catheterization Laboratory, University of Miami Hospital and Clinics, Miami, FL, United States

Francesco Costa Policlinico G. Martino, University of Messina, Messina, Italy

Felix Damas de los Santos Interventional Cardiology Department, National Institute of Cardiology Ignacio Chávez, México City, Mexico

Rustem Dautov Heart and Lung Institute, The Prince Charles Hospital, Brisbane, QLD, Australia; University of Queensland, Brisbane, QLD, Australia

Tony De Martini SIU School of Medicine, Memorial Medical Center, Springfield, IL, United States

Ali E. Denktas Baylor College of Medicine, Houston, TX, United States

Joseph Dens Department of Cardiology, Ziekenhuis Oost Limburg, Genk, Belgium

Zisis Dimitriadis Cardiology I, University Medical Centre, Johannes Gutenberg University Mainz, Mainz, Germany

Anthony Doing University of Colorado Health, Medical Center of the Rockies, Loveland, CO, United States

Mohaned Egred Freeman Hospital, Newcastle University, Newcastle upon Tyne, United Kingdom

Basem Elbarouni University of Manitoba, Winnipeg, MB, Canada; St. Boniface Hospital, Winnipeg, MB, Canada

Ahmed M. El Guindy Department of Cardiology, Aswan Heart Centre, Magdi Yacoub Foundation, Cairo, Egypt

Abdallah El Sabbagh Mayo Clinic, Jacksonville, FL, United States

Panayotis Fasseas Division of Cardiovascular Medicine, Medical College of Wisconsin, Milwaukee, WI, United States

Dmitriy N. Feldman Greenberg Division of Cardiology, Department of Medicine, New York Presbyterian Hospital, Weill Cornell Medical College, New York, NY, United States

Sergey Furkalo National Institute Surgery and Transplantology NAMS, Kiev, Ukraine

Andrea Gagnor Maria Vittoria Hospital, Turin, Italy

Alfredo R. Galassi Department of Health Promotion, Mother and Child Care, Internal Medicine, and Medical Specialties (PROMISE), University of Palermo, Palermo, Italy

Roberto Garbo San Giovanni Bosco Hospital, Turin, Italy

Santiago Garcia Minneapolis Heart Institute and Minneapolis Heart Institute Foundation, Minneapolis, MN, United States

Gabriele L. Gasparini Humanitas Clinical and Research Hospital, Rozzano, Milan, Italy

Anthony H. Gershlick University Hospitals of Leicester (UHL), University of Leicester and Leicester Biomedical Research Centre, Leicester, United Kingdom

Mario Goessl Minneapolis Heart Institute and Minneapolis Heart Institute Foundation, Minneapolis, MN, United States

Luca Grancini Centro Cardiologico Monzino, IRCCS, Milan, Italy

Abdul Hakeem Rutgers Robert Wood Johnson University Hospital, New Brunswick, NJ, United States

Allison B. Hall Eastern Health/Memorial University of Newfoundland, St. John's, NL, Canada

Stefan Harb University Heart Center Graz, Medical University of Graz, Graz, Austria

Raja Hatem Hôpital du Sacré-Coeur de Montréal, Université de Montréal, Montreal, QC, Canada

Jose P.S. Henriques Department of Cardiology, University of Amsterdam, Amsterdam UMC, Amsterdam, The Netherlands

Yangsoo Jang Severance Hospital, Yonsei University College of Medicine, Seoul, Korea

Risto Jussila Helsinki Heart Hospital, Helsinki, Finland

Artis Kalnins Clinic of Cardiovascular diseases, Riga East University hospital, Riga, Latvia

Arun Kalyanasundaram Promed Hospital, Chennai, India

Paul Hsien-Li Kao Cardiovascular Center, Cardiology Division, Department of Medicine, National Taiwan University Hospital, Taipei City, Taiwan

Judit Karacsonyi Division of Invasive Cardiology, Second Department of Internal Medicine and Cardiology Center, University of Szeged, Szeged, Hungary; Minneapolis Heart Institute and Minneapolis Heart Institute Foundation, Minneapolis, MN, United States

Lampros Karagounis Director, A Cardiology Department, European Interbalkan Medical Center, Thessaloniki, Greece

Antonios Karanasos First Department of Cardiology, University of Athens, Hippokration Hospital, Athens, Greece

Dimitri Karmpaliotis Columbia University, New York, NY, United States

Houman Khalili Florida Atlantic University and Delray Medical Center, Delray Beach, FL, United States

Jaikirshan J. Khatri Director of Complex Coronary Intervention, Department of Cardiovascular Medicine, Cleveland Clinic Lerner College of Medicine of Case Western Reserve University, Cleveland, OH, United States

Dmitrii Khelimskii E.N. Meshalkin National Medical Research Center

Byeong-Keuk Kim Division of Cardiology, Severance Cardiovascular Hospital, Yonsei University College of Medicine, Seoul, Korea

Louis P. Kohl Division of Cardiology, Hennepin Healthcare (HCMC) and University of Minnesota Medical School, Minneapolis, MN, United States

Daniel M. Kolansky Hospital of the University of Pennsylvania, Philadelphia, PA, United States

Michalis Koutouzis Red Cross Hospital of Athens, Athens, Greece

Oleg Krestyaninov Meshalkin Novosibrisk Research Institute, Novosibirsk, Russia

Faisal Latif University of Oklahoma, Oklahoma City, OK, United States; SSM Health St. Anthony Hospital, Oklahoma City, OK, United States

Seung-Whan Lee Department of Cardiology, Asan Medical Center, University of Ulsan College of Medicine, Seoul, Korea

Thierry Lefevre ICPS Hôpital Privé Jacques Cartier, Massy, France

Nicholas J. Lembo Columbia University Medical Center, New York-Presbyterian Hospital, Cardiovascular Research Foundation, New York, NY, United States

Ehtisham Mahmud UCSD Cardiovascular Institute-Medicine, University of California, San Diego, CA, United States

Konstantinos Marmagkiolis University of Texas MD Anderson Cancer Center, Houston, TX, United States; HCA Northside Hospital, St. Petersburg, FL, United States

Kambis Mashayekhi University Heart Center Freiburg Bad Krozingen, Bad Krozingen, Germany

Kreton Mavromatis Atlanta VA Health Care System, Emory University School of Medicine, Atlanta, GA, United States

Michael Megaly Minneapolis Heart Institute and Minneapolis Heart Institute Foundation, Minneapolis, MN, United States

Owen Mogabgab Cardiovascular Institute of the South, Houma, LA, United States

Michael R. Mooney Minneapolis Heart Institute and Minneapolis Heart Institute Foundation, Minneapolis, MN, United States

Jeffrey W. Moses Columbia University Medical Center, New York, NY, United States; St. Francis Heart Center, New York, NY, United States

Bilal Murad Minneapolis Heart Institute, United Hospital, St. Paul, MN, United States

Alexander Nap Amsterdam University Medical Center, Amsterdam, The Netherlands

William Nicholson Emory University, Atlanta, GA, United States

Dimitrios N. Nikas 1st Cardiology Department, Ioannina University Hospital, Ioannina Greece

Ilias Nikolakopoulos Minneapolis Heart Institute Foundation, Minneapolis, MN, United States

Goran Olivecrona Department of Cardiology, Lund University/Skåne University Hospital, Lund, Sweden

Mohamed A. Omer Minneapolis Heart Institute and Minneapolis Heart Institute Foundation, Minneapolis, MN, United States

Jacopo Andrea Oreglia De Gasperis Cardio Center, Niguarda Hospital, Milan, Italy

Lucio Padilla Department of Interventional Cardiology and Endovascular Therapeutics, ICBA, Instituto Cardiovascular, Buenos Aires, Argentina

Ioannis Paizis Department of Cardiology, LAIKO General Hospital, Athens, Greece

Carmelo Panetta UMP/University of Minnesota, Minneapolis, MN, United States

Mitul Patel UCSD Cardiovascular Institute, University of California, San Diego, CA, United States

Ashish Pershad Banner-University Medical Center, Phoenix, AZ, United States

Marin Postu IECVD "C.C.Ilescu", Bucharest, Romania

Srini Potluri Cardiac Catheterization Laboratory, Baylor Scott & White Heart Hospital, Plano, TX, United States

Anil Poulose Minneapolis Heart Institute and Minneapolis Heart Institute Foundation, Minneapolis, MN, United States

Stylianos Pyxaras Landshut-Achdorf Hospital, Landshut, Germany

Sunil V. Rao Duke University Medical Center, Durham, NC, United States; Durham VA Medical Center, Durham, NC, United States

Sudhir Rathore Frimley Health NHS Foundation Trust, Surrey, United Kingdom

Amir Ravandi University of Manitoba, Winnipeg, MB, Canada; St. Boniface General Hospital, Winnipeg, MB, Canada; Bergen Cardiac Care Centre, Winnipeg, MB, Canada

Nicolaus Reifart Main Taunus Heart Institute, Bad Soden, Germany

Robert F. Riley Complex Coronary Therapeutics Program, The Christ Hospital Health Network, Cincinnati, OH, United States

Stephane Rinfret McGill University Health Center, Montreal, QC, Canada

Gurpreet S. Sandhu Department of Cardiovascular Medicine, Mayo Clinic, Rochester, MN, United States

Yader Sandoval Department of Cardiovascular Medicine, Mayo Clinic, Rochester, MN, United States

Elias Sanidas Department of Cardiology, LAIKO General Hospital, Athens, Greece

Ricardo Santiago Trinidad PCI Cardiology Group, San Juan, Puerto Rico

Jeffrey M. Schussler Baylor Scott & White Heart and Vascular Hospital, Dallas, TX, United States

Arnold Seto Long Beach VA Medical Center, Long Beach, CA, United States; University of California, Irvine, CA, United States

Alok Sharma Minneapolis VA Medical Center, Minneapolis, MN, United States

Arslan Shaukat Phelps Health, Rolla, MO, United States

Mehdi H. Shishehbor Professor of Medicine, Case Western Reserve University School of Medicine; System Director, Cardiovascular Interventional Center, University Hospitals, Cleveland, Ohio

Evan Shlofmitz MedStar Washington Hospital Center, Washington, DC, United States

Richard Shlofmitz St. Francis Hospital, Roslyn, NY, United States

Paul Sorajja Minneapolis Heart Institute and Minneapolis Heart Institute Foundation, Minneapolis, MN, United States

Anthony Spaedy Cardiac Catheterization Laboratories, Boone Hospital Center, Columbia, MO, United States

Peter Tajti Division of Invasive Cardiology, Second Department of Internal Medicine and Cardiology Center, University of Szeged, Szeged, Hungary

Jacqueline E. Tamis-Holland Division of Cardiology, Department of Medicine, Mount Sinai Mount Sinai Morningside, New York, NY, United States

Aurel Toma Medical University of Vienna, Vienna, Austria

Konstantinos Toutouzas First Department of Cardiology, Athens Medical School, Athens, Greece

Jay H. Traverse Minneapolis Heart Institute Foundation, Abbott Northwestern Hospital, Minneapolis, MN, United States; Cardiovascular Division, University of Minnesota School of Medicine, Minneapolis, MN, United States

Huu Tam Truong Loma Linda VA Healthcare System, Loma Linda, CA, United States; Loma Linda University, Loma Linda, CA, United States

Sotiris Tsalamandris First Department of Cardiology, University of Athens, Hippokration Hospital, Athens, Greece

Ioannis Tsiafoutis Interventional Cardiologist, Red Cross Hospital, Athens, Greece

Imre Ungi University of Szeged, Szeged, Hungary

Emmanouil Vavouranakis First Department of Cardiology, Medical School, University of Athens, Hippokration Hospital, Athens, Greece

Evangelia Vemmou Minneapolis Heart Institute Foundation, Minneapolis, MN, United States

Minh N. Vo Mazankowski Royal Columbian Hospital, BC, Canada

Vassilis Voudris Division of Interventional Cardiology, Chairman Cardiology Department, Onassis Cardiac Surgery Center, Athens, Greece

Yale Wang Minneapolis Heart Institute and Minneapolis Heart Institute Foundation, Minneapolis, MN, United States

Jarosław Wójcik Hospital of Invasive Cardiology "IKARDIA", Nałęczów/Lublin, Poland

Jason Wollmuth Providence Heart and Vascular Institute, Portland, Oregon, United States

Eugene B. Wu Prince of Wales Hospital, Chinese University, Hong Kong

Iosif Xenogiannis Minneapolis Heart Institute Foundation, Minneapolis, MN, United States; Second Department of Cardiology, Attikon University Hospital, National and Kapodistrian University of Athens Medical School, Athens, Greece

Masahisa Yamane Saitama Sekishinkai Hospital, Saitama, Japan

Luiz Fernando Ybarra London Health Sciences Centre, Schulich School of Medicine & Dentistry, Western University, London, ON, Canada

PCI cases online links

Case Number	YOUTUBE link
1	https://www.youtube.com/watch?v = Z4-lZ0GmGbY
2	https://www.youtube.com/watch?v = SSBMDshfnx4
3	https://www.youtube.com/watch?v = R1qMbS8d-f4
4	https://www.youtube.com/watch?v = laYXNLAERKo
5	https://www.youtube.com/watch?v = Yu5KR7gJMNk
6	https://www.youtube.com/watch?v = 5QqwesnusBk
7	https://www.youtube.com/watch?v = 7YXTePUN8A8
8	https://www.youtube.com/watch?v = 9eSeLtQXeYs
9	https://www.youtube.com/watch?v = YsddPIhti-c
10	https://www.youtube.com/watch?v = uzEJi7SWAbc
11	https://www.youtube.com/watch?v = TrSZnlN19BY
12	https://www.youtube.com/watch?v = C9mDQlp6wBc
13	https://www.youtube.com/watch?v = 5x0gJ5hULQ8
14	https://youtu.be/AHteTspH3R8
15	https://www.youtube.com/watch?v = g0IfIsNsops
16	https://www.youtube.com/watch?v = 9ZaHJZKhoC0
17	https://www.youtube.com/watch?v = 9fEVRRVOLME
18	https://www.youtube.com/watch?v = NTNEQzVJINk
19	https://www.youtube.com/watch?v = eV1blkBd2bQ
20	https://www.youtube.com/watch?v = RQiFrT47WXw&t = 3s
21	https://www.youtube.com/watch?v = TDyKZW2fTPA
22	https://www.youtube.com/watch?v = nRNwpgV8ORI&t = 1s
23	https://www.youtube.com/watch?v = lrMbTP0Ou-g
24	https://www.youtube.com/watch?v = -ah-rtCbOfw
25	https://www.youtube.com/watch?v = CX7GS19TTcQ
26	https://www.youtube.com/watch?v = cQMEvu7RavQ
27	https://www.youtube.com/watch?v = x9bB5Hab9A0
28	https://www.youtube.com/watch?v = Wmw8eE8LSuw
29	https://www.youtube.com/watch?v = COqySRTm7h4
30	https://www.youtube.com/watch?v = fxceb437LF0
31	https://youtu.be/cHvPrxoV2-Y
32	https://youtu.be/8EQsasVWJRU
33	https://www.youtube.com/watch?v = 1TwMn7SMhc8
34	https://youtu.be/KRqZDDQ6E3g
35	https://www.youtube.com/watch?v = zDyWd7fNPmw
36	https://www.youtube.com/watch?v = pldQ-P8Ekpc
37	https://www.youtube.com/watch?v = SdRSb8Oh-0Q
38	https://www.youtube.com/watch?v = ARrFtw473ek
39	https://www.youtube.com/watch?v = JNKqsEVl0VE
40	https://www.youtube.com/watch?v = hWVWe9dTqVk
41	https://www.youtube.com/watch?v = eZgX4jMJ158
42	https://www.youtube.com/watch?v = RVcqwU8qhWw
43	https://www.youtube.com/watch?v = tXtTiivM1Kc
44	https://www.youtube.com/watch?v = 4ntheQqS_Lc
45	https://www.youtube.com/watch?v = Ta_qwiFS2-Y
46	https://youtu.be/rQrbOi7-eSE
47	https://www.youtube.com/watch?v = 9HdZn5il-bE
48	https://www.youtube.com/watch?v = _gNLUyKe-Xg
49	https://www.youtube.com/watch?v = l8NIYCW_oXk
50	https://youtu.be/fbVoKuTdVEg

51	https://www.youtube.com/watch?v = NuLm5WUX0ew
52	https://youtu.be/4bZFdJ7POlg
53	https://www.youtube.com/watch?v = qpJnSzpR1kA
54	https://www.youtube.com/watch?v = C-kTk1hv1Nc
55	https://youtu.be/1tCL0HepoqY
56	https://youtu.be/3sGSkQ6hoQ0
57	https://youtu.be/3xTOxrhH3Us
58	https://www.youtube.com/watch?v = 36JzHvjGfLE
59	https://youtu.be/9SL7BnPzFvw
60	https://youtu.be/TNDK5DwtVxo
61	https://www.youtube.com/watch?v = LtWpzX1Mi3g
62	https://youtu.be/mk3HMGICYo4
63	https://youtu.be/mnqzShZ89zE
64	https://youtu.be/z-z76bS3WhY
65	https://youtu.be/qOlOdLF28Gc
66	https://youtu.be/D2QhNbShgj4
67	https://www.youtube.com/watch?v = TsqRwqZju4s
68	https://youtu.be/v3eNPECCPi8
69	https://www.youtube.com/watch?v = ZX-AJEpXeHI
70	https://youtu.be/9sU4RnQbetM
71	https://youtu.be/9qoF_RzfWSc
72	https://youtu.be/6Ne-6W_7h_A
73	https://youtu.be/NndEwAX61hl
74	https://www.youtube.com/watch?v = MTR9XnRyqds
75	https://www.youtube.com/watch?v = XhTQl5mQCdE
76	https://www.youtube.com/watch?v = mUffP1PUD_s
77	https://www.youtube.com/watch?v = MG8W3F3qGa8
78	https://youtu.be/XZoGURFaUkl
79	https://youtu.be/jG01KcZ176w
80	https://www.youtube.com/watch?v = l-kjzOrk8Gg
81	https://www.youtube.com/watch?v = Ql3BHNGVYQA
82	https://youtu.be/lMKovIlh8os
83	https://youtu.be/Wfuh1r4siPQ
84	https://youtu.be/iCpjeqOSXDk
85	https://youtu.be/CBFtofcFw5E
86	https://youtu.be/IzQ-wpYTUMw
87	https://youtu.be/VRqERXKp-rU
88	https://youtu.be/eglCD3rzKLQ
89	https://youtu.be/RNxl_wf9B2Q
90	https://youtu.be/8ezaVo9zSQc
91	https://youtu.be/Zyrf3qbZuxE
92	https://youtu.be/Bd6eqk8aEnw
93	https://youtu.be/U7kaCF14SM4
94	https://youtu.be/_2Yn8td_d9Y
95	https://youtu.be/Q6gLRyyF0jI
96	https://youtu.be/MQO5G8yGy80
97	https://youtu.be/maPX2si7W8k
98	https://youtu.be/uQifXrA_a9w
99	https://youtu.be/qePK1aOyzsc
100	https://youtu.be/zsUIRkRvgHM

CTO PCI cases online links

Case Number	YOUTUBE link
1	https://youtu.be/65ch3syR_6Q
2	https://www.youtube.com/watch?v = l8CavAxZ0fM
3	https://www.youtube.com/watch?v = T1_C4bRPKvQ
4	https://www.youtube.com/watch?v = STjnaCqBWB4
5	https://youtu.be/l1ZJbZPIBoc
6	https://youtu.be/5QqwesnusBk
7	https://www.youtube.com/watch?v = Psln9Ounl1Y
8	https://youtu.be/9eSeLtQXeYs
9	https://www.youtube.com/watch?v = 2s8oMdjgipM
10	https://www.youtube.com/watch?v = p5nnVlbUmTQ
11	https://www.youtube.com/watch?v = rgaCjz26JYQ
12	https://youtu.be/i7e2r7yizng
13	https://www.youtube.com/watch?v = r0-3m3S_Hdg
14	https://www.youtube.com/watch?v = o4aLXc7q-ps
15	https://www.youtube.com/watch?v = 0LssoY7q8WM
16	https://youtu.be/7jMltlY2UBk
17	https://youtu.be/tidtCbQm114
18	https://youtu.be/9ge8x7rwP6Q
19	https://www.youtube.com/watch?v = oedZujPSUGQ
20	https://www.youtube.com/watch?v = xfhkfyxq_I4
21	https://youtu.be/TDyKZW2fTPA
22	https://youtu.be/nRNwpgV8ORI
23	https://www.youtube.com/watch?v = 9ivAdJfB-b8&t = 2s
24	https://youtu.be/-ah-rtCbOfw
25	https://www.youtube.com/watch?v = sFRi-CgeBic
26	https://www.youtube.com/watch?v = rw648MhLzLc
27	https://youtu.be/lmSokEj0kVI
28	https://www.youtube.com/watch?v = MZMGqr3AOho
29	https://www.youtube.com/watch?v = L6lKm8VZy4w
30	https://www.youtube.com/watch?v = 9bLDQT_orbU
31	https://www.youtube.com/watch?v = 2zi7Aw0Wj_4
32	https://youtu.be/8EQsasVWJRU
33	https://www.youtube.com/watch?v = 90t2J_yzfMc
34	https://www.youtube.com/watch?v = 4lthDgjfR-w
35	https://www.youtube.com/watch?v = xXoedns1SG0
36	https://www.youtube.com/watch?v = FM8MagxwF7o
37	https://www.youtube.com/watch?v = xkH5OC_c380
38	https://www.youtube.com/watch?v = u9FljysaPUA
39	https://youtu.be/OKyuSQ_D210
40	https://youtu.be/8Ky6aWpqxkU
41	https://www.youtube.com/watch?v = mRczj1MoxTI
42	https://youtu.be/xGnPiXb_J50
43	https://youtu.be/8gb7uIS2lZI
44	https://www.youtube.com/watch?v = 8o_XFj_WTKg
45	https://youtu.be/Ta_qwiFS2-Y
46	https://youtu.be/zhsZGoKvRW4
47	https://youtu.be/9HdZn5il-bE
48	https://www.youtube.com/watch?v = Jo7-n9GoLsl
49	https://youtu.be/lhco4_iLQKk
50	https://youtu.be/43OqbC8u2-c

51	https://www.youtube.com/watch?v = PBeKhiXqE68
52	https://www.youtube.com/watch?v = Ar6w7Fn7Mq0
53	https://www.youtube.com/watch?v = gHT5CbpPlFg
54	https://www.youtube.com/watch?v = Wq_twcGCq_s
55	https://www.youtube.com/watch?v = CQ2uuHPsPp8
56	https://www.youtube.com/watch?v = ksFrCcrtbJs
57	https://www.youtube.com/watch?v = RIfDfE-e6cM
58	https://www.youtube.com/watch?v = CkyoRDR2Ogc
59	https://www.youtube.com/watch?v = cHClIDaWTZc
60	https://youtu.be/TNDK5DwtVxo
61	https://www.youtube.com/watch?v = 3FyjYP9ckxQ
62	https://youtu.be/mk3HMGICYo4
63	https://www.youtube.com/watch?v = 52wrenRLL1I
64	https://youtu.be/z-z76bS3WhY
65	https://www.youtube.com/watch?v = XfGhPJRgQCE
66	https://www.youtube.com/watch?v = T5TzHmVfqBM
67	https://www.youtube.com/watch?v = 8EVIE9opOy0
68	https://www.youtube.com/watch?v = _fqwBJmvxLU
69	https://www.youtube.com/watch?v = BB9IHRpqJvM
70	https://www.youtube.com/watch?v = VeIAYjBnGrs
71	https://youtu.be/9qoF_RzfWSc
72	https://www.youtube.com/watch?v = LyQUzBOmNiw
73	https://www.youtube.com/watch?v = xAtFXE3hQU8
74	https://www.youtube.com/watch?v = Jh455839rnU
75	https://www.youtube.com/watch?v = qls66s5pfHA
76	https://www.youtube.com/watch?v = MwhcUMGy8iw
77	https://www.youtube.com/watch?v = UXZSq2Sp4a8
78	https://www.youtube.com/watch?v = rGzG5fL7RBQ
79	https://www.youtube.com/watch?v = L2TTclGPUuk
80	https://www.youtube.com/watch?v = F4_qKy5YqdE
81	https://www.youtube.com/watch?v = pHqerc81ZQY
82	https://www.youtube.com/watch?v = MNA2VgMEVjs
83	https://www.youtube.com/watch?v = q4ZpFmmGR50
84	https://www.youtube.com/watch?v = MDGum3jgQVE
85	https://youtu.be/8hOo5Jawjq4
86	https://www.youtube.com/watch?v = Bs0TGvVfQXM
87	https://www.youtube.com/watch?v = FLsKZonQDjA
88	https://www.youtube.com/watch?v = xMcMSepKMoc
89	https://www.youtube.com/watch?v = x5XDGGgdhSI
90	https://www.youtube.com/watch?v = HOJFgzkH3vA
91	https://www.youtube.com/watch?v = _BZ91Uv4k3o
92	https://www.youtube.com/watch?v = cg9HzbLPr3g
93	https://www.youtube.com/watch?v = yRUiHDRB-6w
94	https://www.youtube.com/watch?v = U5jS_YpCio0
95	https://www.youtube.com/watch?v = a84JZGJvzt4
96	https://www.youtube.com/watch?v = IZxNw2C58Gw
97	https://www.youtube.com/watch?v = MMb9gjbweWc
98	https://www.youtube.com/watch?v = ZYg8asrY4bo
99	https://www.youtube.com/watch?v = G013TP0s5Es
100	https://www.youtube.com/watch?v = aeKvhLs4cww
101	https://www.youtube.com/watch?v = JHNJeYyeCJY
102	https://www.youtube.com/watch?v = a3Z8orfDvPU
103	https://www.youtube.com/watch?v = ipHabyQ0GA8
104	https://www.youtube.com/watch?v = 2mYe2Q9_ls0
105	https://www.youtube.com/watch?v = 8dacJb36GzY
106	https://www.youtube.com/watch?v = _Wvej1tV5Sc
107	https://www.youtube.com/watch?v = _jo3zFqt588
108	https://www.youtube.com/watch?v = t_oVUGZKmsI
109	https://www.youtube.com/watch?v = MUEdLGTcj8A
110	https://www.youtube.com/watch?v = CvRFlo-4ClA
111	https://www.youtube.com/watch?v = -MNXJJFMxWo
112	https://youtu.be/xICEE9KOJrI
113	https://www.youtube.com/watch?v = uZ8MjQbKaUE
114	https://www.youtube.com/watch?v = veeK8Nd9fUc

115	https://www.youtube.com/watch?v = q8w-xhW_GX8
116	https://www.youtube.com/watch?v = GeorLGfR3TA
117	https://www.youtube.com/watch?v = 9BzyVbd54pc
118	https://www.youtube.com/watch?v = -bawP8CXlo0
119	https://www.youtube.com/watch?v = ckE99t9akPA
120	https://www.youtube.com/watch?v = 4QrEFV1yj2c
121	https://youtu.be/8iULlNvTBw8
122	https://www.youtube.com/watch?v = Sq2RlA7mp9U
123	https://www.youtube.com/watch?v = WG3LWNtqAwc
124	https://youtu.be/Y2SceKlvTrw
125	https://youtu.be/HXevHVmVGFI
126	https://youtu.be/KV38cR0kgNY
127	https://youtu.be/Mda9W1Nlzms
128	https://youtu.be/Ta1MbCgzgCg
129	https://www.youtube.com/watch?v = 3pBA3EppWpc
130	https://youtu.be/dfoJh8bWtaM
131	https://youtu.be/fBR3p6DBcy4
132	https://youtu.be/8S7bFFfdivM
133	https://youtu.be/91Ty31QL0X8
134	https://youtu.be/7TVoY_HaWXc
135	https://youtu.be/06vvD96kZ7A
136	https://youtu.be/NS719aZYv50
137	https://youtu.be/q6tDMXd8l98
138	https://youtu.be/QANysW9vIE0
139	https://youtu.be/MMkLpsinnis
140	https://youtu.be/4dhjCS9bp1s
141	https://youtu.be/gexEyhnT4Ts
142	https://youtu.be/KABiHnf4G80
143	https://youtu.be/RXdTo1v45MI
144	https://youtu.be/78bi9BATCLQ
145	https://youtu.be/razVovMt8lU
146	https://youtu.be/KUeV7oDLBpg
147	https://youtu.be/leI_YNdil1s
148	https://youtu.be/rlo0VDuu-lY
149	https://youtu.be/657O4WnyN7Q
150	https://youtu.be/XzWU4SptnzQ

Introduction

The goal of percutaneous coronary intervention (PCI) is to restore unimpeded blood flow in epicardial coronary arteries without causing complications.

PCI is performed using the following 14 steps (Fig. 1).

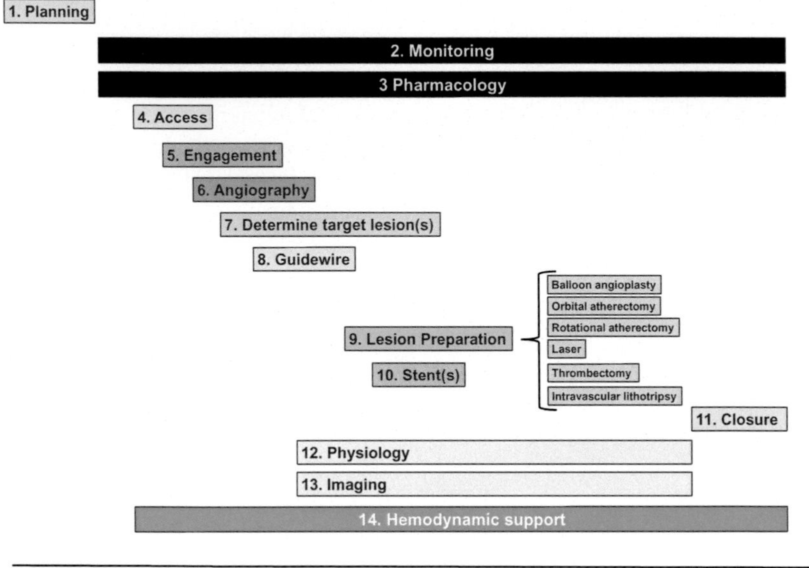

FIGURE 1 The 14 steps of percutaneous coronary intervention.

The following steps are performed in all PCI cases:

- Planning (Chapter 1: Planning).
- Monitoring (Chapter 2: Monitoring).
- Medications (Chapter 3: Medications).
- Access (Chapter 4: Access).
- Engagement (Chapter 5: Coronary and Graft Engagement).
- Angiography (with the exception of the "zero contrast PCI," although the latter still requires a prior angiogram) (Chapter 6: Coronary Angiography).
- Determine target lesion(s) (Chapter 7: Selecting Target Lesion(s)).
- Wiring (Chapter 8: Wiring).
- Vascular closure (Chapter 11: Access Closure).

The following steps are not always performed:

- Lesion preparation (sometimes direct stenting is performed without predilation, although this is generally discouraged) (Chapter 9: Lesion Preparation).

- Stenting (sometimes balloon angioplasty, including drug-coated balloons, or thrombectomy only is performed) (Chapter 10: Stenting).
- Physiology (Chapter 12: Coronary Physiology).
- Imaging (Chapter 13: Coronary Intravascular Imaging).
- Hemodynamic support (Chapter 14: Hemodynamic Support).

The *Manual of PCI* breaks down the PCI procedure into 14 sequential stages. The steps of each stage are then discussed, using the following template:

1. Goal (why?),
2. How?
3. Challenges, and
4. What can go wrong (complications)?

The same format (goal, how, challenges, what can go wrong) is used for the steps of each specialized technique, such as atherectomy and thrombectomy. For each challenge and potential complication, we discuss: (1) potential causes; (2) prevention; and (3) treatment strategies.

The first 14 chapters review in-depth each stage of PCI (part A). The subsequent 10 chapters review performance of those steps in specific clinical and angiographic subgroups (part B). Chapters 25–29, review complications (part C), and Chapter 30 reviews equipment (part D).

Planning is the first stage of any procedure, including PCI, and is a key step. Plans can (and should) change depending on new information that becomes available during the procedure, but creating a plan before starting is invaluable.

The *Manual of PCI* aims to help each operator develop rich, accurate mental representations of what does or can happen during PCI. Developing such mental representations is key to achieving expert performance [1].

The algorithms contained in this book are not the only or necessarily the best algorithms for these procedures. These are algorithms used by the authors, but there will always be room for improvement. Please send feedback on how these algorithms (and the book) can be improved.

Reading this manual (or any book for that matter) will not make you an expert interventionalist. Developing expertise in PCI requires practice—not naïve practice, but deliberate practice (working to improve areas of deficiency with a teacher) [1].

"Always improving" what we do, so that the best possible outcome can be achieved for each patient, is the ultimate goal of this book. We envision a future where all algorithms for all PCI procedures will be freely available to all and continually improved upon.

Reference

[1] Ericsson A, Pool R. Peak: secrets from the new science of expertise. Boston: Houghton Mifflin Harcout Publishing Company; 2016.

Part A

The steps

Chapter 1

Planning

If you fail to plan you are planning to fail.

Benjamin Franklin.

Planning is essential for every procedure, including percutaneous coronary intervention (PCI). Thoughtful planning and appropriate preparation before performing PCI improves the safety, efficiency, outcome, and cost of the procedure.

The following items should be checked, that correspond to each of the 14 steps of the procedure. While planning is in itself the first of the 14 steps, it also serves as a preview of what will occur during each of the subsequent steps (Table 1.1).

1.1 Planning

Consent obtained

- Consent needs to be obtained and documented prior to the procedure. Discussion about the risks and benefits of ad hoc PCI is critical, in patients without a prior angiogram.

History:

- Clinical presentation (stable angina, acute coronary syndromes (ACS), other).
- If stable coronary artery disease, is indication for procedure appropriate? (Review appropriate use criteria [2]).
- Ongoing chest pain?
- Prior cardiac catheterization or other procedure requiring fluoroscopy? If yes, are the prior images and reports available?
- Prior coronary artery bypass graft surgery (CABG)? If yes, is surgical report available?
- Current medications (see Section 1.3).
- Comorbidities
  - Valvular heart disease
  - Congestive heart failure
  - Arrhythmias
  - Peripheral arterial disease (PAD)
  - Renal failure
  - Significant lung disease
  - Obstructive sleep apnea
  - Bleeding disorders
  - Back pain or other musculoskeletal disorders that can affect lying flat on the cardiac catheterization table
  - Diabetes mellitus
  - Advanced age
- Is the patient likely to be noncompliant with medications or require noncardiac surgery in the upcoming 6–12 months? If yes, PCI may be best avoided to minimize the risk of stent thrombosis (due to the surgery and the early discontinuation of dual antiplatelet therapy). Medical therapy only or CABG may be preferred.
- In patients with renal failure or those who are anticoagulated, it may be best to stage non-emergent PCI; ultra low or zero contrast PCI, if feasible, may be beneficial in patients with advanced kidney disease.
- Contrast or latex allergy?

Manual of Percutaneous Coronary Interventions. DOI: https://doi.org/10.1016/B978-0-12-819367-9.00001-9

TABLE 1.1 Preprocedure checklist for cardiac catheterization and PCI.

Patient Name: ____________________ MRN: ____________ Procedure Date:________

Planned Procedure: Diagnostic Cardiac Catheterization ☐
Diagnostic Cardiac Catheterization with possible PCI ☐
Percutaneous Coronary Intervention ☐

1. History

Elective Outpatient Procedures: H&P documented within 30 days? Yes ☐ No ☐

Inpatient Procedures: H&P documented within 24 hours of admission? Yes ☐ No ☐

NPO per institutional protocol prior to procedure*? Yes ☐ No ☐

*NPO GUIDELINE RECOMMENDATIONS (by the American Society of Anesthesiologists (ASA)[1]:

2 hours prior to scheduled procedure time	Clear Liquids, including clear/hard candies and drinks without pulp or dairy
6 hours prior to scheduled procedure time	Light solids, including toast/oatmeal/granola bar, liquids with dairy, hard candies, pulp, and infant formula
8 hours prior to scheduled procedure time	Regular Diet
Chewing tobacco	No chewing tobacco 6 hours prior to procedure

History of prior PCI or CABG: Yes ☐ No ☐ If yes, were reports obtained? Yes ☐ No ☐

Prior radial artery harvesting for CABG? Yes ☐ No ☐

AV fistula for dialysis? Yes ☐ No ☐

Clinical presentation with STEMI? Yes ☐ No ☐

Severe peripheral arterial disease? Yes ☐ No ☐

Prior abdominal aortic aneurysm endograft? Yes ☐ No ☐

Prior iliofemoral surgery/endovascular treatment? Yes ☐ No ☐

Candidacy for stenting:

1. Is there significant anemia (i.e., Hct <30)? Yes ☐ No ☐

If yes, has RBC type and cross been performed? Yes ☐ No ☐

2. Any major surgery performed recently or planned for next year? Yes ☐ No ☐

3. Is there any clinically overt bleeding? Yes ☐ No ☐

4. Is patient on chronic anticoagulation (e.g., warfarin, DOAC)? Yes ☐ No ☐

5. Is there history of medication non-adherence? Yes ☐ No ☐

Allergies

1. Contrast: Yes ☐ No ☐ If yes, was the patient pre-treated? Yes ☐ No ☐

2. Aspirin: Yes ☐ No ☐ If yes, does the patient need desensitization? Yes ☐ No ☐

3. Heparin (HIT) Yes ☐ No ☐ If yes, use alternative anti-thrombotic agents

4. Latex Yes ☐ No ☐ If yes, remove all latex products from procedural use

5. Multiple allergies Yes ☐ No ☐ If yes, consider prednisone pretreatment

(Continued)

TABLE 1.1 (Continued)

Medications

1. Did patient take aspirin within the past 24 hours? Yes ☐ No ☐
2. Did patient take a P2Y12 inhibitor within the past 24 hours? Yes ☐ No ☐
3. Did patient take metformin within the past 24 hours? Yes ☐ No ☐
4. Did patient take sildenafil (or equivalent) within the past 24 hours? Yes ☐ No ☐
5. Did patient receive LMWH within the past 24 hours? Yes ☐ No ☐
   - ☐ If yes for LMWH, dose given and time of last dose ____________________
6. Is patient on immunosuppressants? Yes ☐ No ☐
   - ☐ If yes, avoid use of arterial closure devices
7. Is patient on chronic opioids? Yes ☐ No ☐
   - ☐ If yes, sedation can be challenging

Informed Consent

Was informed consent obtained within 30 days? Yes ☐ No ☐

Was informed consent documented in the medical chart? Yes ☐ No ☐

Is there a healthcare proxy? Yes ☐ No ☐

Is the patient DNR or DNI? Yes ☐ No ☐ Yes, but revoked for procedure ☐

(DNR= do not resuscitate, DNI=do not intubate)

Sedation, Anesthesia and Analgesia

Are ASA and Mallampati Class documented? Yes ☐ No ☐

Is there any contraindication to sedation present? Yes ☐ No ☐

Prior adverse reaction to sedation? Yes ☐ No ☐

Adequate intravenous access? Yes ☐ No ☐

2. Physical Examination

Abnormal cardiac examination? Yes ☐ No ☐

Signs of congestive heart failure? Yes ☐ No ☐

Signs of cardiogenic shock? Yes ☐ No ☐

Distal pulses decreased? Yes ☐ No ☐

Hostile groin (massive obesity, scar from prior procedure, infection, ulcer)? Yes ☐ No ☐

Arteriovenous fistula for dialysis? Yes ☐ No ☐

Severe arm tremor or involuntary movements? Yes ☐ No ☐

Prior radiation skin injury? Yes ☐ No ☐

(Continued)

TABLE 1.1 (Continued)

3. Labs and Imaging

CBC and basic electrolytes within 14 days (outpatient) or 24 hours (inpatient)? Yes □ No □

(CBC=complete blood count): Hemoglobin: ______ WBC: ______ Platelets: ______

Potassium: ______

Creatinine: _____

GFR: _____

Does the patient require pre-procedure hydration? Yes □ No □

INR performed within 24 hours (for patients on warfarin)? Yes □ No □

INR:____

Serum beta hCG for women of childbearing potential? Yes □ No □ NA □

Beta hCG:____

Was **EKG** performed and reviewed? Yes □ No □

If available, **echocardiogram** reviewed? Yes □ No □

If available, coronary **CTA** reviewed? Yes □ No □

If available, prior **angiogram(s)** reviewed? Yes □ No □

If yes, assess the following:

Location of femoral bifurcation: _____

Peripheral arterial disease: _____

Extent of vessel and coronary calcification: _____

Prior stents: _____

Old severe lesions which were untreated: _____

Access issues-crossover: _____

Groin scar: _____

Kissing iliac stents or stenosis: _____

Radial loops: _____

Subclavian stenosis or tortuosity: _____

Need for a long sheath: _____

Anomalous origin of left or right coronary artery: _____

Issues with LIMA or other graft engagement: _____

Diagnostic and guide catheters used [and whether these provided optimal support based on report and angiographic images]: _____

Guidewires used: _____

Issues with stent delivery: _____

Atherectomy was needed: _____

Physical examination:

- Radiation skin injury on the back (Fig. 28.3)? If yes, may need to postpone or modify procedure to avoid repeat radiation of the affected area.
- Cardiovascular examination that includes all pulses in upper and lower extremities.
- Signs of congestive heart failure (pulmonary rales, high jugular venous pressure, lower extremity edema).

Labs:

- Hemoglobin
- White blood cell count
- Platelet count
- International normalized ratio (INR)
- Potassium level
- Creatinine + estimated glomerular filtration rate (GFR) (limit contrast to $\leq 3.7 \times$ GFR for patients at increased risk for contrast nephropathy, such as patients with chronic kidney disease, Section 28.3)
- Pregnancy test (for women of childbearing potential).

Prior imaging:

- Review prior coronary angiograms and PCIs.
- Review noninvasive testing results (echocardiography, magnetic resonance imaging [MRI], stress testing).
- In patients with recent diagnostic angiography or coronary computed tomography angiography (CTA), the target lesion(s) can be determined prior to the procedure.

1.2 Monitoring

- Assess baseline ECG and heart rate.
- Assess patient's baseline vital signs and pulse oximetry.

1.3 Pharmacology

- Allergies?
- Has patient received aspirin?
- For patients with a well-documented aspirin allergy: have they been desensitized?
- For patients allergic to contrast: have they been premedicated (Section 3.3)?
- For planned PCI or for patients with ST-segment elevation acute myocardial infarction (STEMI): have they received a $P2Y_{12}$ inhibitor?
- On metformin: in patients with chronic kidney disease hold metformin the day of the procedure and do not restart until at least 48 hours after the procedure. In patients without chronic kidney disease metformin does not necessarily need to be discontinued; instead renal function can be checked after the procedure and metformin withheld if renal function deteriorates.
- On insulin: reduce insulin to adjust for fasting status before the procedure.
- On warfarin: discontinue 5 days prior to elective procedures and check the INR on the day of the procedure. Radial access is preferred in anticoagulated patients.
- On direct oral anticoagulants (DOAC): discontinue prior to elective procedures, as outlined in Table 1.2.

1.4 Access

History:

- Prior radial artery harvesting for CABG?
- Arteriovenous (AV) fistula for dialysis? Avoid using this arm for cardiac catheterization.
- Access site(s) used for any prior procedures? Has a closure device been used? Consider using contralateral femoral or radial access if an Angioseal was used within 90 days.
- Prior access site complications? If yes, what was the complication and how was it managed? If yes, avoid using the same access site.

TABLE 1.2 How long to stop a DOAC before a cardiac catheterization procedure.

Direct factor Xa inhibitors	Days to hold
Apixaban (Eliquis)	2 days
Edoxaban (Savaysa)	
Creatinine clearance 50–95 mL/min	2 days
Creatinine clearance 15–49 mL/min	3 days
Rivaroxaban (Xarelto)	
Creatinine clearance ≥ 50 mL/min	2 days
Creatinine clearance 15–49 mL/min	3 days
Direct thrombin factor IIa inhibitor	**Days to hold**
Dabigatran (Pradaxa)	
Creatinine clearance > 80 mL/min	2 days
Creatinine clearance 50–79 mL/min	3 days
Creatinine clearance 30–49 mL/min	4 days
Creatinine clearance 15–29 mL/min	5 days

Creatinine clearance calculator: http://www.mdcalc.com/creatinine-clearance-cockcroft-gault-equation/.

- History of PAD? Access through severely diseased or occluded iliofemoral or subclavian arteries should be avoided.
- Clinical presentation: radial access is especially favored in STEMI patients.
- On warfarin or DOAC: radial access is preferred.
- High risk of bleeding: radial access is preferred.
- Patient preference (patients who work extensively with their hands/arms or use them for support may prefer femoral approach).

Physical examination:

- Good distal pulses?
- Morbid obesity? (Favors radial access)

Labs: high INR and low platelet count favor radial access.

Prior imaging:

- Review prior cardiac and/or peripheral catheterization films: disease or tortuosity in aortoiliac and upper extremity vessels?
- Computed tomography (CT) of the chest:
  - Anomalous aortic arch anatomy?
  - Size of iliac/subclavian vessels and presence of disease.
  - Arteria lusoria? (Anomalous origin of right subclavian from the aortic arch.) Arteria lusoria favors use of left radial or femoral access.
- CT of the abdomen/pelvis: location of common femoral artery bifurcation and disease in iliofemoral vessels.
- Ultrasound of peripheral arteries.

Desired outcome: Decide on access site and size/length of the sheath.

1.5 Engagement

- Prior CABG: what is the anatomy (surgical report, prior coronary angiograms)?
- Catheters used in prior coronary angiograms/PCIs? If significant difficulty or inability to engage the coronary arteries well from one access site was encountered, you should switch to a different access site (such as femoral).
- Aortic CT angiography: aortic dilation? Anomalous coronary arteries?
- Aortic stenosis or regurgitation (associated with dilated ascending aorta that may require larger catheters for coronary engagement)?

1.6 Angiography

- Renal failure? If yes:
  - Limit contrast volume, by using biplane angiography if available, limiting cine angiographic projections, using intravascular ultrasound (IVUS), and potentially using contrast savings systems, such as the DyeVert Plus (Osprey Medical) (Section 29.3).
  - Consider using isoosmolar contrast agents (Section 29.3).
  - Administer preprocedural and postprocedural hydration (1−3 mL /kg/h of normal saline).
  - Prior radiation skin injury? If yes: Limit number of cineangiography runs and avoid including the previously affected area within the radiation beam.

1.7 Determine target lesion(s)

History: The presence and severity of symptoms can help determine the need for PCI.
Prior imaging: Prior noninvasive testing can help determine potential culprit lesions. Review of prior angiograms is essential for determining whether any interval changes have occurred.

1.8 Wiring

History:

- Prior challenges wiring the target lesion(s)?

1.9 Lesion preparation

History:

- Prior challenges expanding the target lesion(s)?

Prior imaging:

- Severe calcification: consider atherectomy, laser, and intravascular lithotripsy. if available (Chapter 19: Calcification).
- Large thrombus: consider antithrombotic treatment and thrombectomy (Chapter 20: Acute Coronary Syndromes—Thrombus).

1.10 Stenting

History:

- Able to take dual antiplatelet therapy (DAPT)? (History of bleeding or high risk of bleeding, compliance with medications)

1.11 Access closure

History:

- Active infection or immunocompromised? May be best to avoid use of vascular closure devices to minimize the risk for infection.

1.12 Physiology

History: If symptoms are equivocal and there is no preprocedural noninvasive testing showing ischemia, physiologic coronary assessment can be useful.

- Prior adverse reaction or contraindication to intracoronary vasodilators such as adenosine?

1.13 Imaging

History:

- Prior PCI of the target lesion(s) strongly favors performing intravascular imaging.

1.14 Hemodynamic support

History:

- Congestive heart failure symptoms.
- Low ejection fraction.

Physical examination:

- Elevated jugular venous pressure.
- Lower extremity edema.
- Lung crackles.
- Femoral and radial pulses.

Labs:

- Beta natriuretic peptide (BNP).
- Lactate in patients with cardiogenic shock.

Prior imaging:

- Echocardiography (ejection fraction, left and right ventricular size, valvular abnormalities).
- Access site imaging to determine feasibility of hemodynamic support.

Hemodynamics:

- Right heart catheterization measurements, if available (high right atrial, pulmonary artery, or pulmonary artery capillary wedge pressure, low cardiac output, low cardiac power output, low pulmonary artery pulsatility index [PAPI]).

Consider hemodynamic support in patients with reduced ejection fraction, poor hemodynamics, and/or complex or high-risk planned interventions (Chapter 14: Hemodynamic Support).

References

[1] Practice Guidelines for Preoperative Fasting and the Use of Pharmacologic Agents to Reduce the Risk of Pulmonary Aspiration: Application to Healthy Patients Undergoing Elective Procedures: An Updated Report by the American Society of Anesthesiologists Task Force on Preoperative Fasting and the Use of Pharmacologic Agents to Reduce the Risk of Pulmonary Aspiration. Anesthesiology 2017;126:376–93.

[2] Patel MR, Calhoon JH, Dehmer GJ, et al. ACC/AATS/AHA/ASE/ASNC/SCAI/SCCT/STS 2017 appropriate use criteria for coronary revascularization in patients with stable ischemic heart disease: a report of the American College of Cardiology Appropriate Use Criteria Task Force, American Association for Thoracic Surgery, American Heart Association, American Society of Echocardiography, American Society of Nuclear Cardiology, Society for Cardiovascular Angiography and Interventions, Society of Cardiovascular Computed Tomography, and Society of Thoracic Surgeons. J Am Coll Cardiol 2017;69:2212–41.

Chapter 2

Monitoring

Monitoring the patient should be continually performed from the beginning to the end of the case, so that potential complications are promptly identified and corrected. The following parameters are assessed (Fig. 2.1).

2.1 Patient

1. Patient comfort level: patient discomfort can lead to movement, potentially leading to complications. It can also lead to tachycardia and tachypnea, which may in turn worsen ischemia.
2. Chest pain, abdominal pain, groin pain? Is the pain anticipated based on the procedure or is it unexpected? The pain could be due to ischemia, perforation, or other complications.
3. Level of consciousness and breathing. Is breathing assistance needed (BiPAP or intubation)?
4. Ability to move all extremities (no stroke) or conversely excessive movements that may hinder performance of the procedure.
5. Signs of allergic reactions: skin rash; itching and hives; swelling of the lips, tongue, or throat; hypotension.

2.2 Electrocardiogram

The ECG morphology and heart rate should be evaluated at the beginning of the case, so that subsequent ECG changes can be promptly identified.

Electrocardiographic changes of concern include:

1. New ST segment depression.
2. New ST segment elevation (Fig. 2.2).

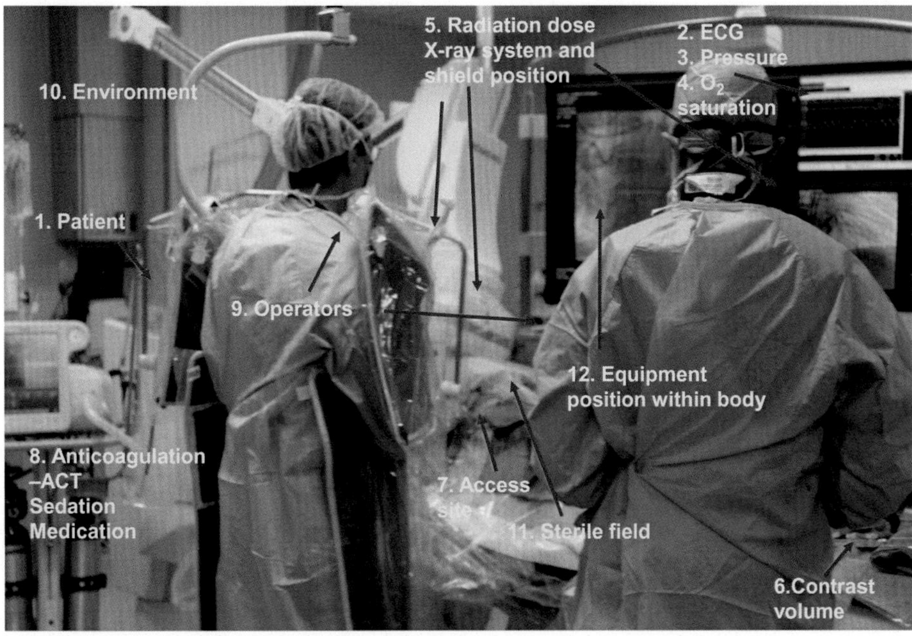

FIGURE 2.1 What to monitor during cardiac catheterization.

Manual of Percutaneous Coronary Interventions. DOI: https://doi.org/10.1016/B978-0-12-819367-9.00002-0

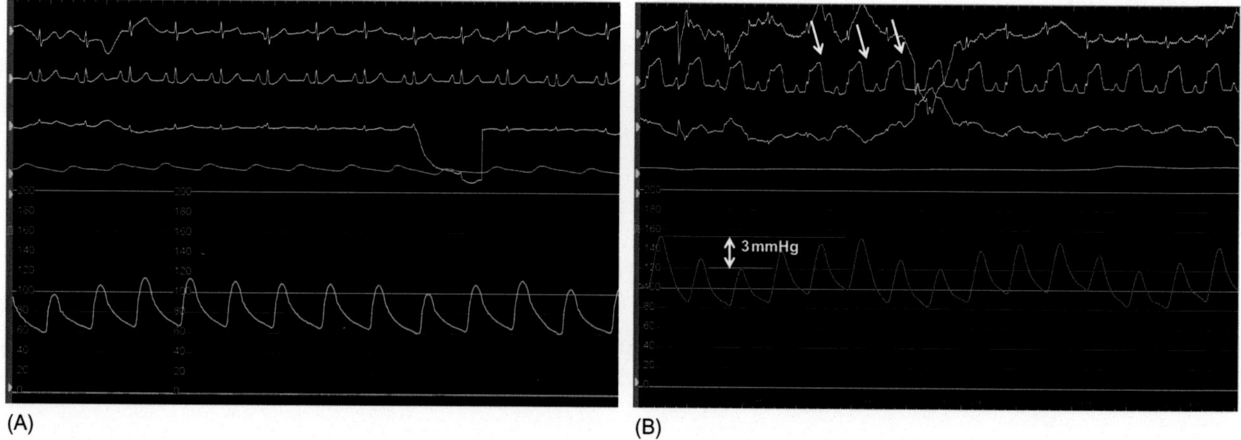

FIGURE 2.2 Electrocardiographic and pressure waveform changes during CTO PCI. (A) Baseline. (B) ST-segment elevation (*arrows*) and development of 35 mmHg pulsus paradoxus after a distal vessel perforation in a patient with prior coronary artery bypass graft surgery. *Reproduced with permission from the* Manual of CTO Interventions, *2nd ed. (Figure 3.35). Copyright Elsevier.*

3. Bradycardia.
4. Tachycardia.
5. QRS widening.
6. Ventricular premature beats during wire manipulations.
7. Ventricular fibrillation.

2.3 Pressure waveform

The arterial pressure should be continuously monitored.

Pressure waveform changes of concern include:

1. Hypotension (see Section 28.1)
2. Pulsus paradoxus (Fig. 2.2)
3. Hypertension
4. Pressure waveform dampening or disappearance (see Section 28.1.1.1.) that may reflect the true aortic pressure, or may be due to:
   a. Deep guide catheter engagement or engagement of coronary arteries with ostial lesions. Injections should not be performed while the pressure waveform is dampened, as they can lead to coronary or aorto-coronary dissection and/or air embolism.
   b. Air entrainment within the guide catheter (e.g., when using the trapping technique for equipment exchange).
   c. Thrombus formation within the catheter. Injecting in such cases can lead to coronary or systemic thromboembolism.
   d. Guide catheter kinking.
   e. Insertion of equipment: for example, inserting an aspiration catheter, such as the Export into a 6 French guide catheter may lead to pressure dampening.
   f. Disconnection of the pressure transducer.

In patients who develop hypotension and in heart failure or shock patients, placement of a Swan Ganz catheter can facilitate decision making regarding hemodynamic support and also help detect any new hemodynamic changes (Section 28.1.1).

2.4 Oxygen saturation

Oxygen desaturation may be due to hypoventilation due to heavy sedation, but it can also be due to early pulmonary edema, artifact, or other causes. Full arterial blood gas can provide more comprehensive information about the patient's oxygenation and metabolic status.

2.5 Radiation dose—X-ray system and shield positioning

The cumulative air kerma and DAP radiation dose should be continuously monitored. Usually the procedure is stopped if the air kerma dose exceeds 5–7 Gray. An air kerma radiation dose higher than 15 Gray is a sentinel event (Section 28.2).

The dose rate is another dynamic parameter that can be tracked.

There are also continuous operator dose monitoring devices (such as the DoseAware, Philips) that can alert to high operator doses in real time, alerting the operator to the need for changes to reduce high radiation dose.

The position of the various shields and the image receptor should be continually monitored and adjusted to minimize patient and operator radiation dose.

2.6 Contrast volume

This can be tracked automatically by some systems (such as the ACIST injector and the DyeVert Plus system). The procedure should generally be stopped before reaching a contrast volume ≥ 3.7 × GFR, although a lower threshold is preferable in patients with chronic kidney disease or single kidney (Section 28.3). Recent contrast administration (for example in patients who had contrast computed tomography) should be taken into consideration when determining the contrast threshold.

2.7 Access site

The pulses at the access site and distally should be assessed at the beginning and the end of the case.

Bleeding and hematoma formation can occur at the access site(s)—continuous inspection and palpation can help in early identification (Chapter 4: Access).

2.8 Medication administration (anticoagulation—ACT, sedation, other medications)

Sedation (Section 3.1) is given in nearly all patients and should be titrated to achieve acceptable patient comfort without compromising respiratory or hemodynamic status.

Anticoagulation (Section 3.4) is achieved with unfractionated heparin in most procedures and monitored using ACT (activated clotting time). Goal ACT (Hemochron device) for PCI is 300–350 seconds for most procedures or >350 seconds for retrograde CTO PCI. When glycoprotein IIb/IIIa inhibitors or cangrelor (Section 3.5) are given, goal ACT is 200–250 seconds.

Other medications may be required, such as vasopressors (Section 3.6), atropine (Section 3.7.2), etc.

2.9 Operator and team performance

Paying attention to the operators' and team's operational state can help identify conditions that may lead to suboptimal outcomes, such as excessive fatigue.

2.10 Cath lab environment

Avoiding excessive noise and distractions is important for better outcomes.

A rule analogous to the "sterile cockpit rule" for flying should be implemented during the critical parts of the procedure. The "sterile cockpit rule" is an informal name for the Federal Aviation Administration regulation stating that pilots shall not require, nor may any *flight* crewmember perform, any duties during a critical phase of *flight*, except those duties required for the safe operation of the aircraft.

2.11 Sterile field and equipment

Keeping the equipment and table organized will facilitate equipment identification and use.

Dried blood and contrast can make the operator gloves and various types of equipment (guidewires, catheters, balloons, stents, etc.) "sticky" and could also create risk of embolism if debris enters the manifold. Regularly wiping the gloves and equipment and flushing the catheters with heparinized saline will facilitate equipment handling and reduce the risk of complications.

2.12 Equipment position within the body

The position of equipment inserted into the body (such as sheaths, guide catheters, guidewires, balloons, stents, etc.) should be constantly monitored for both efficacy and safety.

A classic example is guide disengagement while attempting to deliver balloons and stents, when the operator often focuses on the equipment that needs to be delivered (such as the stent) and does not pay attention to the guide catheter, which may become completely disengaged leading to loss of guide and guidewire position. Conversely, deep guide engagement may result in dissection and acute vessel closure (Section 25.2.1), especially if contrast is injected.

Another example is not monitoring the location of the guidewire tip (especially when collimation is used to minimize radiation dose), which may enter into small branches and result in distal vessel perforation (Section 26.4).

Who is assessing the above parameters:

1. The primary and secondary operators.
2. The cath lab technician (traditionally a technician is constantly monitoring the ECG and pressure tracings).
3. The cath lab RN (who monitors the ECG, pressure, and O_2 saturation). The cath lab RN is usually administering the various medications (sedation, anticoagulation, antiplatelet agents, etc.).

Chapter 3

Medications

In this chapter we discuss the following classes of medications that are commonly used in the cardiac catheterization laboratory:

1. Sedatives and analgesics
2. Vasodilators
3. Contrast media
4. Anticoagulants
5. Antiplatelet agents
6. Vasopressors and inotropes
7. Antiarrhythmics

3.1 Sedatives and analgesics

3.1.1 Goals

- Improve patient comfort.

3.1.2 How?

- Midazolam (Versed): 0.5–1 mg intravenous (IV)—can be repeated. Duration of action: 15–80 minutes.
- Fentanyl: 25–100 mcg IV—can be repeated. Duration of action: 30–60 minutes. Other opioids, such as morphine can also be used.

3.1.3 What can go wrong?

3.1.3.1 Respiratory failure—hypopnea

Causes:

- Excessive sedation may suppress respiratory drive.

Prevention:

- Avoid excessive sedation.
- Monitor oxygen saturation throughout the procedure.

Treatment:

- Stop administering sedation.
- Flumazenil (Romazicon) for reversing midazolam: 0.2 mg IV over 15 seconds. If there is no response after 45 seconds, administer 0.2 mg again over 1 minute. Can repeat at 1-minute intervals up to a total of 1 mg.
- Naloxone (Narcan) for reversing opioids (Fentanyl, morphine, etc.): 0.1–0.2 mg intravenously; can repeat at 2- to 3-minute intervals until the desired degree of reversal is achieved.
- Intubation may be required for severe respiratory depression.

Manual of Percutaneous Coronary Interventions. DOI: https://doi.org/10.1016/B978-0-12-819367-9.00003-2

3.1.3.2 Delayed response to oral P2Y12 inhibitors which may lead to thrombotic complications

Causes:

- Opioids delay gastric empting and slow-down drug adsorption, such as P2Y12 inhibitor absorption.

Prevention:

- Avoid opioids use in STEMI if not deemed necessary.

Treatment:

- Use intravenous antiplatelet agents (e.g., cangrelor or GP IIb/IIIa inhibitors).

3.2 Vasodilators

Medications that cause vasodilation can be categorized into those causing mainly large vessel vasodilation (nitroglycerin) and those causing mainly small vessel vasodilation (nicardipine, nitroprusside, adenosine).

3.2.1 Nitroglycerin

3.2.1.1 Goals

- Dilate coronary arteries (intracoronary nitroglycerin should be routinely administered before coronary angiography, to prevent coronary spasm and allow accurate interpretation of coronary anatomy).
- Treat hypertension.
- Treat pulmonary edema.

3.2.1.2 How?

- Intracoronary/intragraft: 100–300 mcg.
- Intravenous: nitroglycerin drip is usually started at 10 mcg/min and increased by 10 mcg/min at 5-minute intervals until the desired effect is achieved and systolic blood pressure remains above 100 mmHg. Maximum dose is 200 mcg/min.
- Sublingual: 0.4 mg.

3.2.1.3 What can go wrong?

3.2.1.3.1 Hypotension

Causes:

- Excessive dilatation of peripheral veins, reducing blood return to the heart (decreased preload). Also excessive dilatation of peripheral arteries (decreased afterload).
- Coadministration of nitroglycerin and phosphodiesterase type 5 (PDE-5) inhibitors, such as avanafil, sildenafil, vardenafil, and tadalafil.
- Hypertrophic obstructive cardiomyopathy (HOCM): nitroglycerin worsens left ventricular outflow obstruction by decreasing both preload and afterload.

Prevention:

- Avoid high and multiple doses of nitroglycerin.
- Do not administer in patients with hypotension or patients with right ventricular infarction.
- Do not administer in patients who have recently received a PDE-5 inhibitor, such as avanafil (Stendra, within prior 24 hours), sildenafil (Viagra, within prior 24 hours), vardenafil (Levitra, within prior 24 hours), and tadalafil (Cialis, within prior 48 hours).
- Do not administer to patients with hypertrophic obstructive cardiomyopathy (HOCM).

Treatment:

- Do not administer additional doses of nitroglycerin. Nitroglycerin's half-life ranges from 1.5 to 7.5 minutes.
- Administer normal saline.
- Waiting (hypotension often resolves after a few minutes).
- Administer vasopressors (such as norepinephrine or phenylephrine) in cases of extreme or persistent hypotension. If hypotension persists, also assess for other potential causes, such as bleeding.

3.2.1.3.2 Headache, flushing, dizziness

Headache may occur after nitroglycerin administration due to dilation of intracranial arteries. Dizziness can occur due to the hypotensive effect of nitroglycerin.

3.2.1.3.3 Tachycardia

Reflex tachycardia may result from the hypotensive effect of nitroglycerin.

3.2.2 Nicardipine

3.2.2.1 Goals

- Prevent and treat no reflow. Nicardipine is a calcium channel blocker that can be used intracoronary to achieve vasodilation of small arteries. Nicardipine is the preferred agent for treating or preventing no reflow (Section 25.2.3.2), for example, during atherectomy (Section 19.3) and during saphenous vein graft PCI (Section 18.9.2), as it has less hypotensive effect compared with nitroprusside and verapamil and also has shorter duration of action.

3.2.2.2 How?

- Intracoronary: 100–300 mcg.

3.2.2.3 What can go wrong?

3.2.2.3.1 Hypotension

This is treated as described in Section 3.2.1.3.1.

3.2.3 Nitroprusside

3.2.3.1 Goals

- Prevent and treat no reflow.

3.2.3.2 How?

- Intracoronary: 100–300 mcg.

3.2.3.3 What can go wrong?

3.2.3.3.1 Hypotension

This is treated as described in Section 3.2.1.3.1.

3.2.4 Verapamil

3.2.4.1 Goals

- Prevent radial spasm.
- Prevent and treat no reflow.

3.2.4.2 How?

- Radial artery: 2–3 mg.
- Intracoronary: 1 mg intracoronary over 2 minutes.

3.2.4.3 What can go wrong?

3.2.4.3.1 Hypotension

This is treated as described in Section 3.2.1.3.1.

3.2.5 Adenosine

3.2.5.1 Goals

- Prevent and treat no reflow.
- Cause vasodilation during physiologic testing (Section 12.2.6).

3.2.5.2 How?

- Intracoronary: RCA: 50–100 mcg.
- Intracoronary left main: 100–200 mcg—several thousand mcg could be administered (slowly) in case of no reflow.
- Intragraft: 100–200 mcg.
- Intravenous: 140 mcg/kg/min, administered through a central vein or a large peripheral vein.
- Regadenoson or papaverine (papaverine is not available in the United States) can also be administered for inducing vasodilation. Regadenoson is costly and papaverine may cause ventricular fibrillation.

3.2.5.3 What can go wrong?

3.2.5.3.1 Heart block

Causes:

- Adenosine's effect on atrioventricular node.
- This is most likely to occur with injection in the right coronary artery.

Prevention:

- Avoid high doses of adenosine in the right coronary artery.
- Slow adenosine administration.
- Aminophylline administration (250–300 mg intravenously over 10 minutes) may be used to prevent bradycardia [1] during atherectomy of the right coronary artery. Aminophylline is an A1 adenosine receptor antagonist (Section 19.3).

Treatment:

- Watchful waiting (adenosine has short half-life).

3.2.5.3.2 Atrial fibrillation

Atrial fibrillation is the most commonly documented adenosine-induced arrhythmia (2.7% after intravenous administration) [2] and is usually well-tolerated except in patients with accessory pathways [3].

Causes:

- Premature ventricular beats occurring during adenosine administration, sometimes during periods of AV block (Figs. 3.1 and 3.2) [4].

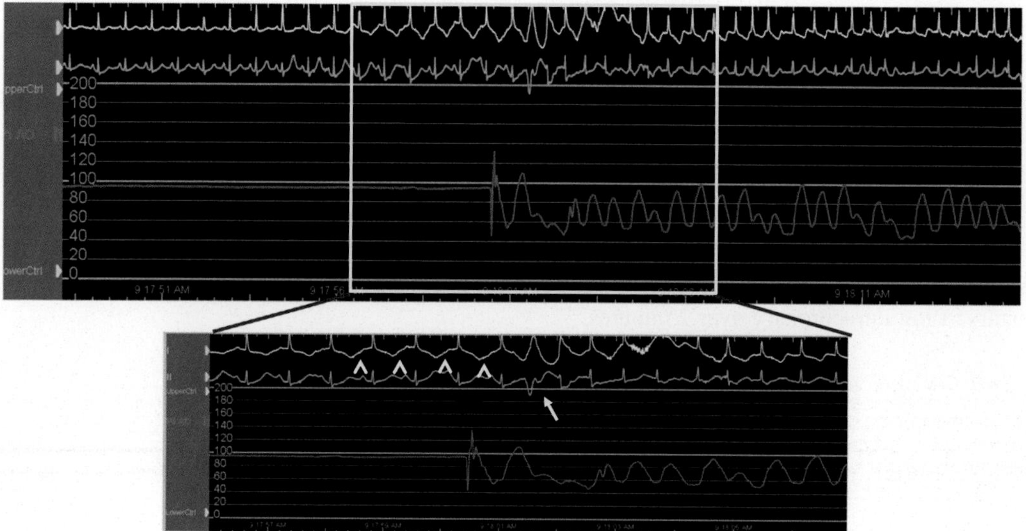

FIGURE 3.1 Atrial fibrillation occurring after adenosine administration. Adenosine caused ST-segment depression (*arrowheads*). A premature ventricular beat (*arrow*) subsequently triggered atrial fibrillation.

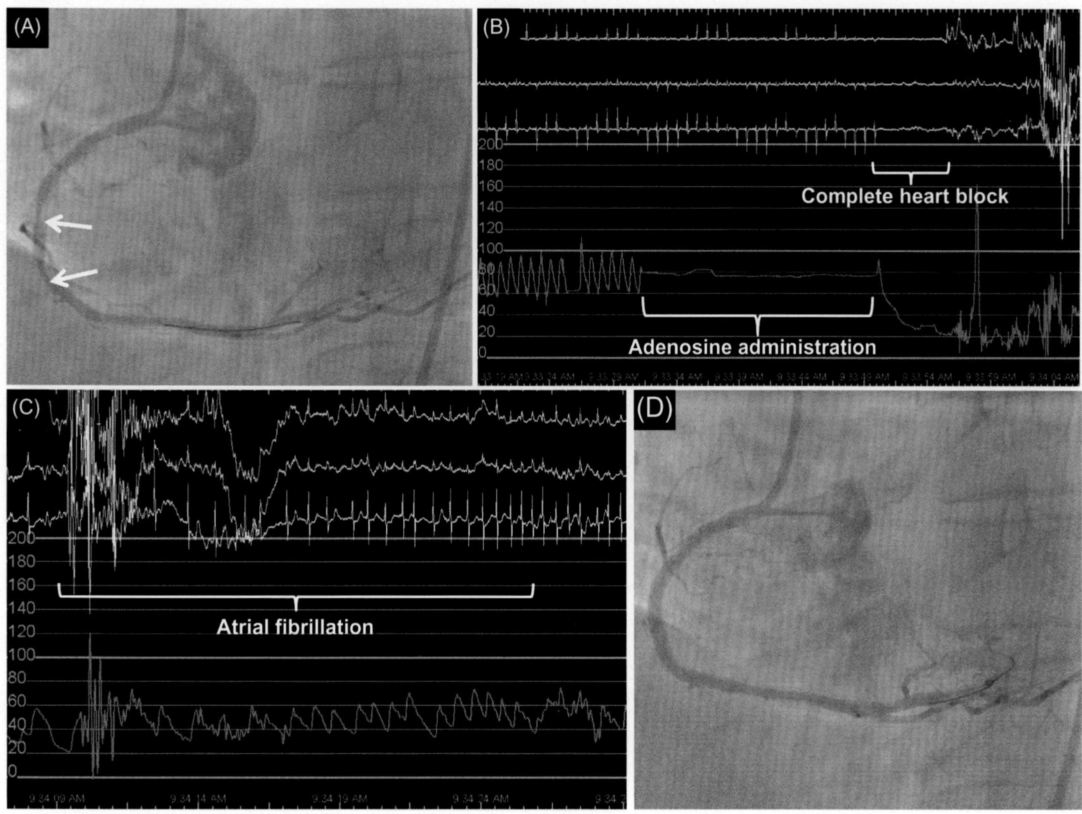

FIGURE 3.2 Coronary angiography demonstrating an in-stent restenotic lesion of the mid right coronary artery (A). Intracoronary adenosine administration through the right coronary artery (40 mcg) resulted in complete heart block (B), followed by development of atrial fibrillation (C). After stenting the right coronary artery lesion resolved. Sinus rhythm was restored with cardioversion at the end of the procedure. *Reproduced with permission from Mahmood A, Papayannis AC, Brilakis ES. Pro-arrhythmic effects of intracoronary adenosine administration. Hellenic J Cardiol 2011;52:352–3 (Figure 2). Copyright Elsevier.*

Prevention:

- Same as for heart block above.

Treatment:

- DC cardioversion. If cardioversion is not desired, antiarrhythmics, such as amiodarone, and AV nodal blocking agents, such as beta blockers or calcium channel blockers could be used.

3.2.5.3.3 Ventricular fibrillation

Causes:

- Torsades des pointes or ventricular fibrillation can be triggered by adenosine administration, usually after a ventricular pause due to the R on T phenomenon (Fig. 3.3), but may also occur without a pause [5].

Prevention:

- Same as for heart block above.

Treatment:

- Ask patient to cough, as forceful coughing could generate sufficient blood flow to the brain to maintain consciousness until definitive treatment (defibrillation) can be administered.
- Defibrillation.

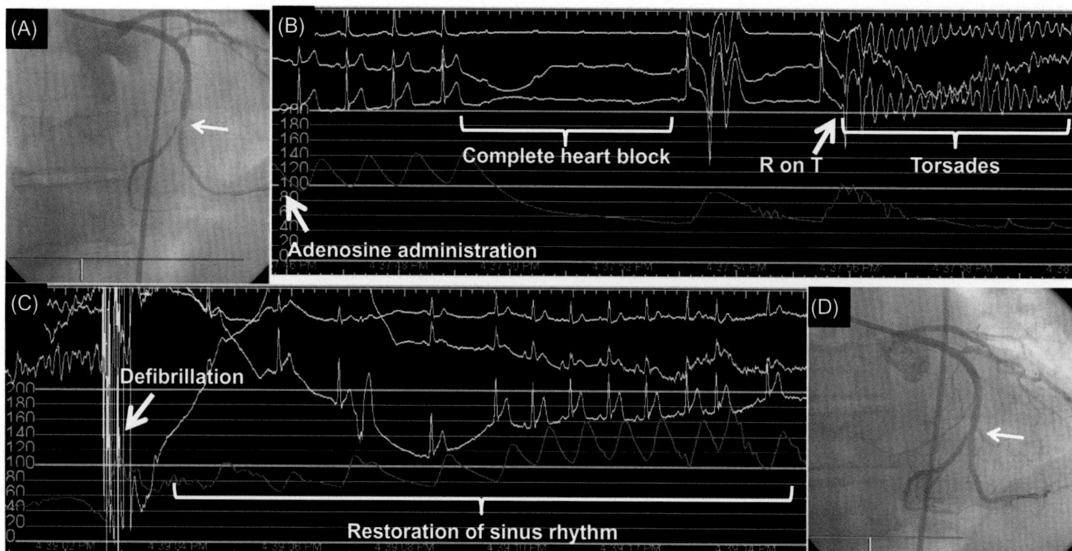

FIGURE 3.3 Coronary angiography demonstrating a bifurcation lesion in the mid circumflex artery (*arrow*, A). Administration of intracoronary adenosine to treat no reflow in the circumflex coronary artery resulted in complete heart block, followed by torsades des pointes due to the R on T phenomenon (B). After defibrillation, sinus rhythm was restored (C). Bifurcation stenting using a "culotte" technique provided an excellent final angiographic result with TIMI 3 flow (D). *Reproduced with permission from Mahmood A, Papayannis AC, Brilakis ES. Pro-arrhythmic effects of intracoronary adenosine administration. Hellenic J Cardiol 2011;52:352–3 (Figure 1). Copyright Elsevier.*

3.3 Contrast media

3.3.1 Goals

- Visualize coronary and peripheral arteries under X-ray.
- Clear blood from the coronary artery to perform OCT (Section 13.3.5).

3.3.2 How?

- Contrast media are injected through the manifold as described in Chapter 6: Coronary Angiography.
- The lowest possible volume of contrast should be administered, as described in Section 28.3.2.
- Isoosmolar contrast media have been associated with lower risk of contrast nephropathy [6].

3.3.3 What can go wrong?

3.3.3.1 Contrast-induced acute kidney injury (discussed in Section 28.3)

Contrast-induced acute kidney injury is a potentially serious complication after coronary angiography and PCI. Patients with advanced chronic kidney disease are at the highest risk of developing this complication. Prevention and treatment of contrast-induced acute kidney injury is discussed in Section 28.3.

3.3.3.2 Allergic reactions

Allergic reaction to contrast agents can occur immediately after administration or be delayed (usually 6–12 hours after contrast administration) and can range in severity from mild (skin rash) to life-threatening (angioedema, anaphylactic shock).

Causes:

- IgE-mediated reactions or direct mast cell degranulation.

Prevention:

- Use of isoosmolar contrast media as compared with ionic, low-osmolar contrast media (ioxaglate) [7]. However, isoosmolar contrast media have higher risk of delayed skin reactions [8,9].

- Pretreatment (for patients known to have allergic reactions to contrast administration):
  - Steroids (usually prednisone 50 mg administered at 13, 7, and 1 hours before the procedure) or 32 mg of methyl-prednisolone administered 12 and 2 hours prior to the procedure.
  - Diphenhydramine (Benadryl) 50 mg administered 1 hour prior to the procedure.
  - Cimetidine 300 mg orally or ranitidine 150 mg administered orally 1 hour prior to the procedure.
- If the patient needs emergency coronary angiography and/or PCI, hydrocortisone sodium succinate (Solu-Cortef) 100 mg should be administered intravenously as soon as possible before the procedure.

Treatment:

- Diphenhydramine (Benadryl) 25–50 mg intravenously.
- Cimetidine 300 mg IV or ranitidine 50 mg IV over 15 minutes.
- Steroids, such as hydrocortisone sodium succinate (Solu-Cortef) 100–400 mg administered intravenously over 1 minute.
- Epinephrine (for anaphylactic shock) (0.3 mg of 1:10,000 solution intravenously - can repeat to total dose of 1 mg).
- Intravenous normal saline.

3.3.3.3 Thyroid dysfunction

Causes:

- Contrast media contain high doses of iodine. Iodine administration to patients with underlying thyroid disease may lead to hypersecretion of thyroid hormones, a phenomenon known as the Jod–Basedow effect (usually occurs 2–12 weeks after iodine administration). Iodine administration can also lead to hypothyroidism.

Prevention:

- Avoid contrast administration in patients with hyperthyroidism or hypothyroidism.

Treatment:

- Referral to endocrinology.

3.4 Anticoagulants

Unfractionated heparin is the most commonly used anticoagulant for PCI [10]. Bivalirudin provides similar outcomes with unfractionated heparin [11,12], hence it is currently used infrequently except in patients with heparin-induced thrombocytopenia (HIT) for whom it is the anticoagulant of choice.

3.4.1 Goals

- Prevent thrombus formation within the coronary artery or within equipment inserted into the body (sheaths, guide catheters, wires, balloon, stents, etc.).
- Reduce the risk of radial artery occlusion (for cases performed via radial access, Section 4.3, step 9).

3.4.2 How?

3.4.2.1 Unfractionated heparin

3.4.2.1.1 Dose

a. Without concomitant glycoprotein IIb/IIIa inhibitor or cangrelor administration: 70–100 units/kg.
b. With concomitant glycoprotein IIb/IIIa inhibitor or cangrelor administration: 50–70 units/kg.
   - Renal failure: no adjustment needed.
   - Half-life: 1–2 hours—the half-life of unfractionated heparin increases with higher heparin doses (from 30 minutes after an intravenous bolus of 25 U/kg, to 60 minutes after a bolus of 100 U/kg, to 150 minutes after a bolus of 400 U/kg) [13].

3.4.2.1.2 Monitoring

Monitoring is performed by checking the activated clotting time (ACT) every 20–30 minutes. Goal ACT:

- Without concomitant glycoprotein IIb/IIIa inhibitor or cangrelor administration:
  - Hemochron device: 300–350 seconds (> 350 seconds is often recommended for retrograde chronic total occlusion interventions) [14].
  - HemoTec device: 250–300 seconds [14].
- With concomitant glycoprotein IIb/IIIa inhibitor or cangrelor administration: 200–250 seconds [14].

3.4.2.1.3 Reversal

Reversal: heparin can be reversed with protamine: 1–1.5 mg of protamine per 100 units of heparin; not to exceed 50 mg. Protamine should be administered slowly to minimize the risk of hypersensitivity reactions or anaphylaxis. Prior NPH insulin or protamine zinc insulin administration [15], fish hypersensitivity, vasectomy, severe left ventricular dysfunction, and abnormal preoperative pulmonary hemodynamics increase the risk of such reactions.

Protamine can cause severe systemic hypotension and relative pulmonary hypertension (due to vasoconstriction of the pulmonary vasculature), usually 5–10 minutes after administration. Methylene blue may help recovery from a protamine reaction [16,17].

3.4.2.2 Bivalirudin

Bivalirudin is a direct thrombin inhibitor that carries no risk of HIT.

3.4.2.2.1 Dose

a. Normal renal function. Bolus: 0.75 mg/kg IV, followed by infusion of 1.75 mg/kg/h IV for up to 4 hours after the procedure [10].
b. Renal failure. Bolus: 0.75 mg/kg IV, followed by infusion. Infusion rate depends on creatinine clearance, as follows:
   - Moderate (CrCl 30–59 mL/min): 1.75 mg/kg/h
   - Severe (CrCl <30 mL/min): 1 mg/kg/h
   - Hemodialysis: 0.25 mg/kg/h

- Half-life: 25 minutes.

3.4.2.2.2 Monitoring

Monitoring is performed using the activated clotting time (ACT). Goal ACT:

- Without concomitant glycoprotein IIb/IIIa inhibitor or cangrelor administration: 300–350 seconds.
- With concomitant glycoprotein IIb/IIIa inhibitor or cangrelor administration: 200–250 seconds.

3.4.2.2.3 Reversal

There is no reversal agent available for bivalidurin. The bivalidurin infusion is discontinued.

3.4.3 Challenges

3.4.3.1 ACT is low after anticoagulant administration

Causes:

- Anticoagulant did not actually reach the circulation (e.g., because the intravenous line malfunctioned).
- Low anticoagulant dose.
- Decreased antithrombin III levels (familial or acquired) may reduce the effect of unfractionated heparin.

Prevention:

- Dose anticoagulants appropriately.
- Ensure intravenous lines are working properly.

Treatment:

- Administer additional anticoagulant doses (potentially through the arterial sheath if there are concerns about the IV line), followed by repeat ACT measurement before proceeding with PCI.

- Avoid inserting devices, such as guidewires, balloons or stents in the coronary arteries until after the ACT is in the therapeutic range to avoid coronary thrombotic complications.

3.4.4 What can go wrong?

3.4.4.1 Bleeding

Causes:

- Excessive anticoagulation.
- Vascular access complication.
- Hypertension.

Prevention:

- Avoid high doses of anticoagulants.
- Monitor anticoagulation level (ACT every 20–30 minutes).
- In patients receiving oral anticoagulation (e.g., for atrial fibrillation), oral anticoagulation should ideally be stopped prior to the procedure (as described in Chapter 1: Planning, Table 1.2), unless the procedure is emergent.

Treatment:

- Reverse anticoagulation with protamine if unfractionated heparin was administered. Reversal should be performed only after all intracoronary equipment has been removed. There is a risk of stent thrombosis if stents were implanted. Protamine can in itself cause hypersensitivity reactions, as described above.
- Further treatment is specific to the site of bleeding.

3.4.4.2 Thrombosis

Causes:

- Low dose of heparin or bivalirudin.
- Administered anticoagulant did not reach circulation (e.g., due to IV line infiltration).
- Defective heparin or bivalirudin batch.
- Heparin-induced thrombocytopenia.
- Resistance to heparin (e.g., in patients with antithrombin III deficiency).

Prevention:

- Assess platelet count prior to the procedure (to ensure that platelet count has not been decreasing, which could be due to HIT).
- Administer correct dose of anticoagulants.
- Ensure that the IV line is working properly.
- Check ACT 5 minutes after anticoagulant administration to ensure that adequate anticoagulation has been achieved before proceeding with PCI.

Treatment:

- Thrombus is managed as described in Chapter 20: Acute Coronary Syndromes—Thrombus, Section 20.9.
- Additional anticoagulant is administered to ensure that the ACT is therapeutic.

3.4.4.3 Heparin-induced thrombocytopenia (HIT)

Heparin administration can cause HIT. There are two types of HIT. Type 1 HIT develops within the first 2 days after heparin administration, and the platelet count normalizes with continued heparin therapy. Type 2 usually develops after 4–10 days from heparin administration and can lead to life-threatening venous and arterial thromboembolism.

Causes:

- Type 1: direct effect of heparin on platelet activation.
- Type 2: immune-mediated disorder.

Prevention:

- Avoid administration of unfractionated heparin, however, as described above, unfractionated heparin is the anticoagulant of choice during PCI.

Treatment:

- Type 1: no specific treatment needed.
- Type 2: discontinuation of all heparin administration and anticoagulation with non-heparin agents (such as bivalirudin or argatroban).

3.5 Antiplatelet agents

Both oral (aspirin and $P2Y_{12}$ inhibitors) and intravenous (glycoprotein IIb/IIIa inhibitors and cangrelor) antiplatelet agents are currently available for use during PCI.

Dual antiplatelet therapy (DAPT) with aspirin and a $P2Y_{12}$ inhibitor is the current standard of care for PCI, except in patients on oral anticoagulants (warfarin or direct oral anticoagulants) in whom aspirin is usually discontinued after the procedure and they only receive clopidogrel in addition to the oral anticoagulant.

3.5.1 Dual antiplatelet therapy

3.5.1.1 Goals

- Prevent stent thrombosis.
- Prevent future acute coronary syndromes.

3.5.1.2 How?

3.5.1.2.1 Medication type

Medication type: all patients (except for those receiving warfarin or direct oral anticoagulants or those who have allergic reactions) should receive aspirin (although recent studies, such as the TWILIGHT study, have suggested that aspirin can often be discontinued after 3 months with the patients only continuing ticagrelor). Clopidogrel is the $P2Y_{12}$ inhibitor of choice for stable CAD patients and prasugrel or ticagrelor for ACS patients, unless contraindicated.

3.5.1.2.2 Dose

- *Aspirin*: 325 mg loading dose, then 81–100 mg daily. Aspirin is commonly stopped in patients who receive oral anticoagulants after an initial periprocedural or longer treatment period (from 1–4 weeks to 6 months after PCI). Patients continue treatment with oral anticoagulant and a $P2Y_{12}$ inhibitor.
- *Clopidogrel*: 600 mg load, ideally given at least 6 hours prior to PCI for stable patients, followed by 75 mg daily thereafter.
- *Ticagrelor*: 180 mg load, followed by 90 mg bid [18]. Ticagrelor is contraindicated in patients with prior intracranial hemorrhage and should be avoided in patients receiving strong CYP3A inhibitors (e.g., ketoconazole, itraconazole, voriconazole, clarithromycin, nefazodone, ritonavir, saquinavir, nelfinavir, indinavir, atazanavir, and telithromycin), strong CYP3A inducers (e.g., rifampin, phenytoin, carbamazepine, and phenobarbital), and simvastatin and lovastatin at doses greater than 40 mg daily.
- *Prasugrel*: 60 mg loading dose, followed by 10 mg daily maintenance dose (5 mg daily dose for patients with body weight <60 kg or patients aged >75 years). Prasugrel is contraindicated in patients with prior transient ischemic attack or stroke [19].

3.5.1.2.3 Pretreatment

Stable CAD patients: not done before diagnostic angiography, but should be given if the patients have planned PCI.
Non-ST-segment elevation ACS: no [20].
STEMI: yes, either before or at the latest at the time of PCI [10], with either ticagrelor 180 mg or prasugrel 60 mg [21].

3.5.1.2.4 Switching between oral $P2Y_{12}$ inhibitors

When to switch [22] (Fig. 3.4).
How to switch [23] (Fig. 3.5).

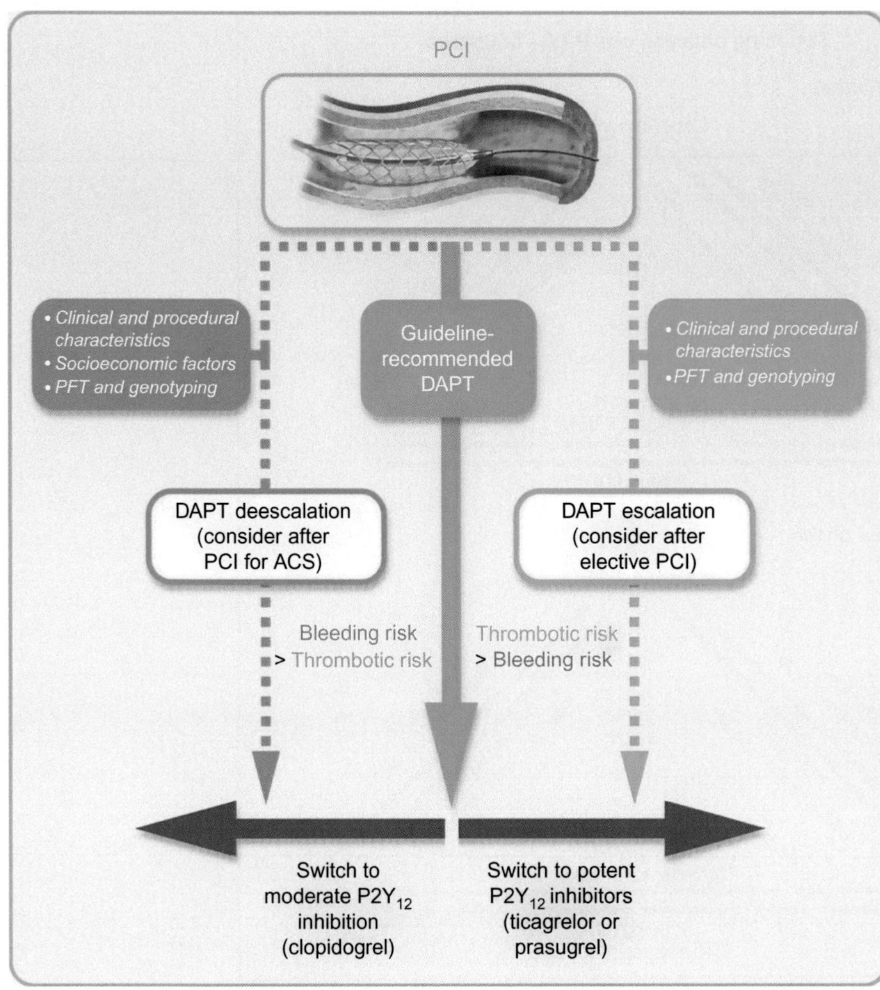

FIGURE 3.4 **When to switch oral P2Y12 inhibitors.** The majority of percutaneous coronary intervention (PCI)–treated patients should be treated with guideline recommended dual antiplatelet therapy (DAPT) (*blue arrow*), which is clopidogrel in stable patients and ticagrelor or prasugrel in patients with ACS (acute coronary syndrome). Alternative strategies may occasionally be considered including a DAPT escalation strategy (*green arrow*) after elective PCI in stable coronary artery disease and a DAPT deescalation strategy (*red arrow*) after PCI for ACS. Escalation strategies may be reasonable when thrombotic risk outweighs bleeding risk, and deescalation strategies may be reasonable when bleeding risk outweighs thrombotic risk. Decision making is guided by clinical and procedural characteristics as well as socioeconomic considerations. Platelet function testing (PFT) and genotyping may be useful to inform guidance of treatment when DAPT escalation or deescalation is desired. *Reproduced with permission from Sibbing D, Aradi D, Alexopoulos D, et al. Updated expert consensus statement on platelet function and genetic testing for guiding P2Y12 receptor inhibitor treatment in percutaneous coronary intervention. JACC Cardiovasc Interv 2019;12:1521–37. Copyright Elsevier.*

3.5.1.2.5 DAPT duration

DAPT duration should be individualized to optimize the balance between benefits (fewer ischemic events) versus risks (bleeding [24]) [25] (Fig. 3.6).

Standard DAPT duration after PCI is 6 months for patients with stable CAD and 12 months in patients who presented with acute coronary syndromes [26]. In patients at high bleeding risk (PRECISE-DAPT score ≥ 25 [27]), DAPT duration can be shortened to 3 and 6 months, respectively, for stable angina and ACS patients [10,28].

Link to online calculator of the PRECISE-DAPT score: http://www.precisedaptscore.com/predapt/webcalculator.html.

Based on the TWILIGHT trial an alternative strategy of aspirin withdrawal after 3 months and continuation with ticagrelor monotherapy, decreases the risk of bleeding without an increase in major adverse events [29].

Prolonged DAPT duration (>12 months) can be considered for patients who do not have high bleeding risk and have high risk for recurrent ischemic events. This can be assessed using the DAPT score [30] (DAPT score ≥ 2 favors >12 month DAPT duration) [10,28].

Link to online calculator of the DAPT score: https://tools.acc.org/DAPTriskapp/#!/content/calculator/.

3.5.1.3 What can go wrong?

3.5.1.3.1 Bleeding

Causes:

- Antiplatelet effect of $P2Y_{12}$ inhibitors and aspirin.
- Concomitant use of anticoagulants (warfarin or DOACs).

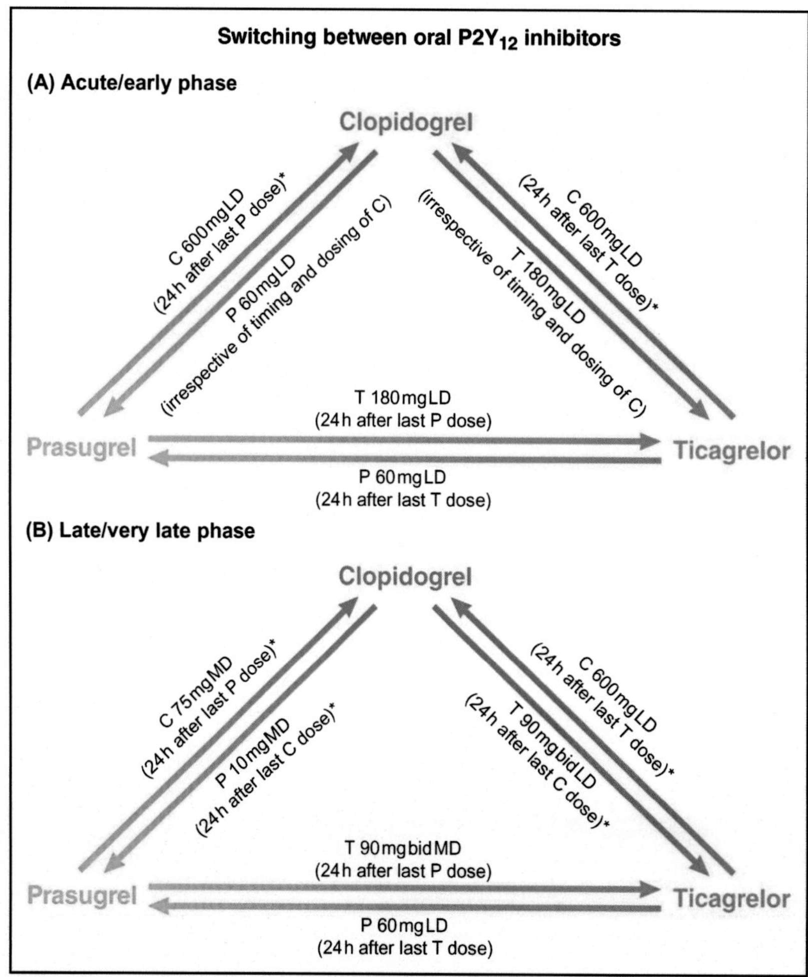

FIGURE 3.5 How to switch between oral $P2Y_{12}$ inhibitors [23]. (A) Switching between oral agents in the acute/early phase. In the acute/early phase (≤30 days from the index event), switching should occur with the administration of a loading dose (LD) in most cases, with the exception of patients who are deescalating therapy because of bleeding or bleeding concerns, in whom a maintenance dose (MD) of clopidogrel (C) should be considered. Timing of switching should be 24 h after the last dose of a given drug, with the exception of when escalating to prasugrel (P) or ticagrelor (T), when the LD can be given regardless of the timing and dosing of the previous clopidogrel regimen.
*Consider deescalation with clopidogrel 75 mg MD (24 h after last prasugrel or ticagrelor dose) in patients with bleeding or bleeding concerns.
(B) Switching between oral agents in the late/very late phase. In the late/very late phase (>30 days from the index event), switching should occur with the administration of a MD 24 h after the last dose of a given drug, with the exception of patients changing from ticagrelor to prasugrel therapy, for whom a LD should be considered. Deescalation from ticagrelor to clopidogrel should occur with administration of a LD 24 h after the last dose of ticagrelor (but in patients in whom deescalation occurs because of bleeding or bleeding concerns, a MD of clopidogrel should be considered).
*Consider deescalation with clopidogrel 75 mg MD (24 h after last prasugrel or ticagrelor dose) in patients with bleeding or bleeding concerns.
Reproduced with permission from Angiolillo DJ, Rollini F, Storey RF, et al. International expert consensus on switching platelet P2Y12 receptor-inhibiting therapies. Circulation 2017;136:1955–75. Permission obtained through copyright.com.

Prevention:

- Reduce aspirin treatment duration in patients receiving warfarin or a DOAC together with a $P2Y_{12}$ inhibitor. Adjust $P2Y_{12}$ inhibitor duration based on the predicted bleeding risk.
- Avoid high potency P2Y12 inhibitors (prasugrel and ticagrelor) in patients at high risk for bleeding, such as patients receiving warfarin or DOACs.
- Routine administration of a proton-pump inhibitor in patients receiving DAPT [31].

Treatment:

- Platelet transfusion in case of life-threatening bleeding and irreversible $P2Y_{12}$ inhibition (i.e., clopidogrel, prasugrel, aspirin).
- Platelet transfusion may not be successful with ticagrelor given its reversible effect on the $P2Y_{12}$ receptor. A specific ticagrelor reversal agent is currently in development [32].

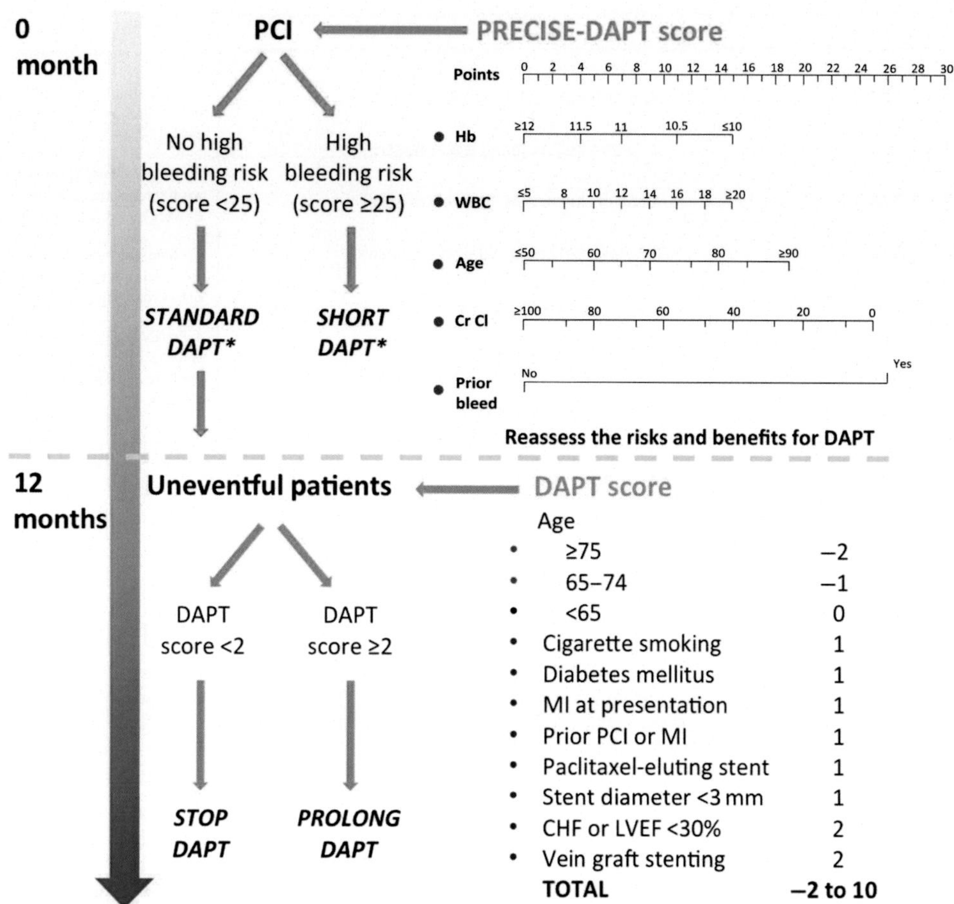

FIGURE 3.6 Determining the optimal duration of dual antiplatelet therapy in acute coronary syndrome patients, using the PRECISE-DAPT and DAPT scores. Variables included in the scores are associated with increased bleeding risk (*red dots*), increased ischemic risk (*blue dots*), or neutral effect (*black dots*). *In the validation study, short-term DAPT consisted of 3–6 months of therapy, and standard DAPT consisted of at least 12-month DAPT duration. *Hb*, hemoglobin; WBC, white blood cell count; *CrCl*, creatinine clearance; *CHF*, congestive heart failure; *DAPT*, dual antiplatelet therapy; *LVEF*, left ventricular ejection fraction; *MI*, myocardial infarction; *PCI*, percutaneous coronary intervention; *PRECISE-DAPT*, Predicting Bleeding Complications in Patients Undergoing Stent Implantation and Subsequent Dual Antiplatelet Therapy. *Reproduced with permission from Valgimigli M, Gargiulo G. DAPT duration after drug-eluting stent implantation: no news is good news. JACC Cardiovasc Interv 2017;10:1211–4. Copyright Elsevier.*

- Deescalate to clopidogrel if the patient was taking ticagrelor or prasugrel.
- Specific measures to control local bleeding.

3.5.1.3.2 Dyspnea

Causes:

- Ticagrelor is more likely to cause dyspnea (likely adenosine-mediated) than other $P2Y_{12}$ inhibitors.
- Other cardiac (such as heart failure) and noncardiac causes (such as primary pulmonary disorders).

Prevention:

- N/A.

Treatment:

- Evaluate for other potential causes of dyspnea.
- Potential treatments (e.g., caffeine [33], aminophylline, theophylline) are currently being studied.
- If dyspnea persists and no other reversible cause is identified, discontinue ticagrelor and replace it with prasugrel or clopidogrel.

Time	Dose
Day 1	
8:00 A.M.	1 spray ketorolac (1 spray in one nostril)
8:30 A.M.	2 sprays ketorolac (1 each nostril)
9:00 A.M.	4 sprays ketorolac (2 each nostril)
9:30 A.M.	6 sprays ketorolac (3 each nostril)
10:30 A.M.	60 mg aspirin
12:00 P.M.	60 mg aspirin
3:00 P.M.	DISCHARGE PATIENT
Day 2	
8:00 A.M.	150 mg aspirin
11:00 A.M.	325 mg aspirin
2:00 P.M.	DISCHARGE PATIENT

To prepare nasal ketorolac:

1. Take ketorolac (60 mg/2 mL) and mix with preservative-free normal saline (2.75 mL).
2. Place combined solution in a nasal spray bottle (one that delivers 100 mL/actuation).
3. Prime with five sprays before use, then each spray actuates 1.26 mg of ketorolac solution.
4. Patient should tilt head down while sprays and should sniff gently to avoid swallowing solution.

FIGURE 3.7 Aspirin desensitization protocol. Nasal ketorolac and oral aspirin challenge protocol. After reactions are treated and resolve, continue the next scheduled ketorolac dose or repeat the oral provoking aspirin dose. Desensitization is complete after 325 mg of aspirin. The patient should take 650 mg of aspirin that evening and then continue 650 mg twice daily as their continuous aspirin dose until further instructed. If no reaction occurs within 3 h after a 325-mg dose, consider it a negative challenge result. *Reproduced with permission from Waldram J, Walters K, Simon R, Woessner K, Waalen J, White A. Safety and outcomes of aspirin desensitization for aspirin-exacerbated respiratory disease: a single-center study. J Allergy Clin Immunol 2018;141:250–6. Copyright Elsevier.*

3.5.1.3.3 Allergic reactions

Causes:

- All medications, including aspirin and $P2Y_{12}$ inhibitors can cause allergic reactions.
- Clopidogrel can very rarely cause thrombotic thrombocytopenic purpura.

Prevention:

- N/A.

Treatment:

- Evaluate whether aspirin or $P2Y_{12}$ inhibitors are causing the reaction—if caused by a $P2Y_{12}$ inhibitor, replace with another $P2Y_{12}$ inhibitor; if caused by aspirin, perform aspirin desensitization (Fig. 3.7) [34].

3.5.2 Intravenous antiplatelet agents

3.5.2.1 Goals

- Prevent thrombus propagation in patients with large intracoronary thrombus or no reflow.
- Prevent stent thrombosis.

3.5.2.2 How?

There are three intravenous antiplatelet agents currently available in the United States: two glycoprotein IIb/IIIa inhibitors, eptifibatide and tirofiban, and one intravenous, short-acting $P2Y_{12}$ inhibitor, cangrelor.

3.5.2.2.1 Cangrelor

Medication	Dose (normal renal function)	Dose in renal failure
Cangrelor	Bolus of 30 μg/kg i.v. followed by 4 μg/kg/min infusion for at least 2 h or duration of procedure, whichever is longer.	No dose adjustment is needed in patients with renal or hepatic insufficiency.

The advantage of cangrelor over GP IIb/IIIa inhibitors is the short half-life, that allows faster return of platelet function in case of bleeding.

Administration of $P2Y_{12}$ inhibitors in patients receiving cangrelor should be as follows (cangrelor inhibits binding of clopidogrel and prasugrel to the platelets):

Ticagrelor: can be administered at any time.

Clopidogrel and prasugrel: administer loading dose immediately after cangrelor discontinuation.

3.5.2.2.2 Glycoprotein IIb/IIIa inhibitors

Medication	Dose (normal renal function)	Dose in renal failure
Eptifibatide	Double bolus of 180 μg/kg i.v. (given at a 10-min interval) followed by an infusion of 2.0 μg/kg/min for up to 18 h.	In patients with creatinine clearance ≤ 60 mL/min the post loading infusion is decreased by 50% to 1 mcg/kg/min.
Tirofiban	Bolus of 25 μg/kg over 3 min i.v., followed by an infusion of 0.15 μg/kg/min for up to 18 h.	In patients with creatinine clearance ≤ 60 mL/min the post loading infusion is decreased by 50% to 0.075 mcg/kg/min.

3.5.2.3 What can go wrong?

3.5.2.3.1 Bleeding

Causes:

- Antiplatelet effect of intravenous antiplatelet agents.
- Thrombocytopenia (caused by glycoprotein IIb/IIIa inhibitors).

Prevention:

- Selective use of cangrelor and GP IIb/IIIa inhibitors: they should be used in patients with large thrombus burden or other thrombotic complications and acceptable bleeding risk.
- Keep ACT low (200–250 seconds).

Treatment:

- Tirofiban/eptifibatide/cangrelor: there is no reversal agent for eptifibatide, tirofiban, or cangrelor. The infusion is stopped followed by a gradual return of platelet reactivity.

3.6 Vasopressors and inotropes

3.6.1 Vasopressors

3.6.1.1 Goals

- Increase blood pressure in case of hypotension.

3.6.1.2 How?

Most IV adrenergic agents provide both vasopressor and inotropic effect, except for phenylephrine that has purely vasopressor effect (as a result it is the vasopressor of choice in patients with hypertrophic cardiomyopathy, in whom an inotropic effect can worsen the intracavitary obstruction). Norepinephrine has mainly a vasopressor with less of an inotropic effect. Vasopressin is also a pure vasopressor. Dopamine at doses >10 mcg/kg/min also has a predominant vasopressor effect.

Vasopressors may be required if mean arterial pressure is <60 mmHg.

Administration through a central line is recommended, if possible.

Dosage:

Phenylephrine: 50–100 mcg IV bolus, followed by infusion at 0.5–1.4 mcg/kg/min.

Norepinephrine: 0.1–0.5 mcg/kg/min IV infusion (norepinephrine is the preferred vasopressor for both cardiogenic and septic shock).

Vasopressin: initial dose: 0.03 U/min IV infusion; can titrate up by 0.005 U/min at 10- to 15-minute intervals up to a maximum dose of 0.1 U/min.

Dopamine: 10–20 mcg/kg/min (at lower doses it mainly has an inotropic and chronotropic effect).

3.6.1.3 What can go wrong?

3.6.1.3.1 Hypertension

Causes:

- Excessive dose of vasopressors.
- Correction of the initial event that caused hypotension.

Prevention:

- Avoid high doses of vasopressors.
- Use mechanical circulatory support.

Treatment:

- The half-life of catecholamines is short, hence blood pressure will likely decrease within a few minutes after discontinuation or dose adjustment.

3.6.2 Inotropes

3.6.2.1 Goals

- Increase myocardial contractility.

3.6.2.2 How?

Epinephrine provides similar vasopressor and inotropic effect. Dopamine at low doses (5-10 mcg/kg/min) has a predominant inotropic effect. Dobutamine has a pure inotropic effect.

Administration through a central line is recommended, if possible.

Dosage:

Epinephrine:

Cardiac arrest: 1 mg IV every 3–5 minutes.

Cardiogenic shock: 2–10 mcg/min IV infusion.

Anaphylactic shock or angioedema: 0.3–0.5 mL of 1 mg/mL solution subcutaneously or IM every 1–2 hours. In case of anaphylaxis epinephrine should be administered immediately without waiting for steroids or other medications to "kick in."

Dopamine: 5–10 mcg/kg/min (at higher doses it acts mainly as a vasopressor).

Dobutamine: 10–40 mcg/kg/min.

3.6.2.3 What can go wrong?

3.6.2.3.1 Arrhythmias

Causes:

- Excessive adrenergic stimulation.

- Underlying cardiac disorders, such as ischemia.

Prevention:

- Use the lowest possible doses of inotropes.
- Use mechanical circulatory support.

Treatment:

- Antiarrhythmics, such as lidocaine and amiodarone.
- Mechanical circulatory support, such as veno-arterial extracorporeal membrane oxygenator (VA-ECMO) may be needed in case of refractory arrhythmias.

3.7 Antiarrhythmics

3.7.1 Amiodarone

3.7.1.1 Goals

- Treat ventricular and atrial tachyarrhythmias.
- Refractory ventricular fibrillation.

3.7.1.2 How?

Dosage:

Non-life-threatening arrhythmias: 150 mg over 10 minutes (may repeat if tachyarrhythmias recur) followed by maintenance infusion of 1 mg/min for 6 hours.

Ventricular fibrillation: 300 mg IV push, followed by maintenance infusion of 1 mg/min for 6 hours.

3.7.1.3 What can go wrong?

3.7.1.3.1 Hypotension

Causes:

- Allergic reaction (if hypotension is due to an allergic reaction the patient may also have other signs of allergic reactions, such as angioedema or urticaria).
- Vasodilation and depression of myocardial contractility; this may be partly due to the solvent, polysorbate 80 or benzyl alcohol, used to assist in dissolving the drug.

Prevention:

- Use caution when administering to patients with low baseline blood pressure.

Treatment:

- Stop amiodarone administration.
- If hypotension is due to an anaphylactic reaction, treat accordingly.
- If hypotension is not due to an anaphylactic reaction: administer vasopressors and inotropes.

3.7.1.3.2 Bradycardia and/or atrioventricular block

Causes:

- Direct action of the medication on conduction through the AV node.

Prevention:

- Do not administer amiodarone to patients with second or third-degree heart block who do not have pacemakers.

Treatment:

- Discontinue amiodarone.
- Temporary pacing may be needed.

3.7.1.3.3 Torsades des pointes or ventricular fibrillation

Causes:

- Prolonged QT interval.
- Administration in patients with preexcitation (Wolff–Parkinson–White syndrome) and concurrent atrial fibrillation.

Prevention:

- Do not administer amiodarone to patients with prolonged QT interval.
- Do not administer amiodarone to patients with preexcitation (Wolff–Parkinson–White syndrome) and concurrent atrial fibrillation.

Treatment:

- Defibrillation.
- Stop amiodarone administration.

3.7.2 Atropine

3.7.2.1 Goals

- Increase heart rate.

3.7.2.2 How?

Dosage: 0.5 mg IV push; may repeat up to a total dose of 3 mg.

3.7.2.3 What can go wrong?

3.7.2.3.1 Rebound tachycardia

Causes:

- Resolution of the underlying cause of bradycardia.
- High atropine dose.

Prevention:

- Use the lowest possible dose of atropine.

Treatment:

- Tachycardia will usually resolve without specific treatment.

3.7.2.3.2 Dry mouth

Atropine reduces salivary gland secretions causing dry mouth.

3.7.2.3.3 Blurry vision

Atropine causes mydriasis, which causes blurry vision. The juice of the Atropa belladonna ("belladonna" means "beautiful woman" in Italian) berries that contains atropine was used during the Renaissance, by women in eyedrops to dilate the pupils and make the eyes appear more seductive.

3.7.2.3.4 Flushing

Atropine can cause dilation of the cutaneous blood vessels, resulting in flushing.

References

[1] Megaly M, Sandoval Y, Lillyblad MP, Brilakis ES. Aminophylline for preventing bradyarrhythmias during orbital or rotational atherectomy of the right coronary artery. J Invasive Cardiol 2018;30:186–9.

[2] Pelleg A, Pennock RS, Kutalek SP. Proarrhythmic effects of adenosine: one decade of clinical data. Am J Ther 2002;9:141–7.

[3] Turley AJ, Murray S, Thambyrajah J. Pre-excited atrial fibrillation triggered by intravenous adenosine: a commonly used drug with potentially life-threatening adverse effects. Emerg Med J 2008;25:46–8.

[4] Mahmood A, Papayannis AC, Brilakis ES. Pro-arrhythmic effects of intracoronary adenosine administration. Hellenic J Cardiol 2011;52:352–3.

[5] Smith JR, Goldberger JJ, Kadish AH. Adenosine induced polymorphic ventricular tachycardia in adults without structural heart disease. Pacing Clin Electrophysiol 1997;20:743–5.

[6] Eng J, Wilson RF, Subramaniam RM, et al. Comparative effect of contrast media type on the incidence of contrast-induced nephropathy: a systematic review and meta-analysis. Ann Intern Med 2016;164:417–24.

[7] Bertrand ME, Esplugas E, Piessens J, Rasch W. Influence of a nonionic, iso-osmolar contrast medium (iodixanol) versus an ionic, low-osmolar contrast medium (ioxaglate) on major adverse cardiac events in patients undergoing percutaneous transluminal coronary angioplasty: a multicenter, randomized, double-blind study. Visipaque in Percutaneous Transluminal Coronary Angioplasty [VIP] Trial Investigators. Circulation 2000;101:131–6.

[8] Schild HH, Kuhl CK, Hubner-Steiner U, Bohm I, Speck U. Adverse events after unenhanced and monomeric and dimeric contrast-enhanced CT: a prospective randomized controlled trial. Radiology 2006;240:56–64.

[9] Sutton AG, Finn P, Grech ED, et al. Early and late reactions after the use of iopamidol 340, ioxaglate 320, and iodixanol 320 in cardiac catheterization. Am Heart J 2001;141:677–83.

[10] Neumann FJ, Sousa-Uva M, Ahlsson A, et al. 2018 ESC/EACTS guidelines on myocardial revascularization. Eur Heart J 2019;40:87–165.

[11] Lincoff AM, Kleiman NS, Kereiakes DJ, et al. Long-term efficacy of bivalirudin and provisional glycoprotein IIb/IIIa blockade vs heparin and planned glycoprotein IIb/IIIa blockade during percutaneous coronary revascularization: REPLACE-2 randomized trial. JAMA 2004;292:696–703.

[12] Kastrati A, Neumann FJ, Mehilli J, et al. Bivalirudin versus unfractionated heparin during percutaneous coronary intervention. N Engl J Med 2008;359:688–96.

[13] Zeymer U, Rao SV, Montalescot G. Anticoagulation in coronary intervention. Eur Heart J 2016;37:3376–85.

[14] O'Gara PT, Kushner FG, Ascheim DD, et al. ACCF/AHA guideline for the management of ST-elevation myocardial infarction: a report of the American College of Cardiology Foundation/American Heart Association Task Force on Practice Guidelines. Circulation 2013;2013:e362–425.

[15] Stewart WJ, McSweeney SM, Kellett MA, Faxon DP, Ryan TJ. Increased risk of severe protamine reactions in NPH insulin-dependent diabetics undergoing cardiac catheterization. Circulation 1984;70:788–92.

[16] Lutjen DL, Arndt KL. Methylene blue to treat vasoplegia due to a severe protamine reaction: a case report. AANA J 2012;80:170–3.

[17] Del Duca D, Sheth SS, Clarke AE, Lachapelle KJ, Ergina PL. Use of methylene blue for catecholamine-refractory vasoplegia from protamine and aprotinin. Ann Thorac Surg 2009;87:640–2.

[18] Wallentin L, Becker RC, Budaj A, et al. Ticagrelor versus clopidogrel in patients with acute coronary syndromes. N Engl J Med 2009;361:1045–57.

[19] Wiviott SD, Braunwald E, McCabe CH, et al. Intensive oral antiplatelet therapy for reduction of ischaemic events including stent thrombosis in patients with acute coronary syndromes treated with percutaneous coronary intervention and stenting in the TRITON-TIMI 38 trial: a subanalysis of a randomised trial. Lancet 2008;371:1353–63.

[20] Montalescot G, Bolognese L, Dudek D, et al. Pretreatment with prasugrel in non-ST-segment elevation acute coronary syndromes. N Engl J Med 2013;369:999–1010.

[21] Schupke S, Neumann FJ, Menichelli M, et al. Ticagrelor or prasugrel in patients with acute coronary syndromes. N Engl J Med 2019;381:1524–34.

[22] Sibbing D, Aradi D, Alexopoulos D, et al. Updated expert consensus statement on platelet function and genetic testing for guiding P2Y12 receptor inhibitor treatment in percutaneous coronary intervention. JACC Cardiovasc Interv 2019;12:1521–37.

[23] Angiolillo DJ, Rollini F, Storey RF, et al. International expert consensus on switching platelet P2Y12 receptor-inhibiting therapies. Circulation 2017;136:1955–75.

[24] Urban P, Mehran R, Colleran R, et al. Defining high bleeding risk in patients undergoing percutaneous coronary intervention. Circulation 2019;140:240–61.

[25] Valgimigli M, Gargiulo G. DAPT duration after drug-eluting stent implantation: no news is good news. JACC Cardiovasc Interv 2017;10:1211–14.

[26] Hahn JY, Song YB, Oh JH, et al. 6-Month versus 12-month or longer dual antiplatelet therapy after percutaneous coronary intervention in patients with acute coronary syndrome (SMART-DATE): a randomised, open-label, non-inferiority trial. Lancet 2018;391:1274–84.

[27] Costa F, van Klaveren D, James S, et al. Derivation and validation of the predicting bleeding complications in patients undergoing stent implantation and subsequent dual antiplatelet therapy (PRECISE-DAPT) score: a pooled analysis of individual-patient datasets from clinical trials. Lancet 2017;389:1025–34.

[28] Levine GN, Bates ER, Bittl JA, et al. ACC/AHA guideline focused update on duration of dual antiplatelet therapy in patients with coronary artery disease: a report of the American College of Cardiology/American Heart Association Task Force on Clinical Practice Guidelines: an update of the 2011 ACCF/AHA/SCAI guideline for percutaneous coronary intervention, 2011 ACCF/AHA guideline for coronary artery bypass graft surgery, 2012 ACC/AHA/ACP/AATS/PCNA/SCAI/STS guideline for the diagnosis and management of patients with stable ischemic heart disease, 2013 ACCF/AHA guideline for the management of ST-elevation myocardial infarction, 2014 AHA/ACC guideline for the management

of patients with non-ST-elevation acute coronary syndromes, and 2014 ACC/AHA guideline on perioperative cardiovascular evaluation and management of patients undergoing noncardiac surgery. Circulation 2016;134:e123–55.

[29] Mehran R, Baber U, Sharma SK, et al. Ticagrelor with or without aspirin in high-risk patients after PCI. N Engl J Med 2019;381:2032–42.

[30] Yeh RW, Secemsky EA, Kereiakes DJ, et al. Development and validation of a prediction rule for benefit and harm of dual antiplatelet therapy beyond 1 year after percutaneous coronary intervention. JAMA 2016;315:1735–49.

[31] Bhatt DL, Cryer BL, Contant CF, et al. Clopidogrel with or without omeprazole in coronary artery disease. N Engl J Med 2010;363:1909–17.

[32] Bhatt DL, Pollack CV, Weitz JI, et al. Antibody-based ticagrelor reversal agent in healthy volunteers. N Engl J Med 2019;380:1825–33.

[33] Lindholm D, Storey RF, Christersson C, et al. Design and rationale of TROCADERO: a TRial Of Caffeine to Alleviate DyspnEa Related to ticagrelOr. Am Heart J 2015;170:465–70.

[34] Waldram J, Walters K, Simon R, Woessner K, Waalen J, White A. Safety and outcomes of aspirin desensitization for aspirin-exacerbated respiratory disease: a single-center study. J Allergy Clin Immunol 2018;141:250–6.

Chapter 4

Access

4.1 Choosing access site

Obtaining arterial access is required for performing diagnostic coronary angiography and percutaneous coronary intervention (PCI).

There is continued controversy about optimal access site selection. Radial (proximal or distal) or ulnar access is associated with significantly fewer access site complications and greater patient satisfaction compared with femoral access [1], but engaging the coronary arteries can be more challenging and guide catheter support may be suboptimal (due to more respiratory motion, subclavian tortuosity, or inability to insert large guide catheters in some patients). Brachial or other access sites, such as transcaval or carotid, are rarely used in clinical practice except for structural heart interventions. Developing expertise in both radial and femoral access is essential for the contemporary interventional cardiologist. The 2018 ESC guidelines state that "Radial access is preferred for any PCI irrespective of clinical presentation, unless there are overriding procedural considerations" [2].

The following algorithm (Fig. 4.1) is recommended for access site selection, assuming expertise in both radial and femoral access:

Step 1. Review prior history, physical examination, noninvasive tests, reports, and films from prior cardiac catheterization procedures. In some patients some access sites are not available or are not suitable for performing cardiac catheterization. Examples include patients with prior radial artery harvesting, occlusion of the common femoral or iliac artery or distal abdominal aorta, extreme iliac or subclavian artery tortuosity, inability to engage the coronary arteries during prior procedures, or dialysis shunts in the ipsilateral arm. Some patients may have a strong preference for a particular access site (e.g., some music players prefer not to use radial, as do some patients who use a walker, whereas the opposite is true for patients who are unable to lie on their back, such as patients with severe back pain or congestive heart failure).

Step 2. If both radial and femoral access are an option, femoral access is usually preferred in patients with prior coronary artery bypass grafting (CABG) (with patent grafts) given increased technical difficulty engaging the grafts via radial access [3]. If radial access is chosen, left radial is preferred to facilitate engagement of the left internal mammary artery (LIMA) graft, which nearly all patients have. If bilateral internal mammary grafts have been used, femoral access is strongly favored.

Step 3. Use of radial access has been associated with lower mortality in ST-segment elevation acute myocardial infarction (STEMI) patients in several, but not all, studies, likely because of use of aggressive anticoagulation and antiplatelet regimens that increase the risk of bleeding.

Similarly, radial access is preferred in patients who are receiving oral anticoagulants (warfarin or direct oral anticoagulants) or are at high risk for bleeding (such as patients with high or low body mass index, renal failure, anemia, or thrombocytopenia) [4].

Step 4. In cases of complex PCI (such as chronic total occlusions, severe calcification and tortuosity, challenging bifurcations, etc.) at least one femoral access (often with a 7 or 8 French sheath, ideally 45 cm long) is often preferred, as it usually provides better support and may improve the efficiency, success, and safety of the procedure. Long (45 cm) sheaths are often used for such procedures.

Step 5. If the initially selected access site is subsequently shown to be suboptimal (failure to advance a wire or catheter to the aortic root, failure to engage the coronary arteries, failure to complete PCI), change to other access sites may facilitate success [5]. However, obtaining femoral access in an anticoagulated patients (radial-to-femoral crossover) has been associated with increased risk of bleeding and should be done with extreme care using meticulous technique [6].

Manual of Percutaneous Coronary Interventions. DOI: https://doi.org/10.1016/B978-0-12-819367-9.00004-4

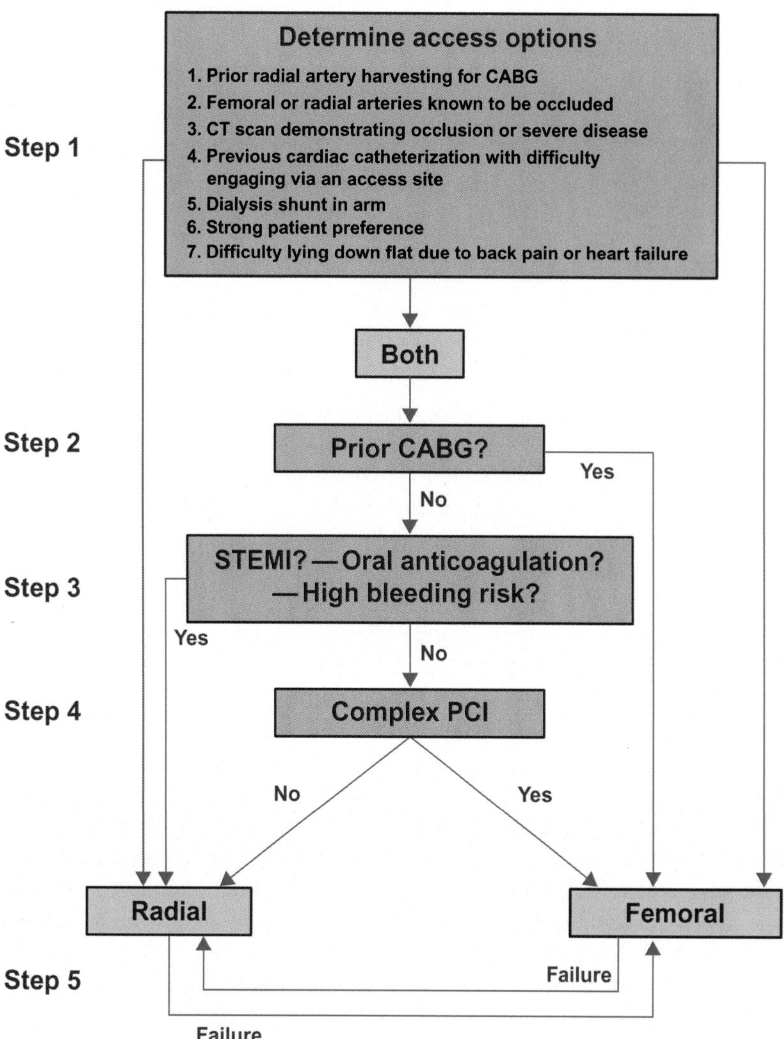

FIGURE 4.1 How to select arterial access site for performing cardiac catheterization and PCI.

4.2 Femoral access

The following sequence is recommended for obtaining femoral access. It uses both fluoroscopy and ultrasound guidance and a micropuncture kit. While there is no definitive proof that use of micropuncture needle (21 gauge, outside diameter: 0.82 mm) is superior to use of a standard needle (18 gauge, outside diameter: 1.27 mm), use of a micropuncture needle is likely to cause less arterial injury (Fig. 4.2). It also allows reattempting the stick (if the initial stick is felt to be suboptimal: too high or too low) without causing significant bleeding. Special caution should be paid to monitoring the position of the 0.018 in. guidewire (that comes with the micropuncture kit) position, as it may be more likely to enter side branches (such as the inferior epigastric artery or lateral circumflex iliac artery) and lead to perforation as compared with a 0.035 in. guidewire [7]. The ideal location for femoral artery puncture depends on the location of the femoral bifurcation.

4.2.1 Step 1. Palpation of the femoral pulse

4.2.1.1 Goals

1. To confirm the presence and location of a femoral pulse.

4.2.1.2 (Fig. 4.3)

1. Right common femoral artery: The operator places his or her left thumb on the right anterior superior iliac spine and the left middle finger on the pubic symphysis. The left second finger of the operator (index finger) should be located over the right common femoral artery above the right femoral head.

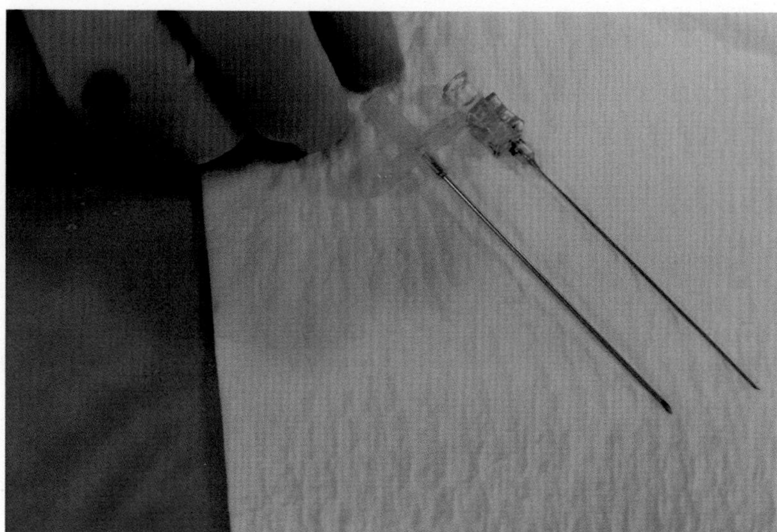

FIGURE 4.2 Standard 18 gauge (*left*) versus micro-puncture 21 gauge (right) needles.

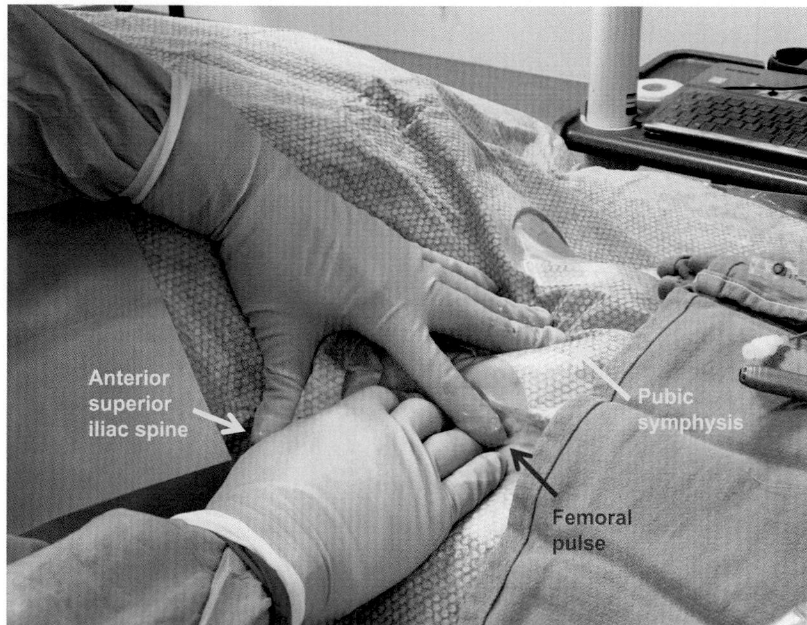

FIGURE 4.3 Palpation for identifying the optimal location for femoral artery puncture.

2. Left common femoral artery: The operator places his or her left thumb on the pubic symphysis and the left middle finger on the left anterior superior iliac spine. The left second finger of the operator (index finger) should be located over the left common femoral artery above the left femoral head.

4.2.1.3 Challenges

1. If the femoral pulse cannot be palpated or is weak, ultrasound should be performed to confirm the presence of a patent femoral artery and its size.

4.2.2 Step 2. Sterile preparation and draping of the groin

4.2.2.1 Goal

To clean the access site and reduce the risk of access site infection.

4.2.2.2 How?

1. The groin areas (ideally both in case there is difficulty in obtaining access in the initially selected groin) are shaved.
2. The groin areas are scrubbed with antiseptic solution, which should be applied from the center (side of puncture) to the periphery making concentric circles, to keep the center sterile.
3. After the antiseptic solution dries up (to allow the adhesive tape to stick), the sterile drape is placed, centering its openings to the groin areas.

4.2.2.3 Challenges

1. In very obese patients it may be challenging to reach the optimal access location due to subcutaneous fat. Taping the pannus may be useful in such cases. There are also specialized disposable systems for pannus retention, such as the Pannus retention system (TZ Medical, Portland, OR).

4.2.2.4 What can go wrong?

1. The groin area can be contaminated (e.g., by the patient's hands). If this happens, scrubbing with antiseptic solution should be repeated.
2. The opening of the drape is positioned too high/low or too lateral/medial. The solution is to reposition the drape, after ensuring that the skin in the new area has been adequately prepared with antiseptic solution.

4.2.3 Step 3. Fluoroscopy of the femoral head

4.2.3.1 Goal

To identify the portion of the femoral artery located above the femoral head. Obtaining access over the femoral head (actually over the pubic bone) is optimal because it facilitates pressing the femoral artery against a hard surface (bone), helping achieve hemostasis after sheath removal. Higher arterial puncture (above the femoral head and above the origin of the inferior epigastric artery) is associated with higher risk of retroperitoneal bleeding. Lower puncture (below the femoral head) is associated with higher risk for pseudoaneurysm (Section 29.1.3) and arteriovenous fistula (Section 29.1.4) formation.

4.2.3.2 How?

1. A hemostat, scissors or other metallic object is placed over the anticipated location of the femoral artery and the femoral head (Fig. 4.4).
2. Fluoroscopy in the AP projection is performed and the hemostat is repositioned until its tip is located at the inferior border of the femoral head (Figs. 4.5 and 4.6).

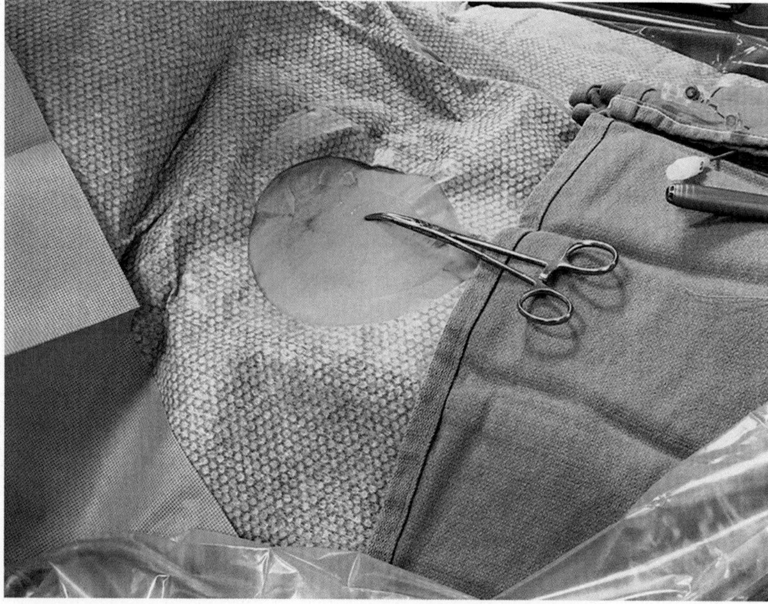

FIGURE 4.4 A hemostat is placed over the area of femoral pulse.

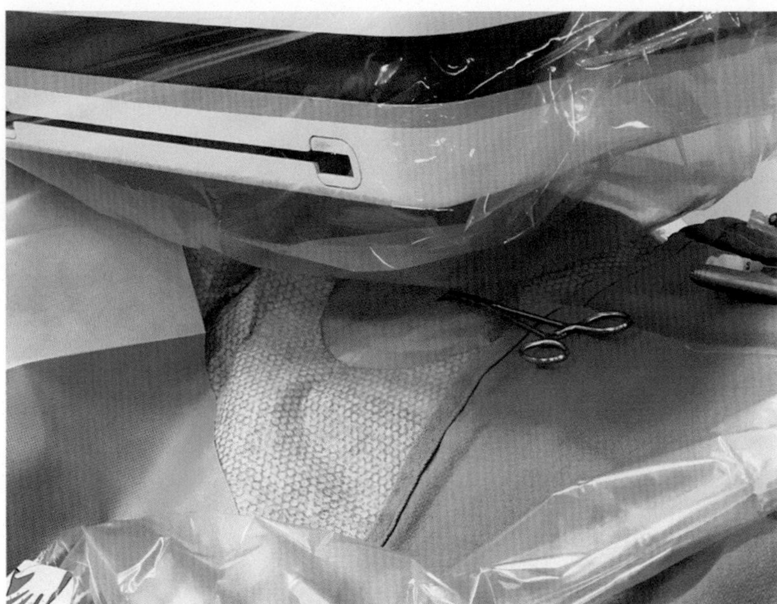

FIGURE 4.5 Fluoroscopy of the hemostat to determine its location relative to the femoral head.

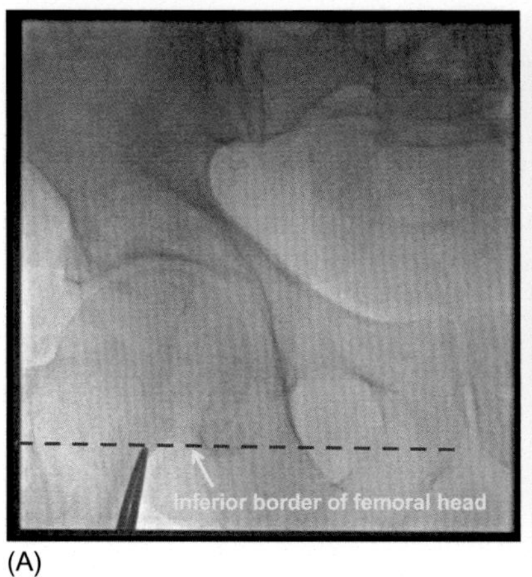

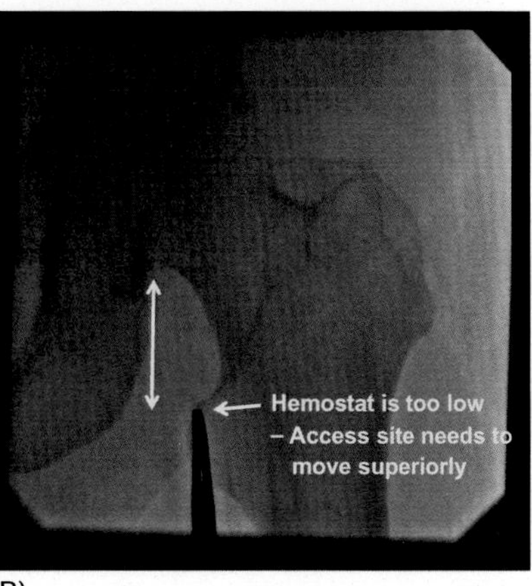

FIGURE 4.6 Localization of the inferior border of the femoral head by fluoroscopy using a hemostat (A). The hemostat position may need to be adjusted if it is too high or too low (B).

3. Using a sterile marker, a line is drawn on the skin at the inferior border of the femoral head (Fig. 4.7). Alternatively a micropuncture needle can be inserted as a marker of the inferior border of the femoral head.

4.2.3.3 Challenges

1. In very obese patients the femoral head may be located under significant abdominal fat. In such cases pushing the overlying tissue up from the femoral crease is preferable than obtaining access through multiple tissue layers.

4.2.3.4 What can go wrong?

1. Do not X-ray the operator's hands! All operator body parts should be removed from the fluoroscopy beam area.
2. Significant anatomical variations exist in the location of the common femoral artery bifurcation. Fluoroscopy alone may not be accurate in identifying the bifurcation of the common femoral artery into the profunda and superficial femoral artery, highlighting the importance of ultrasound (Section 4.2.4).

4.2.4 Step 4. Ultrasound guidance

4.2.4.1 Goal

To identify the optimal segment of the femoral artery for obtaining access. Specifically, access should be obtained above the common femoral artery bifurcation, in a nondiseased arterial segment.

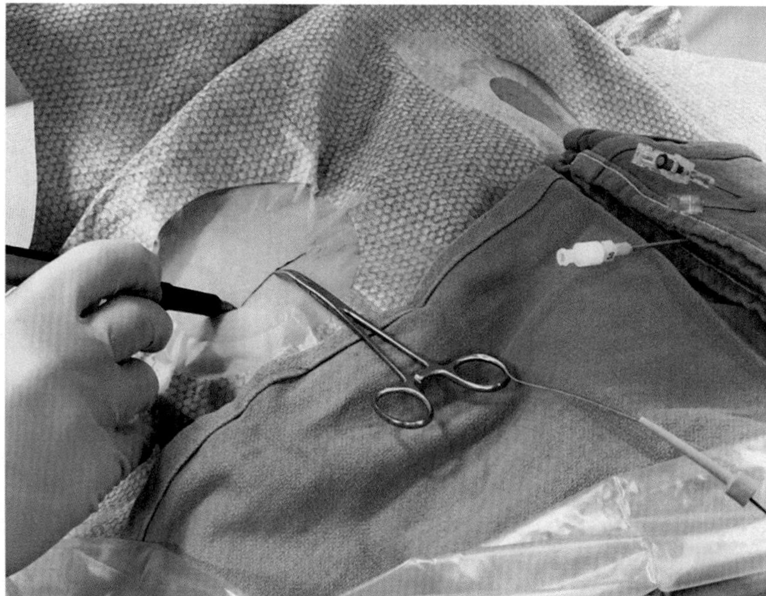

FIGURE 4.7 Marking the location of the inferior border of the femoral head after fluoroscopy is performed.

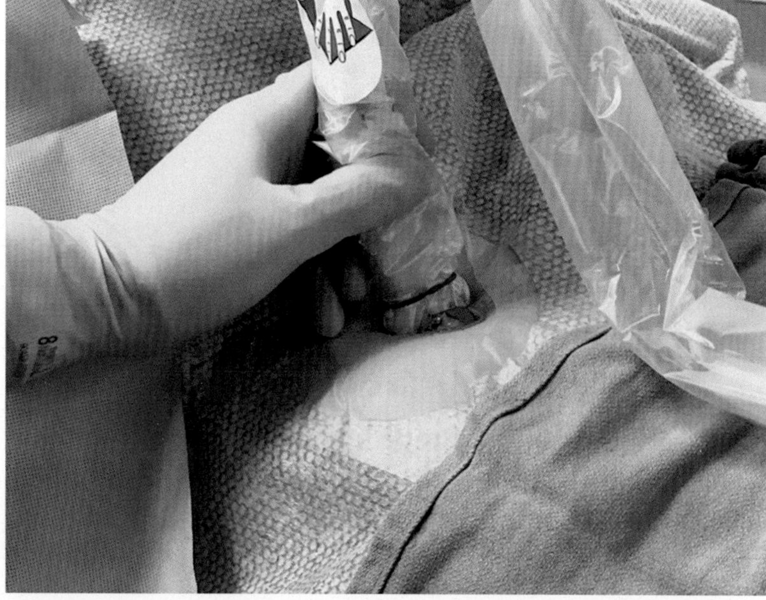

FIGURE 4.8 Ultrasound guidance for obtaining femoral access.

- Routine ultrasound use for obtaining vascular access (arterial and venous) is strongly recommended, as it improves the safety and efficiency of obtaining access, as well as the operator's expertise. It is particularly important for patients with weak pulses, peripheral arterial disease, or obesity, when attempts to obtain access using palpation fail, and when both arterial and venous femoral access is required [8].

4.2.4.2 How?

1. Gel is applied on the ultrasound probe tip (outside the sterile cover—gel should have also been used over the ultrasound probe before inserting it into the sterile cover).
2. The probe is positioned above the lower tip of femoral head line, drawn on step 2.
3. The common femoral artery is identified and scanned up and down (Fig. 4.8) to identify the location of the common femoral artery bifurcation (Fig. 4.9) [9] and the presence of femoral artery calcification or disease. Gain and depth are adjusted to optimize imaging.

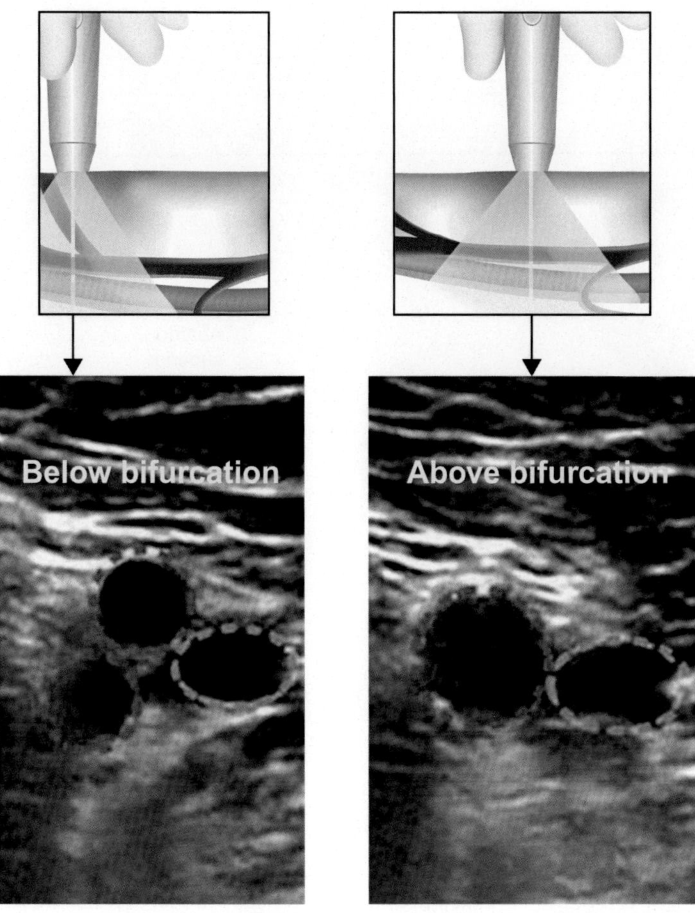

FIGURE 4.9 Illustration of the common femoral artery bifurcation. Imaging above the bifurcation (*right panel*) and below the bifurcation (*left panel*). *Adapted with permission from Sandoval Y, Burke MN, Lobo AS, et al. Contemporary arterial access in the cardiac catheterization laboratory. JACC Cardiovasc Interv 2017;10:2233–41 (Figure 1). Copyright Elsevier.*

- The femoral artery is located lateral to the femoral vein (Fig. 4.10) and, unlike the vein it cannot be collapsed upon compression.

4. The puncture site is selected based on the following criteria:
   - Located above the common femoral artery bifurcation.
   - Away from calcification (Fig. 4.11) [9], intraluminal disease/narrowing, bypass grafts, and previously placed closure devices, such as recently deployed Angioseals, etc.

4.2.4.3 Challenges

1. The femoral artery cannot be identified. Potential causes include obesity, prior surgery, suboptimal drape placement, limited operator experience, or common femoral artery disease.

   Solutions:
   - Use color Doppler to identify flow.
   - Reexamine the X-ray landmarks (Section 4.2.2).
   - Reposition the sterile drape (sometimes the drape is placed too high or too low).
   - Change to an alternative access site (such as contralateral femoral or radial).
   - Change imaging depth or gain of the ultrasound probe.
2. Calcification throughout the length of the common femoral artery. In such cases alternative access sites are preferable.

4.2.5 Step 5. Local anesthetic administration

4.2.5.1 Goal

To eliminate or attenuate the discomfort from obtaining arterial access.

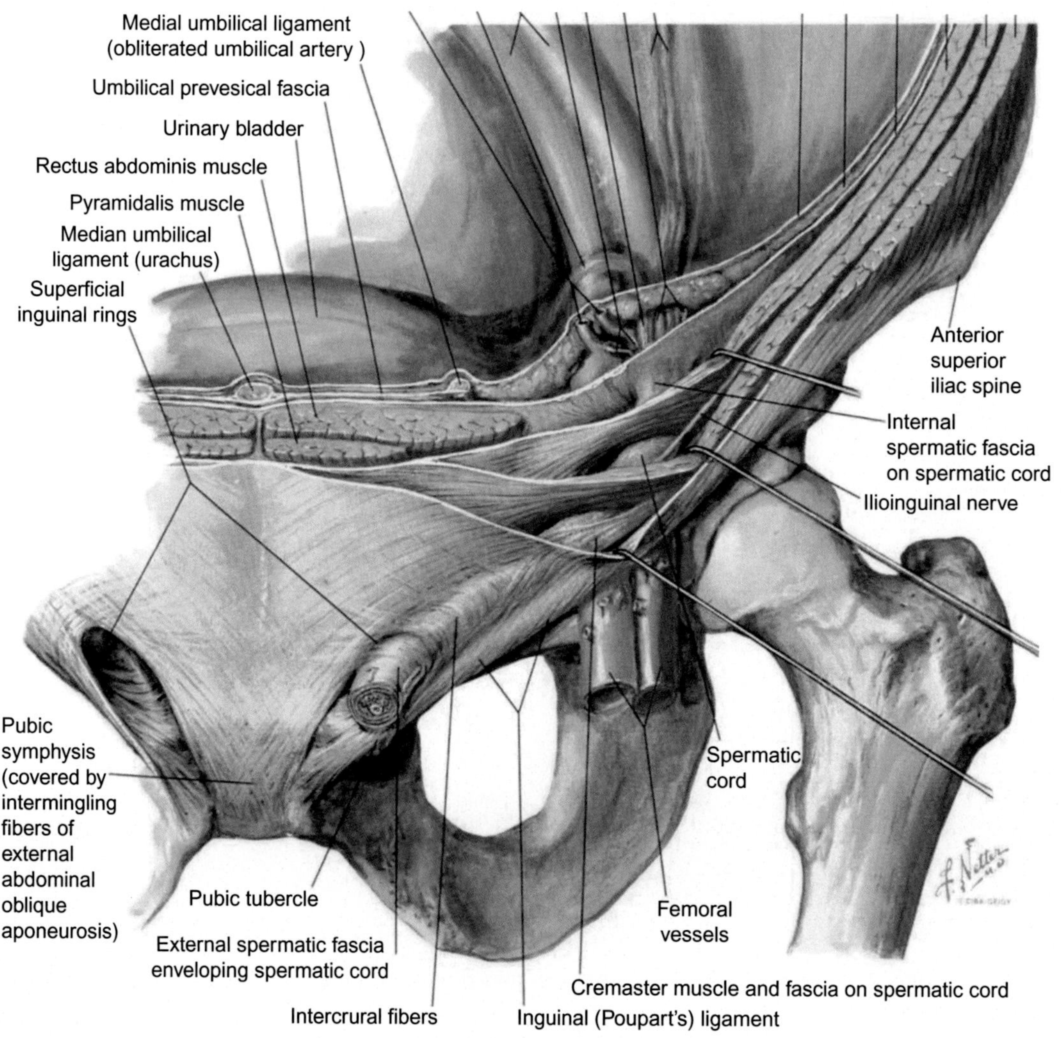

FIGURE 4.10 Anatomic location of the femoral artery and vein. *Reproduced with permission from Netter, Atlas of Anatomy. Copyright Elsevier.*

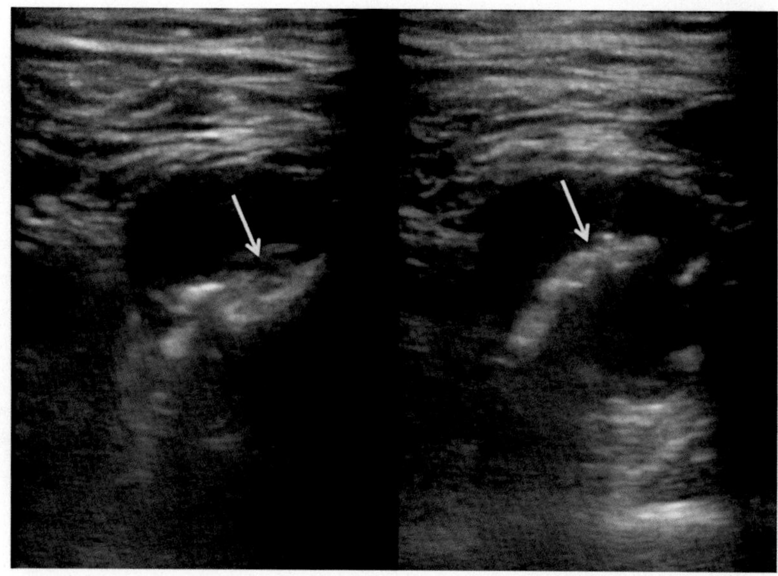

FIGURE 4.11 Severe calcification of the common femoral artery (*arrow*). *Reproduced with permission from Sandoval Y, Burke MN, Lobo AS, et al. Contemporary arterial access in the cardiac catheterization laboratory. JACC Cardiovasc Interv 2017;10:2233–41. Copyright Elsevier.*

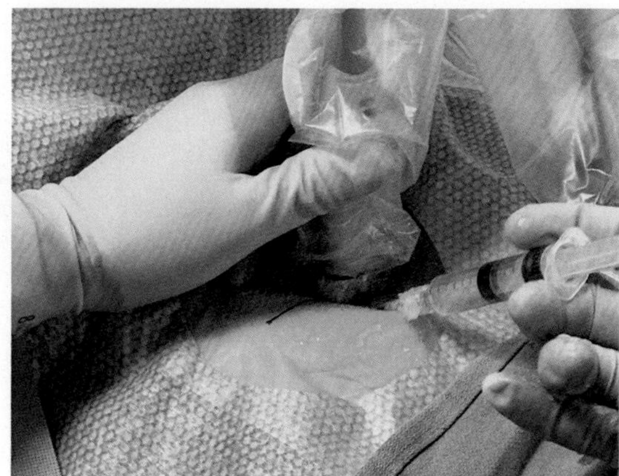

FIGURE 4.12 Local anesthetic administration for obtaining femoral access.

4.2.5.2 How?

1. A syringe (usually 10 cc) is loaded with local anesthetic.
2. The syringe needle (usually 26–31 gauge) is inserted in the skin.
3. Aspiration is performed to confirm that the tip of the needle is not within a vessel.
4. 10–20 cc of local anesthetic are administered in the subcutaneous tissue close to the target femoral artery. To administer >1 syringe of local anesthetic while minimizing the patient's discomfort, the needle can be left in place within the skin, while exchanging the empty syringe for a full one (Fig. 4.12).
5. Ultrasound can help visualize the location of the anesthetic administration, which should be targeted around the common femoral artery target puncture site.

4.2.5.3 Challenges

1. Unable to visualize injection of the local anesthetic in the subcutaneous tissue under ultrasound. In such cases adjustment of the needle angulation is needed to ensure the anesthetic is administered in the area surrounding the planned femoral artery puncture location.

4.2.5.4 What can go wrong?

1. Operator needle injury. It is best to remove the hand that is not holding the local anesthetic syringe from the groin area while administering the local anesthetic (which would be the left hand for right-handed operators). An additional advantage of the ultrasound is that the probe is on the skin over the artery instead of the operator's fingers.
2. Inadvertent puncture of the common femoral artery or vein with the local anesthetic needle. The needle is repositioned before further administration of local anesthetic. If a significant amount of blood enters the local anesthetic syringe, its contents are discarded, and a new syringe loaded with local anesthetic is used.

4.2.6 Step 6. Femoral artery puncture

4.2.6.1 Goal

Puncture the common femoral artery at the desired location and position the access needle tip inside the artery.

4.2.6.2 How?

1. The target femoral artery entry site and the access needle are visualized with ultrasound.
2. The needle entry site and angulation are adjusted to enable puncture at the desired location (Fig. 4.13) [9]. The ultrasound probe should be held perpendicular to the skin surface.
3. The needle is inserted with the bevel facing up.
4. The access needle is advanced under ultrasound visualization until it enters into the common femoral artery at the target entry location. The angulation of the needle is adjusted depending on the depth of the artery (Figs. 4.13 and 4.14).

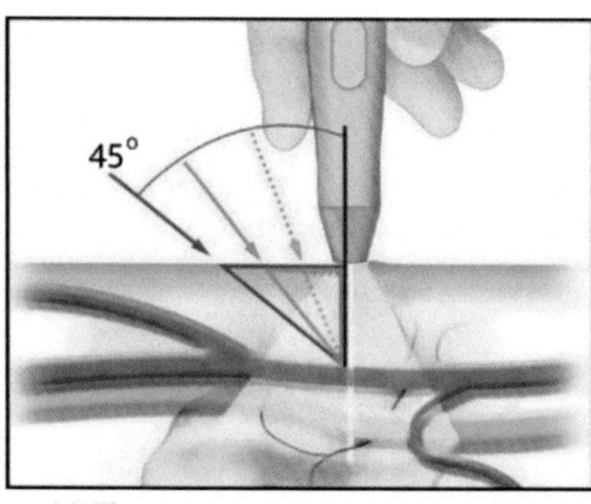

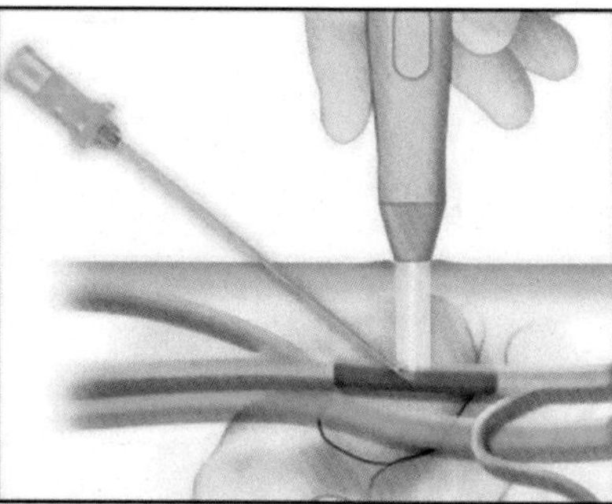

FIGURE 4.13 Illustration of the angulation required to puncture the femoral artery. As the angle of puncture becomes more shallow, the entry point should move further away from the ultrasound probe. *Reproduced with permission from Sandoval Y, Burke MN, Lobo AS, et al. Contemporary arterial access in the cardiac catheterization laboratory. JACC Cardiovasc Interv 2017;10:2233–41 (Figure 1). Copyright Elsevier.*

5. Entry into the artery is confirmed by blood flow through the back end of the access needle.
6. Blood flow is confirmed from the hub of the needle (Figs. 4.15 and 4.16).
7. If a micropuncture needle is used (21 gauge) arterial blood flow is less evident compared with an 18 gauge needle, although pulsatile flow should ideally be present.
8. If an 18 gauge needle is used, the next step is step 9 (insertion of a 0.035 in. guidewire).

4.2.6.3 Challenges

1. Unable to visualize the needle.

   Solutions: (1) change the needle entry point; (2) change the location and angulation of the ultrasound beam; (3) use needles with markings that have enhanced visualization under ultrasound.
2. No blood return.

   This could be due to suboptimal needle tip location (outside or along the wall of the target artery) or due to plugging of the needle.

   Solutions: (1) reposition the needle; (2) remove and flush the needle.
3. Inadvertent puncture of the femoral vein. In this case usually darker nonpulsatile blood flow is seen.

   Solution: The needle is partially withdrawn and redirected laterally.

4.2.6.4 What can go wrong?

1. High puncture increases the risk of retroperitoneal hematoma. It is best to remove the micropuncture needle and obtain access again.
2. Low puncture increases the risk of pseudoaneurysm. It may be preferable to remove the needle and obtain access again.

4.2.7 Step 7. Insertion of a 0.018 in. guidewire

4.2.7.1 Goal

Advance an 0.018 in. guidewire into the target femoral artery.

4.2.7.2 How?

1. The 0.018 in. guidewire (the wire included in the Cook micropuncture kit is 40 cm long) is advanced through the 21 gauge access needle (Figs. 4.17 and 4.18).
2. Fluoroscopy is used to track the wire course, as it might inadvertently enter small branches (e.g., lateral circumflex iliac, inferior epigastric, internal iliac artery) and lead to vessel perforation upon sheath insertion. As is true for all vascular access, the wire should not pushed if any resistance is felt.

4.2.7.3 Challenges

1. Resistance to wire advancement

   Causes:
   - Suboptimal needle position.
   - Inferior direction of wire upon advancement or entry into a small branch.
   - Disease or tortuosity in the femoral or iliac artery.
   - Low puncture (below common femoral artery bifurcation).

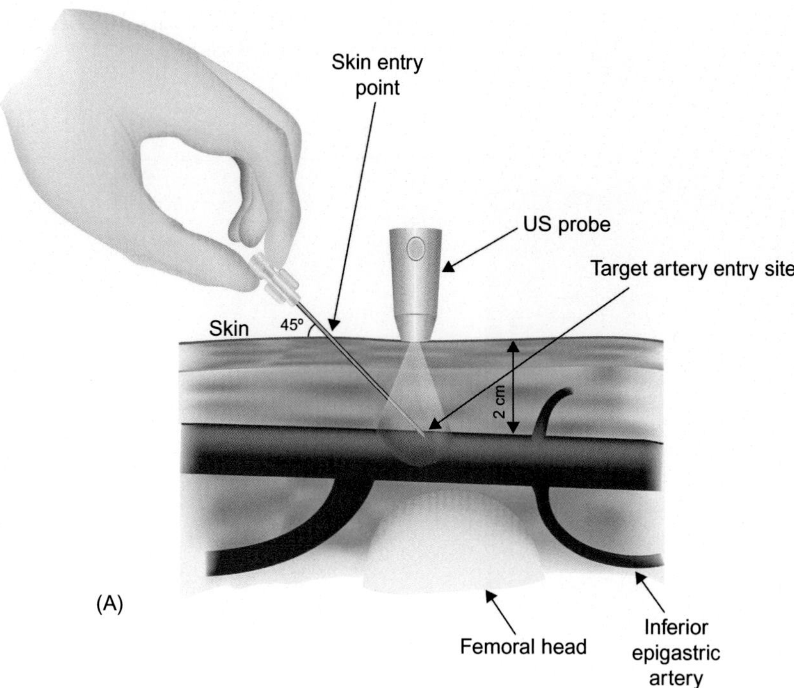

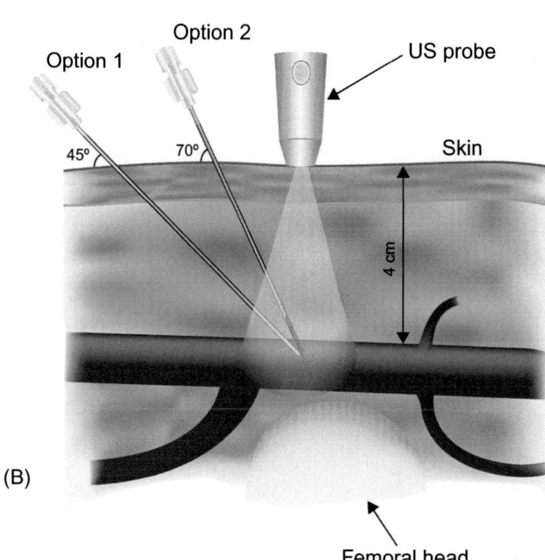

FIGURE 4.14 Adjustment of the needle entry point and the angle of advancement depending on the depth of the femoral artery. For patients in whom the femoral artery is located deeper, either the skin entry point is moved caudally (option 1, preferred) or the needle advancement angle is steeper (option 2).

Solutions:

- Fluoroscopy to assess wire location (see step 8).
- If there is resistance in advancing the wire past the needle tip, the wire is removed and the needle is repositioned until there is good backflow of blood—the guidewire is then reintroduced.
- If the wire is directed caudally, it is withdrawn and redirected under fluoroscopy.
- If the wire is advancing in the anticipated course of the femoral/iliac artery in the presence of tortuosity, advancement continues.

4.2.8 Step 8. Fluoroscopy of the 0.018 in. guidewire

4.2.8.1 Goal

To ensure that the wire is in the anticipated location within the femoral/iliac artery.

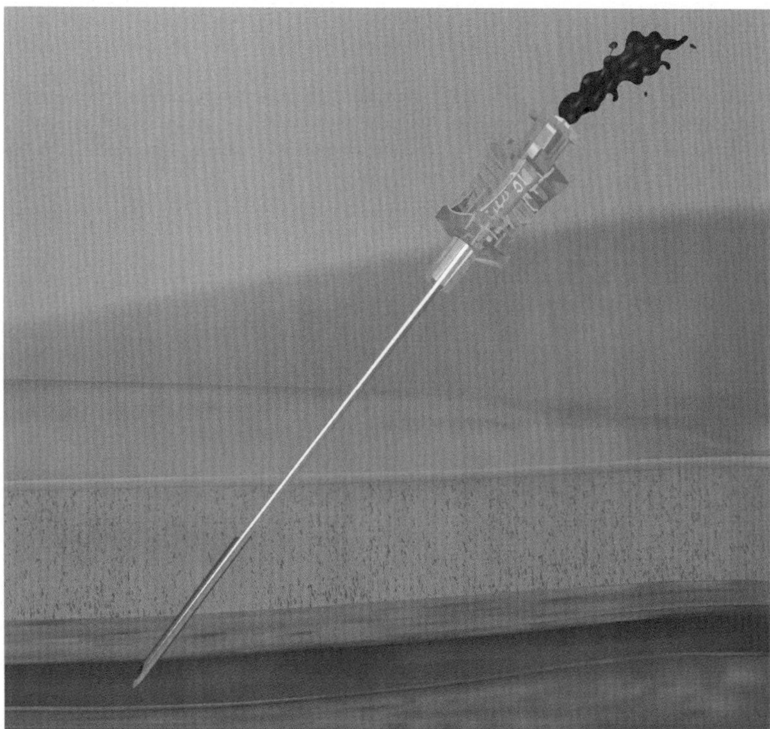

FIGURE 4.15 Pulsatile flow from the access needle upon entering the femoral artery.

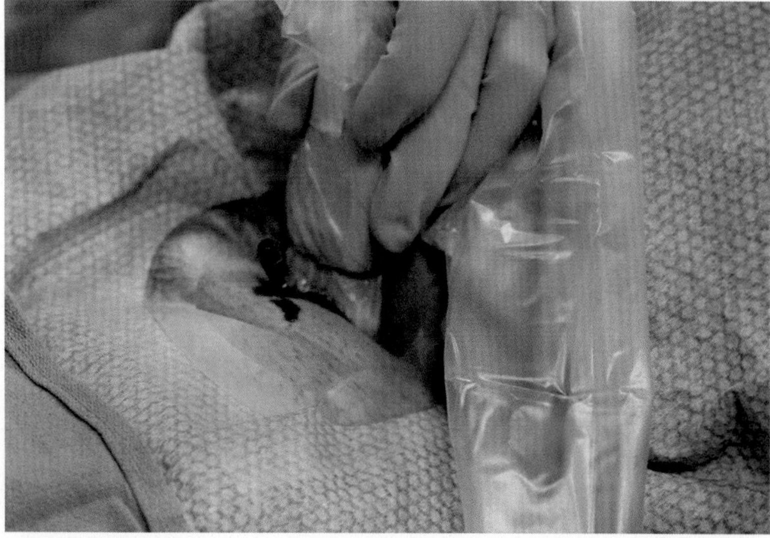

FIGURE 4.16 Pulsatile flow from the access needle upon entering the femoral artery.

4.2.8.2 How?

Fluoroscopy is performed over the groin/iliac artery region. The following are assessed (Fig. 4.19):

1. Guidewire entry site in to the femoral artery—is it within the lower half of the femoral head?
2. Guidewire position—is it following the anticipated course of the iliac/femoral artery?

   If the guidewire position is satisfactory, the needle is removed, holding manual pressure over the guidewire to prevent bleeding.

4.2.8.3 Challenges

1. Guidewire entry point into the femoral artery is too high or too low. If the guidewire position is not acceptable the guidewire and needle are removed, manual pressure is held for 1–2 minutes and puncture is repeated under ultrasound guidance.

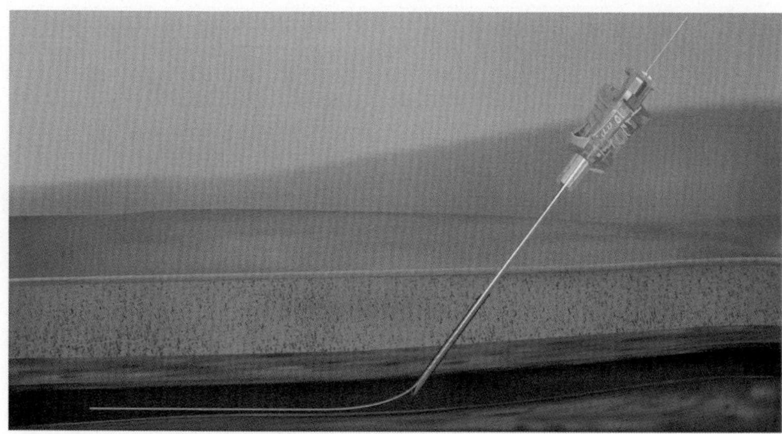

FIGURE 4.17 Guidewire insertion through the puncture needle.

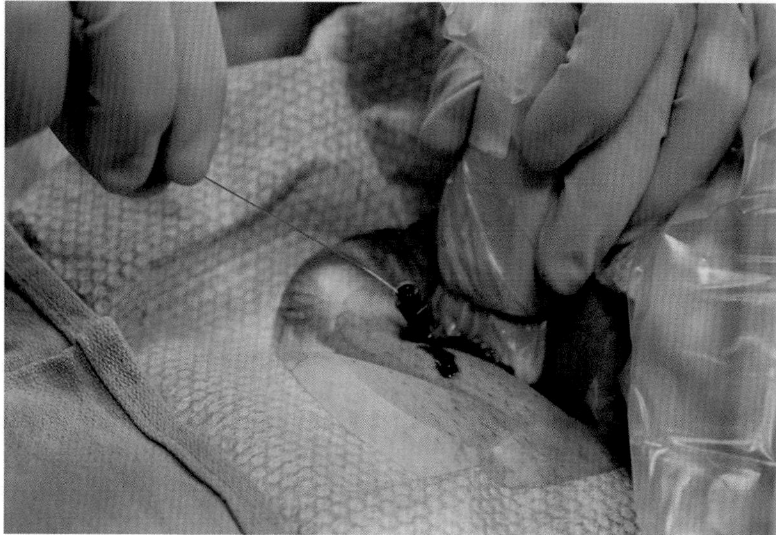

FIGURE 4.18 Guidewire insertion through the puncture needle.

2. Guidewire is coursing too medial (following the course of the inferior epigastric artery), or too lateral (following the lateral circumflex iliac artery, Fig. 4.20), or inferiorly (Fig. 4.21).

   Solution: Guidewire is repositioned under fluoroscopy.

4.2.9 Step 9. Skin nick (optional)

4.2.9.1 Goal

To facilitate advancement of equipment through the skin. Also this is key for delivering a closure device, especially the Perclose device with the associated knot pusher (Section 11.2).

4.2.9.2 How?

1. A scalpel is used to make a nick at the guidewire entry point through the skin. This is best done over the needle, to prevent guidewire damage.
2. Alternatively, a skin nick can be performed after insertion of the micropuncture dilator over the guidewire. Skin nick should, however, not be performed with a sheath in place, since the blade could cut the sheath causing significant bleeding.

4.2.9.3 What can go wrong?

1. Injury of the common femoral artery. The scalpel should not be advanced too deeply.

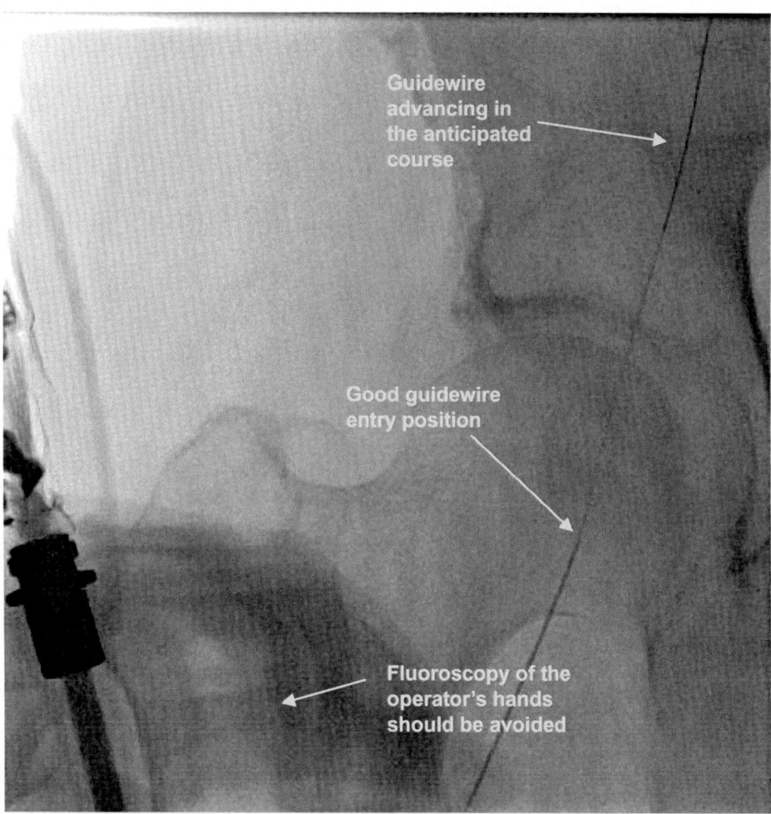

FIGURE 4.19 Fluoroscopy of micropuncture wire after advancement through the micropuncture needle. The entry point of the needle into the common femoral artery is good: just below the middle of the femoral head. Note: fluoroscopy of the operator's hand should be AVOIDED!

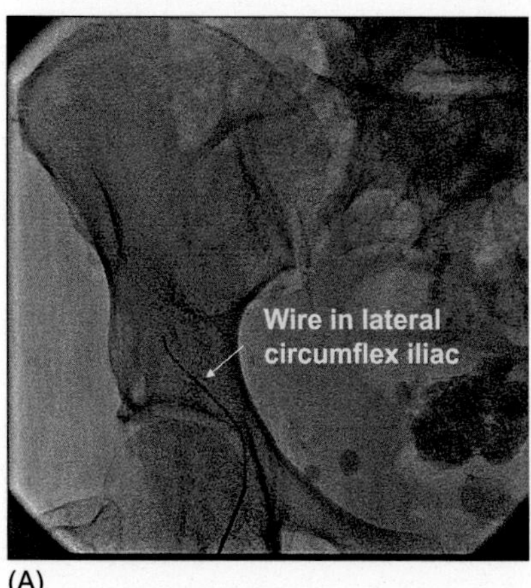

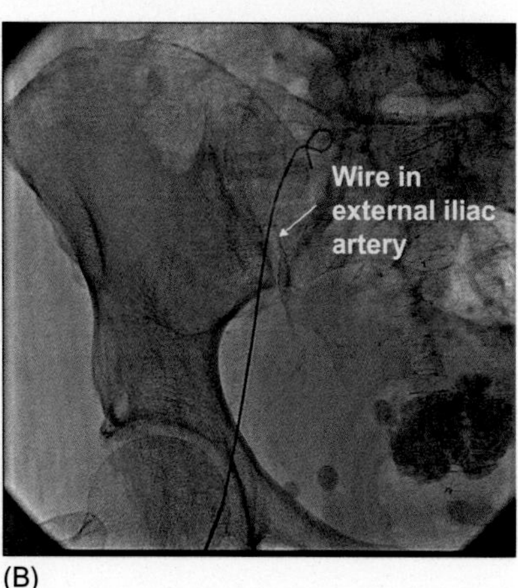

FIGURE 4.20 Guidewire advanced into lateral circumflex iliac artery (*arrow*, panel A). After redirection the guidewire is now located in the external iliac artery (panel B).

2. Subcutaneous bleeding due to excessive cutting. The nick should be of reasonable size to facilitate sheath insertion.
3. Cut of the operator's glove or hand. The nonworking hand should be kept away from the site of the nick.

4.2.10 Step 10. Insertion of the micropuncture-dilator assembly

4.2.10.1 Goal

Insert a dilator assembly into the common femoral artery through which a 0.035 in. guidewire will subsequently be inserted.

4.2.10.2 How?

1. The micropuncture dilator is advanced over the 0.018 in. guidewire (Fig. 4.22).

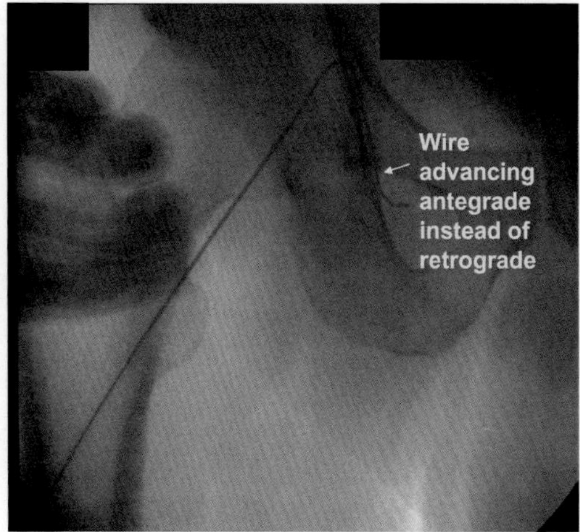

FIGURE 4.21 Guidewire advancing downward instead of upward toward the external iliac artery. The guidewire needs to be withdrawn and redirected. Every effort should be made to avoid performing fluoroscopy of the operator's hands.

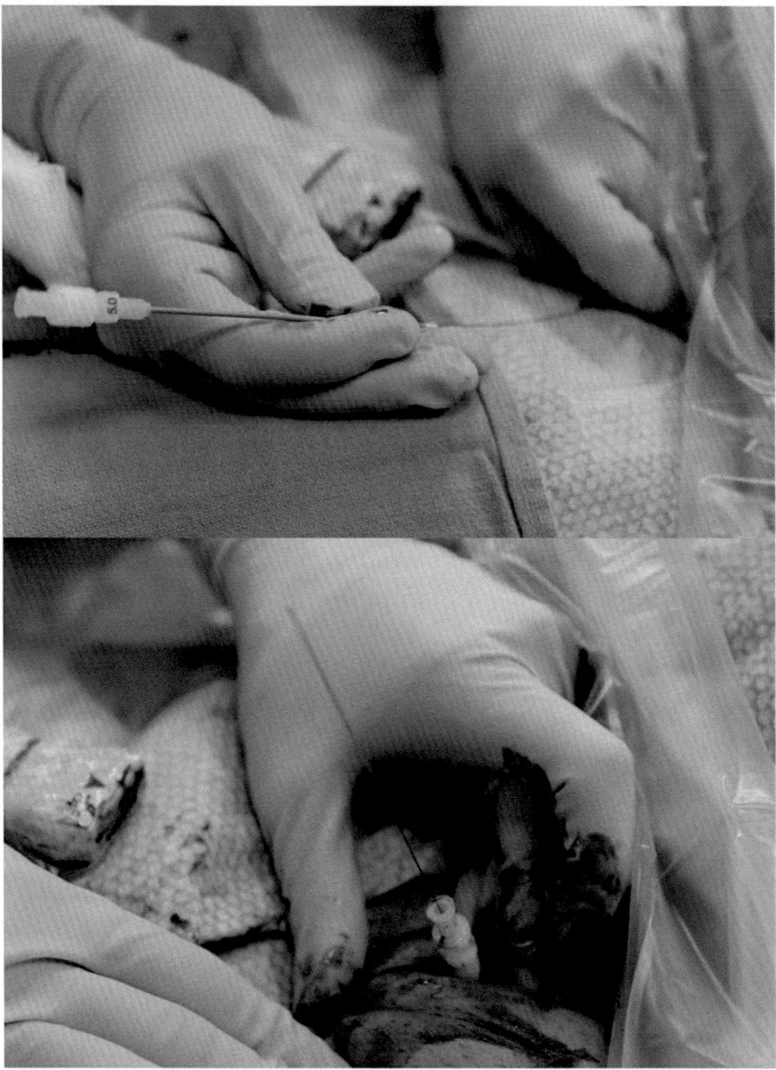

FIGURE 4.22 Inserting the dilator over the micropuncture wire.

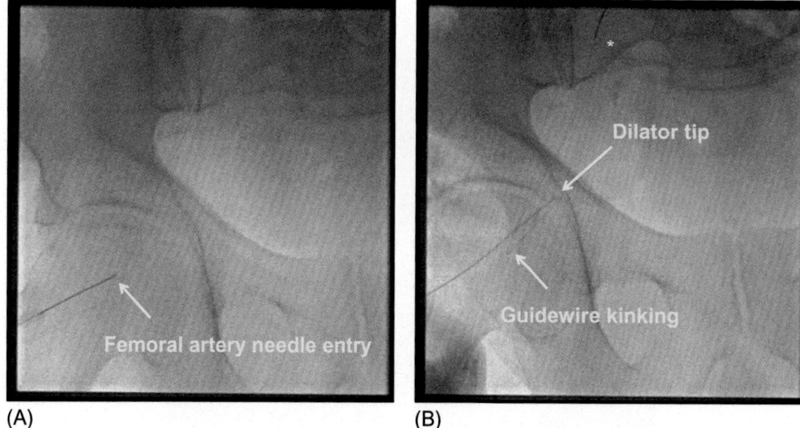

FIGURE 4.23 Example of guidewire kinking during attempts to advance the dilator of the micropuncture kit. After successful puncture of the right common femoral artery over the middle segment of the femoral head (*arrow*, panel A), attempts to advance the micropuncture kit dilator (*superior arrow*, panel B) failed due to guidewire kinking at the femoral artery entry point (*inferior arrow*, panel B). This resulted in partial retraction of the guidewire (*the star* in panel B marks the distal radiopaque portion of the guidewire that was not visible in panel A, in which the guidewire tip was advanced further up in the aorta). Attempts to advance another dilator failed. The guidewire was removed and manual pressure was held for 1 min. Repeat attempt for obtaining arterial access was successful and a 6 French sheath was inserted in the femoral artery.

4.2.10.3 Challenges

1. Unable to advance the dilator or kinking of the dilator in the subcutaneous tissue. Fluoroscopy can help confirm kinking if the diagnosis is not clear.

   Causes: Scar tissue due to multiple prior procedures, calcified plaque on the vessel wall, guidewire kinking which is more likely in obese patients when guidewire is advanced through thick tissue layers (Fig. 4.23).

   Solutions:

   - Perform a skin nick (if not previously performed), or enlarge the skin nick to minimize any resistance to dilator advancement at the skin level.
   - Use a stiffer micropuncture dilator (e.g., S-Mak kit, Merit Medical).
   - If guidewire kinking is the cause, partially withdraw the wire and readvance the dilator. If the dilator tip is flared a new one can be used. If there is enough guidewire length inside the artery the kinked guidewire segment can be cut and removed, followed by insertion of either the standard micropuncture dilator, or a sheath dilator distal segment ("Cutting the Gordian knot technique") [10].
   - Advance the inner dilator of the micropuncture/dilator assembly into the vessel and replace the kinked guidewire with a long 0.018 in. supportive guidewire, such as the Platinum Plus or V-18 (Boston Scientific).

4.2.11 Step 11. Advancement of a 0.035 in. guidewire

4.2.11.1 Goal

To create a rail that will allow insertion of a femoral sheath.

4.2.11.2 How?

1. The 0.018 in. guidewire and the inner portion of the dilator kit is removed, leaving the outer portion of the dilator within the artery.
2. Brisk pulsatile blood flow is confirmed from the back end of the dilator.
3. A 0.035 or 0.038 in. guidewire (Section 30.7.8) is advanced through the dilator to the abdominal aorta (Fig. 4.24). Ideally a long (180 cm wire) should be used, over which the sheath and the catheters will be advanced. Stiff wires (such as the Amplatz Super Stiff, Boston Scientific, Section 30.7.8) are preferred in patients with multiple prior access procedures that may have significant scar tissue.

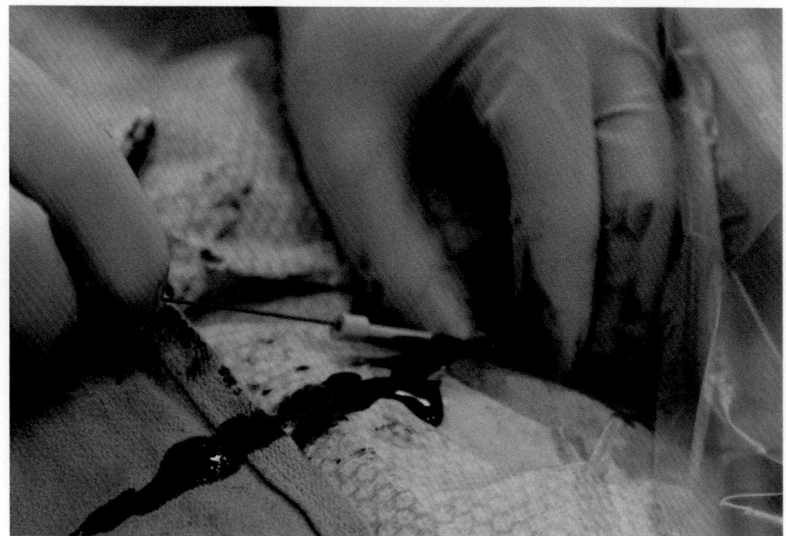

FIGURE 4.24 A 0.035 or 0.038 in. guidewire is advanced through the dilator into the abdominal aorta.

4.2.11.3 Challenges

1. Inability to advance guidewire
   Causes: Dilator kinking, iliac artery tortuosity, wire entry into a side branch, subintimal wire entry.
   Solution: Fluoroscopy of the guidewire (as in Section 4.2.8).

4.2.12 Step 12. Fluoroscopy of the 0.035 in. guidewire

4.2.12.1 Goal

To confirm optimal position of the 0.035 in. guidewire and to guide advancement in case of resistance.

4.2.12.2 How?

Fluoroscopy is performed in the groin area to visualize the guidewire and ensure that it is advancing in the anticipated course.

4.2.13 Step 13. Sheath insertion

4.2.13.1 Goal

To insert a sheath into the femoral artery, through which all catheters will be advanced.

4.2.13.2 How?

1. A sheath is selected for insertion. Short sheaths (10 cm long) are used in most cases, however longer sheaths (25 or 45 cm long) may be needed in cases of iliac tortuosity, when iliac artery stents are present, in obese patients in whom a short sheath may not reach the femoral artery, or if extra back up is needed in complex PCI cases.
2. The sheath is advanced over the 0.035 or 0.038 in. guidewire until the hub reaches the skin (Figs. 4.25 and 4.26).
3. The dilator is removed, leaving the 0.035 in. guidewire in place.

4.2.13.3 Challenges

1. Inability to advance sheath.

   Causes:

   - Poor preparation of the sheath track.
   - Scar from multiple prior procedures.

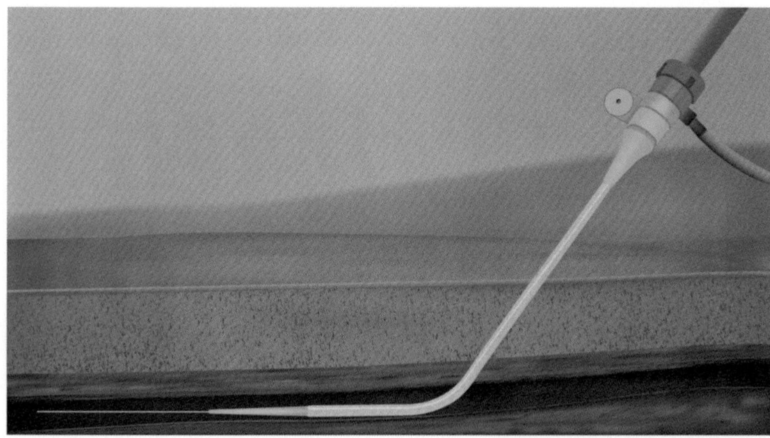

FIGURE 4.25 Insertion of a sheath over the 0.035 or 0.038 in. guidewire.

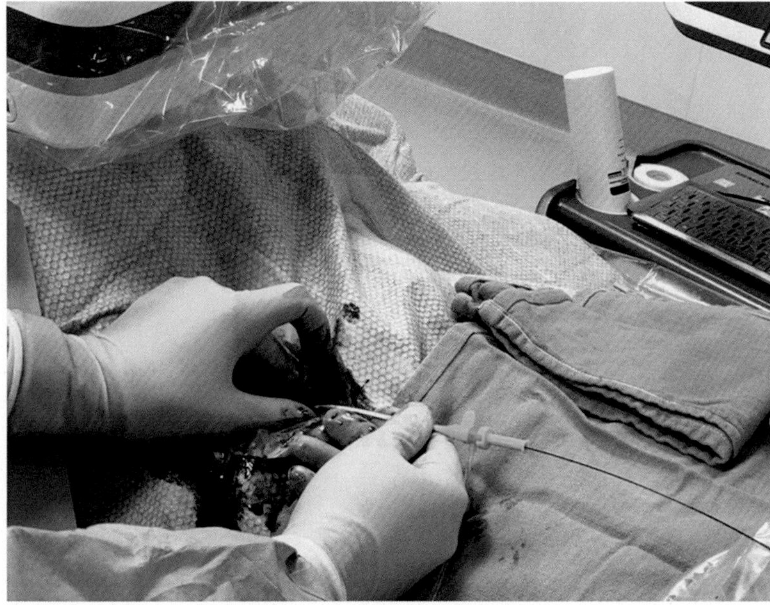

FIGURE 4.26 Insertion of a sheath over the 0.035 or 0.038 in. guidewire.

Solutions:

- Prepare the track better: nick with a scalpel and use the hemostat to dilate the subcutaneous tissue.
- Inspect the sheath: if the tip is damaged, replace with a new sheath.
- Use a smaller sheath.
- Change the existing 0.035 in. guidewire for a stiff 0.035 in. guidewire, such as the Amplatz Super Stiff (Boston Scientific), Supracore (Abbott Vascular), or Lunderquist (Cook Medical) (Section 30.7.8). This assumes that a dilator can be advanced over the existing guidewire and used to perform guidewire exchanges.
- First insert the sheath dilator by itself to create a tract, followed by insertion of the sheath/dilator assembly.

4.2.13.4 *What can go wrong?*

1. Kinking of the 0.035 or 0.038 in. guidewire during attempts to advance the sheath (similar to what can happen when advancing the micropuncture dilator over the micropuncture kit wire in Section 4.2.10). A solution is to remove the sheath, exchange for a stiffer 0.035 or 0.038 in. guidewire (usually over a dilator), followed by repeat attempts to insert the sheath.

4.2.14 Step 14. Sheath aspiration and flushing

4.2.14.1 Goal

To remove any thrombus or plaque from the sheath.

4.2.14.2 How?

1. The side arm of the sheath is aspirated (3–4 mL) and the aspirated blood is discarded (Fig. 4.27).
2. The side arm is flushed with a different syringe that contains heparinized saline.

4.2.15 Step 15. Femoral angiography

4.2.15.1 Goal

To rule out any complications, assess the risk of complications, and determine the feasibility of using a closure device.

4.2.15.2 How?

1. The side arm of the femoral sheath is connected with the manifold.
2. Arterial pressure waveform is confirmed without dampening.
3. If not already left in place a 0.035 or 0.038 in. guidewire is inserted through the sheath into the aorta.

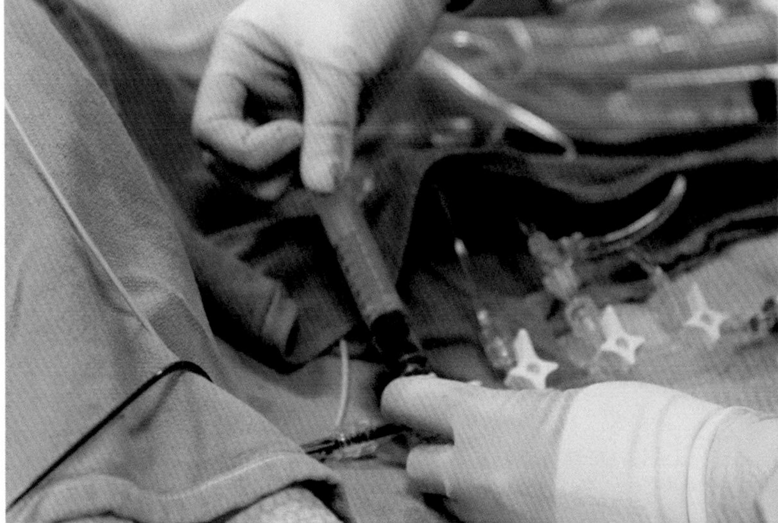

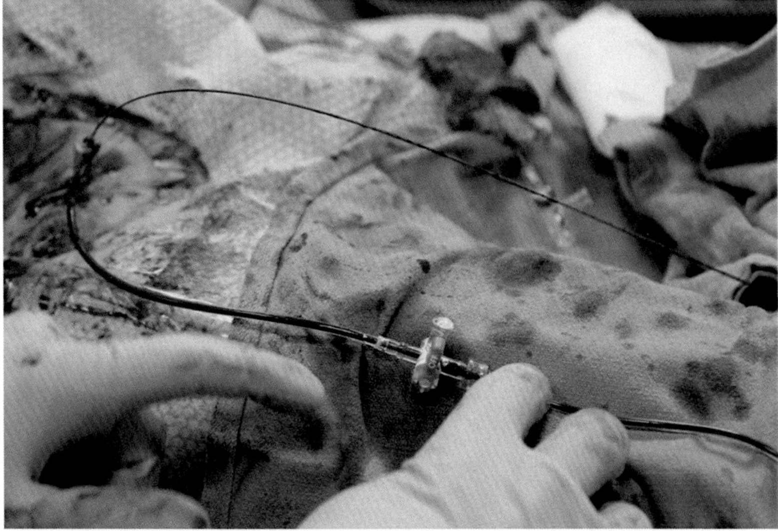

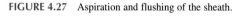

FIGURE 4.27 Aspiration and flushing of the sheath.

4. Fluoroscopy of the groin area is performed and the table is moved so that the femoral sheath is in the center of the image. Low magnification is usually used to allow visualization of the iliac vessels and the common femoral artery bifurcation.
5. The most commonly used projection is 20–30° RAO for the right common femoral artery or 20–30° LAO for the left common femoral artery, as this facilitates visualization of the common femoral artery bifurcation.
6. Cine-angiography (or fluoroscopy using the fluoro-store function) is performed, usually injecting 5–10 mL of contrast. Diluted contrast can be used as well. The sheath is often pulled medially during image acquisition, to maximize the separation of its entry point into the femoral artery.
7. Femoral angiography interpretation—the following parameters are assessed:
   - Perforation.
   - Dissection or intraluminal filling defects (Fig. 4.28).
   - Antegrade flow to the lower extremity.
   - Disease, calcification, and tortuosity of the femoral and iliac vessels.
   - Location of the arterial puncture and size and quality of the femoral artery at the puncture site (to determine if use of a closure device is feasible) (Fig. 4.29).
8. Some operators perform contrast injection through the micropuncture needle before inserting the sheath. The advantage of this technique is that sheath insertion may be avoided if the needle insertion point is suboptimal. The disadvantage is that contrast injection may cause dissection if the needle position is suboptimal.

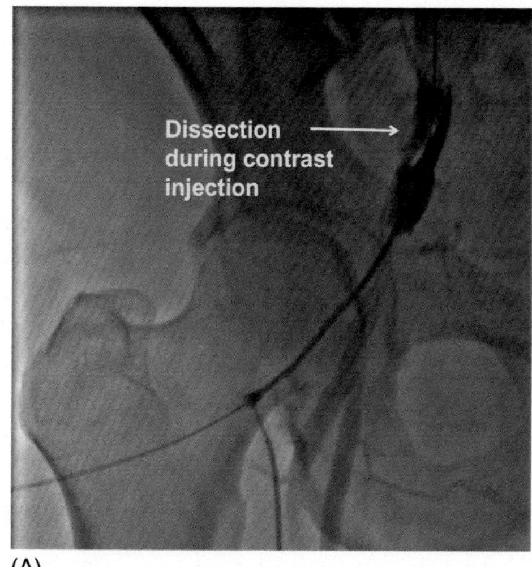

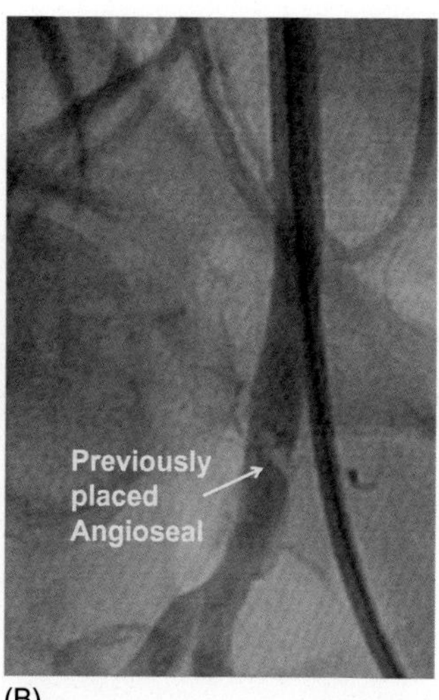

FIGURE 4.28 Aspiration and flushing of the sheath. Iliac dissection during contrast injection (panel A) and filling defect due to a previously placed Angioseal device (panel B).

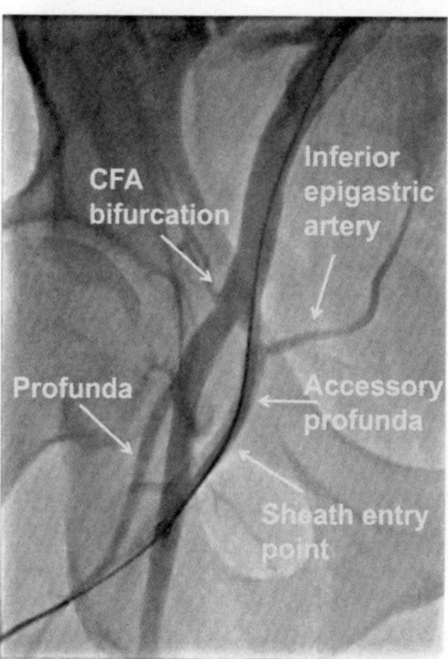

FIGURE 4.29 Sheath insertion into an accessory profunda femoral artery in a patient with high common femoral artery (*CFA*) bifurcation. Given the small size of the accessory profunda artery, a vascular closure device was not used, but manual compression was held instead.

4.2.15.3 What can go wrong?

1. ***Perfortion and bleeding*** **(see also Section 29.1.2).**

   Causes:
   - Guidewire or catheter advancement through small branches.
   - Injury of the arterial wall during sheath insertion.
   - Arteriovenous fistula due to insertion of the venous sheath through the femoral artery.

   Prevention:
   - Advance guidewires (especially micropuncture wires) under fluoroscopy to ensure that they do not enter into a small branch.
   - Never advance wires and sheaths/catheters against resistance.

   Treatment:
   - Bleeding through inferior epigastric/lateral circumflex iliac: embolization with coils or thrombin.
   - Bleeding around sheath: upsize to larger sheath.
   - Covered stent placement.
   - Surgery.
   - In nonemergent cases do not administer anticoagulation and do not perform PCI to minimize risk of recurrent bleeding.

2. ***Femoral or iliac artery dissection*** **(see Section 29.1.1).**

   Causes:
   - Aggressive guidewire advancement against resistance.
   - Iliofemoral atherosclerosis.
   - Contrast injection with the catheter tip against the arterial wall.

   Prevention:
   - Ensure good arterial waveform from the sheath's side arm before contrast injection.
   - Avoid forceful contrast injections.
   - Insert a 0.035 in. guidewire prior to injection, which can help stabilize the sheath, keep the sheath tip away from the arterial wall and also maintain access to the true lumen in case dissection occurs.

   Treatment:
   - Non-flow-limiting dissections: usually no specific treatment is required, as the dissection heals itself.
   - Flow-limiting dissections. There are different treatment options depending on the location of the dissection:
     - Common femoral artery dissection: femoral endarterectomy is the preferred option to avoid stent placement in a location that is prone to stent fracture.
     - Iliac artery dissection: this could be treated by obtaining contralateral femoral arterial access crossing into the true lumen and deploying a self-expanding stent.

3. ***Embolization of air, plaque, thrombus or other foreign material*** **(see Section 29.1.1).**

   Causes:
   - Severe disease in the femoral or iliac arteries.
   - Thrombus formation within the sheath.
   - Poor preparation of the injection syringe or inability to aspirate from the sheath prior to injection.

   Prevention:
   - Ultrasound imaging prior to obtaining access can help avoid accessing severely diseased arterial segments [8].
   - Meticulous aspiration of the sheath before injecting saline.
   - Use separate syringes for aspirating the sheath and flushing the sheath (hence if debris is aspirated it is not reinjected into the femoral artery).

   Treatment:
   - Air embolization: consider 100% oxygen administration.
   - Thrombus or plaque embolization: endovascular or surgical retrieval if lower extremity ischemia occurs.

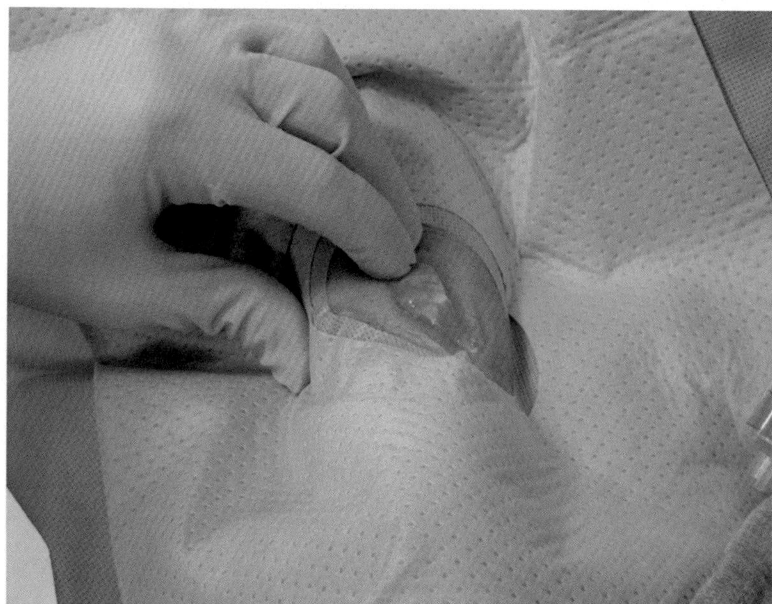

FIGURE 4.30 Palpation of the radial pulse.

4.3 Radial access

4.3.1 Step 1. Palpation of the radial pulse

4.3.1.1 Goal

To confirm the presence of a radial pulse.

4.3.1.2 How?

1. Palpation of the radial artery proximal to the styloid process (Fig. 4.30).
2. In the past the Allen's or Barbeau's test was performed (compression of the radial artery with pulse oximeter placed on the patient's thumb to confirm the presence of arterial waveform, suggesting patent ipsilateral ulnar artery and palmar arch). The RADAR (Predictive Value of Allen's Test Result in Elective Patients Undergoing Coronary Catheterization Through Radial Approach) trial showed feasibility and safety of radial access for the entire spectrum of Allen's test results [11], hence the Allen's test is no longer recommended prior to obtaining radial access.

4.3.2 Step 2. Sterile preparation and draping of the wrist

4.3.2.1 Goals

1. To optimally position the patient's arm to facilitate obtaining access.
2. To clean the access site and reduce the risk of access site infection.

4.3.2.2 How?

1. Right radial: there are dedicated devices for positioning the right arm. Alternatively the arm can be placed alongside the side of the patient.
2. Left radial: the advantage of left radial is less subclavian tortuosity and slightly shorter fluoroscopy times [12], however it can be more challenging and less comfortable for the operator [13]. Left radial should be used in patients with patent LIMA grafts.
   - For left radial access, the left elbow is supported by either a special device, such as the Left Arm Support System (LP Medical, Fig. 4.31), or Cobra board (TZ Medical, Portland, Oregon) or several blankets/pillows and brought over to the patient's abdomen in supination position.
   - Alternatively, finger traps can be used to support the arm, which is partially supinated to allow access to the proximal radial artery. Alternatively a long (16–25 cm) radial sheath is partially inserted in the radial artery obviating the need for supination of the patient's hand (Fig. 4.32).

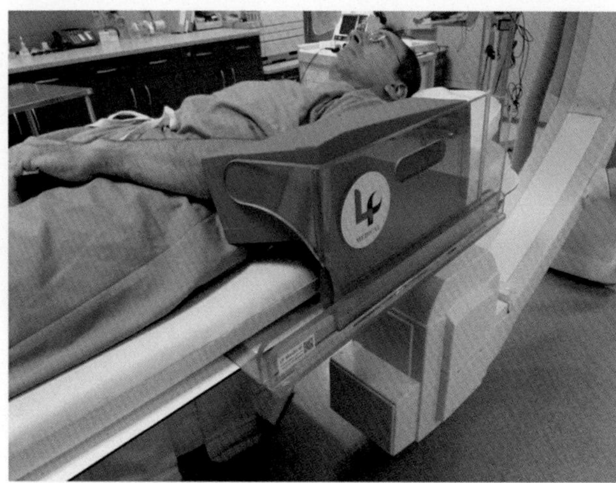

FIGURE 4.31 The left arm support system (LP Medical). *Reproduced with permission from LP Medical LLC.*

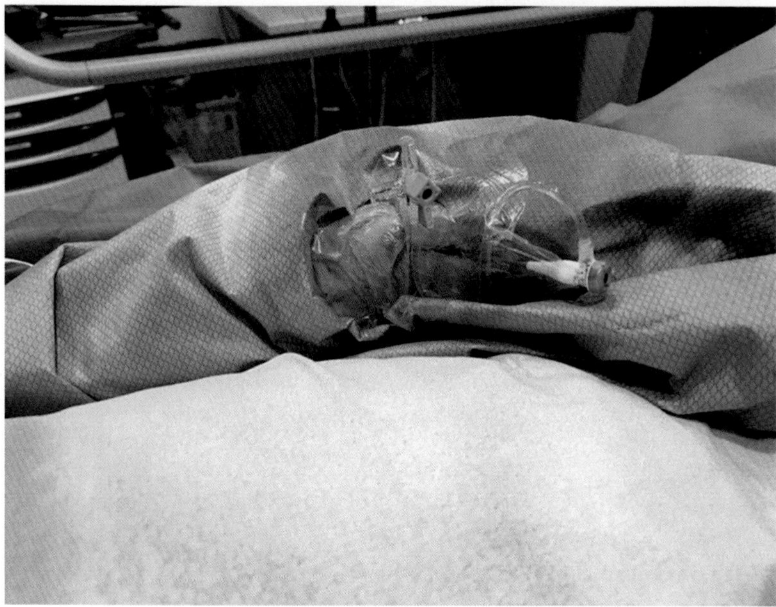

FIGURE 4.32 Use of a long sheath for left radial access. The sheath is partially protruding outside the skin, facilitating catheter insertion and exchanges. *Courtesy of Dr. Stephane Rinfret.*

- Another option is using the Stand Tall radial sheath extension (Radux Medical). Access is obtained from the left or right side of the patient, but the remaining of the procedure is performed with the operator returning to the right side of the patient.
- Left radial access can be obtained in the proximal or distal radial artery. Advantages of distal radial include lower risk for compartment syndrome and radial artery occlusion, whereas disadvantages include smaller artery size limiting the size of sheath that can be inserted [14] and longer time to access (Table 4.1) [14–20].
  Proximal left radial: access is obtained usually on the left side of the patient, as per standard practice.
  Distal left radial: to bring the artery to the surface of the fossa, the patient is asked to grasp his thumb under the other four fingers, with the hand slightly abducted [20].

3. The wrist area is shaved.
4. The wrist is scrubbed with antiseptic solution (both the radial and the ulnar site should be prepared).
5. After the antiseptic solution dries up (to allow the adhesive tape to stick), the sterile drape is placed, centering its opening to the wrist area.

4.3.2.3 *What can go wrong?*

1. The wrist area can be contaminated (e.g., by the patient's hands). In this case, scrubbing with antiseptic solution should be repeated.

TABLE 4.1 Comparison of distal versus proximal radial access for cardiac catheterization [14–17].

		Distal radial	Proximal radial
Success	Obtaining access	No difference	
	Crossover to femoral	No difference	
Efficiency	Time to obtain access		Better
	Difficulty obtaining access		Better
	Able to insert larger sheath [16]		Better
	Time to hemostasis	?	
	Ease of coronary engagement		Better?
Comfort	Operator comfort—right radial	No difference	
	Operator comfort—left radial		Better
	Patient comfort—right radial	No difference	
	Patient comfort—left radial	Better	
Complications	Compartment syndrome	Better	
	Hand ischemia	Better	
	Bleeding	No difference	
	Radial occlusion	Better?	

4.3.3 Step 3. Ultrasound guidance [21]

4.3.3.1 Goal

To locate the radial artery and facilitate obtaining access. Ultrasound guidance has been shown to reduce the number of attempts required to access the radial artery, which can reduce the likelihood of radial spasm [21]. Radial access should be obtained in the segment of the artery above the styloid process.

4.3.3.2 How?

1. Gel is applied on the ultrasound probe tip and the probe is placed into a sterile plastic bag.
2. The probe is positioned above the styloid process of the wrist (Fig. 4.33) or the snuffbox. Gel or normal saline is squirted onto the wrist or snuffbox to optimize imaging.
3. The radial artery is identified (low depth facilitates visualization of the radial artery, which is much smaller and much more superficial than the femoral artery, Fig. 4.34) [9].

4.3.3.3 Challenges

1. The radial artery cannot be identified. Potential causes include obesity, prior harvest of the radial artery, limited operator experience, or radial artery occlusion.
   Solutions:
   - Use of color Doppler to identify flow.
   - Change to an alternative access site (such as ipsilateral ulnar artery, contralateral radial/ulnar artery, or femoral artery).
2. Small radial artery size. This can hinder obtaining access and may also result in spasm, making catheter advancement and manipulations challenging.
   Solution:
   - Change to an alternative access site (such as ipsilateral ulnar artery, contralateral radial/ulnar artery, or femoral artery).

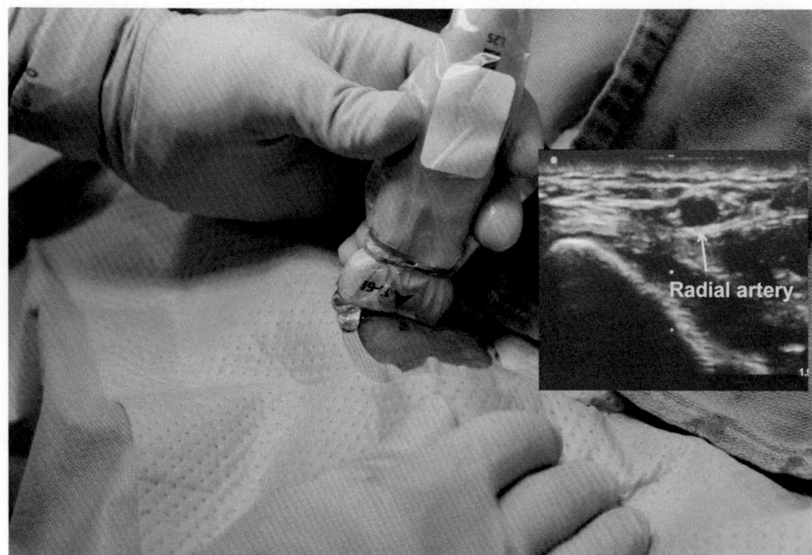

FIGURE 4.33 Ultrasound of the wrist to identify the radial artery.

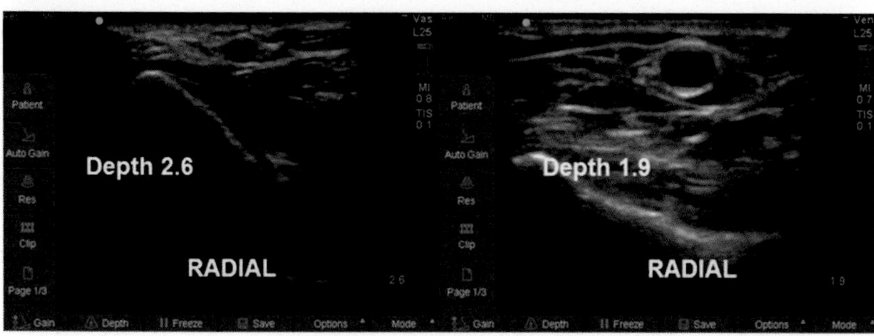

FIGURE 4.34 Identification of the radial artery by ultrasound. Less imaging depth, provides more magnified view and easier identification of the radial artery. *Reproduced with permission from Sandoval Y, Burke MN, Lobo AS, et al. Contemporary arterial access in the cardiac catheterization laboratory. JACC Cardiovasc Interv 2017;10:2233–41. Copyright Elsevier.*

4.3.4 Step 4. Local anesthetic administration

4.3.4.1 Goal

To eliminate or attenuate the discomfort associated with obtaining arterial access.

4.3.4.2 How?

1. A syringe (usually 3 cc) is loaded with local anesthetic.
2. The syringe needle (usually 31 gauge) is inserted in the skin.
3. Aspiration is performed to confirm that the tip of the needle is not within a vessel, followed by injection of approximately 1 cc of local anesthetic (Fig. 4.35). Larger amounts of local anesthetic should be avoided, as they can compress the radial artery and hinder obtaining access.
4. Ultrasound can help visualize the anesthetic as it is being administered and its impact on the radial artery caliber.
5. If radial spasm develops, nitroglycerin can be administered, either sublingually or injected around the radial artery (mixed with lidocaine 1% in a 1:1 ratio) under ultrasound guidance.

4.3.4.3 What can go wrong?

1. Operator needle injury. It is best to remove the hand that is not holding the local anesthetic syringe from the wrist area while administering the local anesthetic.
2. Administration of large volume of local anesthetic that can compress the radial artery and hinder access.
   Solution: Wait for 1–2 minutes or obtain access at a different location of the radial artery.
3. Intraarterial administration of lidocaine should be avoided as it can cause seizures.

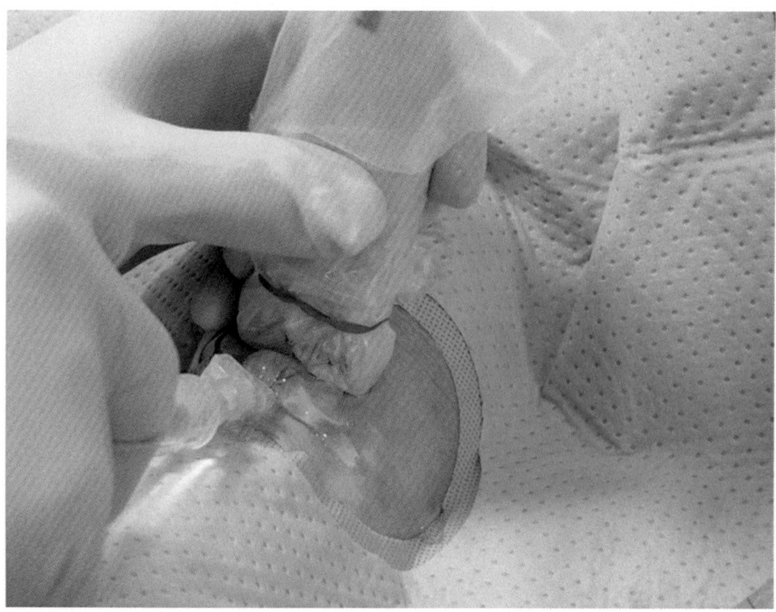

FIGURE 4.35 Local anesthetic administration for radial access.

4.3.5 Step 5. Radial artery puncture

4.3.5.1 Goal

To puncture the radial artery at the desired location and position the access needle tip inside the artery.

4.3.5.2 How?

1. The radial artery entry site and the access needle are visualized with ultrasound.
2. Radial access can be obtained using either a standard micropuncture needle or using an Angiocath. The radial sheath package also contains the needle or Angiocath and a 0.018 in. guidewire.
3. Micropuncture needle: The needle is inserted with the bevel facing up until it enters the radial artery (Figs. 4.36 and 4.37). Puncture can be achieved using two techniques:
   Single wall stick: The needle is inserted into the artery, followed by back bleeding.
   Through and through stick: The needle is advanced through both the front and the back wall of the radial artery, which results in only transient back bleeding.
4. Angiocath: The needle/catheter assembly is inserted with the bevel facing up until there is blood backflow. It is then advanced distally and through the back wall of the artery and the needle removed from the catheter. The catheter is slowly withdrawn. When pulsatile flow appears through the catheter, the 0.018 in. guidewire is inserted into the radial artery.
5. For distal radial access, the artery is punctured under an angle of 30–45 degrees and from lateral to medial. The needle is directed to the point of strongest pulse, proximal in the anatomical snuffbox [20].

4.3.5.3 Challenges

1. Unable to visualize the needle
   Solutions: (1) change the needle angulation or entry point; (2) change the location and angulation of the ultrasound beam; (3) use needles with markings that have enhanced visualization under ultrasound.
2. Failure to puncture the radial artery
   Causes:
   - Small radial artery size.
   - Radial artery spasm.
   - Proximal radial artery occlusion—a pulse may be present due to collateral flow from the palmar arch.
3. Radial artery spasm

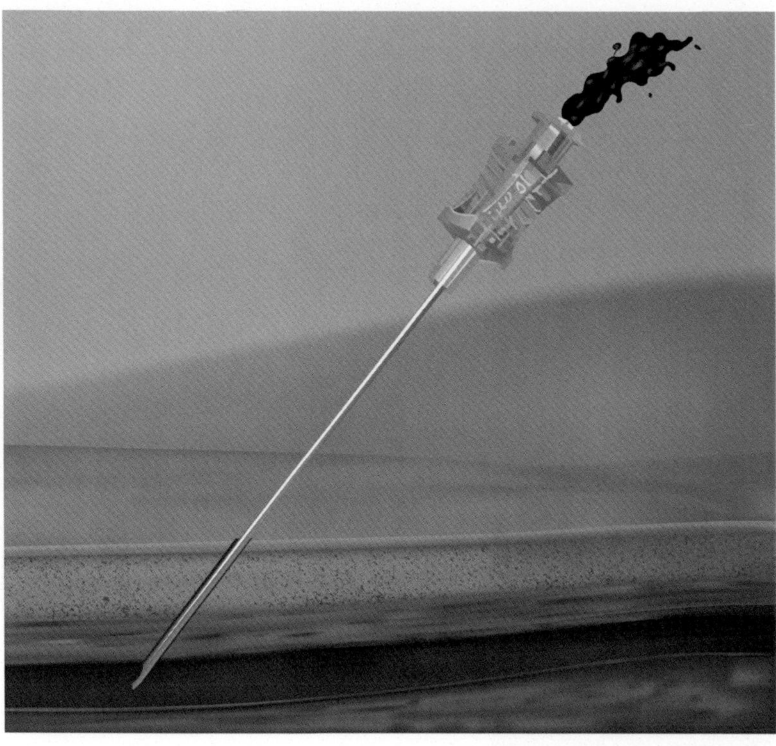

FIGURE 4.36 Radial artery puncture.

- Risk factors:
  - Small radial artery size.
  - Younger patients.
  - Multiple needle passes.
  - Use of intravenous vasopressors or cardiac arrest.
- Prevention
  - Do not use very small radial arteries—potentially use ulnar.
  - Ultrasound guidance for avoiding multiple sticks.
  - Adequate moderate sedation.
  - Sublingual or subcutaneous nitroglycerin administration.
- How to treat
  - Vasodilators (sublingual nitroglycerin, enhanced sedation).
  - Switch to ipsilateral ulnar or to contralateral radial or to femoral access.

4. No blood return

   This is likely due to suboptimal needle tip location (outside or within the wall of the target artery) or due to plugging of the needle.

   Solutions: (1) reposition the needle; (2) remove and flush the needle.

4.3.6 Step 6. Insertion of a 0.018 in. guidewire

4.3.6.1 Goal

To advance a 0.018 in. guidewire into the target radial artery.

4.3.6.2 How?

1. Single wall technique: the 0.018 in. guidewire is advanced through the access needle (Figs. 4.38 and 4.39).

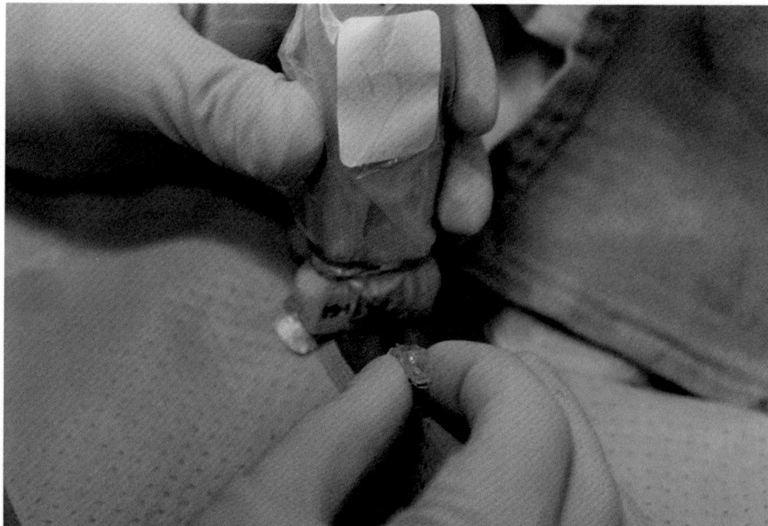

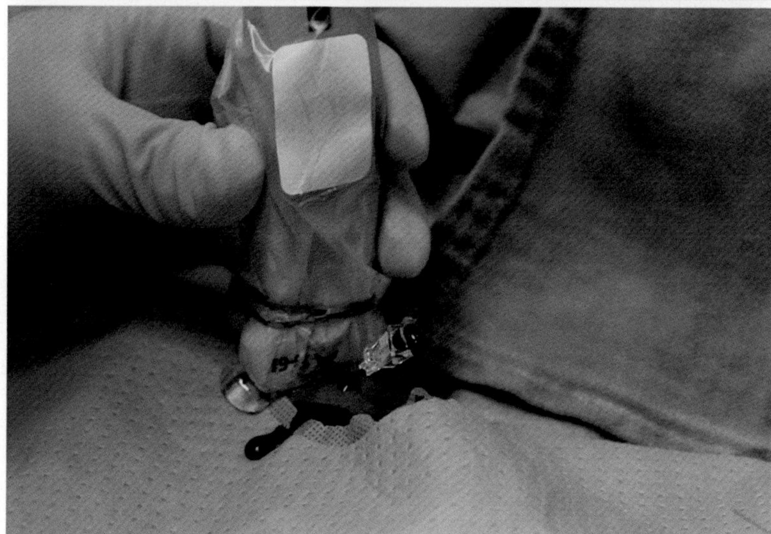

FIGURE 4.37 Radial artery puncture.

2. Through and through technique: the needle is slowly pulled back and when blood flow starts a 0.018 in. guidewire is advanced.

4.3.6.3 Challenges

1. Resistance to wire advancement

   Causes: (1) suboptimal needle position; (2) tortuosity of the radial artery (Fig. 4.40); (3) wire advancement into a side branch; (4) radial artery spasm.

   Solutions:

   - **NEVER push hard**, as this may cause complications (dissection, spasm, perforation).
   - Confirm there is good backflow through the needle hub. If not, the needle tip is repositioned until good blood return is achieved.
   - The wrist is extended, which straightens the radial artery, potentially facilitating wire advancement.
   - If there is good blood backflow but the guidewire cannot advance past the needle tip, the needle angulation is modified, while attempting to advance the guidewire. If the needle was seen entering the radial artery at a side of the artery, the needle can be rotated slightly to align the bevel with the artery lumen, followed by guidewire readvancement.

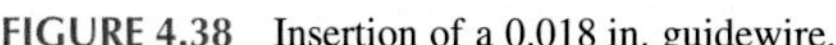

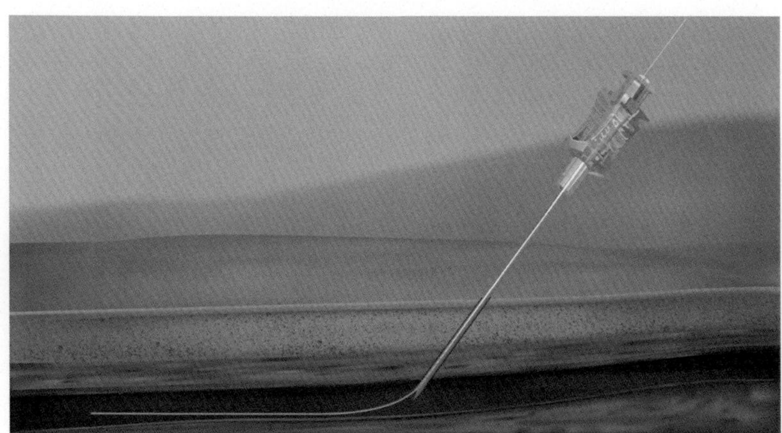

FIGURE 4.38 Insertion of a 0.018 in. guidewire.

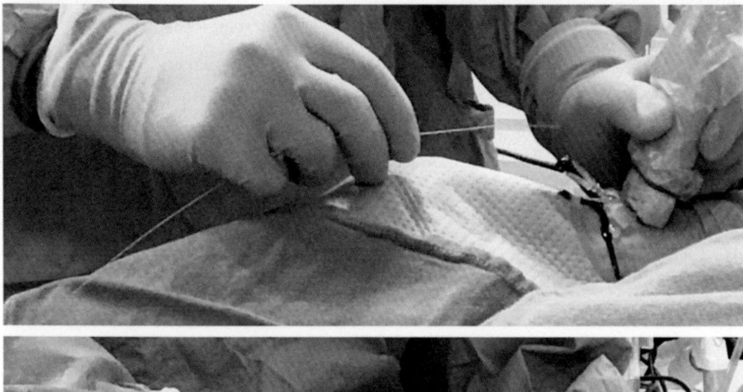

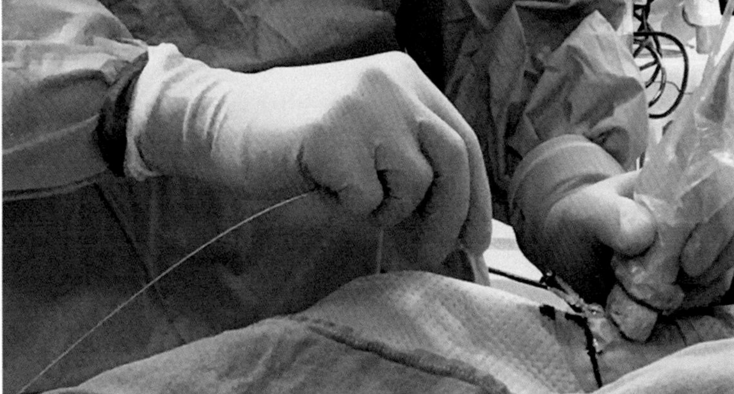

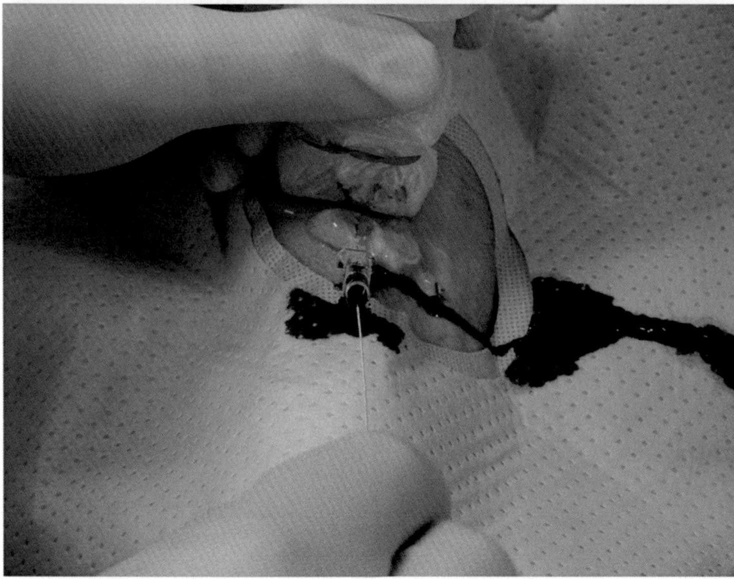

FIGURE 4.39 Insertion of a 0.018 in. guidewire.

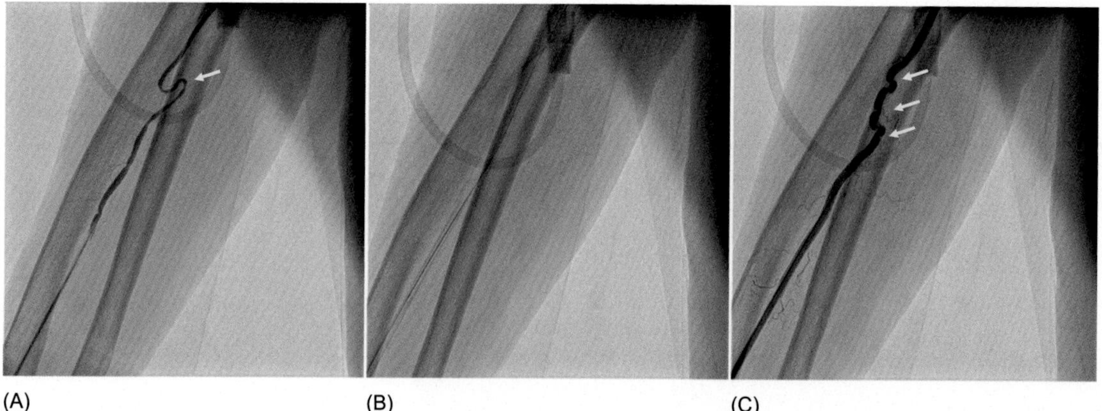

FIGURE 4.40 Example of a radial loop. (Panel A) Spasm developed in a relatively small and tortuous radial artery during attempts for crossing with a 0.035 in. guidewire. (Panel B) The radial loop was crossed with a coronary guidewire followed by sheath insertion over the coronary wire. (Panel C) Pseudolesions due to straightening of the radial loop with the guidewire. After additional vasodilators coronary angiography was completed with 5F Judkins catheters, but PCI could not be performed, requiring crossover to femoral access. *Courtesy of Dr. Ioannis Paizis.*

- If resistance is encountered while advancing the guidewire higher in the forearm, fluoroscopy is used to assess the wire location and the presence of tortuosity.
- A diagnostic catheter is advanced proximal to the wire tip and radial artery angiography is performed to determine the cause of resistance, such as radial artery tortuosity of vasospasm.

4.3.7 Step 7. Skin nick (optional)

4.3.7.1 Goal

To facilitate equipment advancement through the skin.

4.3.7.2 How?

1. The needle is removed.
2. A scalpel is used to make a nick at the guidewire entry point through the skin (Fig. 4.41).

4.3.7.3 What can go wrong?

1. Injury of the radial artery. The scalpel should not be advanced too deeply and should always be advanced in the longitudinal rather than sagittal direction.
2. Injury of the operator. The hand not holding the scalpel should ideally be removed from the sterile field.

4.3.8 Step 8. Sheath insertion

4.3.8.1 Goal

To insert a sheath into the radial artery.

4.3.8.2 How?

1. A sheath is selected for insertion. Sheath size selection depends on the radial artery size (ultrasound can help with artery sizing) and the planned procedure. Using 4 French sheaths for diagnostic angiography or thin wall sheaths (such as the slender sheaths, Terumo and the Prelude Ideal, Merit Medical) are preferred, as they have lower profile and are less likely to cause spasm. Long sheaths are usually avoided, due to risk of radial artery occlusion and spasm.
2. The sheath is dipped in saline solution to activate the hydrophilic coating.
3. The sheath is advanced over the 0.018 in. guidewire until the hub reaches the skin (Figs. 4.42 and 4.43).
4. The dilator and 0.018 in. guidewire are removed.

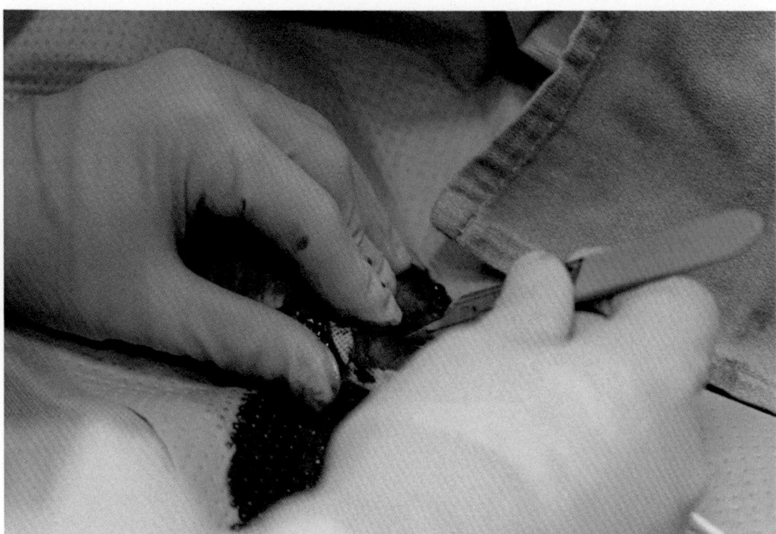

FIGURE 4.41 Skin nick to facilitate sheath insertion.

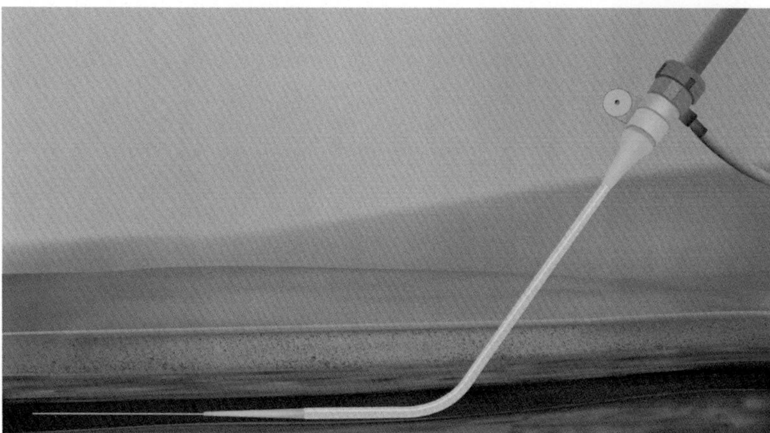

FIGURE 4.42 Radial sheath insertion.

5. Left radial: can use distal radial—or partially insert a long sheath, which will then be extending some distance from the left hand (Fig. 4.32).

4.3.8.3 What can go wrong?

1. Inability to advance sheath
   Causes: skin resistance, spasm, small radial artery size, tortuosity.
   Solutions:
   - Resistance at skin: use scalpel to facilitate insertion.
   - If spasm: administer sedation, intra-arterial verapamil or nitroglycerin, or subcutaneous nitroglycerin.
   - If small radial artery size: use alternative arterial access (ulnar, contralateral radial, or femoral).

4.3.9 Step 9. Sheath aspiration and flushing

4.3.9.1 Goal

To clear the sheath and prevent spasm and radial artery occlusion.

4.3.9.2 How?

1. The side arm of the sheath is aspirated (3–4 mL) and the aspirated blood is discarded.

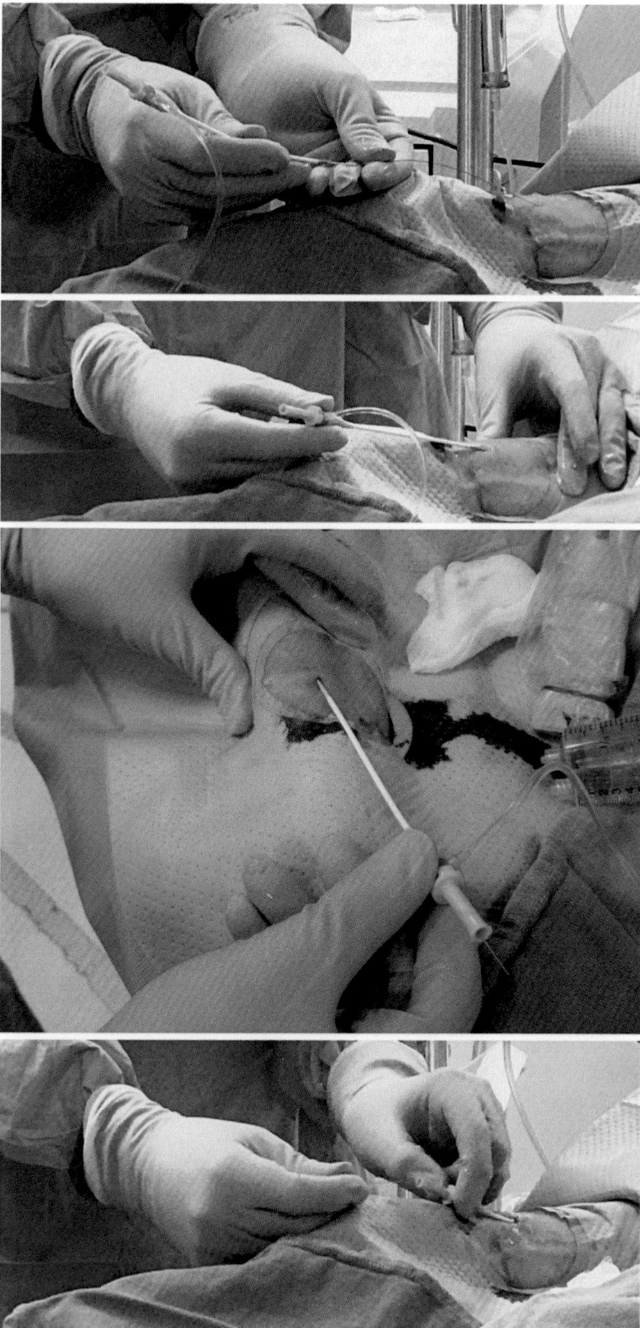

FIGURE 4.43 Radial sheath insertion.

2. One or more vasodilators (such as verapamil 2–3 mg, nicardipine 100–200 mcg, or nitroglycerin 100–200 mcg) are administered (Fig. 4.44). Vasodilator administration prior to sheath removal may also reduce the risk of radial artery occlusion [22].
3. Intravenous heparin is also administered (usually 50 units/kg up to 5000 units for diagnostic catheterization, although a dose of 100 units/kg may be more effective) with the goal to prevent radial artery occlusion. Administration of heparin may be delayed until after the catheter has been advanced to the aortic root, as in some patients with extreme radial or subclavian tortuosity, coronary engagement may be very challenging via radial access and crossover to femoral may be preferable.

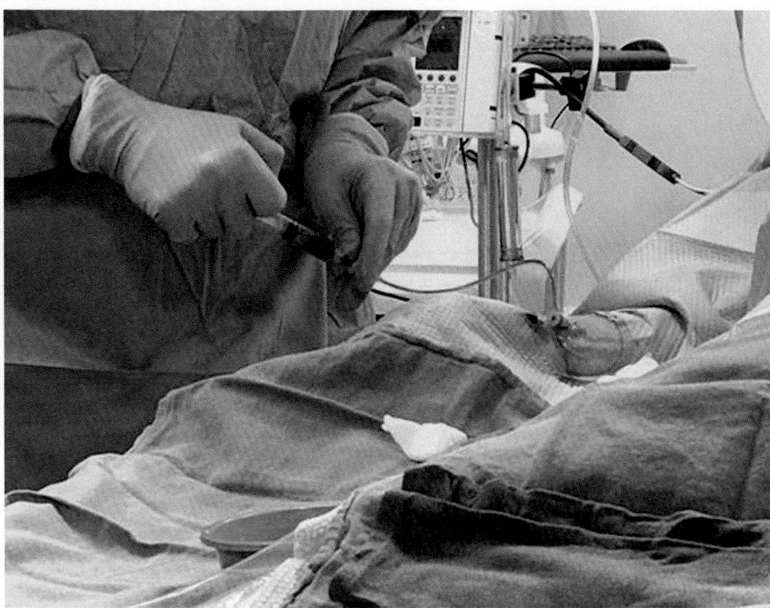

FIGURE 4.44 Vasodilator administration through the radial sheath.

4.3.10 Step 10. Securing the sheath

4.3.10.1 Goal

To prevent sheath movement.

4.3.10.2 How?

1. Movement of the sheath may be prevented using one of several maneuvers:
    - Clip the side arm of the sheath to the drape with a hemostat (Fig. 4.45).
    - Use a plastic adhesive cover, such as a Tegaderm.
    - Suture the sheath to the skin.
    - If the Stand Tall device (RADUX Devices, LLC) is used for left radial access, use the provided clasp that has adhesive backing.

4.4 Other access sites

4.4.1 Ulnar access

Ulnar access is obtained similar to radial access, but can be more challenging because the ulnar artery is located deeper in the wrist and is often located just below the tendon of the muscle flexor carpi ulnaris. Moreover, the ulnar artery is close to the ulnar nerve: puncturing the nerve can be very painful.

4.4.2 Brachial access

Brachial access is used infrequently, due to challenges achieving hemostasis (harder to compress as compared with radial access and compression bands cannot be used). There is also higher risk of brachial nerve injury if there is a hematoma. Access is usually obtained in the antecubital fossa using ultrasound guidance.

4.4.3 Axillary access

Axillary access is sometimes used for insertion of hemodynamic support devices. Detailed description is beyond the scope of this manual [23,24].

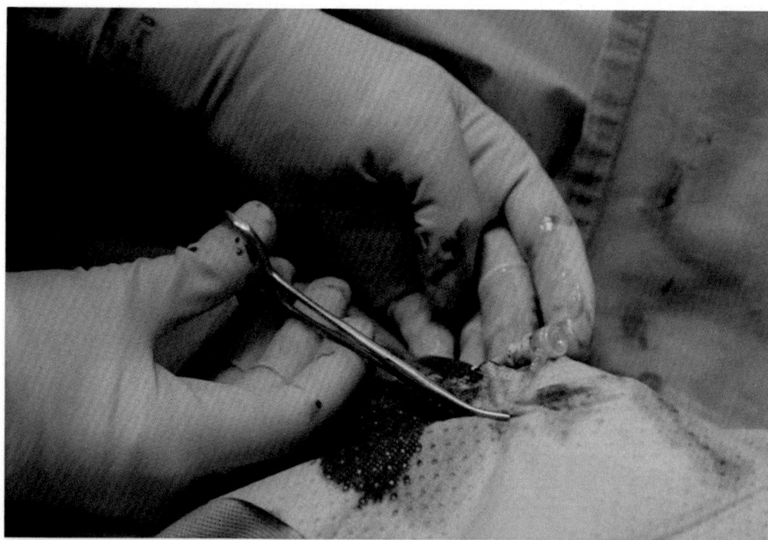
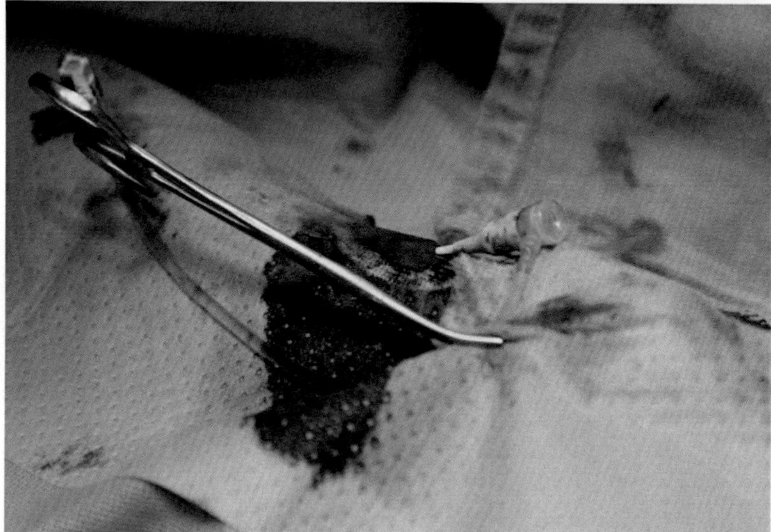

FIGURE 4.45 Securing the radial sheath to the drape with a hemostat.

4.4.4 Transcaval access

Transcaval access if used for insertion of hemodynamic support devices and transcatheter aortic valve replacement. Detailed description of this technique is beyond the scope of this manual.

References

[1] Ferrante G, Rao SV, Juni P, et al. Radial versus femoral access for coronary interventions across the entire spectrum of patients with coronary artery disease: a meta-analysis of randomized trials. JACC Cardiovasc Interv 2016;9:1419–34.

[2] Neumann FJ, Sousa-Uva M, Ahlsson A, et al. 2018 ESC/EACTS guidelines on myocardial revascularization. Eur Heart J 2019;40:87–165.

[3] Michael TT, Alomar M, Papayannis A, et al. A randomized comparison of the transradial and transfemoral approaches for coronary artery bypass graft angiography and intervention: the RADIAL-CABG Trial (RADIAL Versus Femoral Access for Coronary Artery Bypass Graft Angiography and Intervention). JACC Cardiovasc Interv 2013;6:1138–44.

[4] Urban P, Mehran R, Colleran R, et al. Defining high bleeding risk in patients undergoing percutaneous coronary intervention. Circulation 2019;140:240–61.

[5] Cooper L, Banerjee S, Brilakis ES. Crossover from radial to femoral access during a challenging percutaneous coronary intervention can make the difference between success and failure. Cardiovasc Revasc Med 2010;11(266):e5–8.

[6] Azzalini L, Tosin K, Chabot-Blanchet M, et al. The benefits conferred by radial access for cardiac catheterization are offset by a paradoxical increase in the rate of vascular access site complications with femoral access: the Campeau Radial Paradox. JACC Cardiovasc Interv 2015;8:1854–64.

[7] Azzalini L, Jolicoeur EM. The wise radialist's guide to optimal transfemoral access: selection, performance, and troubleshooting. Catheter Cardiovasc Interv 2017;89:399–407.

[8] Seto AH, Abu-Fadel MS, Sparling JM, et al. Real-time ultrasound guidance facilitates femoral arterial access and reduces vascular complications: FAUST (Femoral Arterial Access With Ultrasound Trial). JACC Cardiovasc Interv 2010;3:751–8.

[9] Sandoval Y, Burke MN, Lobo AS, et al. Contemporary arterial access in the cardiac catheterization laboratory. JACC Cardiovasc Interv 2017;10:2233–41.

[10] Vemmou E, Nikolakopoulos I, Xenogiannis I, Tajti P, Hall A, Brilakis ES. The Gordian knot-if you can not solve it, cut it: treating guidewire bending while obtaining arterial access. JACC Cardiovasc Interv 2019;12:892–3.

[11] Valgimigli M, Campo G, Penzo C, et al. Transradial coronary catheterization and intervention across the whole spectrum of Allen test results. J Am Coll Cardiol 2014;63:1833–41.

[12] Norgaz T, Gorgulu S, Dagdelen S. A randomized study comparing the effectiveness of right and left radial approach for coronary angiography. Catheter Cardiovasc Interv 2012;80:260–4.

[13] Kado H, Patel AM, Suryadevara S, et al. Operator radiation exposure and physical discomfort during a right versus left radial approach for coronary interventions: a randomized evaluation. JACC Cardiovasc Interv 2014;7:810–16.

[14] Gasparini GL, Garbo R, Gagnor A, Oreglia J, Mazzarotto P. Feasibility and safety of distal radial access for percutaneous coronary intervention using a 7 FR sheath. Catheter Cardiovasc Interv 2019;94:902.

[15] Aoi S, Htun WW, Freeo S, et al. Distal transradial artery access in the anatomical snuffbox for coronary angiography as an alternative access site for faster hemostasis. Catheter Cardiovasc Interv 2019;94:651–7.

[16] Gasparini GL, Garbo R, Gagnor A, Oreglia J, Mazzarotto P. First prospective multicentre experience with left distal transradial approach for coronary chronic total occlusion interventions using a 7 Fr Glidesheath Slender. EuroIntervention 2019;15:126–8.

[17] Sgueglia GA, Di Giorgio A, Gaspardone A, Babunashvili A. Anatomic basis and physiological rationale of distal radial artery access for percutaneous coronary and endovascular procedures. JACC Cardiovasc Interv 2018;11:2113–19.

[18] Corcos T. Distal radial access for coronary angiography and percutaneous coronary intervention: a state-of-the-art review. Catheter Cardiovasc Interv 2019;93:639–44.

[19] Hadjivassiliou A, Kiemeneij F, Nathan S, Klass D. Ultrasound-guided access of the distal radial artery at the anatomical snuffbox for catheter-based vascular interventions: a technical guide. EuroIntervention 2019. https://dx.doi.10.4244/EIJ-D-19-00555.

[20] Kiemeneij F. Left distal transradial access in the anatomical snuffbox for coronary angiography (ldTRA) and interventions (ldTRI). EuroIntervention 2017;13:851–7.

[21] Seto AH, Roberts JS, Abu-Fadel MS, et al. Real-time ultrasound guidance facilitates transradial access: RAUST (Radial Artery access with Ultrasound Trial). JACC Cardiovasc Interv 2015;8:283–91.

[22] Dharma S, Kedev S, Patel T, Kiemeneij F, Gilchrist IC. A novel approach to reduce radial artery occlusion after transradial catheterization: postprocedural/prehemostasis intra-arterial nitroglycerin. Catheter Cardiovasc Interv 2015;85:818–25.

[23] Cheney AE, McCabe JM. Alternative percutaneous access for large bore devices. Circ Cardiovasc Interv 2019;12:e007707.

[24] Dahle TG, Kaneko T, McCabe JM. Outcomes following subclavian and axillary artery access for transcatheter aortic valve replacement: Society of the Thoracic Surgeons/American College of Cardiology TVT Registry Report. JACC Cardiovasc Interv 2019;12:662–9.

Chapter 5

Coronary and graft engagement

In Chapter 4, Access, we discussed about how to obtain arterial access. In this chapter we describe the steps involved in advancing a catheter over a guidewire from the access site to the coronary ostia and engaging the coronary arteries. Neither diagnostic angiography nor PCI can be performed without engaging the coronary artery ostia. Although the steps involved in engaging the coronary arteries are simple and performed numerous times daily in the cardiac catheterization laboratory, difficulties may be encountered, potentially leading to severe complications.

5.1 Step 1. Catheter selection

5.1.1 Goal

To select a catheter that will most easily and safely engage the target coronary artery in a co-axial orientation while providing optimal support. Guide catheter selection is one of the most critical decisions to ensure procedural success, efficiency, and safety.

5.1.2 How?

Catheter selection is based on arterial access site (radial vs femoral), the target coronary vessel and the size of the aorta, as described in Section 30.2.6. Optimal coronary engagement can be facilitated by optimal (diagnostic or guide) catheter selection.

5.2 Step 2. Advance guidewire to aortic root

5.2.1 Background

Catheter advancement is always performed over a 0.035 or 0.038 in. guidewire for both efficiency and safety (advancing catheters over a wire reduces the risk of arterial injury).

5.2.2 Goal

To advance a 0.035 or 0.038 in. guidewire to the aortic root, which will serve as rail for advancing a catheter to the coronary ostia.

5.2.3 How?

1. A 0.035 or 0.038 in. J-tip guidewire is typically used for advancing the diagnostic or guide catheter.
   FEMORAL access: The guidewire can be inserted to the target vessel through the access needle (or microcatheter if a micropuncture kit is used) and be used to both insert the arterial sheath and allow catheter delivery to the coronary ostia. When cases are performed by a single operator short J-tip guidewires are preferable for sheath insertion.
   RADIAL access: A guidewire with a narrow J-tip is often used, such as the Baby-J wire (Terumo).
2. The guidewire is advanced together with the catheter (the guidewire tip should always stay ahead of the tip of the catheter) under fluoroscopic guidance to the aortic cusps and fixed, usually by the assistant, allowing advancement of the catheter over it.

Manual of Percutaneous Coronary Interventions. DOI: https://doi.org/10.1016/B978-0-12-819367-9.00005-6

5.2.4 Challenges

5.2.4.1 Resistance or failure to advance guidewire (Fig. 5.1)

Step 1. Withdraw the catheter and transduce the side arm of the access sheath. Lack of arterial waveform may mean subintimal sheath position (femoral or radial access) or severe spasm (radial access). In case of subintimal sheath position, it is best to change access site. Retrograde dissections are usually well tolerated (as the blood flow tends to seal the dissection), as long as the dissection length is short and side branches are not affected. Large femoral or iliac dissections may require endovascular repair after obtaining contralateral femoral access (section 29.1.1).

Step 2. Advance a catheter (usually JR4 or multipurpose) over the guidewire close to the resistance point, confirm the presence of arterial waveform, and then perform angiography. Angiography may reveal: (1) occlusion of the vessel, such as iliac artery or subclavian artery; (2) severe lesion; (3) severe tortuosity; (4) side branch that is selectively entered by the guidewire; or (5) aneurysm.

Step 3. For tight lesions, tortuous vessels, or when entering a side branch or crossing an aneurysm, repeat wiring attempts can be performed using various 0.035–0.038 in. guidewires (polymer jacketed or with very soft tip) or 0.014 in. guidewires through a catheter advanced close to the segment of the vessel that is difficult to navigate through. A polymer-jacketed wire should never be inserted through a needle, as the polymer coating may be sheared off when retracting the wire. Occasionally a 0.014 in. guidewire may be needed to cross the diseased segment.

Step 4. Changing access site may provide a solution if crossing attempts fail.

5.2.4.2 Causes of failure to advance guidewire to the aortic root

5.2.4.2.1 Subintimal guidewire position

Prevention:

- Do not force the guidewire, if resistance is felt with advancement through the access needle or during any stage of advancement.

Solutions:

- Withdraw the guidewire and attempt to advance through a different course.
- Change access site.

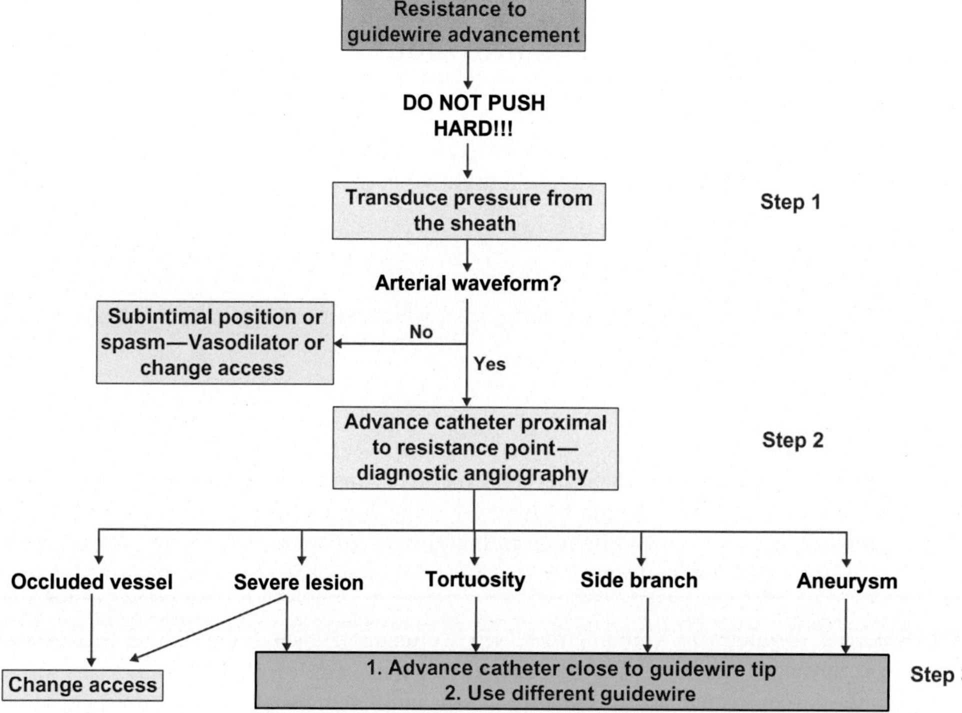

FIGURE 5.1 Resistance advancing a guidewire to the aortic root algorithm.

5.2.4.2.2 Occlusion between the access site and the aortic root

Prevention:

- Preexisting vessel occlusion cannot be prevented, but can be suspected if the pulse in the affected limb is weaker than the pulse in the contralateral limb.
- Forceful guidewire advancement should be avoided to prevent iatrogenic vessel dissection or perforation.

Solution:

- Change access site.

5.2.4.2.3 Tortuosity (iliac artery, aorta, radial, brachial, subclavian artery)

- Sometimes tortuosity is evident by observing the course of the guidewire or visualizing vessel calcification (Figs. 5.2–5.5).

Prevention:

- Tortuosity cannot be prevented, but can be suspected if prior angiograms or computed tomograms are available.

Solutions:

- Use different guidewires, such as the Glidewire (Terumo), Glidewire Advantage (Terumo), Bentson (Cook Medical), and Wholey (Medtronic) for femoral access and the Glidewire or "Baby-J wire" (Terumo) for radial access (Section 30.7.8). A 0.014 in. guidewire may be used in radial access cases to navigate through arterial loops.
- Advance the tip of catheter close to the tip of the guidewire to improve support.

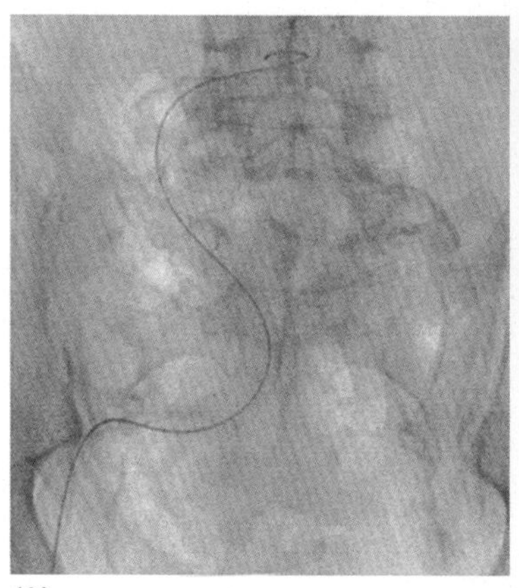
(A)

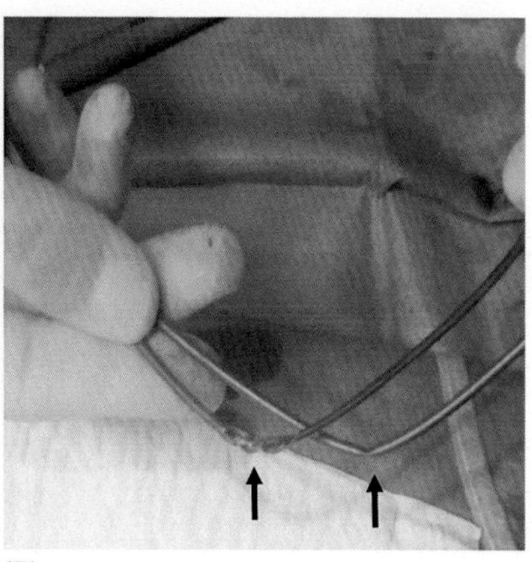
(B)

FIGURE 5.2 Severe iliac tortuosity (panel A) resulting in guide catheter kinking (*arrows*, panel B) during attempts to engage the coronary arteries.

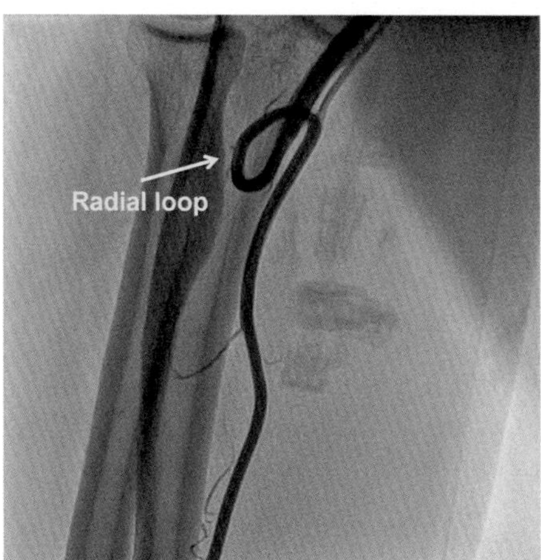

(A)

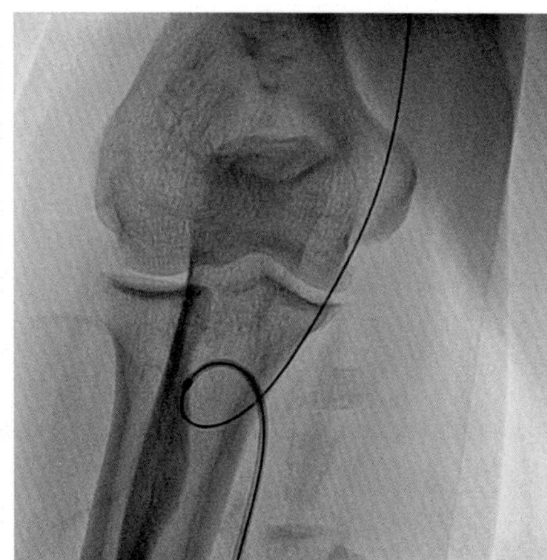
(B)

FIGURE 5.3 Example of a radial loop. (Panel A) Radial loop. (Panel B) Successful crossing with a 0.035 in. Glidewire. *Courtesy of Dr. Ioannis Paizis.*

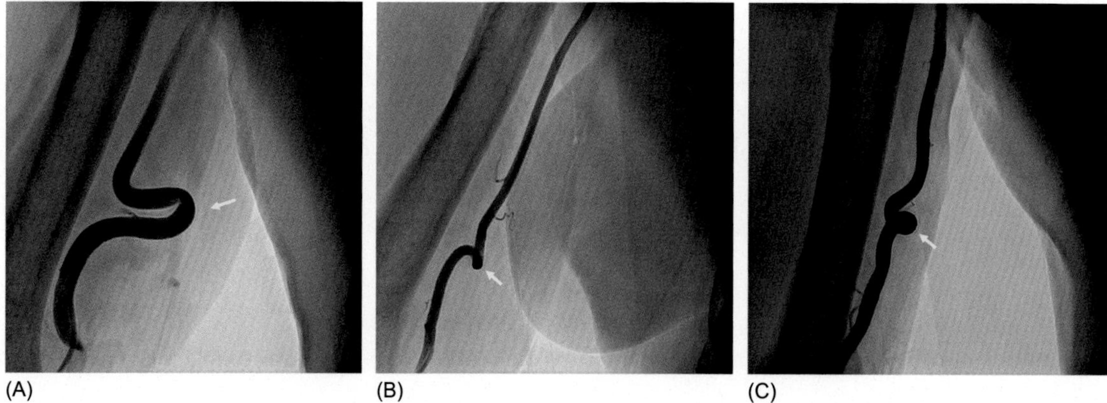

FIGURE 5.4 Example of a brachial loop. *Courtesy of Dr. Ioannis Paizis.*

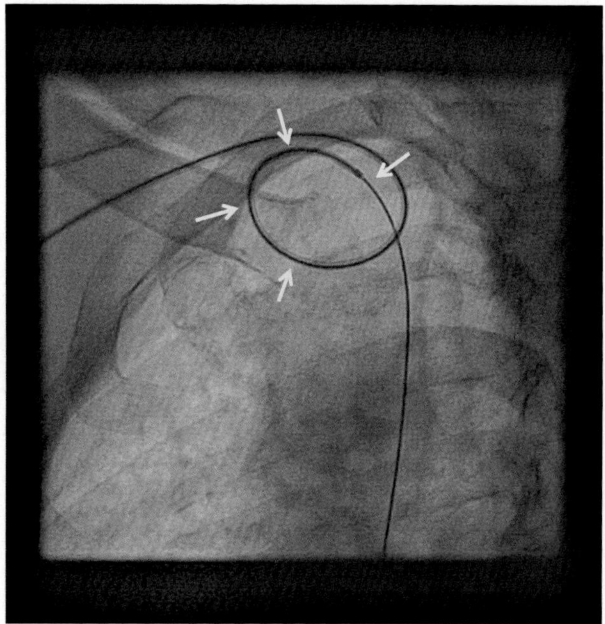

FIGURE 5.5 Subclavian tortuosity. *Courtesy of Dr. Ioannis Paizis.*

- Once crossing is achieved using femoral access, consider inserting a long sheath to prevent recurrent challenges with crossing. In patients with radial access, a 300 cm long 0.035 or 0.038 in. guidewire should be used for all catheter exchanges, keeping the guidewire tip in the aortic root. Moreover, using a long (such as 45 cm) sheath that is at least 1 French larger than the catheter can significantly facilitate catheter manipulation.

5.2.4.2.4 Guidewire enters side branch

This may occur with either femoral or radial access.

Prevention:

- Use of J-tipped guidewires minimizes the likelihood of side branch entry.

Solutions:

- Perform angiography to detect the presence of side branches and determine the vessel course.
- J-shaped wires are less likely to enter side branches. Straight tip wires should not be used for advancing catheters. Use fluoroscopy to detect the wire position and **never push a guidewire when you feel resistance**.

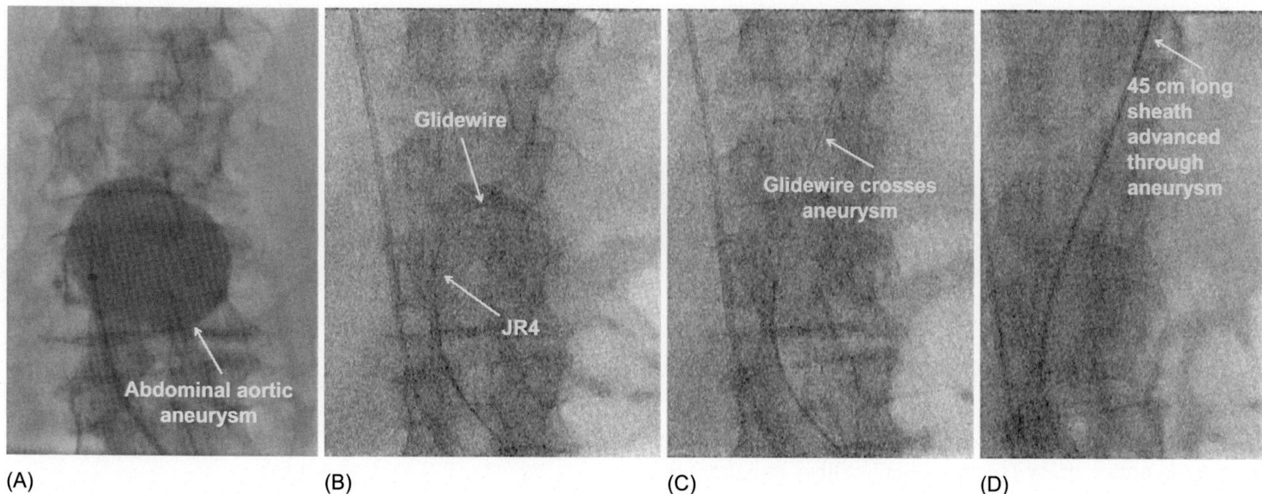

FIGURE 5.6 Crossing an abdominal aortic aneurysm. Initial attempts to cross an abdominal aortic aneurysm (*arrow*, panel A) using a JR4 diagnostic catheter and a stiff Glidewire (panel B) failed. With continued attempts the Glidewire was advanced through the aneurysm (panel C), followed by exchange for a 0.035 in. guidewire and insertion of a 45 cm long sheath across the aneurysm (panel D).

- Use different guidewires, such as the Glidewire (Terumo), Glidewire Advantage (Terumo), Bentson (Cook Medical), and Wholey wire (Medtronic) for femoral access and the Glidewire or "Baby-J wire" (Terumo) for radial access (Section 30.7.8).
- Once crossing is achieved using femoral access, consider inserting a long sheath to prevent recurrent challenges with crossing. In patients with radial access, a 300 cm 0.035 or 0.038 in. guidewire should be used for all catheter exchanges.

5.2.4.2.5 Guidewire enters aneurysm (Fig. 5.6)

Prevention:
- Aneurysms cannot be prevented—awareness of the presence of an aneurysm before cardiac catheterization may facilitate procedural planning (e.g., use radial access in patients with abdominal aortic aneurysms).

Solutions:
- Perform angiography to detect the size and exit point of the aneurysm.
- Use different guidewires, such as the Glidewire, Glidewire Advantage, Wholey, and Bentson for femoral access and the Glidewire or "Baby-J wire" (Terumo) (Section 30.7.8) for radial access
- Once crossing of the aneurysm is achieved using femoral access, consider inserting a long sheath to prevent recurrent challenges with crossing. In patients with radial access, a 300 cm long 0.035 or 0.038 in. guidewire should be used for all catheter exchanges.

5.2.5 What can go wrong?

5.2.5.1 Peripheral artery dissection *(see also Section 29.1.1)*

Attempts to advance a guidewire (and/or the catheter) may cause injury (dissection, perforation) of the peripheral arteries that need to be traversed to reach the coronary arteries.

Causes of dissection:
- Subintimal guidewire advancement.
- Forceful sheath or catheter advancement.
- Advancement of a catheter without a leading guidewire.

Prevention:
- NEVER use force when advancing guidewires and catheters through the peripheral vasculature.
- Use long sheaths in patients with severe iliac tortuosity.

- Perform femoral angiography with a guidewire in place to keep the sheath tip away from the arterial wall.

Diagnosis:
- Perform angiography in cases of challenging catheter advancement. Angiography should only be performed if there is no arterial pressure dampening.

Treatment:
- Most often no specific treatment is required. These are retrograde dissections that are usually sealed off by antegrade blood flow. Large dissections may require endovascular repair, usually with stenting.
- Switch to different access site.

5.2.5.2 Peripheral artery perforation *(see also Section 29.1.2) (Fig. 5.7)*

Causes:
- Forceful guidewire or catheter advancement into a side branch, such as the renal artery.
- Small caliber artery (such as small caliber radial artery).
- Inadvertent advancement of the sheath past the tip of the guidewire.

Prevention:
- Use fluoroscopy while advancing guidewires and guide catheters. Redirect the guidewire if it enters into a side branch.
- Avoid forceful guidewire or catheter advancement.

Diagnosis:
- Perform angiography when the guidewire or catheter follows an unexpected course or when resistance if felt during advancement.

Treatment:
- Balloon occlusion (in case of large artery perforation), such as the iliac artery or the aorta. In cases of iliac artery perforation, obtaining a second access (either femoral or radial) rapidly for balloon occlusion followed by covered stent may prevent or limit the size of a retroperitoneal hematoma.
- Placement of a covered stent to cover the perforation, such as the iCAST (balloon expandable, Atrium Medical) stents that require 6 or 7 French sheaths for delivery, or the Viabahn (self-expanding) or Viabahn VBX (balloon expandable) stents (W.L. Gore) that require a 7 French sheath for delivery (Fig. 29.6). The smallest diameter of both the iCAST and the Viabahn stents is 5 mm.
- Coil (or fat or thrombus) embolization for small branch perforations, such as perforation of the inferior epigastric artery (Section 30.18.2).
- For radial artery, advancement of a catheter over the perforated segment usually suffices to achieve hemostasis.

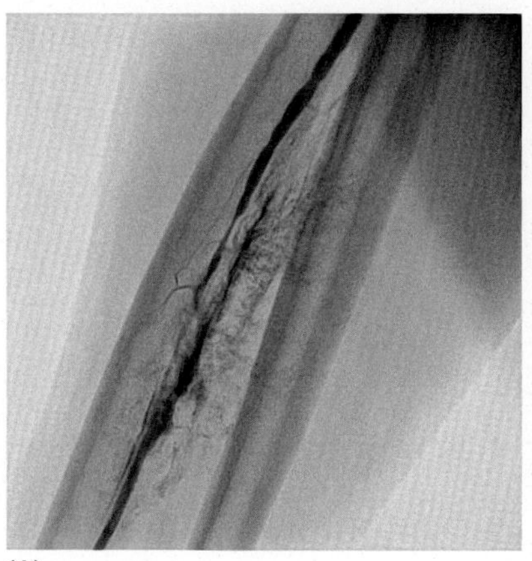
(A)

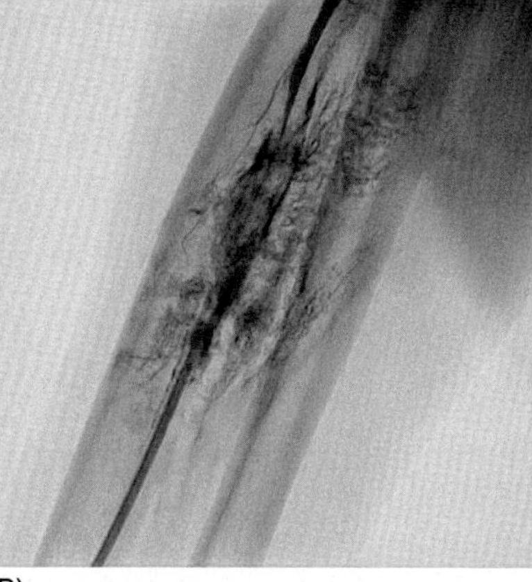
(B)

FIGURE 5.7 Radial artery perforation. *Courtesy of Dr. Ioannis Paizis.*

5.3 Step 3. Advance catheter to aortic root

5.3.1 Goal

To advance the catheter to the aortic root, next to the coronary ostia.

5.3.2 How to?

5.3.2.1 Catheter preparation

- Diagnostic catheters are flushed and loaded over a 0.035 or 0.038 in. guidewire before insertion through the sheath.
- Guide catheters are also advanced over the 0.035 or 0.038 in. guidewire. It is best to connect them to the Y-connector outside the body and flush them prior to insertion through the sheath. Using a hemostatic valve before guide insertion prevents bleeding during guide catheter insertion. Bleeding can be severe through guide catheters, especially large guide catheters (7 Fr and even more 8 Fr). Moreover, especially when using large guide catheters it is important to allow back bleeding, as aortic debris can easily be scraped into the guide catheter during advancement. If the guide is not properly cleared, injection of the debris into a coronary artery or the aorta can be catastrophic (Section 5.4).

5.3.2.2 Catheter advancement

- The catheter is advanced under fluoroscopic guidance to the aortic root, while the 0.035 or 0.038 in. guidewire is fixed. If resistance if felt, do NOT force the guide catheter, as this may cause complications.

5.3.3 Resistance to catheter advancement

Causes:

5.3.3.1 Subintimal guidewire position

Prevention:

- Do not force the guide catheter, if resistance is felt during advancement through the sheath or during any other stage of advancement. In such cases, significant resistance to guidewire advancement may also have been felt earlier.

Solutions:

- Remove the guide and guidewire.
- Change access site and check flow in the affected artery via contralateral injection.

5.3.3.2 Severe stenosis

Prevention:

- Preexisting stenosis cannot be prevented!

Solutions:

- Use a different catheter with less tip angulation.
- Advance a long sheath (usually 45 or 55 cm long) through the stenosis, followed by catheter advancement.
- Dilate the stenosis. Occasionally angioplasty of iliac artery stenoses may be required in case of severe iliac lesions. Stents should generally be avoided prior to completion of the PCI, as they can be dislodged during guide catheter advancement (Fig. 5.8). If iliac stents are placed (or if there are preexisting iliac stents), it is preferable to advance a long sheath through them to minimize the risk of stent deformation or dislodgement.
- For radial access, using a low profile sheath (such as Glidesheath Slender [Terumo] or Prelude Ideal [Merit Medical]) or a sheathless guide catheter may gently dilate the stenosis without injuring the vessel.

5.3.3.3 Tortuosity

Sometimes tortuosity is evident given the course of the guidewire or visualization of vessel calcification.

Prevention:

- Tortuosity cannot be prevented.

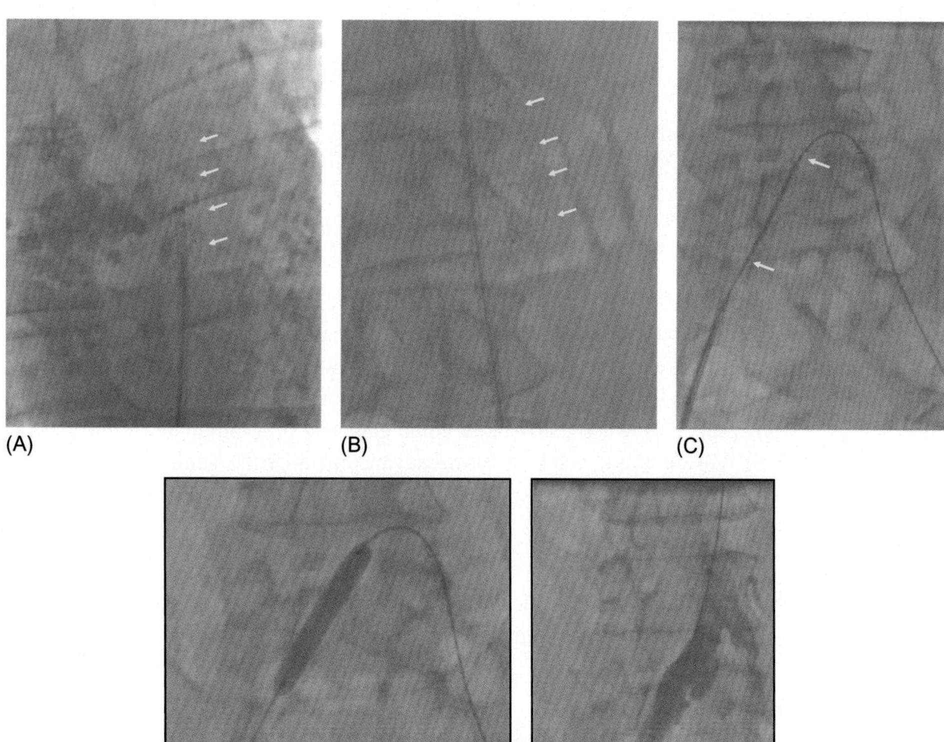

FIGURE 5.8 Peripheral stent dislodgement after guide catheter advancement through the stent (*CTO PCI Manual* Online case 104). A peripheral self-expanding stent (*arrows*, panel A) was caught at the tip of a guide catheter and dislodged into the thoracic aorta. The dislodged peripheral stent (*arrows*, panel B) embolized in the abdominal aorta. The dislodged peripheral stent was snared and withdrawn into the right iliac artery and a balloon expandable stent (*arrows*, panel C) was inserted (*arrows*, panel C) and deployed (panel D), covering the dislodged self-expanding stent (panel E).

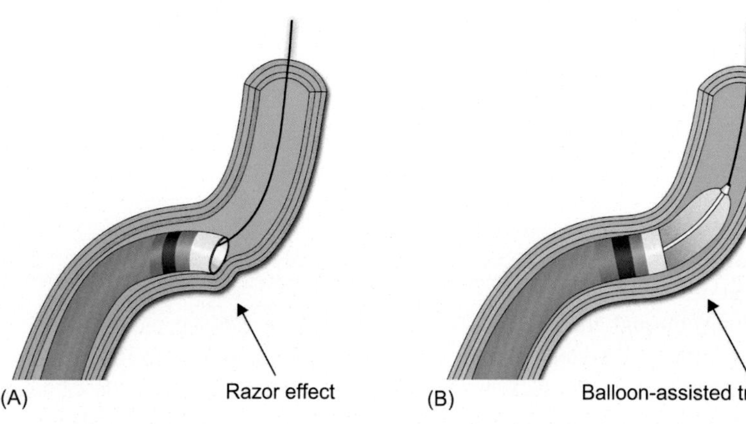

FIGURE 5.9 Illustration of the balloon-assisted tracking technique. A small compliant balloon (sized 1:1 with the guide catheter) is inflated halfway in and halfway out the tip of the guide catheter at low pressure (3–6 atm—the lower the pressure the more flexible the assembly will be, whereas a higher pressure increases pushability). The inflated balloon prevents the catheter from "digging" into the vessel wall (razor effect, panel A).

Solutions:

- Use different guide catheter with less distal angulation.
- Advance a long sheath through the area of tortuosity. Kink-resistant (such as Arrow) sheaths can facilitate catheter advancement and manipulation.
- Parallel sheath technique for femoral access (perform second puncture of the same femoral artery and insert a 4 French sheath, through which a stiff 0.035 in. guidewire is advanced to straighten the iliac tortuosity) [1].
- Balloon-assisted tracking (particularly for radial loops) (Fig. 5.9) [2–4].
- Change access site.

5.3.3.4 Spasm

This is specific to radial access (Fig. 5.10).

Causes:

- Small radial artery size.
- High origin radial artery (Fig. 5.11).
- Radial artery tortuosity.
- Multiple needle passes to obtain access.
- Extensive catheter manipulations.
- Multiple catheter exchanges.

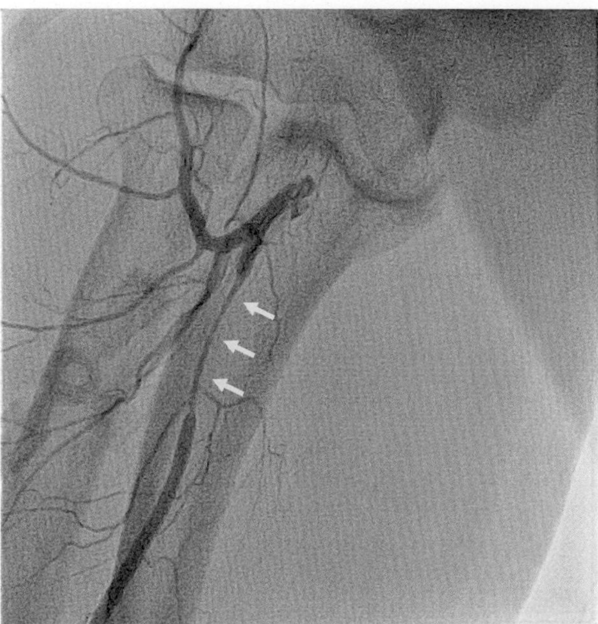

FIGURE 5.10 Radial artery spasm. *Courtesy of Dr. Ioannis Paizis.*

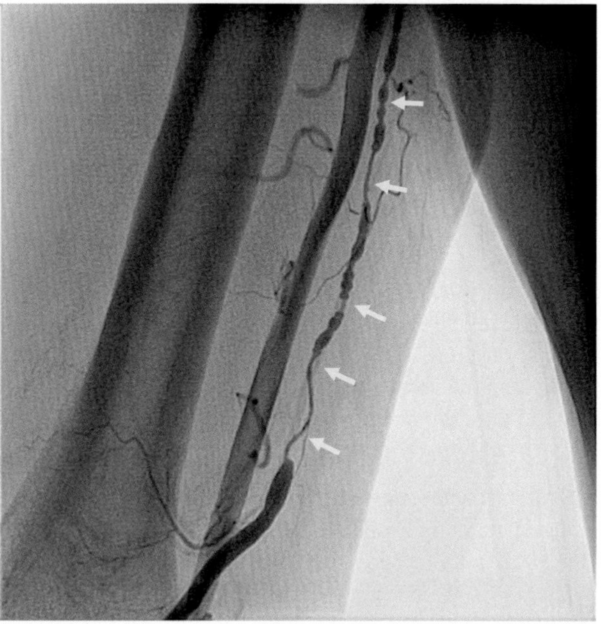

FIGURE 5.11 Spasm in a high origin radial artery. *Courtesy of Dr. Ioannis Paizis.*

- Spasm is more common in women and smokers.
- Inadequate sedation.

Prevention:
- Avoid using small radial arteries: use the ulnar or femoral artery instead. This emphasizes the importance of ultrasound guidance while obtaining radial (and femoral) access.
- Vasodilator administration (such as verapamil 2.5 mg and/or nitroglycerin 100−200 mcg) after sheath insertion.
- Avoid use of large catheters, especially through small radial arteries.
- Adequate conscious sedation prior to obtaining access.

Solutions:
- Administer intra-arterial vasodilators.
- Subcutaneous nitroglycerin injection.
- More sedation.
- Use smaller guide catheters (such as 5 French).
- Use Heartrail (Terumo) guide catheters (they have a dimpled surface that decreases friction and facilitates advancement).
- Use a sheathless guide catheter (such as the Eaucath, Asahi) or the Railway system (Cordis) (or other dilator) with standard guide catheters.
- Use of hydrophilic sheaths and catheters.
- Use of low profile sheaths for radial access.

5.3.3.5 *What to do if there is resistance advancing a catheter to the aortic root (Fig. 5.12)?*

Step 1. Transduce the pressure from the catheter tip. In radial access cases, severe spasm can limit catheter advancement and can be treated with vasodilator administration. For femoral cases, lack of arterial pressure waveform may suggest subintimal sheath position, and it may be best to change access site.

Step 2. Angiography is performed to determine the cause of the resistance to catheter advancement. Angiography may reveal: (1) severe lesion; (2) spasm; (3) or severe tortuosity. It may also reveal a complication, such as perforation or dissection that may require immediate treatment.

Step 3. Treatment is individualized to the cause of resistance to catheter advancement. For severe lesions another catheter can be used, or a long sheath inserted, or the lesion dilated. For severe tortuosity another catheter or a long sheath can be used. For severe spasm, vasodilators are administered.

Step 4. Changing access site may provide a solution if crossing attempts fail.

5.3.4 Failure to reach the aortic root

Causes:
- Very tall patients.
- Severe subclavian or iliac/aortic tortuosity.
- Distal radial access.

Prevention:
- Use of long sheaths in patients with severe iliac or aortic or subclavian tortuosity to possibly shorten the distance from the access point to the coronary ostium.
- Use of a stiff guidewire may straighten the aorta and allow reaching the coronary ostium.
- Use of longer catheters (such as 125 cm long catheters).

Solutions:
- Use longer catheters, however, balloons and stents may not be long enough to reach the target lesion in case PCI is needed (Online case 53). The longest currently available DES is the Xience Sierra (145 cm long shaft), followed by the Promus and Synergy (144 cm long shaft), followed by the Resolute Onyx, Orsiro and Elunir (140 cm long shaft).
- Use guide catheter extensions.
- Parallel sheath technique [1].
- Change access site (femoral to radial and vice versa, although the femoral route is usually shorter).

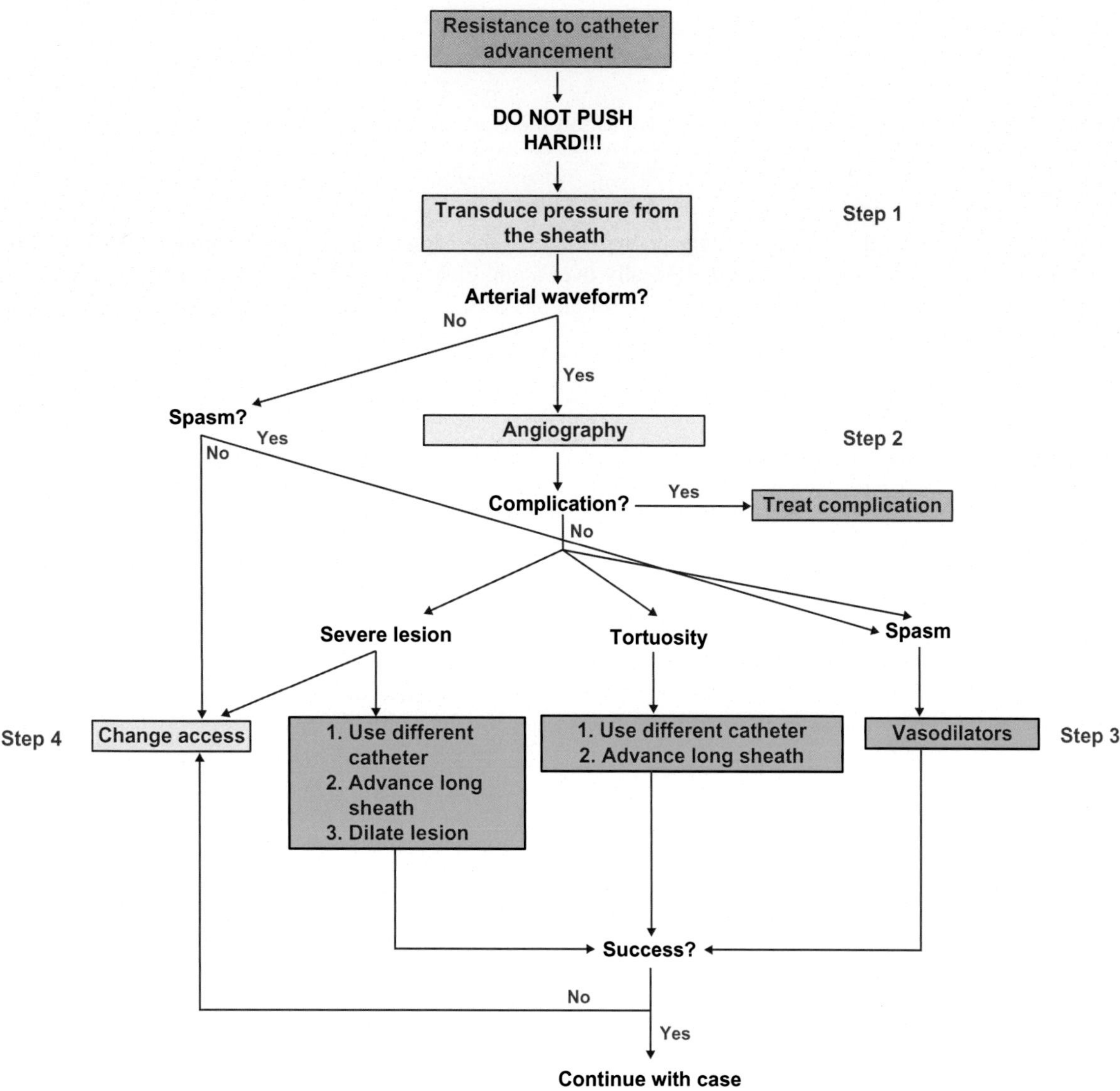

FIGURE 5.12 Algorithm on how to approach resistance advancing a catheter to the aortic root.

5.3.5 What can go wrong?

5.3.5.1 Peripheral artery dissection (as discussed in step 2, section 5.2.5.1)

5.3.5.2 Peripheral artery perforation (as discussed in step 2, section 5.2.5.2)

5.3.5.3 Ventricular arrhythmias: During advancement of the catheter to the aortic root the 0.035 or 0.038 in. guidewire may enter the left ventricle and cause arrhythmias

Causes:

- Entry of the 0.035 in. guidewire into the left ventricle.

Prevention:

- Monitor and adjust the position of the tip of the 0.035 in. guidewire to prevent entry into the left ventricle.

Solution:

- Remove guidewire from the left ventricle.

5.4 Step 4. Aspirate guide catheter

5.4.1 Goal

To clear the catheter of any air or other material (such as thrombus or plaque) before engaging the coronary artery.

5.4.2 How to?

Before attempting to engage the coronary artery or bypass graft, the catheter is aspirated removing 2–3 mL of blood with a syringe. The aspirated blood is discarded, ideally over a white gauze to determine if thrombus (usually from the sheath) or plaque scraped from the aortic wall (which often has iridescent appearance due to cholesterol crystals) [5], has been retrieved (Fig. 5.13).

5.4.3 What can go wrong?

5.4.3.1 Inability to aspirate blood

Causes:

5.4.3.1.1 Catheter is against the aortic or coronary wall, is positioned next to a significant ostial coronary artery lesion, is too large for the size of the coronary artery

Prevention:

- Thoughtful guide catheter selection, based on the size of the aorta and the size and location of the coronary artery ostia.
- Sometimes aspiration is not possible because the vacuum sucks the catheter to the aortic wall or into an ostial stenosis. In this case it is better and safer to back bleed the catheter instead of aspiration.

Solutions:

- Reposition the guide catheter, usually by slight withdrawal.
- Use a different catheter or withdraw the catheter and manually reshape the tip.

5.4.3.2 Catheter contains plaque, thrombus, air, etc.

Prevention:

- Aspirate the sheath before inserting a guide catheter and discard its contents (thrombus may form inside the sheath). Then flush the sheath using another syringe that contains clear heparinized saline.
- Use long femoral sheaths in patients with severe iliac and aortic disease.
- Flush catheters with normal saline before inserting into the body.

Solutions:

- Aspirate the guide catheter.
- If aspiration fails, withdraw slightly the guide catheter and try aspiration again. Forceful aspiration may suck air through the valve of the Y-connector in case of wedged and/or blocked guiding catheter. If this happens, ALL the air must be removed before the next injection.
- If aspiration still fails, remove the guide catheter WITHOUT inserting a guidewire (as the guidewire may push the material contained within the guide into the coronary artery or the aorta).

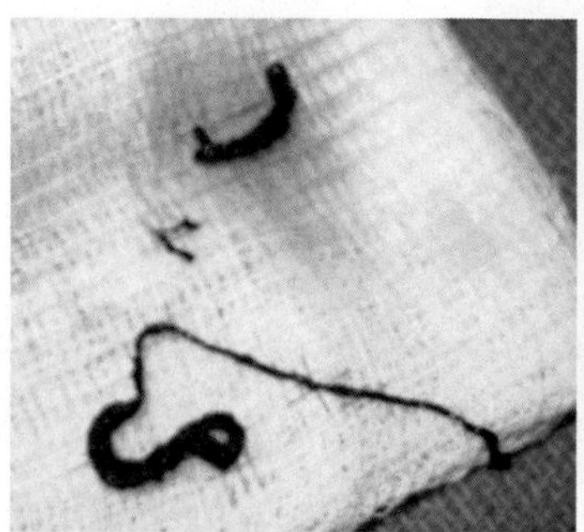

(A)

(B)

FIGURE 5.13 Aspirated material from the guide catheter after advancement to the aortic root. (Panel A) Thrombus. (Panel B) Plaque. *Panel (B) Reproduced with permission from Keeley EC, Grines CL. Scraping of aortic debris by coronary guiding catheters: a prospective evaluation of 1,000 cases. J Am Coll Cardiol 1998;32:1861–5. Copyright Elsevier.*

5.4.3.3 *Catheter is kinked*

Catheter kinking should be suspected if engagement is challenging, when the guide catheter tip is not moving despite turning the hub, or when the pressure waveform is lost.

Prevention:

- Avoid excessive catheter torquing (especially when the catheter tip is not moving proportionally to the torque applied).
- Use long (25–55 cm) femoral sheaths.
- Use different access site (e.g., change from right radial to left radial or to femoral access if there is extreme subclavian tortuosity or arteria lusoria).

Solution:

- Remove guide catheter. Removal of a kinked guide catheter can be challenging and is discussed in the Complications section below.

5.4.3.4 *What to do if you are unable to aspirate blood (Fig. 5.14)?*

Step 1. Partially withdraw the catheter (often a few mm will suffice). Sometimes, the catheter tip is against the aortic wall preventing aspiration. When the catheter is slightly withdrawn or repositioned, aspiration becomes feasible. This can be detected by reappearance of the pressure waveform on the monitor. Do not inject or flush the catheter with dampened or missing pressure waveform due to risk of embolization and/or dissection.

Step 2. Check for catheter kinking, by doing fluoroscopy of the entire catheter length. Kinking should be especially suspected if advancing the catheter was challenging requiring excessive torquing. If the catheter is kinked it should be removed, as outlined in Section 5.6.

Step 3. If you are still unable to aspirate and the catheter is not kinked, the presumed diagnosis should be that the catheter is plugged up with thrombus or plaque.

WARNING: DO NOT INSERT A GUIDEWIRE AND DO NOT FLUSH THE CATHETER IF YOU CANNOT ASPIRATE BACK, AS THIS MAY CAUSE EMBOLIZATION OF THROMBUS/PLAQUE IN THE AORTA, POTENTIALLY CAUSING STROKE OR ACUTE CORONARY OCCLUSION.

The next step is to remove the guide catheter with constant suction through the aspiration syringe.

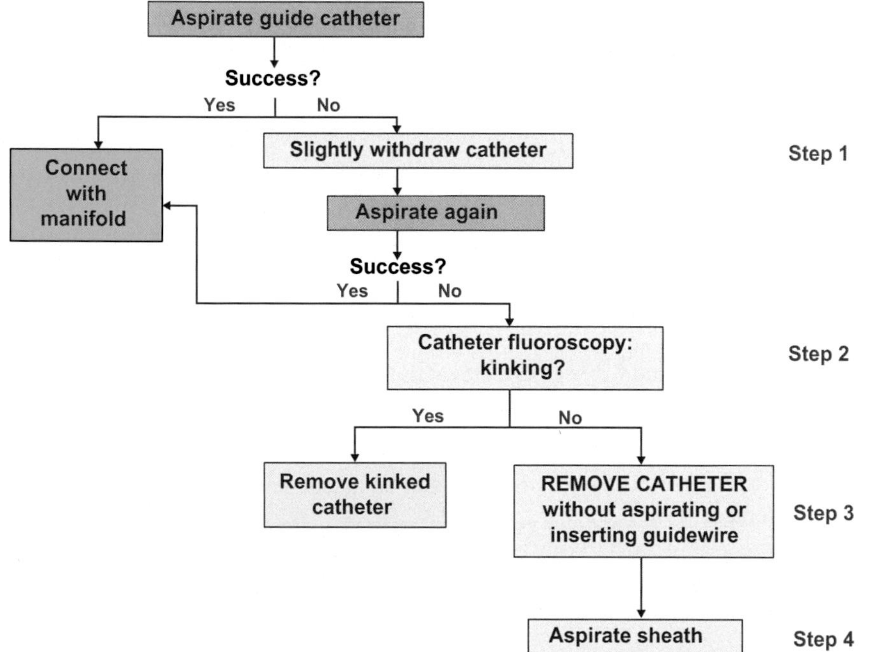

FIGURE 5.14 Failure to aspirate from a catheter algorithm.

Step 4. Aspirating the sheath is important if the catheter was clogged, as this is often due to thrombus forming within the sheath that is "scooped up" by the catheter when it is advanced through the sheath.

5.5 Step 5. Connect with manifold

5.5.1 Goal

To connect the catheter with the manifold, enabling pressure monitoring and contrast injection.

5.5.2 How to?

- The catheter is aspirated, as discussed in step 4.
- The catheter hub is connected with the manifold; the connection is usually performed while the manifold is being flushed (to minimize the risk of entraining air into the catheter).
- Immediately after the connection is completed, flushing stops and blood is aspirated into the manifold syringe.
- If there is any air or visible debris, it is discarded.
- After ensuring there is no air or visible debris, contrast is injected until it exits from the tip of the guide catheter.

5.5.3 What can go wrong?

5.5.3.1 Systemic embolization

Systemic embolization, especially stroke, is one of the most feared complications of cardiac catheterization.

Causes:
- Failure to aspirate (or observe back bleeding from) the guide catheter before attempting coronary artery engagement.
- Plaque dislodgement (in patients with significant aortic atherosclerosis) during catheter advancement.
- Embolization material can be air, plaque, thrombus.

Prevention:
- Aspirate and flush sheath before catheter insertion.
- Aspirate catheter before contrast injection.
- Do not inject if there is dampening of the pressure waveform.
- Use a sheathless guide with an inner dilator that prevents the "razor effect" and does not scrape plaques.

Diagnosis:
- Depends on the area affected. In case of stroke the patients will develop an acute neurologic deficit, such as focal weakness or vision loss. In case of lower extremity embolization, pulses may decrease or disappear and the patient may develop lower extremity discomfort.

Treatment:
- Stroke: emergency head CT and neurology evaluation—many patients may benefit from emergent neurointerventional treatment.
- Other embolization sites: usually conservative management, although sometimes endovascular treatment may be required.

5.5.3.2 Coronary embolization (Section 25.2.3)

Coronary embolization usually manifests with chest pain and ST-segment elevation.

Causes:
- Failure to aspirate the guide catheter before attempting coronary artery engagement.
- Plaque dislodgement (in patients with significant aortic disease) during catheter advancement.
- Embolization material can be air, plaque, thrombus.

Prevention:
- Flush sheath before catheter insertion.
- Aspirate catheter before contrast injection.
- Do not inject if there is dampening of the pressure waveform.

Diagnosis:
- ST-segment elevation is usually the first sign of coronary embolization. Depending on embolus size and target vessel (right coronary artery vs left main) the patient may develop cardiogenic shock or cardiac arrest.
- Coronary angiography confirms decrease or cessation of antegrade coronary flow.

Treatment:
- Air embolism (Section 25.2.3.3): administer 100% oxygen (helps absorb the embolized air)—if antegrade flow stops perform aspiration through guide or thrombectomy catheter. Repetitive forceful saline injections may help "unclog" the occluded vessels. Intracoronary epinephrine (0.05 mg) has also been used in this setting.
- Plaque or thrombus embolism: (1) administer anticoagulation and intravenous antiplatelet agents (glycoprotein IIb/IIIa inhibitors or cangrelor); (2) perform thrombectomy (if there is a large visible thrombus); and (3) administer intracoronary vasodilators, such as nicardipine, nitroprusside, verapamil, and adenosine.
- Hemodynamic support may be needed if the patient develops shock or cardiac arrest.

5.6 Step 6. Ensure there is good pressure waveform

5.6.1 Goal

To ensure that the catheter tip is free and the catheter lumen clear of any foreign material, such as thrombus, plaque, or air.

5.6.2 How to?

The pressure waveform is inspected to ensure that there is no dampening or ventricularization. An example of pressure dampening is shown in Fig. 5.15. Dampening or ventricularization can be mild or severe. In patients with aortic stenosis the waveform may appear "dampened," although it is just "parvus and tardus." Interpretation of the pressure waveform is challenging when using side hole catheters, as the side holes can mask dampening or ventricularization at the tip of the catheter.

5.6.3 Causes and treatment of pressure dampening

5.6.3.1 Catheter is against the aortic or coronary wall, is deeply or subselectively engaged, is positioned next to a significant ostial coronary artery lesion, or is too large for the size of the coronary artery

Prevention:
- Thoughtful catheter selection, based on the size of the aorta and the location and size of the coronary artery ostia.

Solutions:
- Reposition the catheter, usually by slight withdrawal.
- Use a different catheter.

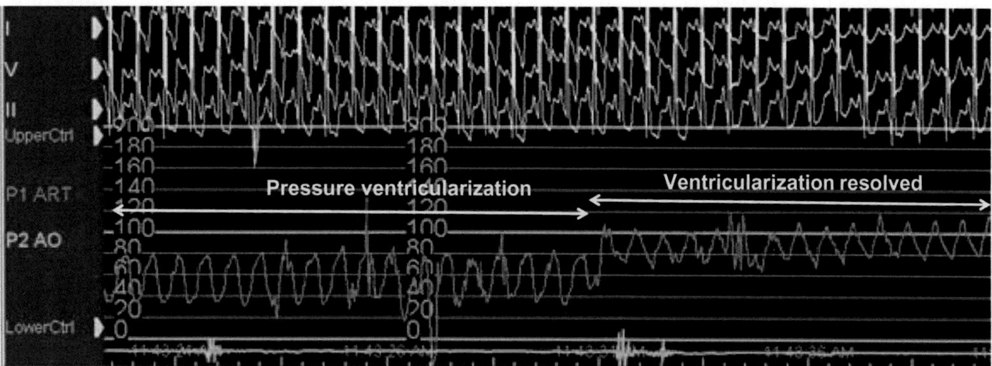

FIGURE 5.15 Illustration of pressure ventricularization upon engaging a coronary artery with an ostial lesion.

5.6.3.2 *Catheter contains plaque, thrombus, air, etc*

Prevention:
- Aspirate the sheath before inserting a guide catheter and discard its contents (thrombus may form inside the sheath). Then flush the sheath using another syringe that contains clear heparinized saline.
- Use long femoral sheaths in patients with severe iliac and aortic disease.
- Flush catheter with normal saline before inserting into the body.
- Use a sheathless guide catheter.

Solutions:
- Aspirate the catheter.
- If aspiration fails, withdraw slightly the guide catheter and try aspiration again.
- If aspiration still fails, remove the guide catheter WITHOUT inserting a guidewire (as the guidewire may push the material contained within the guide catheter into the coronary artery or the aorta).

5.6.3.3 *Catheter is kinked*

Catheter kinking should be suspected if engagement was challenging or when the guide catheter tip is not moving despite turning the hub.

Prevention:
- Avoid excessive catheter torquing. Leaving a 0.035–0.38 in. guidewire inside the guide catheter facilitates catheter manipulations.
- Use long (45–55 cm) femoral sheaths.
- Use a different access site (e.g., change from radial to femoral access if there is extreme subclavian tortuosity or arteria lusoria).

Solutions:
- Remove the catheter. Removal of a kinked catheter can be challenging and is discussed in Section 5.7.4.1.
- Use a different guide catheter.

5.6.3.4 *Pressure transducer malfunction*

Prevention:
- Clear transducer of air bubbles.

Solutions:
- Turn pressure switch to on.
- Use a different transducer.

5.6.3.5 *Algorithmic approach to pressure dampening (Fig. 5.16)*

Step 1. Partially withdraw or reposition the catheter, which should correct the dampening if it is due to the catheter tip being against the aortic or coronary artery wall.

Step 2. Aspirate the catheter, which should correct the dampening if it is due to material introduced inside the catheter.

Step 3. Check the pressure transducer; sometime bubbles can cause pressure dampening and clearing the transducer should correct the dampening.

Step 4. Check for catheter kinking; if the catheter is kinked it should be removed and replaced with a new catheter.

Step 5. If all aforementioned steps fail, the catheter should be removed **without injecting and without advancing a guidewire through it**, while maintaining suction with a syringe.

Step 6. The sheath should be aspirated to ensure that it does not contain thrombus.

5.6.4 What can go wrong?

Failure to recognize pressure dampening followed by contrast injection can cause potentially catastrophic complications, including:

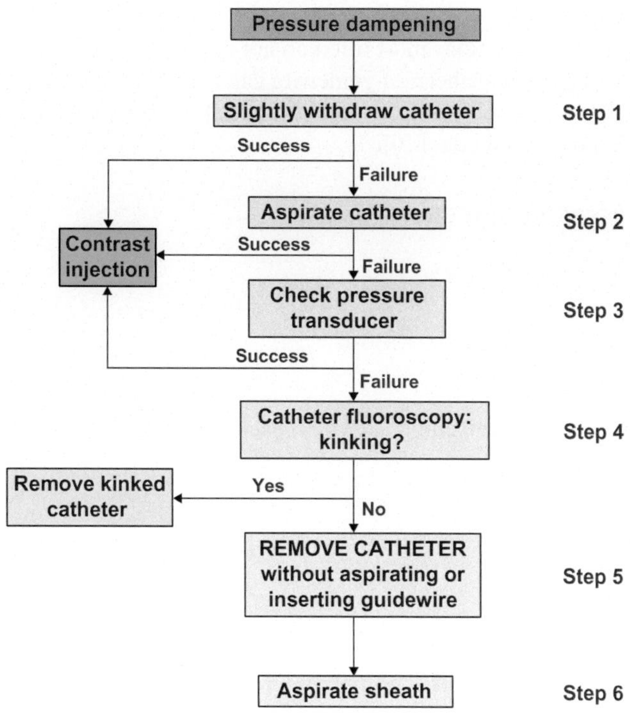

FIGURE 5.16 Algorithmic approach to pressure dampening.

5.6.4.1 *Systemic embolization* (see Section 5.5.3.1)

5.6.4.2 *Coronary embolization* (see Section 5.5.3.2)

5.6.4.3 *Coronary dissection* (PCI Manual Online case 40)

Causes:

- Non-coaxial catheter position.
- Contrast injection despite pressure dampening.
- Ostial lesion.

Prevention:

- **NEVER INJECT** if there is pressure dampening.
- Ensure coaxial catheter position.
- Gentle contrast injection. Use of contrast delivery systems (such as the ACIST system) should be avoided in this setting and manual injections used instead.
- Cautious use of side hole guide catheters, as they mask pressure dampening, but still predispose to ostial vessel dissection. Side hole guides should never be used in unprotected left main arteries.
- Careful catheter manipulation, particularly those with "aggressive" shapes, such as AL catheters (PCI Manual online case 79).

Diagnosis:

- Large coronary dissections will compromise coronary flow and cause ST-segment elevation and may possibly lead to cardiogenic shock or cardiac arrest (especially left main dissections).
- Small coronary dissections may not (at least immediately) impede coronary flow and may be harder to detect. Angiographic filling defects may be visible. The coronary ostium should be assessed during angiography at the end of the procedure to ensure that no complication has occurred. IVUS and especially OCT may help confirm the diagnosis.

Treatment:

- Do NOT lose wire position (if a guidewire is already within the target vessel).

- Stenting for larger coronary dissections.
- Small, non-flow limiting coronary dissections most often do not require specific treatment.
- If a dissection is caused by a diagnostic catheter, a guidewire can be inserted through the diagnostic catheter into the true lumen (if feasible), followed by obtaining a second arterial access and engaging the coronary artery with another guide catheter ("ping pong" technique) [6].

5.6.4.4 Aorto-coronary dissection (Section 25.1.4)

Causes:

- Noncoaxial catheter position.
- Ostial lesion.
- Contrast injection despite pressure dampening.

Prevention:

- **NEVER INJECT** if there is pressure dampening and/or no back bleeding.
- Ensure coaxial catheter position.
- Gentle injections.

Diagnosis:

- Contrast "staining" within the aortic wall.

Treatment:

- **Do NOT inject**.
- Stent the coronary artery ostium to seal off the contrast entry point. High-pressure flaring of the stent is often required.
- Use of a covered stent may be needed if standard stents fail to seal the area of contrast entry into the aortic wall, but they carry increased risk of restenosis and thrombosis.

5.7 Step 7. Manipulate catheter to engage coronary ostia

After the catheter is advanced to the aortic root, it is aspirated to clear it from any thrombus or debris, connected to the manifold and good pressure waveform is confirmed (steps 1–6 as described above). The catheter is then ready to engage the target coronary artery.

5.7.1 Goal

To insert the tip of the catheter into the target coronary artery. The following are characteristics of optimal catheter engagement:

1. No pressure dampening.
2. Coaxial orientation with the vessel (need orthogonal projections to verify).
3. 2–3 mm engagement depth. Deep guide catheter engagement provides additional support, but may also lead to vessel dissection and/or pressure dampening and ischemia.

5.7.2 How to?

- Various maneuvers are performed to engage the coronary ostia depending on the target coronary artery and catheter shape, as described below:

5.7.2.1 Engaging the left main

FEMORAL access

Using the JL, XB, or EBU catheters: The 0.035 in. guidewire is looped over the aortic valve. The catheter tip is advanced 2–3 cm higher than the bottom of the right coronary cusp at the upward reflection of the J wire. The guidewire is then withdrawn, allowing the catheter to fold into the left coronary cusp (Fig. 5.17).

If the catheter tip is advanced too far into the right cusp, subsequent advancement of the catheter would not allow the tip to move into the left cusp, but would drive it deeper into the right cusp instead (Fig. 5.18).

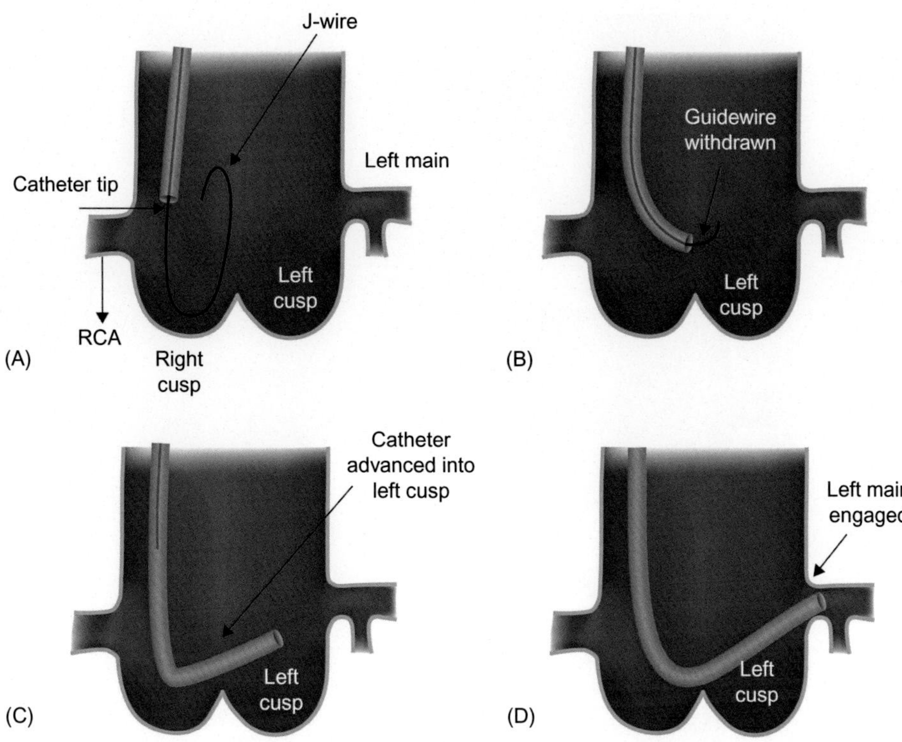

FIGURE 5.17 Engaging the left main coronary artery. The catheter tip is advanced 2–3 cm above the upward deflection of the J wire (panel A). The J wire is then withdrawn (panel B) while simultaneously allowing the catheter into the left cusp (panel C) and then into the left main ostium (panel D).

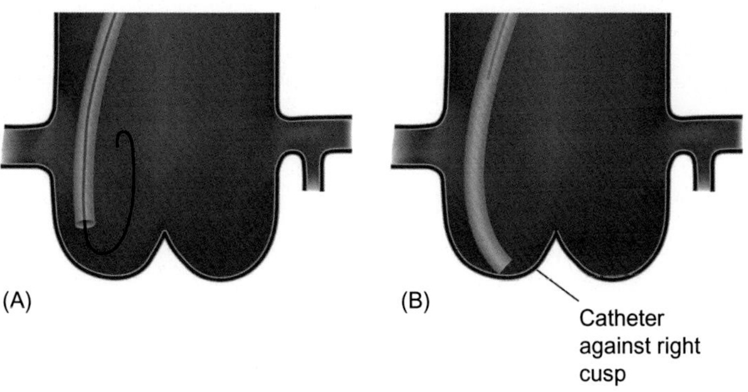

FIGURE 5.18 What to avoid when trying to engage the left main. Advancing the catheter tip too deep into the right coronary cusp (panel A), prevents it from entering the left coronary cusp (panel B).

Sometimes, some rotation is needed for the catheter tip to reach the left main ostium. Counterclockwise rotation may be required, as the left main ostium usually lies posteriorly. Smaller curve guide catheters (such as EBU 3.0 or 3.5) facilitate treating left anterior descending artery lesions and larger curve guide catheters (such as EBU 4.0 or EBU 4.5) facilitate treating circumflex lesions.

RADIAL access

- *Universal shape catheters (such as the Jacky or Tiger) are often used to engage both the left main and the RCA via radial access*: The 0.035 in. guidewire is looped over the aortic valve and the catheter is advanced while withdrawing the guidewire until the tip enters the left main ostium. The catheter is then usually withdrawn to allow coaxial alignment.
- *Standard catheters (JL, XB or EBU) can also be used via radial access*, especially left radial access.

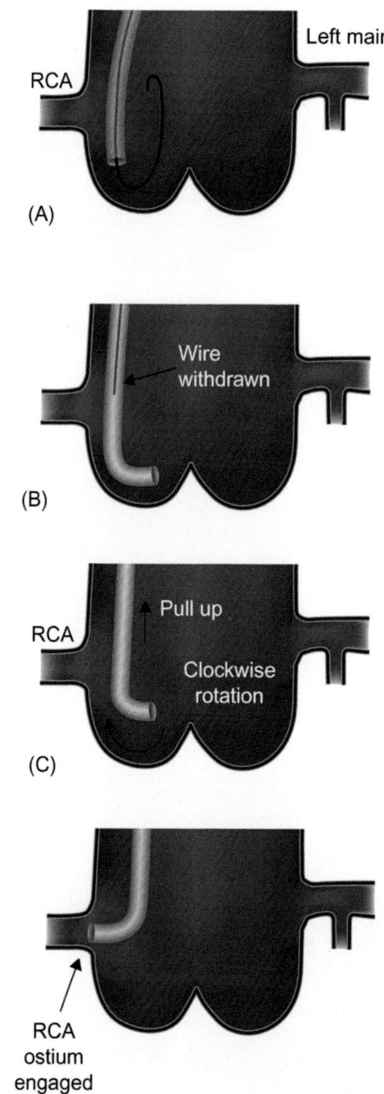

FIGURE 5.19 Engaging the right coronary artery with a JR4 catheter.

5.7.2.2 Engaging the right coronary artery

FEMORAL access

- *JR4 catheter*: The catheter is advanced until it reaches the bottom of the right coronary cusp (Fig. 5.19, panel A) (in contrast to catheters used to engage the left main that should be advanced up to 2–3 cm above the aortic cusp). After withdrawing the guidewire (Fig. 5.19, panel B) the JR4 catheter is retracted a few centimeters and rotated clockwise using the right hand while the left hand is performing back and forth movements (Fig. 5.19, panel C) to transmit the rotation from the hub to the tip of the catheter and engage the right coronary ostium (Fig. 5.19, panel D).
- *Williams right*: This is a preshaped catheter, designed to engage the RCA with limited or no catheter manipulation. Occasionally, clockwise rotation may be needed similar to the JR4 catheter. It can be particularly useful for treating aorto-ostial RCA lesions.
- *AL1*: The AL catheter is advanced until the distal tip is facing up and then rotated until it engages the RCA ostium. AL guide catheters provide the strongest support and are preferred for complex RCA interventions, such as RCA chronic total occlusions.

RADIAL access

- *Universal shape* diagnostic (such as the Jacky or Tiger) or guide catheters can be used for engaging the RCA.

- *AL* guide catheters are preferred for RCA PCI via radial access, as JR guide catheters often provide poor support.

5.7.2.3 *Engaging left-sided bypass grafts (see Section 18.5)*

- The catheter is advanced up and down using the left hand (every movement covers a distance of several cm from the aortic cusp to the origin of the brachiocephalic trunk) with gentle rotation after each up and down movement using the right hand.

5.7.2.4 *Engaging right-sided bypass grafts (see Section 18.5)*

These are engaged in the LAO projection. The catheter (usually multipurpose or RCB) is rotated to face the right side of the aorta in the LAO view and is then advanced up and down until the tip engages the bypass graft ostium.

5.7.2.5 *Engaging IM grafts (see Section 18.5)*

The catheter (IM or occasionally, VB1, JR4, Williams, Bartorelli-Cozzi, or JIM) is advanced over a 0.035 in. guidewire past the origin of the IM graft and is then withdrawn with simultaneous counterclockwise rotation, preferably in the RAO projection.

- Difficulty in catheter manipulation may be due to severe proximal arterial tortuosity that could also predispose to catheter kinking during manipulation attempts. Insertion of long sheaths may improve catheter handling in such cases.

5.7.3 Causes of failure to engage

5.7.3.1 *Anomalous coronary artery origin*

Solutions:

- Perform cusp angiography (or ascending aortography), which can help clarify the presence of an anomalous coronary artery and/or the location of the coronary ostia (PCI Manual Online case 56).
- Use a different angiographic projection (RAO instead of the more commonly used LAO projection).
- Based on the information obtained from the aortography or the cusp angiography use different guide or diagnostic catheters, more suitable to reach the ostium of the target vessel.
- Once the ostium is identified, a different catheter may be needed (such as an AL catheter).
- Coronary CT angiography can be also performed to clarify anatomy in patients with coronary anomalies.

5.7.3.2 *Ostial coronary artery occlusion*

Solutions:

- Perform cusp angiography (or ascending aortography) which can help clarify the presence of an anomalous coronary artery and/or the location of the coronary ostia.
- If the coronary artery ostium has a flush occlusion, engagement may be impossible. Ostial occlusion can also be confirmed by detecting collateral flow to the supplied myocardial territory from another coronary artery.

5.7.3.3 *Inability to manipulate catheter due to peripheral tortuosity*

Solutions:

- Manipulate catheter after inserting a 0.035 or 0.038 in. guidewire. The guidewire increases the stiffness of the catheter, facilitating manipulations and decreasing the likelihood of catheter kinking.
- Inserting the back end of a 0.035 in. guidewire (without exiting the guide catheter) will straighten the guide catheter tip and help with engagement, especially when using radial access.
- Insert a long (such as 45 cm long) sheath.
- If there is still difficulty manipulating the catheter, try using a catheter that this slightly smaller than the sheath (e.g., 7 French guide within an 8 Fr sheath).
- Use a diagnostic catheter to engage a coronary artery that cannot be engaged by a guide catheter. It is easier to manipulate diagnostic catheters (that have thicker walls) than guide catheters (that have thinner walls) (Section 30.2.1). After coronary engagement with a diagnostic catheter a 300 cm long 0.014 in. supportive guidewire (such

as the Iron Man, Grand Slam, or Mailman, Section 30.7.4) can be advanced into the coronary artery and used to remove the diagnostic catheter and insert a guide catheter.

- Spasm (Section 29.2.3) may hinder catheter manipulation when radial access is used. Vasodilator administration and use of smaller diameter catheters may help.
- Change access site, for example from radial to femoral.

5.7.3.4 *Suboptimal catheter size*

Solutions:

- Use a smaller curve (when catheter curve is too big for the size of the aorta) or larger curve (when catheter curve is too small for the size of the aorta) catheter.
- Use a guide catheter extension.
- In very dilated aortic roots with a hard to reach left main artery, a 6 French AL3 guide catheter can be used through which a 4 French diagnostic catheter is advanced (in a telescoping fashion) to reach the left main artery.

5.7.3.5 *Suboptimal catheter shape*

Solutions:

- Try different guide catheter shapes.
- If the guide catheter tip is close to the ostium, inserting a 0.014 in. guidewire to the target coronary artery may be feasible and can facilitate catheter engagement. This is often done for engaging IMA grafts.

5.7.3.6 *What do to if the catheter fails to engage (Fig. 5.20)*

Step 1. If the location of the coronary artery or bypass graft ostia is not understood, performing nonselective cusp angiography (nonselective injection in the coronary cusps) or aortography (usually 60 mL injection over 3 seconds through a pigtail catheter), may allow visualization of the coronary and graft ostia and facilitate subsequent engagement attempts. It may also demonstrate ostial occlusion of the target artery, obviating the need for selective engagement.

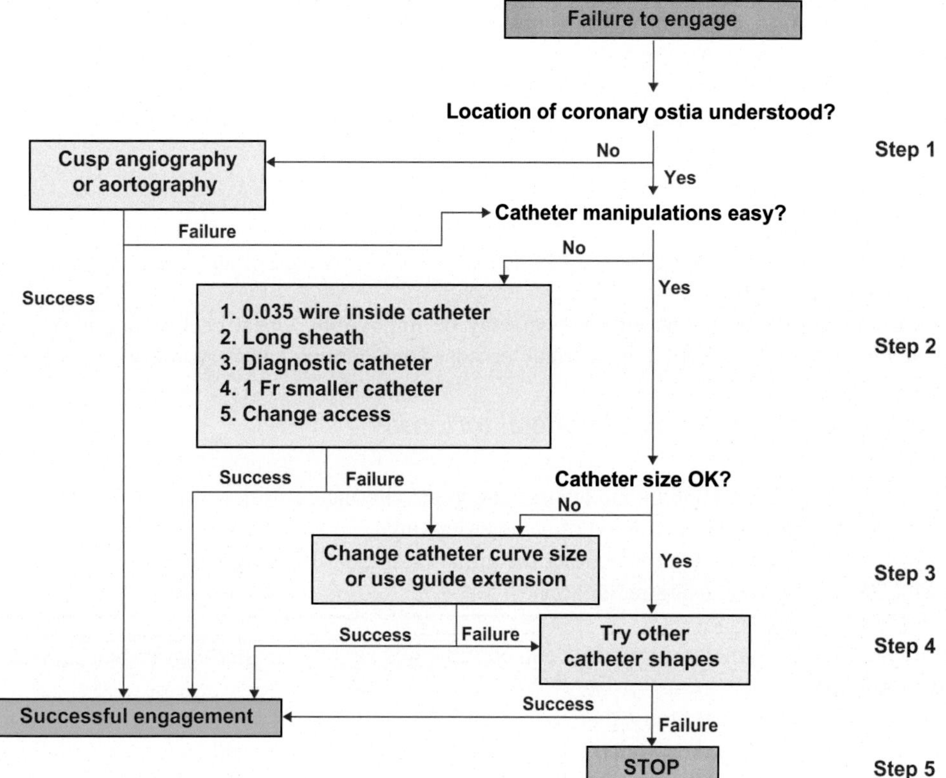

FIGURE 5.20 Failure to engage the coronary artery or bypass graft ostium algorithm.

Step 2. Failure to engage may be due to difficulty manipulating the catheter due to peripheral tortuosity. Several techniques can increase the maneuverability of the catheters, such as inserting a 0.035 or 0.038 in. guidewire in the catheter, using a long sheath, using a diagnostic catheter instead of a guide, using a catheter 1 French smaller than the arterial sheath, or possibly even changing access site.

Step 3. If the catheter curve size is suboptimal, a larger or smaller curve catheter, or a guide catheter extension can be used.

Step 4. Different catheter shapes can be tried to identify one that is successful in engaging the coronary or bypass graft ostium.

Step 5. If all aforementioned steps fail, consider stopping the procedure, especially if large volume of contrast is administered and there is no emergency in determining coronary anatomy (e.g., in patients without STEMI or cardiogenic shock). Coronary computed tomography could then be used to clarify coronary and bypass graft anatomy.

5.7.4 What can go wrong?

5.7.4.1 Catheter kinking

Causes:
- Severe iliac/aortic/subclavian tortuosity.
- Difficulty engaging the coronary ostia requiring extensive catheter manipulation.

Prevention:
- Use long sheaths.
- Avoid excessive catheter rotation.
- Insert a 0.035 or 0.038 in. guidewire inside the catheter during catheter manipulations.

Diagnosis:
- Catheter tip is not moving when rotating the catheter hub.
- Arterial pressure waveform is dampened or lost.
- Fluoroscopy will confirm the location of the kinking.

Treatment (Fig. 5.21) [7]:

Step 1: Gently untwist the catheter (turn in opposite direction than during engagement attempts) and advance a 0.035 in. guidewire through the kinked catheter segment—this can be challenging depending on the severity of kinking. The back end of a guidewire can be advanced (although this carries a risk of perforating the kinked catheter). Alternatively, a 0.069 in. guidewire may be used.

Step 2. Obtain another arterial access (radial or femoral) and snare the tip of the kinked catheter. This can allow straightening of the kinked segment and facilitate "untwisting". If the kink occurs with radial access and it is below the brachial artery, a blood pressure cuff can be inflated at the brachial artery to "fix" the kinked catheter and facilitate untwisting and removal.

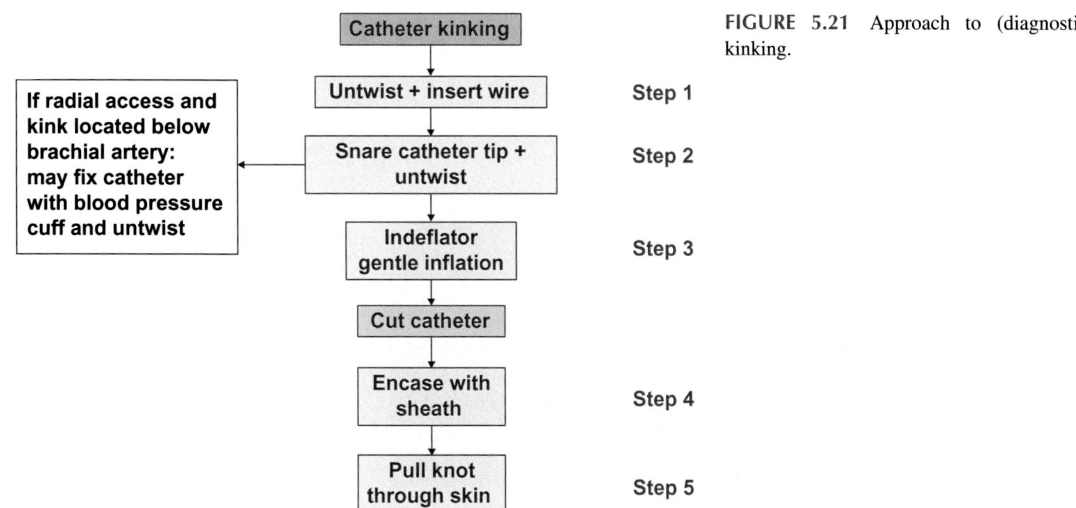

FIGURE 5.21 Approach to (diagnostic or guide) catheter kinking.

Step 3. An indeflator may be attached to the hub of the kinked catheter and inflated gently to 3–4 atm, which may help untwist the kink.

Step 4. The catheter is cut at the hub and a sheath (either of same size or 1–2 French larger than the catheter size) is advanced over the catheter in an attempt to encase the kinked catheter segment and allow withdrawal. However, advancing a sheath over a catheter without a dilator may cause vessel trauma.

Step 5. The kinked catheter segment is pulled through the arteriotomy (provided that there is a long enough catheter segment to allow insertion of a guidewire), followed by insertion of a guidewire (PCI Manual Online case 48).

If the kinked catheter fractures during retrieval attempts, snaring of the catheter fragments can allow their removal without the need for surgical intervention [8].

Contrast injection during catheter engagement manipulations may lead to any of the following complications:

5.7.4.2 Systemic embolization (see complications under step 5, section 5.5.3.1)

5.7.4.3 Coronary embolization (see complications under step 5, section 5.5.3.2)

5.7.4.4 Coronary dissection (see complications under step 6, section 5.6.4.3)

5.7.4.5 Aorto-coronary dissection (see complications under step 6, section 5.6.4.4)

5.8 Step 8. Ensure there is good pressure waveform

This is done as described in step 6.

Additional causes of pressure dampening after catheter engagement include:

5.8.1 Ostial lesions

Solutions:
- Keep catheter disengaged from the coronary ostium. This may be facilitated by advancing a 0.014 in. guidewire into the target coronary vessel. The catheter may be intermittently engaged when necessary to advance equipment or inject contrast (Fig. 5.22).
- Use a different catheter, either of smaller French size or with different shape which is less likely to deeply engage the target coronary artery.
- Manually reshape the tip of the catheter.

5.8.2 Small size of target vessel

Solutions:
- Keep catheter disengaged from the coronary ostium. This may be facilitated by advancing a 0.014 in. guidewire into the target coronary vessel.
- Use a smaller French size catheter.

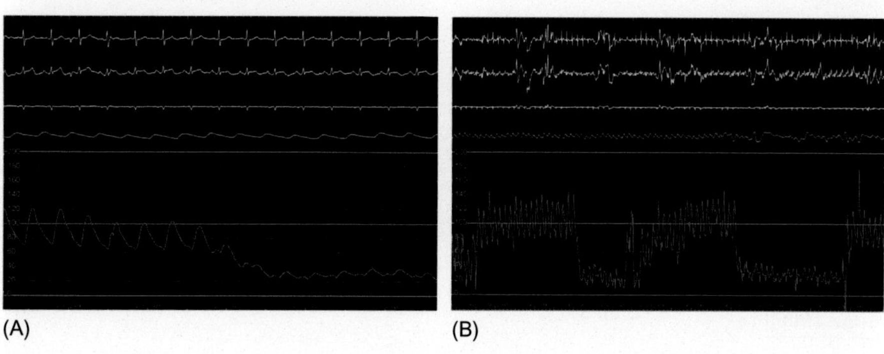

FIGURE 5.22 Pressure dampening upon engagement requiring intermittent engagement and disengagement of the catheter.

5.9 Step 9. Proceed with contrast injection as described in Chapter 6

References

[1] Reifart N, Sotoudeh N. The parallel sheath technique in severe iliac tortuosity: a simple and novel technique to improve catheter manoeuvrability. EuroIntervention 2014;10:231–5.

[2] Agelaki M, Koutouzis M. Balloon-assisted tracking for challenging transradial percutaneous coronary intervention. Anatol J Cardiol 2017;17:E1.

[3] Patel T, Shah S, Pancholy S. Balloon-assisted tracking of a guide catheter through difficult radial anatomy: a technical report. Catheter Cardiovasc Interv 2013;81 E215–8.

[4] Patel T, Shah S, Pancholy S, Rao S, Bertrand OF, Kwan T. Balloon-assisted tracking: a must-know technique to overcome difficult anatomy during transradial approach. Catheter Cardiovasc Interv 2014;83:211–20.

[5] Keeley EC, Grines CL. Scraping of aortic debris by coronary guiding catheters: a prospective evaluation of 1,000 cases. J Am Coll Cardiol 1998;32:1861–5.

[6] Brilakis ES, Grantham JA, Banerjee S. "Ping-pong" guide catheter technique for retrograde intervention of a chronic total occlusion through an ipsilateral collateral. Catheter Cardiovasc Interv 2011;78:395–9.

[7] Ben-Dor I, Rogers T, Satler LF, Waksman R. Reduction of catheter kinks and knots via radial approach. Catheter Cardiovasc Interv 2018;92:1141–6.

[8] Michael TT, Banerjee S, Brilakis ES. Percutaneous retrieval of a fractured guide catheter using contralateral snaring. J Invasive Cardiol 2012;24 E176–8.

Chapter 6

Coronary angiography

Although adjunctive modalities (stress testing, noninvasive imaging including coronary computed angiography, coronary physiology, and intracoronary imaging) can help evaluate coronary anatomy, coronary angiography remains the most commonly used technique for assessing the presence and severity of coronary artery stenoses and for planning coronary revascularization (surgical or percutaneous).

For coronary angiography to provide accurate information about coronary anatomy, it should be performed using meticulous technique, which can help maximize the accuracy of the imaging, while minimizing the potential risks associated with the procedure.

6.1 Step 1. Ensure there is good pressure waveform

6.1.1 Goal

To ensure that the catheter tip is free and coaxial with the coronary artery ostium, not obstructing coronary flow or engaged against the arterial wall and that the catheter lumen is clear of any foreign material, such as thrombus, plaque, or air.

This is performed as described in step 6 of Chapter 5: Coronary and Graft Engagement (Section 5.6).

6.2 Step 2. Test contrast injection

6.2.1 Goal

To confirm that the tip of catheter is engaged in the coronary artery (or bypass graft) ostium. Optimally, the catheter should engage the ostium coaxially to avoid tenting which could lead to dissection.

Sometimes test injection may not be necessary if the operator is certain (based on catheter movement and "jump" motion of the catheter) that the coronary ostium is engaged.

6.2.2 How to?

1. 1–2 mL of contrast is injected through the catheter (that has been cleared of any thrombus, air, or debris as described in Section 5.6.

6.2.3 What can go wrong?

6.2.3.1 Coronary artery dissection (Section 5.6.4.3 and Section 25.2.1 and PCI Manual Online case 40)

Causes:

- Suboptimal catheter engagement (e.g., deep catheter intubation, subintimal or noncoaxial intubation, or catheter positioned at the edge of or under an eccentric plaque, especially of the left main coronary artery).
- Ostial coronary lesion.
- Forceful contrast injection (usually when performed in the setting of suboptimal catheter engagement as described above).
- Dampened arterial waveform.

Prevention:

- Ensure nondampened pressure waveform (step 1 in this chapter) before injecting contrast.
- Ensure coaxial catheter placement.
- Inject contrast gently (if manual injection is used).

Manual of Percutaneous Coronary Interventions. DOI: https://doi.org/10.1016/B978-0-12-819367-9.00006-8

Treatment (Section 25.2.1):
- STOP injecting contrast.
- Gentle catheter retraction.
- Wire through dissected coronary artery with a workhorse guidewire into the true lumen, or reenter distally using chronic total occlusion techniques. It may be necessary to select a guide catheter with a slightly different distal curve than the one that caused the dissection.
- Stent area of dissection. Intravascular imaging is strongly recommended to ensure complete coverage of the ostial dissection by the stent.
- Treat patient symptoms as appropriate.

6.2.3.2 Coronary embolization (Sections 5.5.3.2 and 25.2.3)

6.3 Step 3. Administer intracoronary nitroglycerin

PCI Manual Online case 56 (spasm of anomalous circumflex—resolved after nitroglycerin administration)

6.3.1 Goal

To prevent and/or correct coronary spasm that could be interpreted as fixed coronary lesions and to optimize balloon and stent sizing.

6.3.2 How?

Check systolic blood pressure and administer nitroglycerin (50–200 mcg intracoronary) unless the patient has severe hypotension:
- $>$ 120 mmHg: administer 200 mcg nitroglycerin.
- 100–120 mmHg: administer 100 mcg nitroglycerin.
- $<$ 100 mmHg: 50 mcg or no nitroglycerin (due to concern for worsening hypotension).

Nitroglycerin can be administered through the manifold or directly through the Y-connector using the introducer needle ("nitro on a stick").

Alternatively, a sublingual nitroglycerin (0.4 mg) can be given.

Wait 1–2 minutes prior to coronary angiography, as it can take time (up to 2 minutes) for vasodilation to occur [1]. A drop in systemic blood pressure is common.

As an alternative, in cases of suspected spasm that will not recede following administration of nitroglycerin, use of OCT or IVUS may help differentiate between coronary spasm and true atherosclerotic stenosis.

6.3.3 What can go wrong?

6.3.3.1 Hypotension (see Section 28.1.1)

Causes:
- Vasodilatory effect of nitroglycerin.
- Vasodilators administered prior to angiography to facilitate radial access, such as verapamil.
- Acute inferior myocardial infarction with right ventricular infarction.
- Hypovolemia (which is possible in patients who are kept NPO before cardiac catheterization if they are not given intravenous fluids).
- Recent use of phosphodiesterase type 5 (PDE5) inhibitors, such as sildenafil (Viagra, within prior 24 hours), vardenafil (Levitra, within 24 hours), avanafil (Stendra, within 24 hours) and tadalafil (Cialis, within 48 hours).
- Vasovagal reaction to nitrates.
- Transducer or connection issues (pseudohypotension).

Prevention:
- Do not administer nitroglycerin if systemic pressure is low or if the patient is suspected to be hypovolemic, has right ventricular infarction, or has recently taken a phosphodiesterase type 5 (PDE5) inhibitor.

Treatment:
- Administer normal saline (often done prophylactically after nitroglycerin administration).
- Observe until blood pressure recovers.
- In severe hypotension cases, administer vasopressors, such as phenylephrine.

6.4 Step 4. Optimally position patient, image receptor, shields, and operator

6.4.1 Goal

To optimize the position of the X-ray image receptor and the patient to obtain excellent cineangiography images with the lowest possible radiation dose for both the patient and the operator.

6.4.2 How?

Patient position: the table should be placed as high as possible, while remaining comfortable and ergonomic for the operator.

Image receptor position: the image receptor should be placed as close as possible to the patient.

Image receptor angulation:

Left main: usual initial projection is AP (anteroposterior) to assess the left main ostium.

Left anterior descending artery: RAO (right anterior oblique) with some caudal and AP cranial.

Right coronary artery: usual initial projection is LAO (left anterior oblique) with cranial angulation.

Saphenous vein and radial grafts: usual initial projection is LAO.

IM grafts: usual initial projection is AP.

Shielding position: placed between patient and operator to reduce scatter radiation to the operator.

Operator position: as far as possible from the patient and X-ray tube. This can be facilitated by using a tubing extension for the manifold.

6.4.3 Challenges

6.4.3.1 Poor image quality

Causes:
- Large patient size.
- Excessive angulation.
- Image receptor too far from the patient.
- Differences in density of surrounding structures.

Prevention:
- Use shallower angulation, which facilitates X-ray beam penetration.
- Collimate to the area of interest and use filters.

Treatment:
- Use shallower angulation, which facilitates X-ray beam penetration.
- Change X-ray settings (such as frame rate and pulse rate to increase radiation dose and penetration).

6.4.4 What can go wrong?

6.4.4.1 Patient injury

Causes:
- Excessive movement of the image receptor.

Prevention:
- Caution while moving image receptor.
- Some X-ray systems have sensors that prevent the image receptor from hitting the patient.

Treatment:
- Depends on the type of injury.

6.4.4.2 *Operator injury*

Causes:
- Failure to move away from the X-ray system trajectory.

Prevention:
- Move away from the X-ray system moving parts.

Treatment:
- Depends on the type of injury—usually no specific treatment is required.

6.5 Step 5. Assess pressure and ECG

6.5.1 Goal

To ensure that there is no pressure dampening or new ECG changes before performing angiography.

6.5.2 How?

1. Assess the pressure waveform, to verify that there is no dampening, ventricularization, or new hypotension. This is performed as outlined in Section 5.6.
2. Assess the ECG, to confirm that there are no new ECG changes (contrast administration may sometimes cause ST segment elevation or depression or T-wave inversion).

6.5.3 Challenges

1. Poor-quality ECG signal. This may require repositioning of the electrodes.
2. Poor-quality pressure signal. The transducer is flushed and the connections checked.

6.5.4 What can go wrong?

1. Coronary ischemia due to catheter occluding or partially occluding antegrade coronary flow, which can lead to hypotension and arrhythmias, even ventricular fibrillation, if not corrected expeditiously. Remedy by withdrawing the catheter, and reengaging more carefully or by using a smaller French size catheter.
2. Coronary dissection due to contrast injection despite ventricularization of the pressure waveform.

6.6 Step 6. Perform cineangiography

6.6.1 Goal

To record high-quality cineangiographic videos of each coronary artery in orthogonal projections, minimizing overlap.

6.6.2 How?

1. The patient and image receptor are positioned optimally to allow coronary artery visualization with minimal or no panning. For example, for imaging the left coronary system in the RAO cranial projection the catheter tip should be at the left upper corner of the screen. When performing the LAO caudal (spider) projection, the catheter should be in the middle and center of the screen.
2. The operator presses the cineangiography pedal.
3. After 1–2 seconds (to allow visualization of any prior stent, calcium or contrast clearing from the prior injection) the assistant or operator (using a syringe or an automated system, if available) injects contrast to fully opacify the coronary artery, ensuring there is contrast reflux into the aorta. The syringe is held vertical to prevent potential bubble embolization.
4. Panning is performed (if needed) to ensure imaging of all the branches and the distal portion of the injected coronary artery, as well as potential collateral filling of other vessels However, panning is best avoided as it degrades the quality of the image, especially when performing dual injection for chronic total occlusion interventions.

5. Image acquisition is continued until all branches (and collateral branches in cases of contralateral or ipsilateral vessel occlusion) are opacified. Long acquisitions should be avoided except when delayed collateral filling is present.
6. At the end of the injection the syringe is refilled with contrast, to facilitate the subsequent injection.
7. Repeat until orthogonal projections are obtained for all coronary arteries and bypass grafts:
   a. Left main: AP, RAO caudal, RAO cranial, LAO cranial, LAO caudal (spider).
   b. LAD: AP or RAO cranial, RAO caudal.
   c. RCA: LAO, LAO cranial or AP cranial, RAO.
   d. Saphenous vein and radial grafts: LAO, RAO.
   e. IM: AP, lateral, AP cranial or RAO cranial.

6.6.3 Challenges

6.6.3.1 *Inability to fill the coronary artery*

Complete filling of the coronary artery is critical for accurate interpretation of the coronary angiography. If complete filling cannot be achieved, it is best to declare that the angiogram was nondiagnostic, than provide a potentially erroneous interpretation.

Causes:
- Small catheter (especially 4 French).
- Side hole catheters.
- Large or ectatic coronary arteries.
- Poor engagement of the coronary artery.
- High coronary flow (e.g., in patients with severe aortic stenosis, hypertrophic obstructive cardiomyopathy, dilated cardiomyopathy, arteriovenous fistula).
- High contrast viscosity (isoosmolar contrast agents are more viscous than low-osmolar contrast agents).
- Weak injection.
- High left ventricular end-diastolic pressure.

Prevention and treatment:
- Use a catheter with larger lumen (e.g., a guide catheter instead of a diagnostic catheter or a larger French size diagnostic catheter).
- Avoid side hole catheters.
- Stronger injection, possibly using an automated injector, although manual injection allows more tailoring of the force of injection: if there is slight pressure dampening manual injection is preferred.

6.6.3.2 *Catheter disengages coronary artery during injection*

Causes:
- Poor engagement.
- Strong contrast injection.
- Insufficient catheter support relative to the required contrast injection force.
- Patient taking deep breaths, especially during radial procedures.

Prevention:
- Ensure good guide engagement with coaxial alignment prior to contrast injection.
- Start injection softly and increase force while injecting.

Treatment:
- Reengage catheter and retry.
- Use another catheter.
- Insert guidewire into coronary artery to anchor catheter.
- Use a guide catheter extension to better engage the coronary artery.
- Slowly ramp up the injection pressure.

6.6.3.3 *Unable to optimally visualize some coronary segments*

Causes:
- Poor engagement.

- Weak contrast injection.
- Overlap of coronary segments.
- Patient size (larger patients are more difficult to image as the X-ray has to penetrate through thicker tissue planes).
- Not enough angiographic views.

Prevention:
- Optimize coronary artery filling.
- Use various projections to minimize overlap of various coronary segments.

Treatment:
- Improve engagement (if poor visualization is due to poor filling).
- Use larger catheter (if poor visualization is due to poor filling).
- Use different projections (at least two orthogonal projections).

6.6.3.4 What can go wrong (see also Chapter 25: Acute Vessel Closure)?

1. Spasm (that is why nitroglycerin should be administered before coronary angiography unless there is a contraindication, such as hypotension) (Section 25.2.5).
2. Air embolization (see Chapter 5: Coronary and Graft Engagement, step 5 Section 5.5 and Section 25.2.3.3).
3. Thrombus or other debris embolization (see Chapter 5: Coronary and Graft Engagement, step 5 Section 5.5 and Section 25.2.3).
4. Dissection (Section 25.2.1).

6.7 Step 7. Assess pressure and ECG

6.7.1 Goal

To ensure that there is no pressure dampening, new hypotension, or new ECG changes after each angiography run.

6.7.2 How?

1. Assess the pressure waveform, to verify that there is no dampening or new hypotension.
2. Assess the ECG, to confirm that there are no new ECG changes (ST segment or rhythm changes) that require treatment, such as ventricular fibrillation.

6.8 Step 8. Angiogram interpretation

6.8.1 Goal

To determine coronary flow and the presence and severity of coronary stenoses in the acquired cineangiography runs.

6.8.2 How?

The following parameters are assessed for each artery and lesion in each cineangiographic run [2].

6.8.2.1 Coronary flow

Coronary flow usually assessed using the TIMI (Thrombolysis in Myocardial Infarction) flow grade as follows (Fig. 6.1): ***Grade 0*** (no perfusion): There is no antegrade flow beyond the point of occlusion.

Grade 1 (penetration without perfusion): The contrast material passes beyond the area of obstruction but "hangs up" and fails to opacify the entire coronary bed distal to the obstruction for the duration of the cineangiographic filming sequence.

Grade 2 (partial perfusion): The contrast material passes through the obstruction and opacifies the coronary bed distal to the obstruction. However, the rate of entry of contrast material into the vessel distal to the obstruction or its rate of clearance from the distal bed (or both) is perceptibly slower than its entry into or clearance from comparable areas not perfused by the previously occluded vessel (e.g., the opposite coronary artery or the coronary bed proximal to the obstruction).

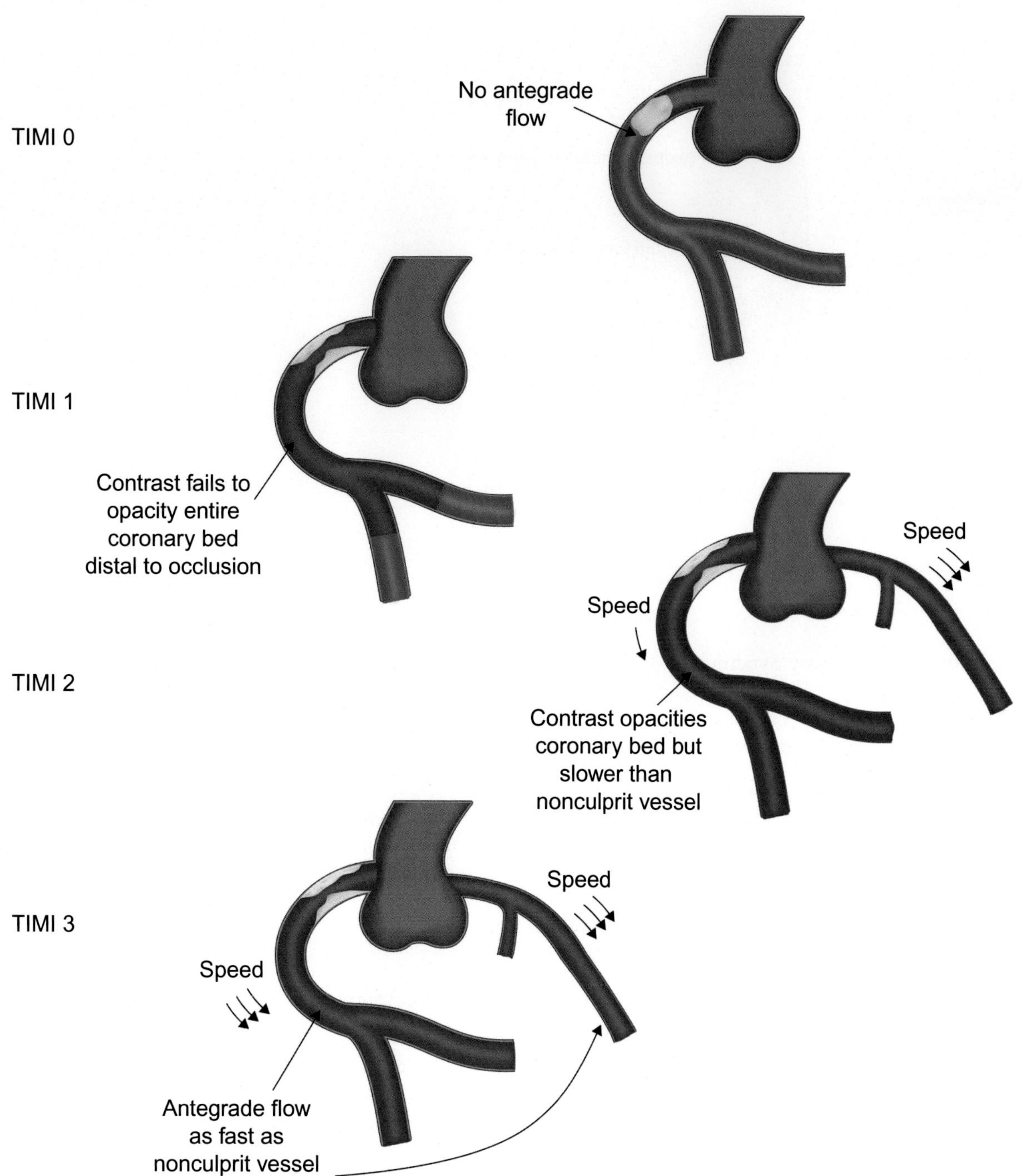

FIGURE 6.1 Thrombolysis In Myocardial Infarction (TIMI) flow grade classification.

Grade 3 (complete perfusion): Antegrade flow into the bed distal to the obstruction occurs as promptly as antegrade flow into the bed proximal to the obstruction and is as rapid as clearance from an uninvolved bed in the same vessel or the opposite artery.

6.8.2.2 Presence of coronary lesions

Presence of coronary lesions is assessed as percent diameter stenosis by comparing them with the adjacent angiographically normal coronary artery segment (Fig. 6.2). Estimating lesion stenosis at or beyond major side branches can be challenging due to uncertain change in diameter before and after the bifurcation. Occlusive (100%) lesions both acute

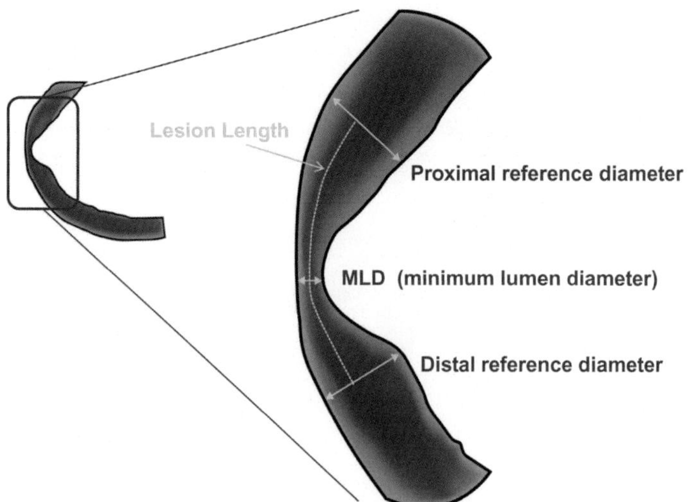

$$\text{Average reference diameter} = \frac{\text{Proximal + Distal reference diameter}}{2}$$

$$\%\ \text{Diameter stenosis} = 100 \times \left[1 - \frac{\text{MLD}}{\text{Average reference diameter}}\right]$$

FIGURE 6.2 Estimation of percent diameter stenosis.

(Chapter 20: Acute Coronary Syndromes—Thrombus) and chronic (Chapter 21: Chronic Total Occlusions) are harder to recanalize.

Orthogonal projections of each arterial segment are needed to assess each lesion. The projection in which the stenosis appears to be the worst and least foreshortened is used (Fig. 6.3).

- $\geq 70\%-90\%$ diameter stenosis for non-left main lesions and $\geq 50\%$ diameter stenosis for left main lesions is considered significant; however, % diameter stenosis alone has limitations when assessing the hemodynamic severity of each lesion, as illustrated in Fig. 6.4 [3]. Physiologic assessment (as described in Chapter 12: Coronary Physiology) is preferred for determining whether a lesion should be intervened upon, although it should not be used in culprit lesions for acute coronary syndromes.

6.8.2.3 Lesion length (estimated in mm)

Lesion length is usually measured in mm from the proximal to the distal "normal" reference segment (Fig. 6.2).

6.8.2.4 Thrombus (Chapter 20: Acute Coronary Syndromes—Thrombus)

Coronary angiography has limited capacity to detect thrombus, which is seen as an intraluminal filling defect with or without contrast staining and can be confused with a calcific nodule. The most commonly used thrombus classification is the TIMI classification that divides angiographic thrombus into five categories [4]:

- TIMI thrombus grade 1: possible thrombus is present, with such angiography characteristics as reduced contrast density, haziness, irregular lesion contour, or a smooth convex "meniscus" at the site of total occlusion, suggestive but not diagnostic of thrombus.
- TIMI thrombus grade 2: definite thrombus, with greatest dimensions $\leq 1/2$ the vessel diameter.
- TIMI thrombus grade 3: definite thrombus but with greatest linear dimension $>1/2$ but <2 vessel diameters.
- TIMI thrombus grade 4: definite thrombus, with the largest dimension ≥ 2 vessel diameters.
- TIMI thrombus grade 5: total occlusion.

6.8.2.5 Calcification (Chapter 19: Calcification)

Severe calcification is usually defined as radioopacities noted without cardiac motion before contrast injection, involving both sides of the arterial wall. Moderate calcification is defined by densities involving $<50\%$ of the

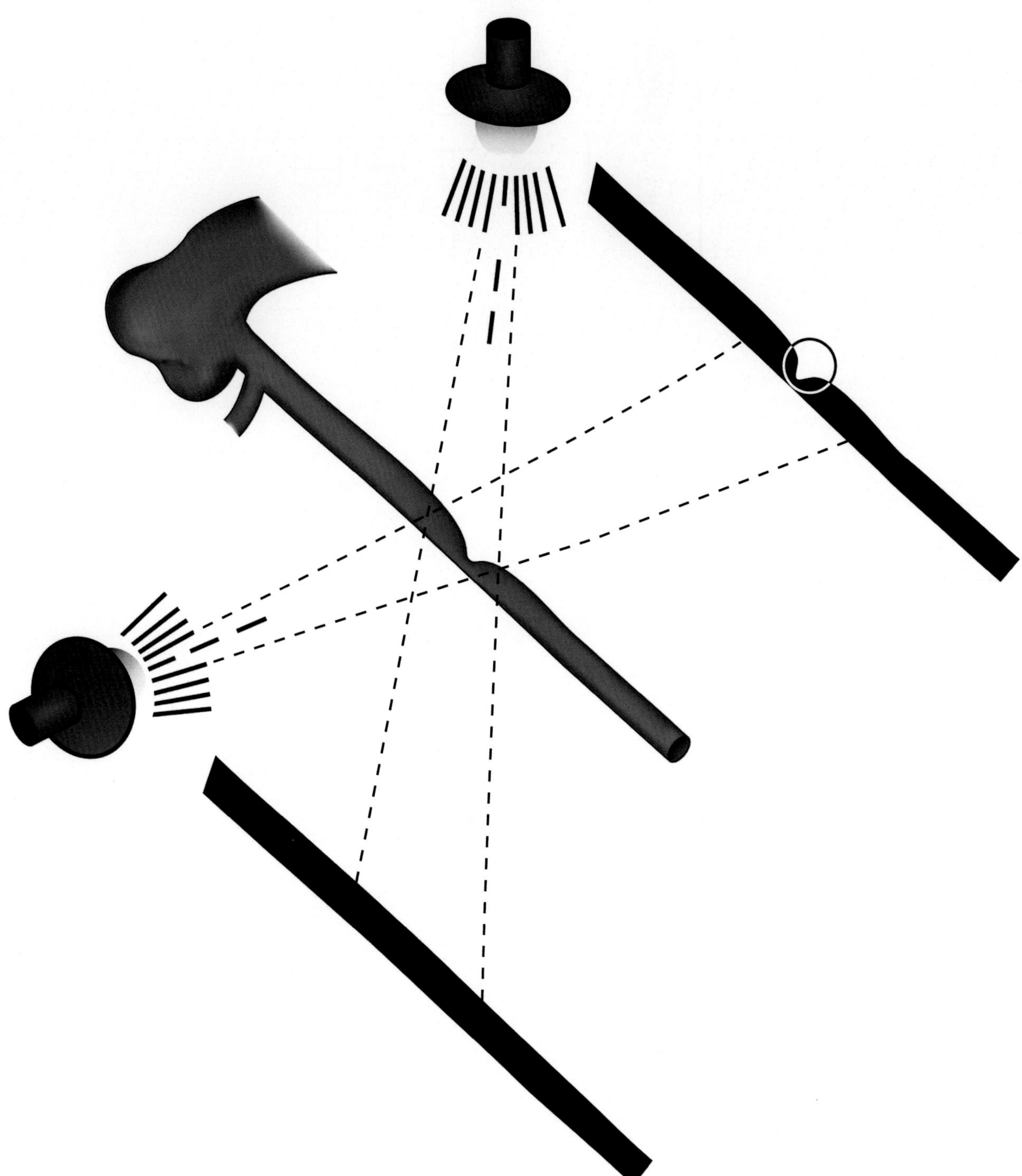

FIGURE 6.3 Illustration of the importance of orthogonal projections for assessing eccentric coronary lesions.

reference lesion diameter, often noted only during cardiac movement before contrast injection. Coronary angiography underestimates the presence and severity of coronary calcium as compared with intravascular imaging [5].

6.8.2.6 Dissection (Section 25.2.1)

Coronary artery dissections resulting from percutaneous transluminal coronary angioplasty are categorized according to the National Heart, Lung, and Blood Institute classification system (Figs. 6.5 and 6.6) [6]. Although the system was developed in the balloon angioplasty era it remains useful in the current stent era.

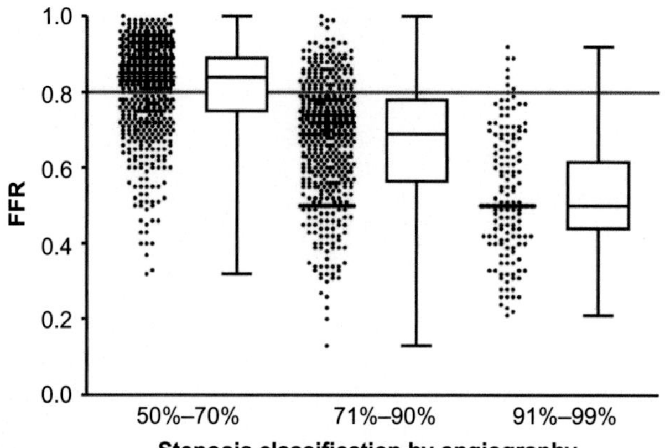

FIGURE 6.4 Angiographic versus functional severity of coronary artery stenoses. *Reproduced from Tonino PAL, Fearon WF, De Bruyne B, et al. Angiographic versus functional severity of coronary artery stenoses in the FAME study: fractional flow reserve versus angiography in multivessel evaluation. J Am Coll Cardiol 2010;55:2816–21. Copyright Elsevier.*

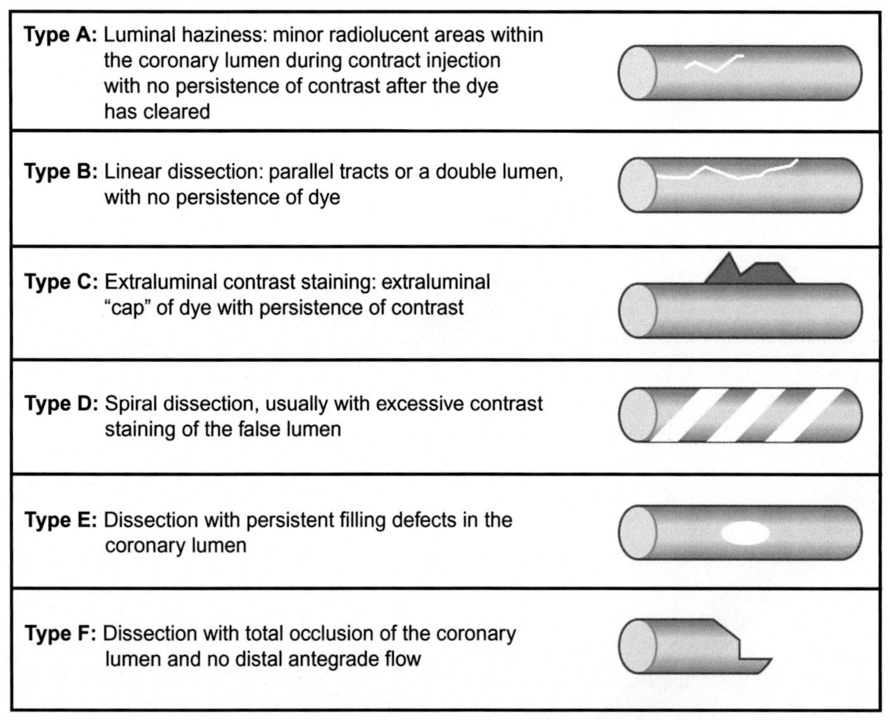

FIGURE 6.5 The National Heart, Lung and Blood Institute (NHLBI) grading of coronary artery dissections [6].

Some patients have spontaneous coronary artery dissection (Section 22.1). In some other patients, aortic dissection may involve the coronary ostia causing an acute coronary syndrome [7].

6.8.2.7 Branch involvement (bifurcation lesions, Chapter 16: Bifurcations)

According to the European Bifurcation Club a bifurcation lesion is ''a coronary artery narrowing occurring adjacent to, and/or involving, the origin of a significant side branch''. PCI of bifurcation lesions is discussed in detail in Chapter 16.

6.8.2.8 Ostial location (aorto-ostial or branch ostial, Chapter 15: Ostial Lesions)

Ostial lesions are lesions located within 3-5 mm of the vessel origin. There are 2 types of ostial lesions: (a) aorto-ostial lesions, and (b) branch ostial lesions (Figure 15.1). Aorto-ostial lesions are located at the ostia of the right coronary

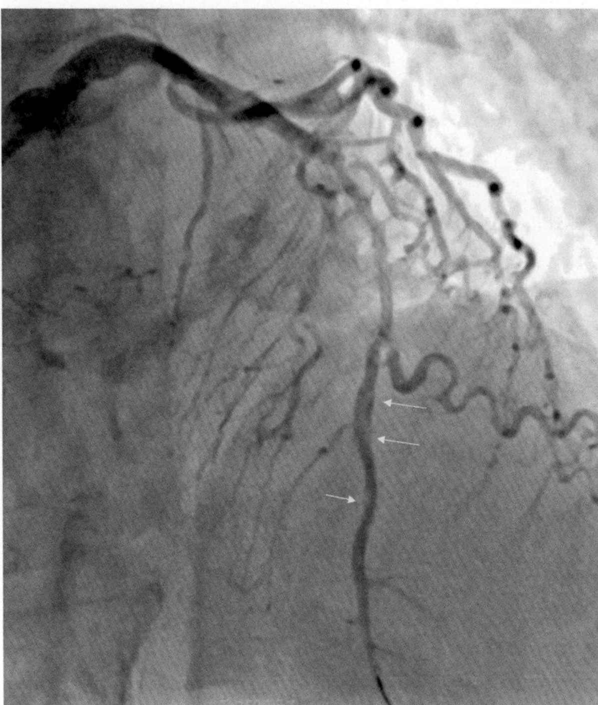

FIGURE 6.6 Type D dissection following balloon angioplasty characterized by a spiral filling defect (*arrows*) in the mid and distal LAD. *Courtesy of Dr. Ivan Chavez.*

artery, left main, and aorto-coronary bypass grafts (both saphenous vein grafts and arterial grafts). Branch ostial lesions are located at the ostia of branches of the coronary vessels. PCI of ostial lesions is described in detail in Chapter 15.

6.8.2.9 Proximal and within lesion tortuosity

Proximal tortuosity is usually defined as *moderate* in the presence of at least 2 bends >70° or 1 bend >90° and as *severe* if there are 2 bends >90° or 1 bend >120° proximal to the target lesion. Within lesion bend >45° is considered significant for chronic total occlusions. Chapter 8.6.10 part C describes various techniques about how to advance a guidewire through tortuosity.

6.8.2.10 Aneurysm

Coronary artery aneurysm is defined as a localized dilation of a portion or diffuse segments of the coronary artery with a diameter of more than 1.5 times the adjacent normal coronary artery (Fig. 6.7). PCI Manual online case 80.

6.8.2.11 Collateral flow

Collateral filling of a coronary artery from another coronary artery is suggestive of acute or chronic occlusion or high-grade stenosis of the collateralized coronary artery. Collateral filling of a vessel is categorized using the Rentrop classification (Fig. 6.8). The size of collaterals is assessed using the Werner classification (Fig. 6.9) [8].

6.8.2.12 Perforation

Perforation is coronary veseel injury resulting in bleeding through the vessel wall. Coronary perforations are usually classified according to severity using the Ellis classification (Figure 26.3, Chapter 26: Perforation) [9]:

- Class 1: A crater extending outside the lumen only in the absence of linear staining angiographically suggestive of dissection.
- Class 2: Pericardial or myocardial blush without a ≥1 mm exit hole.
- Class 3: Frank streaming of contrast through a ≥1 mm exit hole.

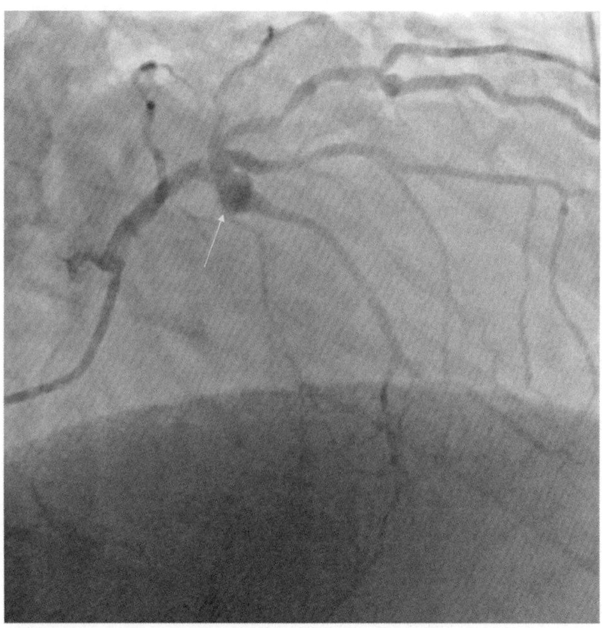

FIGURE 6.7 Example of a coronary artery aneurysm (*arrow*). *Courtesy of Dr. Ivan Chavez.*

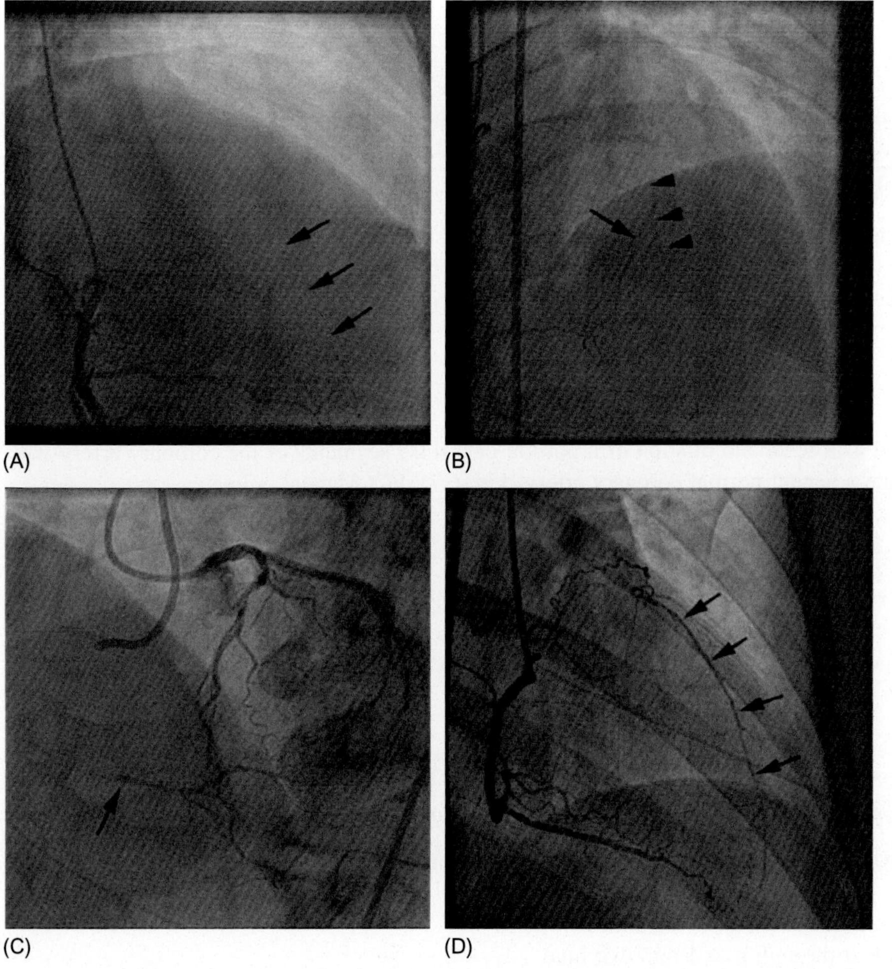

FIGURE 6.8 Rentrop classification of collateral filling. (A) Example of Rentrop Grade 0 left anterior descending artery (LAD) chronic total occlusion (CTO) with no visible collaterals from the right coronary artery (RCA) donor vessel (*arrows* indicate LAD not well seen). (B) Rentrop Grade 1 collateral (*arrow*) from RCA to LAD (*arrowheads*) only filling a side branch. (C) Rentrop Grade 2 collateral from LAD to RCA resulting in partial filling of RCA CTO (*arrow*). (D) Rentrop Grade 3 from RCA resulting in complete filling of LAD CTO (*arrows*). *Reproduced with permission from Vo MN, Brilakis ES, Kass M, Ravandi A. Physiologic significance of coronary collaterals in chronic total occlusions. Can J Physiol Pharmacol 2015;93:867–71.*

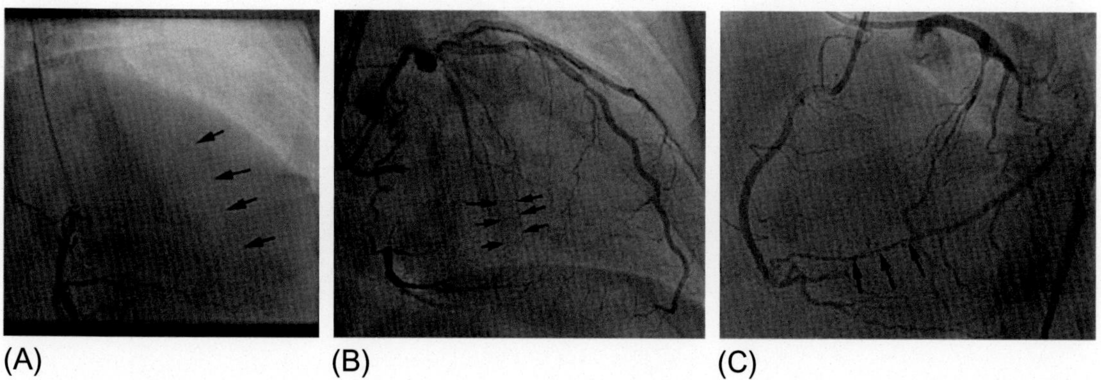

FIGURE 6.9 Werner classification of collateral size. (A) CC0 (no continuous connection) from the right coronary artery (RCA) to the left anterior descending artery (LAD) that is not well seen (*arrows*). (B) CC1 (threadlike continuous) connections (*arrows*) from the LAD to the RCA. (C) Epicardial CC2 (side-branch like) connection (*arrows*) from the RCA to the left circumflex. *Reproduced with permission from Vo MN, Brilakis ES, Kass M, Ravandi A. Physiologic significance of coronary collaterals in chronic total occlusions. Can J Physiol Pharmacol 2015;93:867–71.*

- Class 3-cavity spilling: Perforation into an anatomic cavity chamber, such as the coronary sinus, the right ventricle, etc.

  Coronary perforations should also be classified according to location as follows: (a) large vessel perforations, (b) distal vessel perforations, and (c) collateral vessel perforations (Figure 26.1 and 26.2). Management of coronary perforations is discussed in detail in Chapter 26.

6.8.2.13 Fistulae

Coronary fistulae are abnormal direct connections between a coronary artery and a systemic vein, the pulmonary artery, a pulmonary vein, a cardiac chamber or the coronary sinus. Most fistulae do not require treatment, which is reserved for:

1. Large amount of blood flow that may result in right- or left-sided volume overload.
2. Ischemia due to the steal phenomenon (preferential flow of blood through the low-pressure fistula instead of through the coronary capillaries).

References

[1] Ahmed B, Martinez JD, Schevchuck A, et al. Appropriate timing of nitroglycerin prior to intravascular ultrasound. J Invasive Cardiol 2012;24:422–6.

[2] Cannon CP, Battler A, Brindis RG, et al. American College of Cardiology key data elements and definitions for measuring the clinical management and outcomes of patients with acute coronary syndromes. A report of the American College of Cardiology Task Force on Clinical Data Standards (Acute Coronary Syndromes Writing Committee). J Am Coll Cardiol 2001;38:2114–30.

[3] Tonino PAL, Fearon WF, De Bruyne B, et al. Angiographic versus functional severity of coronary artery stenoses in the FAME study: fractional flow reserve versus angiography in multivessel evaluation. J Am Coll Cardiol 2010;55:2816–21.

[4] Gibson CM, de Lemos JA, Murphy SA, et al. Combination therapy with abciximab reduces angiographically evident thrombus in acute myocardial infarction: a TIMI 14 substudy. Circulation 2001;103:2550–4.

[5] Mintz GS, Popma JJ, Pichard AD, et al. Patterns of calcification in coronary artery disease. A statistical analysis of intravascular ultrasound and coronary angiography in 1155 lesions. Circulation 1995;91:1959–65.

[6] Huber MS, Mooney JF, Madison J, Mooney MR. Use of a morphologic classification to predict clinical outcome after dissection from coronary angioplasty. Am J Cardiol 1991;68:467–71.

[7] Lentini S, Perrotta S. Aortic dissection with concomitant acute myocardial infarction: from diagnosis to management. J Emerg Trauma Shock 2011;4:273–8.

[8] Vo MN, Brilakis ES, Kass M, Ravandi A. Physiologic significance of coronary collaterals in chronic total occlusions. Can J Physiol Pharmacol 2015;93:867–71.

[9] Ellis SG, Ajluni S, Arnold AZ, et al. Increased coronary perforation in the new device era. Incidence, classification, management, and outcome. Circulation 1994;90:2725–30.

Chapter 7

Selecting target lesion(s)

The following decisions need to be made after coronary angiography is performed.

1. Is coronary revascularization needed?
   Like every other procedure, coronary revascularization should be done when the anticipated benefits exceed the potential risks. Potential benefits are improving symptoms and improving prognosis. This is discussed separately for patients with stable angina (Section 7.1) and for patients with acute coronary syndromes (ACS) (Section 7.2).
2. If yes, should it be done with percutaneous coronary intervention (PCI) or coronary artery bypass graft surgery (CABG)?
   PCI and CABG have advantages and disadvantages: PCI is generally easier to perform and carries lower upfront risk, but is associated with higher need for repeat revascularization compared with CABG. The choice of revascularization modality is discussed in Sections 7.1 and 7.2.
3. If PCI is selected, which lesions should be treated, in-which sequence and with what techniques?
   Target lesion selection depends on clinical presentation (e.g., culprit lesions should be treated first in ACS patients), lesion location and lesion complexity. This is discussed in Section 7.3.

Algorithms for determining the need, modality, and sequence (in case of PCI) of coronary revascularization are discussed in the following sections.

7.1 Stable angina - Chronic Coronary Syndromes

(Fig. 7.1)

7.1.1 Symptoms?

The goal of coronary revascularization is to improve symptoms (help patients feel better) or improve prognosis (live longer or reduce risk of subsequent unwanted events, such as myocardial infarction).

The following symptoms are often caused by coronary artery disease (CAD):

1. Chest pain that is provoked by exertion and relieved by rest. Chest pain at rest is the hallmark of acute coronary syndromes and is discussed in Section 7.2.
2. Dyspnea, which is a frequent "anginal equivalent."

Symptoms can only be improved if they are present at baseline! In other words, asymptomatic patients cannot feel better after coronary revascularization. Some patients, however, deny symptoms in part because they gradually limit their activities. Obtaining information from their family and performing an exercise stress test can help determine if a patient is truly asymptomatic or not.

Coronary revascularization is more effective in relieving angina than medical therapy [1], reduces or eliminates the need for antianginal medications and improves quality of life, but is not associated with improved prognosis (lower risk of myocardial infarction or death) in stable CAD patients [1,2,3] except possibly in patients with high-risk coronary anatomy [4] as described in Sections 7.1.2 and 7.1.7.

7.1.2 CAD extent

The extent of CAD is a key determinant of whether coronary revascularization is needed and of the optimal type of coronary revascularization (PCI or CABG).

Manual of Percutaneous Coronary Interventions. DOI: https://doi.org/10.1016/B978-0-12-819367-9.00007-X

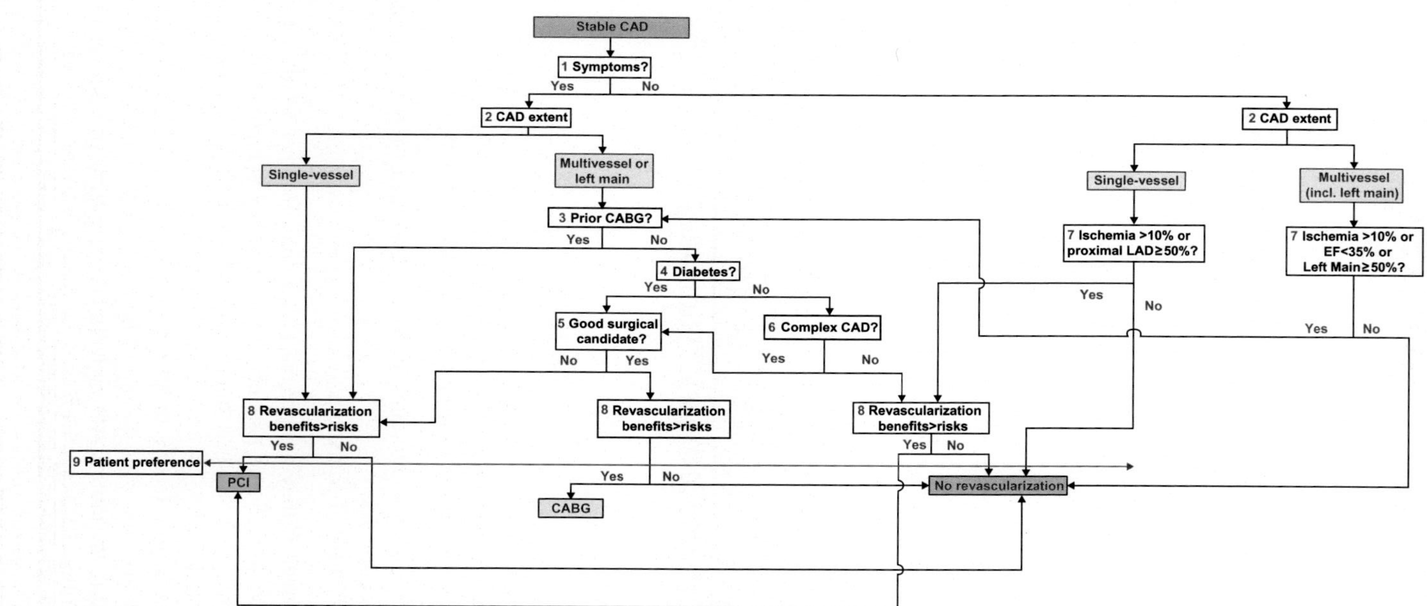

FIGURE 7.1 Algorithm for deciding about coronary revascularization in stable angina patients.

With the exception of highly stenotic coronary lesions (diameter stenosis ≥90%), that are almost always hemodynamically significant, coronary physiology (Chapter 12: Coronary physiology) can help more accurately determine the severity of CAD [5].

Patients with single-vessel disease are in most cases treated with PCI.

Patients with multivessel (or left main) coronary artery disease can be treated with either PCI or CABG:

- PCI is preferred for prior CABG patients, poor surgical candidates, and patients with less complex CAD.
- CABG is preferred for patients with diabetes, reduced ejection fraction, and patients with complex multivessel CAD, as discussed in more detail below.

7.1.3 Prior CABG

Due to increased risk of death and complications, redo CABG is performed infrequently in patients who have already had CABG [6]. However, redo CABG could be considered in patients with severely diseased or occluded bypass grafts, native vessels not amenable to PCI, depressed left ventricular ejection fraction and absence of patent arterial grafts. Internal mammary artery grafts should be used, if feasible, in patients undergoing reoperation [7].

7.1.4 Diabetes mellitus

Patients with diabetes mellitus and multivessel CAD had better outcomes with CABG (including lower mortality) in several trials [8–12] (although most, but not all [12], such studies included first generation drug-eluting stents [DES]). This is likely due to higher risk of restenosis and disease progression in non-stented coronary segments or vessels in diabetic patients and has been linked to the use of left internal mammary artery to left anterior descending (LIMA-LAD) grafts. Hence, CABG is generally preferred for diabetic patients with multivessel CAD, who are good surgical candidates [3].

7.1.5 Good surgical candidate?

The risk of CABG depends on the patients' cardiac status and noncardiac comorbidities.

This is often assessed using the Society of Thoracic Surgeons (STS) score (http://riskcalc.sts.org/stswebriskcalc/) that predicts in-hospital or 30-day mortality and in-hospital morbidity, or the Euroscore II that predicts in-hospital mortality (http://www.euroscore.org/calc.html).

The following factors are assessed:

Cardiac

1. Recent acute coronary syndrome
2. Urgency of the operation
3. Angina severity
4. Left ventricular function
5. Congestive heart failure
6. Pulmonary hypertension
7. Valvular heart disease
8. Active endocarditis
9. Prior cardiac surgery
10. Porcelain aorta
11. Atrial fibrillation
12. Heart block
13. Cardiogenic shock

Noncardiac

1. Age
2. Gender
3. Height and weight
4. Mobility
5. Renal disease

6. Liver disease
7. Chronic lung disease
8. Peripheral arterial disease
9. Cerebrovascular disease
10. Diabetes mellitus
11. Immunocompromised status
12. Prior mediastinal radiation
13. Severe chest deformation or scoliosis
14. Illicit drug-use
15. Cancer
16. Hematocrit
17. Platelet count
18. Current medications ($P2Y_{12}$ inhibitors, glycoprotein IIb/IIIa inhibitors, etc.)
19. Frailty

7.1.6 Complex CAD?

The complexity of CAD is most commonly assessed using the Syntax score (http://www.syntaxscore.com/) [13], which incorporates extent of disease and angiographic characteristics, such as lesion location, diameter stenosis, presence of bifurcations or trifurcations, aorto-ostial location, severe tortuosity, lesion length, calcification, thrombus, and vessel size (Fig. 7.2, Table 7.1) [4].

7.1.7 Factors potentially associated with improved prognosis with coronary revascularization

Coronary revascularization in stable CAD patients is mainly done to improve symptoms as discussed in Section 7.1.1, but could also improve prognosis in the following patient subgroups [4]:

- Left main disease with stenosis >50%.
- Proximal LAD stenosis >50%.
- Two- or three-vessel disease with stenosis >50% and impaired LV function (LVEF ≤35%).
- Large area of ischemia detected by functional testing (> 10% LV) or abnormal invasive FFR.
- Single remaining patent coronary artery with stenosis >50%.
- Sudden cardiac death due to ventricular fibrillation or ventricular tachycardia.

7.1.8 Benefits versus risks

(Fig. 7.3)

Before proceeding with PCI the risks and benefits need to be assessed to ensure that benefits outweigh the risks. Estimation of the risks and benefits can be based on the following four areas:

1. Impact of CAD

   The more severe the impact of CAD on the patient's clinical condition (e.g., severe angina even with minimal exertion), the greater the potential benefit of PCI.
2. Operator competence

   Greater operator experience is associated with higher success and lower complication rates. Experienced operators may also be more adept in managing complications should they occur. Moreover, the operator and staff condition (e.g., well rested and not-sleep deprived, etc.) can affect the outcome of the procedure [15–17].
3. Lesion(s) complexity

   More complex lesions (such as CTOs, heavily calcified lesions, bifurcations, etc.) can be more challenging to recanalize and carry increased risk of complications.
4. Comorbidities

   More comorbidities may increase the risk of the procedure and decrease the benefits (e.g., coronary revascularization should not be performed with the goal to improve prognosis in patients with noncardiac terminal disease, such as cancer).

Steps	Variable assessed	Description
Step 1	Dominance	The weight of individual coronary segments varies according to coronary artery dominance (right or left). Codominance does not exist as an option in the Syntax score.
Step 2	Coronary segment	The diseased coronary segment directly affects the score as each coronary segment is assigned a weight depending on its location, ranging from 0.5 (i.e., the posterolateral branch) to 6 (i.e., left main in case of left dominance). 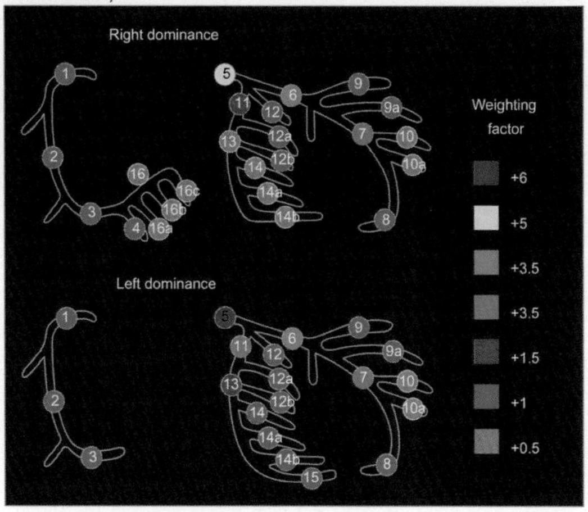 Right dominance Left dominance Weighting factor: +6, +5, +3.5, +3.5, +1.5, +1, +0.5 ©ESC 2018
Step 3	Diameter stenosis	The score of each diseased coronary segment is multiplied by two in case of a stenosis 50%–99% and by five in case of total occlusion. In case of total occlusion, additional points will be added as follows: Age >3 months or unknown +1 Blunt stump +1 Bridging +1 First segment visible distally +1 per nonvisible segment Side branch at the occlusion +1 if <1.5 mm diameter +1 if both <1.5 mm and ≥1.5 mm diameter +0 if ≥1.5 mm diameter (i.e., bifurcation lesion)
Step 4	Trifurcation lesion	The presence of a trifurcation lesion adds additional points based on the number of diseased segments: 1 segment +3 2 segments +4 3 segments +5 4 segments +6
Step 5	Bifurcation lesion	The presence of a bifurcation lesion adds additional points based on the type of bifurcation according to the Medina classification[126]: Medina 1,0,0–0,1,0–1,1,0 +1 Medina 1,1,1–0,0,1–1,0,1–0,1,1 +2 Moreover, the presence of a bifurcation angle <70° adds one additional point
Step 6	Aortoostial lesion	The presence of aortoostial lesion segments adds one additional point
Step 7	Severe tortuosity	The presence of severe tortuosity proximal of the diseased segment adds two additional points
Step 8	Lesion length	Lesion length >20mm adds one additional point
Step 9	Calcification	The presence of heavy calcification adds two additional points
Step 10	Thrombus	The presence of thrombus adds one additional point
Step 11	Diffuse disease/ small vessels	The presence of diffusely diseased and narrowed segments distal to the lesion (i.e., when at least 75% of the length of the segment distal to the lesion has a vessel diameter <2mm) adds one point per segment number

FIGURE 7.2 Calculation of the Syntax score. *Reproduced with permission from Neumann FJ, Sousa-Uva M, Ahlsson A, et al. 2018 ESC/EACTS guidelines on myocardial revascularization. Eur Heart J 2019;40:87–165.*

TABLE 7.1 Revascularization decision making based on Syntax score.

No. of diseased coronary vessels	Syntax score	Revascularization modality [4,14]
2		PCI or CABG
3	0–22	PCI or CABG
3	>22	CABG
Left main	0–32	PCI or CABG
Left main	>32	CABG

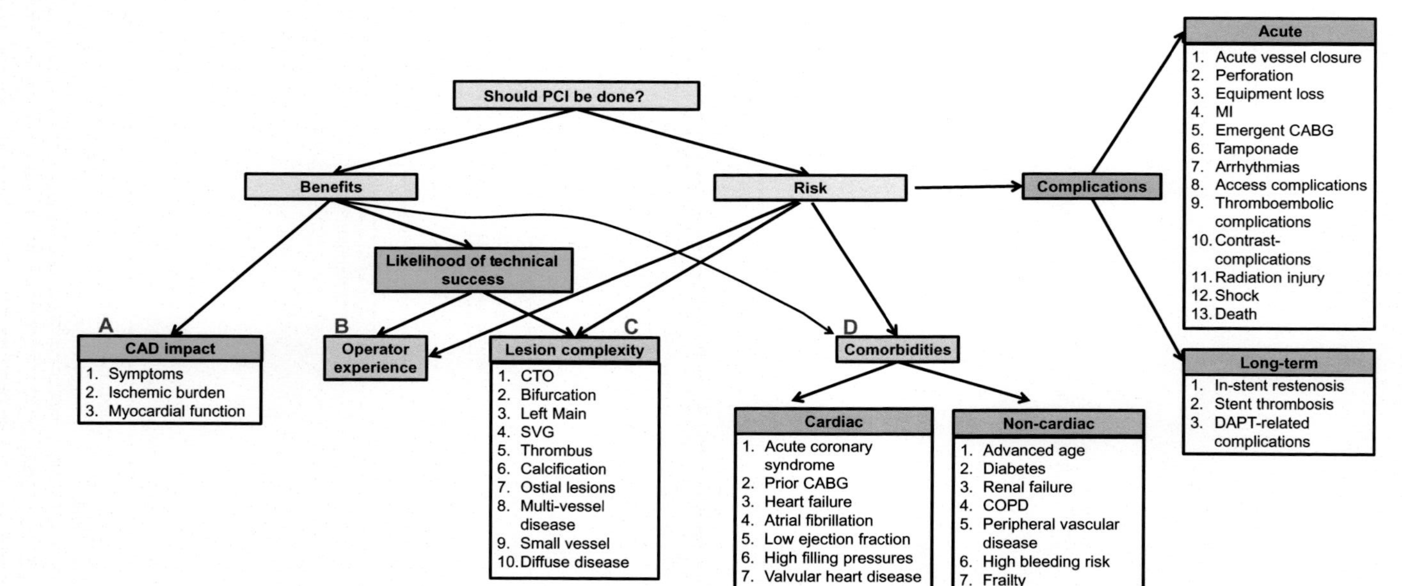

FIGURE 7.3 Assessing risks and benefits of PCI.

7.1.9 Patient preference

The final decision should always be the patient's. The interventionalist's and the heart team's role is to educate the patient and help him/her in the decision making process.

7.2 Acute coronary syndromes

(Fig. 7.4)

7.2.1a Culprit lesion evident?

Treatment of the culprit lesion takes priority in ACS patients. A culprit lesion is the stenosis that is causing the patient's symptoms and clinical syndrome.

Sometimes the culprit lesion is easy to identify (vessel occlusion with evident thrombus), but sometimes no culprit lesion is evident (see Section 7.2.10) or multiple possible culprit lesions are present. The following criteria can help determine the culprit lesion(s):

1. Artery supplies myocardium that corresponds to area of ST-elevation on the electrocardiogram
2. Artery supplies myocardium that is hypokinetic by left ventriculography or echocardiography
3. Thrombus (Chapter 20: Acute coronary syndromes—thrombus)
4. Contrast staining
5. Poor or no collaterals
6. Ease of wiring

7.2.1b Ongoing ischemia?

The classic example of ongoing ischemia is ST-segment elevation myocardial infarction, when "time is muscle." Another example is patients with non-ST segment elevation ACS and ongoing chest pain. Immediate PCI of the culprit lesion(s) is indicated in such cases.

7.2.2 CAD extent

Same concepts apply as discussed in Section 7.1.2.

7.2.3 Prior CABG

Same concepts apply as discussed in Section 7.1.3.

7.2.4 Diabetes mellitus

Same concepts apply as discussed in Section 7.1.4.

7.2.5 Good surgical candidate?

Same concepts apply as discussed in Section 7.1.5. Patients with STEMI are usually preloaded with oral $P2Y_{12}$ inhibitors that increase the risk of bleeding if emergency CABG is performed.

7.2.6 Complex CAD?

Same concepts apply as discussed in Section 7.1.6.

7.2.7 CAD extent

Same concepts apply as discussed in Section 7.1.7.

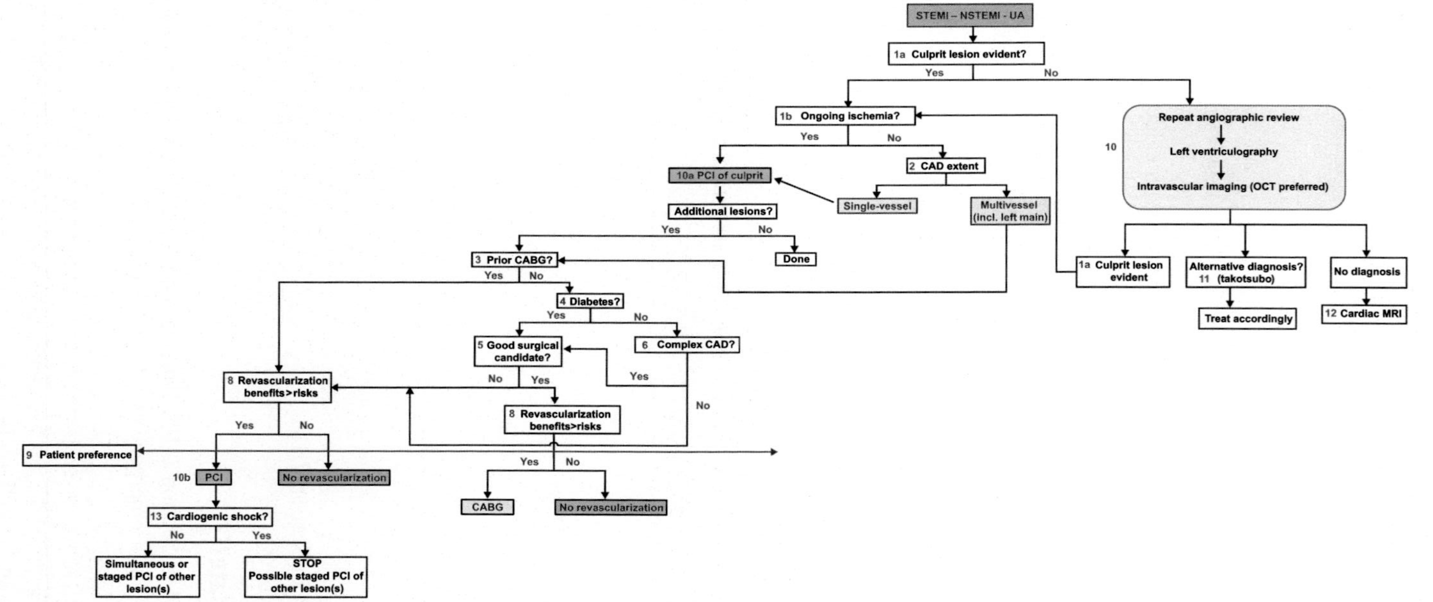

FIGURE 7.4 Algorithm for deciding about coronary revascularization in patients with acute coronary syndromes.

7.2.8 Revascularization benefits and risks

Same concepts apply as discussed in Section 7.1.8. However, the urgency for performing PCI is often higher in ACS patients.

7.2.9 Patient preference

Same concepts apply as discussed in Section 7.1.9. However, the urgency for making a decision is higher in ACS patients, given the potential adverse consequences of delayed revascularization.

7.2.10 Assessment for culprit lesion(s)

When there is no immediate culprit lesion(s) on initial angiographic assessment, the following steps are recommended:

1. Repeat careful angiographic review: sometimes the angiographic findings, such as "staining" or thrombus can be very subtle.
2. Left ventriculography can be diagnostic of related conditions (such as takotsubo cardiomyopathy) or mechanical complications of myocardial infarction (such as papillary muscle rupture or free wall rupture). Left ventriculography can also demonstrate areas of myocardial hypokinesis that can help determine the culprit lesion(s). Alternatively, a transthoracic echocardiogram can be performed in the cardiac catheterization laboratory.
3. Intravascular imaging. Due to its high spatial resolution, optical coherence tomography (OCT) is the intravascular imaging modality of choice for determining the presence (or confirming the absence) of culprit lesion(s). OCT findings of culprit lesions are thrombus and plaque rupture (see Section 13.1.1.2).

7.2.11 Alternative diagnoses (such as takotsubo)

Left ventriculography can clarify the area of myocardial infarction or suggest alternative diagnoses (such as myocarditis).

7.2.12 Cardiac MRI

Cardiac MRI can clarify the area of myocardial infarction or suggest alternative diagnoses (such as myocarditis).

7.2.13 Cardiogenic shock

Based on the results of the CULPRIT-SHOCK trial [18,19], PCI of the culprit lesion only is recommended in the acute setting in ACS patients presenting with cardiogenic shock. Staged revascularization of non-culprit lesions can be performed if neccssary.

7.3 PCI timing

7.3.1 Ad hoc versus delayed PCI

PCI can be performed "ad hoc" (i.e., immediately following diagnostic coronary angiography) or can be performed at a later time.

The advantage of ad hoc PCI is avoidance of a second procedure with any concomitant risks. Disadvantages of ad hoc PCI include higher contrast and radiation dose, less planning, and less time devoted to a detailed discussion about the risks and benefits of the procedure.

Ad hoc PCI is needed in most cases in ACS patients with active ischemia, given potential delays associated with CABG. In stable CAD patients or ACS patients without active ischemia and complex anatomy, delayed PCI is favored in the following scenarios [20]:

1. Complex or high-risk disease (such as unprotected left main, chronic total occlusions, multivessel disease, bifurcations, etc.)
2. Excessive contrast or radiation during diagnostic procedure or anticipated during percutaneous coronary intervention.

3. Site of service (e.g., facility without onsite surgery when the patient risk or lesion risk is high or facility lacking necessary interventional equipment).
4. Inadequate informed consent (e.g., diagnostic catheterization identifies anatomy for which the risk of PCI is significantly higher than was discussed before the diagnostic angiography).
5. Uncertainty regarding extent of symptoms in patients with stable ischemic heart disease.
6. Lack of evidence of ischemia and unavailability of fractional flow reserve or intravascular ultrasound.
7. Complication during diagnostic catheterization (e.g., stroke or access site bleeding).
8. Operator or patient fatigue after diagnostic catheterization.
9. Scheduling challenges (e.g., a new patient presents with ST-segment elevation acute myocardial infarction while ad hoc PCI is being considered for a patient with stable ischemic heart disease).
10. Inadequate pretreatment (e.g., no aspirin before diagnostic catheterization, inadequate trial of antianginal therapy, and inadequate hydration).
11. Patient preference.

7.3.2 Immediate versus staged multivessel/multi-lesion revascularization

In patients with STEMI and multivessel coronary artery disease, complete revascularization (either immediate or staged [21]) improves hard outcomes (death or myocardial infarction) [22]. Better outcomes with complete revascularization have also been demonstrated in observational studies of stable coronary artery disease patients [23].

Factors favoring staged PCI in patients with multivessel CAD include:

1. High complexity of the remaining lesion(s).
2. No active ischemia caused by the remaining lesion(s) in ACS patients.
3. Excessive contrast or radiation dose during the first part of PCI or anticipated during PCI of additional lesions.
4. Complications.
5. Operator fatigue.
6. Patient fatigue.

7.4 PCI lesion sequence selection

In ACS patients with multivessel CAD, the culprit lesion(s) should be treated first, especially if they are causing active ischemia (e.g., the STEMI culprit lesion).

In stable patients with multiple lesions in the same coronary artery, the more distal lesions are generally treated first, unless there is severe pressure dampening or ischemia caused by more proximal or ostial lesion or inability to advance devices through a tight proximal lesion. In general, all significant lesions within the same target vessel should be treated during the same procedure.

In stable patients with multivessel disease, parameters to consider for choosing the sequence of PCI include:

1. Lesion complexity: for example, treating the less complex lesions first may facilitate subsequent treatment of the more complex lesions.
2. Contrast and radiation dose: in patients who have received high radiation or contrast dose, treating the easier lesion(s) first and deferring the more complex lesions may be preferable.
3. Alternative revascularization options: for example, in a patient with an LAD chronic total occlusion and non-occlusive lesions in the RCA and circumflex, some operators may choose to attempt LAD CTO PCI first and refer the patient for CABG if the attempt is unsuccessful. However, other operators may choose to treat the non-occlusive RCA and circumflex lesions first to increase the safety of the LAD CTO PCI attempt, since the success of CTO PCI is 85–90% at experienced centers [24].

References

[1] Al-Lamee R, Thompson D, Dehbi HM, et al. Percutaneous coronary intervention in stable angina (ORBITA): a double-blind, randomised controlled trial. Lancet 2018;391:31–40.

[2] Boden WE, O'Rourke RA, Teo KK, et al. Optimal medical therapy with or without PCI for stable coronary disease. N Engl J Med 2007;356:1503–16.

[3] Maron DJ, Hochman JS, Reynolds HR, et al. Initial invasive or conservative strategy for stable coronary disease. N Engl J Med 2020;382:1395–407.

[4] Neumann FJ, Sousa-Uva M, Ahlsson A, et al. 2018 ESC/EACTS Guidelines on myocardial revascularization. Eur Heart J 2019;40:87–165.

[5] Tonino PAL, Fearon WF, De Bruyne B, et al. Angiographic versus functional severity of coronary artery stenoses in the FAME study: fractional flow reserve versus angiography in multivessel evaluation. J Am Coll Cardiol 2010;55:2816–21.

[6] Morrison DA, Sethi G, Sacks J, et al. Percutaneous coronary intervention versus repeat bypass surgery for patients with medically refractory myocardial ischemia: AWESOME randomized trial and registry experience with post-CABG patients. J Am Coll Cardiol 2002;40:1951–4.

[7] Levine GN, Bates ER, Blankenship JC, et al. 2011 ACCF/AHA/SCAI guideline for percutaneous coronary intervention: a report of the American College of Cardiology Foundation/American Heart Association Task Force on Practice Guidelines and the Society for Cardiovascular Angiography and Interventions. Circulation 2011;124:e574–651.

[8] Farkouh ME, Domanski M, Sleeper LA, et al. Strategies for multivessel revascularization in patients with diabetes. N Engl J Med 2012;367:2375–84.

[9] Farkouh ME, Domanski M, Dangas GD, et al. Long-term survival following multivessel revascularization in patients with diabetes: The FREEDOM follow-on study. J Am Coll Cardiol 2019;73:629–38.

[10] Investigators B. The final 10-year follow-up results from the BARI randomized trial. J Am Coll Cardiol 2007;49:1600–6.

[11] Kamalesh M, Sharp TG, Tang XC, et al. Percutaneous coronary intervention versus coronary bypass surgery in United States veterans with diabetes. J Am Coll Cardiol 2013;61:808–16.

[12] Park SJ, Ahn JM, Kim YH, et al. Trial of everolimus-eluting stents or bypass surgery for coronary disease. N Engl J Med 2015;372:1204–12.

[13] Sianos G, Morel MA, Kappetein AP, et al. The SYNTAX Score: an angiographic tool grading the complexity of coronary artery disease. EuroIntervention 2005;1:219–27.

[14] Mohr FW, Morice MC, Kappetein AP, et al. Coronary artery bypass graft surgery versus percutaneous coronary intervention in patients with three-vessel disease and left main coronary disease: 5-year follow-up of the randomised, clinical SYNTAX trial. Lancet 2013;381:629–38.

[15] Lobo AS, Sandoval Y, Burke MN, et al. Sleep deprivation in cardiology: a multidisciplinary survey. J Invasive Cardiol 2019;31:195–8.

[16] Sandoval Y, Lobo AS, Somers VK, et al. Sleep deprivation in interventional cardiology: implications for patient care and physician-health. Catheter Cardiovasc Interv 2018;91:905–10.

[17] Iverson A, Stanberry L, Garberich R, et al. Impact of sleep deprivation on the outcomes of percutaneous coronary intervention. Catheter Cardiovasc Interv 2018;92:1118–25.

[18] Thiele H, Akin I, Sandri M, et al. One-year outcomes after PCI strategies in cardiogenic shock. N Engl J Med 2018;379:1699–710.

[19] Thiele H, Akin I, Sandri M, et al. PCI strategies in patients with acute myocardial infarction and cardiogenic shock. N Engl J Med 2017;377:2419–32.

[20] Blankenship JC, Gigliotti OS, Feldman DN, et al. Ad hoc percutaneous coronary intervention: a consensus statement from the Society for Cardiovascular Angiography and Interventions. Catheter Cardiovasc Interv 2013;81:748–58.

[21] Wood DA, Cairns JA, Wang J, et al. Timing of Staged Nonculprit Artery Revascularization in Patients With ST-Segment Elevation Myocardial Infarction: COMPLETE Trial. J Am Coll Cardiol 2019;74:2713–23.

[22] Mehta SR, Wood DA, Storey RF, et al. Complete revascularization with multivessel PCI for myocardial infarction. N Engl J Med 2019;381:1411–21.

[23] Farooq V, Serruys PW, Bourantas CV, et al. Quantification of incomplete revascularization and its association with five-year mortality in the synergy between percutaneous coronary intervention with taxus and cardiac surgery (SYNTAX) trial validation of the residual SYNTAX score. Circulation 2013;128:141–51.

[24] Tajti P, Burke MN, Karmpaliotis D, et al. Update in the percutaneous management of coronary chronic total occlusions. JACC Cardiovasc Interv 2018;11:615–25.

Chapter 8

Wiring

Goal: To advance a guidewire through and distal to the target coronary lesion in the most efficient way and without causing a complication and to provide an adequate platform to support balloon, stent and other equipment delivery.

Similar to use of a guide catheter, wire insertion is essential for performing PCI and/or lesion assessment through coronary imaging or physiology. Obtaining optimal guide catheter support, as outlined in Sections 9.5.8.1.1-9.5.8.1.6, can facilitate wiring.

Wiring occurs in nine steps.

8.1 Step 1. Determine whether a microcatheter is needed

8.1.1 Goal:

To facilitate guidewire advancement to the desired coronary location, as well as guidewire exchanges.

8.1.2 How?

A microcatheter can significantly facilitate guidewire handling and should be considered in any complex lesion, such as highly tortuous lesions, near-occlusive lesions and chronic total occlusions (CTOs). If a decision is made to use a microcatheter, the guidewire should be inserted together with the microcatheter, bypassing the need to use the introducer needle (step 4 of this section).

There are five main microcatheter categories: large, small, angulated, dual-lumen, and plaque modification, as discussed in detail in Section 30.6. Use of a microcatheter for navigating through tortuosity is discussed in Section 8.6, for use in CTOs in Chapter 21: Chronic Total Occlusions, and for wiring the side branch of bifurcation lesions in Chapter 16: Bifurcations (Section 16.2.8).

If a microcatheter is not available an over-the-wire balloon could be used instead, but it has limitations (balloons are stiffer and more likely to kink and the marker of small balloons is at the middle, hence the location of the balloon tip can be hard to determine). On the other hand, over-the-wire balloons cost significantly less than microcatheters.

8.2 Step 2. Guidewire selection

8.2.1 Goal:

To select a guidewire capable of advancement to the desired location while minimizing the risk of complications. The various guidewire types are discussed in Section 30.7.

8.2.2 How?

1. Noncomplex lesions: workhorse guidewire (Section 30.7.1).
2. Tortuous lesions:
   - Option 1: use advanced workhorse guidewires, such as composite core (such as Sion blue), dual core (such as Samurai), or nitinol (such as Runthrough, TurnTrac, Versaturn, Minamo) guidewires.
   - Option 2: use soft tip guidewires with hydrophilic tip coating, such as the Sion, Suoh 03, and Samurai RC.
   - Option 3: use soft, nontapered polymer-jacketed guidewire (such as Sion black, Pilot 50, Fielder FC, or Whisper)
   - Guidewires with higher tip load and hydrophilic coating or polymer jacket carry higher risk of complications, such as dissection or perforation, and should only be used if standard workhorse guidewires fail to advance to

Manual of Percutaneous Coronary Interventions. DOI: https://doi.org/10.1016/B978-0-12-819367-9.00008-1

the desired coronary segment. Such guidewires should subsequently be exchanged for a workhorse guidewire for balloon and stent delivery. If polymer jacketed guidewires are used for equipment delivery the position of their tip should be monitored constantly to minimize the risk of distal wire perforation.

3. CTOs are discussed in Chapter 21: Chronic total occlusions and in the *Manual of CTO Interventions*. CTOs are different from other coronary lesions, as the initial wire choice is usually a polymer-jacketed, tapered tip guidewire. Stiff, highly penetrating guidewires (both polymer-jacketed and nonpolymer jacketed guidewires) are also often used.
4. More than one guidewire may be needed for crossing a lesion: for example, a workhorse guidewire may be needed for reaching the target lesion and a polymer-jacketed guidewire for crossing it. If it is anticipated that >1 guidewire will be required, use of a microcatheter is recommended to facilitate wire exchanges and enhance wire manipulation.
5. After a lesion is crossed with a polymer-jacketed or stiff tip guidewire, the guidewire should be changed (using a microcatheter or over-the-wire balloon) for a workhorse guidewire, a highly supportive guidewire (such as Grand Slam, Iron Man, Wiggle wire), or an atherectomy guidewire such as the Rotawire Floppy or Rorawire extra support for rotational atherectomy (Sections 19.9.5.3 and 30.10.1.1.3) or ViperWire Advance or ViperWire Advance Flex Tip for orbital atherectomy (Sections 19.9.6.2 and 30.10.2.3) if atherectomy is planned. Guidewire exchanges using a microcatheter are best performed using the trapping technique (Section 8.9.2.1).

8.3 Step 3. Shape the guidewire tip

8.3.1 Goal:

To shape the wire tip in the optimal way for advancing it to the desired coronary artery location.

8.3.2 How?

Some wires are preshaped, obviating the need for shaping the tip, but at the same time limiting the options for customizing the guidewire tip shape.

The shape of the tip depends on the target vessel size and on lesion and vessel angulation: for example, big bends should be used for big vessels and small bends for small vessels (Fig. 8.1). In general the distal bend should be smaller than the diameter of the target vessel. Sometimes different bends may be required for wiring a complex lesion, such as CTOs: for example, one bend may be needed to reach to the target lesion and another bend to cross the lesion. Sometimes placing 2 bends (a proximal bend that is usually larger and a distal bend that is usually smaller) can improve the guidewire reach in larger vessels and vessels with angulated ostia).

Shaping the guidewire tip is best performed by advancing the guidewire tip through the introducer and bending it with a finger from the other hand (Fig. 8.2, panel A). Another technique is to compress the guidewire between a finger and the introducer needle and stretch it (Fig. 8.2, panel B). Advancing the guidewire through the introducer allows formation of small bends with high accuracy, however may also lead to fracture of the wire core, especially with nitinol guidewires, such as the Runthrough that should only be shaped with the needle that it comes packaged with.

8.3.3 What can go wrong?

1. Excessive guidewire bending

   Causes:
   - Forceful or excessive wire manipulation.
   - Removing the guidewire from its hoop by pulling it from the tip.

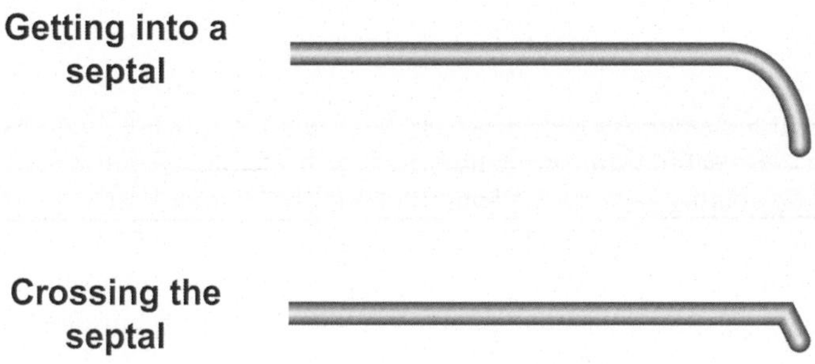

Getting into a septal

Crossing the septal

FIGURE 8.1 Different wire bends are required for different tasks. For example, during CTO PCI, entering a septal branch requires a large bend (panel A), whereas crossing the septal requires a small bend (panel B).

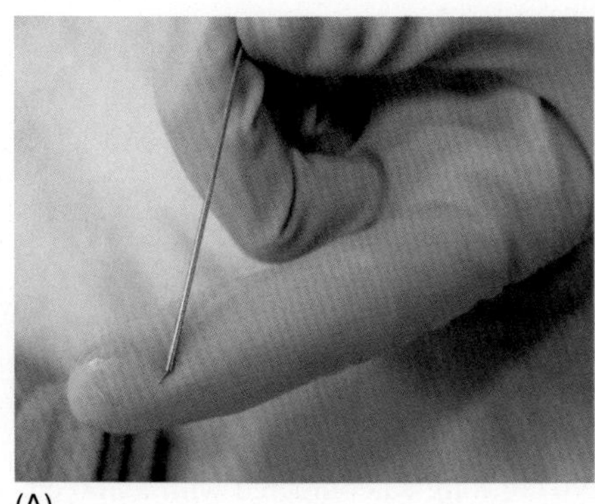
(A)

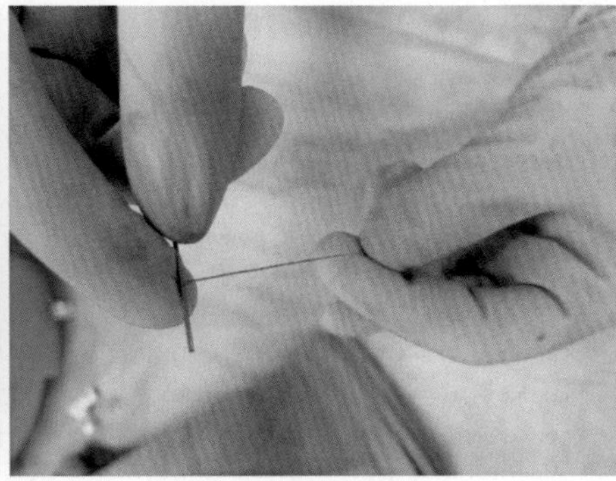
(B)

FIGURE 8.2 How to shape the guidewire tip. Placing small bends is best achieved by advancing the guidewire through the introducer and bending it at the tip (panel A). Large bends can be formed by "sliding" the side of the introducer against the wire (panel B), but this may cause wire deformation. *Reproduced with permission from Brilakis ES. Manual of coronary chronic total occlusion interventions. A step-by-step approach. 2nd ed. Cambridge, MA: Elsevier; 2017 (Figure 4.7).*

Prevention:
- Use the introducer for shaping wire tip.
- Do not hold the tip of the guidewire while removing it from the hoop.
- Initially create a small bend and subsequently change it if it fails to advance. It is always possible to create additional bend(s) on the guidewire, but it can be hard to remove them.

Treatment:
- If the wire cannot be reshaped, it may need to be discarded and another guidewire used.

8.4 Step 4. Insert the guidewire into the guide catheter

8.4.1 Goal:

To insert the guidewire into the guide catheter through the hemostatic valve of the Y-connector.

8.4.2 How?

The guidewire tip is withdrawn into the introducer needle. The introducer needle is inserted into the hemostatic valve, followed by guidewire advancement. If the guidewire tip has a large bend, it may need to be back loaded into the introducer.

Alternatively, the guidewire can be preloaded into a microcatheter or over-the-wire balloon, the tip of which is then inserted through the hemostatic valve.

8.4.3 What can go wrong?

1. Guidewire tip deformation

Causes:
- Guidewire tip not fully withdrawn into the introducer needle.
- Guidewire enters side arm of the Y-connector.
- Guidewire comes in contact with devices in the guide catheter, such as guide catheter extensions, balloons or stents.

Prevention:
- Ensure that guidewire tip is not protruding from the tip of the introducer needle or the microcatheter.
- Do not force the guidewire against resistance.
- Ensure that the introducer is advanced all the way through the Y-connector.
- Remove other devices from the guide catheter whenever possible.
- Use fluoroscopy when advancing guidewires through the proximal collar of guide catheter extensions.

Treatment:
- Exchange the guidewire for a new one (if the damaged guidewire cannot be reshaped).

8.5 Step 5. Advance the guidewire to the tip of the guide catheter

8.5.1 Goal:

To advance the guidewire to the tip of the guide catheter

8.5.2 How?

The guidewire is advanced to the tip of the guide catheter. Fluoroscopy is used to check the guidewire position.

Some guidewires, such as the BMW (Abbott Vascular), have length markers on their shaft at 90 and 100 cm from the tip, that can be used to minimize use of fluoroscopy during wire advancement. It is important, however, to know the length of the guide catheter (the guidewire will exit sooner when used in a 90 cm long as compared with a 100 cm long guide catheter). Caution should be used when wiring through side hole guide catheters, as the guidewire can exit through the side hole, instead of the catheter tip (Fig. 8.3).

8.5.3 What can go wrong?

8.5.3.1 Inadvertent advancement into the coronary artery

The guidewire may be advanced into the coronary artery inadvertently or during contrast or saline injections without fluoroscopy guidance, which may lead to dissection, perforation, or loss of guide catheter position.

Causes:
- Too distal guidewire advancement.

Prevention:
- Careful monitoring of the position of the guidewire tip.
- Some guidewires (such as the BMW), have a proximal marker that can help prevent too distal advancement.
- The torquer can be tightened on the guidewire at approximately 90 cm from the guidewire tip to prevent excessive guidewire advancement.

Treatment:
- Do NOT remove the guidewire (sometime the guidewire enters the intended branch).
- Contrast injection to check guidewire position, followed by guidewire redirection, if needed.

8.5.3.2 Guidewire tip deformation

Causes:
- Advancement through a guide catheter extension.
- Advancement past balloons/stent previously inserted in the guide catheter.
- Advancement through side holes of the guide catheter (Fig. 8.3).

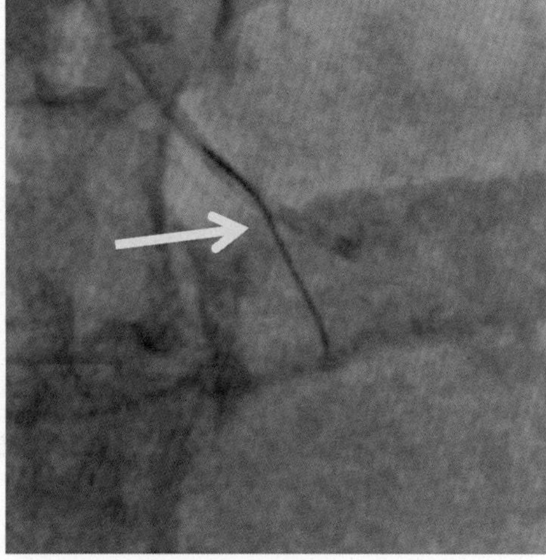

FIGURE 8.3 Wire advancement through the side hole of a multipurpose catheter (*arrow*).

Prevention:

- Remove guide extensions and/or balloons or stents before inserting another guidewire.
- If this is not feasible, guidewire advancement through the collar of the guide extension or past balloons/stents should be done under fluoroscopy without forcing the wire.
- Balloons can be advanced into the coronary artery while advancing additional wires through the guide catheter; the baloon is then retracted into the guide catheter once wire advancement is completed.
- Use a dual lumen microcatheter: the monorail lumen of the dual lumen microcatheter is advanced over the initially placed guidewire, followed by insertion of the new guidewire through the over-the-wire lumen. Alternatively, an aspiration thrombectomy catheter can be used (Sections 20.9.6 and 30.12).The trapping technique (Section 8.9.2.1) is then used to remove the dual lumen microcatheter.

Treatment:

- Attempt to reshape the guidewire tip.
- If reshaping fails, the guidewire is discarded and a new guidewire is used.

8.6 Step 6. Advance the guidewire from the tip of the guide catheter to the target lesion

8.6.1 Goal:

To advance the guidewire from the tip of the guide catheter to the target lesion.

8.6.2 How?

The guidewire is advanced from the guide tip to the lesion under fluoroscopic guidance with intermittent contrast injections. A still frame image of the coronary anatomy can be used as reference to assist with guidewire advancement and minimize the need for contrast injections. Another way to guide wiring is the dynamic roadmap that is available in some X-ray systems.

Advancing the wire: general principles (they apply to steps 6–8)

8.6.2.1 View selection

For wiring the LAD or circumflex, a caudal view is used initially to enter from the left main into the LAD or circumflex. Subsequent wiring of the LAD is easier using the RAO cranial or AP cranial view, whereas wiring of the circumflex/obtuse marginal is easier in the RAO caudal or LAO caudal views.

For wiring the RCA the LAO view is used to advance to the distal vessel, followed by LAO or AP cranial to wire into the PDA or the right posterolateral branch.

8.6.2.2 Use of a torquer

Use of a torquer improves wire control, but may not be necessary for simple lesions. This is a matter of preference and experience.

8.6.2.3 Pushing versus turning

The two key movements of a guidewire are pushing/withdrawing and rotation. Various combinations of push and turn are employed, depending on the vessel and lesion morphology.

8.6.2.4 Underhand versus overhand and using both hands

This is a matter of personal preference, but the authors prefer underhand holding of the wire as it allows easier wire manipulation. For complex lesions (such as CTOs) many (right-handed) operators advance or withdraw the wire using the left hand and rotate it using the right hand.

8.6.2.5 Keep tip free

Keeping the wire tip free is important for both (1) succeeding in advancing the wire to the desired location, and (2) minimizing the risk of vessel injury. If the tip goes "under a plaque," aggressive advancement may lead to vessel dissection and possibly tip entrapment. If the wire tip movement becomes restrained, the wire should be withdrawn and readvanced. Alternatively, the wire can be left in place with a second wire advanced next to it (parallel wiring).

8.6.2.6 *Reference image, intermittent injections and orthogonal views to visualize wiring progress*

Having a stored reference image can facilitate guidewire advancement. Moreover, intermittent contrast injections are key for determining the wire position and need for adjustment. Injecting a small amount of contrast usually suffices to visualize the wire position. For chronic total occlusion interventions injecting from the contralateral coronary artery or bypass graft is often required to determine guidewire position and is one of the key principles of CTO crossing. Orthogonal projections are also important to confirm that the guidewire is in the desired location, especially when complex lesions (such as CTOs, tortuous lesions, overlapping side branches, etc.) are being wired.

8.6.2.7 *Looping the wire tip is OK*

Forming a loop at the wire tip can be both an effective and safe wiring strategy, especially after crossing the area of stenosis. A knuckled guidewire may in some cases advance more easily through tortuous coronary segments without entering into side branches. A knuckled guidewire is also less likely to cause perforation from inadvertent too distal advancement, but if it does occur, the perforation will be larger as compared with guidewire tip exit only. The knuckle should be kept small and not allowed to extend beyond the radiopaque portion of the wire. Knuckles should be avoided at the tip of highly supportive guidewires, such as the Grand Slam and Iron Man (Section 30.7.4).

8.6.2.8 *Change if tip damaged*

Sometimes the guidewire tip may get deformed, especially if advanced through tortuosity or caught in a small branch. In such cases continued manipulation of the damaged guidewire is less likely to be successful and carries higher risk of complications. The guidewire should be changed, ideally using a microcatheter and the trapping technique (Section 8.9.2.1).

8.6.2.9 *Escalate–de-escalate*

This is a standard CTO crossing technique, but also applies to non-CTO lesions. If non-workhorse guidewires are used to cross the lesion (e.g., polymer-jacketed guidewires, such as Fielder FC, Sion black, or Pilot 200) they should be exchanged for a workhorse guidewire after lesion crossing before proceeding with balloon, stent and other equipment delivery.

8.6.2.10 *If multiple guidewires are used—keep them organized*

If multiple guidewires are used, for example, when treating bifurcation lesions, it is important to keep track of which wire is where. One approach is to separate them with towels, with the wire in the most superior branch as visualized in

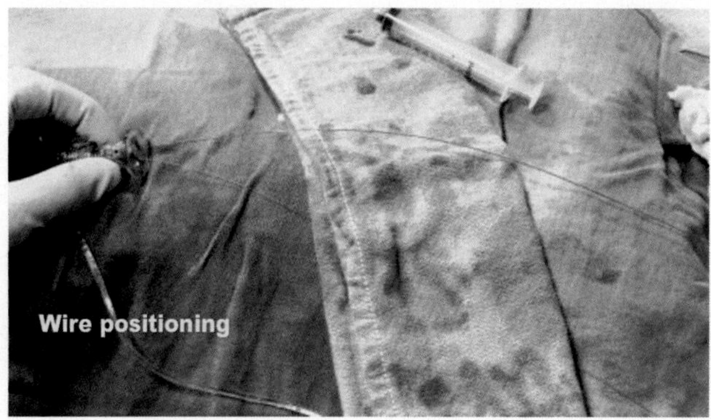

FIGURE 8.4 Use of a towel to separate two guidewires used simultaneously during PCI. Typically, the wire on top of the screen is placed on top of the towel, whereas the wire located lower in the screen is placed under the towel.

the X-ray screen being in the top and the wire in the most inferior branch in the bottom (Fig. 8.4). Another option is to use different color guidewires, slightly bend the back end of the side-branch guidewire, or use a torquer on the proximal end of one of the wires.

8.6.3 Challenges

8.6.3.1 Difficulty advancing the guidewire through the tip of the guide catheter

Causes:
- Suboptimal guide engagement.

Prevention:
- Optimal guide engagement (coaxial, not too deep, no pressure waveform dampening).

Solutions:
- Reposition the guide catheter tip to obtain optimal vessel engagement.
- Change the shape of the guidewire tip.
- Leave the original guidewire in place and wire with a second guidewire.

8.6.3.2 Guidewire inadvertently enters a side branch

Causes:
- Challenging coronary anatomy (tortuosity).
- Suboptimal guidewire tip shaping.

Prevention:
- Optimal shaping of guidewire tip.
- Use guidewires that provide 1:1 torque response.
- Use a microcatheter to facilitate torque transmission and allow reshaping of the guidewire tip.

Treatment:
- Reposition the guide catheter tip. For example, if the guidewire keeps entering into a conus branch in the right coronary artery, the guide may be withdrawn followed by attempts to obtain coaxial engagement.
- Change the shape of the guidewire tip.
- Change guide catheter: for example, if a guidewire keeps entering the circumflex instead of the LAD, using a smaller guide catheter will facilitate wire advancement into the LAD. Alternatively a guidewire is left into the initially wired vessel stabilizing the guide catheter, followed by insertion of a second guidewire into the target vessel.

8.6.3.3 Guidewire cannot be advanced through tortuosity

Causes:
- Challenging coronary anatomy (tortuosity).
- Suboptimal guidewire tip shaping.

Prevention:
- Optimal shaping of guidewire tip.
- Use guidewires that provide 1:1 torque response.
- Use hydrophilic or polymer jacketed guidewires (which, however, may increase the risk of dissection and perforation).

Treatment (Fig. 8.5):
- **Solution 1**: Use a different guidewire (with hydrophilic coating or a soft-polymer jacketed guidewire) and change the shape of the guidewire tip. Some guidewires, such as the Pilot family are less likely to prolapse compared with other guidewires.
- **Solution 2**: Use a microcatheter (with straight or angulated tip, Section 30.6) to facilitate advancement. The microcatheter also allows guidewire exchanges without the need to rewire from the ostium of the vessel. A dual lumen microcatheter can also be used, wiring the angulated vessel through the over-the-wire lumen.
- **Solution 3**. Reversed (also called "hairpin") guidewire technique (*CTO Manual* Online case 71) [1–3]. In this technique a polymer-jacketed wire is bent approximately 3 cm from the wire tip and the bend is inserted through the hemostatic valve of the Y-connector (Fig. 8.6).

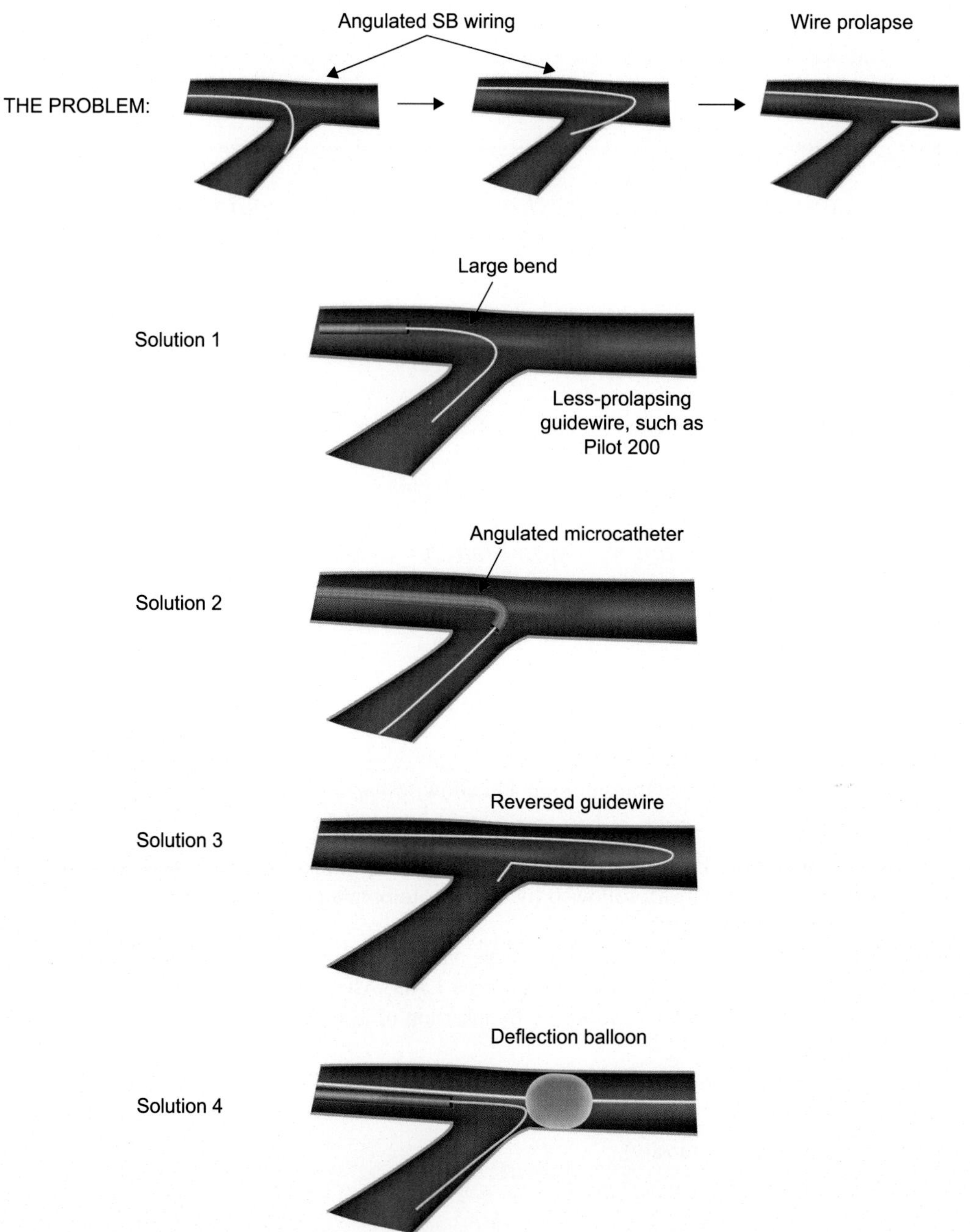

FIGURE 8.5 Strategies for wiring through tortuosity.

The hairpin wire is then advanced into the main vessel (Fig. 8.7, panel A), and pulled back (Fig. 8.7, panel B) entering the main branch (Fig. 8.7, panel C).

Alternatively, the hairpin wire can be advanced through a dual lumen microcatheter (Section 30.6.4, Fig. 30.71).

In a variation of this technique called "streamlined reverse wire technique" [4] the polymer-jacketed guidewire with a bend 3 cm from the tip is advanced through the over-the-wire lumen of a dual lumen microcatheter that is already placed in the coronary vessel until it enters into another side branch. The dual-lumen microcatheter and guidewire are then advanced, creating the hairpin at the tip of the wire, which is then pulled back in order to enter the angulated target branch.

What can go wrong?

Use of the "hairpin wire" technique may cause vessel dissection. Also after the "hairpin" enters the main vessel, further advancement may be challenging due to the bend in the wire.

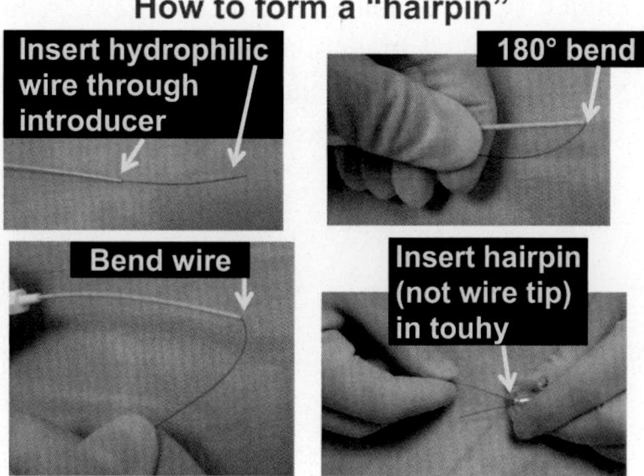
How to form a "hairpin"
Insert hydrophilic wire through introducer
180° bend
Bend wire
Insert hairpin (not wire tip) in touhy

FIGURE 8.6 How to form a reversed (also called "hairpin") guidewire.

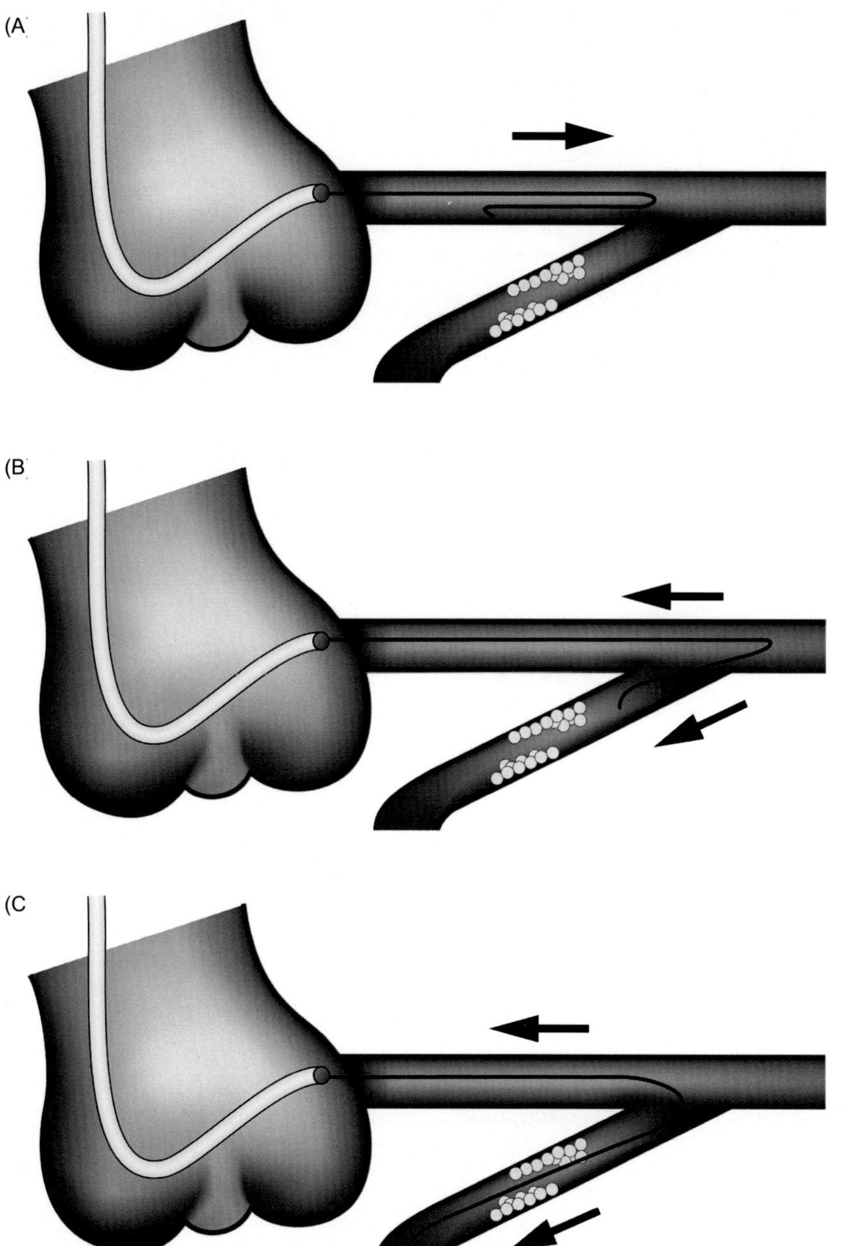
(A)
(B)
(C)

FIGURE 8.7 Illustration of the reversed guidewire (also called "hairpin" guidewire) technique.

- **Solution 4**: Deflection balloon technique (CTO Manual online case 127). A balloon is inflated in the main vessel immediately distal to the takeoff of the angulated branch, allowing wires and microcatheters to be "deflected" into the side branch. The disadvantages of this technique include ischemia (since the main vessel is occluded for the duration of the maneuver) and risk of vessel injury, such as dissection.

8.6.3.4 What can go wrong?

1. ***Coronary dissection***

   Causes:
   - Severe or ulcerated lesions.
   - Use of stiff or polymer-jacketed guidewires.
   - Aggressive guidewire manipulation.

   Prevention:
   - Avoid using aggressive guidewires in severe lesions.
   - Gentle guidewire manipulation.

   Solutions:
   - Advance another wire into the distal true lumen, followed by stenting (see Section 25.2.1).
   - Use of CTO techniques, such as the Stingray balloon can facilitate reentry into the distal true lumen, but significant expertise is needed for successful implementation of these techniques in case of acute vessel closure due to dissection.

8.7 Cross lesion with the guidewire

8.7.1 Goal:

To advance the guidewire through the target lesion.

8.7.2 How?

The guidewire is advanced through the lesion under fluoroscopic guidance and intermittent contrast injections.

8.7.3 Challenges

1. *Wire fails to go through the lesion*

   Causes:
   - Highly stenotic and/or calcific lesion.
   - Suboptimal guidewire tip shape.
   - Tortuosity.
   - Poor guide catheter support.

   Prevention:
   - Optimize guidewire tip shape.
   - Obtain optimal guide catheter support.
   - Obtain optimal guidewire support and control by using a microcatheter.
   - Use hydrophilic-coated guidewires (Section 30.7.1) or polymer-jacketed guidewires (Section 30.7.2).

   Treatment:
   - Change the shape of the guidewire tip.
   - Use a different guidewire (e.g., polymer-jacketed instead of workhorse guidewire).
   - Use a microcatheter (or over-the-wire balloon) for support.

8.7.4 What can go wrong?

1. ***Acute vessel closure***

This is more likely to occur in very stenotic lesions. The most common cause is dissection. However, in very tortuous vessels crossing with a guidewire may cause vessel straightening and pseudolesion formation (Fig. 8.8) (Section 25.2.6) that may mimic dissection.

Prevention:

- Avoid forceful advancement of the guidewire: when tip deflection is observed pull back and rotate slightly before readvancing.
- Cautious use of polymer-jacketed guidewires.
- Partial guidewire withdrawal in tortuous lesions, so that the soft part of the wire is just distal to the lesion, can help clarify whether any apparent lesions are actually pseudolesions.

Solutions:

- Attempt to advance the same or different guidewire through the lesion. It is critical to confirm distal true lumen wire position before ballooning or stent deployment.
- Intravascular ultrasound can be used to determine wire position.
- Distal injections through a microcatheter should be avoided, as they may extend the dissection if the wire tip is in the subintimal space.
- If the guidewire is subintimal and attempts to wire the distal true lumen fail, reentry using the Stingray system (or various penetrating guidewires) can be attempted [5,6].

8.8 Advance guidewire distal to the target lesion

8.8.1 Goal:

To advance the guidewire as far distally as safely possible from the target lesion to facilitate subsequent equipment advancement.

8.8.2 How?

Guidewire is advanced under fluoroscopic guidance and intermittent contrast injection through the lesion into a distal vessel/branch. Using a stored reference image in the same projection (or the dynamic coronary roadmap feature of Philips X-ray systems) can facilitate wire advancement and limit contrast injections.

8.8.3 What can go wrong?

1. ***Wire advanced too distally causing perforation***

Prevention:

- Monitor distal guidewire position and prevent distal advancement. This is especially important when multiple wires are used.

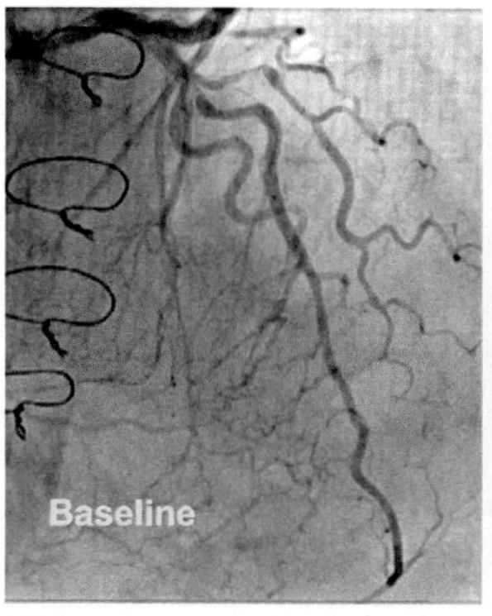

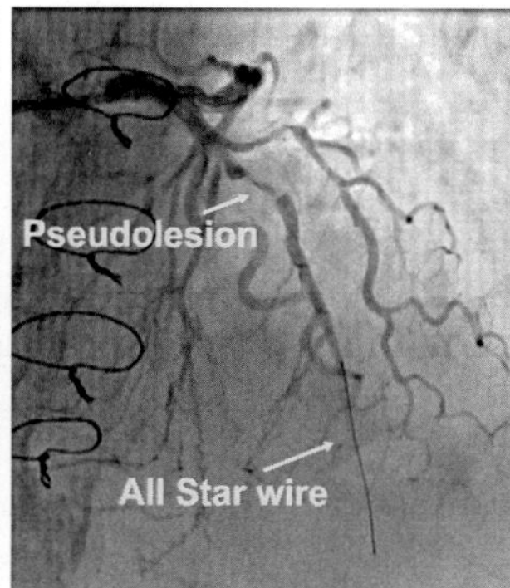

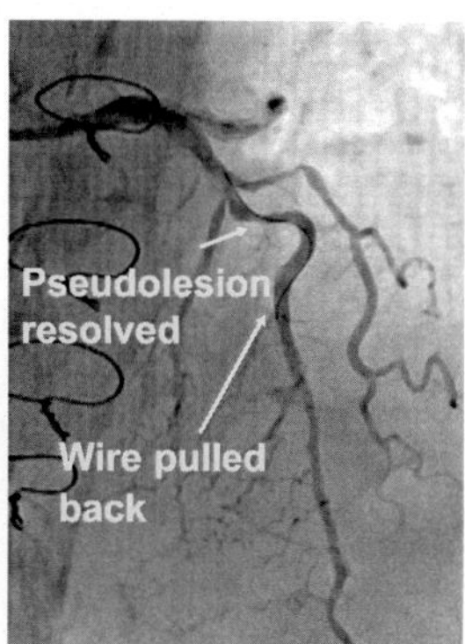

FIGURE 8.8 Example of pseudolesion after insertion of a supportive guidewire (All Star) through a tortuous mid LAD lesion. *Courtesy of Michael Megaly, MD and Ivan Chavez, MD.*

- If a stiff or polymer-jacketed guidewire is used to cross the lesion, exchange it for a workhorse guidewire immediately after crossing.

Treatment:
- Start with balloon inflation to stop bleeding into the pericardium, followed by fat or coil embolization if extravasation continues. This is described in detail in Section 26.4.

8.9 Remove microcatheter (if used)

8.9.1 Goal:

To remove the microcatheter used for lesion wiring while maintaining distal wire position without wire movement.

8.9.2 How?

There are four techniques for removing or exchanging microcatheters or any over-the-wire system when using a short (180–190 cm) guidewire [7]:

1. Trapping (preferred technique).
2. Hydraulic exchange.
3. Use of a guidewire extension.
4. "Circumcision" of the over-the-wire system.

8.9.2.1 Trapping

Trapping is the best technique for removing or exchanging microcatheters or any over-the-wire system when using a short (180–190 cm) or even a long (300 cm) guidewire, because it:

1. Minimizes guidewire movement, which can result in guidewire position loss, distal vessel injury, and/or perforation.
2. Minimizes radiation exposure (as fluoroscopy is only needed during the initial phase of catheter withdrawal).

The small inner diameter of a 6 Fr guide catheter prevents utilization of the trapping technique for over-the-wire balloons, the CrossBoss catheter, and the Stingray balloon. However, 6 Fr guides with 0.071 in. inner diameter, such as the Medtronic Launcher line of guide catheters, allow trapping of low profile microcatheters such as Turnpike, SuperCross, Corsair, Caravel, and Finecross.

The TrapLiner catheter (Section 30.3, Fig. 30.20) combines guide extension functionality with a built-in trapping balloon to pin short guidewires. The TrapLiner is available in 6–8 Fr sizes. The advantages of this system include reduced requirement for fluoroscopy to position a trapping balloon and reduced equipment use, as trapping balloons can be difficult to reinsert through the Y-connector after multiple inflations during long procedures. The TrapLiner catheter works best with short (135 cm) microcatheters, whereas long microcatheters (150 cm) are often too long to expose the guidewire for trapping with the TrapLiner balloon.

Trapping technique (Fig. 8.9):

***Step 1*:** Withdraw the over-the-wire balloon or microcatheter into the guide catheter just proximal to the position where the trapping balloon will be inflated (Fig. 8.9, panel B)

***Step 2*:** Insert the trapping balloon through the Y-connector, next to (but not over) the guidewire. Either standard compliant balloons or dedicated "trapping" balloons can be used. Keeping the stylet in the end of the balloon catheter can assist in the advancement through the hemostatic valve of the Y-connector; the stylet should then be removed before advancing the balloon further.

Type of balloon: Monorail balloons are preferred to over-the-wire balloons that have larger profiles compared with rapid exchange balloons and may not fit into the guide along with other equipment, such as the Stingray balloon and the Venture catheter. Dedicated trapping balloons (Section 30.9.4) have smaller profiles and can trap equipment in smaller guides.

Size of the trapping balloon: 2.5 mm for 6 or 7 Fr guide catheters, 3.0 mm for 7 and 8 Fr guide catheters.

Length of the trapping balloon: Ideally ≥ 20 mm (longer balloon length provides more area of contact and "pins" the wire better, which is especially important for trapping polymer-jacketed wires, which are more slippery).

Caveats: A previously used trapping balloon may get deformed and be difficult to reinsert through the Y-connector on subsequent attempts. In such cases the trapping balloon should be exchanged for a new one or the stylet used to

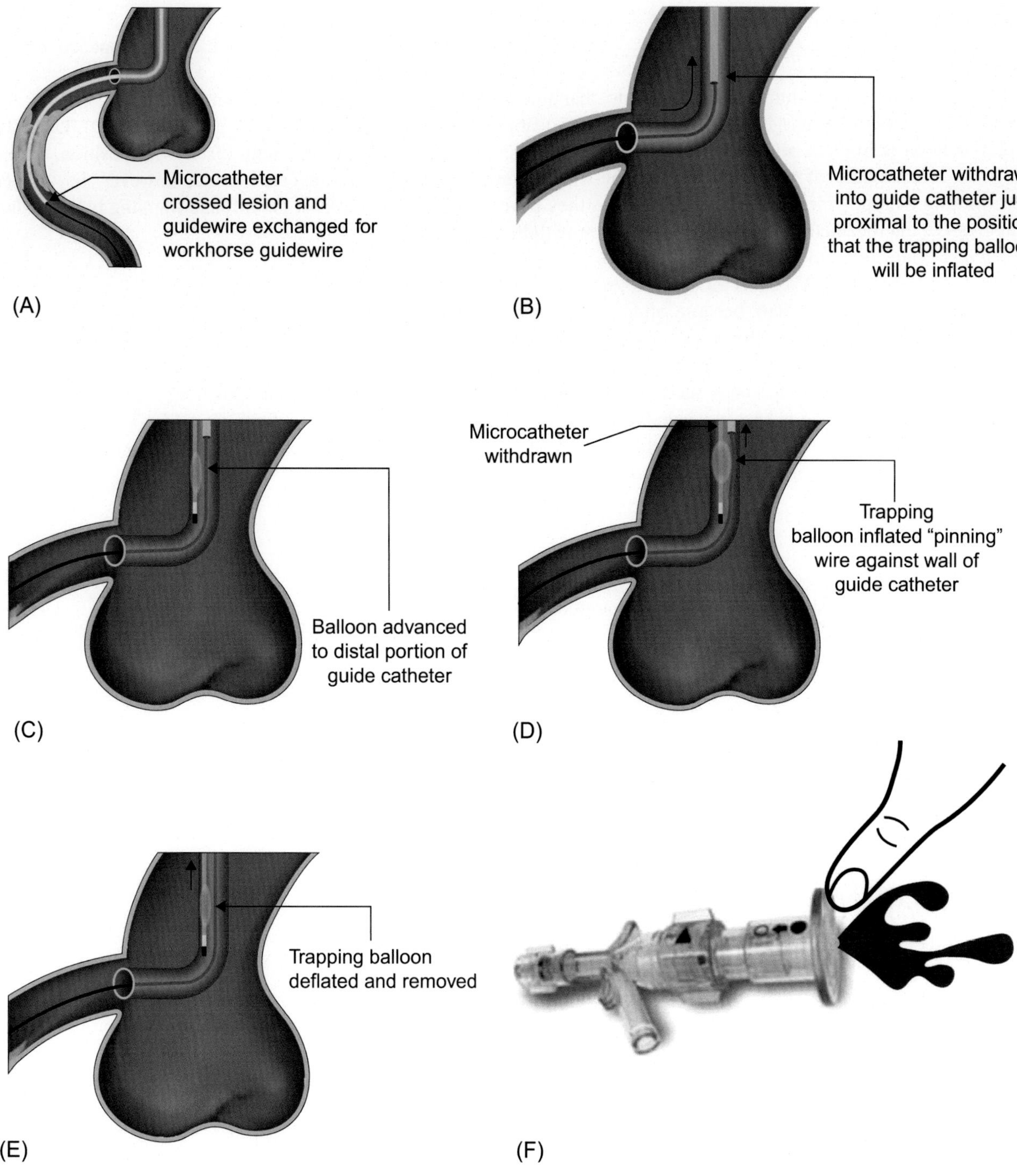

FIGURE 8.9 Illustration of the trapping technique.

reinsert the balloon though the hemostatic valve. Attempting to trap multiple devices within a single guide catheter is more likely to fail, resulting in loss of wire position.

Step 3**:** Advance the trapping balloon to the distal portion of the guide catheter beyond the distal marker of the microcatheter or over-the-wire balloon, usually at or near the primary curve (Fig. 8.9, panel C).

Caveat: when short or shortened guide catheters are used, balloon shaft markers cannot be relied upon to prevent inadvertent advancement of the trapping balloon into the target vessel.

***Step 4*:** Inflate the trapping balloon (Fig. 8.9, panel D).

Pressure: 15–20 atm to provide adequate "trapping." The guide catheter pressure should be dampened during trapping.

Caveat: Balloon may rupture if inflated next to "handmade" side holes in the guide catheter.

***Step 5*:** Withdraw the over-the-wire balloon or microcatheter from the guide catheter (Fig. 8.9, panel D). Once the trapping balloon is in place and inflated, fluoroscopy is generally not required to maintain distal wire position, except during the initial phase of microcatheter/over-the-wire balloon withdrawal. Extra care should, however, be utilized when working with polymer-jacketed guidewires, as they have a tendency to slip past the inflated trapping balloon, during aggressive microcatheter or over-the-wire balloon withdrawal.

***Step 6*:** Deflate and remove the trapping balloon (Fig. 8.9, panel E).

***Step 7*:** Back bleed the Y-connector (Fig. 8.9, panel F).

This is a very important step, because often air is entrained into the guide catheter during trapping. If contrast is injected without back bleeding, coronary air embolization is likely to occur. An alternative technique to prevent air embolization is to use continuous saline flushing when removing the catheters while the trapping balloon is inflated.

Step 8: The microcatheter has been successfully removed while maintaining distal wire position.

Tip: if an anchor balloon is kept in a side-branch and its size is appropriate i.e. ≥2.5 mm in a 6 or 7 Fr guide catheter and ≥3.0 mm in an 8 Fr guide catheter, this balloon can be deflated and temporarily withdrawn into the guide catheter for trapping. This maneuver saves time and cost.

What can go wrong?

1. *Air embolization*

   Prevention:
   - Back bleed the guide catheter after using the trapping technique.

   Treatment:
   - Treatment of air embolism is described in Section 25.2.3.3.
2. *Loss of wire position*

   Prevention:
   - Apply trapping technique with meticulous attention to detail.

   Treatment:
   - Rewire the vessel.

8.9.2.2 Hydraulic exchange

(Fig. 8.10)

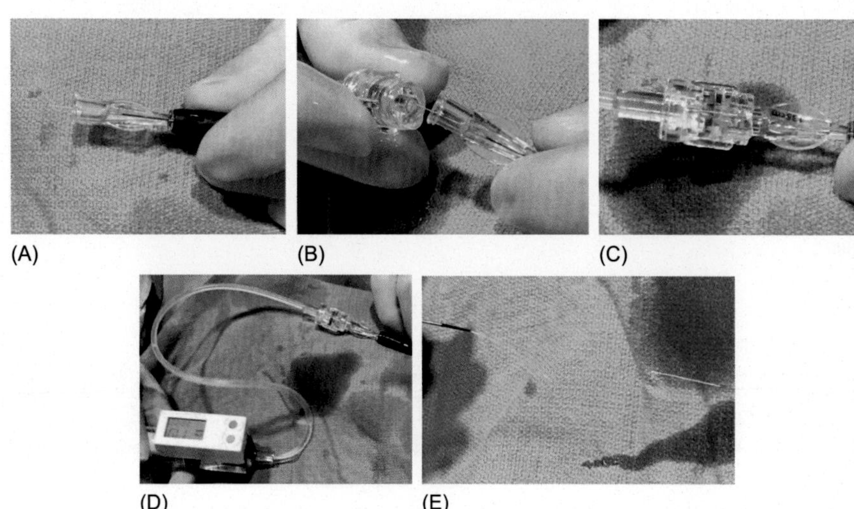

FIGURE 8.10 Illustration of the hydraulic exchange technique. *Courtesy Dr. William Nicholson. Reproduced with permission from Brilakis ES. Manual of coronary chronic total occlusion interventions. A step-by-step approach. 2nd ed. Cambridge, MA: Elsevier; 2017 (Figure 3.32).*

Hydraulic exchange (also called "jet exchange" or "Nanto technique") is easier to perform, but less reliable than trapping and may not maintain the distal position of the guidewire [8,9]. Hydraulic exchange should not be used with stiff tip or polymer-jacketed guidewires to minimize the risk of distal perforation. Hydraulic exchange can be performed through any size guide catheter as follows:

***Step 1*:** Fill an inflation device with normal saline (alternatively a standard mixture of contrast and saline can be used). The inflation device should be filled with the maximum volume possible. Alternatively, a 3 cc luer lock syringe can be used, however an inflation device is preferred.

Step 2: Withdraw the microcatheter or over-the-wire balloon as far as possible until the back end of the short guidewire is at the hub of the microcatheter/over-the-wire balloon (Fig. 8.10, panel A).

Step 3: Connect the saline-filled inflation device to the hub of the microcatheter/over-the-wire balloon (wet to wet connection) (Fig. 8.10, panel B and C).

Step 4: Rotate the indeflator arm until a pressure of 14–20 atm is reached, while performing fluoroscopy (Fig. 8.10, panel D).

***Step 5*:** Under fluoroscopy, when the inflation device pressure reaches 14–20 atm withdraw the microcatheter/over-the-wire balloon (Fig. 8.10, panel E), while maintaining the indeflator pressure. It is very important that the inflation device pressure is maintained at 14–20 atm as the microcatheter is withdrawn. Otherwise, if the pressure is lost, the wire may be withdrawn. If partial guidewire withdrawal occurs, slowing down the speed of microcatheter withdrawal while maintaining the inflation device pressure can often correct/salvage the situation.

***Caveats*:** Use of the hydraulic exchange should be avoided in CTO PCI in general and especially when using stiff, tapered-tip guidewires or stiff polymer-jacketed guidewires in order to minimize the risk of inadvertent guidewire advancement and distal guidewire perforation. Hydraulic exchange should also be avoided for removing over-the-wire balloon catheters and the Stingray balloon because increased friction between the inner surface of the catheter and the guidewire may result in loss of guidewire position.

8.9.2.3 Use of a guidewire extension

(Fig. 8.11)

How?

By inserting the back end of the short guidewire into the guidewire extension. A guidewire extension can be used with any size guide catheter [7].

***Caveats*:**

1. Each guidewire type has specific guidewire extension that should be available in the catheterization laboratory. For example:
   a. Abbott guidewires: "DOC" guidewire extension (145 cm long).
   b. Asahi guidewires: "Asahi" guidewire extension (150/165 cm long).
   c. Cordis guidewires: "Cinch" guidewire extension (145 cm long).
2. The connection should be tightened as much as possible to minimize the risk of the guidewire and guidewire extension coming apart during withdrawal of the microcatheter or over-the-wire balloon.
3. Be careful to avoid kinking or bending of the guidewire when tightening the connection.

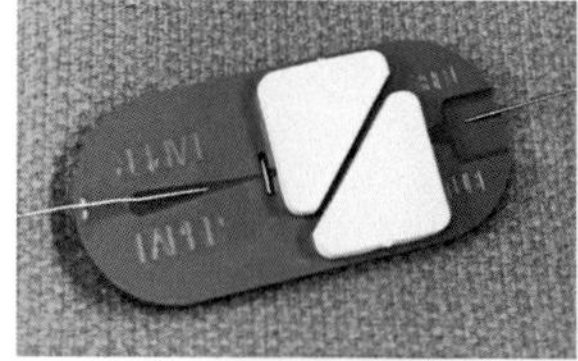
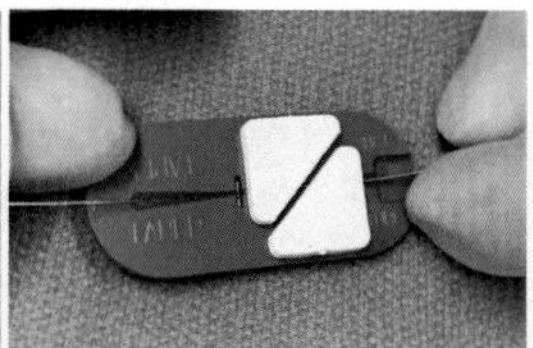
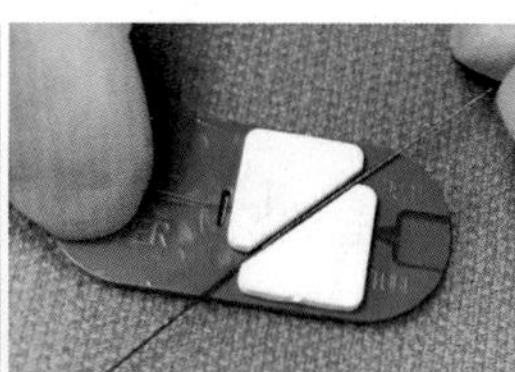
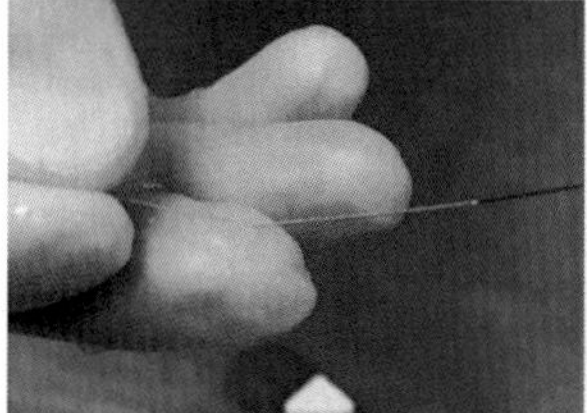
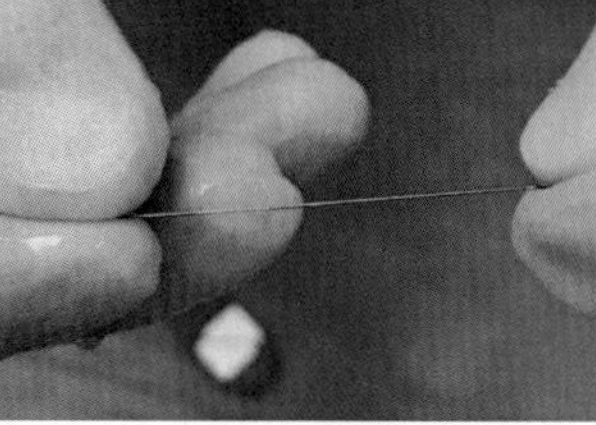

FIGURE 8.11 Using a guidewire extension to remove a microcatheter over a short guidewire. *Courtesy of Dr. William Nicholson. Reproduced with permission from Brilakis ES. Manual of coronary chronic total occlusion interventions. A step-by-step approach. 2nd ed. Cambridge, MA: Elsevier; 2017 (Figure 3.33).*

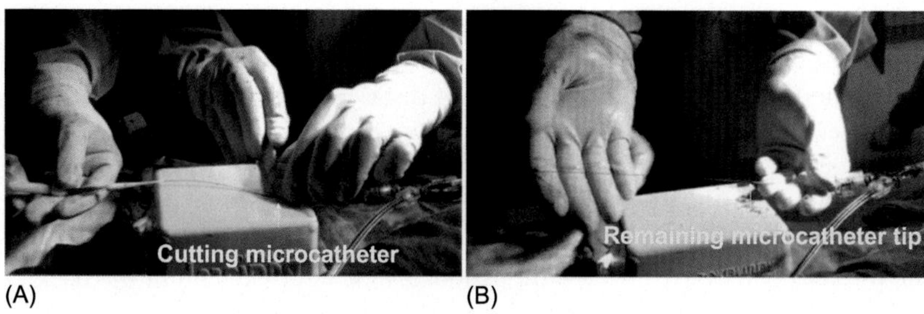

FIGURE 8.12 Illustration of the circumcision technique for removing a microcatheter over a short guidewire. (Panel A) A scalpel is used over a hard surface (Backstop disposal system [Merit Medical] in this case) to cut the proximal part of the microcatheter. (Panel B) After the microcatheter is cut, the proximal part of the microcatheter is removed and the process is repeated until the entire microcatheter has been removed. *Reproduced with permission from Brilakis ES. Manual of coronary chronic total occlusion interventions. A step-by-step approach. 2nd ed. Cambridge, MA: Elsevier; 2017 (Figure 3.34) also online video: Microcatheter "circumcision" technique.*

8.9.2.4 Circumcision technique

(Fig. 8.12)

This is an advanced and cumbersome technique, but can be very useful, especially in cases where the guidewire becomes entrapped within the microcatheter/over-the-wire balloon or when appropriate extension guidewires are not available and trapping is not possible. The circumcision technique can be performed with any size guide catheter.

***Step 1*:** Withdraw the microcatheter or over-the-wire balloon as far as possible until the back end of the short guidewire is at the hub of the microcatheter/over-the-wire balloon.

***Step 2*:** Using a scalpel and a hard surface, a circumferential cut is made as close to the Y-connector as possible. (Fig. 8.12A)

***Step 3*:** Once circumferential cutting is complete, the microcatheter/over-the-wire balloon fragment is removed. (Fig. 8.12B)

***Step 4*:** Steps 1–3 are repeated until the entire microcatheter/over-the-wire balloon is removed. Cutting two to three segments is usually necessary to remove the entire microcatheter or over-the-wire balloon.

Caveats:

1. Cutting should be done very carefully to minimize the risk of injuring the patient or the operator, and to avoid damaging the guidewire making it difficult, if not impossible, to pass other equipment over it.
2. Consider using a hard surface, such as an upside down saline bowl to perform the circumcision on.
3. The microcatheter/over-the-wire balloon will be destroyed during this maneuver and hence cannot be reused.

8.10 Monitor wire position

8.10.1 Goal:

To avoid inadvertent guidewire withdrawal or distal guidewire advancement.

8.10.2 How?

The tip of the guidewire is monitored. If the tip position changes the guidewire position is adjusted.

8.10.3 What can go wrong?

8.10.3.1 Perforation

Perforation can occur if the guidewire tip is advanced too distally. The risk is increased if polymer jacketed or stiff and/or tapered tip guidewires are used.

Prevention:

- Continuous or at least intermittent monitoring of the guidewire position. This is especially important if collimation is used not allowing continuous monitoring of the guidewire tip position and when there is significant difficulty with equipment delivery or withdrawal that often results in excessive guidewire movement.

- Another prevention option is use of the ControlRad system that allows high image resolution in part of the screen, with low image resolution in the remaining portion of the screen (including the location of the guidewire).

Treatment:
- Full discussion of distal guidewire perforation is discussed in Section 26.4. The first step is to inflate a balloon proximal to the perforation site to prevent further bleeding into the pericardium, while preparations are made for definitive treatment, that usually involves fat or coil embolization.

8.10.3.2 Guidewire withdrawal proximal to the target lesion

This could result in inability to treat acute vessel closure, if dissection occurs at the lesion site.

Prevention:
- Continuous monitoring of the guidewire position.
- Good lesion preparation.
- Avoid very aggressive attempts to advance equipment through a lesion. For example, if a stent cannot cross a lesion, repeat balloon angioplasty (or sometimes atherectomy) can facilitate crossing.

References

[1] Ide S, Sumitsuji S, Kaneda H, Kassaian SE, Ostovan MA, Nanto S. A case of successful percutaneous coronary intervention for chronic total occlusion using the reversed guidewire technique. Cardiovasc Interv Ther 2013;28:282–6.

[2] Kawasaki T, Koga H, Serikawa T. New bifurcation guidewire technique: a reversed guidewire technique for extremely angulated bifurcation—a case report. Catheter Cardiovasc Interv 2008;71:73–6.

[3] Suzuki G, Nozaki Y, Sakurai M. A novel guidewire approach for handling acute-angle bifurcations: reversed guidewire technique with adjunctive use of a double-lumen microcatheter. J Invasive Cardiol 2013;25:48–54.

[4] Hasegawa K, Yamamoto, W, Nakabayashi S, Otsuji O. Streamlined reverse wire technique for the treatment of complex bifurcated lesions, Catheter Cardiovasc Interv 2019; doi: 10.1002/ccd.28656.

[5] Martinez-Rumayor AA, Banerjee S, Brilakis ES. Knuckle wire and stingray balloon for recrossing a coronary dissection after loss of guidewire position. JACC Cardiovasc Interv 2012;5:e31–2.

[6] Shaukat A, Mooney M, Burke MN, Brilakis ES. Use of chronic total occlusion percutaneous coronary intervention techniques for treating acute vessel closure. Catheter Cardiovasc Interv 2018;92:1297–300.

[7] Brilakis ES. Manual of coronary chronic total occlusion interventions. A step-by-step approach. 2nd ed Cambridge, MA: Elsevier; 2017.

[8] Nanto S, Ohara T, Shimonagata T, Hori M, Kubori S. A technique for changing a PTCA balloon catheter over a regular-length guidewire. Cathet Cardiovasc Diagn 1994;32:274–7.

[9] Feiring AJ, Olson LE. Coronary stent and over-the-wire catheter exchange using standard length guidewires: jet exchange (JEX) practice and theory. Cathet Cardiovasc Diagn 1997;42:457–66.

Chapter 9

Lesion preparation

9.1 Goal

To adequately prepare the target lesion to facilitate stent delivery and expansion.

9.2 When is lesion preparation needed?

(Fig. 9.1)

Lesion preparation (in most cases with balloon angioplasty) should be performed in nearly all lesions because it:

1. Facilitates stent delivery and decreases the risk of stent loss.
2. Helps determine optimal stent diameter and length (especially when no intracoronary imaging is used and when there is poor flow of contrast distal to the target lesion).
3. Facilitates stent expansion and helps determine the need for additional lesion modification (e.g., with atherectomy in heavily calcified lesions) to prevent "stent-regret."

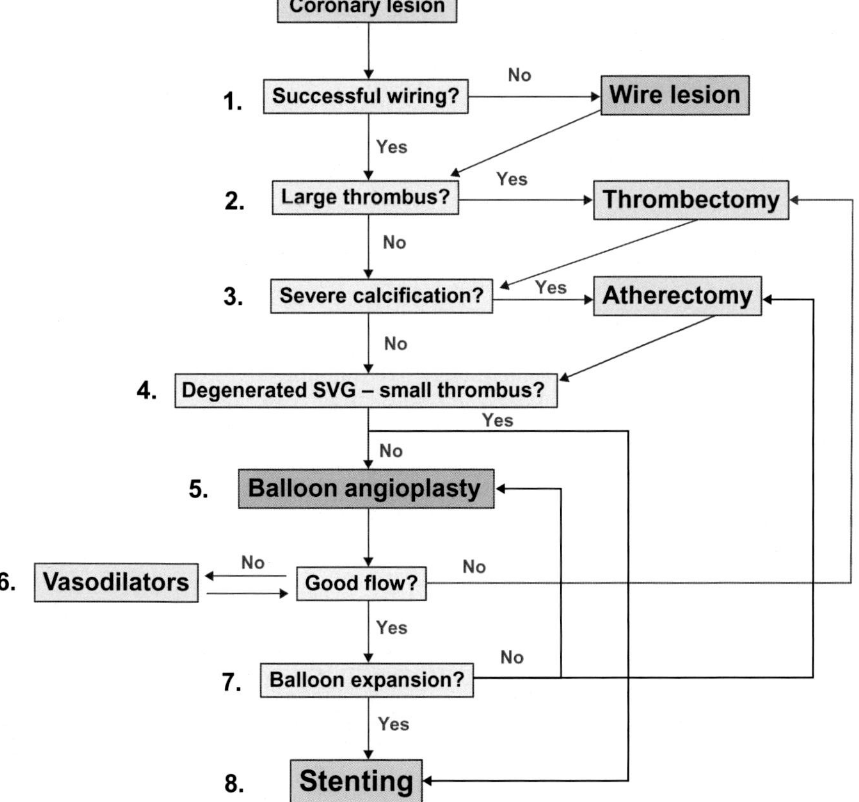

FIGURE 9.1 Lesion preparation algorithm.

Manual of Percutaneous Coronary Interventions. DOI: https://doi.org/10.1016/B978-0-12-819367-9.00009-3

9.3 Confirm successful wiring

Coronary lesion preparation should not be performed until after successful wiring of the lesion (as described in Chapter 8: Wiring). Optimal distal wire position should be confirmed before balloon angioplasty (and/or stenting).

9.4 Large thrombus

If large thrombus is present, coronary thrombectomy is often performed, as described in Chapter 20: Acute coronary syndromes—thrombus.

9.5 Severe calcification

If severe calcification is present, atherectomy or intracoronary lithotripsy is often performed, as described in Chapter 19: Calcification.

9.6 Degenerated SVG or thrombus

Primary stenting (i.e., stenting without balloon angioplasty or other lesion preparation) may be preferred in few lesion types, such as degenerated saphenous vein grafts with friable lesions, or lesion with thrombus (when no thrombectomy is planned or when thrombectomy fails).

9.7 Balloon angioplasty

9.7.1 Step 1. Confirm that the guidewire is optimally positioned through the target lesion

9.7.1.1 Goal

To ensure that the guidewire is optimally positioned (through the target lesion and advanced several centimeters distally, but not into small distal branches that could lead to perforation).

9.7.1.2 How?

Contrast injection is performed to confirm that the guidewire is in optimal position. If position is suboptimal, the guidewire is repositioned.

9.7.1.3 What can go wrong?

If the guidewire is suboptimally positioned, complications can happen at the time of balloon inflation, as follows:

9.7.1.3.1 Perforation

Causes:
- Wire entered small distal branch.

Prevention:
- Ensure optimal distal guidewire position (wire should not be in a small branch to minimize the risk of distal coronary perforation; this is, especially important for branches located close to the occlusion, as the balloon may enter that branch and perforate it while dilating the lesion).

Treatment:
- Treatment of perforation is discussed in Chapter 26: Perforation.

9.7.2 Step 2. Confirm that the guide catheter is aspirated and flushed

9.7.2.1 Goal

To ensure there is no air or debris inside the guide catheter.

9.7.2.2 How?

1. Back bleed the Y-connector.
2. Aspirate the guide catheter, then fill with contrast.

9.7.2.3 What can go wrong?

9.7.2.3.1 *Distal embolization* (Section 25.2.3)

Causes:

- Air sucked into the guide catheter during balloon withdrawal. This is more likely when withdrawing large balloons.
- Thrombus or plaque entry into the guide catheter while withdrawing a balloon (may be more likely with bulky large diameter balloons, or plaque modification balloons, such as the Angiosculpt, Chocolate, and cutting balloon, Section 30.9.3).

Prevention:

- Guide aspiration followed by flushing, as described above.

Treatment:

- Air embolization is treated with administration of 100% oxygen, possibly aspiration, and administration of intracoronary epinephrine in case of cardiac arrest. Air embolization usually resolves with supportive measures without needing additional intervention. Air embolization is described in detail in Section 25.2.3.3.
- Embolization of plaque or thrombus is usually treated with thrombectomy, as described in Chapter 20: Acute coronary syndromes—thrombus.

9.7.3 Step 3. Select balloon type and size

9.7.3.1 Goal

To choose optimal balloon size and type.

9.7.3.2 How?

9.7.3.2.1 Balloon diameter

For achieving lesion expansion: for most lesions the balloon diameter is chosen to match the distal reference vessel diameter (1:1 ratio). The goal is to determine whether the balloon fully expands and therefore no additional lesion modification is needed before stenting. This is particularly important for calcified and in-stent restenotic lesions.

For crossing: for very tight lesions that are hard to cross, small balloons ($\leq$1.5 mm, Section 30.9.2) are initially used to modify the lesion entry point and allow subsequent delivery of larger balloons to further modify the lesion before stent placement.

9.7.3.2.2 Balloon length

Balloon length should be shorter than the estimated lesion length (to avoid injury of coronary segments proximal or distal to the target lesion that will require implantation of longer length stents).

9.7.3.2.3 Balloon compliance

Noncompliant balloons are preferred, because they can also be used for postdilation of the target lesion after stenting. However, they are less deliverable than compliant balloons.

9.7.3.2.4 Balloon delivery system

There are two balloon delivery systems: monorail and over-the-wire. Monorail balloons should be used in the vast majority of cases, as they are simpler to use, do not require use of long guidewires, and are easier to deliver because they have a stiffer shaft.

9.7.3.2.5 Plaque modification balloons

Some balloons have nitinol wires (such as the Angiosculpt and the Chocolate) or cutting blades (cutting balloon) aiming to modify the lesion and facilitate expansion (Section 30.9.3). Plaque modification balloons are harder to deliver (due to larger profile and lesser flexibility). The cutting balloon also requires slow inflation and deflation. Plaque modification balloons can, in some cases, facilitate expansion of "balloon undilatable lesions" (Section 23.2). The SIS OPN balloon (Section 30.9.7) can be very useful for dilating resistant lesions, but is not currently available in the United States.

9.7.4 Step 4. Prepare the balloon

9.7.4.1 *Goal*

Remove the air from the balloon, fill the balloon with a contrast solution, and connect to an indeflator.

9.7.4.2 *How?*

1. The balloon is removed from its packaging.
2. A luer-lock syringe is filled usually with a 50% contrast/50% saline solution. Lower concentration of contrast will speed the inflation/deflation of the balloon but will decrease the balloon visibility.
3. The syringe is connected with the balloon proximal hub.
4. Negative suction is performed with the syringe plunger positioned up.
5. The balloon is connected with the indeflator and negative suction is applied.

9.7.4.3 *Challenges*

1. *Operator finger injury*. The balloon/stent should be slowly removed from the container hoop to reduce the risk of the balloon/stent stylet injuring the operator's hands.

9.7.4.4 *What can go wrong?*

9.7.4.4.1 Inability to visualize balloon when inflated

Causes:

- Inadvertent filling of the balloon with saline instead of contrast solution.
- Poor balloon preparation with significant amount of air remaining in the balloon.

Prevention:

- Ensure that diluted contrast (and not pure saline) is used to prepare and inflate the balloon. The syringe containing the contrast/saline solution should be appropriately labeled.
- Ensure that air is removed during balloon preparation.

Treatment:

- If balloon cannot be visualized once inflated, it should be prepared again using a contrast solution and optimal preparation technique.

9.7.5 Step 5. Load the balloon on the guidewire

9.7.5.1 *Goal*

To load the balloon on guidewire to allow subsequent advancement to the target coronary lesion.

9.7.5.2 *How?*

1. The balloon tip is advanced over the back end of the guidewire.
2. This maneuver is easier to perform by holding the guidewire with the thumb and index finger of the left hand with the tip of the guidewire resting on the middle finger at an angle. The balloon tip is held by the right hand and its tip is advanced over the guidewire tip (Fig. 9.2 panel A).

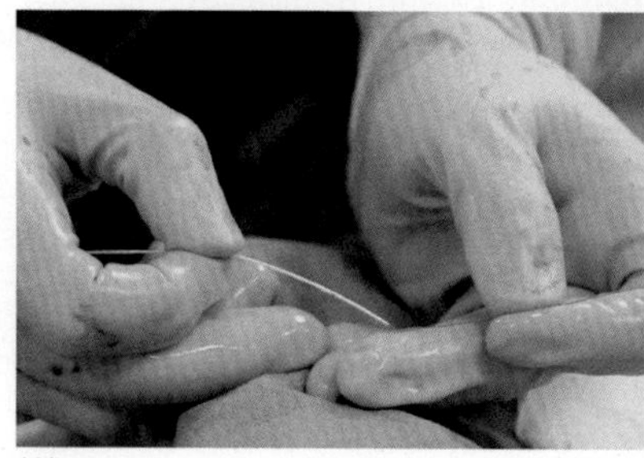
(A)

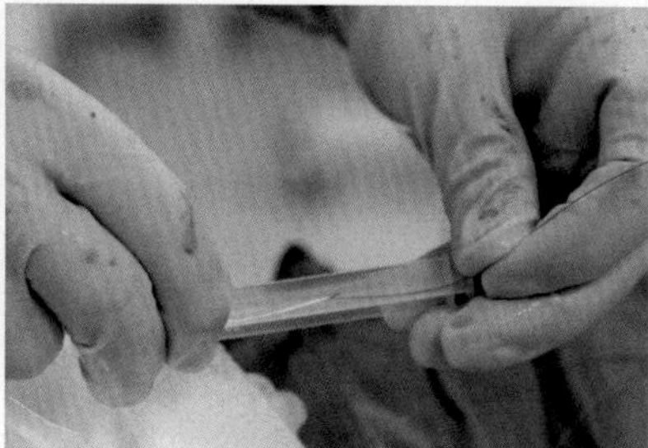
(B)

FIGURE 9.2 Loading a balloon over a guidewire. (Panel A) Holding the guidewire with the thumb and index finger of the left hand with the tip of the guidewire resting on the middle finger at an angle. (Panel B) Using the groove of a syringe plunger.

3. A variation of the above maneuver to facilitate advancing the balloon over the guidewire is to place them both in the groove of a syringe plunger (Fig. 9.2, panel B).

9.7.6 Step 6. Advance the balloon monorail segment through the Y-connector

9.7.6.1 Goal

To insert the balloon through the Y-connector into the guide catheter.

9.7.6.2 How?

1. The back end of the guidewire is fixed by an assistant or another operator (or the other hand of the operator if there is no assistant or second operator) (Fig. 9.3).
2. For standard Y-connectors: the Y-connector is opened up followed by advancement of the balloon. The Y-connector is subsequently closed.
3. For Y-connectors with automated hemostatic valves: there is no need to open the hemostatic valve. The balloon is advanced through the valve.

9.7.6.3 What can go wrong?

9.7.6.3.1 Balloon deformation/damage

Causes:

- Forceful balloon advancement through too tightly closed hemostasis valve.
- Cutting balloons may be at higher risk for damage.

Prevention:

- Open hemostasis valve before inserting balloon or use automated hemostasis devices, such as the Copilot, Guardian, Watchdog, and OKAY II (section 30.5).

Treatment:

- Replace the damaged balloon with a new one.

9.7.7 Step 7. Advance the balloon to the tip of the guide catheter

9.7.7.1 Goal

To advance the balloon to the tip of the guide catheter.

9.7.7.2 How?

1. The left hand is holding the Y-connector. The thumb and index finger of the left hand is fixing the guidewire.
2. The balloon is advanced without fluoroscopy until the first marker (for 90 cm long guide catheters) or the second marker (for 100 cm long guide catheters) is at the Y-connector.
3. The balloon is advanced under fluoroscopy to the tip of the guide catheter.

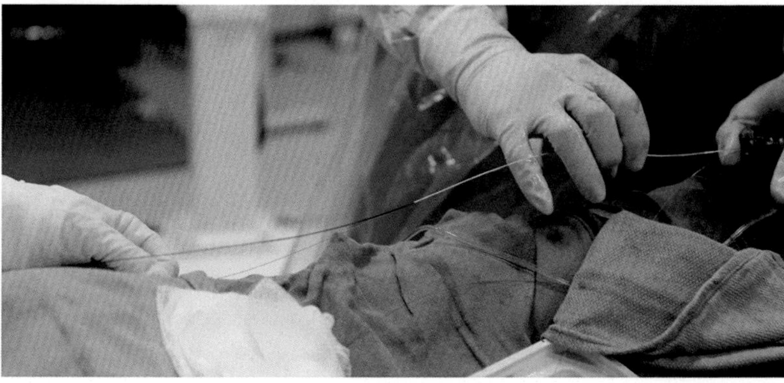

FIGURE 9.3 Inserting a balloon through the Y-connector.

9.7.7.3 What can go wrong?

9.7.7.3.1 Resistance to balloon advancement

Causes:
- Guide catheter extensions, when the guidewire wraps around the delivery rod.
- Guide catheter kinking.
- Too much equipment inside the guide catheter (too many balloons, wires, stents, etc.) (Section 30.2.2). Deciding the needed guide catheter size based on the types of equipment planned to be used can be estimated using the "Complex PCI Solutions" app (Section 30.2.2, Fig. 30.7).
- Wrapped guidewires.

Prevention:
- If guide catheter extensions are used, place the push rod of the guide extension in a towel to avoid wrapping with the guidewire.
- If there are additional guidewires and balloons/stents inside the guide, remove them before advancing the balloon.
- The "Complex PCI Solutions" app (Section 30.2.2) can be used to determine equipment compatibility within various sizes (6, 7, 8 Fr) guide catheters.

Treatment:
- If the resistance to balloon advancement is at the level of entry through a guide extension cylinder, then the guide extension is removed and reinserted (to correct wire wrapping around the guide catheter extension).
- If the resistance is due to too much equipment, the balloon is removed and reinserted after some of the equipment is removed.
- If the resistance is due to guide catheter kinking, the guide catheter is replaced with another guide catheter.

9.7.7.3.2 Kinking of the balloon shaft

Causes:
- Forceful balloon advancement.
- Resistance to balloon advancement.

Prevention:
- Avoid forceful advancement.
- Advance balloon by a short distance with each movement ("small bites").

Treatment:
- For small kinks the procedure can usually continue without any changes. The operator should not attempt to "straighten the kink" before inserting the balloon, as this may weaken the shaft and potentially lead to fracture.
- For large kinks, the balloon should be removed and discarded, as there is risk of balloon shaft fracture (Section 27.3.1.2).

9.7.7.3.3 Inadvertent advancement of the balloon into the coronary artery

Causes:
- Poor visualization, especially in obese patients.
- Lack of attention to balloon shaft markers.

Prevention:
- Meticulous attention to balloon shaft markers (Fig. 9.4).
- Do not advance the balloon shaft past the markers without fluoroscopic guidance.
- Do not push hard against resistance, unless the balloon location is clearly visualized.

Treatment:
- Withdraw balloon and perform angiography to determine whether distal dissection or perforation has occurred.

9.7.8 Step 8. Advance the balloon to the target lesion

9.7.8.1 Goal

To advance the balloon to the target lesion.

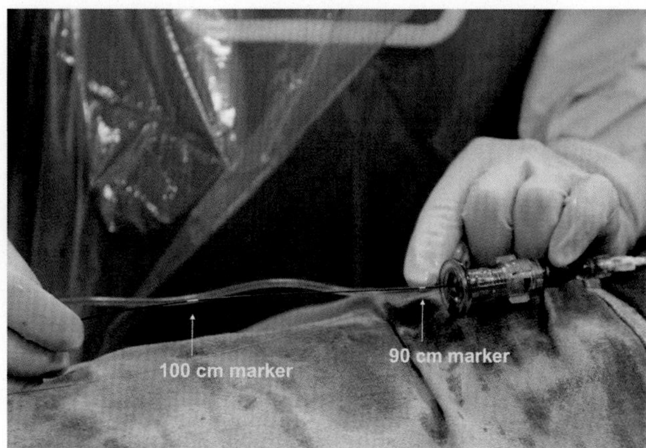

FIGURE 9.4 The 90 cm (distal) and 100 cm (proximal) marker on the balloon shaft.

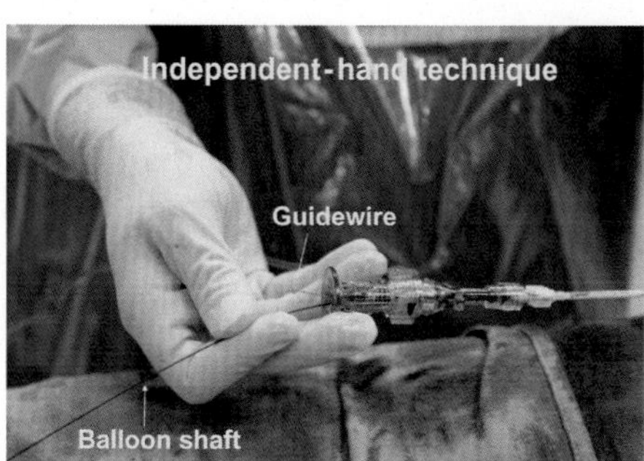

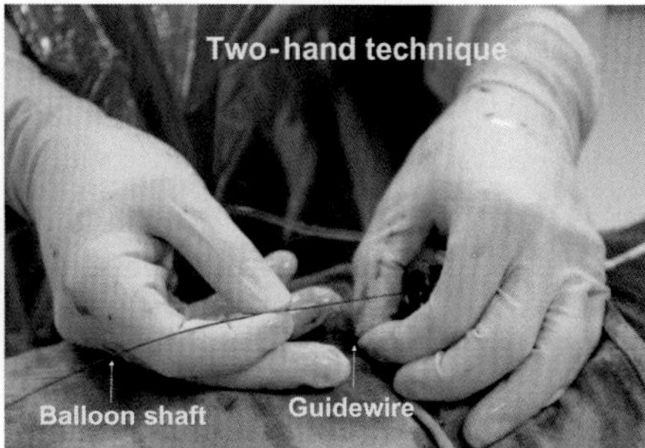

FIGURE 9.5 The independent-hand and the two-hand techniques for advancing a balloon.

9.7.8.2 *How?*

1. The balloon is advanced toward the lesion using the "independent-hand technique" or the "two-hand technique" (Fig. 9.5).

   "Independent-hand technique". The right hand holds the Y-connector with the middle and ring finger and the balloon shaft with the thumb and index finger. The guidewire is placed between the ring and little "pinky" finger of the right hand. This allows the operator to push the balloon with one hand (the right hand) while holding and simultaneously advancing the guide catheter with the other hand (left hand) to increase support.

   "Two-hand technique": The left hand holds the Y-connector and the guidewire, and the right hand advances the balloon shaft.
2. The position of the balloon is confirmed using X-ray landmarks (such as clips, previous stents, etc.) and/or contrast injection. Small diameter ($\leq$1.5 mm) balloons have only one marker in the middle, hence they cannot be used to assess the length of the lesion. Larger balloons ($\geq$2.0 mm) have two markers and can thus be used to determine the length of the lesion.

   The StentBoost (Philips) and ClearStent (Siemens) stent imaging enhancement tools can facilitate balloon placement inside previously placed stents (Section 22.2.6).

9.7.8.3 *Challenges*

9.7.8.3.1 Failure to reach the lesion with the balloon (balloon shaft is not long enough) [1]

Causes:

- Very distal lesion location, especially in patients with prior coronary bypass grafts and lesions distal to bypass graft anastomoses.

Prevention:

- Use short (or shortened) guide catheters when treating very distal coronary lesions.
- Use balloons with the longest available shaft.

Treatment:

- Use short (or shortened) guide catheters when treating very distal coronary lesions.
- Use balloons with the longest available shaft.

9.7.8.3.2 Failure to reach the lesion with the balloon (due to disease/tortuosity in proximal vessel) or cross it with a balloon (balloon uncrossable lesion)

Causes:

- Significant coronary lesions proximal to the target lesion.
- Severe proximal or within lesion tortuosity.
- Severe proximal or within lesion calcification.
- Highly stenotic (especially when also calcified and/or tortuous) target lesion.
- Use of high-profile, noncompliant, or already used balloons.

Prevention and treatment:

- (Fig. 9.6)

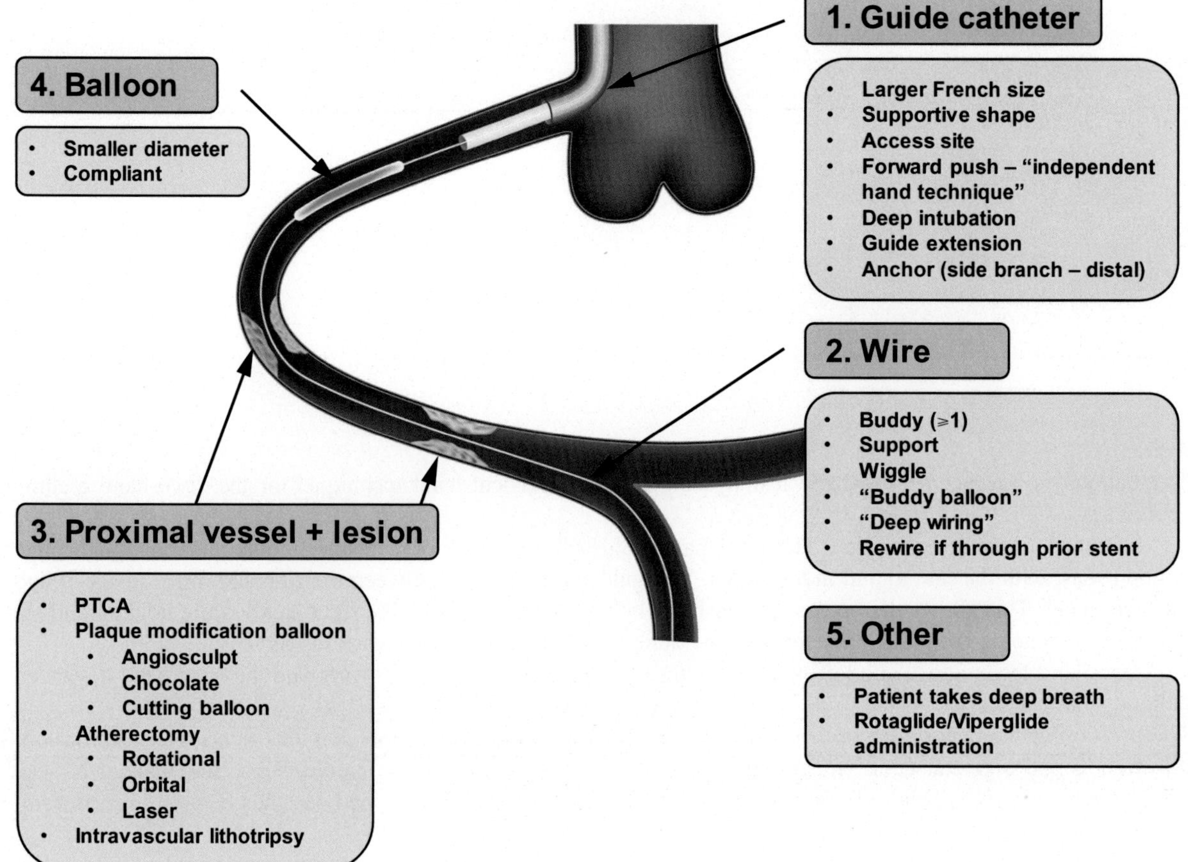
1. Guide catheter
• Larger French size
• Supportive shape
• Access site
• Forward push – "independent hand technique"
• Deep intubation
• Guide extension
• Anchor (side branch – distal)
2. Wire
• Buddy (≥1)
• Support
• Wiggle
• "Buddy balloon"
• "Deep wiring"
• Rewire if through prior stent
3. Proximal vessel + lesion
• PTCA
• Plaque modification balloon
• Angiosculpt
• Chocolate
• Cutting balloon
• Atherectomy
• Rotational
• Orbital
• Laser
• Intravascular lithotripsy
4. Balloon
• Smaller diameter
• Compliant
5. Other
• Patient takes deep breath
• Rotaglide/Viperglide administration

FIGURE 9.6 How to achieve balloon delivery to the target lesion.

9.7.8.3.2.1 Guide catheter

Strong guide catheter support is critical for balloon delivery, especially in complex lesions.

9.7.8.3.2.1.1 Larger diameter guide catheters (such as 7 or 8 Fr) provide stronger support than smaller ones (5 or 6 Fr).

9.7.8.3.2.1.2 Supportive guide catheter shapes (e.g., EBU guides provide stronger support than JL4 guides for left coronary interventions, and AL1 guide catheters provide stronger support than JR4 guide catheters for RCA interventions).

9.7.8.3.2.1.3 Access site. Femoral access provides stronger support than radial access. Long sheaths increase guide catheter support. Left radial access provides stronger support for PCI of the right coronary artery, while right radial provides stronger support for PCI of the left coronary artery.

9.7.8.3.2.1.4 Forward push. Forward push of the guide catheter can increase support, but may also sometimes lead to guide prolapse and disengagement. The "independent-hand" technique (the left-hand pushes the guide, while the right hand both fixes the guidewire and advances the balloon, Fig. 9.5) allows a single operator to simultaneously manipulate the guide and advance the balloon or stent or other equipment.

9.7.8.3.2.1.5 Deep intubation. Deep intubation of the target vessel with the guide catheter increases support, but at the same time may cause dissection. Deep intubation is usually done in the RCA or in SVGs and rarely in the left coronary artery.

9.7.8.3.2.1.6 Guide catheter extensions. (Guide catheter extensions Section 30.3) are some of the most commonly used tools for facilitating balloon and stent delivery. They not only increase guide catheter support, but may also effectively favorably "modify" the proximal vessel if advanced distally, close to (or through) the target lesion (the balloon or stent is then advanced through the guide extension without coming in contact with the wall of the target vessel).

9.7.8.3.2.1.7 Anchor techniques. (side branch anchor and distal anchor) can also provide strong support for equipment delivery. The most commonly used is the *side-branch anchoring* technique (Fig. 9.7), in which a balloon is inflated in a proximal side branch of the target vessel [2–11]. Using a long balloon increases support by increasing the "grip" of the vessel, especially if the hydrophilic coating is wiped off the surface of the balloon with a wet gauze before insertion.

The *distal anchor* technique (Fig. 9.8) is similar to the side-branch anchor, except that the balloon is inflated distal to or at the occlusion within the target artery [12–14]. Two guidewires are required for the distal anchor technique, one to deliver the anchor balloon and a second guidewire (which is pinned by the anchor balloon against the vessel wall), for delivering equipment, such as microcatheters, balloons, stents and guide catheter extensions to the lesion [5,7,15–17]. The distal anchor technique requires at least a 6 Fr guide catheter, but is better performed with a 7 or 8 Fr guide catheter.

The *buddy wire stent anchor* technique (Fig. 9.9) can be used if the proximal vessel needs stenting: a buddy wire can be inserted and a stent deployed over the main guidewire, effectively "trapping" the buddy wire, which then provides strong guide catheter support.

Limitations of the anchor techniques:

1. Injury at the site of the anchor balloon inflation, which is usually inconsequential in small side branches [12]. The risk can be minimized by sizing the balloon 1:1 to the side branch and inflating the anchor balloon at relatively low pressures (4–8 atm).
2. Distal dissection can occur with the distal anchor technique, requiring extensive stenting.
3. Larger guide catheters (at least 7 Fr) are often needed for delivering a distal anchor balloon and other equipment [7,16,17].
4. Use of a side branch anchor may modify the proximal vessel anatomy and hinder antegrade wiring.

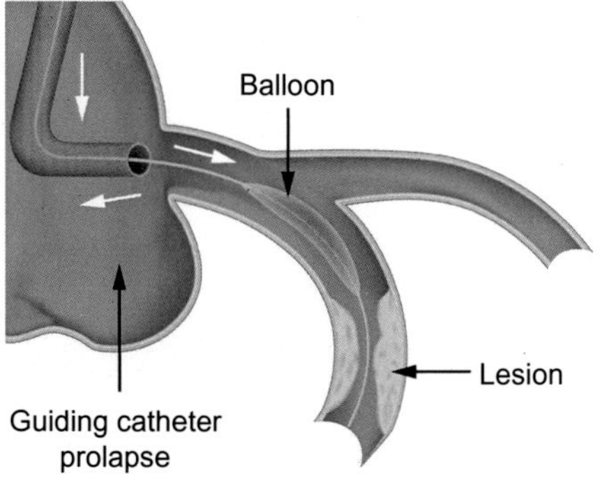

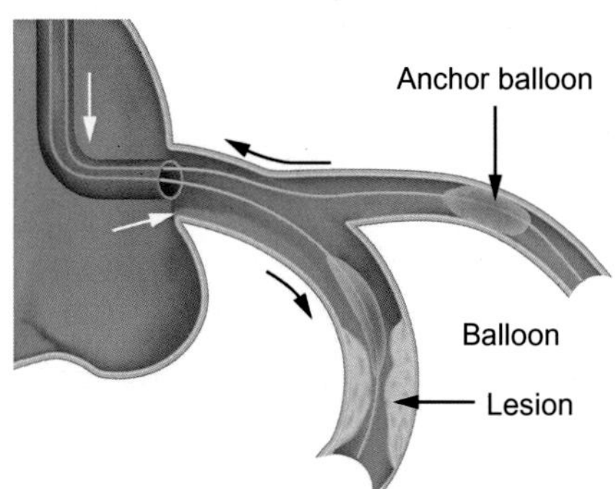

FIGURE 9.7 The side branch anchor technique.

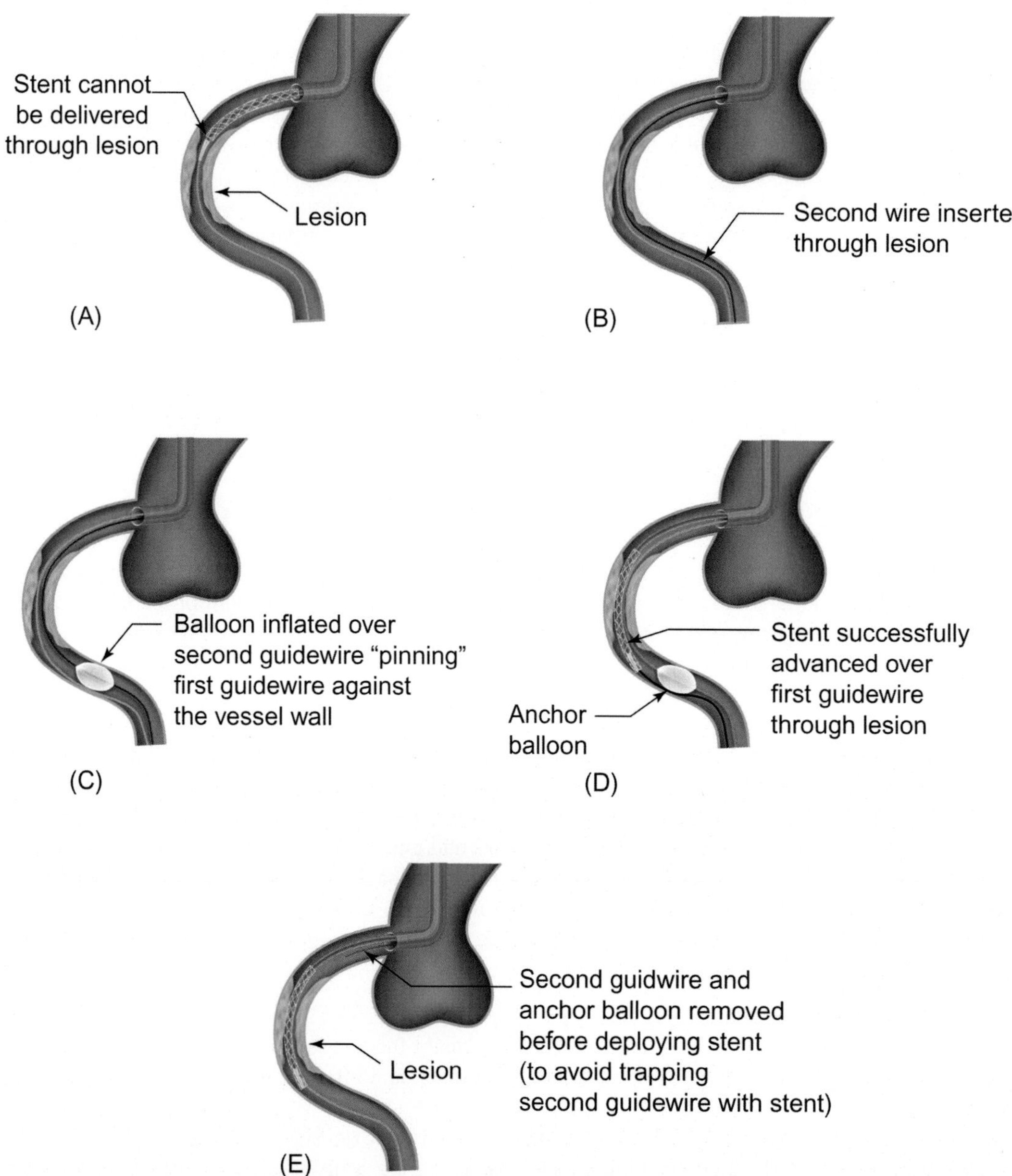
Stent cannot be delivered through lesion
Lesion
(A)
Second wire inserted through lesion
(B)
Balloon inflated over second guidewire "pinning" first guidewire against the vessel wall
(C)
Anchor balloon
Stent successfully advanced over first guidewire through lesion
(D)
Second guidwire and anchor balloon removed before deploying stent (to avoid trapping second guidewire with stent)
Lesion
(E)

FIGURE 9.8 Illustration of the distal anchor technique. The distal anchor technique can be useful when difficulty is encountered delivering a stent (or other equipment) through a lesion (panel A). A second guidewire is inserted next to the initial guidewire (panel B) and a balloon (which is usually easier to deliver than a stent) is delivered distal to the lesion and inflated, "pinning" the initial guidewire against the vessel wall (panel C) and enabling stent delivery over the first guidewire (panel D). The second guidewire (and balloon) are subsequently withdrawn before stent deployment (panel E). *Modified with permission from Brilakis ES. Manual of coronary chronic total occlusion interventions. A step-by-step approach. 2nd ed. Cambridge, MA: Elsevier; 2017 (Figure 3.29).*

5. Rarely, ischemia can occur if large branches are used for anchoring, such as diagonal branches. Some patients experience arrhythmias and/or hemodynamic instability with prolonged balloon occlusion of the conus branch of the proximal right coronary artery when used for anchoring. Intermittent deflation of the anchoring balloon may be necessary to relieve ischemia in such cases.

9.7.8.3.2.2 Wire

One of the simplest ways to facilitate delivery is to use one (or more) *buddy wire(s)*. The buddy wire increases guide support and also straightens the target vessel.

Another option is to use *support* guidewires (such as the Iron Man, Grand Slam, Mailman, or the *Wiggle* wire, Section 30.7.4).

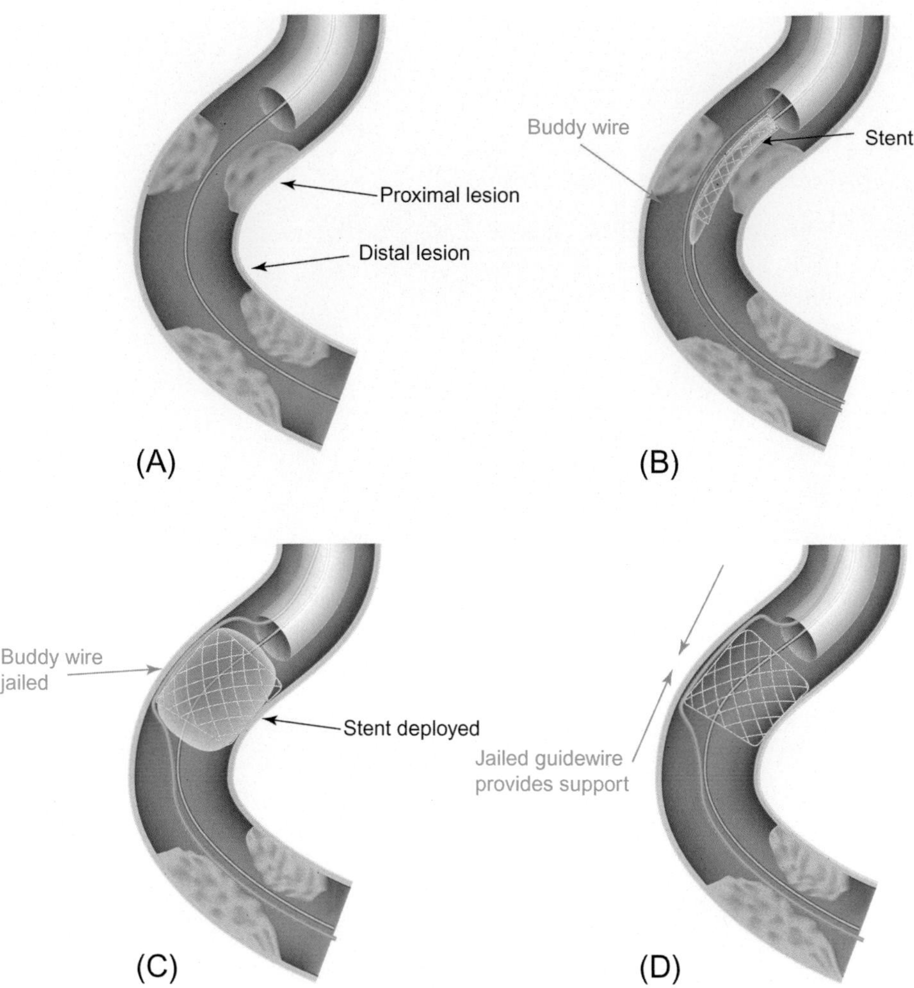

FIGURE 9.9 Illustration of the buddy wire stent anchor technique.

Partial withdrawal of the wire while advancing the stent may facilitate advancement.

Another option is to *advance a balloon* over a buddy wire across the area of resistance to stent advancement, to "deflect" the stent during advancement.

If there is resistance to balloon delivery at the location of a previously deployed stent, it is possible that the wire has crossed under a stent strut (Fig. 9.10). *Rewiring through the stent*, ideally after knuckling the wire tip more proximally, can solve this problem.

9.7.8.3.2.2.1 "Deep wire crossing" technique. In the "deep wiring" technique a polymer-jacketed guidewire is advanced through the lesion then into the ventricle/aorta through arterioluminal communications, which increases wire support. The obvious risk of this technique is coronary perforation.

How?

(Fig. 9.11)

The operator gently advances a polymer jacketed guidewire through the selected distal coronary artery branches relying on tactile feedback under fluoroscopic control without contrast, similar to the "surfing technique" described by Sianos for septal collateral crossing during retrograde recanalization of chronic total occlusions.

What can go wrong?

1. Unable to advance a guidewire through the selected branch into a ventricular cavity. In this case, the operator must carefully pull the wire back and redirect it to another channel.
2. Distal perforation. This is more likely to happen when the operator continues wire manipulations when the wire is meeting resistance.

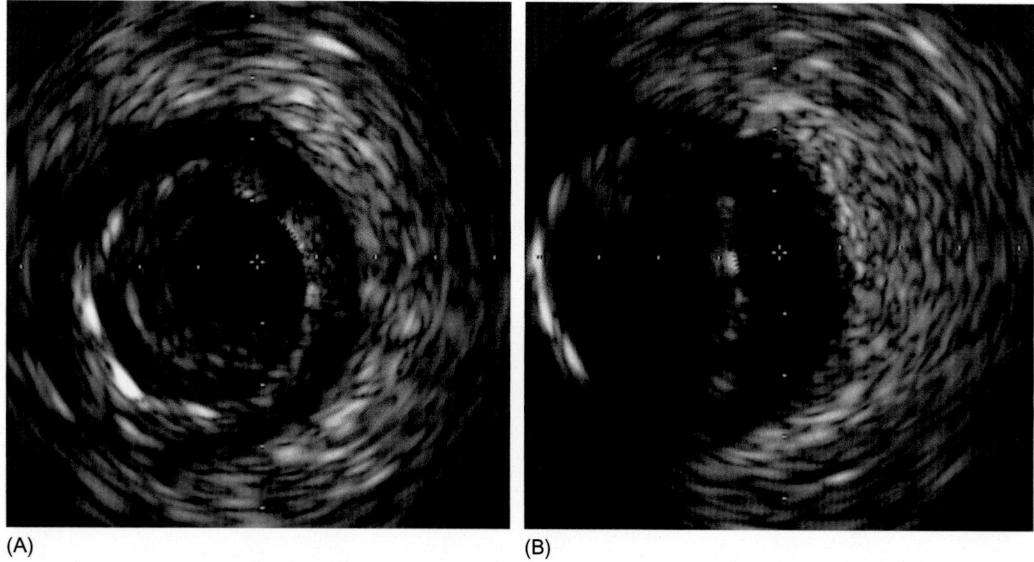

FIGURE 9.10 Failure to deliver a stent through a previously deployed stent due to substent wire advancement. IVUS images taken from guidewire which passed coaxially through the proximal edge of a previously deployed stent (panel A) and subsequently through a side strut into the substent space (panel B). *Courtesy of Dr. Jaikirshan Khatri.*

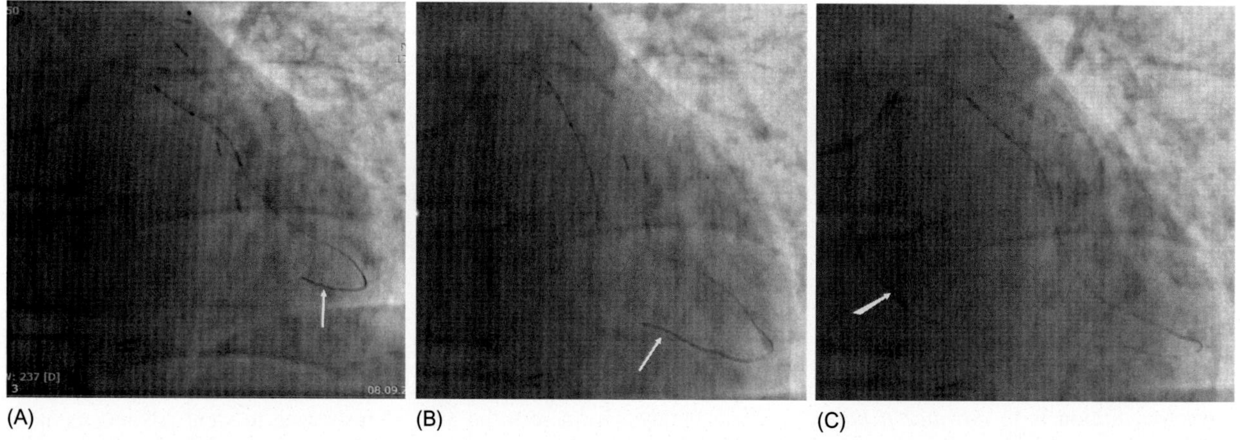

FIGURE 9.11 Illustration of stages in "deep wire crossing" technique. (A) Guidewire is in the distal septal branch. (B) Advancement of wire distally with intensive rotational movements. (C) Guidewire is in the cavity of the ventricle. *Courtesy of Dr. Oleg Krestyaninov and Dr. Dmitrii Khelimskii.*

3. Ventricular exrasystoles or tachyarrhythmias. These usually resolve spontaneously without requiring additional treatment.

9.7.8.3.2.3 Proximal vessel preparation

If the balloon cannot reach the target lesion due to proximal disease (especially in the setting of tortuosity and calcification), preparation of the proximal vessel (with balloon angioplasty, atherectomy, or intravascular lithotripsy) could facilitate balloon delivery.

9.7.8.3.2.4 Balloon

Small diameter balloons (such as the 1.0 mm Sapphire Pro, the Threader (Section 30.9.2), or other 1.0–1.50 mm balloons are more likely to cross highly resistant lesions. The marker is usually located in the middle of the balloon for 1.0–1.5 mm balloons, hence longer length is preferred.

Compliant balloons are more flexible, have lower profile and are thus more deliverable compared with noncompliant balloons.

Used balloons are less deliverable—changing for a *new* balloon may allow crossing.

9.7.8.3.2.5 Other

Asking the patient to *take a deep breath* could facilitate balloon or stent delivery, possibly by straightening the coronary arteries or changing the seating of the guide catheter.
Rotaglide or Viperglide administration during balloon or stent advancement might reduce friction and facilitate delivery.

The algorithm for approaching "balloon uncrossable" lesions is discussed in Section 23.1.

9.7.8.4 What can go wrong?

9.7.8.4.1 Wire and guide catheter position loss

Causes:

- Forceful balloon advancement through areas of resistance.

Prevention:

- Avoid forceful balloon advancement, if possible.
- Always monitor the guide catheter and the distal wire position and adjust the effort to advance the balloon accordingly.
- Optimal guide catheter selection and coaxial engagement to optimize support as described in Chapter 5: Coronary and graft engagement.

Treatment:

- Reengagement with the guide catheter and rewiring of the lesion or readvancement of the guidewire.

9.7.8.4.2 Loss of distal vessel visualization

Causes:

- Balloon advancement through a tight lesion.
- Dissection.
- Distal embolization.

Prevention:

- Avoid use of oversized balloons.

Treatment:

- If the reason for loss of distal vessel visualization is balloon advancement through a tight lesion, withdraw the balloon, inject contrast until it reaches distal to the lesion, and advance balloon, which then causes contrast to be entrapped distal to the lesion and allows visualization of the distal vessel.
- If the reason for loss of distal vessel visualization is dissection, stenting should be performed (Section 25.2.1).
- If the reason for loss of distal vessel visualization is distal embolization, thrombectomy should be performed and vasodilators should be administered (Section 25.2.3).

9.7.9 Step 9. Inflate balloon

9.7.9.1 Goal

To inflate the balloon to modify the target lesion.

9.7.9.2 How?

1. The indeflator arm is rotated until the target pressure is reached under fluoroscopy.
2. Inflation speed: inflation is usually achieved in 3–5 seconds. For *cutting balloon* inflation should be much slower (5 seconds for each atm, not to exceed 12 atm), to prevent the blades from puncturing the balloon and reduce the risk of blade entrapment in calcified spicules.
3. Inflation pressure: depends on balloon type and balloon diameter in relation to the target vessel diameter.
    - For noncompliant balloons sized $\leq$1:1 to the target vessel diameter, inflation pressures are usually 12–20 atm.
    - For compliant balloons sized $\leq$1:1 to the target vessel diameter, inflation pressures are usually 12–14 atm.
    - For cutting balloons: 6–8 atm, not to exceed 12 atm.
    - For the SIS OPN balloon: 30–35 atm.
4. Inflation duration: 10–20 seconds.

5. If fluoroscopy shows suboptimal lesion expansion, longer balloon inflations or higher pressures (up to 20–26 atm for noncompliant balloons) can be used.

9.7.9.3 *What can go wrong?*

9.7.9.3.1 Balloon rupture

Causes:

- Complex coronary lesions (such as highly calcified lesions).
- High inflation pressures (> 20 atm).

Prevention:

- Consider atherectomy or intravascular lithotripsy prior to balloon angioplasty in heavily calcified lesions, as described in Chapter 19: Calcification.
- Avoid high inflation pressures (> 20 atm).

Diagnosis:

- The diagnosis should be suspected if there is sudden indeflator pressure drop.
- Fluoroscopy can confirm balloon rupture.

Treatment:

- If indeflator pressure suddenly drops, suction should be immediately applied to prevent contrast from forcefully exiting through the ruptured balloon, potentially causing a perforation.
- The ruptured balloon is removed from the guide catheter.
- The guide catheter is aspirated to remove any air that may have entered into the guide catheter.
- Coronary angiography is performed to confirm that no perforation has occurred. In case of perforation a new balloon should be immediately delivered to the perforation site and inflated, followed by additional treatments as described in Chapter 26: Perforation.

9.7.9.3.2 Watermelon seeding (balloon slipping forward or backward during inflation)

Causes:

- Severe lesion.
- Severe angulation.
- Excessive guide catheter movement.
- Short balloon length.
- Rapid balloon inflation.

Prevention:

- Gentle traction of the balloon during inflation if slipping forward, or gentle push of the balloon if slipping backward.
- Slow inflation speed.
- Longer balloon length, if possible.

Treatment:

- Immediately deflate balloon to minimize risk of injuring not-target coronary segments.
- Repeat inflation attempts at lower pressures and with slower inflation speed.
- Use one (or more) buddy wires.
- Use a plaque modification balloon (Section 30.9.3).
- Use longer balloons.

9.7.9.3.3 Failure to expand (balloon undilatable lesion)

Causes:

- Severe calcification.
- In-stent restenosis lesions.

Prevention:

- Consider atherectomy prior to balloon angioplasty in heavily calcified lesions, as described in Chapter 19: Calcification.

Diagnosis:

- A waist remains on the balloon during inflation on fluoroscopy.

Treatment:

- Treatment of balloon undilatable lesions is described in detail in Section 23.2.
- Intravascular imaging can help characterize the lesion and determine optimal treatment.

9.7.10 Step 10. Deflate the balloon

9.7.10.1 Goal

To deflate the balloon to prepare it for removal.

9.7.10.2 How?

1. The indeflator arm is rotated to decrease the inflation pressure and then pulled back and locked to create vacuum and remove the contrast/saline solution from the balloon.
2. Deflation can take several seconds, especially for longer and larger balloons that require larger contrast volume as compared with smaller balloons.

9.7.10.3 What can go wrong?

9.7.10.3.1 Balloon fails to deflate [18]

Causes (Fig. 9.12):

- Lesion or stent recoil on the balloon.
- Strangulation of the balloon in the guide catheter.
- Damage of the balloon hypotube.

Prevention:

- Avoid kinking the balloon shaft. If shaft kinking occurs, promptly replace the balloon for a new one.

Treatment:

- *Treatment of balloon entrapment is discussed in* Sections 27.3.3.11-27.3.3.14.

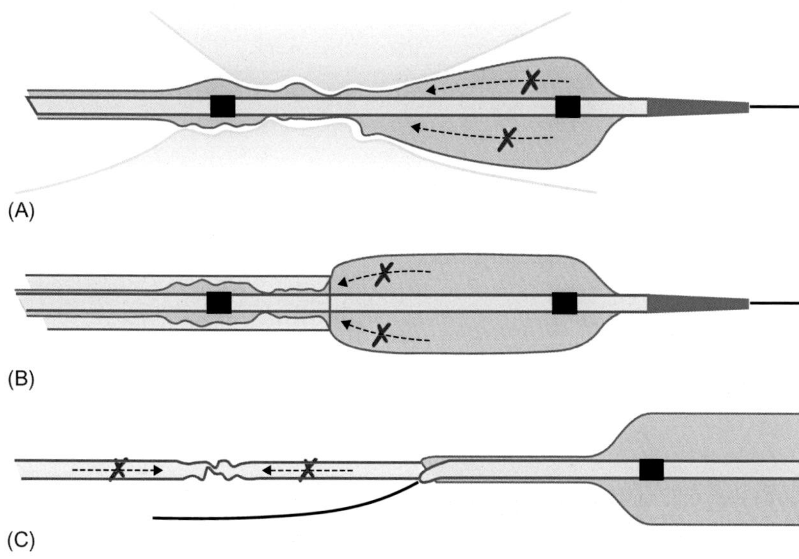

FIGURE 9.12 Mechanisms of contrast media entrapment. Several potential mechanisms can result in the contrast media being trapped inside the balloon. (A) Acute recoil of calcified lesion or stent and (B) strangulation of the balloon in the guide catheter. (C) Damage to the hypotube (kink or break). *Dotted arrows* and *red X* indicate compromised contrast media flow from the balloon into the hypotube [18]. *Reproduced with permission from Leibundgut G., Degen C., Riede F. Transcutaneous puncture of an undeflatable coronary angioplasty balloon catheter. Case Rep Cardiol 2018;2018:6252809 (Figure 3).*

9.7.11 Step 11. Withdraw balloon into the guide catheter

9.7.11.1 Goal

To withdraw the balloon from the lesion into the guide catheter.

9.7.11.2 How?

1. Confirm by fluoroscopy that the balloon has fully deflated.
2. Withdraw the balloon while monitoring the position of the guide catheter and guidewire.
3. If the guide catheter is "sucked into the vessel," balloon withdrawal should stop, followed by repositioning of the guide catheter and guidewire. It may be necessary to exert traction on the guide catheter to prevent it from being "sucked into the vessel."

9.7.11.3 What can go wrong?

9.7.11.3.1 Balloon entrapment

Balloon entrapment is discussed in detail in Section 27.3.3.

9.7.11.3.2 Guide-induced dissection (Section 24.2.1)

Causes:
- Forceful balloon withdrawal (sometimes before full balloon deflation) resulting in the guide catheter tip being "sucked" inside the target coronary artery.

Prevention:
- Avoid forceful guide catheter withdrawal.
- Ensure that balloon is fully deflated before attempting to withdraw it.
- Partially disengage the guide catheter from the vessel ostium before removing the balloon.
- Use the independent-hand technique (Fig. 9.5) in which the left hand pulls the guide catheter while the right hand retracts the balloon.

Treatment:
- Ensure that guidewire position is maintained.
- Stenting.

9.7.12 Step 12. Remove balloon from the guide catheter

9.7.12.1 Goal

To remove the balloon from the guide catheter.

9.7.12.2 How?

1. The operator's left hand fixes the wire just outside the Y-connector, and the balloon is withdrawn using the right hand until it reaches the left hand fingers.
2. The wire is then fixed with the right hand while the left hand gradually withdraws the balloon until it is completely removed from the guidewire (Fig. 9.13).

9.7.12.3 What can go wrong?

9.7.12.3.1 Loss of guidewire position

Causes:
- Suboptimal fixing of the guidewire.
- High-pressure balloon inflations ($>$ 20 atm) that can make the balloons sticky.
- "Sticky" guidewire.
- Guidewire kinking.
- Guidewire wrapping if more than one guidewires are used.

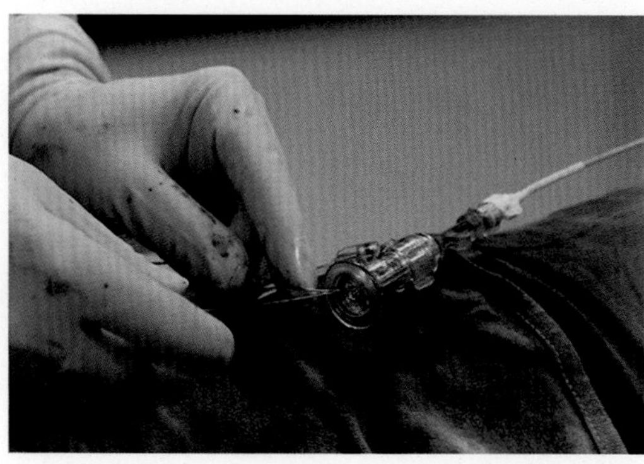
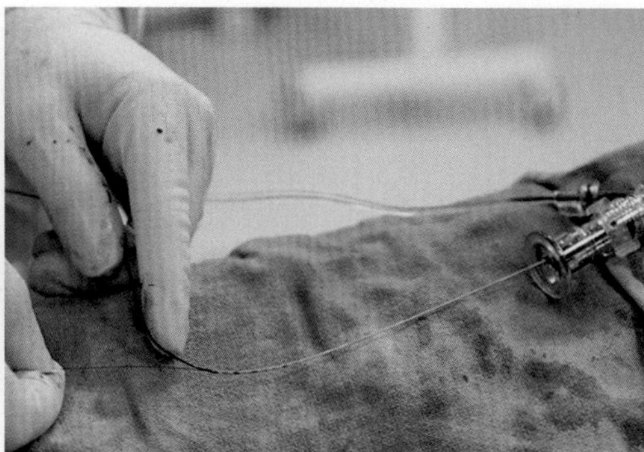

FIGURE 9.13 Removing the balloon while maintaining guidewire position.

Prevention:
- Wipe the guidewire segment outside the guide with a wet gauze to ensure it is clean and slippery.
- Make small balloon movements.
- Avoid guidewire kinking. If the guidewire becomes kinked it should be removed after inserting another intact guidewire.
- If balloon movement is challenging, consider fluoroscopy of the guidewire tip to ensure that the guidewire is not moving.
- Trapping the guidewire can prevent guidewire tip movement.
- If more than one guidewires are used, ensure that they are well separated and wiped prior to balloon removal.

Treatment:
- Rewire the lesion.

9.7.13 Step 13. Check the balloon angioplasty result

9.7.13.1 Goal

To assess the result achieved by balloon angioplasty and the occurrence of any complications.

9.7.13.2 How?

1. Following balloon removal, the guide catheter is aspirated and flushed.
2. Optimal guide engagement is confirmed, as the guide may have become disengaged or deeply engaged during balloon removal.
3. Contrast is administered to evaluate the balloon angioplasty result.
4. Alternatively, intravascular imaging can be performed to assess the procedural result and minimize the use of contrast.

9.7.13.3 Balloon angioplasty—special scenarios

9.7.13.3.1 Multiple lesions

In cases of multiple lesions or long lesions, the more distal lesions or segments should generally be treated first, followed by treatment of the more proximal lesions. This is done because more distal delivery may fail if the balloon is first inflated in the more proximal lesions.

9.7.13.3.2 Advancing balloons through stents

In case of difficulty advancing balloons through a stent, wiring under the stent struts should be suspected. Ideally the wire should be withdrawn and a knuckled guidewire advanced through the stent. Forceful advancement may lead to longitudinal stent deformation [19–22].

9.8 Good flow?

In case of poor antegrade flow, vasodilators are administered and thrombectomy may be performed (unless the poor flow is considered to be due to dissection, in which case stenting is needed) (Chapter 25: Acute vessel closure).

9.9 Balloon expansion

Before stenting, good balloon expansion should be confirmed (especially in heavily calcified lesions), to prevent stent underexpansion.

9.10 Stenting

Stenting is performed, as described in Chapter 10: Stenting.

References

[1] Tajti P, Sandoval Y, Brilakis ES. "Around the world"—how to reach native coronary artery lesions through long and tortuous aortocoronary bypass grafts. Hellenic J Cardiol 2018;59:354–7.

[2] Fujita S, Tamai H, Kyo E, et al. New technique for superior guiding catheter support during advancement of a balloon in coronary angioplasty: the anchor technique. Catheter Cardiovasc Interv 2003;59:482–8.

[3] Hirokami M, Saito S, Muto H. Anchoring technique to improve guiding catheter support in coronary angioplasty of chronic total occlusions. Catheter Cardiovasc Interv 2006;67:366–71.

[4] Kirtane AJ, Stone GW. The Anchor-Tornus technique: a novel approach to "uncrossable" chronic total occlusions. Catheter Cardiovasc Interv 2007;70:554–7.

[5] Matsumi J, Saito S. Progress in the retrograde approach for chronic total coronary artery occlusion: a case with successful angioplasty using CART and reverse-anchoring techniques 3 years after failed PCI via a retrograde approach. Catheter Cardiovasc Interv 2008;71:810–14.

[6] Fang HY, Wu CC, Wu CJ. Successful transradial antegrade coronary intervention of a rare right coronary artery high anterior downward takeoff anomalous chronic total occlusion by double-anchoring technique and retrograde guidance. Int Heart J 2009;50:531–8.

[7] Lee NH, Suh J, Seo HS. Double anchoring balloon technique for recanalization of coronary chronic total occlusion by retrograde approach. Catheter Cardiovasc Interv 2009;73:791–4.

[8] Saito S. Different strategies of retrograde approach in coronary angioplasty for chronic total occlusion. Catheter Cardiovasc Interv 2008;71:8–19.

[9] Surmely JF, Katoh O, Tsuchikane E, Nasu K, Suzuki T. Coronary septal collaterals as an access for the retrograde approach in the percutaneous treatment of coronary chronic total occlusions. Catheter Cardiovasc Interv 2007;69:826–32.

[10] Surmely JF, Tsuchikane E, Katoh O, et al. New concept for CTO recanalization using controlled antegrade and retrograde subintimal tracking: the CART technique. J Invasive Cardiol 2006;18:334–8.

[11] Rathore S, Katoh O, Matsuo H, et al. Retrograde percutaneous recanalization of chronic total occlusion of the coronary arteries: procedural outcomes and predictors of success in contemporary practice. Circ Cardiovasc Interv 2009;2:124–32.

[12] Di Mario C, Ramasami N. Techniques to enhance guide catheter support. Catheter Cardiovasc Interv 2008;72:505–12.

[13] Mahmood A, Banerjee S, Brilakis ES. Applications of the distal anchoring technique in coronary and peripheral interventions. J Invasive Cardiol 2011;23:291–4.

[14] Brilakis ES. Manual of coronary chronic total occlusion interventions. A step-by-step approach. 2nd ed. Cambridge, MA: Elsevier; 2017.

[15] Christ G, Glogar D. Successful recanalization of a chronic occluded left anterior descending coronary artery with a modification of the retrograde proximal true lumen puncture technique: the antegrade microcatheter probing technique. Catheter Cardiovasc Interv 2009;73:272–5.

[16] Mamas MA, Fath-Ordoubadi F, Fraser DG. Distal stent delivery with Guideliner catheter: first in man experience. Catheter Cardiovasc Interv 2010;76:102–11.

[17] Fang HY, Fang CY, Hussein H, et al. Can a penetration catheter (Tornus) substitute traditional rotational atherectomy for recanalizing chronic total occlusions? Int Heart J 2010;51:147–52.

[18] Leibundgut G, Degen C, Riede F. Transcutaneous puncture of an undeflatable coronary angioplasty balloon catheter. Case Rep Cardiol 2018;2018:6252809.

[19] Lee HH, Hsu PC, Lee WH, et al. Longitudinal stent deformation caused by retraction of the looped main branch guidewire. Acta Cardiol Sin 2016;32:616–18.

[20] Rhee TM, Park KW, Lee JM, et al. Predictors and long-term clinical outcome of longitudinal stent deformation: insights from pooled analysis of Korean multicenter drug-eluting stent cohort. Circ Cardiovasc Interv 2017;10.

[21] Kobayashi N, Hata N, Okazaki H, Shimizu W. Longitudinal stent deformation as a cause of very late stent thrombosis: optical coherence tomography images. Int J Cardiol 2016;202:601–3.

[22] Guler A, Guler Y, Acar E, et al. Clinical, angiographic and procedural characteristics of longitudinal stent deformation. Int J Cardiovasc Imaging 2016;32:1163–70.

Chapter 10

Stenting

Goal: To deliver and adequately expand a stent, completely covering the target lesion.

10.1 When to stent?

Stenting (in most cases with drug-eluting stents [DES]) is performed in the vast majority of coronary lesions because stenting:

1. Prevents vessel recoil and reduces the risk of acute closure, especially in lesions with dissection, rupture, and thrombus.
2. Reduces the risk of restenosis.

Stenting should *not* be performed in the following scenarios:

1. Poor antegrade flow (unless poor flow is caused by a dissection). Stenting in the setting of no reflow will worsen it.
2. Inability to expand the target lesion with a balloon.
3. High risk of compromising an important coronary branch that cannot be protected.
4. Very small target vessel and target lesion [although 2.0 mm stents are currently available (Resolute Onyx, Medtronic) that could be used in some of those lesions].

Stenting should not be performed until after successful wiring of the lesion (as described in Chapter 8: Wiring) and (in most cases) successful lesion preparation (as described in Chapter 9: Lesion Preparation). Optimal distal wire position should be confirmed prior to stenting.

10.2 How to stent

Several of the stenting steps are similar to the balloon angioplasty steps, which are described in Chapter 9: Lesion Preparation.

10.2.1. Step 1. Confirm that a guidewire is advanced through the target lesion and optimally positioned distally (Section 9.5.1).

10.2.2. Step 2. Confirm that the guide catheter is aspirated and flushed (Section 9.5.2).

10.2.3. Step 3. Select stent type and size.

10.2.3.1. Goal: To choose optimal stent type and size.

10.2.3.2. How?

10.2.3.2.1. Stent type

DES vs. bare metal stents (BMS): Newer generation drug-eluting stents are currently used in the vast majority of patients, due to better efficacy (less restenosis) and safety (stent thrombosis) as compared with bare metal stents with the exception of saphenous vein grafts, where DES and BMS have similar outcomes [1,2].

Stent implantation (DES or BMS) should be avoided if possible, if noncardiac surgery is needed within 6–12 months [3].

Stent brand: Thinner strut stents have higher deliverability, which is often achieved at the cost of lower visibility. Stent brand selection depends on the associated clinical studies results, the target lesion characteristics and local availability and cost.

Manual of Percutaneous Coronary Interventions. DOI: https://doi.org/10.1016/B978-0-12-819367-9.00010-X

10.2.3.2.2. Stent size

***Stent diameter*:**

Stent diameter can be selected using coronary angiography and/or intravascular imaging (Sections 13.3.6 and 13.4.4.2). Intravascular imaging significantly facilitates accurate stent size selection and can also confirm the adequacy of lesion preparation.

The stent diameter should match the distal reference vessel diameter (as assessed by angiography and/or intravascular imaging). Sizing stents based on the proximal vessel may result in distal vessel dissection or perforation, as the proximal vessel is usually larger than the distal vessel due to normal vessel tapering; it can also lead to side branch occlusion due to carina shift. After implantation, the proximal portion of the stent should be postdilated with a larger balloon to match the proximal reference vessel diameter [proximal optimization technique (POT), Section 16.1.10.1.3]. When the proximal vessel diameter is much larger than the distal vessel diameter (e.g., in left main lesions or in aneurysmal coronary vessels) it is important to know the limits of expansion of the various stents (Fig. 10.1). Peripheral stents can be used in very large (≥6 mm) vessels.

When intravascular imaging is used for stent sizing, it can be based on distal reference vessel diameter, or if the media is visible, on media to media measurements (selected stent diameter should be 85% of media to media dimension) (Section 13.1.2).

***Stent length*:** the stent should be long enough to cover the entire target lesion, including coronary artery segments proximal or distal to the lesion that were injured with balloon angioplasty. Failure to cover the entire lesion/predilated segment is called "geographic miss" and increases the risk of restenosis. Long stents (40 mm in the United States and 60 mm in Europe) are currently available.

The desired stent length can be estimated angiographically by using a known length balloon to predilate the lesion. It can also be measured using intravascular imaging with automated pullback (IVUS or OCT).

Long stents can be challenging to deliver, especially through tortuous and calcified segments. In such lesions, delivering >1 shorter stents may be easier. Alternatively better lesion preparation, and increased guide catheter support can be used to facilitate delivery.

10.2.3.3. What can go wrong?

Poor stent size selection may not result in complications until the time of stent delivery and deployment, as follows:

1. Inability to deliver stent to the target lesion (see step 8).
2. Perforation (if stent diameter is too big for the target vessel) (see step 13).

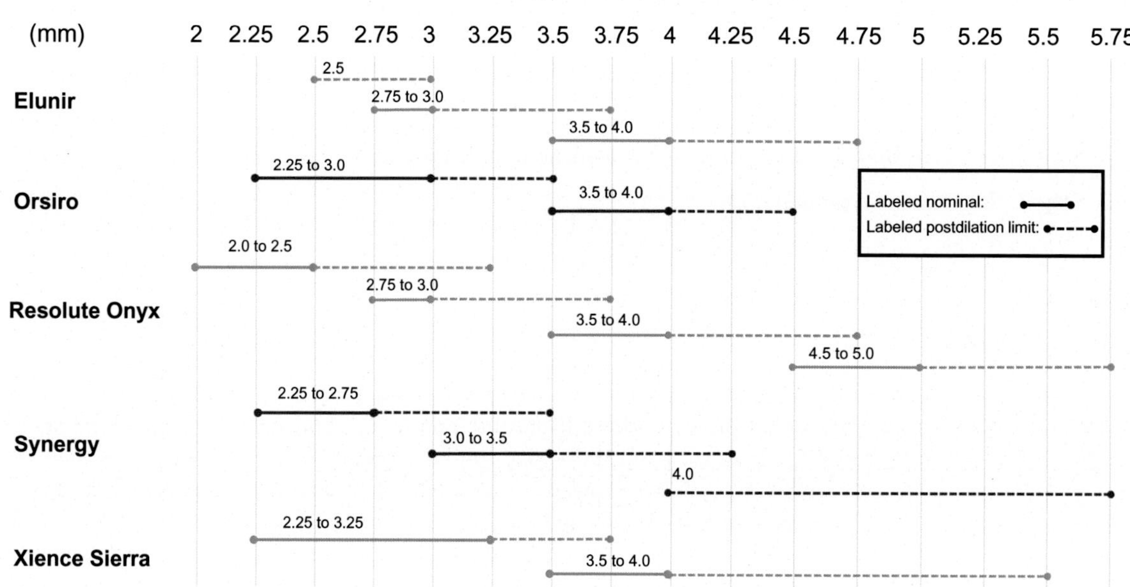

FIGURE 10.1 Labeled stent postdilatation limits.

3. Geographic miss (if stent is too short for the target lesion) (see step 13).

10.2.4. Step 4. Prepare stent balloon.

10.2.4.1. Goal**:** To remove the air from the stent balloon, fill the stent balloon lumen with a contrast solution, and connect to an indeflator.

10.2.4.2. How?

1. The stent is removed from its packaging.
2. A luer-lock syringe is filled usually with 50% contrast/50% saline solution. Lower concentration of contrast will speed up the inflation/deflation of the balloon but will also decrease the stent balloon visibility.
3. The syringe is connected with the stent balloon proximal hub.
4. Suction is performed with the syringe positioned upside down to remove the air from the balloon.
5. The stent balloon is connected with the indeflator.
6. Many operators do NOT apply suction through the indeflator (to minimize the risk of stent dislodgement from the delivery balloon). However, with contemporary stents, the risk of dislodgement and stent loss is very low, and suction with the indeflator can usually be applied.

10.2.5. Step 5. Load the stent on the guidewire (Section 9.5.5).

10.2.6. Step 6. Advance the stent balloon monorail segment through the Y-connector (Section 9.5.6).

10.2.7. Step 7. Advance the stent to the tip of the guide catheter (Section 9.5.7).

10.2.8. Step 8. Advance the stent to the target lesion.

10.2.8.1. Goal**: To advance the stent to target lesion.**

10.2.8.2. How?

1. The stent is advanced toward the target lesion using the "independent-hand" or "two-hand" technique (Fig. 9.5).
   "Two-hand technique". The left hand fixes the guidewire while the right hand pushes the balloon.
   "Independent technique" (preferred): The right hand is holding the touhy, the stent balloon shaft and the guidewire. This frees the left hand that can push the guide catheter forward to increase the guide catheter support.
2. The position of the stent is confirmed using X-ray landmarks (such as clips, previous stents, etc.) and/or contrast injection (Fig. 10.2). Contrast injections should be minimized in patients with chronic kidney disease. Angiographic coregistration with intravascular imaging or physiology systems can be utilized to assist in optimal stent positioning.

10.2.8.3. *Challenges*

10.2.8.3.1. ***Failure to reach the lesion with the stent (stent shaft is not long enough)***

Causes:

- Very distal lesion location, especially in patients with prior coronary bypass grafts and lesions distal to bypass graft anastomoses.

Prevention:

- Use short (or shortened) guide catheters when treating very distal coronary lesions.
- Use stents with the longest available shaft (Xience: 145 cm, Promus/Synergy: 144 cm, Resolute/Elunir/Orsiro: 140 cm).
- Use guide extensions that may allow safer deep intubation and better align the guide catheter.

Treatment:

- Use short (or shortened) guide catheters when treating very distal coronary lesions.
- Use stents with the longest available shafts.

10.2.8.3.2. ***Failure to reach the lesion or cross the lesion with the stent***

Causes:

- Proximal vessel: significant disease, tortuosity, and/or calcification.
- Severe lesion tortuosity.
- Severe lesion calcification.
- Tight lesion.
- Poor lesion preparation.

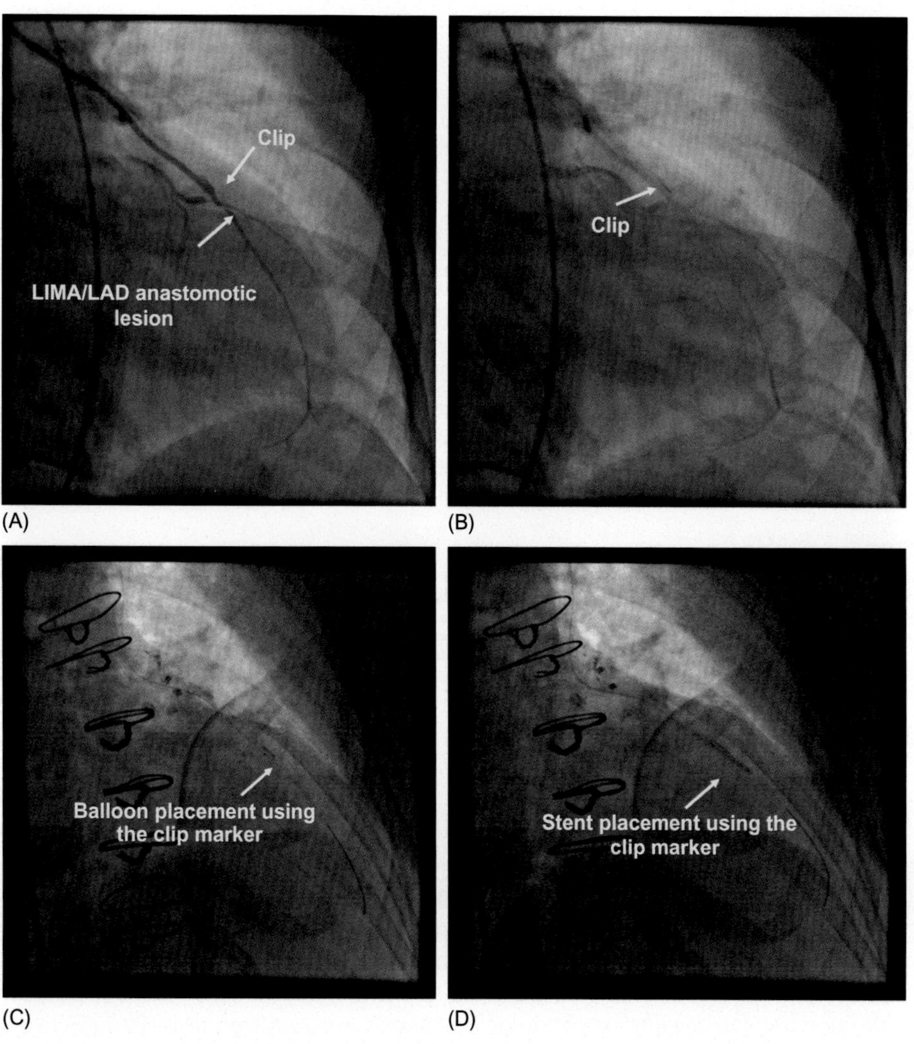

FIGURE 10.2 Use of a surgical clip (panels A and B) as a landmark for balloon (panel C) and stent (panel D) positioning.

Prevention:

- If there is proximal disease, treat the proximal vessel.
- Good lesion preparation.

Treatment:

(Fig. 10.3)

1. **Guide catheter**

Strong guide catheter support is critical for (balloon and) stent delivery, especially in complex lesions.

Larger diameter guide catheters (such as 7 or 8 Fr), *supportive shapes* (e.g., EBU guides provide much stronger support than JL4 guides for left coronary interventions, and AL1 or 3D Right guide catheters provide stronger support than JR4 guide catheters for RCA interventions), and *femoral access* (as compared with radial) provide stronger support.

For truly complex lesions, many operators increase the likelihood of success by using femoral access with 8 Fr, supportive shape guide catheters.

Forward push of the guide catheter can increase support, but may also in some cases lead to guide catheter prolapse and disengagement. The "independent-hand" technique (the left hand pushes the guide, while the right hand both fixes the guidewire and advances the stent, Fig. 9.5) allows a single operator to simultaneously manipulate the guide catheter and advance the stent.

Deep intubation of the guide catheter increases support, but at the same time carries risk of causing dissection. Deep intubation is usually done in the RCA or in SVGs and rarely in the left coronary artery.

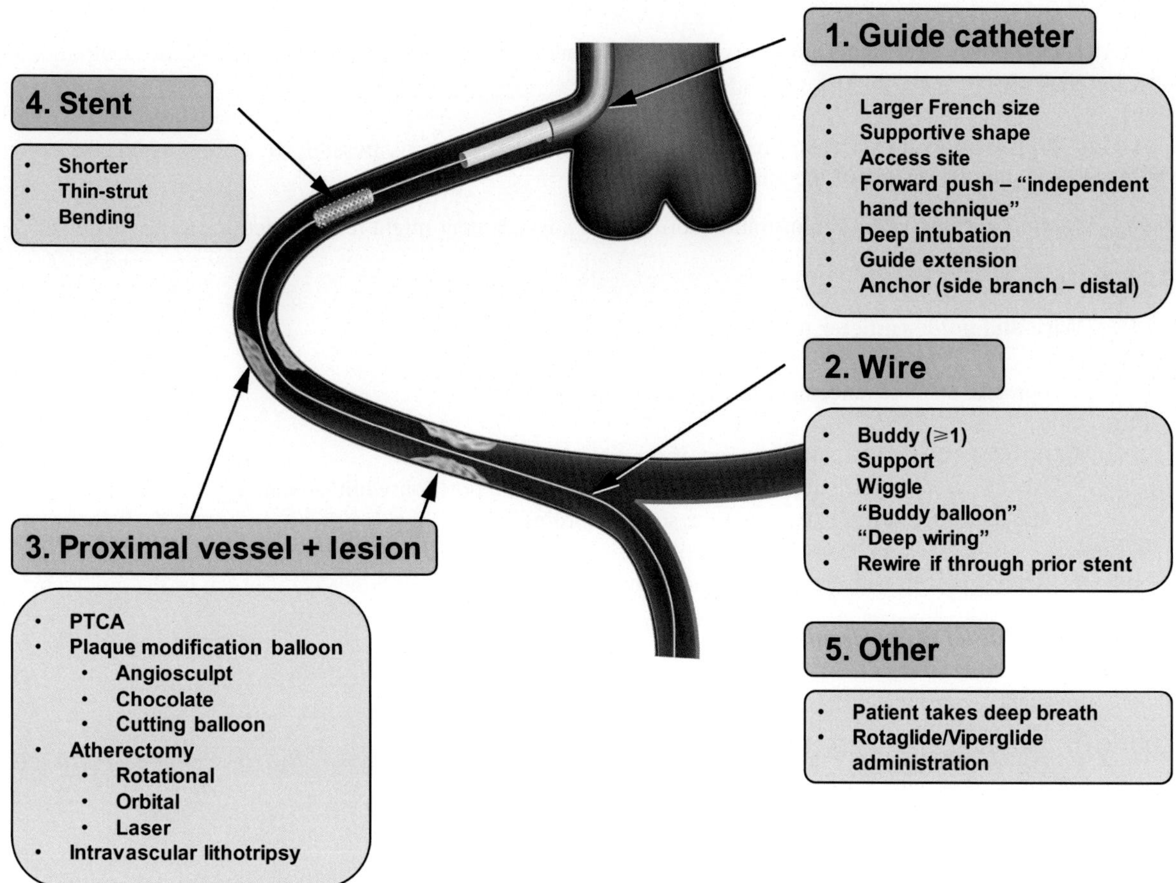
1. Guide catheter
Larger French size
Supportive shape
Access site
Forward push – "independent hand technique"
Deep intubation
Guide extension
Anchor (side branch – distal)
2. Wire
Buddy (≥1)
Support
Wiggle
"Buddy balloon"
"Deep wiring"
Rewire if through prior stent
3. Proximal vessel + lesion
PTCA
Plaque modification balloon
Angiosculpt
Chocolate
Cutting balloon
Atherectomy
Rotational
Orbital
Laser
Intravascular lithotripsy
4. Stent
Shorter
Thin-strut
Bending
5. Other
Patient takes deep breath
Rotaglide/Viperglide administration

FIGURE 10.3 How to facilitate stent delivery to the target lesion.

Guide catheter extensions (Section 30.3) are among the most commonly used tools for facilitating stent delivery. They not only increase guide catheter support, but may also effectively "modify" the proximal vessel if advanced distally close to (or through) the target lesion (the stent is then advanced through the guide extension without coming in contact with the wall of the target vessel).

Anchor techniques (side branch anchor and distal anchor) can also provide strong support for equipment delivery and are discussed in detail in Section 9.5.8.1.7.

2. **Wire**

One of the simplest ways to facilitate delivery is to use one (or more) buddy wire(s). The buddy wire increases guide support and also straightens the target vessel.

Another option is to use support guidewires (such as the Iron Man, Grand Slam, or the Wiggle wire (Section 30.7.4).

Partially withdrawal of the wire while advancing the stent may facilitate advancement.

Another option is to advance a balloon over a second guidewire across the area of resistance to stent advancement, in order to "deflect" the stent during repeat advancement attempts.

In the "deep wiring" technique a polymer-jacketed guidewire is advanced through the lesion then into the ventricle/aorta through arterioluminal communications, which increases wire support [4]. The obvious risk of this technique is coronary perforation.

If there is resistance to stent delivery at the location of a previously deployed stent, it is possible that the wire has crossed under a stent strut. Rewiring through the stent, ideally after forming a knuckle at the wire tip more proximally can solve this problem.

3. **Proximal vessel + lesion preparation**

Stent delivery can be challenging, especially through proximal lesions (especially tortuous and calcified) and through severe, calcified lesions. Adequate lesion preparation with balloon angioplasty (standard and modified balloons) and/or atherectomy or intravascular lithotripsy can facilitate both stent delivery and stent expansion, and minimize the risk of stent loss.

4. **Stent**

   Shorter and thinner-strut stents are easier to deliver. Bending the stent may facilitate advancement through tortuosity, but may also increase the risk of stent loss.
5. **Other**

   Asking the patient to take a deep breath could facilitate stent delivery, possibly by straightening the coronary arteries or changing the seating of the guide catheter.

Finally, Rotaglide or Viperglide administration during stent advancement might reduce friction and facilitate delivery

10.2.8.4. What can go wrong?

10.2.8.4.1. **Wire and guide catheter position loss**

Causes:
- Forceful stent advancement through areas of resistance.

Prevention:
- Avoid forceful stent advancement, if possible.
- Always monitor the guide catheter and the distal wire position while advancing (or withdrawing stents) and adjust the effort to advance the stent accordingly.

Treatment:
- Reengagement with the guide catheter and rewiring of the lesion or advancement of the guidewire.

10.2.8.4.2. ***Loss of distal vessel visualization***

Causes:
- Stent advancement through a tight lesion.
- Dissection.
- Distal embolization.
- Spasm.

Prevention:
- Adequate lesion preparation.
- Nitroglycerin administration.

Treatment:
- If the reason is stent advancement through a tight lesion, withdraw the stent, inject contrast until it reaches distal to the lesion, and then advance the stent, which will entrap contrast distal to the lesion and allow visualization of the distal vessel.
- If the reason is dissection, stenting should be performed.
- If the reason is spasm, intracoronary nitroglycerin should be administered.

10.2.8.4.3. ***Stent loss***

Causes:
- Forceful stent advancement through areas of resistance.
- Stent deformation during advancement or withdrawal attempts.

Prevention:
- Meticulous lesion preparation.
- Avoid forceful stent advancement, if possible.
- Avoid pushing the proximal part of the stent outside the guide catheter if the stent is not advancing through the target coronary artery.
- Maintain coaxial guide catheter alignment when attempting to withdraw the stent into the guide catheter.

Treatment:
- See Section 27.1.

10.2.9. Step 9. Inflate balloon and deploy stent.

***10.2.9.1. Goal*:** To inflate the stent balloon and deploy the stent across the target lesion.

10.2.9.2. How?

1. The indeflator arm is rotated until the target pressure is reached under fluoroscopy.

2. Inflation speed: Inflation is usually achieved in 3–5 seconds.
3. Inflation pressure: Stent deployment is usually done at 12–14 atm, followed in most cases by high-pressure (≥20 atm) postdilation with a shorter noncompliant balloon. Exceptions to the need for high-pressure postdilation are excellent stent expansion confirmed by intravascular imaging, concerns about thrombus or plaque embolization, or concerns about compromising a large jailed branch.
4. Inflation duration: The longer the duration of stent balloon inflation, the better the stent expansion. One way to assess for continued expansion is to monitor the pressure of the stent balloon. If the pressure continues to decrease, this signifies continued stent expansion, and inflation should be continued.
5. If fluoroscopy (often using the "StentBoost" or "CLEARstent" technology) or intravascular imaging show suboptimal lesion expansion, longer duration balloon inflations [5] or higher pressures (up to 20 atm for semi-compliant balloons; up to 26-28 atm for noncompliant balloons) could be attempted.

10.2.9.3. What can go wrong?

10.2.9.3.1. **Suboptimal stent placement**

Causes:

- Patient or operator inadvertent movement.
- Significant stent movement due to cardiac motion.

Prevention:

- Deploy stent only when optimal position has been achieved.
- In case of excessive stent movement due to cardiac motion, rapid pacing during stent deployment can minimize the risk of deployment in an unintended location.

Treatment:

- Additional stent placement may be required if the target lesion is incompletely covered.
- In case of ostial stenting if a long portion of the stent is protruding into the aorta, snaring and removal can be performed.

10.2.10. Step 10. Deflate stent balloon (Section 9.5.10).

10.2.11. Step 11. Withdraw stent balloon into the guide catheter (Section 9.5.11).

10.2.12. Step 12. Remove stent balloon from guide catheter (Section 9.5.12).

10.2.13. Step 13. Check the stenting result.

***10.2.13.1. Goal*:** To determine the stenting result achieved by balloon angioplasty and the occurrence of any complications.

10.2.13.2. How?

1. Following balloon removal, the catheter is aspirated and flushed.
2. Optimal guide engagement is confirmed, as the guide may have become disengaged or deeply engaged during balloon removal.
3. Nitroglycerin may be administered to minimize the risk of spasm.
4. Contrast is administered to evaluate the balloon angioplasty result.
5. The "StentBoost" or "CLEARstent" X-ray modes can help evaluate stent expansion.
6. Alternatively, intravascular imaging with IVUS or OCT (Section 13.3.6.3) or physiology (Section 12.1.2) can be performed to assess the procedural result and minimize the use of contrast.

10.2.13.3. What can go wrong?

10.2.13.3.1. ***Failure to expand (balloon undilatable lesion)***

Causes:

- Severe calcification
- In-stent restenosis
- Suboptimal lesion preparation

Prevention:

- Meticulous lesion preparation, especially for heavily calcified lesions, as described in Chapter 9: Lesion Preparation and Chapter 19: Calcification.

Diagnosis:
- A waist remains on the balloon during inflation on fluoroscopy.
- Intravascular imaging.
- Use of the "StentBoost" or "CLEARstent" X-ray imaging techniques.

Treatment:
- See balloon undilatable algorithm (Section 23.2)

10.2.13.3.2. *Failure to fully cover the target lesion*

Causes:
- Shorter than needed stent length.
- Excessive vessel injury during lesion preparation.

Prevention:
- Optimal stent length selection based on balloon length during predilation and/or based on intravascular imaging with automated pullback.
- Avoid predilation outside the target coronary lesion.

Diagnosis:
- Residual stenosis at proximal or distal edge of the stent.

Treatment:
- Implantation of additional stent(s) to fully cover the target coronary lesion.

10.2.13.3.3. *Distal edge dissection*

Causes:
- Stent oversizing and/or very high pressure inflations.
- Incomplete lesion coverage, especially when the stent edge is implanted over a lipid rich plaque.

Prevention:
- Avoid stent oversizing and very high-pressure stent balloon inflation.
- Optimal stent length selection based on balloon size during predilation and/or based on intervascular imaging with automated pullback.

Diagnosis:
- Angiographic lucency at the stent edges.
- If dissection is significant, it may lead to slow flow or no flow.
- Flap by intravascular ultrasound or by optical coherence tomography (Section 13.3.6.3, 4. Edge dissection)

Treatment:
- Additional stent implantation (for significant dissections).

10.2.13.3.4. *Side branch occlusion*

Causes:
- Jailing of the side branch ostium.
- Plaque shift and carina shift [6] after stenting.

Prevention:
- Place guidewire in important side branches before placing a stent.
- Balloon or stent side branch prior to stenting the main branch.
- Perform the proximal optimization technique (POT, Section 16.1.10.1.3) proximal to the carina.

Diagnosis:
- Slow flow or no flow in side branch.

Treatment:
- Balloon angioplasty and/or stenting of the side branch.

References

[1] Patel NJ, Bavishi C, Atti V, et al. Drug-eluting stents versus bare-metal stents in saphenous vein graft intervention. Circ Cardiovasc Interv 2018;11:e007045.

[2] Elgendy IY, Mahmoud AN, Brilakis ES, Bavry AA. Drug-eluting stents versus bare metal stents for saphenous vein graft revascularisation: a meta-analysis of randomised trials. EuroIntervention 2018;14:215–23.

[3] Banerjee S, Angiolillo DJ, Boden WE, et al. Use of antiplatelet therapy/DAPT for post-PCI patients undergoing noncardiac surgery. J Am Coll Cardiol 2017;69:1861–70.

[4] D. Khelimskii, A. Badoyan, O. Krestyaninov , The deep-wire crossing technique: a novel method for treating balloon-uncrossable lesions, J Invasive Cardiol 2019;31:E362–E368.

[5] Saad M, Bavineni M, Uretsky BF, Vallurupalli S. Improved stent expansion with prolonged compared with short balloon inflation: a meta-analysis. Catheter Cardiovasc Interv 2018;92:873–80.

[6] Kang SJ, Mintz GS, Kim WJ, et al. Changes in left main bifurcation geometry after a single-stent crossover technique: an intravascular ultrasound study using direct imaging of both the left anterior descending and the left circumflex coronary arteries before and after intervention. Circ Cardiovasc Interv 2011;4:355–61.

Chapter 11

Access closure

11.1 Femoral access

11.1.1 Femoral access closure algorithm

The following algorithm (Fig. 11.1) reflects the experience and current practice of the authors; other vascular closure devices can be incorporated in the algorithm depending on local availability and expertise. Use of vascular closure devices is favored for shortening the time to ambulation and potentially reducing the risk of complications, although the latter remains controversial [1].

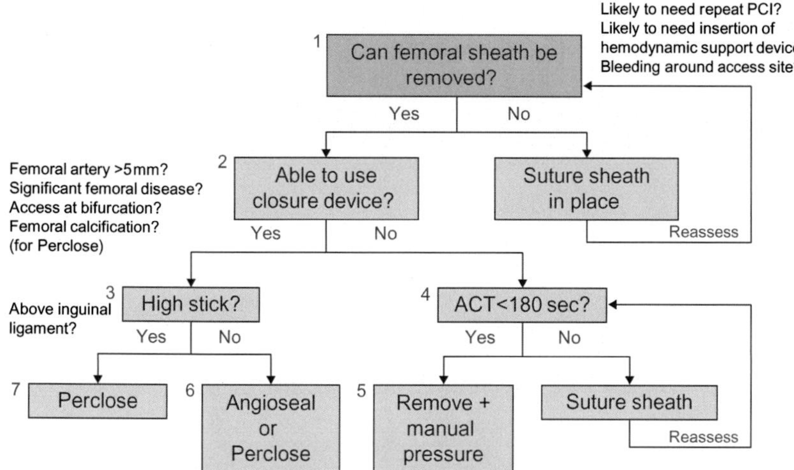

FIGURE 11.1 Algorithm for achieving hemostasis in patients with femoral access.

11.1.2 Can the femoral sheath be safely removed?

In some patients the femoral sheath may need to remain in place, for example, in patients who are planned to have staged PCI within a few hours or patients who are likely to need a hemodynamic support device in the next few hours.

Conversely, the femoral sheath may need to be removed immediately, for example, if there is continued bleeding around the sheath (an alternative is to insert a larger sheath if the oozing is due to an arteriotomy size that is larger than the femoral sheath size).

If the femoral sheath needs to stay in place it is secured with sutures. If long sheaths were used they are usually replaced with short (10–13 cm) sheaths.

11.1.3 Able to use a vascular closure device?

Use of vascular closure devices improves patient's comfort, speeds up ambulation, and may reduce access complications (although vascular closure devices may cause complications as well).

Femoral angiography is critical for determining the feasibility of vascular closure. Use of vascular closure devices requires:

- Large femoral artery (> 5 mm) without significant disease
- Access above the common femoral artery bifurcation.

Manual of Percutaneous Coronary Interventions. DOI: https://doi.org/10.1016/B978-0-12-819367-9.00011-1

The most commonly used vascular closure devices are the Angioseal (Terumo) and the Perclose (Abbott Vascular), and these are discussed in more detail later in this chapter. Use of the Perclose should be avoided in heavily calcified femoral arteries and use of Angioseal should be avoided in cases of high femoral artery puncture [2].

There are other vascular closure devices, such as the Starclose (Abbott Vascular), Mynx (Cardinal Health), Exoseal (Cordis), Vascade (Cardiva), and Celt (Vasorum) that can be used depending on local availability and expertise.

11.1.4 High-stick?

Use of the Angioseal (Section 30.15.1.1) should be avoided when arterial stick is high (above the inguinal ligament) due to difficulty advancing the collagen plug all the way to the artery [2]. Use of the Perclose (Section 30.15.1.2) is preferred in such cases.

11.1.5 ACT < 180 seconds?

Vascular closure devices can be used even when the ACT is high, although it may be best to not use them if the ACT is > 300 seconds.

Removing the femoral sheath and holding manual pressure should be delayed until after ACT is <180 seconds to minimize the risk of bleeding.

Protamine (Section 3.4) can be administered in some patients to reverse anticoagulation and allow for earlier sheath removal.

11.1.6 Manual compression

11.1.6.1 Goal

To achieve hemostasis without complications.

11.1.6.2 How?

1. Patient should be on telemetry with noninvasive blood pressure monitoring.
2. Personnel should be available to administer atropine or fluids if needed for a vasovagal reaction.
3. Sterile gloves should be used.
4. The operator's hands are placed above the femoral puncture site.
5. The sheath is removed while applying gentle pressure with small back bleed.
6. Firm pressure is applied confirming hemostasis.
7. Duration: approximately 2–3 minutes × the sheath size, i.e., 12–18 minutes for 6 French and 16–24 minutes for 8 French sheaths.
8. The access site is checked for hematoma.
9. Distal pulses are checked.
10. A clear sterile dressing (such as Tegaderm) is applied over the puncture site.

11.1.6.3 Challenges

1. Manual compression can be challenging in morbidly obese patients.

11.1.6.4 What can go wrong?

11.1.6.4.1 Bleeding (Section 29.1.2)

Causes:

- Suboptimal placement of the operator hands (too low or to the side of the arterial entry point).
- Inadequate pressure.
- Anticoagulation.
- Hypertension.

- Obesity.
- Hematoma formation.

Prevention:
- Optimal hand positioning.
- Application of firm pressure.
- ACT is checked prior to sheath removal to ensure patient is not anticoagulated.
- If patient is hypertensive, medications are given to lower blood pressure.

Treatment:
- Reposition hands.
- Firm pressure.
- Endovascular intervention or emergency surgery may be needed if hemostasis cannot be achieved.

11.1.6.4.2 Lower extremity ischemia (Section 29.1.1)

Causes:
- Iliac or femoral artery dissection.
- Femoral artery thrombosis.
- Thrombus or plaque embolization.

Prevention:
- Use meticulous technique while obtaining access (Chapter 4: Access).

Treatment:
- Emergency angiography followed by endovascular or surgical intervention.

11.1.6.4.3 Hypotension (Section 28.1)

Causes:
- Bleeding.
- Cardiac causes (tamponade, ischemia, arrhythmias, valvular regurgitation, and left- or right-ventricular failure).
- Vasovagal reaction or anaphylactic reaction.

Prevention:
- Optimal access technique.
- Normal saline administration prior to removing the sheath.
- Local anesthetic administration prior to sheath insertion to minimize the risk for pain-induced vasovagal reactions.

Treatment:
- Bleeding: resuscitation with normal saline and blood transfusion if needed, followed by endovascular or surgical intervention, depending on type of bleeding.
- Cardiac failure: treat the underlying cardiac cause: for example pericardiocentesis in case of tamponade, vasopressors, inotropes, and hemodynamic support devices in case of heart failure.
- Vasovagal reaction: normal saline administration—atropine may be used in case of bradycardia.

11.1.7 Angioseal VIP

The Angioseal VIP device is described in Section 30.15.1.1.

11.1.7.1 Step 1. Prepare the device for use

11.1.7.1.1 How?

1. Open the Angioseal sterile package.
2. Remove the dilator and the sheath from the package. Insert the dilator into the sheath, ensuring that the two pieces snap together securely (Fig. 11.2). There is a reference indicator on the dilator that should align with the indicator on the sheath.

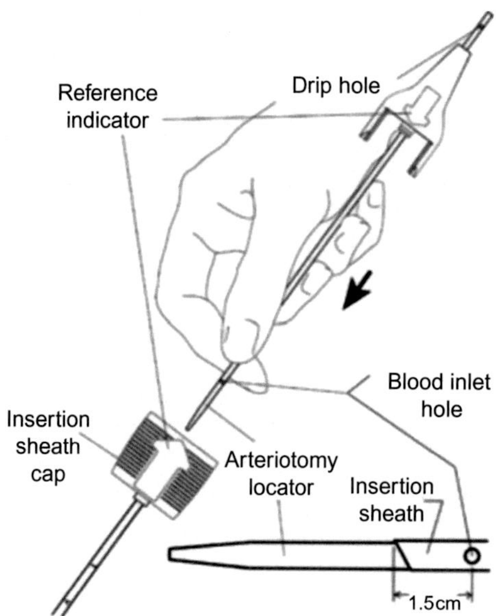

FIGURE 11.2 Preparation of the Angioseal device. ©*2020 Terumo Medical Corporation. All rights reserved.*

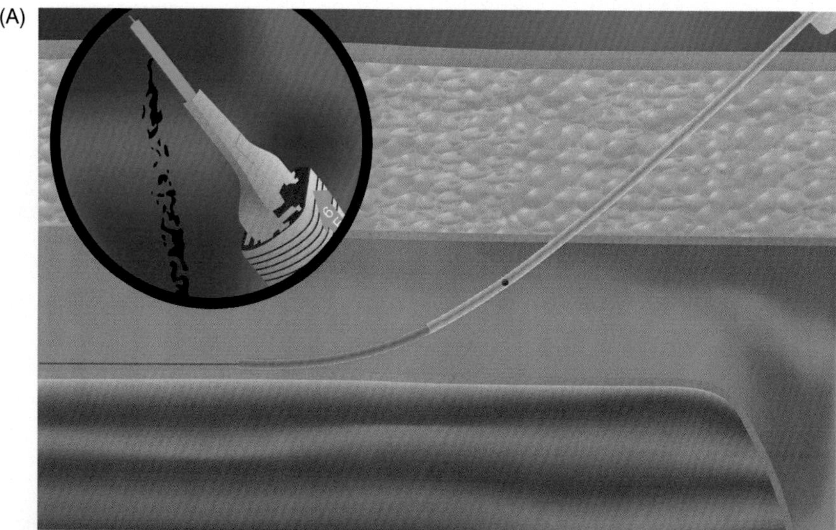

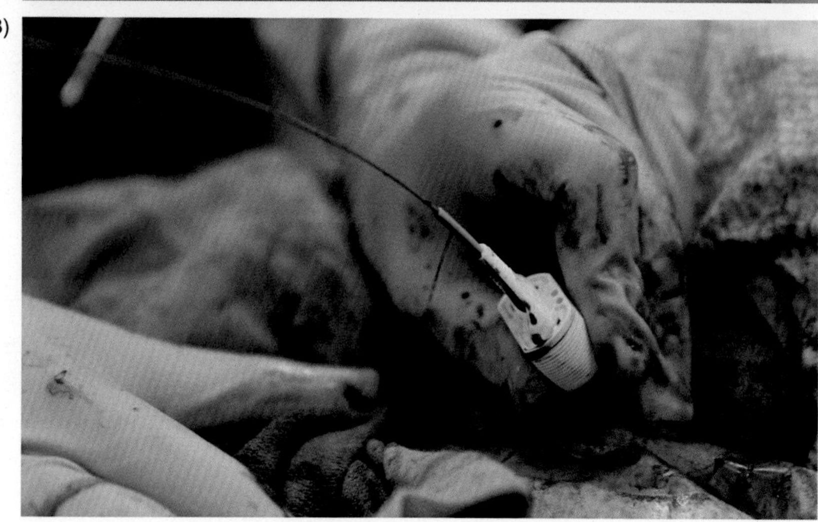

FIGURE 11.3 (A) Insertion of the Angioseal device into the femoral artery. (B) Insertion of the Angioseal and bleeding through the drip hole.

11.1.7.2 Step 2. Advancement of a 0.035 in. guidewire and removal of the femoral sheath

11.1.7.2.1 How?

1. Advance a 0.035 in. guidewire through the femoral sheath. The wire provided with the Angioseal device is adequate for sheaths up to 25 cm in length. Longer sheaths will require a longer 0.035 in. wire. A 0.038 in. guidewire can be used for the 8 French but not the 6 French Angioseal.
2. Perform fluoroscopy to confirm that the guidewire is in the aorta.
3. Remove the femoral sheath leaving the 0.035 in. guidewire in place.

11.1.7.3 Step 3. Insertion of the Angioseal sheath

11.1.7.3.1 How?

1. Advance the Angioseal sheath/dilator assembly into the vessel until flow of blood is observed through the drip hole (Fig. 11.3).
2. Slowly withdraw the Angioseal sheath/dilator assembly until blood flow from the drip hole stops (Fig. 11.4). This suggests that the blood inlet holes of the insertion sheath have just exited the artery.
3. Readvance the Angioseal sheath/dilator assembly until blood begins to flow from the drip hole of the dilator.

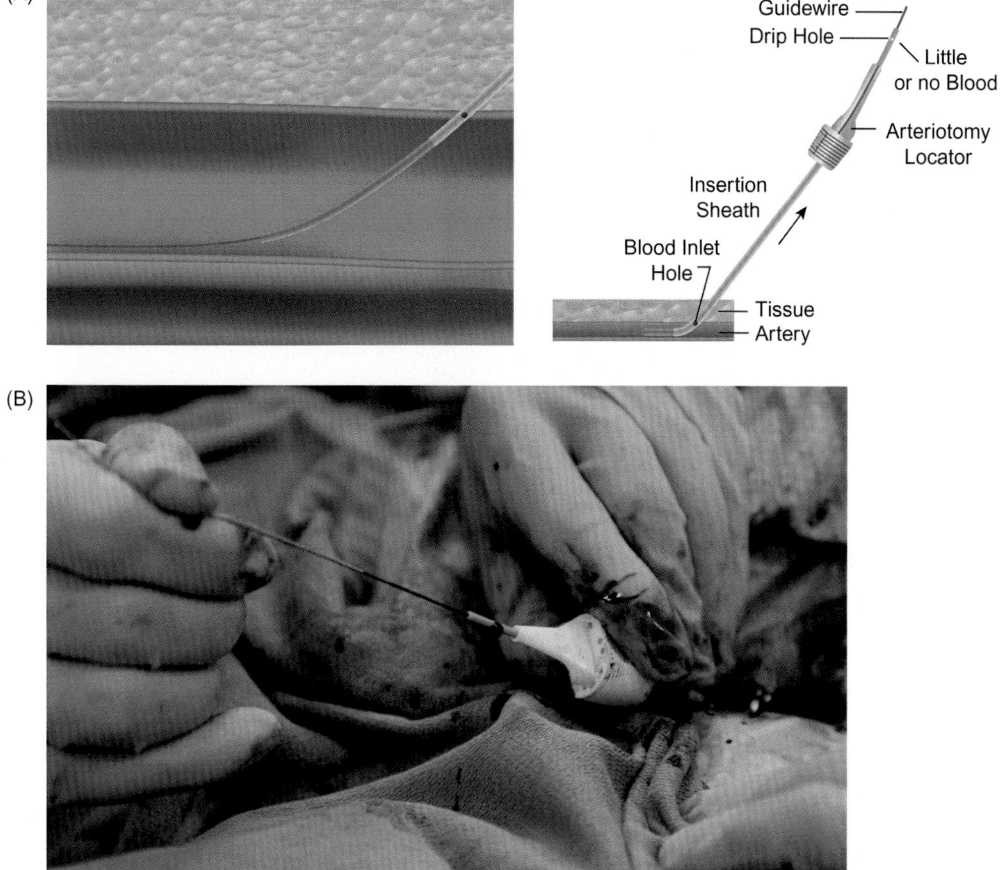

FIGURE 11.4 Withdrawal of the Angioseal device until blood flow from the drip hole stops. *Panel A: ©2020 Terumo Medical Corporation. All rights reserved.*

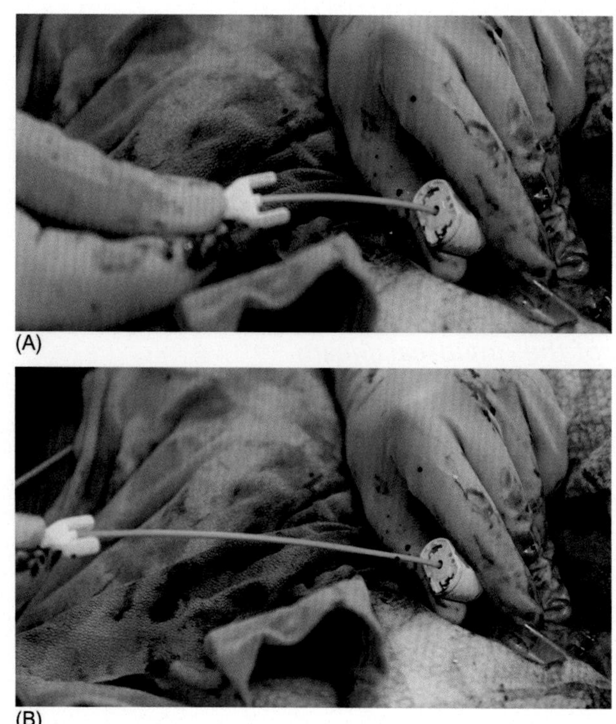

FIGURE 11.5 Removal of the dilator and the guidewire.

11.1.7.4 Step 4. Removal of the guidewire and dilator

11.1.7.4.1 How?

1. Fix the tip of the Angioseal sheath in place using the operator's left hand. There should be no movement of the sheath.
2. Flex the dilator tip upward to separate it from the sheath.
3. Remove the dilator and guidewire (ensuring that the sheath does not move, Fig. 11.5).

11.1.7.5 Step 5. Insertion of the Angioseal device

11.1.7.5.1 How?

1. Hold the Angioseal device close to its tip.
2. Ensure that Angioseal device reference indicator is facing up.
3. Insert the Angioseal device tip through the hemostatic valve of the Angioseal sheath.
4. Advance the Angioseal device until it is completely inserted into the Angioseal sheath and the devices snap together (Fig. 11.6).

11.1.7.6 Step 6. Deployment of the anchor

11.1.7.6.1 Goal

Deploy the anchor inside the artery (Fig. 11.7).

11.1.7.6.2 How?

1. The Angioseal sheath cap continues to be held firmly during this maneuver. There should be no movement of the sheath during anchor deployment.

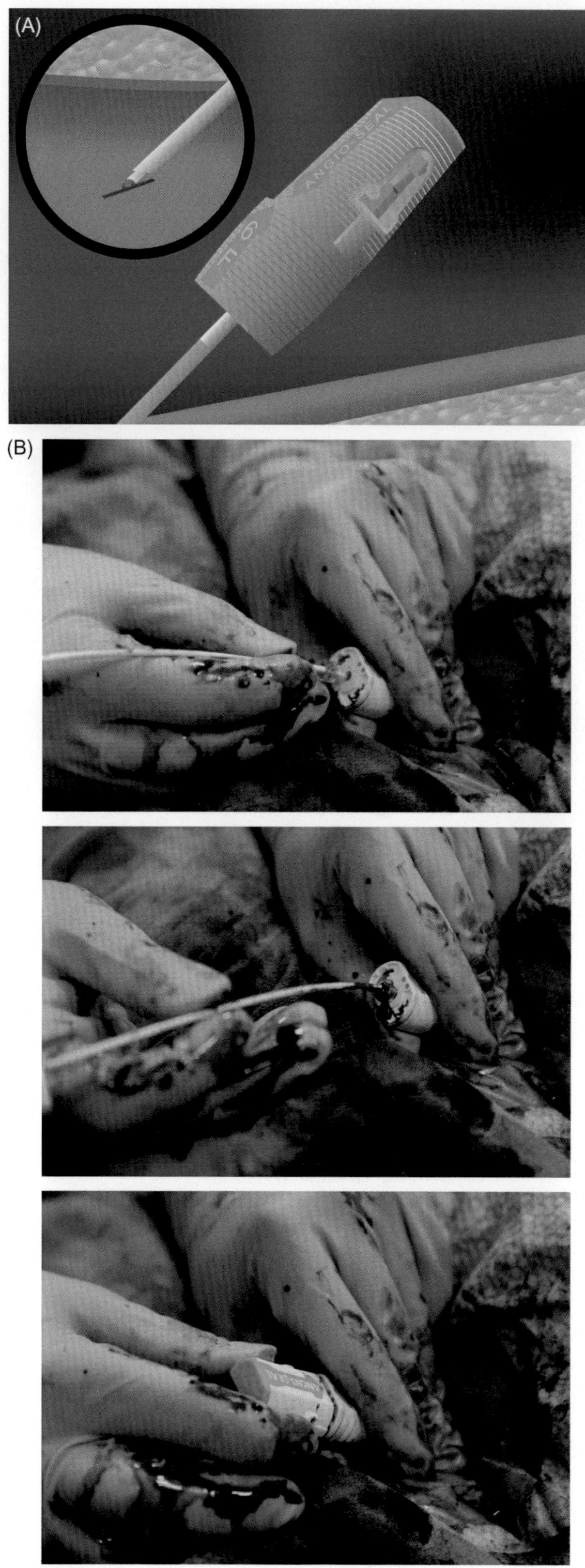

FIGURE 11.6 Advancement of the Angioseal device into the Angioseal sheath until they snap together.

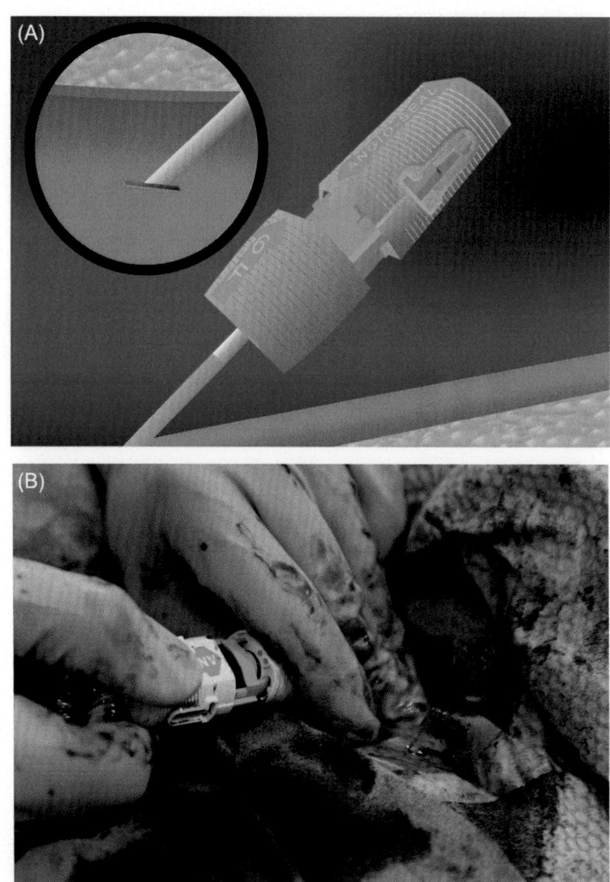

FIGURE 11.7 Deployment of the Angioseal anchor.

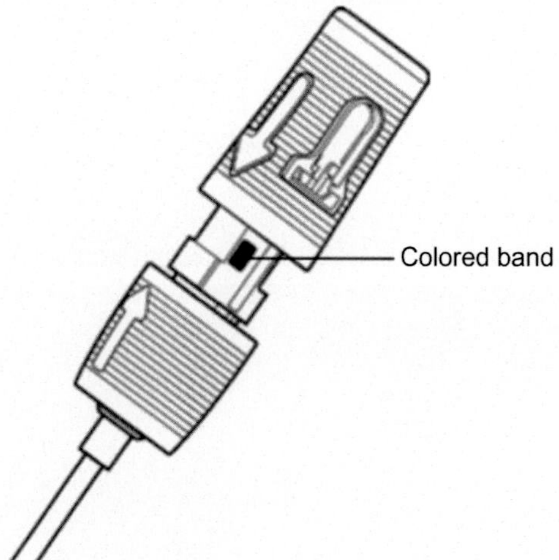

FIGURE 11.8 Location of the colored band on the Angioseal device after deploying the anchor. *©2020 Terumo Medical Corporation. All rights reserved.*

2. Slowly pull back the Angioseal device until resistance is felt. The resistance is due to the Angioseal anchor catching the distal tip of the sheath.
3. The Angioseal device sleeve colored band should be completely visible (Fig. 11.8).

FIGURE 11.9 Anchor leg catches sheath. *©2020 Terumo Medical Corporation. All rights reserved.*

11.1.7.6.3 What can go wrong?

11.1.7.6.3.1 Incomplete/improper anchor deployment

Causes:

- Incomplete withdrawal of the Angioseal device.
- Anchor leg catches sheath (Fig. 11.9).

Prevention:

- Ensure that device is pulled back completely before proceeding with subsequent steps.

Treatment:

- If the anchor catches prematurely, readvance the device into the sheath, and then pull back again.

11.1.7.7 Step 7. Device withdrawal

11.1.7.7.1 How?

1. Pull back the Angioseal device/sheath assembly at an angle of approximately 45° with the skin. This movement pulls the anchor against the vessel wall (Fig. 11.10).
2. After the sheath is removed from the skin a tamper tube will appear. Gently advance the tamper forward while maintaining tension on the device suture (Fig. 11.11). Pushing down with the tamper compacts the collagen outside the arteriotomy. **It is critical to not advance the tamper until after resistance if felt on the back end of the Angioseal device, which signifies that the anchor is against the vessel wall.**
3. The backward tension on the device and forward pushing with the tamper should continue for 8–10 seconds. In most cases a black marker will appear on the device suture (Fig. 11.12).
4. Check that hemostasis is achieved.

11.1.7.7.2 What can go wrong?

11.1.7.7.2.1 Angioseal anchor is pulled through arteriotomy

Causes:

- Poor device selection (e.g., 6 French Angioseal for an 8 French sheath).
- Too forceful withdrawal of the Angioseal device/sheath assembly.

Prevention:

- Ensure correct Angioseal size is selected (6 French Angioseal for 6 or 7 French sheaths and 8 French Angioseal for 8 French sheaths; the 8 French Angioseal has also been used for sealing 9 French sheaths with good results [3]).
- Avoid very forceful device withdrawal.

Treatment:

- Manual pressure. If hemostasis cannot be achieved anticoagulation may need to be reversed and angiography performed to determine whether significant arterial injury has occurred.

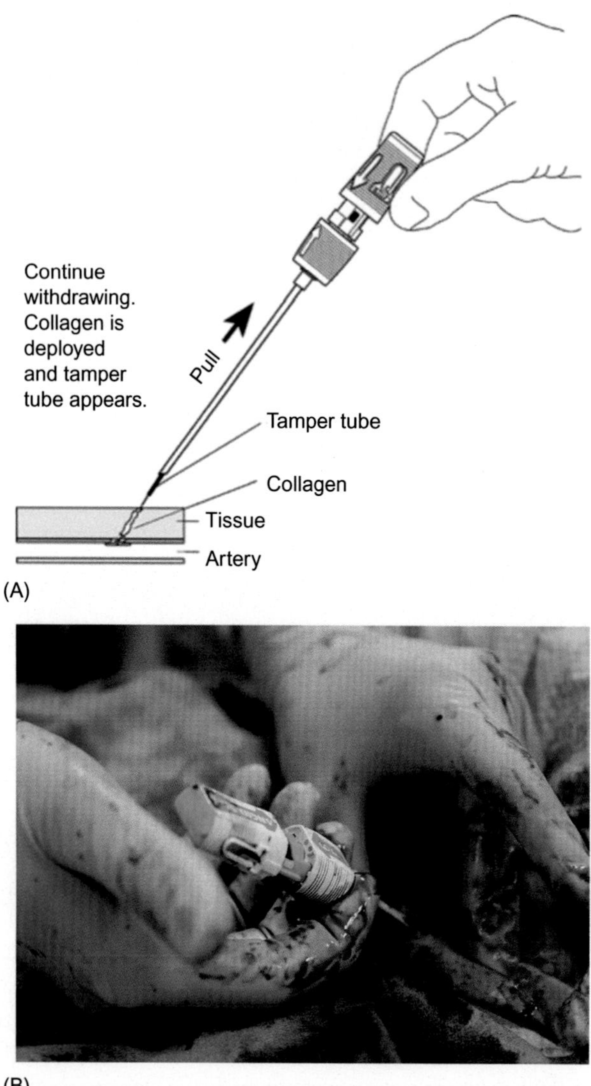
Continue
withdrawing.
Collagen is
deployed
and tamper
tube appears.
Pull
Tamper tube
Collagen
Tissue
Artery
(A)
(B)

FIGURE 11.10 Withdrawal of the Angioseal device. *Panel A: ©2020 Terumo Medical Corporation. All rights reserved.*

11.1.7.7.2.2 Angioseal deployed in the artery

Causes:

- If the sheath is advanced too far into the femoral artery, the footplate may get caught on atherosclerosis in the external iliac or proximal common femoral artery and the operator may falsely feel that the plate has reached the arteriotomy site. The plug is then deployed in the common femoral artery. This can be avoided by not using Angioseal in small or diseased arteries or visualizing the footplate on ultrasound during deployment.
- The operator pushes the plug down into the arteriotomy before feeling the footplate appose on the femoral wall. As mentioned above, it is critical to not advance the tamper until after resistance if felt on the back end of the Angioseal device, which signifies that the anchor is against the vessel wall.
- The arteriotomy is either in the superficial femoral artery or profunda (rather than the common femoral) and the footplate becomes entrapped at the bifurcation instead of the arteriotomy. The plug is then deployed in the artery.

Treatment:

- Angioseal deployment inside the femoral artery can cause acute limb ischemia and should be promptly treated. This will often require surgical removal of the plug, although endovascular treatments have been described.

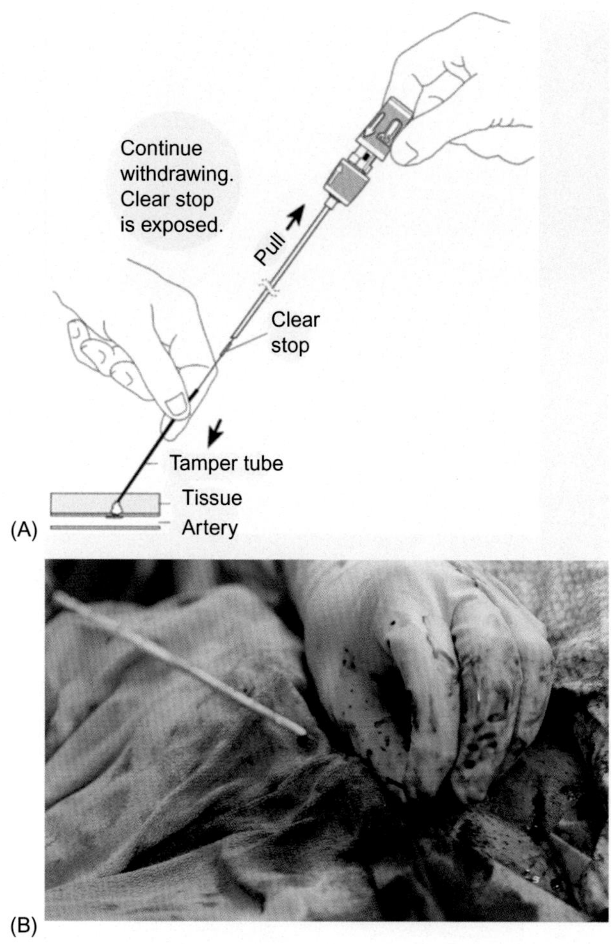

FIGURE 11.11 Advancing the tamper along the suture while maintaining tension on the suture. *Panel A: ©2020 Terumo Medical Corporation. All rights reserved.*

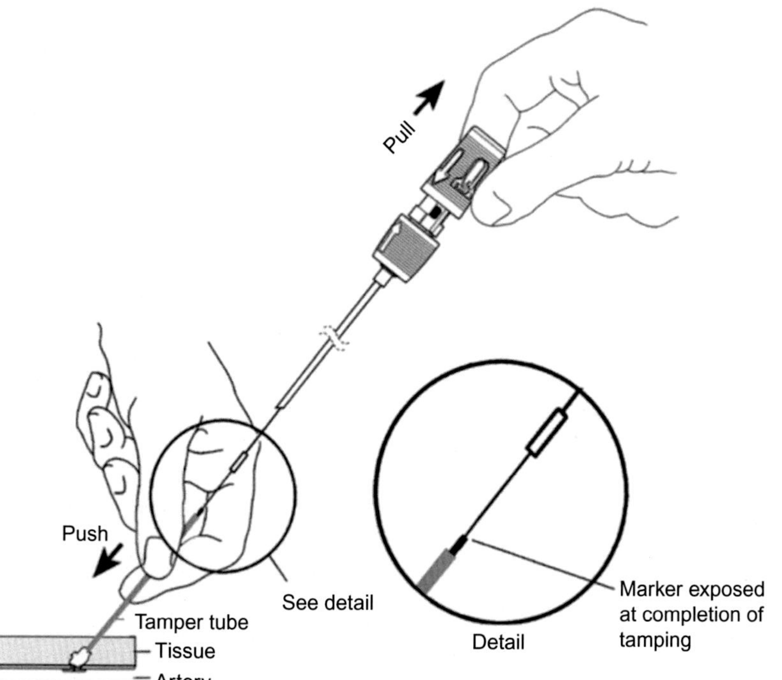

FIGURE 11.12 Illustration of the black marker on the suture device. *©2020 Terumo Medical Corporation. All rights reserved.*

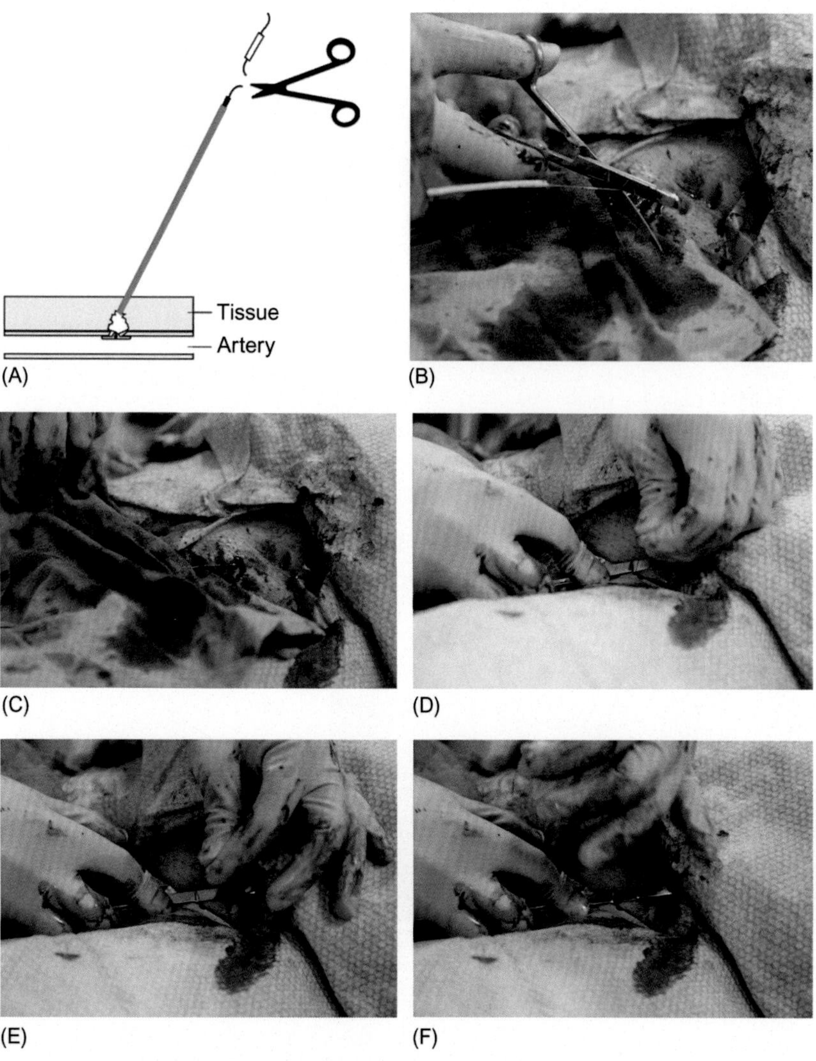

FIGURE 11.13 Cutting the Angioseal suture.

11.1.7.8 Step 8. Cutting the suture

11.1.7.8.1 How?

1. The suture is cut below the clear stop (Fig. 11.13).
2. The tamper tube is removed.
3. The suture is pulled and cut below the skin level. The cut should be below the black compaction marker, which is not absorbable. In obese patients it may be necessary to push down on the tissue around the suture to expose the suture for cutting.

11.1.7.8.2 What can go wrong?

11.1.7.8.2.1 Retroperitoneal **hematoma** *(Section 29.1.2)*

Causes:

- Diseased common femoral artery.
- High arterial stick.
- Suboptimal Angioseal deployment technique.

Prevention:

- Avoid use of Angioseal (or other arterial closure devices) in diseased femoral vessels and in patients with high arterial sticks.
- Use optimal Angioseal deployment technique.

Treatment:

- If the patient becomes hemodynamically unstable, treat accordingly with blood transfusions and fluid resuscitation.
- Reverse anticoagulation.
- Surgical or endovascular treatment may be required in case of continued active bleeding.
- In patients with high suspicion for retroperitoneal hematoma, direct transfer back to the cardiac catheterization laboratory for endovascular treatment may be preferable instead of noninvasive imaging that may delay treatment.

11.1.7.8.2.2 Infection

Causes:

- Systemic infection.
- Immunocompromised patients.
- Imperfect sterile technique.

Prevention:

- Avoid use of the Angioseal (or other arterial closure devices) in patients with active infection and in immunocompromised patients.
- Use sterile technique—repeat groin aseptic preparation immediately before deployment of closure device.
- Apply a sterile dressing over the site immediately after deployment.

Treatment:

- Antibiotics.
- Surgical debridement and reconstruction may be needed.

11.1.8 Perclose

The Perclose device is described in Section 30.15.1.2.

11.1.8.1 Step 1. Device preparation

11.1.8.1.1 How?

1. Open the Perclose sterile package.
2. Flush the marker lumen with saline, until the saline exits through the marker port.

11.1.8.2 Step 2. Advancement of a 0.035–0.038 in. guidewire and removal of the femoral sheath

11.1.8.2.1 How?

1. Advance a 0.035 or 0.038 in. guidewire through the femoral sheath.
2. Perform fluoroscopy to confirm that the guidewire is in the distal aorta.
3. Remove the femoral sheath, leaving the 0.035–0.038 in. guidewire in place.

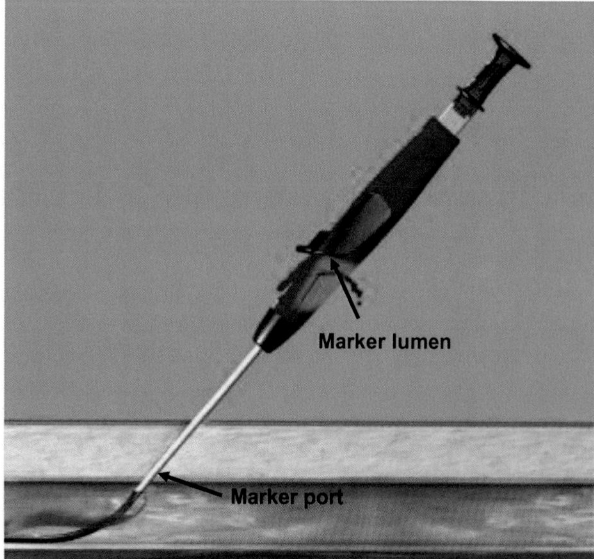

FIGURE 11.14 Advancement of the Perclose device until pulsatile blood flow starts from the marker lumen. *Courtesy Abbott Vascular. ©2019 Abbott. All Rights Reserved.*

11.1.8.3 Step 3. Tissue track preparation

11.1.8.3.1 How?

1. Dilate the tissue tract along the guidewire using a hemostat to facilitate subsequent delivery of the Perclose device.

11.1.8.4 Step 4. Insertion of the Perclose device

11.1.8.4.1 How?

1. Advance the Perclose device over the 0.035–0.038 in. guidewire until the wire exit port is at the skin level.
2. Remove the 0.035–0.038 in. guidewire.
3. Further advance the Perclose device until pulsatile blood flow starts from the marker lumen (Fig. 11.14).

11.1.8.4.2 Challenges

11.1.8.4.2.1 Unable to advance device

Causes:

- Groin scar tissue formation from prior arteriotomies.
- Suboptimal femoral track creation.
- Wire kinking.

Prevention:

- Meticulous track preparation to facilitate subsequent device advancement.
- Use of a stiff 0.035 or 0.038 guidewire, such as the Amplatz super-stiff.

Treatment:

- In case of poor track preparation: the Perclose device is removed and the tissue track is prepared again.
- In case of wire kinking: the wire is exchanged for another (ideally stiffer) wire and the tissue track is prepared again.

11.1.8.5 Step 5. Perclose foot deployment

(Fig. 11.15)

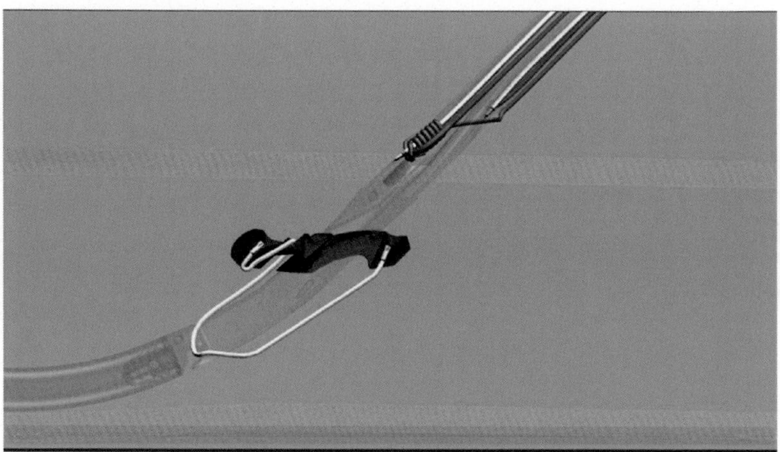

FIGURE 11.15 Deployment of the Perclose foot. *Courtesy Abbott Vascular. ©2019 Abbott. All Rights Reserved.*

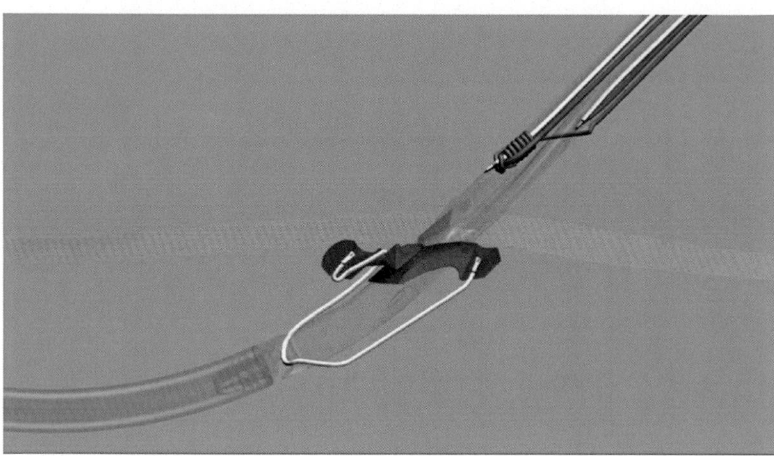

FIGURE 11.16 Withdrawal of the Perclose suture until resistance if felt against the vessel wall. *Courtesy Abbott Vascular. ©2019 Abbott. All Rights Reserved.*

11.1.8.5.1 How?

1. Hold the Perclose device at a 45° angle.
2. Confirm continued pulsatile flow from the marker lumen.
3. Lift the lever, deploying the foot of the Perclose device.

11.1.8.6 Step 6. Perclose suture deployment

11.1.8.6.1 How?

1. Place the device between the middle and index finger of the right hand.
2. Pull back the device until resistance is felt against the vessel wall (Fig. 11.16).
3. Blood flow from the marker lumen should stop or significantly decrease.
4. Hold the device body with the left hand, while the right hand pushes the plunger, deploying the suture (Fig. 11.17A–E).
5. Using the thumb as fulcrum pull back the plunger, disengaging the needles and the plunger from the body of the device.
6. One suture (anterior) will continue to be connected with the needle. Cut this suture using the cutter in the front of the Perclose device.

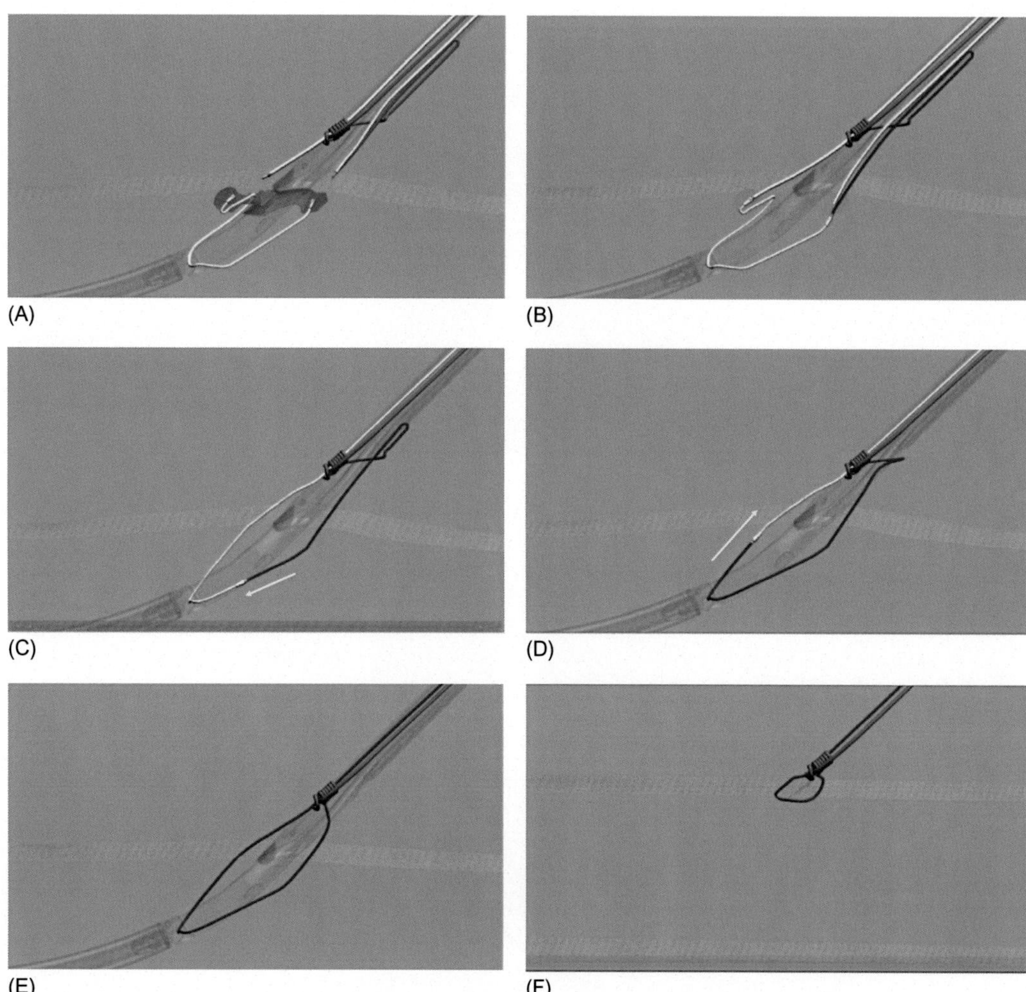

FIGURE 11.17 Deployment of the Perclose suture. *Courtesy Abbott Vascular. ©2019 Abbott. All Rights Reserved.*

11.1.8.6.2 Challenges

11.1.8.6.2.1 Suture is cut or loose

Causes:

- Device failure.
- Forceful pulling of the device.
- Deployment not made in a 45° angle.

Prevention:

- Careful selection of which arteriotomies to seal.
- Gentle handling of the Perclose device.
- Stabilize device at a 45° angle before suture deployment.

Treatment:

- Reinsert a 0.035–0.038 in. guidewire into the artery through the Perclose device, remove the used Perclose device and deploy another Perclose device.

11.1.8.7 Step 7. Perclose foot retraction

11.1.8.7.1 How?

1. Relax tension on the device.
2. Push the Perclose lever down, retracting the foot to its original closed position.

11.1.8.8 Step 8. Perclose device retraction

11.1.8.8.1 How?

1. Retract the Perclose device until the guidewire exit port is visible above the skin line. At this point the 0.035 in. guidewire can be reinserted into the vessel to maintain access to the vessel, which is important when closing large arteriotomies.

11.1.8.8.2 What can go wrong?

11.1.8.8.2.1 Unable to retract device

Causes:
- Device malfunction.
- Poor tissue track preparation.
- Severe calcification.

Prevention:
- Good track preparation prior to Perclose device deployment.

Treatment:
- Gentle rotation and withdrawal movement in both directions.
- If unable to remove device, vascular surgery needs to be consulted for further evaluation and treatment.

11.1.8.9 Step 9. Perclose device removal and suture tightening

11.1.8.9.1 How?

1. Remove the sutures from the Perclose device.
2. The blue suture limb is the longer of the two and is used to advance the knot, whereas the white tip suture is used to lock the knot.
3. Advance the knot pusher over the blue suture limb.
4. Wrap the blue suture limb around the operator's left index finger close to the skin.
5. If there is concern for device failure, the 0.035 in. wire can be readvanced through the Perclose device and the device removed over the wire to maintain wire access.
6. Remove the Perclose device, while applying gentle back tension on the blue suture.
7. Advance the knot pusher (or the suture trimmer) to push the knot down to the arteriotomy site (Fig. 11.17F).
8. If there is bleeding additional pushing of the knot may be required.
9. If the 0.035 in. wire was left in placc, it should be removed once hemostasis is achieved.

11.1.8.10 Step 10. Suture cutting

11.1.8.10.1 How?

1. Inspect the site to determine that hemostasis has been achieved.
2. Pull back the white-tipper wire limb to lock the knot.
3. Remove the knot pusher.
4. Insert the knot trimmer over both sutures, which are then cut, followed by removal of the trimmer.

11.2 Radial access

Manual compression (Fig. 11.18) can be used for achieving radial artery hemostasis and it achieves good results, however mean compression time can be long (mean manual compression time was 22 minutes in the Memory trial) [4].

Currently, a radial compression device (Section 30.15.2) is used at most centers for achieving hemostasis after radial catheterization, as follows:

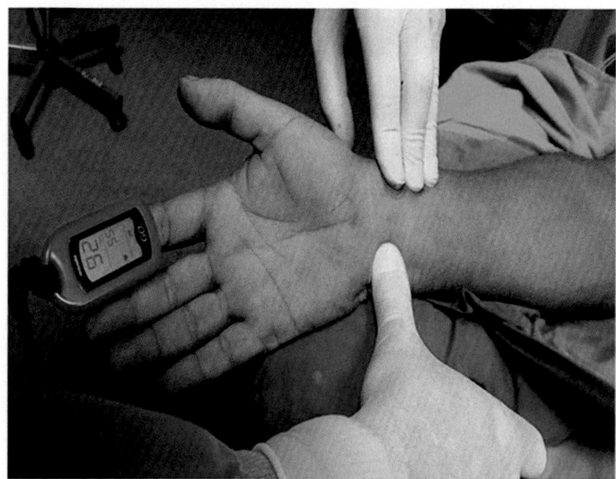

FIGURE 11.18 For manual compression, apply pressure with three fingers over the puncture site and centrally to establish hemostasis. The operator is trying to achieve patent hemostasis, and patency was checked every 1–2 min by transient manual compression of the ulnar artery and evaluation of the status of radial artery patency by plethysmography. *Reproduced from Petroglou D, Didagelos M, Chalikias G, et al. Manual versus mechanical compression of the radial artery after transradial coronary angiography: the MEMORY multicenter randomized trial. JACC Cardiovasc Interv 2018;11:1050–8. Copyright Elsevier.*

11.2.1 Step 1. Determine that the radial sheath can be removed

11.2.1.1 How?

Determine that the procedure is completed and there is no more need for additional PCI or hemodynamic monitoring through the radial artery.

Unlike femoral sheaths, radial sheaths should be removed immediately after the procedure, even when the ACT is high and even if the patient is receiving intravenous antiplatelet agents, such as glycoprotein IIb/IIIa inhibitors or cangrelor. If the radial sheath is not removed immediately, it must be carefully flushed and connected to a pressure sensor to prevent thrombosis and embolization.

11.2.2 Step 2. Vasodilator administration

11.2.2.1 Goal

To prevent or relieve radial artery spasm and facilitate removal of the radial sheath.

11.2.2.2 How?

1. Administer a vasodilator (such as verapamil 2–3 mg and/or nitroglycerin 200–500 mcg) through the side arm of the radial sheath. The vasodilator is typically is mixed 1:1 with arterial blood that has just been withdrawn through the sheath to minimize patient discomfort.

11.2.3 Step 3. Placement of a radial artery compression device

11.2.3.1 Goal

To prevent radial artery bleeding.

11.2.3.2 How?

1. Place a radial compression device, such as the TR band (Terumo) on the patient's wrist immediately above the radial puncture site.

11.2.4 Step 4. Removal of the radial sheath

11.2.4.1 Goal

To remove the radial artery sheath.

11.2.4.2 How?

1. Remove the radial sheath and inflate the cuff of the radial compression device with 10–15 cc of air.
2. Slowly release air until the arteriotomy starts to bleed, after which reinflate with 2–3 cc.

11.2.4.3 What can go wrong?

11.2.4.3.1 Inability to remove the radial artery sheath

Causes:

- Radial artery spasm (Section 29.2.3).

Prevention:

- Vasodilator administration at the beginning of the case.
- Avoid radial sheath oversizing.
- Minimize catheter manipulations.

Treatment:

- Vasodilators (verapamil, nitroglycerin). Vasodilators can be administered intravenously if the operator is unable to aspirate through the side port.
- Subcutaneous nitroglycerin injection around the radial artery.
- More sedation—consider propofol.
- Warm blankets.
- Ischemia-induced vasodilation (inflate a blood pressure cuff in the arm above systolic pressure × 5 minutes which often results in ischemia-induced vasodilation).
- General anesthesia or nerve block.

11.2.4.3.2 Bleeding (Section 29.2.1)

Causes:

- Suboptimal compression device placement.

Prevention:

- Careful placement of the radial compression device and rapid adjustment in case of failure to achieve hemostasis.

Treatment:

- Reposition the radial compression device.
- Manual pressure and arm elevation.

11.2.5 Step 5. Radial artery compression

11.2.5.1 Goal

To achieve hemostasis while minimizing the risk for radial artery occlusion.

11.2.5.2 How?

The patent hemostasis technique [5] is preferred as it minimizes the risk of radial artery occlusion. The goal is to maintain flow within the radial artery during compression (Figs. 11.19 and 11.20) to minimize the risk of thrombus formation and subsequent radial artery occlusion.

1. Place a pulse oximeter sensor over the index finger.
2. Occlude the ipsilateral ulnar artery.
3. The radial compression device is loosened until the plethysmographic signal returns (confirming radial artery patency) or bleeding occurs.

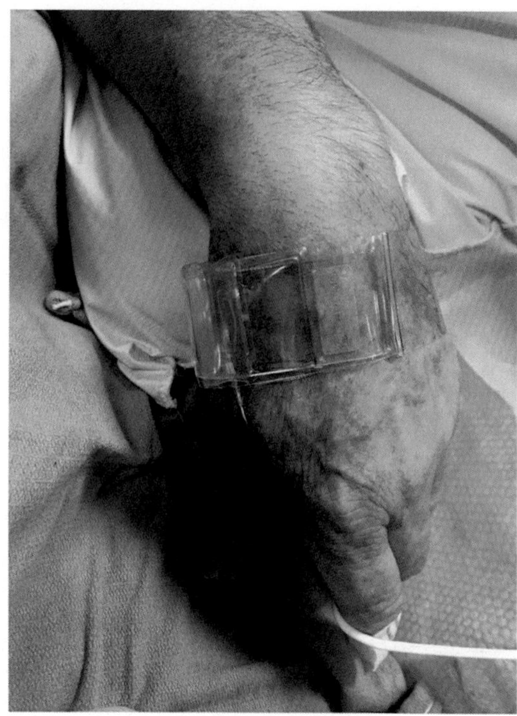

FIGURE 11.19 Example of compression device used for hemostasis with distal radial access.

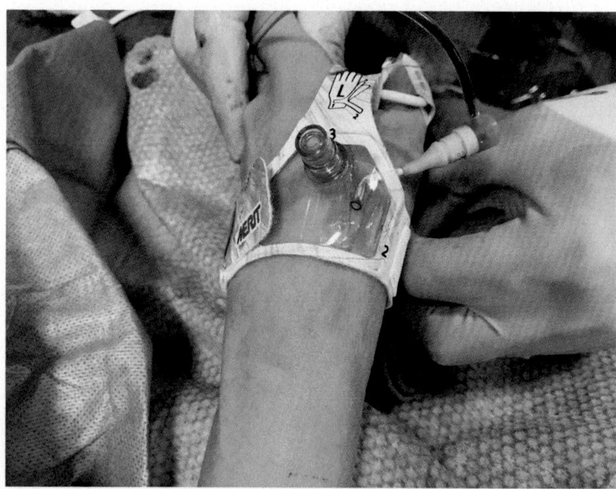

FIGURE 11.20 Example of dedicated compression device for achieving hemostasis of distal radial access.

4. If bleeding occurs at the pressure required to maintain patency, manual compression is used.
5. Duration of compression: The radial compression band is left in place for 1–2 hours. Longer duration should be avoided as it increases the risk of radial artery occlusion [6].
6. Simultaneous compression of the ulnar artery further reduces the risk of radial artery occlusion [7,8].

References

[1] Robertson L, Andras A, Colgan F, Jackson R. Vascular closure devices for femoral arterial puncture site haemostasis. Cochrane Database Syst Rev 2016;3:CD009541.

[2] Ellis SG, Bhatt D, Kapadia S, Lee D, Yen M, Whitlow PL. Correlates and outcomes of retroperitoneal hemorrhage complicating percutaneous coronary intervention. Catheterization and Cardiovascular Interventions 2006;67:541–5.

[3] Janssen H, Killer-Oberpfalzer M, Lange R. Closure of large bore 9 F arterial puncture sites with the AngioSeal STS device in acute stroke patients after intravenous recombinant tissue plasminogen activator (rt-PA). J Neurointerv Surg 2019;11:28–30.

[4] Petroglou D, Didagelos M, Chalikias G, et al. Manual versus mechanical compression of the radial artery after transradial coronary angiography: the MEMORY multicenter randomized trial. JACC Cardiovasc Interv 2018;11:1050–8.

[5] Pancholy S, Coppola J, Patel T, Roke-Thomas M. Prevention of radial artery occlusion-patent hemostasis evaluation trial (PROPHET study): a randomized comparison of traditional versus patency documented hemostasis after transradial catheterization. Catheter Cardiovasc Interv 2008;72:335–40.

[6] Pancholy SB, Patel TM. Effect of duration of hemostatic compression on radial artery occlusion after transradial access. Catheter Cardiovasc Interv 2012;79:78–81.

[7] Pancholy SB, Bernat I, Bertrand OF, Patel TM. Prevention of radial artery occlusion after transradial catheterization: the PROPHET-II randomized trial. JACC Cardiovasc Interv 2016;9:1992–9.

[8] Koutouzis MJ, Maniotis CD, Avdikos G, Tsoumeleas A, Andreou C, Kyriakides ZS. ULnar artery Transient compression facilitating Radial Artery patent hemostasis (ULTRA): a novel technique to reduce radial artery occlusion after transradial coronary catheterization. J Invasive Cardiol 2016;28:451–4.

Chapter 12

Coronary physiology

12.1 When should coronary physiology be used?

12.1.1 Before PCI

1. Determine significance of intermediate coronary lesions.

   Several studies have shown that PCI of lesions with adenosine fractional flow reserve (FFR) > 0.80 (FFR > 0.75 was used in earlier studies [1]) or lesions with nonhyperemic indices that do not show ischemia (such as instantaneous wave free ratio—iFR > 0.89 [2,3]) can be safely deferred without increasing the incidence of adverse outcomes. This also applies to left main lesions [4]. In addition to iFR, several other nonhyperemic indices have been developed (such as the resting full cycle ratio [RFR], the diastolic hyperemia-free ratio [DFR], and the diastolic pressure ratio [DPR]) and provide nearly identical measurements with the iFR.
2. Determine which lesions require revascularization in patients with multivessel CAD.

Use of adenosine FFR (usually in lesions with <90% angiographic diameter stenosis) during multivessel PCI is associated with treatment of fewer lesions and better clinical outcomes as compared with angiography-guided PCI [5–7]. Physiologic assessment should not be performed for clinical decision making in culprit or suspected culprit lesions in patients presenting with an acute coronary syndrome (ACS). Intracoronary imaging should be used to assess ACS culprit lesions [8,9].

12.1.2 During PCI

1. Post-PCI adenosine FFR is emerging as a tool to determine functionally optimal PCI results [10]. Post-PCI adenosine FFR > 0.91 has been associated with optimal outcomes.

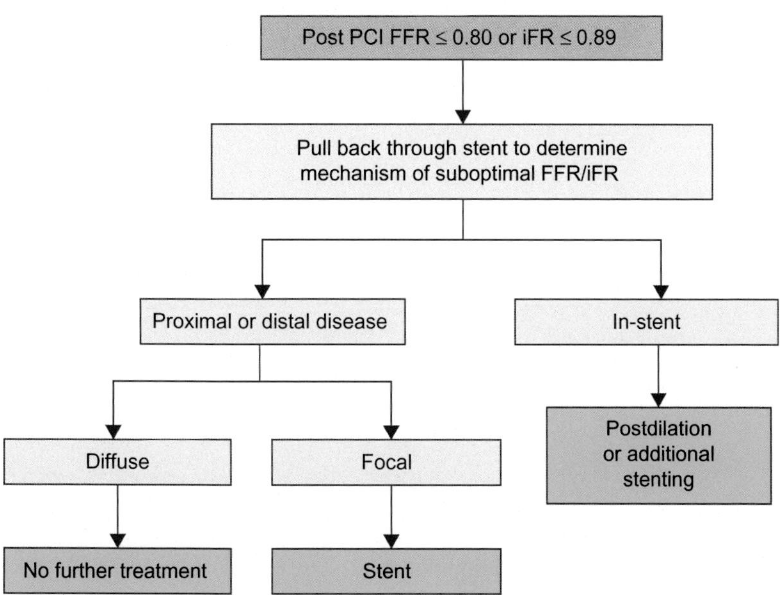

FIGURE 12.1 Use of post-PCI FFR to optimize PCI results.

Manual of Percutaneous Coronary Interventions. DOI: https://doi.org/10.1016/B978-0-12-819367-9.00012-3

Causes of suboptimal post-PCI adenosine FFR (Fig. 12.1) [10]:

- Geographic miss of culprit lesion.
- Inadequate lesion coverage (not covering from healthy to healthy segment).
- Suboptimal stent selection and deployment (undersized—underexpanded).
- Diffuse disease.
- Stent edge dissection.
- Serial lesions (another intermediate lesion becoming hemodynamically manifest after treating the culprit lesion).
- Coronary vasospasm.
- Extreme vessel tortuosity (causing wire bias and accordioning effect).
- Large collateral supply by the treated vessel to a CTO-supplied territory.
- Myocardial bridging.
- Pressure wire-related technical issues (such as drift or wire malfunction).

2. To assess the side branch during provisional bifurcation stenting.

Angiographic assessment of jailed side branches during bifurcation stenting can be challenging. Intracoronary nitroglycerin may help reverse coronary spasm. Adenosine FFR of the jailed side branch can help guide further treatment [11–13]:

FFR ≤ *0.80*: further treatment of the side branch is needed (assuming that the vessel diameter is ≥ 2 mm), as there is increased risk of subsequent adverse events.

FFR > *0.80*: no further treatment of the side branch is needed.

12.2 How to do coronary physiologic assessment?

Starting point: The guide catheter is engaged in the target coronary vessel. The guide pressure waveform should not be dampened, as this can result in underestimation of the lesion severity (artificially increasing the FFR or nonhyperemic indices) [14–16].

12.2.1 Step 1: Flush and zero pressure guidewire outside the body

12.2.1.1 Goals

To prepare the pressure guidewire for use.

12.2.1.2 How?

Flush the pressure guidewire container with normal saline, connect to the patient interface module (PIM), and zero it. The wire is shaped as described in Chapter 8 (Wiring), paying attention to not damage the pressure sensor (that is usually located 3 cm from the tip).

12.2.1.3 Challenges

1. Unable to zero pressure wire.

   Causes:
   - Poor pressure wire connection.
   - Defective pressure wire.

   Prevention:
   - Check all connections prior to connecting guidewire.

   Treatment:
   - Check all connections.
   - Replace pressure wire with a new one.

12.2.1.4 What can go wrong?

2. Inadequate calibration: if this is not appreciated prior to insertion of the guidewire in the guide catheter, it will delay the procedure, as it will require removal of the guidewire and recalibration.
3. Wire damage that may require discarding the pressure wire and using a new one.

12.2.2 Step 2: Insert pressure guidewire in the guide catheter through the Y-connector and advance to the tip of the guide catheter (Fig. 12.2)

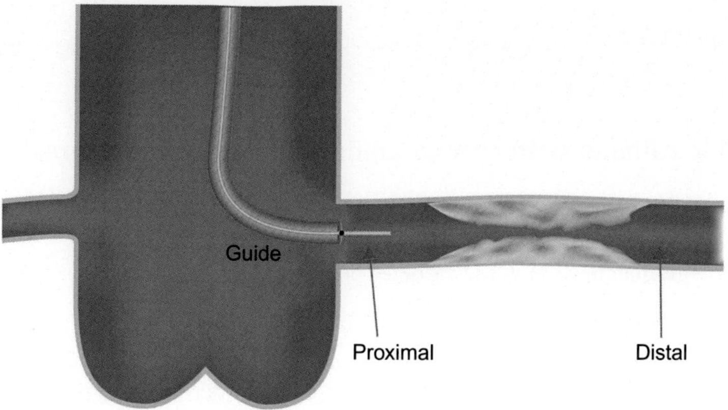

FIGURE 12.2 Advancing guidewire with the proximal end of the radiopaque portion of the guidewire at the tip of the guide catheter.

12.2.2.1 Goals

To deliver the pressure guidewire to the guide catheter tip.

12.2.2.2 How?

Insert the pressure guidewire using an introducer through the hemostatic valve of the catheter-connector and advance it until it reaches the tip of the guide catheter. Partially advance the pressure wire into the coronary artery until the proximal end of the radiopaque portion of the wire is at the tip of the guide catheter (Fig. 12.2).

12.2.2.3 Challenges

1. Inability to advance the pressure wire within the target vessel
   Causes:
   - Suboptimal guide catheter engagement.
   - Suboptimal guide catheter choice (for example using a too large EBU guide catheter may result in the wire preferentially entering the circumflex instead of the LAD and vice versa).
   - Suboptimal shape of the guidewire tip.

   Prevention:
   - Ensure excellent guide catheter engagement using an appropriately sized guide catheter as outlined in Chapter 5, Coronary and Graft Engagement.
   - Appropriate shaping of the guidewire tip.

   Treatment:
   - Change guide catheter for a better fitting guide catheter.
   - Modify the shape of the guidewire tip.

12.2.2.4 What can go wrong?

2. Wire deformation
   Causes:
   - Suboptimal wire shaping prior to insertion into the guide catheter.
   - Wire protruding through the introducer tip while being inserted.
   - Other equipment (such as guide catheter extensions, balloons, and stents) present inside the guide catheter.

   Prevention:
   - Ensure the guidewire is fully retracted within the introducer while inserting through the hemostatic valve of the Y-connector.
   - If there is resistance to wire advancement, do NOT push hard!

- If additional equipment is present within the guide catheter, either remove it prior to inserting the pressure guidewire, or advance thepressure guidewire under fluoroscopy.

Treatment:
- Reshape pressure guidewire if feasible, or use a new one.
- Follow all precautions outlined under "Prevention" above.

12.2.3 Step 3: Remove introducer, flush guide catheter with normal saline, and equalize pressures (Fig. 12.3)

12.2.3.1 Goals

To ensure that the pressure read by the pressure guidewire matches the guide catheter pressure.

12.2.3.2 How?

The introducer is removed, the guide catheter flushed with normal saline, and the guide catheter-derived and pressure wire-derived pressures are equalized.

12.2.3.3 Challenges

1. Unable to perform pressure equalization (sometimes the system may equalize the mean pressures, but the pressure waveforms may not overlap introducing error in the measurement of the hyperemic FFR).
   Causes:
   - Failure to remove the wire introducer.
   - Failure to flush the guide catheter.
   - Pressure guidewire failure.
   - Suboptimal guidewire tip position (too deep into the target vessel).
   - Air in the aortic pressure line.

   Prevention:
   - Ensure that the introducer is removed from the tuohy.
   - Ensure that the guide catheter is flushed.
   - Ensure that the systemic pressure line is flushed and properly zeroed.
   - Ensure that the proximal portion of the radiopaque segment of the pressure guidewire is at the guide catheter tip.

   Treatment:
   - Perform steps described under "Prevention" above.
   - If pressure equalization still fails, change pressure guidewire.

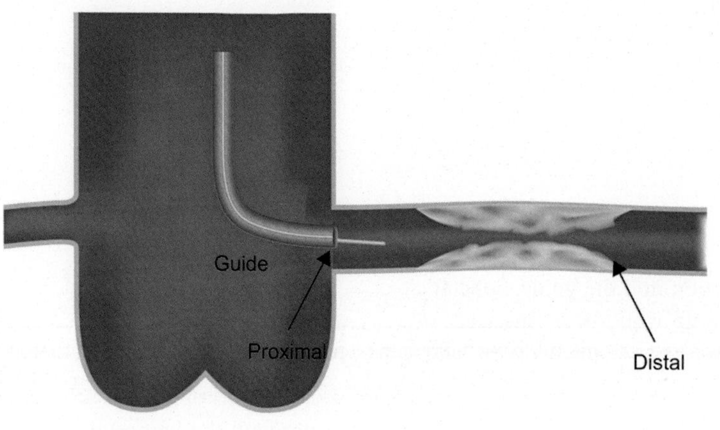

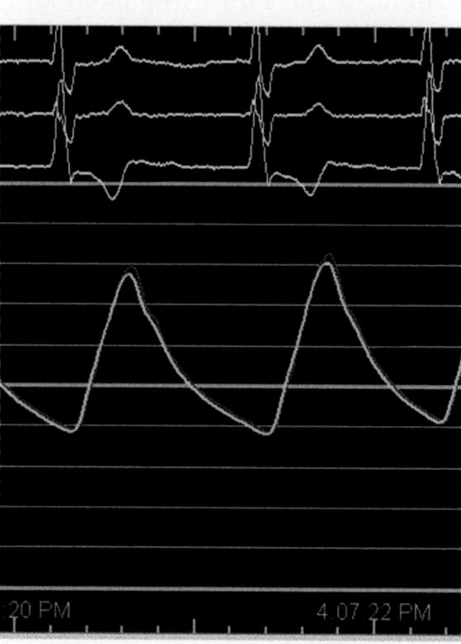

FIGURE 12.3 Pressure equalization.

12.2.4 Step 4: Advance pressure wire distal to the target lesion(s) (Fig. 12.4)

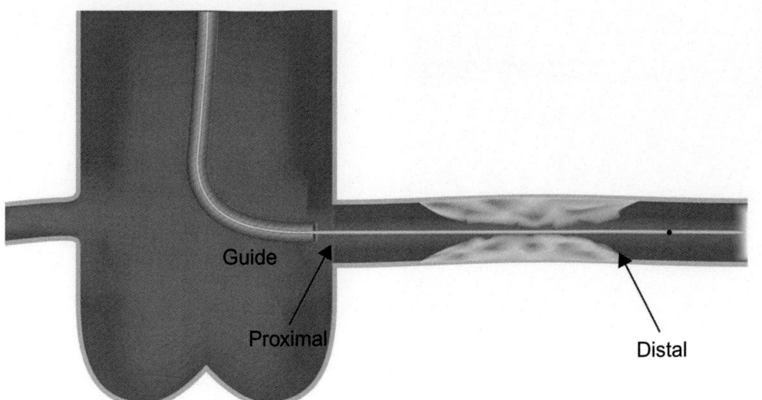

FIGURE 12.4 Advancing the pressure guidewire distal to the target lesion(s).

12.2.4.1 Goal

To advance the pressure guidewire distal to the target lesion(s).

12.2.4.2 How?

The introducer is reinserted (optional) and the guidewire is advanced through and distal to the target lesion(s) and ideally as distal in the coronary artery as possible to assess flow through the entire artery [16].

12.2.4.3 Challenges

1. Unable to advance guidewire past lesion.

   See Sections 8.6–8.8.

12.2.4.4 What can go wrong?

12.2.4.4.1 Target vessel injury

See Section 8.7.

12.2.4.4.2 Pressure drift

Causes:

- Technical problem with the pressure wire.
- Loose or malfunctioning pressure transducer.

Prevention:

- Ensure that pressures are equalized prior to advancing the pressure guidewire through the lesion.
- Avoid disconnecting the pressure guidewire connection during wire manipulations.

Treatment:

- Retract pressure guidewire so that the proximal end of the radiopaque portion of the pressure guidewire is at the tip of the guide catheter, reequalize pressures, readvance the pressure guidewire through the lesion, and repeat physiologic assessment.

12.2.5 Step 5: Measure a resting physiologic index (Fig. 12.5)

12.2.5.1 Goal

Assess the hemodynamic significance of the target lesion without vasodilator administration.

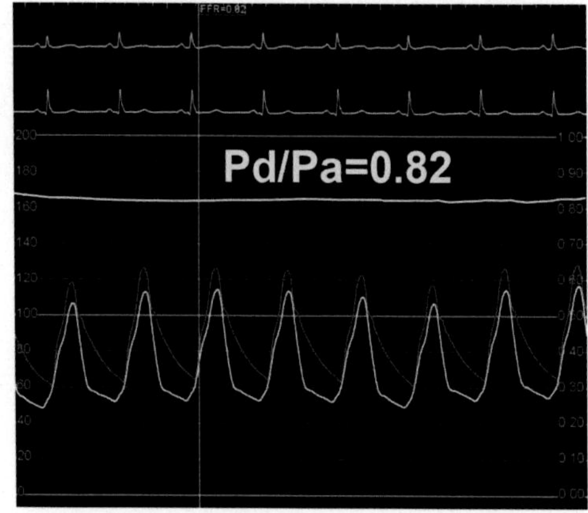

FIGURE 12.5 Resting (nonhyperemic) physiologic index measurement.

TABLE 12.1 Resting indices for physiologic lesion assessment.

Resting index	Abbreviation explanation	Cutoff
Pd/Pa	Pressure distal to lesion/ Pressure in the guide	≤0.91
iFR	Instantaneous wave-free ratio	≤0.89
dPR	Diastolic pressure ratio	≤0.89
dFR	Diastolic hyperemia-free ratio	≤0.89
RFR	Resting full-cycle ratio	≤0.89

12.2.5.2 How?

The pressure from the pressure guidewire (Pd) and the pressure from the guide catheter (Pa) are simultaneously recorded, followed by calculation of a resting index.

There are several nonhyperemic indices, all of which are strongly correlated (Table 12.1); however only the iFR has clinical trial data validation [2,3].

Advantages of resting physiologic indices over hyperemic indices:

1. Shorter procedure time.
2. Lower cost.
3. No complications associated with administration of hyperemic agents.

12.2.5.3 What can go wrong?

12.2.5.3.1 Pressure drift

Resting indices are much more influenced by pressure drift than hyperemic indices [17]. Approaching pressure drift is discussed in Section 11.2.8.

12.2.6 Step 6: Measure hyperemic FFR (may not be needed in many cases in which resting indices suffice for clinical decision making) (Fig. 12.6)

12.2.6.1 Goal

Determine the hemodynamic significance of the target lesion with vasodilator administration.

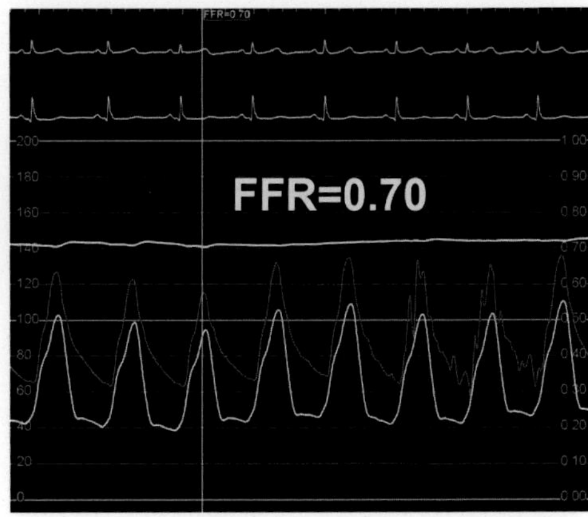

FIGURE 12.6 Adenosine FFR measurement.

12.2.6.2 How?

1. Administer a vasodilator (adenosine is most commonly used; regadenoson can also be used but is costly; papaverine can be used outside the US). Contrast (usual volume administered is 8 mL) also has some vasodilatory effect: the ischemic threshold for contrast-FFR is 0.83.
   - Adenosine dose:
     - Intracoronary in the RCA: 50–100 mcg
     - Intracoronary in the left main: 200 mcg; up to 600 mcg have been used in some studies [18]
     - Intravenous: 140–180 mcg/kg/min
   - Papaverine dose (usually avoided due to risk of ventricular arrhythmias – it is not currently available in the US):
     - Intracoronary in the RCA: 8 mg
     - Intracoronary in the left main: 12 mg
2. If intracoronary adenosine is administered, the guide catheter must be flushed afterwards to ensure delivery of all the adenosine administered into the target coronary artery.
3. One-two seconds after intracoronary adenosine administration, the pressure from the pressure wire (Pd) and the pressure from the guide catheter (Pa) are simultaneously recorded: the lowest value is the adenosine FFR.

   With intravenous adenosine the adenosine FFR is measured 1–3 minutes after initiation of the infusion when stable hyperemia has been achieved. Usually a systolic blood pressure drop of 10 mm Hg and a heart rate increase by 5–10 beats/minute during adenosine infusion is a good surrogate of adenosine efficacy in the circulation (hyperemia).
4. Interpretation: although adenosine FFR and contrast FFR are often interpreted as a binary variable (yes/no) based on the 0.80 cutoff (for adenosine FFR) or 0.83 cutoff (for contrast FFR) (Table 12.2), they are actually continuous parameters. Ischemia and likewise FFR exist as a continuum. There is no major difference between a FFR of 0.80 and 0.81 and this should be taken into account for clinical decision making. The potential benefit of coronary revascularization is higher in lesions with lower FFR [19].

12.2.6.3 Challenges

12.2.6.3.1 Artifacts

Causes:
- Wire artifact.
- Measurement started too early, before the guide pressure waveform returned to baseline after intracoronary adenosine administration.
- Dampening of the guide catheter pressure.

Prevention:
- Visual inspection of FFR pressure tracings to ensure lack of artifact (and of guide catheter pressure dampening) before FFR recording is performed.

Treatment:
- Repeat adenosine administration and FFR measurement, ensuring no artifact is present.

TABLE 12.2 Cutoffs of hyperemic physiologic indices.

Hyperemic index	Abbreviation explanation	Cutoff
Adenosine FFR	Fractional flow reserve	≤0.80
Contrast FFR	Contrast FFR	≤0.83

12.2.6.3.2 Failure to induce adequate vasodilation

Causes:

- Guide catheter is disengaged prior to intracoronary adenosine administration.
- Erroneous administration of normal saline instead of adenosine.
- Intravenous line is not working (e.g., due to IV infiltration).
- Administration of very low adenosine dose.
- Recent caffeine intake (caffeine prevents adenosine binding to the adenosine receptor) [20].

Prevention:

- Ensure good guide catheter engagement prior to vasodilator administration.
- Administer appropriate vasodilator doses.
- Ensure patient has not received caffeine prior to the procedure.

Treatment:

- Repeat adenosine administration and FFR measurement, ensuring good guide catheter engagement and appropriate adenosine dose.

12.2.6.4 What can go wrong?

12.2.6.4.1 Bradycardia—atrioventricular (AV) block

Causes:

- Adenosine effect on the AV node.

Prevention:

- Do not administer high doses (> 100 mcg) of intracoronary adenosine in the right coronary artery.

Treatment:

- Because of the short half-life of adenosine, bradycardia usually resolves after a few seconds.
- Ask patient to cough (self-CPR).
- If AV block without escape persists, transient CPR and temporary pacing may be needed.

12.2.6.4.2 Ventricular fibrillation (Fig. 3.3) [21]

Causes:

- R on T phenomenon

Prevention:

- Do not administer high doses (> 100 mcg) of intracoronary adenosine in the right coronary (or a dominant circumflex) artery.

Treatment:

- Defibrillation.

12.2.6.4.3 Atrial fibrillation (Fig. 3.2) [21]

Causes:

- Adenosine-induced AV block.
- Adenosine-induced atrial fibrillation.

Prevention:

- Do not administer high doses (> 100 mcg) of intracoronary adenosine in the right coronary (or a dominant circumflex) artery.

Treatment:

- Cardioversion (if the patient does not spontaneously convert to sinus rhythm).

12.2.7 Step 7: Perform pressure wire pullback while recording pressure

12.2.7.1 Goal

Determine the location of pressure drop(s) that suggest hemodynamically significant lesion(s). This is particularly important in patients with multiple lesions or diffuse disease.

12.2.7.2 How?

1. Guidewire is withdrawn under fluoroscopy with simultaneous pressure measurements. Pullback can be performed without vasodilator administration while measuring resting indices or during intravenous adenosine administration (adenosine FFR pullback).
2. Pressure wire pullback should stop when the radiopaque portion of the wire reaches the tip of the guide catheter.
3. Interpretation (Fig. 12.7):

In patients with multiple lesions, although FFR reflects the summed result of stenoses in a series, FFR does not apply to individual lesions, as both upstream and downstream lesions limit maximal flow across the other lesion. The stenosis with the biggest pressure gradient should be treated first. FFR of the remaining lesion(s) should then be repeated, and if it remains abnormal, these lesion(s) should also be stented. With iFR however, pre-angioplasty pullback provides virtual intervention and predicts hemodynamic outcome for serial or diffuse lesions [22,23]. Newer systems can coregister angiography and the pullback measurements and provide a map of the pressure drop across various coronary segments (Fig. 12.8) [24].

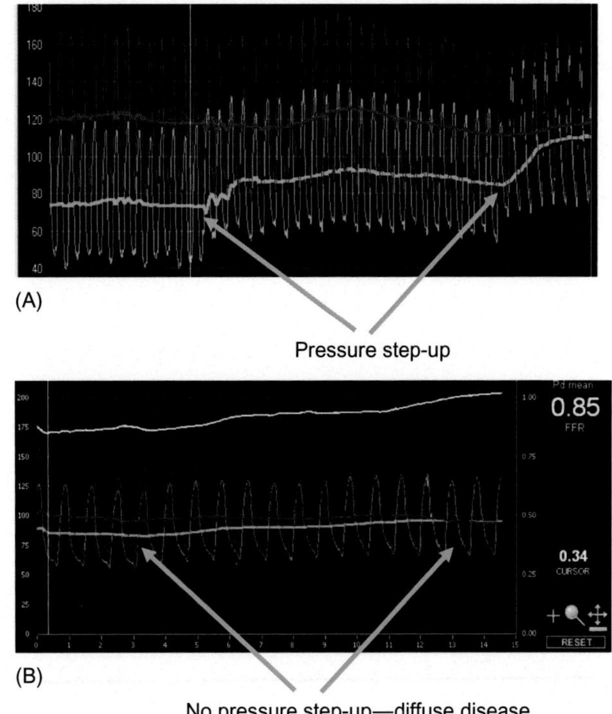

FIGURE 12.7 Illustration of various types of pressure step-up. (A) Focal coronary lesions with clear pressure step-ups. (B) Diffuse coronary disease without pressure step-ups. PCI is challenging in the latter setting, as long stent length would be required. Coronary bypass graft surgery is preferable, if feasible, for such patients.

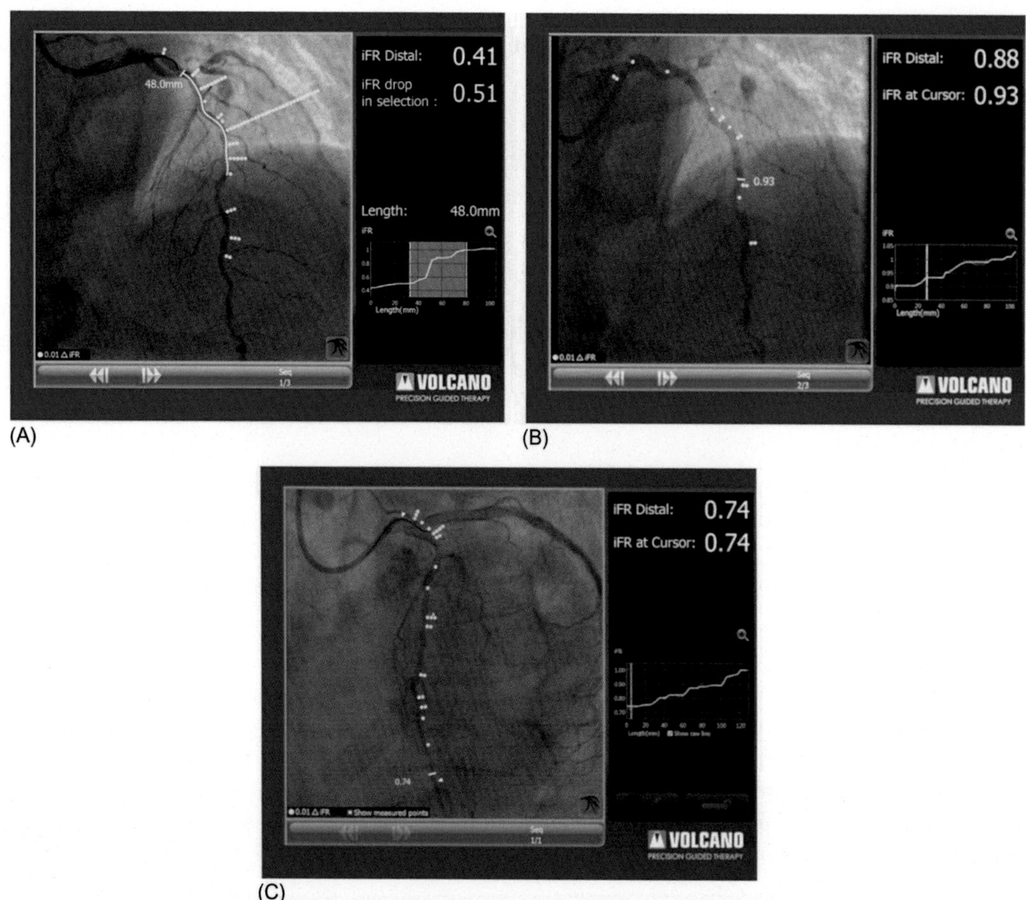

FIGURE 12.8 Angiography-physiology coregistration properties and use. (A) Using the joystick, the operator delineates the "big step" over the trend line in the bottom right (*light blue rectangle*). The system automatically outlines a *white line* over the corresponding angiogram roadmap of the left anterior descending artery on the left and measures its length (48 mm). The physiology values in the upper right shows the initial low distal instantaneous flow reserve (iFR) (0.41) and the estimated increase in iFR (0.51 + 0.41 = 0.92) if a proper angioplasty (i.e., 48-mm stent deployment) covers the entire length delineated on the roadmap. (B) After deployment of a 48-mm stent in the left anterior descending artery, which covers exactly the length and site of the *white line*, the iFR distal to the stent increased to 0.93. (C) No significant "step-up" can be detected over the trend line, so stenting is not practical in this case. *Reproduced with permission from Frimerman A, Abu-Fane R, Levi Y, et al. Novel method for real-time coregistration of coronary physiology and angiography by iFR. JACC Cardiovasc Interv 2019;12:692–4. Copyright Elsevier.*

12.2.8 Step 8: Confirm that no pressure drift has occurred (Fig. 12.9)

12.2.8.1 Goal

Confirm that no pressure drift has occurred.

12.2.8.2 How?

1. Assess the Pd/Pa after the radiopaque segment of the pressure wire is withdrawn to the tip of the guide catheter.
2. Pd/Pa > 1.02 or <0.98 suggests significant pressure drift: the pressure wire and guide pressures should be reequalized and steps 4–7 repeated.

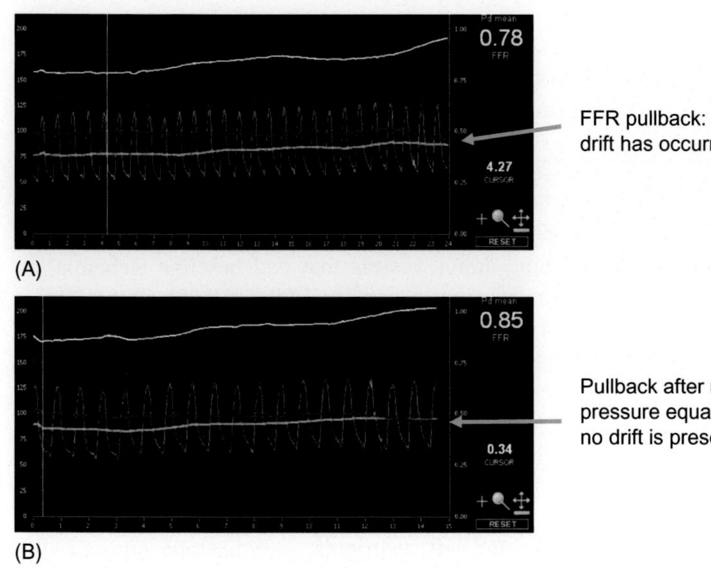
0.78
FFR
4.27
CURSOR
RESET
(A)
FFR pullback: pressure drift has occurred
0.85
FFR
0.34
CURSOR
RESET
(B)
Pullback after repeat pressure equalization: no drift is present

FIGURE 12.9 **Impact of pressure drift on FFR measurements.** (A) Adenosine FFR pullback suggesting a hemodynamically significant lesion is present (FFR = 0.78). However, inspection of the pressures shows that pressures have drifted, hence the measurements are not valid. (B) Repeat adenosine FFR pullback after pressures were reequalized at the guide catheter, showing no pressure drift at the end of the pullback. The lesion is, therefore, not hemodynamically significant (FFR = 0.85).

12.3 Physiologic assessment in various lesion and patient subsets

12.3.1 Aorto-ostial lesions

The key challenge in aorto-ostial lesions is guide catheter pressure dampening, which can artificially increase the physiologic indices and as a result underestimate the severity of the stenosis. To prevent this, physiologic assessment of aorto-ostial lesions should be performed after disengaging the guide catheter from the coronary artery ostium. Adenosine can either be administered intracoronary followed by guide catheter disengagement, or administered intravenously.

12.3.2 Left main

Both hyperemic and resting physiologic indices can be used in left main lesions.

For ostial left main lesions, the guide catheter should be disengaged to prevent pressure dampening, as should be done in all aorto-ostial lesions.

Physiologic assessment should be performed in both the LAD and the circumflex with pullback to demonstrate whether the step-up occurs in the left main. FFR $\geq$ 0.80 in one vessel (such as the LAD) but not the other (such as the circumflex) suggests that the left main stenosis is not significant.

12.3.3 Saphenous vein grafts

Physiologic assessment of SVG lesions is subject to several limitations. First, FFR of a SVG is the result of SVG flow, flow through the native coronary artery (unless the latter is occluded) and flow via collaterals; hence FFR may be normal, even when a SVG has severe stenoses if there is competitive flow via the native coronary artery. Second, the basic underpinning of functional assessment of native coronary artery lesions is slow rate of disease progression in nonsignificant lesions. This assumption does not hold true for SVG lesions, since intermediate SVG lesions have high rates of progression [25–28] and also additional lesions may develop at other locations within the same SVG. We, therefore, recommend against performing physiologic assessment of SVG lesions.

12.3.4 Acute coronary syndromes

Physiologic assessment should not be performed in myocardial infarction culprit vessels due to microcirculatory dysfunction [8,9,29]. However, nonculprit lesions can be assessed using coronary physiology [30,31].

12.3.5 Chronic total occlusions

The hemodynamic significance of a lesion in the donor vessel that collateralizes the CTO vessel may change after successful CTO recanalization. In a study by Sachdeva et al., six of nine donor vessels that had baseline ischemia, as assessed by adenosine FFR measurement (FFR ≤ 0.80) reverted to nonischemic adenosine FFR (>0.80) after successful CTO recanalization [32]. The mean increase in donor vessel adenosine FFR after CTO recanalization was 0.098 ± 0.04. In another study the mean increase in donor vessel adenosine FFR after CTO recanalization was 0.03 [33]. Therefore, physiologic assessment of a donor vessel supplying collaterals to a CTO territory is limited and should ideally be performed after CTO recanalization.

12.3.6 Congestive heart failure

In patients with congestive heart failure and high right atrial pressure and left ventricular end-diastolic pressure, the adenosine FFR measurement will be impacted. Although a correction can be made by incorporating right atrial pressure in the FFR calculation (FFR = (Pd − Pv)/(Pa − Pv) it is best to avoid functional assessment of coronary lesions in such patients.

12.3.7 Discrepancy between resting and hyperemic indices

Discordance between resting and hyperemic indices is noted in approximately 14% of cases [34] and is particularly notable in large coronary arteries (e.g., left main, proximal LAD) and in patients with diabetes [35]. In case of discrepancy between resting and hyperemic indices, decision making should be based on clinical presentation.

References

[1] Zimmermann FM, Ferrara A, Johnson NP, et al. Deferral vs. performance of percutaneous coronary intervention of functionally non-significant coronary stenosis: 15-year follow-up of the DEFER trial. Eur Heart J 2015;36:3182–8.

[2] Davies JE, Sen S, Dehbi HM, et al. Use of the instantaneous wave-free ratio or fractional flow reserve in PCI. N Engl J Med 2017;376:1824–34.

[3] Gotberg M, Christiansen EH, Gudmundsdottir IJ, et al. Instantaneous wave-free ratio versus fractional flow reserve to guide PCI. N Engl J Med 2017;376:1813–23.

[4] Hamilos M, Muller O, Cuisset T, et al. Long-term clinical outcome after fractional flow reserve-guided treatment in patients with angiographically equivocal left main coronary artery stenosis. Circulation 2009;120:1505–12.

[5] Tonino PAL, De Bruyne B, Pijls NHJ, et al. Fractional flow reserve versus angiography for guiding percutaneous coronary intervention. N Engl J Med 2009;360:213–24.

[6] Pijls NH, Fearon WF, Tonino PA, et al. Fractional flow reserve versus angiography for guiding percutaneous coronary intervention in patients with multivessel coronary artery disease: 2-year follow-up of the FAME (Fractional Flow Reserve Versus Angiography for Multivessel Evaluation) study. J Am Coll Cardiol 2010;56:177–84.

[7] van Nunen LX, Zimmermann FM, Tonino PA, et al. Fractional flow reserve versus angiography for guidance of PCI in patients with multivessel coronary artery disease (FAME): 5-year follow-up of a randomised controlled trial. Lancet 2015;386:1853–60.

[8] Hakeem A, Edupuganti MM, Almomani A, et al. Long-term prognosis of deferred acute coronary syndrome lesions based on nonischemic fractional flow reserve. J Am Coll Cardiol 2016;68:1181–91.

[9] Masrani Mehta S, Depta JP, Novak E, et al. Association of lower fractional flow reserve values with higher risk of adverse cardiac events for lesions deferred revascularization among patients with acute coronary syndrome. J Am Heart Assoc 2015;4:e002172.

[10] Hakeem A, Uretsky BF. Role of postintervention fractional flow reserve to improve procedural and clinical outcomes. Circulation 2019;139:694–706.

[11] Koo BK, Park KW, Kang HJ, et al. Physiological evaluation of the provisional side-branch intervention strategy for bifurcation lesions using fractional flow reserve. Eur Heart J 2008;29:726–32.

[12] Kang SJ, Ahn JM, Kim WJ, et al. Functional and morphological assessment of side branch after left main coronary artery bifurcation stenting with cross-over technique. Catheter Cardiovasc Interv 2014;83:545–52.

[13] Koo BK, Kang HJ, Youn TJ, et al. Physiologic assessment of jailed side branch lesions using fractional flow reserve. J Am Coll Cardiol 2005;46:633–7.

[14] Aminian A, Dolatabadi D, Lefebvre P, et al. Importance of guiding catheter disengagement during measurement of fractional flow reserve in patients with an isolated proximal left anterior descending artery stenosis. Catheter Cardiovasc Interv 2015;85:595–601.

[15] Patel KS, Christakopoulos GE, Karatasakis A, et al. Prospective evaluation of the impact of side-holes and guide-catheter disengagement from the coronary ostium on fractional flow reserve measurements. J Invasive Cardiol 2016;28:306–10.

[16] Toth GG, Johnson NP, Jeremias A, et al. Standardization of fractional flow reserve measurements. J Am Coll Cardiol 2016;68:742–53.

[17] Cook CM, Ahmad Y, Shun-Shin MJ, et al. Quantification of the effect of pressure wire drift on the diagnostic performance of fractional flow reserve, instantaneous wave-free ratio, and whole-cycle Pd/Pa. Circ Cardiovasc Interv 2016;9:e002988.

[18] Lopez-Palop R, Carrillo P, Frutos A, et al. Comparison of effectiveness of high-dose intracoronary adenosine versus intravenous administration on the assessment of fractional flow reserve in patients with coronary heart disease. Am J Cardiol 2013;111:1277–83.

[19] Johnson NP, Toth GG, Lai D, et al. Prognostic value of fractional flow reserve: linking physiologic severity to clinical outcomes. J Am Coll Cardiol 2014;64:1641–54.

[20] Nakayama M, Chikamori T, Uchiyama T, et al. Effects of caffeine on fractional flow reserve values measured using intravenous adenosine triphosphate. Cardiovasc Interv Ther 2018;33:116–24.

[21] Mahmood A, Papayannis AC, Brilakis ES. Pro-arrhythmic effects of intracoronary adenosine administration. Hellenic J Cardiol 2011;52:352–3.

[22] Kikuta Y, Cook CM, Sharp ASP, et al. Pre-angioplasty instantaneous wave-free ratio pullback predicts hemodynamic outcome in humans with coronary artery disease: primary results of the international multicenter iFR GRADIENT registry. JACC Cardiovasc Interv 2018;11:757–67.

[23] Nijjer SS, Sen S, Petraco R, et al. Pre-angioplasty instantaneous wave-free ratio pullback provides virtual intervention and predicts hemodynamic outcome for serial lesions and diffuse coronary artery disease. JACC Cardiovasc Interv 2014;7:1386–96.

[24] Frimerman A, Abu-Fane R, Levi Y, et al. Novel method for real-time coregistration of coronary physiology and angiography by iFR. JACC Cardiovasc Interv 2019;12:692–4.

[25] Abdel-Karim AR, Da Silva M, Lichtenwalter C, et al. Prevalence and outcomes of intermediate saphenous vein graft lesions: findings from the stenting of saphenous vein grafts randomized-controlled trial. Int J Cardiol 2013;168:2468–73.

[26] Rodes-Cabau J, Bertrand OF, Larose E, et al. Comparison of plaque sealing with paclitaxel-eluting stents versus medical therapy for the treatment of moderate nonsignificant saphenous vein graft lesions. The moderate VEin graft LEsion stenting with the Taxus stent and Intravascular ultrasound (VELETI) pilot trial. Circulation 2009;120:1978–86.

[27] Rodes-Cabau J, Bertrand OF, Larose E, et al. Five-year follow-up of the plaque sealing with paclitaxel-eluting stents vs medical therapy for the treatment of intermediate nonobstructive saphenous vein graft lesions (VELETI) trial. Can J Cardiol 2014;30:138–45.

[28] Rodes-Cabau J, Jolly SS, Cairns J, et al. Sealing intermediate nonobstructive coronary saphenous vein graft lesions with drug-eluting stents as a new approach to reducing cardiac events: a randomized controlled trial. Circ Cardiovasc Interv 2016;9:e004336.

[29] Cuculi F, De Maria GL, Meier P, et al. Impact of microvascular obstruction on the assessment of coronary flow reserve, index of microcirculatory resistance, and fractional flow reserve after ST-segment elevation myocardial infarction. J Am Coll Cardiol 2014;64:1894–904.

[30] Ntalianis A, Sels J-W, Davidavicius G, et al. Fractional flow reserve for the assessment of nonculprit coronary artery stenoses in patients with acute myocardial infarction. J Am Coll Cardiol Intv 2010;3:1274–81.

[31] Engstrom T, Kelbaek H, Helqvist S, et al. Complete revascularisation versus treatment of the culprit lesion only in patients with ST-segment elevation myocardial infarction and multivessel disease (DANAMI-3-PRIMULTI): an open-label, randomised controlled trial. Lancet 2015;386:665–71.

[32] Sachdeva R, Agrawal M, Flynn SE, Werner GS, Uretsky BF. Reversal of ischemia of donor artery myocardium after recanalization of a chronic total occlusion. Catheter Cardiovasc Interv 2013;82:E453–8.

[33] Ladwiniec A, Cunnington MS, Rossington J, et al. Collateral donor artery physiology and the influence of a chronic total occlusion on fractional flow reserve. Circ Cardiovasc Interv 2015;8.

[34] Cook CM, Jeremias A, Petraco R, et al. Fractional flow reserve/instantaneous wave-free ratio discordance in angiographically intermediate coronary stenoses: an analysis using doppler-derived coronary flow measurements. JACC Cardiovasc Interv 2017;10:2514–24.

[35] Chacko Y, Fearon WF. Should we just go with the flow? JACC Cardiovasc Interv 2017;10:2525–7.

Chapter 13

Coronary intravascular imaging

13.1 When to do coronary intravascular imaging?

Coronary intravascular imaging can be performed before, during, and after PCI to determine the need for coronary revascularization, and help plan and optimize the result of PCI, as described below. Intravascular ultrasound (IVUS) and optical coherence tomography (OCT) are the currently available modalities for coronary intravascular imaging.

Although some operators have advocated imaging of all coronary lesions undergoing PCI [1], the benefit of intravascular imaging appears greater in more complex lesions, such as [2]:

1. Left main lesions.
2. Long lesions [3].
3. Chronic total occlusions [4].
4. Stent failure (in-stent restenosis or stent thrombosis).
5. Severely calcified lesions.
6. Complex bifurcations.
7. Acute coronary syndrome lesions.
8. To minimize contrast administration (with IVUS) [5].

13.1.1 Before PCI

Is revascularization needed?

1. *Determine significance of left main coronary lesions*

   Left main lesions with minimum lumen area (MLA) $> 6.0\ mm^2$ by IVUS or $>5.4\ mm^2$ by OCT do not require revascularization [6,7]. However, assessment of ostial left main lesions can be challenging with OCT and there are no outcomes data with OCT of the left main. Moreover, the left main minimal lumen area cut off for determining that a lesion is functionally significant differs between various populations, hence caution is needed when interpreting those measurements.

   In non-left main lesions, intravascular imaging should not be used for determining their hemodynamic severity and need for revascularization.
2. *Determine the culprit lesion in patients with acute coronary syndromes (ACS)*

   The presence of plaque ulceration, erosion, thrombus, or a calcified nodule can help determine the presence and location of culprit lesion(s) in ACS patients. OCT is preferred over IVUS due to higher resolution (Fig. 13.1) [8].
3. *Evaluate angiographically ambiguous lesions, such as suspected dissection, thrombus, and calcified nodule.*
4. *Determine cause of stent failure (in-stent restenosis and stent thrombosis).*
5. *Predict risk of distal embolization: lesions with large lipid core plaque* (Section 25.2.3.2) *or large thrombus* (Section 20.9.6) *are at increased risk.*

13.1.2 During PCI

1. Determine the need for lesion preparation before stenting.

   Atherectomy or intracoronary lithotripsy may be required in heavily calcified lesions (Sections 13.3.6.2.3 and 19.13).
2. Assist with chronic total occlusion crossing (IVUS).
3. Select stent landing zone free of atherosclerosis to avoid geographic miss.
4. Choose balloon and stent diameter and length by measuring the lesion length and reference vessel diameter.

Manual of Percutaneous Coronary Interventions. DOI: https://doi.org/10.1016/B978-0-12-819367-9.00013-5

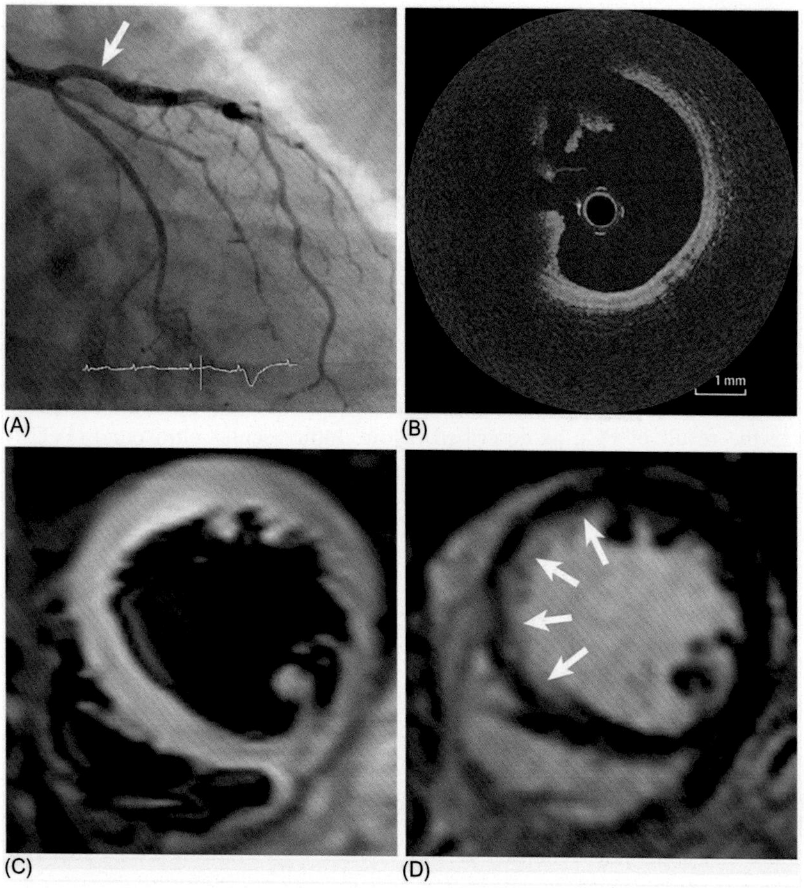

FIGURE 13.1 Use of OCT to determine the presence and location of the culprit lesion in a patient with STEMI without an angiographically obvious culprit lesion. Coronary angiography demonstrated a mild-to-moderate lesion in the proximal left anterior descending artery (*arrow*, panel A) in a 44-year-old woman with ST-segment elevation myocardial infarction. OCT showed a ruptured fibrous cap with red thrombus in the proximal LAD (panel B). Cardiac magnetic resonance revealed T2 signal hyperintensity (panel C) along with subendocardial late gadolinium enhancement (*arrows*, panel D) in the antero-septal wall. *Reproduced with permission from Opolski MP, Spiewak M, Marczak M, et al. Mechanisms of myocardial infarction in patients with nonobstructive coronary artery disease: results from the optical coherence tomography study. JACC Cardiovasc Imaging 2019;12: 2210–21.*

5. Evaluate result after stenting, as follows:
   - Stent expansion.
   - Stent strut apposition.
   - Tissue protrusion.
   - Edge dissection.
   - Geographic miss (residual disease).
6. Use of IVUS for evaluating the result of PCI can reduce the volume of contrast required and therefore the risk for contrast-induced acute kidney injury (Section 28.3), especially in high-risk patients [9,10].

13.1.3 After PCI

1. Determine the mechanism of stent failure (stent thrombosis and in-stent restenosis).

13.2 Imaging modality selection

13.2.1 Goals

Choose the optimal imaging modality to achieve the desired goal.

13.2.2 How?

There are two major intravascular imaging modalities, IVUS and optical coherence tomography (OCT), with important differences as outlined in Table 13.1. OCT has 10-fold higher resolution (10–15 μm compared with 100 μm for IVUS),

TABLE 13.1 Comparison of IVUS and OCT.

	IVUS	OCT
Requires contrast administration	No	Yes
Image resolution	+/++	+++
Tissue penetration	+++	+
Speed of pullback	+	+++
Ease of image interpretation	++	+++
Need for predilation of severe lesions	+	++
Imaging thrombus	+	+++
Imaging calcium	++	+++
Imaging aorto-ostial lesions	+++	+
Plaque morphology	++	+++
Imaging stents	++	+++

TABLE 13.2 Intravascular imaging modality selection depending on the goal of imaging.

GOAL	Preferred coronary imaging modality	Comment
Baseline assessment		
Determine significance of left main lesions	IVUS	OCT can be used for non aorto-ostial lesions
Determine culprit lesion for ACS	OCT	
Evaluate suspected dissection or thrombus	OCT	OCT has higher resolution than IVUS but may cause extension of the dissection due to contrast injection
Determine cause of stent failure	IVUS or OCT	
Assess presence, extent, and composition of coronary plaque	IVUS or OCT—IVUS is preferred for assessing plaque volume	
During PCI		
Determine need for lesion preparation before stenting	IVUS or OCT	
Facilitate CTO crossing	IVUS	
Select stent landing zone	IVUS or OCT	
Choose balloon and stent diameter and length	IVUS or OCT	
Evaluate stent expansion	IVUS or OCT	
Evaluate stent apposition	OCT	
Evaluate stent edge dissections	OCT	
Minimize contrast utilization	IVUS	

but requires blood clearing which is usually achieved by contrast injection, and has low penetration. OCT is superior to IVUS in detecting thrombus, dissection, and assessing plaque morphology and stent strut coverage and apposition.

Imaging modality selection depends on:

1. The goal of imaging (Table 13.2).
2. Patient and lesion characteristics (such as lesion location and chronic or acute kidney disease).

3. Local availability and expertise in acquiring and interpreting intravascular images.
4. Chronic kidney disease: IVUS is preferred in patients with chronic kidney disease (as it does not require contrast administration) to minimize the risk of acute kidney injury (Section 28.3). Dextran can be used instead of contrast for OCT imaging but has received limited study.

13.3 OCT step-by-step

Starting point: The guide catheter is well engaged in the target coronary vessel and a guidewire has been advanced across the coronary area of interest. Moreover, unless contraindicated, intracoronary nitroglycerin has been administered.

Usually a ≥6 French guide catheter is needed for adequate contrast injection to create a blood-free field with manual injection. Injection of contrast through a 5 French guide catheter would require a power injector to adequately fill the lumen and allow high-quality OCT acquisition.

13.3.1 Step 1: Prepare OCT catheter for use

13.3.1.1 Goal

Prepare the OCT catheter for use.

13.3.1.2 How?

1. The OCT catheter is removed from the sterile packaging.
2. The OCT catheter is flushed with 100% contrast.
3. The OCT pullback device is inserted in a dedicated sterile bag.
4. The OCT catheter is connected to the OCT pullback device.

13.3.1.3 What can go wrong?

13.3.1.3.1 Failure of OCT catheter to image

Causes:

- Defective catheter.
- Kinking of the catheter shaft.
- Damage of the pullback device (for example by improper disconnection of an OCT catheter from the pullback device).
- Damage of the imaging lens during wire insertion on the short monorail segment.

Prevention:

- Gentle and careful equipment handling.

Treatment:

- Catheter malfunction: replace with new catheter.
- Pullback device malfunction: call for repair.

13.3.2 Step 2: Advance OCT catheter past target coronary segment

13.3.2.1 Goal

Advance the OCT catheter distal to the target coronary vessel segment.

13.3.2.2 How?

After distal wiring of the target vessel using a 0.014 in. coronary guidewire, the OCT catheter is advanced distal to the coronary area of interest. The OCT catheter has 3 markers: the most distal is at the catheter tip, the middle is at the wire exit port, and the most proximal is 50 mm from the imaging lens. To ensure that the lesion of interest is imaged, the middle marker should be positioned at least 5–7 mm distal to the target site for imaging.

The two proximal imaging markers can help determine the optimal mode for image acquisition (high-resolution vs survey mode, see step 5 below), as they demarcate the approximate length of the high-resolution imaging run. The two proximal markers are integrated within the imaging core, hence they track with the movement of the imaging lens during pullback.

13.3.2.3 Challenges

13.3.2.3.1 Failure to advance OCT catheter through the lesion

Causes:

- Severely stenotic lesion.
- Severe calcification.
- Severe tortuosity.
- Poor guide catheter support.

Prevention:

- Severely stenosed or calcified lesions may need balloon predilation, to allow: (1) advancement of the OCT catheter and (2) contrast penetration through the lesion for visualization.
- Obtaining good guide catheter support, as outlined in Section 9.5.8.
- Consider using a supportive coronary guidewire in severely tortuous anatomy.

Treatment:

- Remove OCT catheter and predilate the lesion with a small balloon.
- Increase guide catheter support.

13.3.2.4 What can go wrong?

13.3.2.4.1 Kinking of the OCT catheter

Causes:

- Forceful OCT catheter advancement.

Prevention:

- Lesion preparation for highly stenotic lesions.
- Avoid forceful OCT catheter advancement.

Treatment:

- Replace catheter with a new one if malfunction persists after straightening the OCT catheter.

13.3.3 Step 3: Flush OCT catheter

13.3.3.1 Goal

Remove any blood from within the OCT catheter to enable accurate vessel visualization.

13.3.3.2 How?

The OCT catheter is flushed with 100% contrast followed by confirmation of clearance of blood (Fig. 13.2) and air from within the catheter. This step is important to prevent shadow (from air) or attenuation (from blood) artifacts during OCT image acquisition.

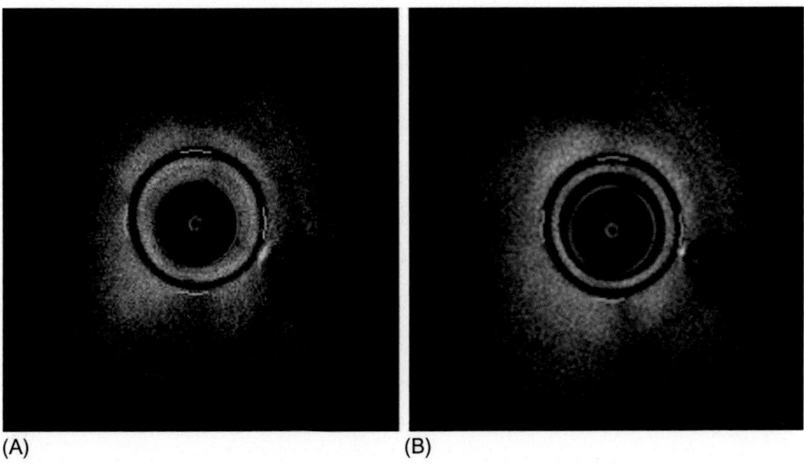

FIGURE 13.2 OCT cross section with blood within the catheter (panel A). OCT cross section following flushing with contrast to purge the blood from the catheter (panel B). *Courtesy of Drs Evan and Richard Shlofmitz.*

13.3.4 Step 4: Calibration

13.3.4.1 Goal

Calibrate the OCT catheter to allow automated measurements.

13.3.4.2 How?

Push the calibration button. To ensure accurate calibration, do not calibrate while purging the catheter.

13.3.5 Step 5: Perform OCT imaging run

13.3.5.1 Goal

Obtain high-quality OCT images of the target coronary segment.

13.3.5.2 How?

1. The desired pullback length is selected (54 mm for the high-resolution mode vs 75 mm for the survey mode) for the Illumien OPTIS system (Abbott Vascular) (Table 13.3). The OCT survey mode provides a rapid pullback of 75 mm over 2.1 seconds (pullback speed 36 mm/sec) (i.e. 5 frames/mm) and is the standard mode for pre-PCI imaging. The high-resolution mode images a 54 mm segment at 10 frames per mm over 3 seconds, providing twice the frame density as the survey mode and is the optimal acquisition method when high detail is desired.
2. Good guide catheter engagement is confirmed with a small test contrast injection.
3. The "enable" button is pressed.
4. Contrast injection is performed, usually under cineangiography to allow simultaneous angiographic visualization of the coronary segment being examined. Automated injection facilitates filling of the coronary vessel and is usually done as follows:
   - Left coronary artery: 4 mL/sec × 4 seconds.
   - Right coronary artery: 3 mL/sec × 3 seconds.
5. Pullback is automatically triggered when the system detects a blood-free lumen after contrast injection (manual activation of the pullback is also feasible).

13.3.5.3 Challenges

13.3.5.3.1 Unable to initiate pullback due to poor contrast penetration

Causes:

- Poor contrast penetration through the imaged coronary segment.
- Suboptimal contrast injection.
- Guide disengagement.

Prevention:

- Verify coaxial guide catheter engagement prior to injection.
- Use automated contrast injector device or thin syringe for contrast injection.
- Verify that contrast goes through lesion.

TABLE 13.3 Comparison of the survey mode and the high-resolution mode for OCT image acquisition.

	Survey mode	High-resolution mode
Image acquisition speed	180 frames/sec	180 frames/sec
Pullback length	75 mm	54 mm
Pullback speed	36 mm/sec	18 mm/sec
Pullback duration	2.1 sec	3.0 sec
Total frames recorded	375	540
Frame density	5 frames/mm	10 frames/mm

Treatment:
- Engage guide catheter well and/or use guide catheter extension.
- Powerful contrast injection (infusion rate of at least 3 mL/s for the right coronary artery and 4 mL/s for the left main).
- If the lesion remains highly stenotic: repeat balloon dilation prior to OCT run.
- Utilize manual trigger for pullback.
- Increase the rate or duration of automated injection; consider nonlinear rise if the guide position is stable.

13.3.5.3.2 Artifacts

Causes:
- Suboptimal contrast clearance from the vessel.
- Suboptimal OCT catheter preparation.
- One or more buddy wires.

Prevention:
- Fill vessel well with contrast ensuring good guide catheter engagement; filling of the vessel can be facilitated by using a power injector.
- Remove buddy wire(s), if applicable.

Treatment:
- Repeat OCT image acquisition after correcting the underlying problem.

13.3.5.4 What can go wrong?

13.3.5.4.1 Distal migration of OCT catheter

Causes:
- Forceful contrast injection without holding the OCT catheter (and coronary guidewire) in place.

Prevention:
- Fix the OCT catheter and coronary guidewire during contrast injection.

Treatment:
- Assess for distal vessel injury or perforation and treat accordingly.

13.3.6 Step 6: Image interpretation

13.3.6.1 Goal

Accurately analyze the OCT images obtained.

13.3.6.2 How?

1. Adjust visualization depth to include entire vessel in the image.
2. Review images as follows:

13.3.6.2.1 Identify OCT artifacts (Fig. 13.3)

13.3.6.2.2 Baseline lesion assessment

13.3.6.2.2.1 OCT Image Assessment

OCT can detect abnormalities in both the lumen and the wall of the coronary artery. The signal intensity and attenuation characteristics helps characterize pathological morphologies identified on OCT. A simplified algorithm for analyzing the OCT images is shown in Fig. 13.4 [11] and common morphologies identified by OCT are shown in Fig. 13.5 [11].

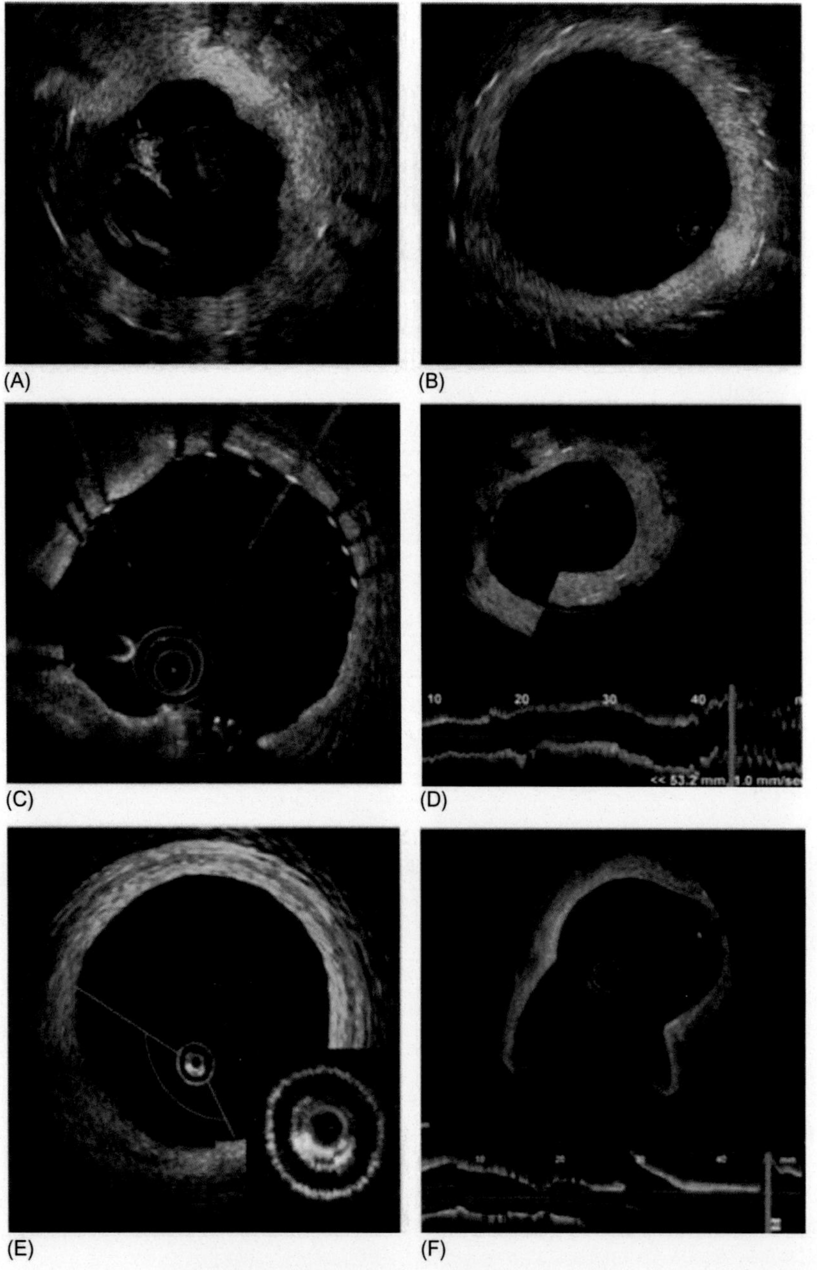

FIGURE 13.3 Most frequent OCT artifacts. (A) incomplete blood displacement, resulting in light attenuation. (B) Eccentric image wire can distort stent reflection orientation: the struts align toward the imaging wire "sunflower effect" and are elongated "merry-go-round." (C) Saturation artifact, some scan lines have a streaked appearance. (D) Sew-up artifact: result of rapid wire or vessel movement along 1 frame formation, resulting in misalignment of the image. (E) Air bubbles, inside the catheter, produce an attenuated image along the corresponding arc. Detail reveals the bubbles, bright structures, between 5 and 9 o'clock. (F) Fold over artifact (Fourier-domain optical coherence tomography system), the longitudinal view demonstrates that the cross section is located at the level of a side branch (*blue line*). *Reproduced with permission from Bezerra HG, Costa MA, Guagliumi G, Rollins AM, Simon DI. Intracoronary optical coherence tomography: a comprehensive review clinical and research applications. JACC Cardiovasc Interv 2009;2:1035–46 [19]. Copyright Elsevier.*

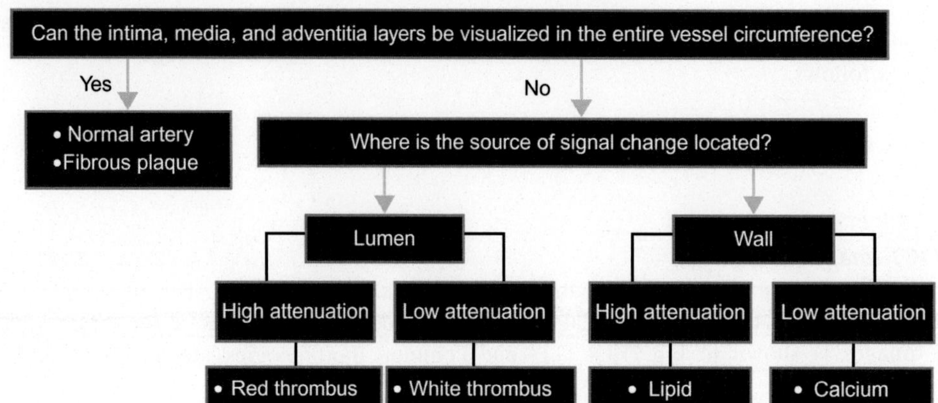

FIGURE 13.4 Simplified algorithm for interpretation of optical coherence tomographic images in native coronary arteries. This algorithm is useful in describing native coronary lesions, with the understanding that many lesions have a mixed appearance and contain more than one pathological morphology mentioned in the schematic. *Reproduced with permission from Ali ZA, Karimi Galougahi K, Maehara A, et al. Intracoronary optical coherence tomography 2018: current status and future directions. JACC Cardiovasc Interv 2017;10:2473–87. Copyright Elsevier.*

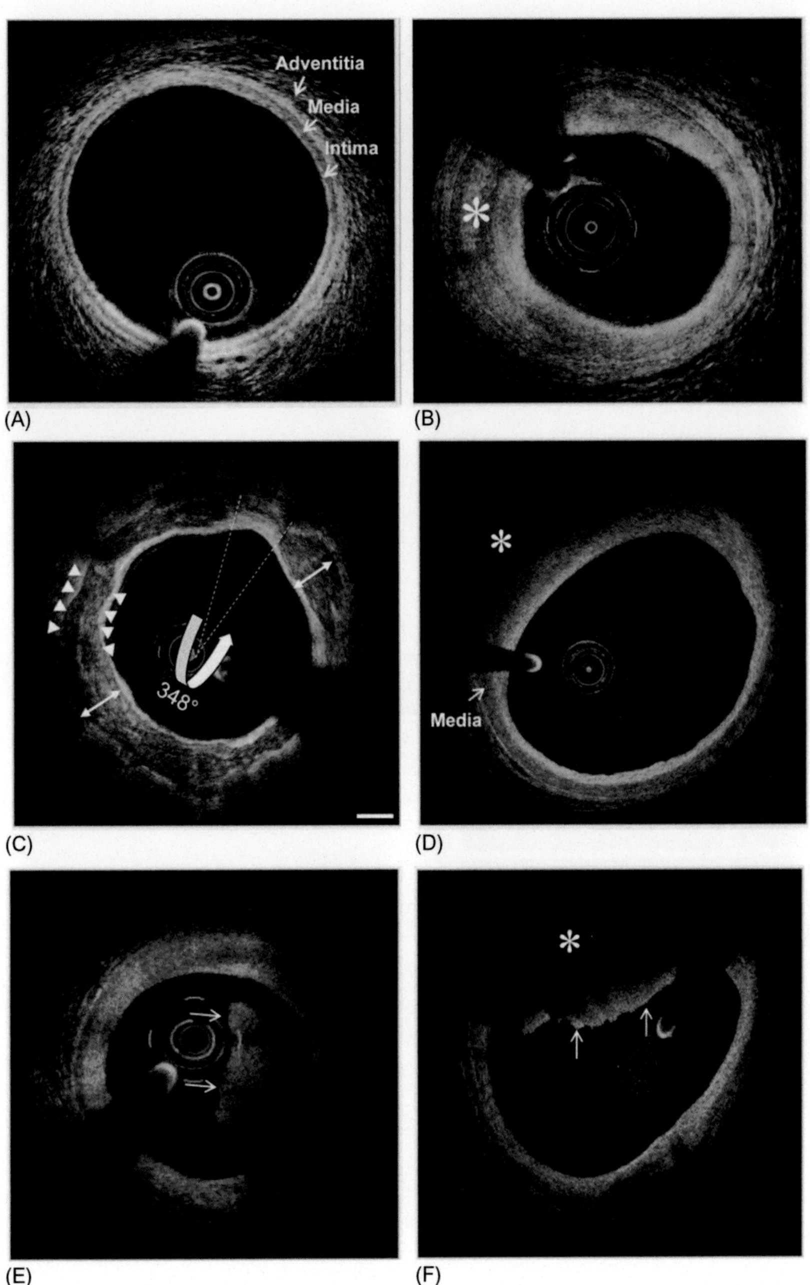

FIGURE 13.5 Common morphologies detected on intracoronary optical coherence tomography. (A) Normal artery: the bright-dark-bright three-layered appearance corresponding to intima, media, and adventitia is visualized in the entirety of the vessel circumference. (B) Fibrous plaque: note homogeneous, signal-rich regions (*asterisk*). (C) Calcific plaque: characterized as a signal-poor region with sharply delineated borders (*arrowheads*). Analysis of calcification depth (*double arrows*) and arc (measured at 348°) is feasible in most cases with optical coherence tomography. (D) Lipid-rich plaque: characterized as a signal-poor region with poorly defined borders (*asterisk*). Note that because of light attenuation, the media is not visible beyond the lipid content of the plaque. (E) White thrombus: mass floating within the lumen (*arrows*) that is platelet and white blood cell rich with minimal optical coherence tomographic signal attenuation. (F) Red thrombus: mass floating/attached to the luminal surface (*arrows*) that is rich in red blood cells and therefore highly attenuates the optical coherence tomographic signal, casting a shadow on the vessel wall behind the mass (*asterisk*). *Reproduced with permission from Ali ZA, Karimi Galougahi K, Maehara A, et al. Intracoronary optical coherence tomography 2018: current status and future directions. JACC Cardiovasc Interv 2017;10:2473–87.*

13.3.6.2.2.2 Should PCI be done?

OCT can be used to: (1) determine whether PCI (or coronary revascularization in general) should be done; and (2) how to perform PCI.

OCT can help determine the need for PCI as follows:

1. *Hemodynamic significance*
   Minimum lumen area should not be used for determining the hemodynamic significance of non-left main lesions. For left main lesions a minimum lumen area cutoff of 5.4 mm^2 has been proposed for detecting hemodynamically significant lesions, but has not been prospectively validated [12].
2. *Culprit lesion for ACS*
   The *presence* of intracoronary thrombus, plaque erosion or ulceration, or calcified nodule can help identify the culprit lesion in patients with acute coronary syndromes in whom the culprit lesion is unclear.

3. *Spontaneous coronary artery dissection*
   OCT can help with the diagnosis of spontaneous coronary artery dissection (SCAD) (Fig. 13.6) [13], but concerns exist over whether wiring and contrast injection may lead to hydraulic extension of the dissection flap and acute vessel closure.

13.3.6.2.2.3 How should PCI be done?

Once a decision is made to perform PCI, OCT can help plan it by focusing on the following three aspects:

1. Need for atherectomy.
2. Stent length.
3. Stent diameter.
   a. *Determine need for atherectomy*
      Using OCT, calcium appears as low signal (dark) and low attenuation (allows deep penetration) areas with sharp borders (Fig. 13.5).
      Severe calcification can hinder stent delivery and stent expansion. The following three OCT findings have been associated with stent under-expansion: (1) maximum calcium angle >180°; (2) maximum calcium thickness >0.5 mm; and (3) calcium length >5.0 mm (Figs. 13.7 and 13.8) [14]. Presence of all three criteria was

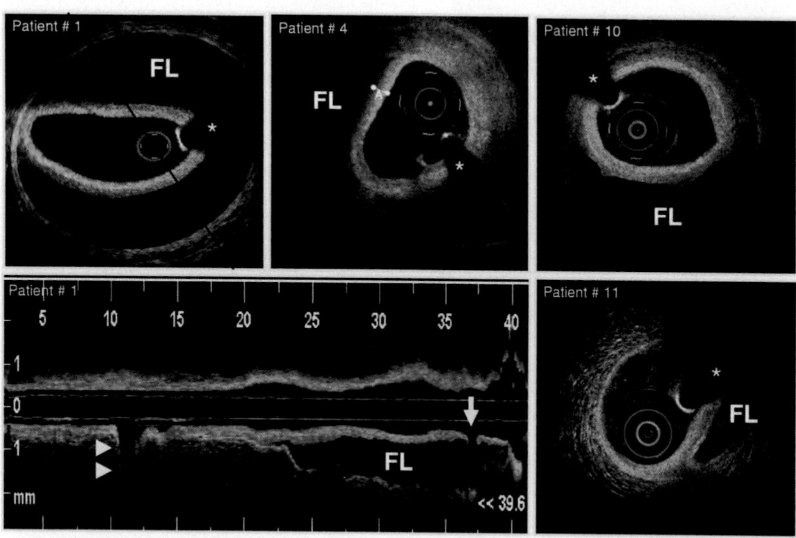

FIGURE 13.6 OCT examples of spontaneous coronary dissection. *FL*, false lumen. *Reproduced with permission from Alfonso F, Paulo M, Gonzalo N, et al. Diagnosis of spontaneous coronary artery dissection by optical coherence tomography. J Am Coll Cardiol 2012;59:1073–79. Copyright Elsevier.*

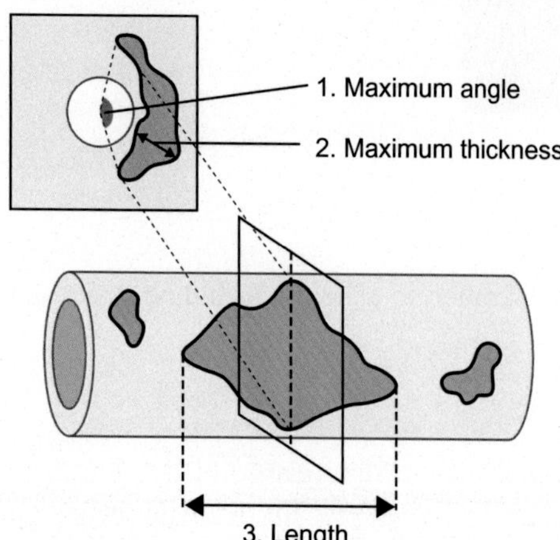

OCT-based calcium score			
1. Maximum calcium angle (°)	≤180°	→	0 point
	>180°	→	2 points
2. Maximum calcium thickness (mm)	≤0.5 mm	→	0 point
	>0.5 mm	→	1 point
3. Calcium length (mm)	≤5.0 mm	→	0 point
	>5.0 mm	→	1 point
Total score	**0 – 4 points**		

FIGURE 13.7 Optical coherence tomography-based calcium scoring system. OCT-based calcium score was composed of three parameters that were derived from prestent OCT: (1) maximum calcium angle; (2) maximum thickness; and (3) calcium length. One or two points were assigned for each parameter, and the total score (0–4) calculated. *OCT*, optical coherence tomography. *Reprinted from Fujino A, Mintz GS, Matsumura M, et al. A new optical coherence tomography-based calcium scoring system to predict stent underexpansion. EuroIntervention 2018;13:e2182–9. Copyright (2018), with permission from Europa Digital & Publishing.*

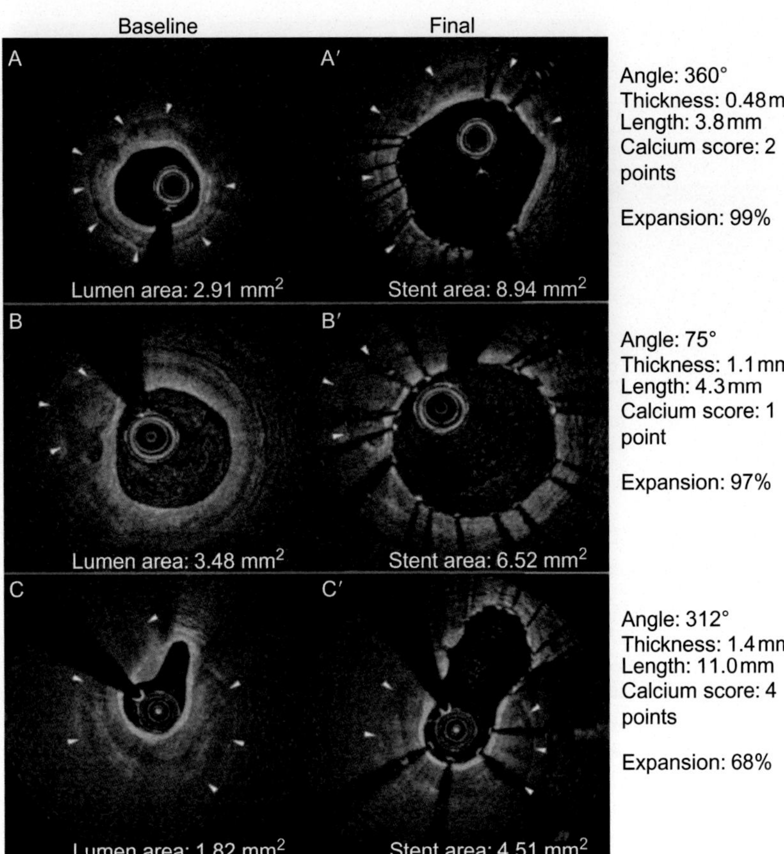

FIGURE 13.8 Association between calcium score and stent expansion. Case A. Calcium score of 2 (large-angled and thin calcium) and excellent stent expansion. Case B. Calcium score of 1 (small-angled and thick calcium) and excellent stent expansion. Case C. Calcium score of 4 with stent underexpansion. *Reprinted from Fujino A, Mintz GS, Matsumura M, et al. A new optical coherence tomography-based calcium scoring system to predict stent underexpansion. EuroIntervention 2018;13:e2182–9. Copyright (2018), with permission from Europa Digital & Publishing.*

associated with 22% likelihood of stent underexpansion [14] and favors use of plaque modification strategies, such as atherectomy (Chapter 19: Calcification) or intravascular lithotripsy. Deep calcified lesions may respond to balloon angioplasty, whereas superficial calcified lesions are more likely to require ablative techniques.

b. *Identify landing zone and select stent length*
OCT assessment can help identify the proximal and distal landing zone and determine the optimal stent length (Fig. 13.9). The landing zones should have <50% plaque burden, and not contain lipid-rich plaque, especially thin-cap fibroatheromas (TCFA), as these are prone to edge dissections. However, in the presence of heavy plaque burden assessment of the external elastic membrane (EEM) can be challenging by OCT due to limited penetration in contrast to IVUS, especially through lipid plaques.
OCT coregistration can help with accurate stent positioning in various anatomic subgroups including bifurcations [15].
c. *Select stent diameter*
The stent diameter should match the distal reference diameter, that can be assessed by either measuring the distal lumen (recommended by the authors) [2] or the external elastic membrane (EEM) to EEM distance (Fig. 13.9) [11]. Stents should NOT be sized based on the proximal reference diameter as this may lead to distal vessel dissection (because coronary vessels taper). If the proximal reference diameter is larger than the distal reference diameter, the mid and proximal portions of the stent are subsequently postdilated with a larger diameter, but shorter length, balloon. If stent sizing is based on lumen diameter, the stent diameter is up-rounded by 0.25 mm. If stent sizing is based on EEM it is down-rounded by 0.25 mm. EEM-based sizing is not always possible due to poor EEM visualization. EEM-based sizing was possible in 77% of cases in the ILUMIEN III study [16].

13.3.6.2.3 Post stent assessment

The following four key parameters are assessed after stenting (Fig. 13.10) [2]:

1. Stent expansion.
2. Stent strut malapposition.

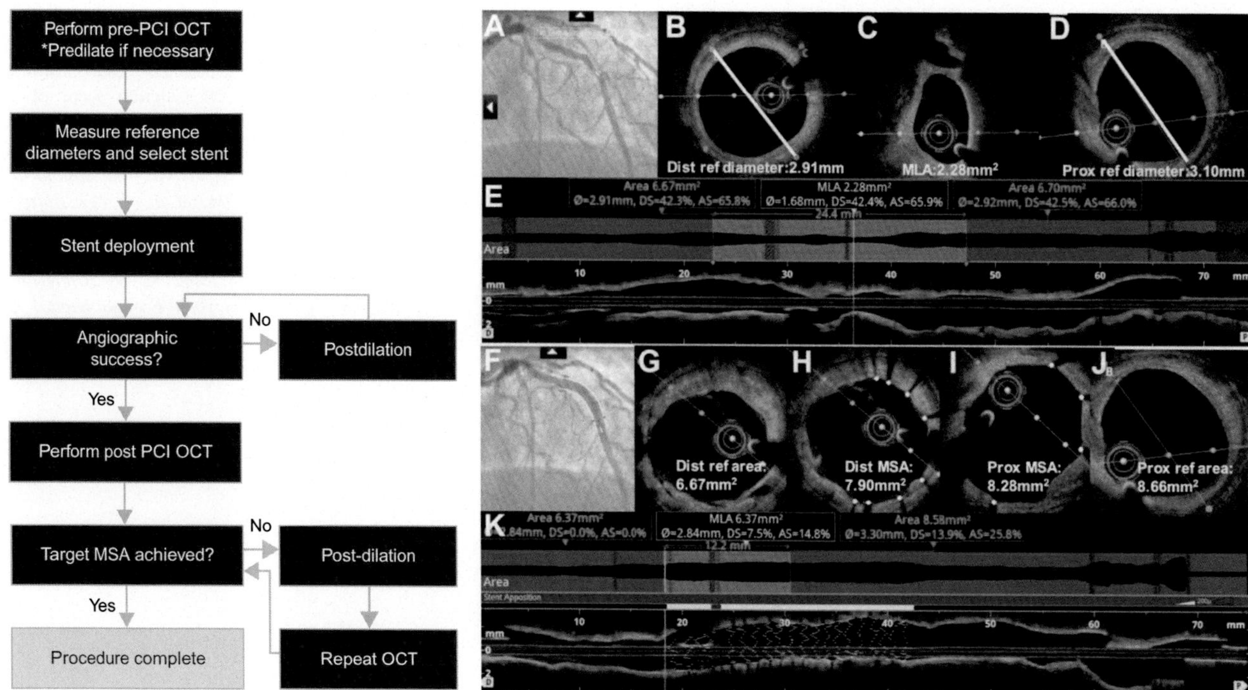

FIGURE 13.9 Example of optical coherence tomography–guided percutaneous coronary intervention. Coregistration of angiography and optical coherence tomography (OCT) (A), including longitudinal (E) and cross-sectional optical coherence tomographic frames (B–D), along with longitudinal luminal automated measures that provides minimal, mean, and reference diameter measurements as well as area and diameter stenosis along the pull-back. An example of OCT-guided PCI in a patient with angina is illustrated on the coregistered angiographic (A) and optical coherence tomographic (B–E) images. On the basis of the external elastic lamina–based measurements of the distal (B, 2.91 mm) and proximal (D, 3.10 mm) reference segments, and the distance on optical coherence tomographic automation (24.4 mm) (E), a 3.0 × 24 mm drug-eluting stent was directly deployed without predilatation because of the presence of fibroatheromatous plaque without calcification. On post-PCI OCT (F–K), the minimum stent area (MSA) in both the distal (G, H) and proximal (I, J) halves of the stent met protocol criteria for adequate expansion ($7.90 > 90\% \times 6.67$ mm^2 distally and $8.28 > 90\% \times 8.66$ mm^2 proximally); thus, postdilatation was deemed unnecessary. *Reproduced with permission from Ali ZA, Karimi Galougahi K, Maehara A, et al. Intracoronary optical coherence tomography 2018: current status and future directions. JACC Cardiovasc Interv 2017;10:2473–87. Copyright Elsevier.*

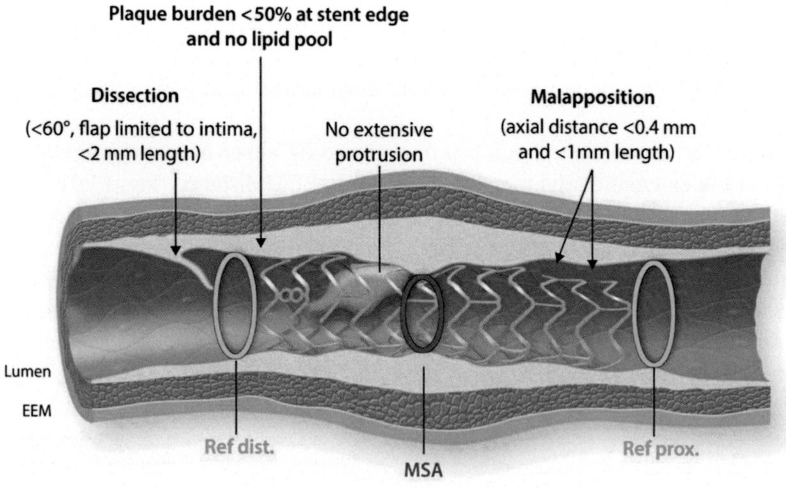

FIGURE 13.10 Summary of post-PCI optimization targets. The most relevant targets to be achieved following stent implantation in non-left main lesions are shown. These include optimal stent expansion (absolute as well as relative to reference lumen diameter); avoidance of landing zone in plaque burden >50% or lipid rich tissue; avoidance of large malapposition regions, irregular tissue protrusion, and dissections. Thresholds provided reflect the consensus of the European Association of Percutaneous Cardiovascular Interventions writing group. Some are based on consistent and robust prospective data (e.g., stent expansion, landing zone) and others are less established (e.g., malapposition). *Reproduced with permission from Raber L, Mintz GS, Koskinas KC, et al. Clinical use of intracoronary imaging. Part 1: guidance and optimization of coronary interventions. An expert consensus document of the European Association of Percutaneous Cardiovascular Interventions. EuroIntervention 2018;14:656–77.*

3. Tissue protrusion.
4. Edge dissection.
   a. *Stent expansion*
   Stent expansion refers to the size of the stent as determined by minimal stent area. Stent underexpansion increases the risk of in-stent restenosis and stent thrombosis. For non-left main lesions goal minimum stent lumen area is $>4.5\ mm^2$ (vs $\geq 5.5\ mm^2$ if measured with $IVUS^2$) or >80% of the average reference lumen area. For left main lesions, goal minimum stent lumen area is $>8.0\ mm^2$ [2]. Alternatively, goal minimal stent area is >90% of the proximal reference for proximal MSA, and distal reference for distal MSA. Treatment of acute stent underexpansion should begin with high inflation pressures (>18 atm) with a noncompliant balloon sized at least equal to the stent diameter.
   b. *Stent malapposition*
   Malapposition refers to lack of contact of the stent struts with the vessel wall. Although acute malapposition (separation of at least one stent strut from the intimal layer by 200 μm or more) has not been associated with increased risk of stent failure, extensive stent malapposition (maximum malapposed distance ≥400 μm or maximum malapposed length ≥1 mm) should be corrected with balloon postdilation, usually with a semicompliant balloon inflated at low pressures.
   c. *Extensive tissue protrusion*
   Tissue protrusion has not been associated with adverse outcomes [17]. Postdilation may be performed in such cases, but is typically necessary only when the resultant effective lumen area is inadequate.
   d. *Edge dissection*
   OCT is more sensitive than IVUS in detecting edge dissections. Small, nonflow limiting dissections do not require treatment. Larger dissections (>60° arc and/or >3 mm in length), especially in the distal stent edge, may be best treated with additional stent implantation [2].

13.4 IVUS step-by-step

Starting point: guide catheter is engaged in the target coronary vessel and a guidewire has been advanced across the coronary area to be imaged.

13.4.1 Step 1: Select and prepare IVUS catheter

There are two major categories of IVUS catheters, the phased array, such as the Eagle Eye (Philips, 20 MHz) and the rotational, such as the Refinity (Philips, 45 MHz), OptiCross HD (Boston Scientific, 60 MHz), Kodama (ACIST, 60 MHz) and DualPro (InfraRedx, 35–65 MHz; this catheter also allows near-infrared detection of lipid core plaque) (Table 13.4).

The phased array IVUS does not require any preparation, in contrast to the rotational IVUS catheters that require flushing prior to insertion into the body. Moreover, in the solid state IVUS the lumen is black without blood speckle and there is the option of color Doppler for easier identification of the lumen. On the other hand rotational IVUS catheters have lower-profile and as a result they can cross severe coronary lesions more easily, and provide higher resolution images.

TABLE 13.4 Comparison of the solid state and rotational IVUS catheters.

	Phased array	Rotational
Need for flushing before insertion?	No	Yes
Blood speckle in lumen	No	Yes
Color Doppler for visualizing flow	Yes	No
Crossing severe lesions	Harder	Easier
Resolution	Lower	Higher (especially with high-frequency 60 MHz catheters)
Use for CTO crossing	Preferred as imaging occurs closer to the tip	

13.4.2 Step 2: Advance IVUS catheter past target coronary segment

13.4.2.1 Goal

Advance the IVUS catheter distal to the coronary vessel segment that needs imaging.

13.4.2.2 How?

The IVUS catheter is advanced over the guidewire distal to the coronary vessel segment that needs imaging. Before advancing through the lesion, the solid-state IVUS catheter needs to go to a large portion of the vessel or in the aorta, followed by pressing the "clear view" button.

13.4.2.3 Challenges + What can go wrong?

Same as challenges + what can go wrong described in Section 13.3.2 for OCT.

13.4.3 Step 3: Perform IVUS pullback

13.4.3.1 Goal

Obtain high-quality IVUS images of the target coronary segment.

13.4.3.2 How?

1. Imaging depth is adjusted to ensure that the entire vessel wall is included.
2. Imaging is started to ensure that there are no artifacts.
3. The IVUS catheter is pulled back while recording the images, either using manual pullback, or using automated pullback speed (usually 1–10 mm/sec depending on the catheter).

13.4.3.3 Challenges

13.4.3.3.1 Catheter "jumps"

Causes:

- Tight coronary lesions.
- Severe coronary tortuosity.
- Severe calcification.
- Imaging across stent struts, such as in bifurcation stenting.
- This only happens with the solid state IVUS catheter, as the rotational catheter pullback occurs within the catheter.

Prevention:

- Lesion pretreatment
- Slow, steady pullback.

Treatment:

- Use rotational IVUS catheter.
- Additional lesion pretreatment.
- Forward imaging (instead of imaging during pullback).

13.4.3.4 What can go wrong?

This is similar to what is described in Section 13.3.2 for OCT.

13.4.4 Step 4: IVUS Image interpretation

13.4.4.1 Goal

Accurately analyze the IVUS image obtained.

13.4.4.2 How?

1. Adjust visualization depth to include entire vessel in the image.
2. Review images as follows.

13.4.4.2.1 IVUS artifacts (Fig. 13.11)

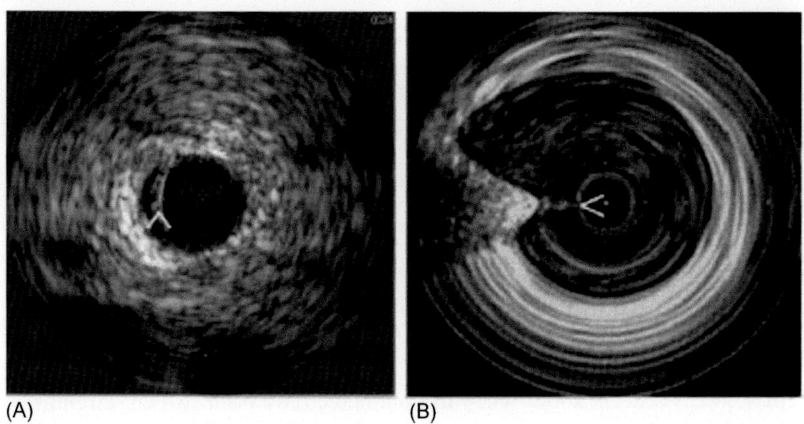

FIGURE 13.11 Common IVUS artifacts. (Panel A) Ring down artifact (arrowhead). (Panel B) Non-uniform rotational distortion (*arrowhead—can only occur with rotational IVUS systems*).

13.4.4.2.2 Baseline lesion assessment

13.4.4.2.2.1 Part A. IVUS image assessment

IVUS can detect calcification and attenuation (Fig. 13.12) [18]. Attenuation in the absence of calcification is highly suggestive of lipid core plaque.

IVUS can be used to determine: (1) whether PCI (or coronary revascularization in general) should be done; and (2) how PCI should be done.

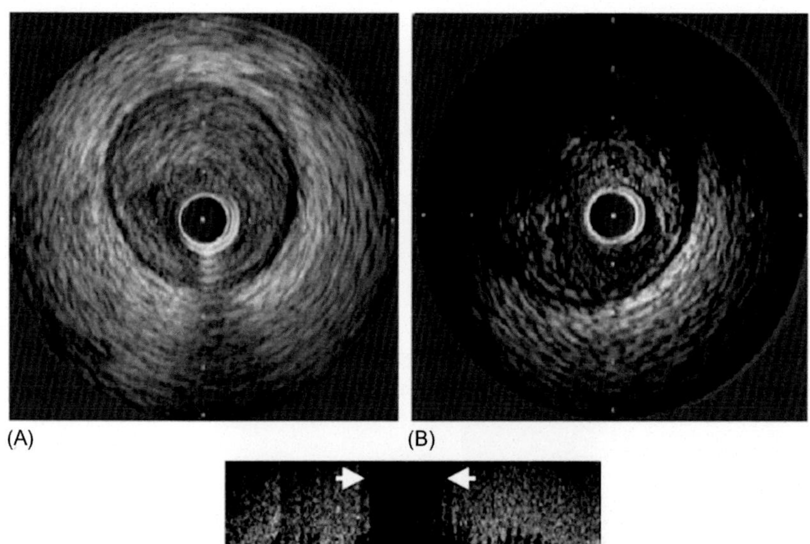

FIGURE 13.12 Cross-sectional intravascular ultrasound (IVUS) images of plaque without attenuation (panel A) and plaque with attenuation without calcification, highly suggestive of lipid-rich plaque (panel B) in cross-sectional images and longitudinal reconstruction (panel C, between the *white arrows*). Ultrasound attenuation occurred despite the absence of calcium. *Reproduced with permission from Lee SY, Mintz GS, Kim S-Y, et al. Attenuated plaque detected by intravascular ultrasound: clinical, angiographic, and morphologic features and post-percutaneous coronary intervention complications in patients with acute coronary syndromes. J Am Coll Cardiol Intv 2009;2:65–72. Copyright Elsevier.*

13.4.4.2.2.2 Part B. Should PCI be done?

1. Hemodynamic significance

Minimum lumen area should not be used for determining the hemodynamic significance of non-left main lesions. For left main lesions minimum lumen area of 6.0 mm^2 has been proposed for detecting hemodynamically significant lesions [12] whereas a cutoff of 5.4 mm^2 is recommended for OCT measurements (although the OCT cutoff does not have prospective validation) [2].

13.4.4.2.2.3 Part C. How should PCI be done?

Once a decision is made to perform PCI, IVUS can help plan the procedure by focusing on the following three aspects:

1. Need for atherectomy (when severe calcification is present).
2. Stent length.
3. Stent diameter.
   a. *Determine need for atherectomy*
   Although there are no commonly accepted IVUS criteria about when to do atherectomy prior to PCI, circumferential and superficial calcification by IVUS favors performance of atherectomy (or intravascular lithotripsy, if available).
   b. *Identify landing zone and select stent length*
   IVUS assessment can help determine the needed stent length by identifying the proximal and distal landing zone (Fig. 13.10). The landing zones should have <50% plaque burden, and not contain lipid-rich plaque. Lesion length can only be assessed using an automated pullback device and is more accurate when using rotational IVUS.
   c. *Select stent diameter*
   This is done as described in the corresponding OCT section.

13.4.4.2.3 Poststent assessment

The following four key parameters are assessed after stenting (Fig. 13.10):

1. Stent expansion
   The IVUS-defined criteria for optimal stent deployment include [1]:
   a. MLA in the stented segment ≥ 5.5 mm^2 for non-left main lesions or ≥ 8.0 mm^2 for left main lesions [2] (Fig. 13.13).
   b. Plaque burden <50% at the proximal and distal stent edge.
2. Stent strut malapposition (Fig. 13.14).

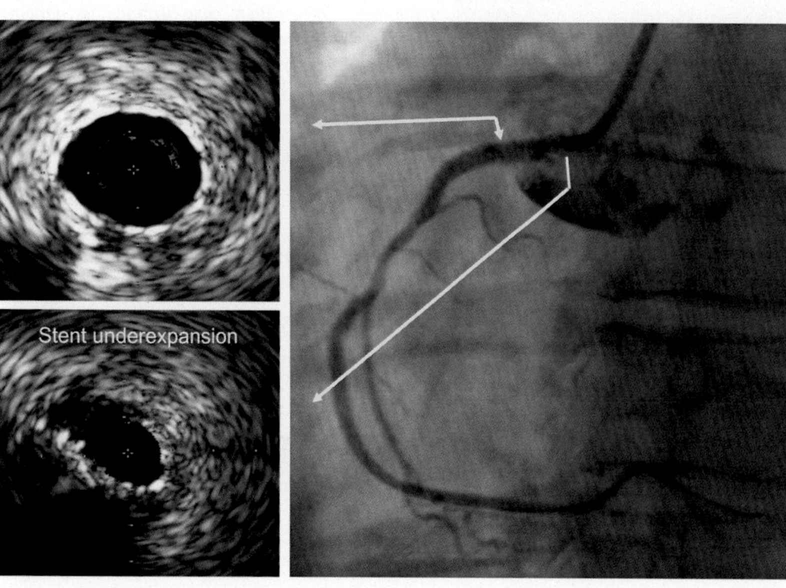

FIGURE 13.13 IVUS example of stent underexpansion in the ostium of the right coronary artery in a patient with prior chest radiation therapy.

3. Edge dissection.
   There should be no proximal or distal edge dissection that involves the media with a length of >3 mm (Fig. 13.15) [1].
4. Intramural hematoma (Fig. 13.16).

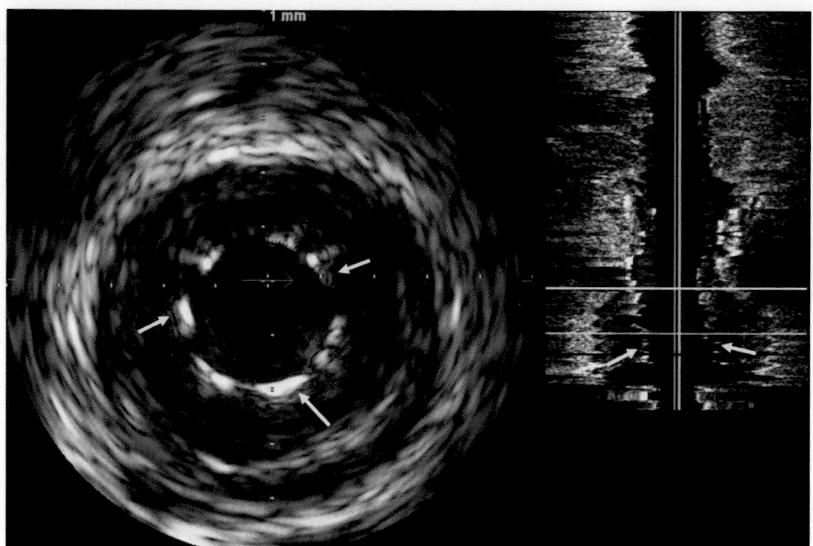

FIGURE 13.14 Example of severe stent malapposition and underexpansion (due to stent undersizing) imaged by phased array intravascular ultrasound with color Doppler.

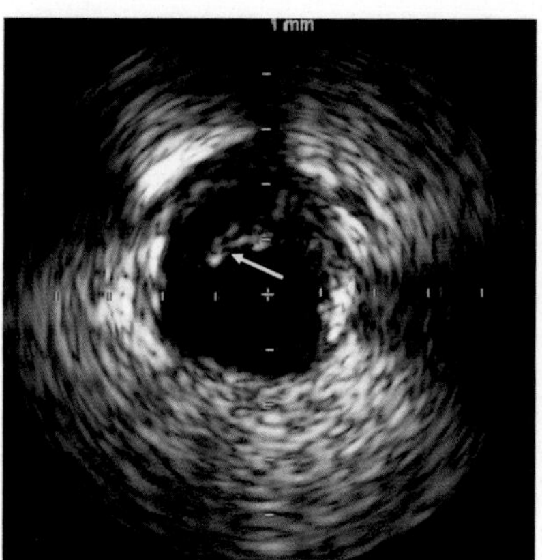

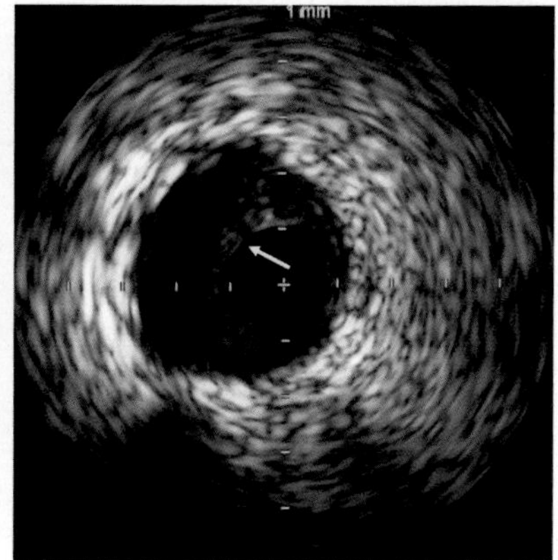

FIGURE 13.15 Example of coronary dissection imaged by phased array intravascular ultrasound with color Doppler, showing flow behind the intimal flap (*arrow*).

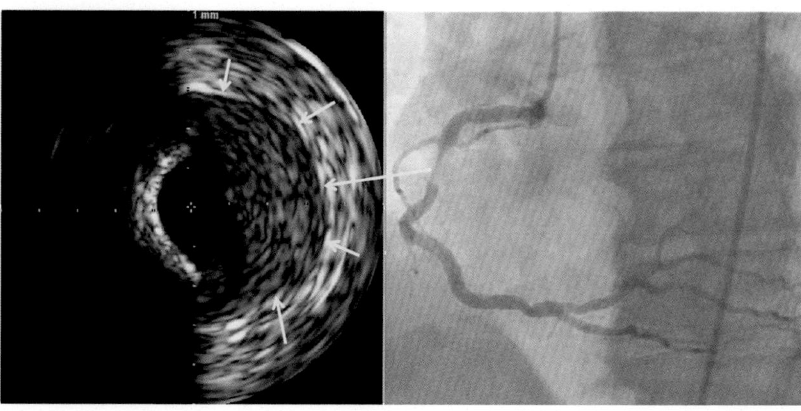

FIGURE 13.16 Intramural hematoma after stenting, imaged using phased array IVUS.

References

[1] Zhang J, Gao X, Kan J, et al. Intravascular ultrasound versus angiography-guided drug-eluting stent implantation: the ULTIMATE trial. J Am Coll Cardiol 2018;72:3126–37.

[2] Raber L, Mintz GS, Koskinas KC, et al. Clinical use of intracoronary imaging. Part 1: guidance and optimization of coronary interventions. An expert consensus document of the European Association of Percutaneous Cardiovascular Interventions. EuroIntervention 2018;14:656–77.

[3] Hong SJ, Kim BK, Shin DH, et al. Effect of intravascular ultrasound-guided vs angiography-guided everolimus-eluting stent implantation: the IVUS-XPL randomized clinical trial. Jama 2015;314:2155–63.

[4] Kim BK, Shin DH, Hong MK, et al. Clinical impact of intravascular ultrasound-guided chronic total occlusion intervention with zotarolimus-eluting versus biolimus-eluting stent implantation: randomized study. Circ Cardiovasc Interv 2015;8:e002592.

[5] Escaned J, Collet C, Ryan N, et al. Clinical outcomes of state-of-the-art percutaneous coronary revascularization in patients with de novo three vessel disease: 1-year results of the SYNTAX II study. Eur Heart J 2017;38:3124–34.

[6] Jasti V, Ivan E, Yalamanchili V, Wongpraparut N, Leesar MA. Correlations between fractional flow reserve and intravascular ultrasound in patients with an ambiguous left main coronary artery stenosis. Circulation 2004;110:2831–6.

[7] de la Torre Hernandez JM, Hernandez Hernandez F, Alfonso F, et al. Prospective application of pre-defined intravascular ultrasound criteria for assessment of intermediate left main coronary artery lesions results from the multicenter LITRO study. J Am Coll Cardiol 2011;58:351–8.

[8] Opolski MP, Spiewak M, Marczak M, et al. Mechanisms of myocardial infarction in patients with nonobstructive coronary artery disease: results from the optical coherence tomography study. JACC Cardiovasc Imaging 2019;12:2210–21.

[9] Mariani Jr. J, Guedes C, Soares P, et al. Intravascular ultrasound guidance to minimize the use of iodine contrast in percutaneous coronary intervention: the MOZART (Minimizing cOntrast utiliZation With IVUS Guidance in coRonary angioplasTy) randomized controlled trial. JACC Cardiovasc Interv 2014;7:1287–93.

[10] Azzalini L, Uretsky B, Brilakis ES, Colombo A, Carlino M. Contrast modulation in chronic total occlusion percutaneous coronary intervention. Catheter Cardiovasc Interv 2019;93:E24–9.

[11] Ali ZA, Karimi Galougahi K, Maehara A, et al. Intracoronary optical coherence tomography 2018: current status and future directions. JACC Cardiovasc Interv 2017;10:2473–87.

[12] D'Ascenzo F, Barbero U, Cerrato E, et al. Accuracy of intravascular ultrasound and optical coherence tomography in identifying functionally significant coronary stenosis according to vessel diameter: a meta-analysis of 2,581 patients and 2,807 lesions. Am Heart J 2015;169:663–73.

[13] Alfonso F, Paulo M, Gonzalo N, et al. Diagnosis of spontaneous coronary artery dissection by optical coherence tomography. J Am Coll Cardiol 2012;59:1073–9.

[14] Fujino A, Mintz GS, Matsumura M, et al. A new optical coherence tomography-based calcium scoring system to predict stent underexpansion. EuroIntervention 2018;13:e2182–9.

[15] Shlofmitz E, Sosa F, Goldberg A, et al. Bifurcation and ostial optical coherence tomography mapping (BOOM)—case description of a novel bifurcation stent technique. Cardiovasc Revasc Med 2018;19:47–9.

[16] Ali ZA, Maehara A, Genereux P, et al. Optical coherence tomography compared with intravascular ultrasound and with angiography to guide coronary stent implantation (ILUMIEN III: OPTIMIZE PCI): a randomised controlled trial. Lancet 2016;388:2618–28.

[17] Qiu F, Mintz GS, Witzenbichler B, et al. Prevalence and clinical impact of tissue protrusion after stent implantation: an ADAPT-DES intravascular ultrasound substudy. JACC Cardiovasc Interv 2016;9:1499–507.

[18] Lee SY, Mintz GS, Kim S-Y, et al. Attenuated plaque detected by intravascular ultrasound: clinical, angiographic, and morphologic features and post-percutaneous coronary intervention complications in patients with acute coronary syndromes. J Am Coll Cardiol Intv 2009;2:65–72.

[19] Bezerra HG, Costa MA, Guagliumi G, Rollins AM, Simon DI. Intracoronary optical coherence tomography: a comprehensive review clinical and research applications. JACC Cardiovasc Interv 2009;2:1035–46.

Chapter 14

Hemodynamic support

14.1 Hemodynamic support: when and what device

Maintaining adequate tissue perfusion is critical for survival. PCI may result in decreased cardiac output and tissue hypoperfusion or may be performed in the setting of decreased or absent perfusion (such as in cardiogenic shock or cardiac arrest). Although no randomized trials have demonstrated a decrease in in-hospital mortality or major complications with use of hemodynamic support devices [1,2], such devices may increase the safety of high-risk PCI and potentially improve the outcomes of cardiogenic shock or cardiac arrest. There are 6 hemodynamic support devices currently available [intraaortic balloon pump (IABP), Impella, veno-arterial extracorporeal membrane oxygenator (VA-ECMO), Tandem Heart, Protek Duo, and Impella RP]. The first section of this chapter will discuss when such devices should be used, and the second will provide step-by-step instruction for use of IABP, Impella, VA-ECMO, and Tandem Heart.

14.1.1 Hemodynamics

The key parameters that determine the need for hemodynamic support are the patient's ejection fraction and hemodynamics before, during or after PCI (Fig. 14.1).

***Before PCI*:**

The most important factor for determining the need for hemodynamic support are the baseline hemodynamics. In some settings the indication is clear. For example, in cardiac arrest, hemodynamic support with VA-ECMO is essential to prevent death (VA-ECMO supports both circulation and oxygenation). Similarly, in many patients with cardiogenic shock hemodynamic support is needed. In stable patients undergoing elective PCI, determining whether hemodynamic support is needed can be challenging and depends heavily on the anticipated risk of PCI (Section 14.1.2).

Right heart catheterization can help determine the patient's hemodynamic status and is strongly recommended in patients with low ejection fraction undergoing complex PCI or patients with unclear hemodynamic status. High pulmonary capillary wedge pressure (>18 mmHg) is suggestive of left ventricular failure and low pulmonary artery pulsatility index (PaPi <0.9) [3] is suggestive of right ventricular failure. Patients with severely elevated pulmonary capillary wedge pressure (>25 mmHg) and low mixed venous oxygen saturation ($<50\%$) are at high risk of periprocedural pulmonary edema and shock.

The Swan Ganz catheter can be left in place during the procedure to monitor hemodynamics (pulmonary artery pressure) during PCI and determine the need for escalation or de-escalation of support.

***During and after PCI*:**

Hemodynamic support may be needed during or after PCI if the patient develops hemodynamic instability (such as hypotension or acute pulmonary edema). Sometimes the need for hemodynamic support may be anticipated, whereas in other cases it may be unexpected (such as acute vessel closure or left main dissection).

14.1.2 Risk of PCI

The risk of PCI depends on location and morphology of the target lesion(s) and anticipated PCI techniques.

1. Lesion location:
   a. Unprotected left main (especially in patients with occluded or diseased right coronary artery).

Manual of Percutaneous Coronary Interventions. DOI: https://doi.org/10.1016/B978-0-12-819367-9.00014-7

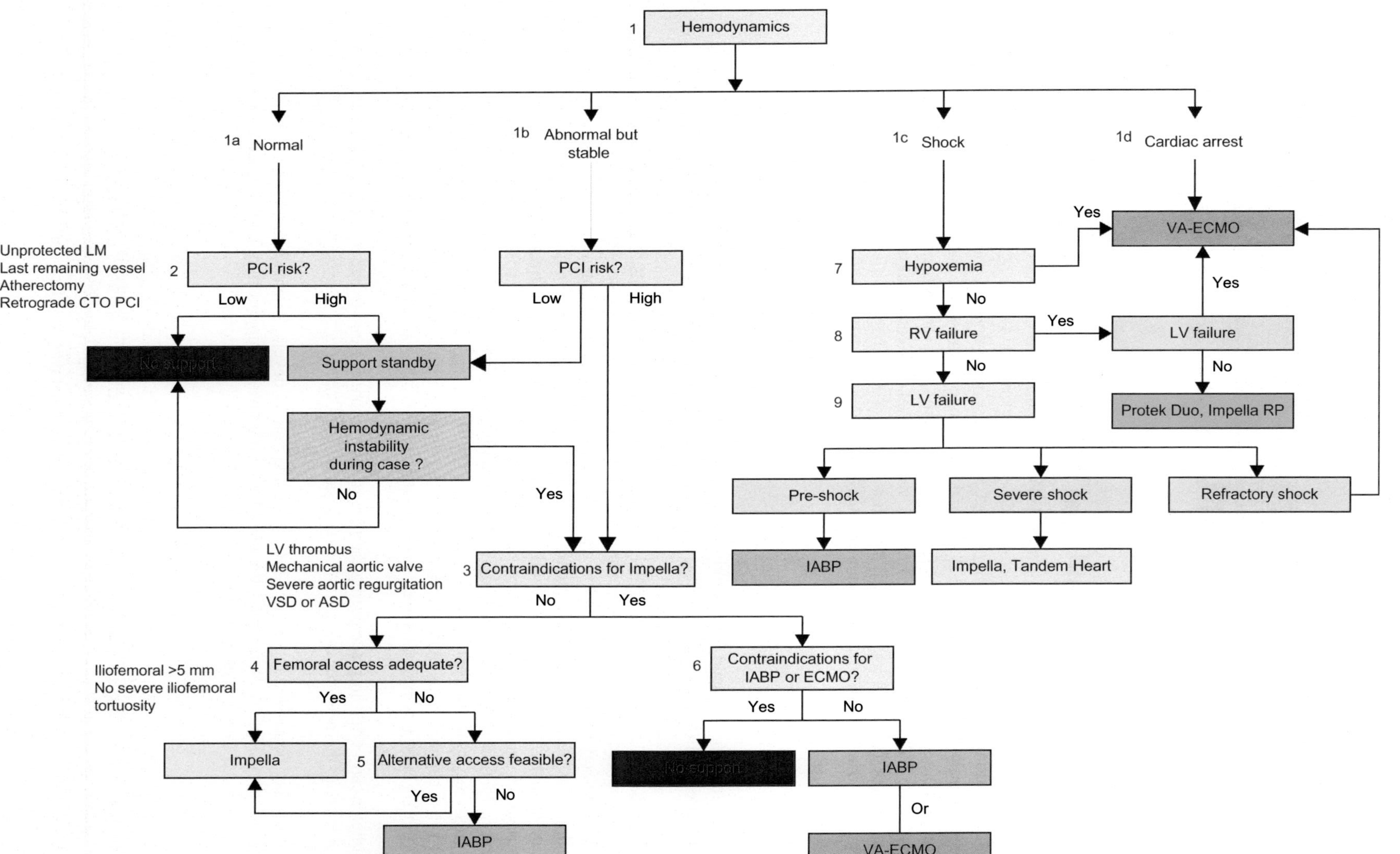

FIGURE 14.1 Determining whether hemodynamic support is needed and optimal device selection.

  b. Last remaining vessel.
2. Lesion morphology:
  a. Severe calcification and tortuosity.
  b. High-grade baseline stenosis that carries high risk of acute vessel occlusion during wiring attempts.
  c. Bifurcation lesion with large side branches with high-grade stenoses that may be challenging to protect.
  d. Large intracoronary thrombus.
3. Procedural plan (complex, prolonged procedures resulting in significant ischemia):
  a. Atherectomy.
  b. Retrograde CTO PCI.

Prophylactic hemodynamic support is in most cases recommended (if feasible) in patients who have *both* abnormal hemodynamics and are need PCI of high-risk lesions [4].

Hemodynamic support is *not used prophylactically* in most patients with (1) abnormal hemodynamics undergoing a low-risk procedure or (2) normal hemodynamics undergoing high-risk PCI, but *should be immediately available for use* in case of hemodynamic decompensation. Many operators insert a 4 French femoral sheath so that they can rapidly insert a hemodynamic support device (IABP, Impella, VA-ECMO) if the need arises during the procedure. Prophylactic upfront hemodynamic support may be considered in some of these patients.

14.1.3 Contraindications for Impella

Where available, the most commonly used device for hemodynamic support in high-risk PCI is the Impella CP device [1].

Contraindications to Impella device use are: (1) mechanical aortic valve, (2) left ventricular thrombus, (3) severe aortic regurgitation, and (4) atrial or ventricular septal defect.

14.1.4 Femoral access

The Impella CP device requires insertion of a 14 French sheath, which in turn requires iliofemoral arteries >5 mm in diameter without severe calcification, tortuosity, or obstructive lesions. Iliac stenting can sometimes be performed to facilitate insertion of the Impella device (CTO Manual case 145) Intravascular lithotripsy can also be used to facilitate insertion of large bore sheaths and catheters through calcified iliofemoral arteries [5].

14.1.5 Alternative access sites

In case of suboptimal femoral access or in case of failure to advance the Impella CP device, alternative access sites can be considered, such as axillary and transcaval. Both require expertise with large bore access and closure techniques.

14.1.6 Contraindications for VA-ECMO or IABP

1. Severe aortic regurgitation.
2. Aortic dissection.
3. Uncontrolled bleeding (although VA-ECMO and IABP can be used without anticoagulation for a short period of time in patients with active bleeding).

14.1.7 Hypoxemia

In patients with cardiogenic shock and hypoxemia, VA-ECMO is used as it can support both oxygenation and circulation.

14.1.8 Right ventricular failure

Right ventricular function can be assessed using echocardiography (may not be immediately available in the cath lab) or using invasive hemodynamics (right atrial pressure >12 mmHg or Pulmonary Artery Pulsatility Index [PAPI] < 0.9) [3].

$$\text{Pulmonary Artery Pulsatility Index (PAPI):} = \frac{\text{sPAP} - \text{dPAP}}{\text{RA}}$$

sPAP = systolic pulmonary artery pressure
dPAP = diastolic pulmonary artery pressure

RA = right atrial pressure

In patients with cardiogenic shock and isolated RV failure without LV failure, right ventricular support devices may be used, such as the Protek Duo and Impella RP.

In patients with cardiogenic shock and biventricular failure, VA-ECMO is usually used, although a combination of a left and a right ventricular support device can also be used.

14.1.9 Left ventricular failure

In preshock patients (cardiac index >2.0 L/min/m^2), use of IABP may suffice, although continuous hemodynamic monitoring, ideally with a Swan Ganz catheter, is needed to determine the need for escalating support.

In shock patients with isolated severe LV failure the Impella CP device is most commonly used (Tandem Heart can be used at centers experienced in its use).

In patients with LV failure and refractory shock (continued hypoperfusion) despite Impella CP or Tandem Heart use, escalation to VA-ECMO or Impella 5.0 may be required. The Impella is often left in place to provide left ventricular unloading (LV venting) during VA-ECMO support.

14.2 Hemodynamic support: device comparison

Four devices are currently available in the US for providing percutaneous left ventricular hemodynamic support: the intraaortic balloon pump (IABP), the Impella (2.5, CP, and 5.0, Abiomed Inc, Danvers, Massachusetts), the Tandem Heart (Liva Nova, Pittsburgh, PA), and veno-arterial extracorporeal membrane oxygenator (VA-ECMO) (Fig. 14.2, Table 14.1) [6]. Two devices are available for right ventricular support, the Protek Duo (Liva Nova) and the Impella RP (Abiomed).

Intraaortic balloon pump (IABP) is the smallest device but also supplies the least hemodynamic support. IABP is usually inserted percutaneously through a 7–8 French femoral arterial sheath. The IABP mechanism of action is inflation of a balloon with helium in the aorta during diastole, increasing coronary perfusion, displacing blood peripherally and increasing cardiac output, while reducing left ventricular end-diastolic pressure and reducing afterload.

The Impella is a non-pulsatile axial flow pump that is advanced through the aortic valve and moves blood from the left ventricle into the aorta. Use of the Impella results in left ventricular unloading with reduction in left ventricular end-diastolic pressure and volume. The Impella is available in two percutaneous types (2.5 and CP), and one type that requires surgical cut down (5.0).

The Tandem Heart is a centrifugal pump that propels blood from the left atrium into the femoral artery. It uses a 21 French cannula placed through a transseptal puncture and a 15–19 French arterial cannula. With an appropriately sized arterial cannula, full perfusion can be achieved. The Tandem Heart requires continuous monitoring to prevent displacement of the transseptal cannula into the right atrium. In isolation, the Tandem Heart does not oxygenate the blood, but an oxygenator can be placed into the circuit to allow full cardiopulmonary support.

VA-ECMO consists of a centrifugal, non-pulsatile pump that circulates the blood and has a membrane oxygenator. Venous blood is aspirated through a venous cannula, advanced through the oxygenator and returned to the patient through an arterial cannula. Similar to Tandem Heart, VA-ECMO requires large size cannulae and a perfusionist to manage the system. VA-ECMO can generate adequate flows to maintain systemic perfusion, but also increases myocardial oxygen demand due to increase of left ventricular end diastolic pressure and volume, that could adversely impact myocardial recovery [7].

The Protek Duo dual-lumen cannula (Fig. 14.3, available in 29 and 31 French sizes) contains two lumens. The inflow lumen has several inflow vents across the superior vena cava and the right atrium and drains blood into an extracorporeal centrifugal pump that delivers blood through the outflow lumen into the main pulmonary artery bypassing the right ventricle.

The Impella RP device can provide right ventricular support by pumping blood from the inferior vena cava to the pulmonary artery and consists of a 22 French motor mounted on an 11 French catheter.

14.3 IABP insertion: step-by-step

14.3.1 Step 1: Obtain femoral access

Goal: Safely obtain arterial access to insert the IABP

How? As described in Chapter 4: Access. The femoral sheath is provided with the IABP kit.

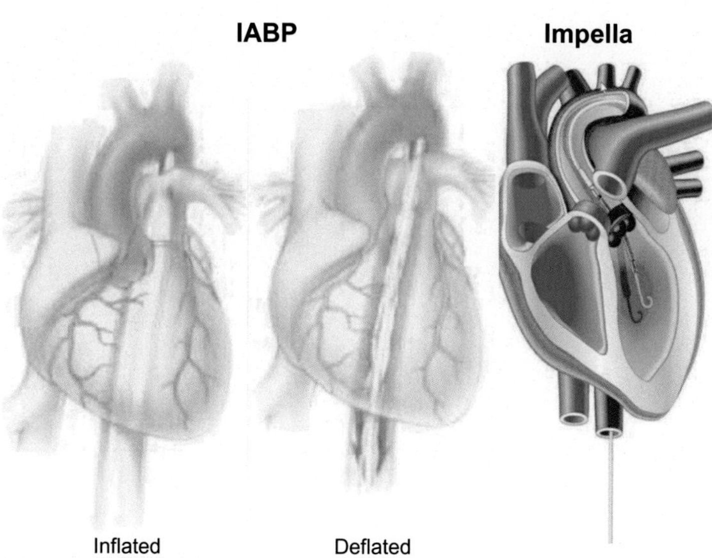

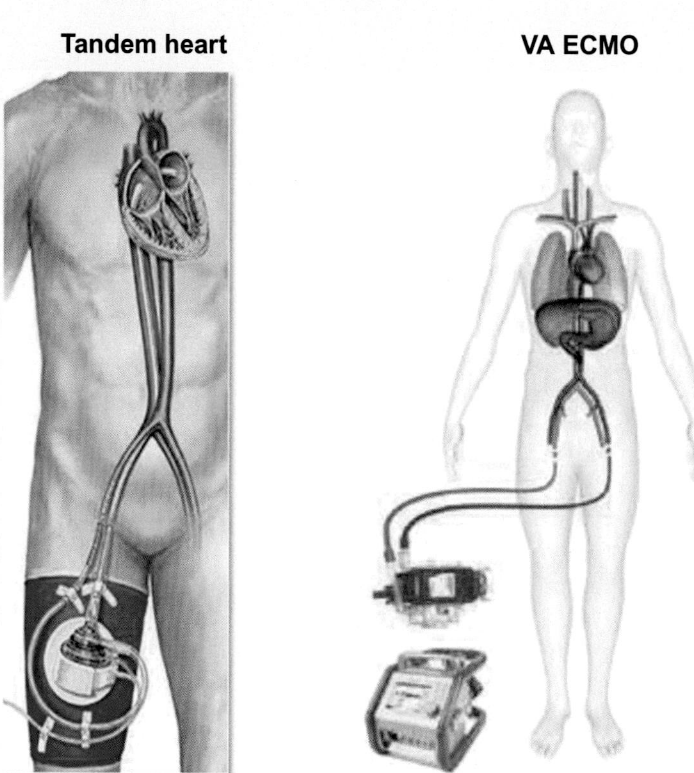

FIGURE 14.2 Devices that provide left ventricular hemodynamic support. *Reproduced with permission from Brilakis ES. Manual of coronary chronic total occlusion interventions. A step-by-step approach. 2nd ed. Cambridge, MA: Elsevier; 2017 (Figure 2.81).*

14.3.2 Step 2: Prepare IABP for use

***Goal*:** Prepare IABP for insertion and use.

***How*?**

1. Select IABP size (usually done by height, Table 14.2).
2. Attach one-way valve to the gas lumen (do not remove until IABP is in position in the aorta).
3. Use the provided syringe to apply vacuum off the IABP helium lumen. The syringe is then removed, while keeping the one-way valve in place.
4. Remove the IABP catheter from the tray (immediately prior to insertion).
5. Remove stylet from IABP catheter wire lumen.

TABLE 14.1 Left ventricular hemodynamic support devices.

	IABP	Impella	Tandem Heart	VA-ECMO
Feasibility				
Availability	+++	++	+	+
Arterial access size required	7–8 French	12 French (Impella 2.5) 14 French (Impella CP) 21 French (Impella 5.0)	15–19 French arterial 21 French venous	15–19 French arterial 21–25 French venous
Contraindications	• High bleeding risk • Severe aortic regurgitation • Thoracic or abdominal aorta aneurysm	• High bleeding risk • Severe aortic regurgitation[a] • Severe PAD[b] • Left ventricular thrombus • Mechanical aortic valve • Ventricular septal defect[a]	• High bleeding risk • Severe aortic regurgitation • Severe PAD[b] • Left atrial thrombus	• High bleeding risk • Severe aortic regurgitation • Severe PAD[b]
Efficacy				
Cardiac output increase (L/min)	0.3–0.5	≈2.5 (Impella 2.5) ≈3.5 (Impella CP) ≈5.0 (Impella 5.0)	4–5[c]	4–5[c]
Affected by arrhythmias	Yes	Yes	Yes	No
Requires adequate right ventricular function	Yes	Yes	Yes	No
Can correct respiratory failure	No	No	Yes[d]	Yes
Complications				
Risk of lower limb ischemia	+	++	+++	+++
Transseptal puncture required	No	No	Yes	No
Risk of bleeding	+	++	++	++
Risk of hemolysis	+	++	++	++
Risk of stroke	++	++	++	+++
Difficulty of insertion	+	++	++++	++

PAD, peripheral arterial disease.
[a]*Except as bridge to surgery.*
[b]*Transcaval access can be used for placing the arterial cannula in case of severe peripheral arterial disease.*
[c]*Depending on arterial cannula size.*
[d]*Adding an oxygenator to the Tandem Heart circuit.*

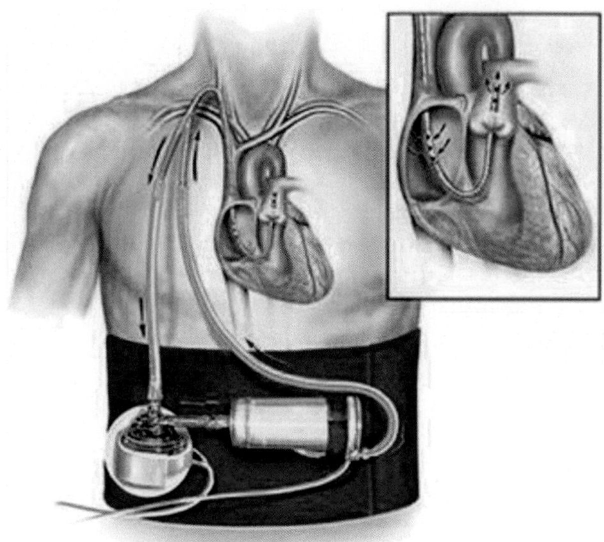

FIGURE 14.3 The Protek Duo right ventricular support device. *Reproduced with permission from LivaNova.*

TABLE 14.2 Intraaortic balloon pump size selection.

	30 cc	40 cc	50 cc
Height	147–162 cm 4′10″–5′4″	162–182 cm 5′4″–6′0″	>182 cm >6′0″

6. Flush IABP wire lumen with normal saline.
7. Connect fiber-optic cable to the console and zero fluid transducer.

14.3.3 Step 3: Advance guidewire to aortic arch

***Goal*:** Advance IABP guidewire (0.025 inch, provided in the IABP kit), over which the IABP will be inserted.

***How*?** IABP wire is advanced under fluoroscopy to the aortic arch.

Challenges and troubleshooting for advancing the guidewire can be overcome as described in Chapter 5: Coronary and Graft Engagement.

14.3.4 Step 4: Advance IABP catheter to aortic arch

***Goal*:** Advance IABP catheter to optimal position.

***How*?** The IABP is advanced under fluoroscopy until the tip of the catheter is at the level of the tracheal bifurcation.

14.3.5 Step 5: Start IABP function

***Goal*:** Initiate IABP function.

***How*?**

1. The guidewire is removed and the wire lumen is connected to an arterial pressure monitoring system.
2. The helium port is connected with the IABP console.
3. Counterpulsation is initiated.

14.3.6 Monitor IABP function

***14.3.6.1. Goal*:** Ensure IABP optimal functioning.

14.3.6.2. How (Fig. 14.4)?

1. Use 1:2 augmentation and ensure that timing of balloon inflation is optimal.
2. Monitor systemic pressure
3. Monitor IABP position.
4. Ensure therapeutic anticoagulation is administered. If anticoagulation cannot be administered, 1:1 augmentation should be used to minimize the risk of thrombus formation on the IABP balloon.

14.3.6.3. Challenges

*14.3.6.3.***1. *Poor augmentation (mean pressure not increasing)***

Causes:

- Hypovolemia.
- Low balloon position.
- Small size balloon for the patient's aorta.
- Low systemic vascular resistance.
- Improper timing.
- Kinked catheter.
- IABP delivered volume is too low.

Prevention:

- Optimal IABP size selection and placement in the aorta, as well as optimal inflation volume and timing.

Treatment:

- Treat underlying cause(s).

*14.3.6.3.***2. *Poor afterload reduction***

Causes:

- Balloon not inflated to full volume.
- Compliant aortic wall.
- Improper balloon position.
- Partial obstruction of gas lumen.
- Improper timing.

Prevention:

- Optimal IABP size selection, optimal placement, optimal inflation timing and optimal inflation volume.

Treatment:

- Treat underlying cause(s)

*14.3.6.3.***3. *IABP migration***

Causes:

- Catheter inadvertently pulled back.

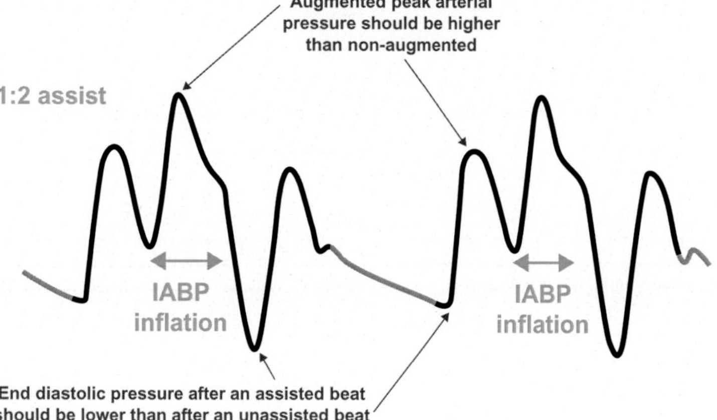

FIGURE 14.4 How to assess IABP function.

Prevention:

- Suture IABP after insertion.

Treatment:

- Reposition and then secure IABP catheter.

14.3.6.4. What can go wrong?

14.3.6.4.1. ***Access site complications, including limb ischemia and retroperitoneal bleeding.***

Causes:

- Suboptimal access technique.
- Anticoagulation.

Prevention:

- Optimal access technique as outlined in Chapter 4: Access.

Treatment:

- Tailored to the type of complication: an ischemic limb will usually recover after removing the IABP and its sheath with manual hemostasis, however it may sometimes require surgical or endovascular treatment; retroperitoneal bleeding may require cessation of anticoagulation. If the anticoagulation is discontinued during IABP use, 1:1 augmentation should be used to minimize the risk of thrombus formation on the IABP.

14.3.6.4.2. *Aortic dissection.*

14.3.6.4.3. *Thromboembolism (due to thrombus formation on the IABP).*

14.3.6.4.4. *IABP catheter entrapment.*

14.4 Impella insertion: step-by-step

14.4.1 Step 1: Obtain femoral arterial access + perform angiography

***Goal*:** Safely obtain arterial access to insert the Impella device.

How?

1. Access is obtained as described in Chapter 4: Access.
2. Usually a 5–6 French arterial sheath is inserted first and femoral angiography is performed to ensure that the artery is large enough and without significant disease to allow insertion of the Impella sheath.

14.4.2 Step 2: Preclose

***Goal*:** To deploy two Perclose sutures to allow hemostasis upon removal of the Impella sheath.

How?

1. Two Perclose sutures are deployed (one at 10 o'clock and the other at 2 o'clock orientation) as described in Chapter 11: Access Closure.
2. The sutures are wrapped into gauze or inserted into insulin syringes without tightening them up.

14.4.3 Step 3: Insert Impella sheath

***Goal*:** Insert the Impella sheath.

How?

1. A stiff guidewire (provided with the Impella kit) is inserted through the femoral artery.
2. Sequentially insert and remove the 8 Fr, 10 Fr, and 12 French dilators.
3. The Impella peel-away sheath [there are two lengths, short (13 cm) and long (25 cm)] is inserted.
4. Alternatively a 14 French Cook sheath can be used.
5. Femoral angiography is performed after sheath insertion to ensure that the Impella sheath is not causing limb ischemia.

14.4.4 Step 4: Anticoagulation

***Goal*:** Prevent thrombus formation on the Impella catheter.

How?

1. Anticoagulation is administered, usually with unfractionated heparin.

2. Recommended ACT is >250 seconds.
3. Once ACT is >250 seconds, remove the Impella sheath dilator and aspirate and flush the sheath.

14.4.5 Step 5: Cross aortic valve and insert an 0.018 stiff guidewire into the left ventricle

14.4.5.1. ***Goal*****:** Cross the aortic valve with a pigtail catheter.
14.4.5.2. ***How?***

1. A pigtail catheter is advanced to the aortic root. A pigtail catheter is preferred over an Amplatz or multipurpose catheter because it is less likely to advance behind a papillary muscle.
2. The pigtail catheter is advanced through the aortic valve into the left ventricle.
3. A 260 cm long stiff 0.014 or 0.018 in. guidewire (such as the Platinum Plus) is then advanced through the pigtail catheter into the left ventricle.
4. The pigtail catheter is removed leaving the 0.018 in. guidewire into the left ventricle.

14.4.5.3. Challenges

14.4.5.3.1. ***Difficulty advancing the pigtail catheter through the aortic valve***

Causes:

- Aortic stenosis.

Prevention:

- Balloon dilation of a stenotic aortic valve may be needed.

Treatment:

- Use 0.035 in. J-wire to stiffen the pigtail catheter and enable advancement through the aortic valve.

14.4.6 Step 6: Prepare Impella device for use

Goal**:** Prepare the Impella device so that it is ready for insertion.
How?

1. The catheter is connected with the Impella controller and purged, as described in the Impella instructions for use.
2. Preparation of the Impella is usually done by the cath lab technician while the interventionalist is obtaining arterial access and preclosing the femoral artery.

14.4.7 Step 7: Load Impella device on the 0.018 in. guidewire

14.4.7.1. Goal**:** Load the Impella device on the 0.018 in. guidewire to allow delivery into the left ventricle.
14.4.7.2. How?

1. The 0.018 in. guidewire is inserted into the red EasyGuide lumen at the tip of the pigtail, after straightening the tip.
2. The guidewire is advanced until it exits from the red EasyGuide lumen near the label.
3. The red EasyGuide lumen is removed.

14.4.7.3. Challenges

14.4.7.3.1. ***Unable to use red EasyGuide lumen***

Causes:

- Red EasyGuide lumen inadvertently removed.
- Impella device required removal and reinsertion.

Treatment:

- The Impella catheter is backloaded on the back end of the 0.018 in. guidewire.
- The guidewire should exit the outlet area on the inner radius of the Impella cannula aligned with the straight black line on the catheter.

14.4.7.4. What can go wrong?

*14.4.7.4.*1. ***EasyGuide red tube not completely removed*** (Fig. 14.5)

Causes:

- Improper preparation.
- Fracture of the red EasyGuide lumen.

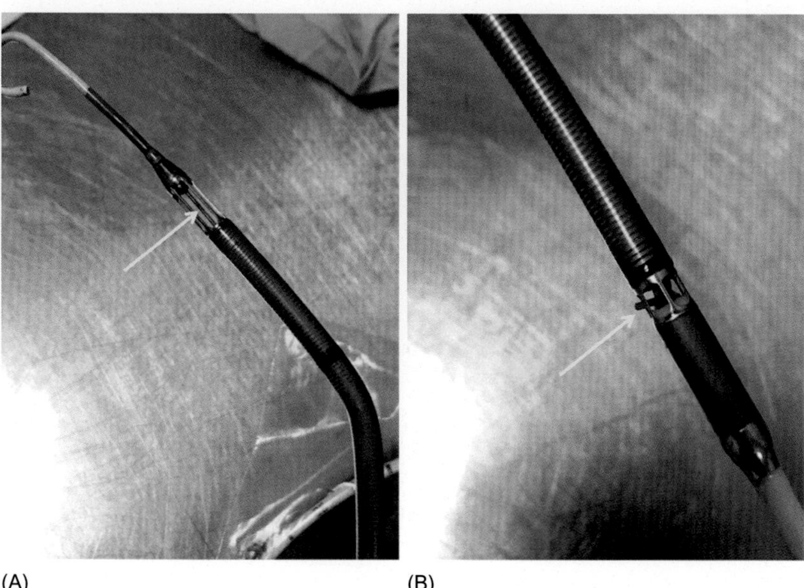

FIGURE 14.5 Example of failure of the Impella device to function due to fracture and retention of the red EasyTube lumen within the device.

Prevention:

- Ensure that red EasyGuide tube is completely removed prior to insertion of the Impella device.

Treatment:

- Remove Impella device and remove red EasyGuide tube.
- Use a new Impella device.

14.4.8 Step 8: Insert Impella into the left ventricle

14.4.8.1. Goal**:** To insert the Impella device in the left ventricle over the 0.018 in. guidewire.

14.4.8.2. How?

1. The 0.018 in. guidewire is fixed by the assistant
2. The Impella device is advanced through the sheath. This is usually done with fast, small movements, as significant bleeding occurs while the inlet area is inside the sheath and the outlet area is outside.
3. The Impella is advanced through the iliac arteries and the aortic valve into the left ventricle under fluoroscopic guidance. The inlet area should be approximately 3.5 cm below the aortic valve. If unsure about the location of the aortic valve, a pigtail catheter can be placed from a second access point in one of the coronary cusps.
4. The 0.018 in. guidewire is removed.

14.4.8.3. Challenges

14.4.8.3.1. ***Unable to advance Impella device through the iliac arteries***

Causes:

- Iliofemoral lesions.
- Iliofemoral tortuosity and calcification.

Prevention:

- Insert the Impella device through the least diseased femoral artery.

Treatment:

- Use the long Impella peel-away sheath (sometimes paradoxically using the short sheath may be advantageous in cases of iliofemoral tortuosity or disease).
- Balloon angioplasty of the iliac arteries.
- Intravascular lithotripsy of heavily calcified iliac arteries [8].

14.4.9 Step 9: Initiation of Impella support

Goal**:** Start the Impella pump.

How?

1. Confirm position of the Impella device within the left ventricle, with the inlet being approximately 3.5 cm below the aortic valve.
2. Confirm that the 0.018 in. guidewire is removed.
3. Confirm that aortic waveform is displayed in the Impella controller.
4. Press the "Start Impella" button.
5. Monitor Impella flow. Usually the "Auto" support option is selected in the cath lab. After 3 hours of support the device automatically defaults to P9 support.

14.4.10 Step 10. Monitor Impella device function

***14.4.10.1. Goal*:** Ensure proper functioning of the Impella device during PCI.

14.4.10.2. How?

1. Monitor flow.
2. Monitor pressure waveform (an aortic waveform should be present—if a ventricular waveform is present the catheter should be pulled back).

14.4.10.3. What can go wrong?

14.4.10.3.1. ***Impella pulled back into aorta***

Causes:
- Patient movement.
- Imperfect suturing of the Impella.

Prevention:
- Good patient sedation.
- Secure Impella device in the groin.

Treatment:
- Sometimes the Impella device can be advanced until it reenters into the left ventricle—this is ideally done using fluoroscopic guidance.
- If reinsertion into the left ventricle is not possible, the Impella device is removed and steps 5–10 are repeated.

14.4.10.3.2. ***Suction alarm***

Causes:
- Hypovolemia.
- Right ventricular failure.
- Impella cannula positioned under the papillary muscles.

Prevention:
- Optimize volume status (while maintaining right atrial pressure <10 mmHg).
- Minimize Impella device movement.
- Cross the aortic valve using a pigtail catheter to minimize the likelihood of insertion under the papillary muscles.

Treatment:
- Reduce P-level by 1 or 2 P-levels.
- If hypovolemia: administer fluids.
- Check and correct the position of the device.
- If needed, support the right ventricle.

14.4.10.3.3. ***Impella stops***

Causes:
- Device malfunction.
- Battery runs out.
- Thrombus formation.

Prevention:
- Ensure that the console is plugged into the electric outlet.
- Proper anticoagulation.

Treatment:

- Attempt to restart device.
- If it does not restart: remove from the ventricle to avoid aortic regurgitation.

14.4.11 Step 11. Insert sheath and guide through Impella sheath

***14.4.11.1. Goal*:** To use the Impella sheath to insert the guide catheter. The Impella shaft is only 9 French, whereas the Impella sheath is 14 French, hence there is space for inserting up to a 7 French sheath next to the Impella shaft. This technique obviates the need for a second arterial puncture for inserting the guide catheter [9].

14.4.11.2. How (Fig. 14.6)?

1. A micropuncture needle is used to puncture the Impella sheath valve at the center of one of the two superior quadrants (Fig. 14.7).
2. After inserting a 0.018 and then a 0.035 in. guidewire, a 7 French sheath (such as Pinnacle destination sheath) is inserted through the Impella valve.
3. At the end of the procedure the 7 French sheath is removed. The valve maintains hemostasis.

14.4.11.3. *What can go wrong*?

*14.4.11.3.*1. ***Guide does not reach the coronary ostia***

Causes:

- The 7 French 45-cm long sheath cannot be advanced all the way into the body.
- Use of 90 cm long guide catheters.
- Tall patients or patients with highly tortuous aorta.

Prevention:

- Use 100 cm (or 125 cm if available) long guide catheters.
- Use short 7 French sheaths.

*14.4.11.3.*2. ***Impella device forward movement during sheath insertion through the Impella sheath the may lead to left ventricular injury***

Causes:

- Suboptimal fixation of the Impella device during sheath insertion.

Prevention:

- Hold the Impella device while advancing a sheath through the Impella sheath. Also monitor the position of the Impella device in the left ventricle using fluoroscopy.

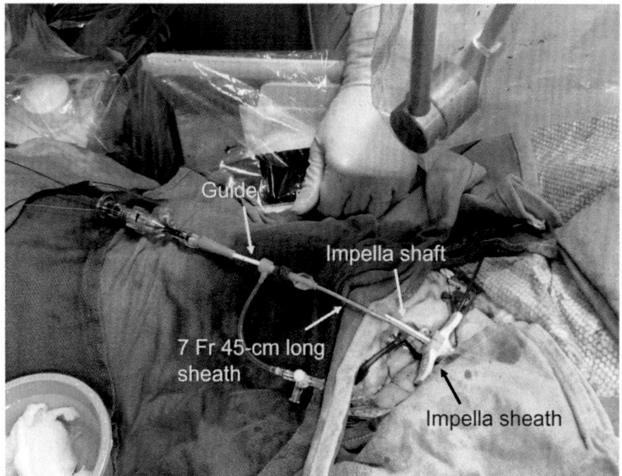

FIGURE 14.6 Inserting a 7 French sheath through the Impella sheath next to the Impella shaft.

FIGURE 14.7 Puncture of the Impella sheath to insert another sheath.

14.4.12 Step 12: Remove Impella (if support is no longer needed) or remove the peel away sheath (if continued support is planned) at the end of PCI

***Goal*:** Remove the device (if support is not needed after PCI) or remove the peel-away sheath (if support is needed).

How?

Remove Impella

1. Impella support is slowly decreased to P2 levels to ensure that patient tolerates weaning.
2. If support is no longer needed, the Impella device is withdrawn into the aorta.
3. The Impella is turned off and the catheter disconnected from the Impella controller.
4. The Impella is removed through the peel-away sheath. Removal should be quick as there will be severe bleeding when the Impella outlet is out of the sheath and the inlet is still in the sheath.
5. The Impella sheath is removed and the Perclose sutures tightened to achieve hemostasis (a balloon can be inserted into the iliac artery from the contralateral femoral artery to minimize bleeding and facilitate hemostasis).

Remove peel-away sheath

1. If continued support is desired at the end of PCI, the peel-away sheath can be removed while advancing the repositioning sheath into the artery.
2. While an assistant holds manual pressure on the groin, the peel-away sheath is retracted until it is *completely outside the body* (if it remains partially inside the femoral artery peeling of the sheath can cause arterial injury and difficult to control bleeding).
3. The two wings of the peel-away sheath are bend back and peeled away until the sheath is completely separated from the Impella catheter (Fig. 14.8).
4. The side arm of the repositioning sheath is flushed.
5. The repositioning sheath is advanced into the femoral artery.
6. The Perclose sutures are partially tightened to minimize bleeding around the repositioning sheath.
7. A gauze can be placed underneath the catheter to create a 45° entry angle.
8. The repositioning sheath and the Impella device are sutured and secured on the skin to minimize the risk of Impella device movement.

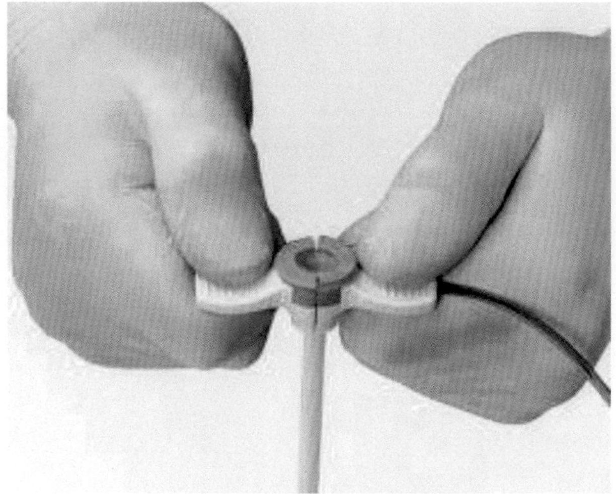

FIGURE 14.8 Splitting the Impella peel-away sheath. Reproduced with permission by Abiomed.

14.5 VA-ECMO insertion: step-by-step

14.5.1 Step 1: Obtain femoral arterial and venous access + perform angiography

***Goal*:** Obtain arterial and venous access in order to allow insertion of the arterial and venous cannulae.

***How*?**

1. Access is obtained as described in Chapter 4: Access.
2. Venous access is usually obtained in the right common femoral vein (it is easier to advance the venous cannula to the right atrium) and arterial access in the left common femoral artery.
3. Venous and arterial access should not be placed in the same groin to avoid venous congestion from the femoral vein being collapsed by the adjacent arterial cannula.
4. Angiography of the femoral arterial access is performed (except for emergency cases) to ensure that the artery is large enough and without significant disease to allow insertion of the arterial cannula.

14.5.2 Step 2: Obtain antegrade arterial access

***Goal*:** To preserve limb perfusion in the limb supplied by the femoral artery used for insertion of the arterial cannula (the large arterial cannula may cause lower extremity ischemia) (Fig. 14.9).

***How*?**

1. It is best to perform this step before inserting the arterial cannula, which may interfere with obtaining access. However, in urgent cases this step can be performed after extracorporeal circulation has started.
2. Antegrade access is obtained (ideally using ultrasound guidance and a micropuncture needle) in the common femoral or superficial femoral artery.
3. Six French 24-cm long Super Arrow-flex sheaths (Teleflex) are usually used for antegrade perfusion, as they are less likely to kink.
4. An antegrade perfusion catheter may not be needed when smaller (such as 15–17 French) arterial cannulas are used.
5. The need for an antegrade perfusion catheter can be assessed by near infrared reflectance spectroscopy (NIRS). If NIRS tissue saturations is $<50\%$ or if there is $>15\%$ difference compared with the uncannulated contralateral limb, then an antegrade perfusion catheter should be placed [10].

14.5.3 Step 3: Anticoagulation

***Goal*:** Prevent thrombus formation within the cannulae and the ECMO circuit.

***How*?**

1. Unfractionated heparin is administered (70–100 Units/kg).

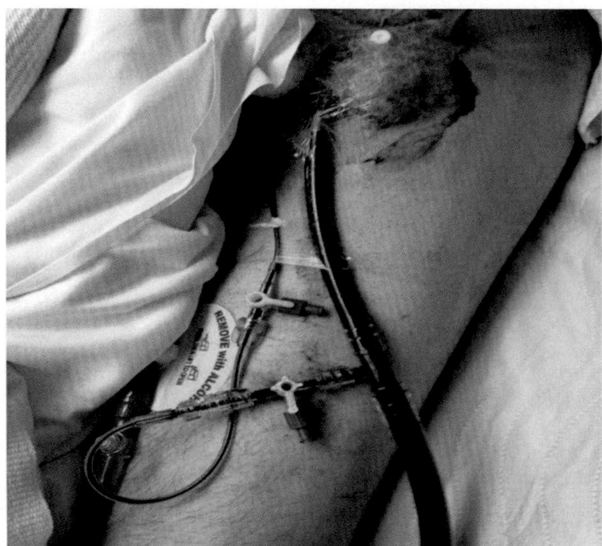

FIGURE 14.9 Placement of a sheath in the SFA for maintaining limb perfusion during VA-ECMO support. *Courtesy of Dr. Khaldoon Alaswad.*

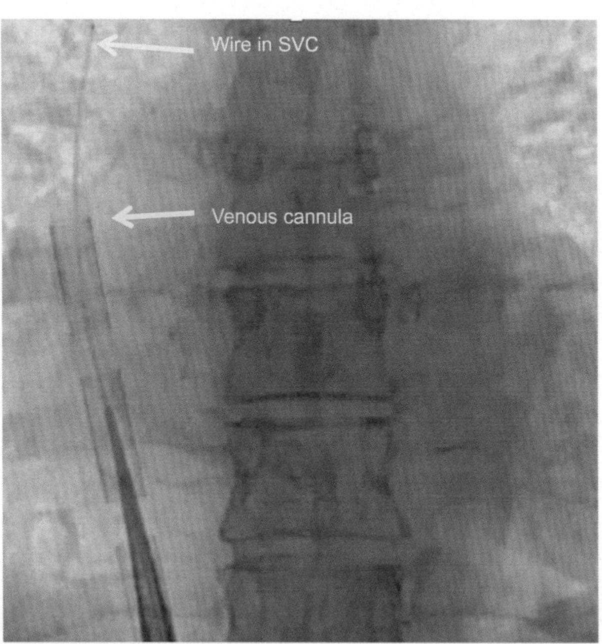

FIGURE 14.10 For venous cannula insertion the 0.035 in. stiff guidewire should be inserted in the SVC and the venous cannula tip should be at the SVC-right atrial junction. *Courtesy of Dr. Chavez.*

14.5.4 Step 4: Insert venous cannula

***14.5.4.1. Goal*:** Insert venous cannula.

14.5.4.2. How?

1. A stiff guidewire, such as a Lunderquist (Cook Medical) or Amplatz Super Stiff (Boston Scientific) is inserted from the right common femoral vein into the SVC. It is important to ensure that the guidewire is in a distal SVC position to avoid migration of the wire back into the right atrium. Migration of the wire into the right atrium may lead to advancement of the cannula against an intracardiac structure, potentially leading to perforation.
2. The skin entry point is dilated with a hemostat.
3. The venous cannula (25 French is used in most patients) is inserted until it reaches the right atrium. The optimal position of the venous cannula tip is at the superior vena cava-right atrial (SVC-RA) junction (Fig. 14.10).

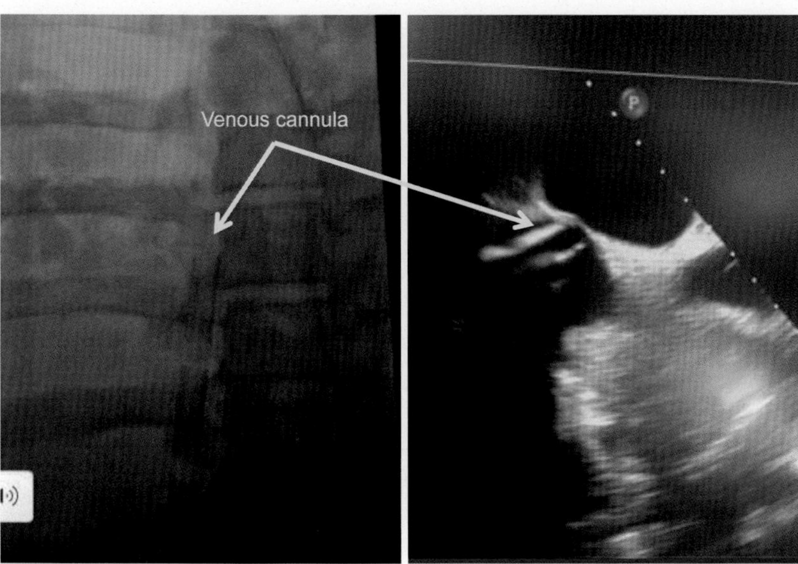

FIGURE 14.11 Improper positioning of the venous cannula against the intraatrial septum, likely as a result of poor distal wire positioning. *Courtesy of Dr. Ivan Chavez.*

4. The back end of the cannula is clamped.

   14.5.4.3. Challenges

   1. Difficult cannula advancement, which is usually due to resistance at the entry site. Treatment is with additional skin dilation before cannula insertion.
   2. Patients with venous thrombosis. In such cases the venous cannula may be placed in the contralateral common femoral vein or the internal jugular vein.
   3. Patients with an inferior vena cava filter. In such patients the venous cannula can be placed through the internal jugular vein.

   14.5.4.4. What can go wrong?

   1. Lower extremity ischemia.
   2. Vessel laceration during cannula advancement. The venous cannula should be advanced under fluoroscopy after ensuring that the guidewire is in the SVC. Advancement should stop if any resistance is felt.
   3. Improper cannula positioning (Fig. 14.11).

14.5.5 Step 5: Insert arterial cannula

***14.5.5.1. Goal*:** Insert arterial cannula.

14.5.5.2. How?

1. A stiff guidewire, such as Lunderquist (Cook Medical) or Amplatz Super Stiff (Boston Scientific) is inserted via the common femoral artery into the thoracic aorta.
2. The skin entry point is dilated with a hemostat.
3. The arterial cannula (15 French is used for most women and 17 French for most men, although larger size cannulae may occasionally be needed for larger patients) is inserted. The proximal wider portion of the cannula should not be advanced through the skin.
4. The back end of the cannula is clamped.

   14.5.5.3. Challenges

   1. Difficulty advancing the cannula. Ensure that the guidewire is in the correct position and that there is no kinking or prolapsing. Cannula insertion may cause perforation.

   14.5.5.4. What can go wrong?

   1. Arterial cannula displacement that can lead to massive bleeding or hematoma.
   2. Bleeding around the cannula.
   3. Vessel perforation. The cannula should be inserted under fluoroscopy over a stiff guidewire.

14.5.6 Step 6: Connect cannulae with ECMO circuit

***14.5.6.1. Goal*:** To connect the cannulae with the tubes connecting it with the centrifugal pump.

14.5.6.2. How?

1. The cannulae and the ECMO tubes are brought next to each other.
2. Saline is injected above the connection point by an assistant to completely fill the cannula and the tube and prevent air from entering the circuit.
3. The connections are checked to ensure that there are no bubbles.

14.5.6.3. Challenges

14.5.6.3.1. ***Bubbles within the tubing***

Causes:

- Imperfect connection.

Prevention:

- Connect tubing with cannula during continuous saline dribbling.
- Avoid oversqueezing the tubing while connecting, as air may be introduced upon release.
- Do not tighten connections until all air is removed.

Treatment:

- Bubbles in arterial cannula: remove bubbles through side port.
- Bubbles in venous cannula: small bubbles will be removed by the ECMO circuit. For bigger bubbles disconnect from the pump tube and reconnect.

***14.5.6.4.** What can go wrong?*

1. *Thrombus embolism.*
2. *Air embolism if connections are not secured and sealed.*
3. *Circuit clotting.*

14.5.7 Step 7: Clamps removed and pump started

***Goal*:** Start extracorporeal circulation.

How?

1. Clamps are removed.
2. Pump is started.
3. Pump flows are monitored.

14.5.8 Step 8: Connect side port of antegrade arterial sheath to ECMO circuit

***Goal*:** Prevent lower extremity ischemia.

How?

1. The circuit needs to be clamped to prevent bleeding.
2. The arterial sheath is allowed to back bleed to remove any air from the line.
3. The sheath is connected to the side port of the arterial cannula using a male to male connector.
4. The clamps are removed and VA-ECMO is restarted.

14.6 Tandem Heart: step-by-step

Insertion of the Tandem Heart is similar to insertion of a VA-ECMO circuit except for the venous sheath that is introduced into the left atrium (LA) to aspirate oxygenated blood which is pumped by a centrifugal pump through a 15, 17, or 19 French arterial cannula.

14.6.1. How to insert a Tandem Heart

1. Obtain arterial and venous access.
2. Venous access is usually placed in the right groin to facilitate transseptal puncture.

Transseptal Cannula Placement:

3. A Brockenbrough needle is most commonly used to perform transseptal puncture under fluoroscopic, intracardiac echocardiography, or transesophageal echocardiography guidance.
4. Anticoagulation with heparin should be administered immediately after entry into the LA.
5. An extra-support wire is placed into the left atrium.
6. The skin entry is gradually dilated, followed by use of a 21 French atrial septal dilator to dilate the transseptal puncture.
7. The LA cannula is introduced into the left atrium, ensuring that the fenestrated end is fully inside the left atrium.
8. The rest of the Tandem Heart placement including the arterial access and connection of the cannulae to the pump is similar to the VA-ECMO circuit placement.

14.6.2. Challenges

14.6.2.1. Anatomic variations that could lead to inability to obtain a transseptal access or inadvertent entry into the aorta or puncture of the atrial wall.

Prevention:

- Use of transesophageal echocardiography or intracardiac echocardiography guidance.
- Verify left atrial access before introducing the sheath into the left atrium.
- Perform transseptal puncture under positive intrathoracic pressure and lower the proximal end of the cannula below the level of the heart.

Treatment:

- Inadvertent puncture into the aorta or other heart structure is usually well tolerated if the transseptal sheath is not advanced.

14.7 Mechanical chest compression systems

There are three systems for automated performance of chest compressions, the LUCAS device (Physio-Control Inc./Jolife AB, Lund, Sweden), the LifeStat (Michigan Instruments, Grand Rapids, MI) and the Autopulse (Zoll, Chelmsford, MA). Use of these devices can facilitate patient management in case of cardiac arrest while in the cardiac catheterization laboratory [11], but also carries risk of complications such as rib fracture and even cardiac laceration [12].

References

[1] O'Neill WW, Kleiman NS, Moses J, et al. A prospective, randomized clinical trial of hemodynamic support with Impella 2.5 versus intra-aortic balloon pump in patients undergoing high-risk percutaneous coronary intervention: the PROTECT II study. Circulation 2012;126:1717–27.

[2] Perera D, Stables R, Clayton T, et al. Long-term mortality data from the balloon pump-assisted coronary intervention study (BCIS-1): a randomized, controlled trial of elective balloon counterpulsation during high-risk percutaneous coronary intervention. Circulation 2013;127:207–12.

[3] Korabathina R, Heffernan KS, Paruchuri V, et al. The pulmonary artery pulsatility index identifies severe right ventricular dysfunction in acute inferior myocardial infarction. Catheter Cardiovasc Interv 2012;80:593–600.

[4] Rihal CS, Naidu SS, Givertz MM, et al. 2015 SCAI/ACC/HFSA/STS clinical expert consensus statement on the use of percutaneous mechanical circulatory support devices in cardiovascular care: endorsed by the American Heart Association, the Cardiological Society of India, and Sociedad Latino Americana de Cardiologia Intervencion; Affirmation of Value by the Canadian Association of Interventional Cardiology-Association Canadienne de Cardiologie d'intervention. J Am Coll Cardiol 2015;65:e7–e26.

[5] Di Mario C, Goodwin M, Ristalli F, et al. A prospective registry of intravascular lithotripsy-enabled vascular access for transfemoral transcatheter aortic valve replacement. JACC Cardiovasc Interv 2019;12:502–4.

[6] Brilakis ES. Manual of coronary chronic total occlusion interventions. A step-by-step approach. 2nd ed. Cambridge, MA: Elsevier; 2017.

[7] Tomasello SD, Boukhris M, Ganyukov V, et al. Outcome of extracorporeal membrane oxygenation support for complex high-risk elective percutaneous coronary interventions: a single-center experience. Heart Lung 2015;44:309–13.

[8] Riley RF, Corl JD, Kereiakes DJ. Intravascular lithotripsy-assisted Impella insertion: a case report. Catheter Cardiovasc Interv 2019;93:1317–19.

[9] Wollmuth J, Korngold E, Croce K, Pinto DS. The Single-access for Hi-risk PCI (SHiP) technique. Catheter Cardiovasc Interv 2019.

[10] Patton-Rivera K, Beck J, Fung K, et al. Using near-infrared reflectance spectroscopy (NIRS) to assess distal-limb perfusion on venoarterial (V-A) extracorporeal membrane oxygenation (ECMO) patients with femoral cannulation. Perfusion 2018;33:618–23.

[11] Yadav K, Truong HT. Cardiac arrest in the catheterization laboratory. Curr Cardiol Rev 2018;14:115–20.

[12] Elison D, Don C, Nakamura K. Coronary artery and right ventricular perforation due to mechanical CPR trauma. JACC: Case Rep 2019;1:407–10.

Part B

Complex lesion subsets

Complexity during PCI may be due to:

(a) Coronary lesion **location**:
  Chapter 15: Ostial Lesions.
  Chapter 16: Bifurcations.
  Chapter 17: Left Main Percutaneous Coronary Intervention.
  Chapter 18: Bypass Graft—Prior CABG Patients.

(b) Coronary lesion **morphology**:
  Chapter 19: Calcification.
  Chapter 20: Acute Coronary Syndromes—Thrombus.
  Chapter 21: Chronic Total Occlusions.
  Chapter 22: Other Complex Lesion Types.

(c) Response to **therapy**:
  Chapter 23: Balloon Uncrossable and Balloon Undilatable Lesions.

(d) **Patient characteristics**:
  Chapter 24: Complex Patient Subgroups.

Chapter 15

Ostial lesions

CTO Manual Online cases: 2, 10, 12, 20, 37, 58, 65, 69, 70, 78, 83, 84, 90, 91, 92, 102, 110, 118, 130, 135, 136, 148, 149
PCI Manual Online cases: 18, 28, 29, 30, 31, 50, 52, 56, 57, 68, 70, 71, 79, 83, 85, 86, 92

Ostial lesions are lesions located within 3−5 mm of the vessel origin [1,2].

There are two types of ostial lesions: (1) aorto-ostial lesions, and (2) branch ostial lesions (Fig. 15.1) [1].

1. Aorto-ostial lesions: involve the ostia of the right coronary artery, left main, and aorto-coronary bypass grafts (both saphenous vein grafts and arterial grafts, such as left internal mammary artery grafts [LIMA]).
2. Branch ostial lesions: involve the ostia of branches of the coronary vessels, such as left anterior descending artery LAD), ramus, and circumflex (branches of the left main), diagonals (branches of the LAD), obtuse marginals (branches of the circumflex) the posterior descending and posterolateral (branches of the right coronary artery [RCA] in right dominant coronary circulation and of the circumflex in left dominant coronary circulation), and ostial lesions of Y-grafts (graft attached to another graft, such as free right internal mammary artery [RIMA] attached to LIMA).

15.1 Aorto-ostial lesions

15.1.1 Planning

Aorto-ostial lesions can be challenging to engage and severe ischemia can occur upon engagement. It is best to have a guidewire preloaded within the guide catheter before attempting to engage, so that the wire can be immediately advanced into the coronary artery upon engagement, allowing subsequent guide disengagement to prevent ischemia ("hit and run"). Alternatively, the wire can be advanced into the vessel with non-selective guide engagement, minimizing the risk of ischemia and ostial trauma.

Aorto-ostial lesions can be fibrotic, calcified, rigid, and prone to recoil. As a result, they can be challenging to dilate, sometimes requiring atherectomy.

15.1.2 Monitoring

Pressure dampening is common when treating aorto-ostial lesions, requiring constant attention to the guide pressure waveform (Section 5.6).

15.1.3 Medications

Medications are administered as described in Chapter 3, Medications. No special medications are required for treatment of aorto-ostial lesions.

15.1.4 Access

- Arterial access is obtained as described in Chapter 4, Access.
- Radial or femoral access are both acceptable for aorto-ostial lesion stenting. In highly complex cases femoral access and large (i.e., 7 or 8 French) guide catheters may provide better support, but are more likely to cause pressure dampening upon engagement.

Manual of Percutaneous Coronary Interventions. DOI: https://doi.org/10.1016/B978-0-12-819367-9.00015-9

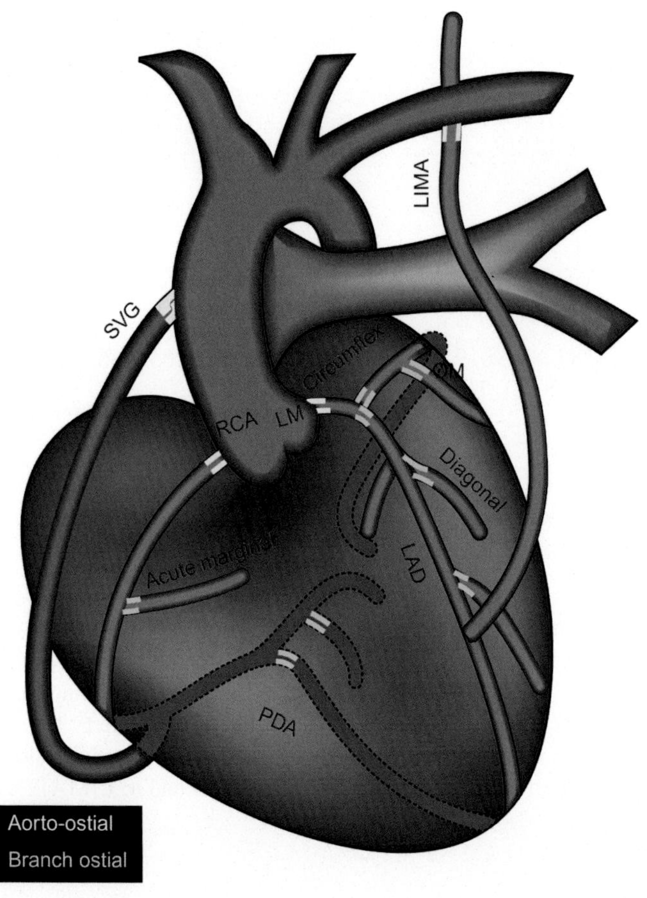

FIGURE 15.1 Types of ostial lesions: aorto-ostial (*yellow*) and branch ostial (*green*).

15.1.5 Engagement

A common challenge associated with aorto-ostial lesion PCI is pressure dampening upon engagement (Fig. 15.2), that may lead to ischemia and hemodynamic compromise [3].

Other causes of pressure dampening (such as intracatheter thrombus or debris, catheter kinking, non-coaxial alignment, or transducer malfunction) should be considered and excluded as discussed in Chapter 5, Coronary and Graft Engagement, Section 5.6, Step 6.

Prevention of pressure dampening:

- Use small (such as 6 French) guide catheters.
- Avoid deep guide engagement.
- Use guide catheters with side holes (Section 30.2.5). However, side-hole guides may provide a false sense of security, as pressure dampening and ischemia may not be appreciated. They should not be used in an unprotected left main coronary artery, given the large area at risk of ischemia, but can be used in the right coronary artery, bypass grafts, and a protected left main coronary artery. Similarly, they should not be used in the right coronary artery if it is the donor vessel to a totally occluded vessel in the left system. Disadvantages of side-hole guide catheters include contrast exit though the side holes that may hinder visualization of the lesion, and lead to use of larger contrast volume for completing the PCI.
- Use of the Ostial Pro device (Section 30.13.1) for keeping the guide catheter outside the ostium.
- Place a second guidewire through the guide catheter into the aortic root ("floating wire") [4] to keep the guide catheter off the coronary ostium (Fig. 15.3).
- Use a small guide catheter extension to engage the coronary artery and "back out" the guide catheter.

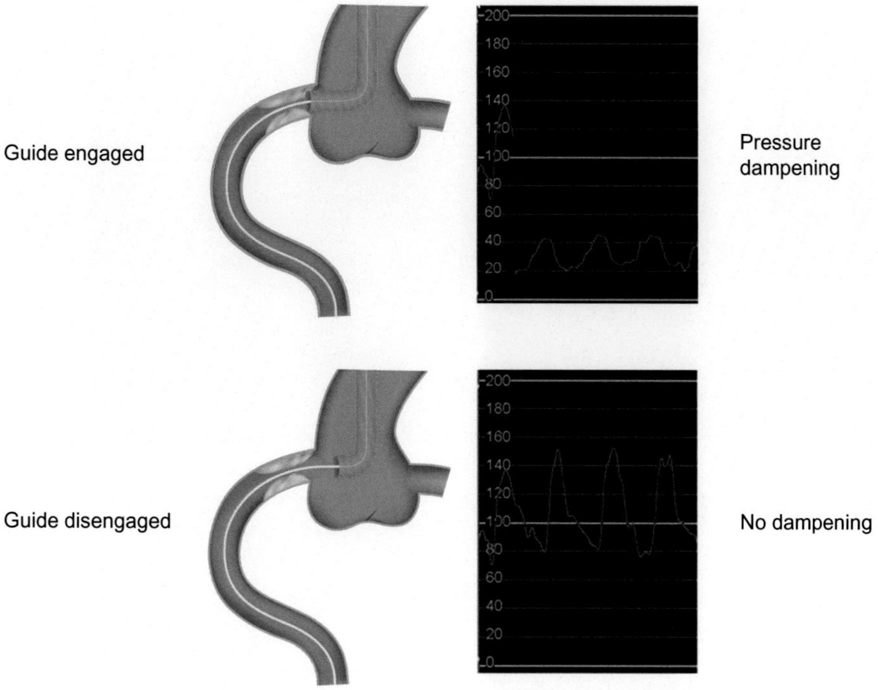

FIGURE 15.2 Pressure dampening upon engagement of a coronary artery that has an aorto-ostial lesion.

Solutions:

- A guidewire should be inserted and advanced to the tip of the guide catheter prior to engagement, to facilitate immediate wiring upon engagement, as described under "Planning". Use of extra support guidewires (such as Grand Slam, Iron Man, Wiggle, etc.) can facilitate guide engagement and disengagement as well as coaxial alignment.
- Use smaller (such as 6 French) guide catheters as larger guide catheters are more likely to cause pressure dampening upon engagement.
- Less aggressive guide catheters, such as a JR4 (and sometimes IM catheters for shepherd's crook RCA) are preferred for isolated aorto-ostial disease, as they are easier to intermittently engage and disengage. These catheters, however, may not provide enough support for more diffusely diseased or tortuous vessels compared with more aggressive guide catheters, such as AL shapes.
- Avoid contrast injections while pressure is dampened to minimize the risk of coronary ostial dissection that can propagate down the coronary artery or in the ascending aorta.
- Intermittently disengage the guide catheter to alleviate ischemia. Disengagement is best done by torquing the guide to a noncoaxial position as opposed to pulling it back, as the latter may lead to loss of distal wire position.
- Place a second guidewire through the guide catheter into the aortic root ("floating wire") [4] to keep the guide catheter off the coronary ostium.

Contrast injection while pressure is dampened can lead to aorto-coronary dissection.

***Aorto-coronary dissection*:**

Prevention:

- Use small guide catheters.
- Avoid deep guide engagement.
- Use gentle manual injection—avoid automated injectors or use a gentle injection preset.
- Avoid injection while pressure is dampened.
- If possible, wire the vessel prior to contrast injection.
- Use a small guide catheter extension to engage the coronary artery, keeping the guide catheter outside the ostium.

Solutions:

- Maintain guidewire position within the target vessel.
- STOP INJECTING contrast.

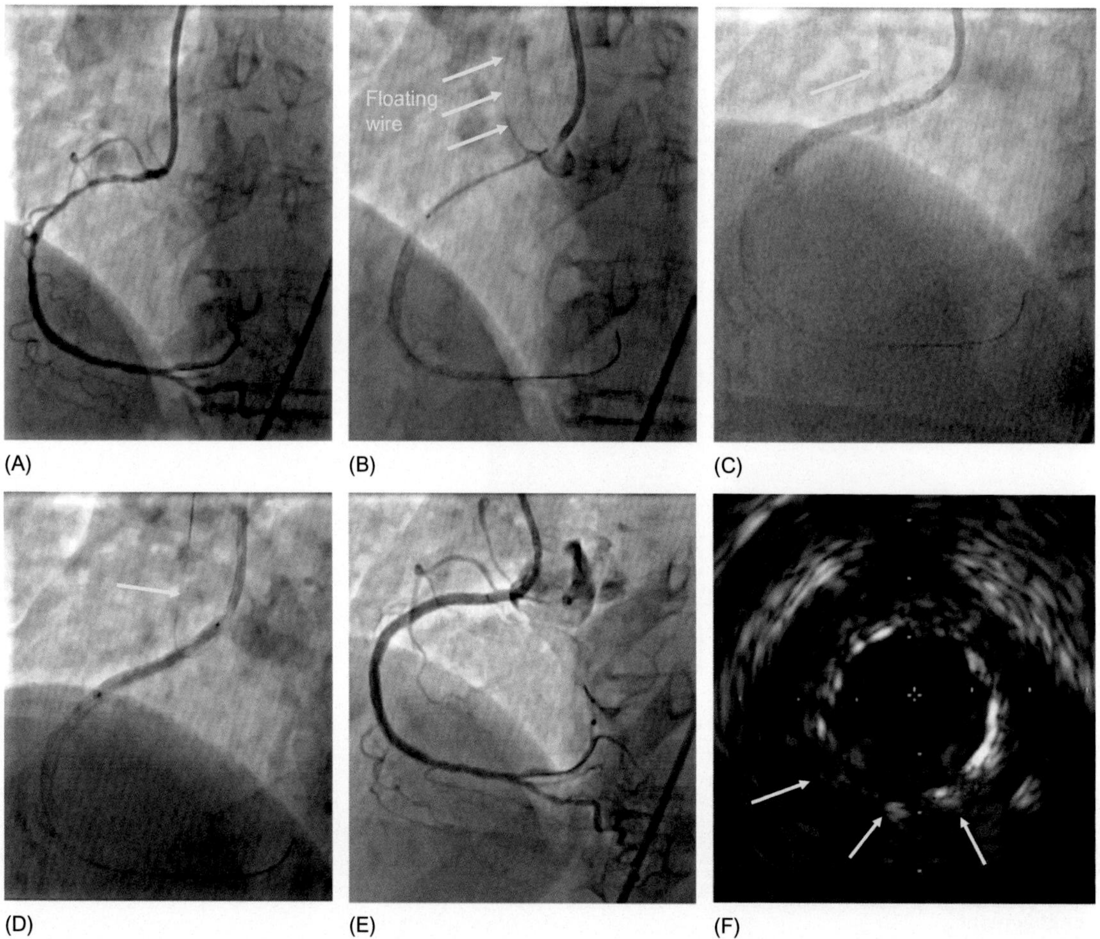

FIGURE 15.3 The floating wire technique for aorto-ostial stenting. While treating an aorto-ostial and proximal right coronary artery lesion (panel A) a second floating guidewire (*arrows*, panel B) was inserted in the aorta, serving as a marker of the ostium guiding stenting implantation (panel C). After postdilation of the ostium by withdrawing the stent balloon (panel D) a nice final result was achieved (panel E). Intravascular ultrasound confirmed that the RCA ostium was covered with stent struts protruding into the aorta (*arrows*, panel F).

- If vessel needs wiring, confirm intraluminal position with IVUS (without contrast injections).
- Stent vessel ostium to seal entry into the dissection flap.

Guide engagement in patients with protruding stents in the aorta.

Engaging vessels with prior aorto-ostial stenting can be difficult due to protruding stent struts, and may result in stent deformation. Sometimes, it can also be challenging to engage the guide catheter in a coaxial manner. In such cases, the vessel can often be wired with the guide catheter disengaged (and ideally knuckling the wire to avoid wire advancement through stent struts), followed by guide engagement over the wire. In the RCA, wiring through previously deployed aorto-ostial stents is easier with a JR4 guide catheter. If a more supportive guide shape is needed, guide exchange can be performed using a microcatheter and a 300 cm supportive guidewire (Chapter 5: Coronary and Graft Engagement, Section 5.7, Step 7).

15.1.6 Angiography

Performing coronary angiography is described in detail in Chapter 6, Coronary Angiography.

- Nitroglycerin should be administered before angiography to exclude coronary spasm.
- For aorto-ostial lesions with severe dampening upon engagement, nonselective angiography can reduce the risk of dissection, but may provide suboptimal quality images.

- The "independent-hand" technique for holding the guide catheter (Fig. 15.4) can significantly facilitate all steps of aorto-ostial lesion stenting.
- Multiple angiographic acquisitions should be avoided to minimize the risk of aorto-ostial lesion dissection; intravascular imaging can help determine the presence and severity of aorto-ostial lesions and response to treatment.
- If the contrast is not allowed to clear the coronary artery after injection, ventricular fibrillation or ventricular tachycardia may occur. Failure of the contrast to clear is likely due to occlusion of the ostium from the catheter. The catheter should be disengaged, to allow contrast clearance.
- Determining the optimal view to define the true ostium is essential to avoid geographic miss during stenting. If a patient has a prior coronary CT angiography, the optimal angle can be noted prior to the procedure. In general, the following projections offer the least foreshortening of the ostium:
  - RCA: steep LAO +/− caudal.
  - Left main: LAO caudal and LAO cranial.
  - Right-sided SVGs: LAO.
  - Left-sided SVGs: RAO.
  - LIMA and RIMA: AP with shallow cranial.

15.1.7 Determining culprit lesion(s)

If the severity of an aorto-ostial lesion is questionable, nitroglycerin should be administered to exclude spasm and coronary physiology (or intravascular imaging for ostial left main lesions) should be used (Section 15.1.13). Because of the slanted nature of coronary takeoffs, two-dimensional IVUS area measurements should be interpreted with care in aorto-ostial lesions, especially when making therapeutic decisions about the left main.

15.1.8 Wiring

Performing coronary wiring is described in detail in Chapter 8, Wiring.

Tips and tricks:

1. The guidewire should be advanced to the tip of the guide catheter prior to engagement, to facilitate immediate wiring upon engagement.
2. In difficult to wire vessels, a hydrophilic-coated or polymer-jacketed guidewire can be advanced through a microcatheter to the distal vessel and then exchanged for a safer guidewire. Wiring should be done with great caution to minimize the risk of subintimal tracking and acute vessel closure.
3. Once the vessel is wired, the original guidewire can be exchanged for a supportive guidewire (such as Iron Man, Grand Slam, Mailman, and Wiggle, Section 30.7.4) which can facilitate engagement and disengagement of the guide catheter.

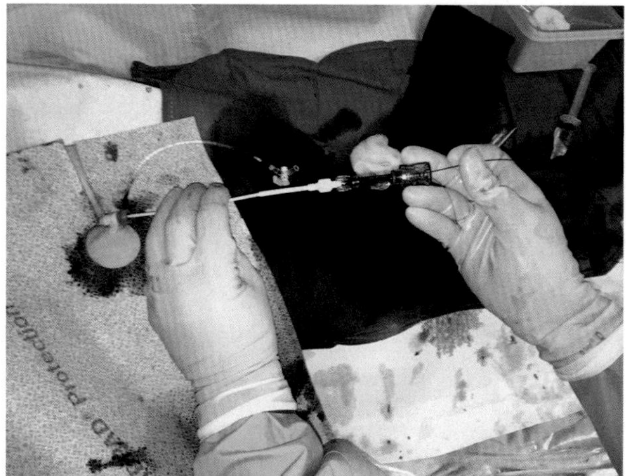

FIGURE 15.4 The "independent-hand" technique. The left hand advances and withdraws the guide catheter while the right hand advances or withdraws the balloon or stent or other equipment.

15.1.9 Lesion preparation

Lesion preparation is described in detail in Chapter 9, Lesion Preparation.

Aorto-ostial lesions are often resistant to dilatation and prone to recoil due to the greater thickness of muscular and elastic tissue in the aortic wall. These lesions are also often heavily calcified and "balloon undilatable" (Section 23.2), hence excellent preparation is important prior to stenting.

Balloon angioplasty: Balloons should be placed partially into the vessel and partially into the aorta to ensure adequate lesion preparation. Noncompliant and longer balloons (such as 30 mm long) are useful. Use of plaque modification balloons (Angiosculpt, Chocolate, and cutting balloon, Section 30.9.3) can facilitate lesion expansion and prevent excessive balloon movement.

Challenges

1. **Watermelon seeding** (excessive movement of the balloon either distally into the vessel or proximally into the aorta).

   Prevention:

   - Slow balloon inflation (1 atm increase at a time, while observing the balloon position with gentle retraction of the balloon catheter).
   - Use a buddy wire.
   - Use a plaque modification balloon (such as Angiosculpt, Chocolate, and cutting balloon, Section 30.9.3).
   - Use longer balloons (such as 30 mm long balloons).
   - Use a winged balloon.
   - Use intravascular lithotripsy (Section 29.9.8) or the very-high-pressure SIS OPN balloon (Sections 30.9.7 and 23.2.10).

   Treatment:

   - Same as for prevention above.

2. **Balloon undilatable** lesions (discussed in detail in Section 23.2)

   Prevention:

   - Ensure full balloon expansion prior to placing stent. The balloon should be sized 1:1 with the target vessel.
   - Intravascular imaging to ensure complete lesion expansion.

   Treatment:

   - As outlined in the balloon undilatable algorithm (Section 23.2).

Atherectomy**:** Atherectomy (Chapter 19: Calcification) can be very useful in severely calcified balloon undilatable aorto-ostial lesions. Both rotational (Section 30.10.1) and orbital (Section 30.10.2) atherectomy can be used.

Rotational atherectomy should ideally be done using the Rotawire extra support. Placing the guide catheter coaxial to the artery is important to decrease the risk of perforation or aortic dissection and facilitate maximal debulking of the lesion. Often, the burr needs to be activated at the end of the guide catheter with tension off the burr to prevent it from "jumping" through the lesion that could increase the risk of burr entrapment. Larger burr sizes may be required to achieve sufficient aorto-ostial lesion modification.

Orbital atherectomy of aorto-ostial lesions can be challenging, as activation of the device that is not engaged into the vessel ostium can result in unconstrained device orbit that could cause vessel injury and/or perforation [5]. If the atherectomy crown can cross the lesion (sometimes with the help of GlideAssist), the crown should be advanced distal to the lesion and the aorto-ostial lesion treated while withdrawing the crown [5]. If the aorto-ostial lesion is too severe for the crown to cross, make sure the guide catheter is coaxial and place the nose cone of the device into the ostium keeping the crown 5 mm from the lesion. This will constrain the orbit and allow the lesion to be treated in a more controlled manner [5].

Intracoronary lithotripsy (Section 3.9.8) is another emerging option for dilating heavily calcified and hard to expand aorto-ostial lesions.

15.1.10 Stenting

When there are distal lesions in addition to the aorto-ostial lesion, the more distal lesions should ideally be stented first, as advancing equipment through an ostial stent may result in stent deformation and/or equipment loss.

***Aorto-ostial lesion stenting—step-by-step*:**

***Baseline*:** The aorto-ostial lesion is pretreated and guide catheter is engaged (Fig. 15.5).

15.1.10.1 Step 1: The stent is advanced into the target vessel (Fig. 15.6)

Challenges

1. Loss of guide catheter (and possibly guidewire) position (can occur during any stage of aorto-ostial lesion stenting).

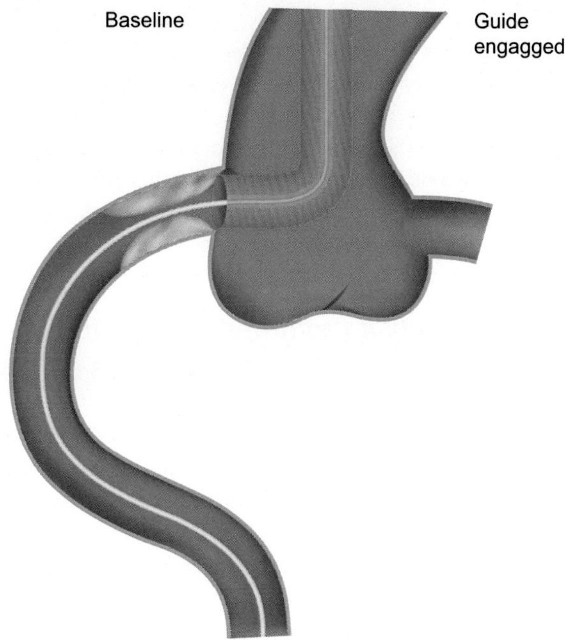

FIGURE 15.5 Guide catheter is engaged in a vessel with an aorto-ostial lesion.

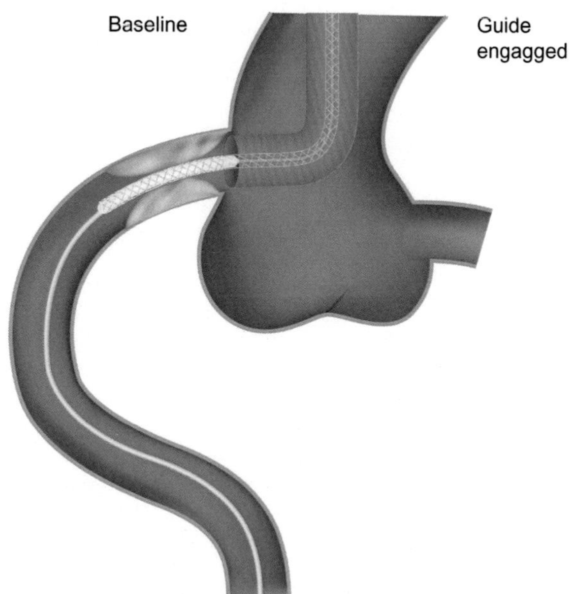

FIGURE 15.6 Advancing a stent through the aorto-ostial lesion.

Prevention: Use stiff, supportive guidewires or buddy wire(s).

Treatment: Re-engage with the same or a different guide catheter. If a guidewire has been advanced through the lesion it can facilitate catheter advancement to engage the ostium. Use of a guide catheter extension can be very helpful.

15.1.10.2 Step 2: The guide catheter is disengaged

(Fig. 15.7)

Challenges

1. Loss of guide catheter (and possibly guidewire) position (can occur during any stage of aorto-ostial lesion stenting).
2. The withdrawal movement to disengage the guide catheter may cause the stent move forward past the lesion; vice-versa, the withdrawal of the stent may cause the guide catheter to deeply re-engage the vessel.

   Solutions:
   - Readjust the stent and guide catheter to account for the counter forces.
   - Position the stent just outside the guide catheter and maintain this geometry by moving both as a single unit.
   - Use extra support guidewires or buddy wires. Controlled disengagement can be achieved by pushing on the guidewire rather than pulling on the guide catheter.

15.1.10.3 Step 3: Stent is withdrawn until it protrudes 1–2 mm in the aorta, as confirmed in multiple projections

(Fig. 15.8)

The guide catheter is placed just outside the ostium far enough out to not obstruct the ostium or interfere with stent placement, yet close enough to allow adequate visualization of the ostium during contrast injection.

LAO projections are often used for the right coronary artery ostium and anteroposterior with cranial angulation for the left main.

Challenges

1. Positioning the stent often requires pushing and pulling the equipment across the aorto-ostial lesion, which may lead to stent damage or dislodgement.

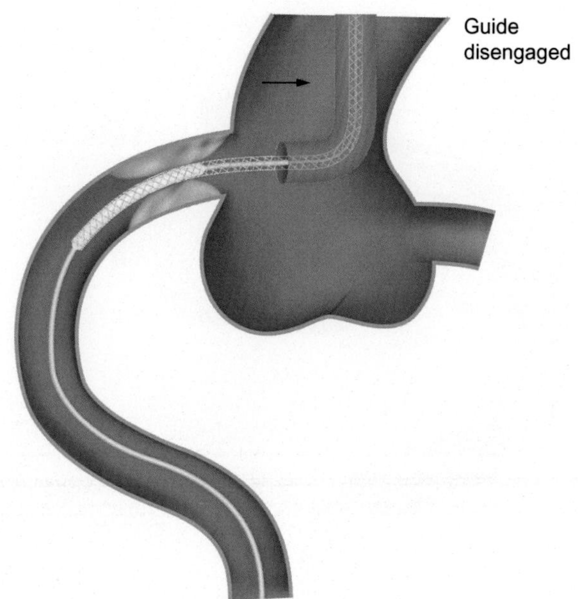

FIGURE 15.7 Disengaging the guide catheter before aorto-ostial lesion stenting.

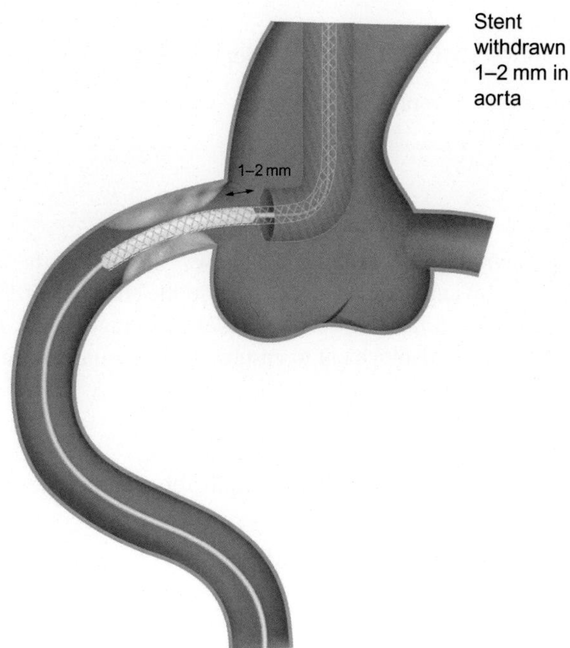

FIGURE 15.8 Optimal stent placement for treating an aorto-ostial lesion.

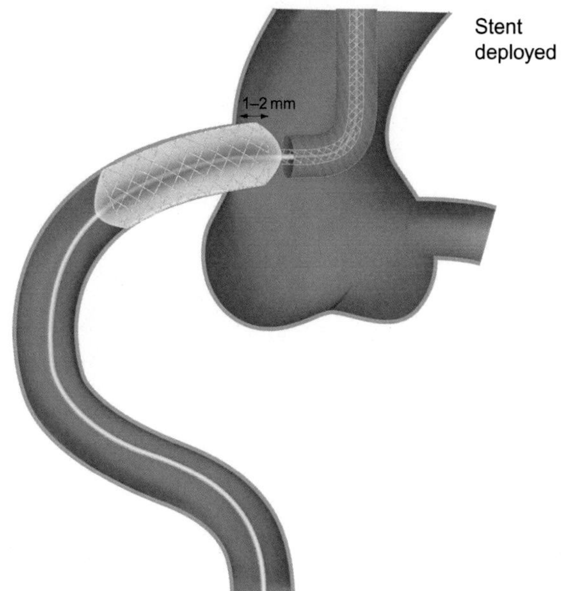

FIGURE 15.9 Stent deployment for treating an aorto-ostial lesion.

Prevention: avoid advancing the stent past the aorto-ostial lesion inside the vessel, especially if severe calcification or tortuosity is present (discussed in detail in Section 27.1.2).

Treatment: Remove the damaged stent if possible. If this is not possible or if the stent has been lost see Section 27.1.3 for management options.

15.1.10.4 Step 4: The stent is deployed

(Fig. 15.9)

Challenges

The main causes of suboptimal aorto-ostial stent placement/deployment are the following (Fig. 15.10):

1. Missing the ostium (incomplete ostium coverage by the stent).
2. Excessive stent overhang into the aorta.
3. Stent underexpansion.
4. Acute stent recoil.

Challenge 1. Missing the ostium (stent does not protrude all the way into the aorta, incompletely covering the lesion). Geographic miss is often underdiagnosed by angiography alone because the shoulder of the stent balloon will treat the true ostium, improving the stenosis and providing a false sense of complete coverage. Thus, intracoronary imaging (preferably with IVUS) is strongly recommended post stent deployment to ensure optimal ostial coverage (and stent expansion).

Prevention:

- Use multiple angiographic projections before deploying the stent.
- Use longer than 8 mm stents to ensure adequate stent length for complete lesion coverage and also reduce the risk of stent embolization.

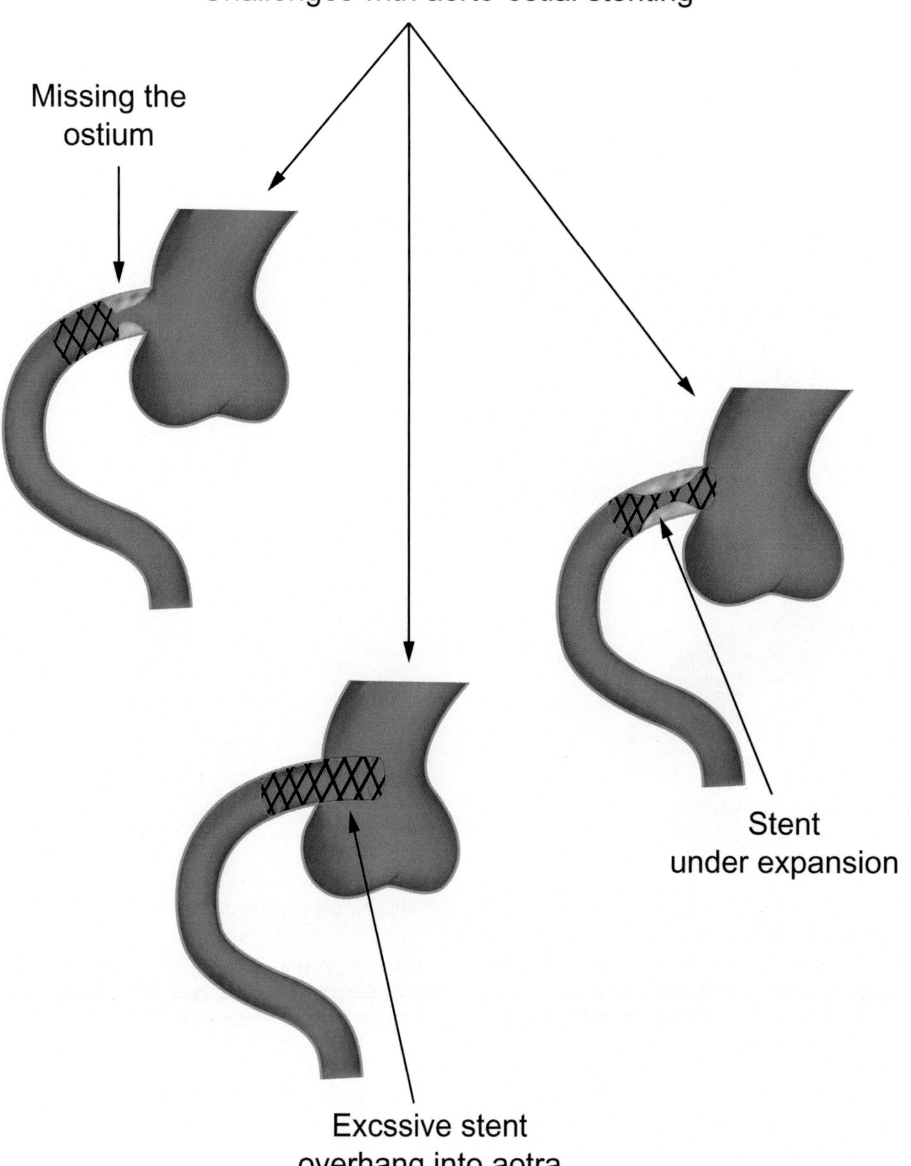

FIGURE 15.10 Challenges of aorto-ostial lesion stenting.

- Ask the patient to hold his/her breath during stent deployment to minimize movement.
- Understand the location of the stent edge relative to the radiopaque markers. In most drug-eluting stent platforms the stent is located just inside of the radiopaque markers, except for the Xience stent (Abbott Vascular, Abbott Park, IL), which is crimped at the center of the radiopaque marker.
- Use a second guidewire advanced into the aorta (floating wire) as marker of the vessel ostium.
- Use the Ostial Pro device (Section 30.13.1) which demarcates the ostium of the vessel.
- Use the Szabo technique (Fig. 15.11): a second guidewire is threaded through one of the most proximal stent struts and then advanced into the aorta; the stent is advanced in the usual fashion over the first guidewire but the second wire guide prevents deep advancement of the stent into the lesion, ensuring ostial coverage. However, the Szabo technique may lead to stent deformation or stent loss [6–11].
- Use of IVUS to determine the location of the true ostium in relation to another radiopaque structure in the chest (such as coronary calcification) can facilitate ostium localization (regardless of view selected) and minimize contrast injections.
- Rapid pacing can minimize stent movement that may be excessive in some patients.

Solutions:

- Deploy another stent to fully cover the ostium.
- If an 8 French guide catheter is being used, IVUS guided deployment may be considered.

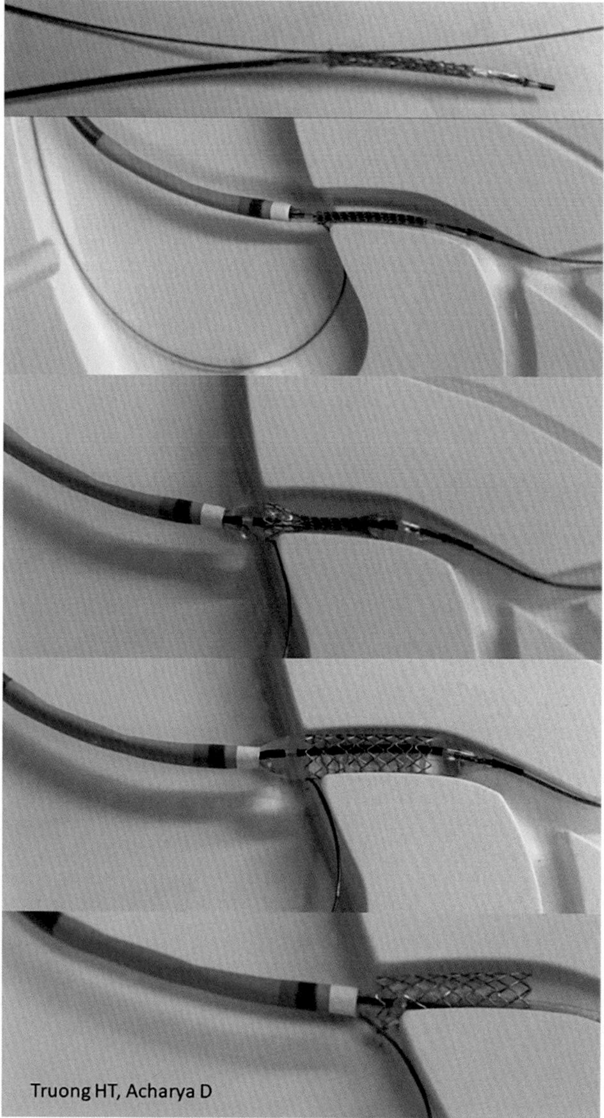

FIGURE 15.11 Illustration of the Szabo technique. (1) A second guidewire is threaded through the most proximal strut of the stent. (2) The stent is advanced and the second guidewire prevents deep stent engagement. (3) Stent is deployed. (4) Second guidewire is removed. *Courtesy of Dr. Tam Truong.*

Challenge 2. Stent protruding too far into the aorta

Prevention:

- Use multiple angiographic projections before deploying the stent.
- Ask the patient to hold his or her breath while deploying the stent.
- Use the Ostial Pro device (Section 30.13.1) or the Szabo technique (Fig. 15.11).
- The indeflator should be prepared and connected to the stent, so that inflation can occur immediately when the stent position is optimal.

Solutions:

- Inflate a short (6–8 mm length) oversized balloon to flare the overhanging portion of the stent (Fig. 15.12).
- Use the Ostial Flash balloon (Section 30.13.2) to flare the overhanging struts.
- Consider prolonged dual antiplatelet therapy (DAPT).
- If the stent protrudes more than 50% out of the ostium it is sometimes possible to snare and remove it, although this maneuver carries the risk of coronary injury and/or stent embolization.

Challenge 3. Failure to expand the stent

Aorto-ostial lesions can often be hard to expand due to fibrosis and calcification; hence primary stenting should be avoided in most cases.

Prevention:

- Meticulous lesion preparation, as described in Chapter 9, Lesion Preparation.

Treatment:

- As for other balloon undilatable lesions, as described in Section 23.2.

Challenge 4. Acute stent recoil

Sometimes a stent may not have enough radial strength to prevent recoil.

Prevention:

- Use stents with strong radial strength in aorto-coronary lesions.

Treatment:

- Postdilate the stent again and wait for a few minutes. If recoil recurs, a stent with strong radial strength (such as the Herculink peripheral stent, Abbott Vascular) can be implanted within the recoiled stent (Fig. 15.13).

15.1.10.5 Step 5: The stent balloon is withdrawn partially into the aorta, followed by high-pressure balloon inflation

(Fig. 15.14)

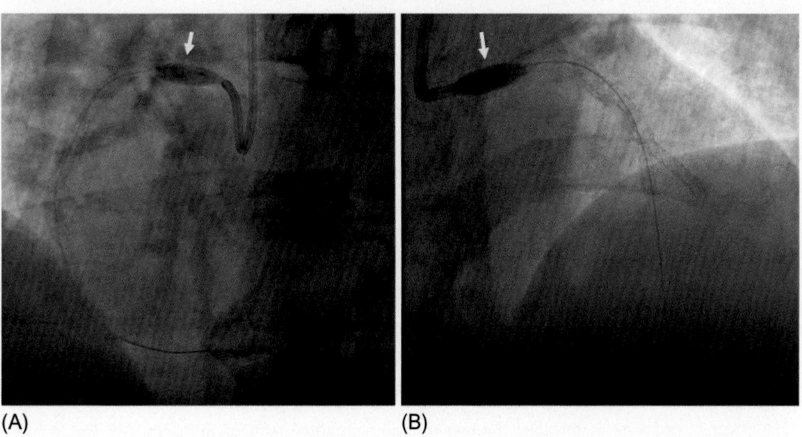

FIGURE 15.12 Inflation of a 6 mm long balloon to flare the ostium of the right coronary artery (panel A) and left main (panel B). *Courtesy of Dr. Alok Sharma.*

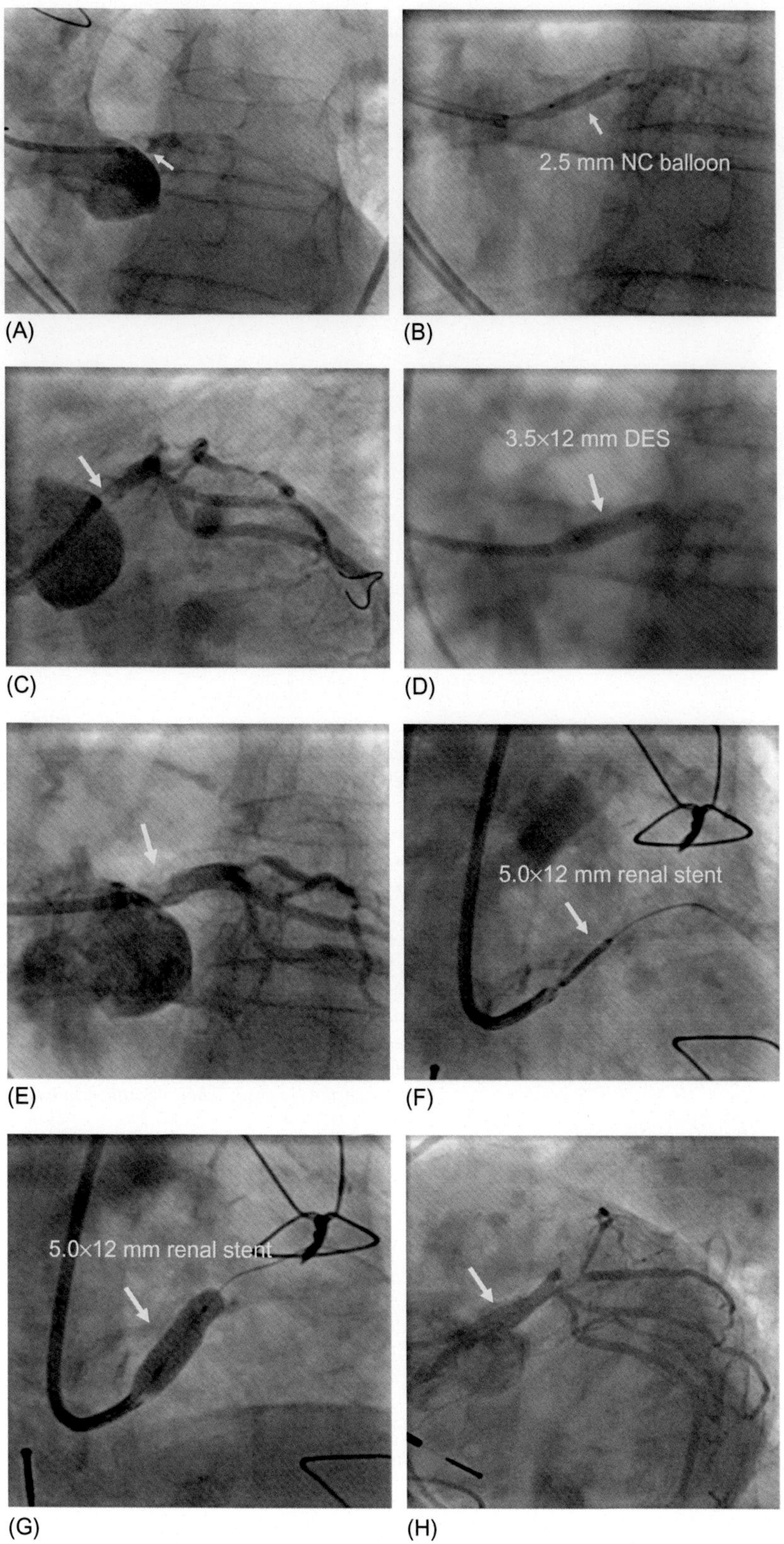

FIGURE 15.13 Ostial left main stent recoil. (Panel A) Heavily calcified ostial left main lesion (*arrow*) with TIMI 2 flow. (Panel B) Predilation with a 2.5 mm noncompliant balloon. (Panel C) Improved antegrade flow after predilation. (Panel D) Deployment of a 3.5 × 12 mm DES. (Panel E) Stent recoil (*arrow*). (Panel F) Placement of a renal stent inside the DES. (Panel G) Deployment of the renal stent. (Panel H) Expansion of the ostial left main lesion (*arrow*). *Courtesy of Dr. Joachim Büttner.*

The goal of this step is to flare the few millimeters of overhanging stent struts, to facilitate re-engagement without stent distortion.

An alternative way to flare the proximal edge of the stent is by using the Ostial Flash balloon (Section 30.13.2) (Fig. 15.15) [12].

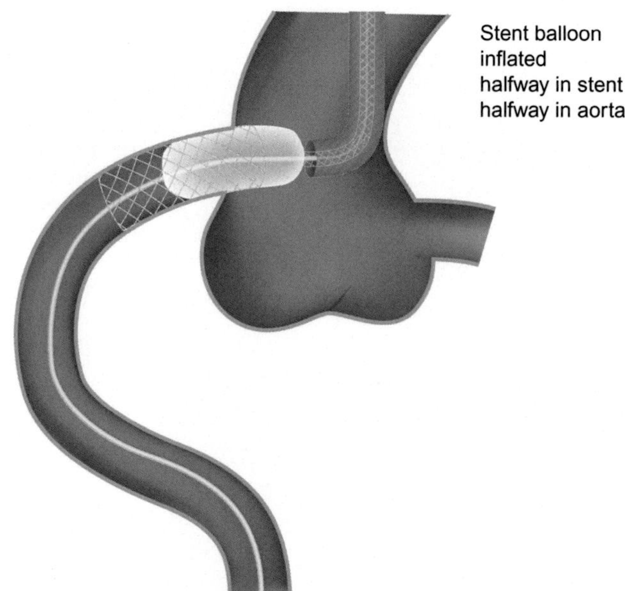

FIGURE 15.14 Stent balloon inflation after partial withdrawal into the aorta to flare the ostium.

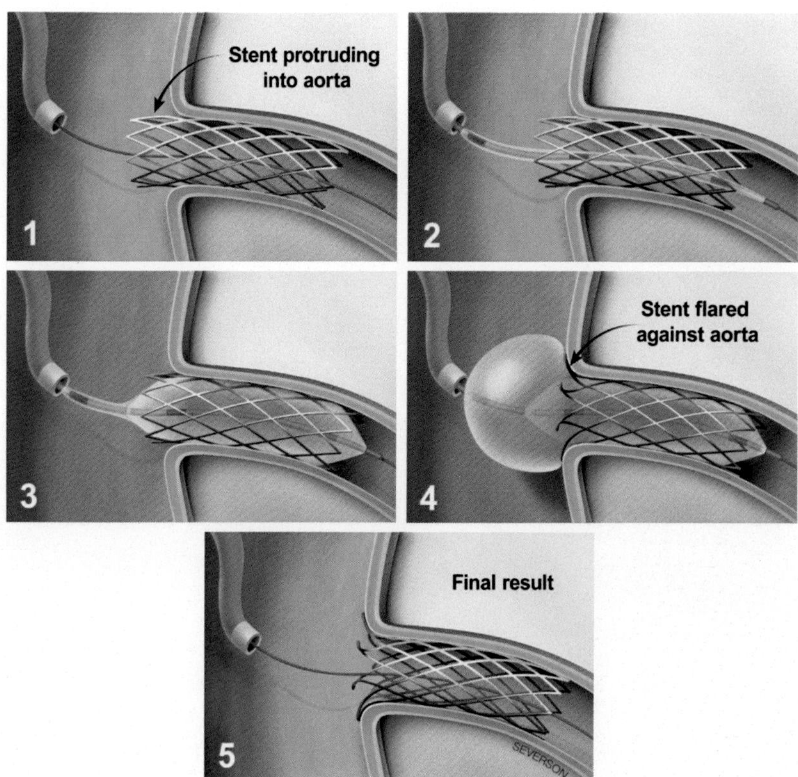

FIGURE 15.15 Illustration of the Ostial Flash balloon. Panel 1: The stent is positioned approximately 1–2 mm from the vessel ostium. Panel 2: The Ostial Flash balloon is positioned with the proximal marker in the aorta outside the guide catheter, middle marker at the ostium, and distal marker inside the stent. Panel 3: The distal balloon is inflated inside the stent with an inflating device. Panel 4: The proximal balloon is then inflated with a 1 cc syringe to flare the stent struts against the aorta. Panel 5: Final result. *Reproduced with permission from Nguyen-Trong PJ, Martinez Parachini JR, Resendes E, et al. Procedural outcomes with use of the flash ostial system in aorto-coronary ostial lesions. Catheter Cardiovasc Interv 2016;88:1067–74.*

15.1.10.6 Step 6: The guide catheter is advanced over the stent balloon to re-engage the target coronary artery

(Fig. 15.16)

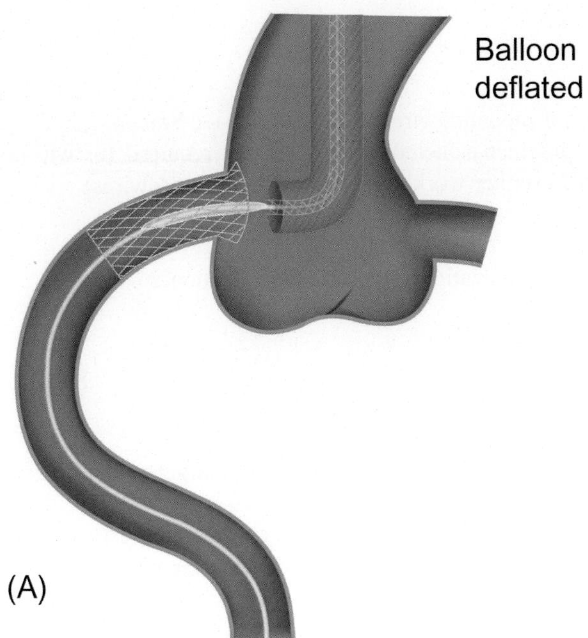
Balloon
deflated
(A)

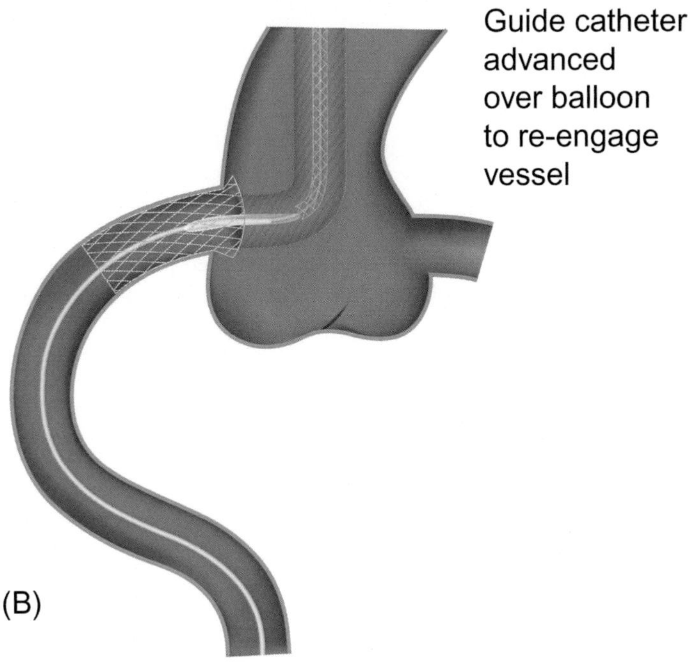
Guide catheter
advanced
over balloon
to re-engage
vessel
(B)

FIGURE 15.16 Guide re-engagement after aorto-ostial lesion stenting.

Challenges

1. ***Stent deformation upon guide catheter advancement***

   Prevention:
   - Ensure that the guide catheter is properly aligned before engagement.
   - Avoid forceful guide (or guide catheter extension) advancement against resistance.

   Solutions:
   - If there is significant stent deformation, additional balloon dilation or implantation of an additional stent may be required.

2. ***Distal wire perforation***

Prevention:

- Closely monitor the distal guidewire position and reposition if the guidewire enters a small side branch.
- Do not use polymer-jacketed guidewires for stenting. If polymer-jacketed guidewires are required for wiring they should be exchanged for soft, non-coated tip, guidewires (either workhorse, or stiff body guidewires).

Solutions:

- If distal wire perforation occurs, treatment is with fat or coil embolization as described in Section 26.4.

15.1.10.7 Step 7: Final assessment

(Fig. 15.17)

The final result can be assessed in multiple ways:

1. Pressure waveform: pressure dampening should no longer be present, except in cases where the target vessel diameter is of similar diameter or smaller than the guide catheter.
2. Angiography: there should be brisk and wide blood flow around the vessel ostium. Inject enough contrast to allow reflux into the aorta, otherwise residual ostial stenosis may be missed. To prevent catheter disengagement from the coronary ostium it is helpful to leave a short segment of the guidewire within the vessel.
3. Physiology: FFR or a non-hyperemic index index can be measured.
4. Imaging: IVUS can confirm full ostium coverage and adequate stent expansion, as discussed below.

15.1.11 Closure

Performed as described in Chapter 11, Access Closure.

15.1.12 Physiology

Baseline assessment

Pressures should be equalized with the guide catheter disengaged from the ostium to avoid dampened pressure tracings. This can be done by having both the pressure guidewire or catheter and the guide catheter in the ascending aorta. Pressure

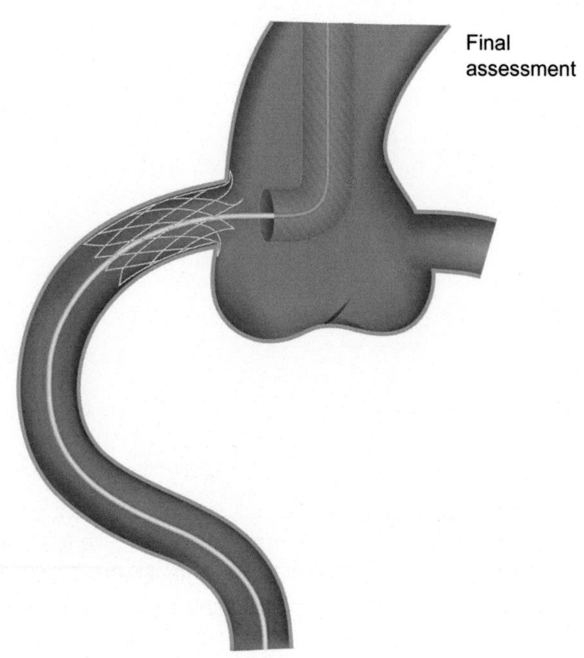

FIGURE 15.17 Final assessment of aorto-ostial stenting.

dampening can result in artificially higher FFR/resting indices and underestimation of the lesion severity [13]. Additionally, the guide catheter should be disengaged when performing adenosine FFR or nonhyperemic index measurements, as pressure dampening can result in artificially higher FFR/resting indices and underestimation of the lesion severity [13].

Final assessment

Pressure dampening should be detected and corrected by disengaging the guide catheter prior to physiologic assessment of aorto-ostial lesions post stenting, as described above.

15.1.13 Imaging

IVUS is preferred over OCT for assessment of aorto-ostial lesions, because obtaining optimal OCT images requires excellent replacement of the blood column with contrast that can be challenging with aorto-ostial lesions.

Baseline assessment

IVUS or OCT can help determine the reference vessel diameter and the presence of severe calcification, which may necessitate aggressive lesion preparation.

Final assessment

IVUS can confirm good expansion and complete ostium coverage with protrusion of the stent into the aorta (Fig. 15.18).

15.1.14 Hemodynamic support

Hemodynamic support is used as per standard practice as described in Chapter 14, Hemodynamic Support. Although PCI of ostial left main lesions rarely requires hemodynamic support, hemodynamic support could be considered if there is a high likelihood of hemodynamic compromise.

15.2 Branch ostial lesions

Branch ostial lesions are essentially Medina 0.0.1 or 0.1.0. bifurcation lesions, hence the relevant sections from Chapter 16 apply.

15.2.1 *Planning*

15.2.2 *Monitoring*

15.2.3 *Medications*

15.2.4 *Access*

15.2.5 *Engagement*

15.2.6 *Angiography*

Angiography in various projections can help better delineate the branch ostial lesion and facilitate balloon angioplasty and stent placement.

15.2.7 *Selection of target lesion(s)*

15.2.8 *Wiring*

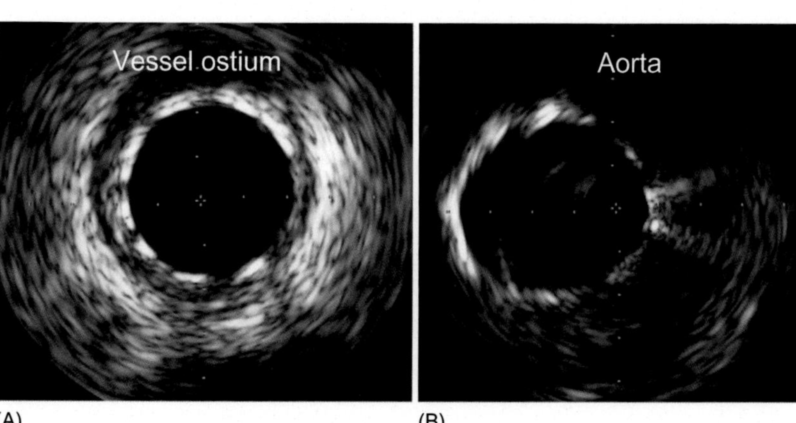

FIGURE 15.18 IVUS demonstrating good expansion of the vessel ostium (panel A) and protrusion into the aorta (panel B). *Courtesy of Dr. Alok Sharma.*

Wiring can be challenging depending on the angulation and the severity of the branch ostial stenosis. Techniques used for wiring through tortuosity, such as use of microcatheters, especially dual lumen microcatheters, and hydrophilic-coated or polymer-jacketed guidewires can facilitate wiring (Section 8.6).

Wiring both the target branch vessel and the main vessel is advised to maintain access to both vessels during side branch (SB) stenting.

15.2.9 *Lesion preparation*

Preparation of the branch ostial lesion should be done while avoiding injury of the main vessel. This is especially important if there is mild to moderate disease of the main vessel near the bifurcation. Watermelon seeding of the balloon back to the main vessel may be less important due to usually smaller branch vessel size (and hence smaller balloon), but injury may still occur in diseased vessels (see Section 15.1.9).

Use of plaque modification balloons (such as Angiosculpt, Chocolate, and cutting balloon) can minimize movement of the balloon and the risk of dissection of the side branch ostium.

15.2.10 *Stenting* (Fig. 15.19)

15.2.10.1 **T-stenting**

15.2.10.1.1 Both branches are wired (Fig. 15.20).

15.2.10.1.2 The SB is predilated (Fig. 15.21).

15.2.10.1.3 The stent (sized according to the SB diameter) is advanced into the SB with the proximal stent marker placed at the SB ostium (Fig. 15.22).

15.2.10.1.4 The stent is deployed and the SB balloon is removed (Fig. 15.23).

15.2.10.1.5 Final assessment (Fig. 15.24).

The operator should ascertain that no injury occurred to the main vessel (MV) by obtaining multiple orthogonal angiographic views and using intravascular imaging.

15.2.10.2 **Mini-crush (step-by-step)**

15.2.10.2.1 Both branches are wired (Fig. 15.25).

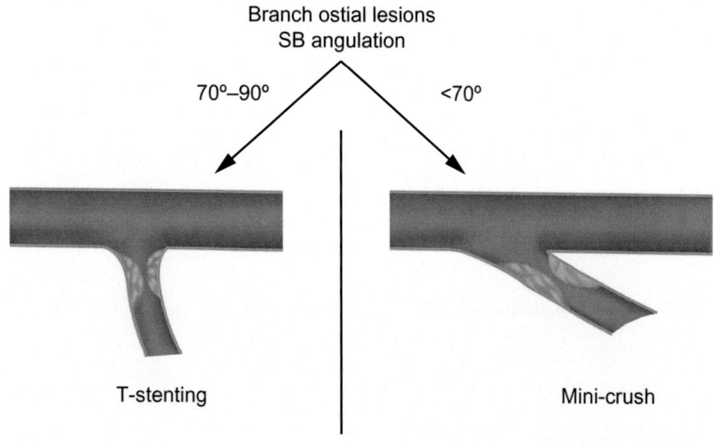

FIGURE 15.19 Stenting of branch ostial lesions depends on the SB angulation. For 70- to 90-degree angulation, T-stenting is usually performed, as described below. For <70-degree angulation, mini-crush is often performed, because significant prolapse of the SB stent into the main vessel (MV) is needed to fully cover the SB ostium.

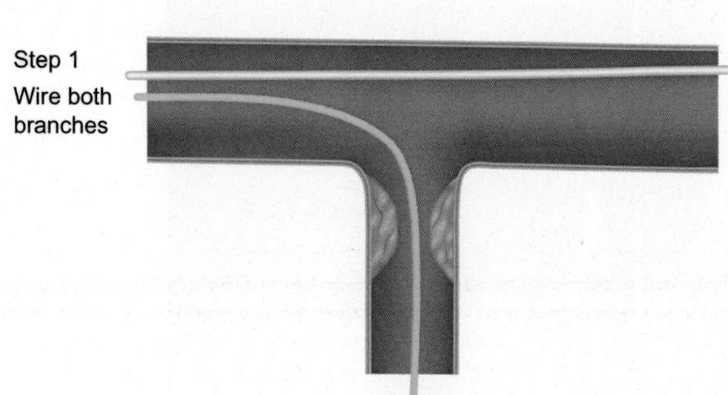

FIGURE 15.20 T-stenting: wiring both the main vessel and the side branch of the bifurcation.

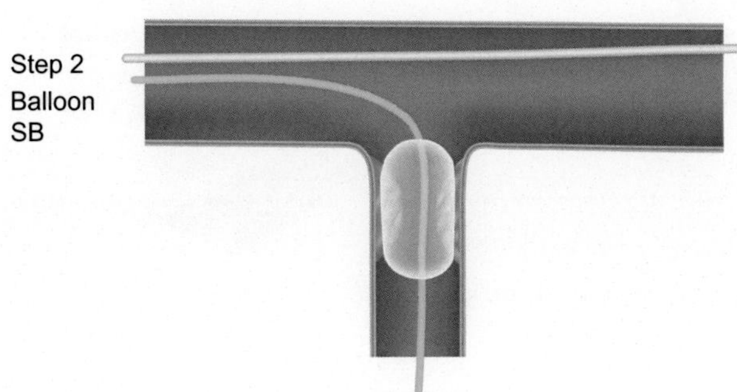

FIGURE 15.21 T-stenting: side branch predilation.

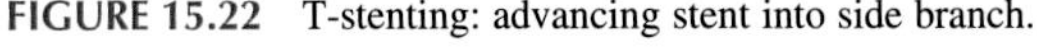

FIGURE 15.22 T-stenting: advancing stent into side branch.

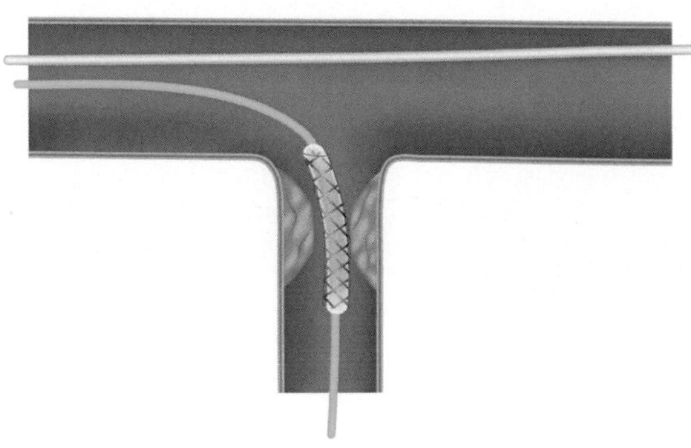

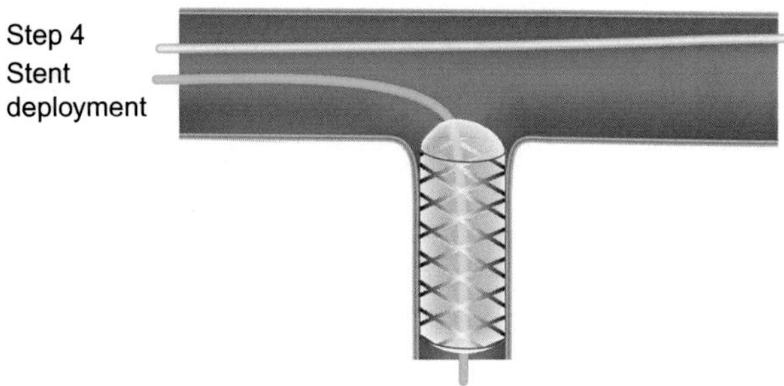

FIGURE 15.23 T-stenting: stent deployment into side branch.

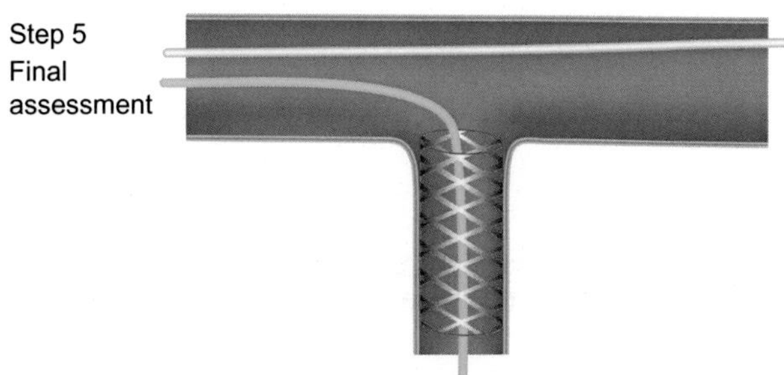

FIGURE 15.24 T-stenting: final assessment after branch ostial lesion stenting.

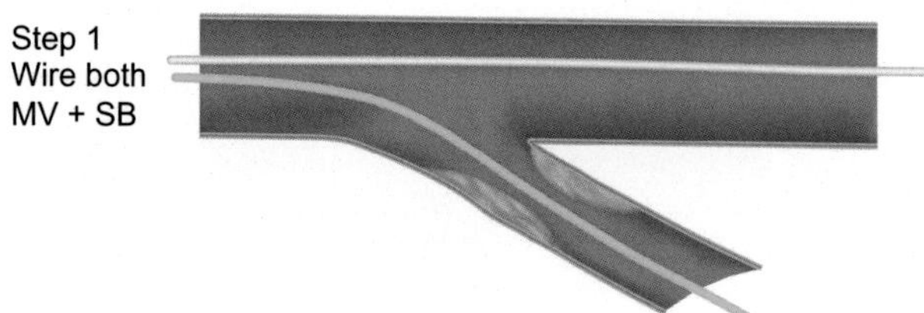

FIGURE 15.25 Mini-crush: wiring both the main vessel and the side branch.

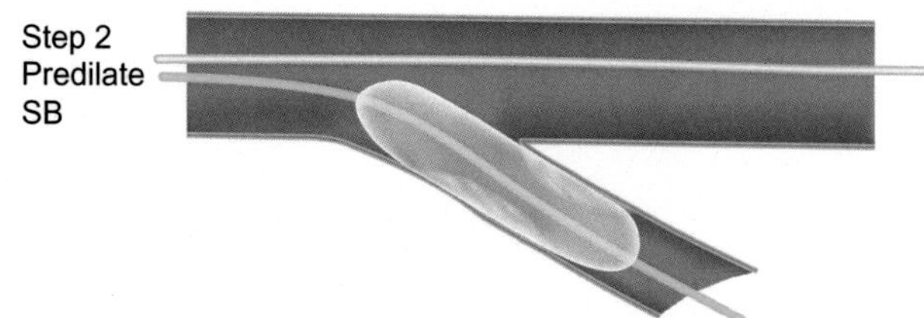

FIGURE 15.26 Mini-crush: side branch predilation.

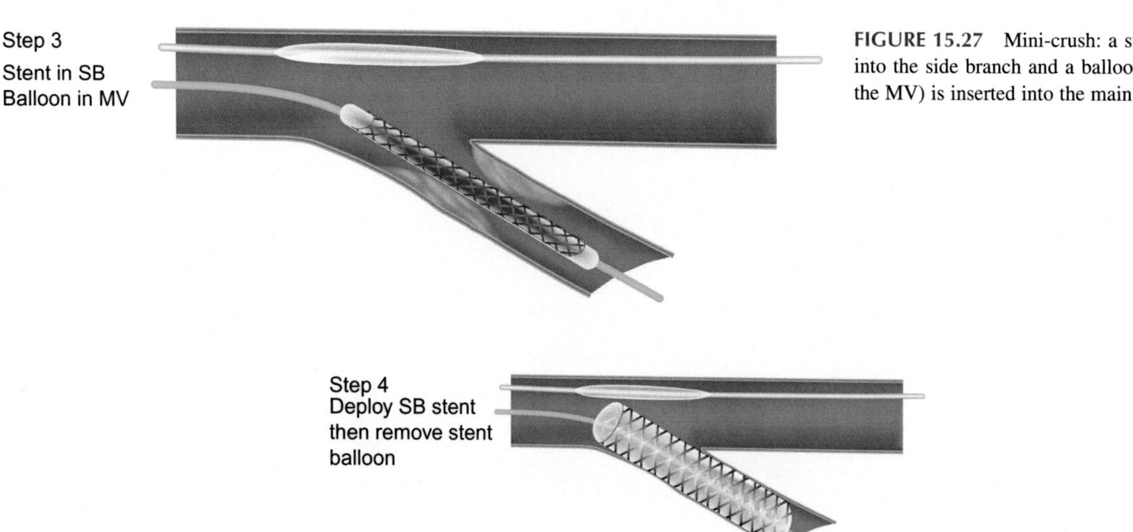

FIGURE 15.27 Mini-crush: a stent is inserted into the side branch and a balloon (sized 1:1 to the MV) is inserted into the main vessel.

Step 4
Deploy SB stent
then remove stent
balloon

FIGURE 15.28 Mini-crush: the side branch stent is deployed, followed by removal of the side branch stent balloon.

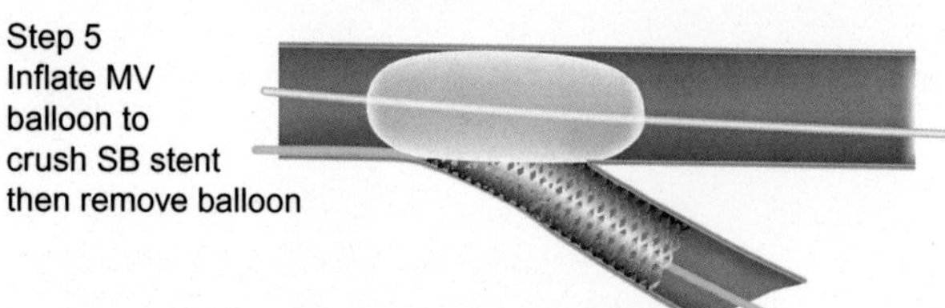

FIGURE 15.29 Mini-crush: the main vessel balloon is inflated, crushing the portion of the side branch stent that was protruding into the main vessel.

15.2.10.2.2 The SB is predilated (Fig. 15.26).

15.2.10.2.3 A stent (sized according to the SB diameter) is advanced into the SB (protruding in to the MV enough to cover the proximal and distal aspects of the side branch ostium). A balloon is advanced in the MV and positioned across from the SB ostium (Fig. 15.27).

15.2.10.2.4 Deploy the SB stent, then remove the stent balloon (Fig. 15.28).

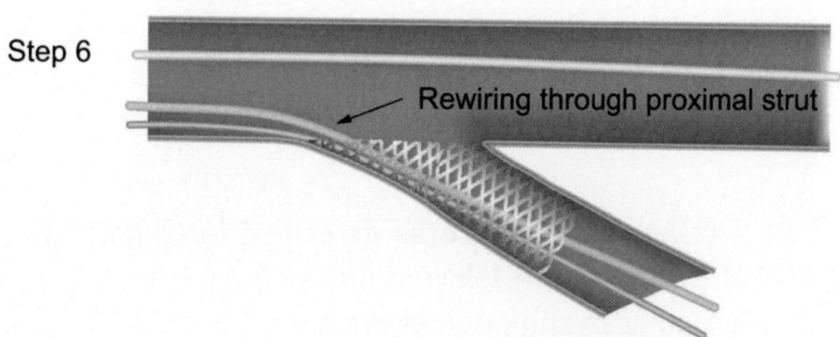

FIGURE 15.30 Mini-crush: the side branch is rewired through a proximal cell to ensure that the SB balloon will push the crushed stent to the carina (similar to DK crush). If the SB is rewired from a distal strut there is a risk upon balloon inflation of moving the crushed stent further away from the carina. If the proximal MV is significantly larger than the distal MV, the proximal optimization technique (POT) can be performed. If MV dissection occurs, it may require stenting using the double kissing crush (DK crush) technique.

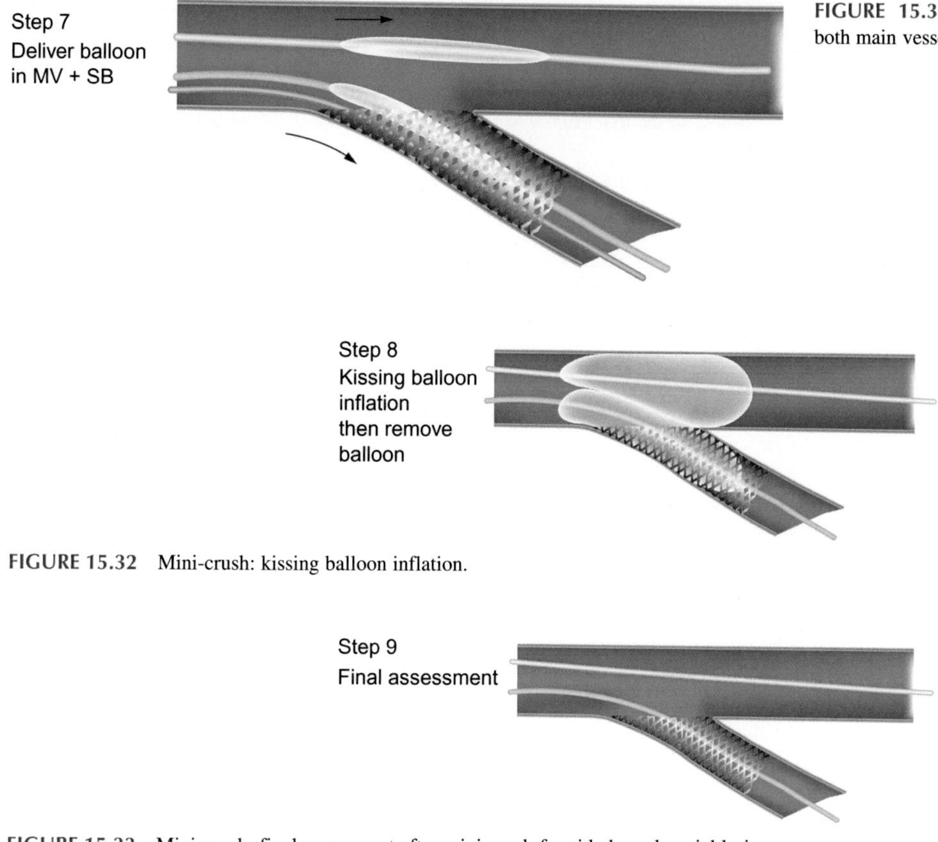

FIGURE 15.31 Mini-crush: balloon delivery into both main vessel and side branch.

FIGURE 15.32 Mini-crush: kissing balloon inflation.

FIGURE 15.33 Mini-crush: final assessment after mini-crush for side branch ostial lesions.

15.2.10.2.5 The MV balloon is inflated to crush the protruding portion of the SB stent, then the MV balloon is removed (Fig. 15.29).

15.2.10.2.6 The SB stent is rewired with another wire through a proximal strut (Fig. 15.30).

15.2.10.2.7 Balloons are delivered both the SB and the MV, in preparation for kissing balloon inflation (Fig. 15.31) and the jailed side branch wire is removed (the jailed wire acts as an anchor facilitating balloon delivery).

15.2.10.2.8 Kissing balloon inflation, then both balloons are removed (Fig. 15.32).

15.2.10.2.9 Final assessment (Fig. 15.33).

If there is significant plaque shift into the main branch, a stent may need to be placed in the main vessel as described in Chapter 16, Bifurcations.

15.2.11 *Closure*

15.2.12 *Physiology*

15.2.13 *Imaging*

Intravascular imaging can help determine stent expansion, adequate ostial coverage, and protrusion of the SB stent into the main vessel.

15.2.14 *Hemodynamic support*

Hemodynamic support is used as per standard practice as described in Chapter 14, Hemodynamic Support. It is a rare scenario to need support for PCI of a Medina 0.0.1 lesion unless the territory at risk is large (e.g., LAD), atherectomy is required, and significant left ventricular dysfunction is present.

References

[1] Jokhi P, Curzen N. Percutaneous coronary intervention of ostial lesions. EuroIntervention 2009;5:511–14.

[2] Tajti P, Burke MN, Karmpaliotis D, et al. Prevalence and outcomes of percutaneous coronary interventions for ostial chronic total occlusions: insights from a multicenter chronic total occlusion registry. Can J Cardiol 2018;34:1264–74.

[3] Alame A, Brilakis ES. Best practices for treating coronary ostial lesions. Catheter Cardiovasc Interv 2016;87:241–2.

[4] Katoh O, Reifart N. New double wire technique to stent ostial lesions. Cathet Cardiovasc Diagn 1997;40:400–2.

[5] Shlofmitz E, Martinsen BJ, Lee M, et al. Orbital atherectomy for the treatment of severely calcified coronary lesions: evidence, technique, and best practices. Expert Rev Med Devices 2017;14:867–79.

[6] Kern MJ, Ouellette D, Frianeza T. A new technique to anchor stents for exact placement in ostial stenoses: the stent tail wire or Szabo technique. Catheter Cardiovasc Interv 2006;68:901–6.

[7] Vaquerizo B, Serra A, Ormiston J, et al. Bench top evaluation and clinical experience with the Szabo technique: new questions for a complex lesion. Catheter Cardiovasc Interv 2012;79:378–89.

[8] Gutierrez-Chico JL, Villanueva-Benito I, Villanueva-Montoto L, et al. Szabo technique versus conventional angiographic placement in bifurcations 010-001 of Medina and in aorto-ostial stenting: angiographic and procedural results. EuroIntervention 2010;5:801–8.

[9] Ferrer-Gracia MC, Sanchez-Rubio J, Calvo-Cebollero I. Stent dislodgement during Szabo technique. Int J Cardiol 2011;147:e8–9.

[10] Kwan TW, Chen JP, Cherukuri S, et al. Transradial Szabo technique for intervention of ostial lesions. J Interv Cardiol 2012;25:447–51.

[11] Schrage B, Schwarzl M, Waldeyer C, et al. Percutaneous coronary intervention for ostial and bifurcation lesions using the Szabo technique: a single center experience. Minerva Cardioangiol 2017;65:331–5.

[12] Nguyen-Trong PJ, Martinez Parachini JR, Resendes E, et al. Procedural outcomes with use of the flash ostial system in aorto-coronary ostial lesions. Catheter Cardiovasc Interv 2016;88:1067–74.

[13] Patel KS, Christakopoulos GE, Karatasakis A, et al. Prospective evaluation of the impact of side-holes and guide-catheter disengagement from the coronary ostium on fractional flow reserve measurements. J Invasive Cardiol 2016;28:306–10.

Chapter 16

Bifurcations

16.1 Bifurcation algorithm

(Fig. 16.1)

16.1.1 Does the side branch need to be preserved?

This depends on the size of the side branch (SB) (usually branches <2 mm in diameter do not need to be preserved) and the supplied myocardial territory. If a decision is made to preserve the SB, it must be decided whether to simply wire the SB to help preserve patency during provisional stenting of the main vessel (MV), or whether to perform an upfront dedicated two-stent bifurcation percutaneous coronary intervention (PCI) strategy. To help decide this, it is necessary to assess the likelihood of SB occlusion after MV stenting.

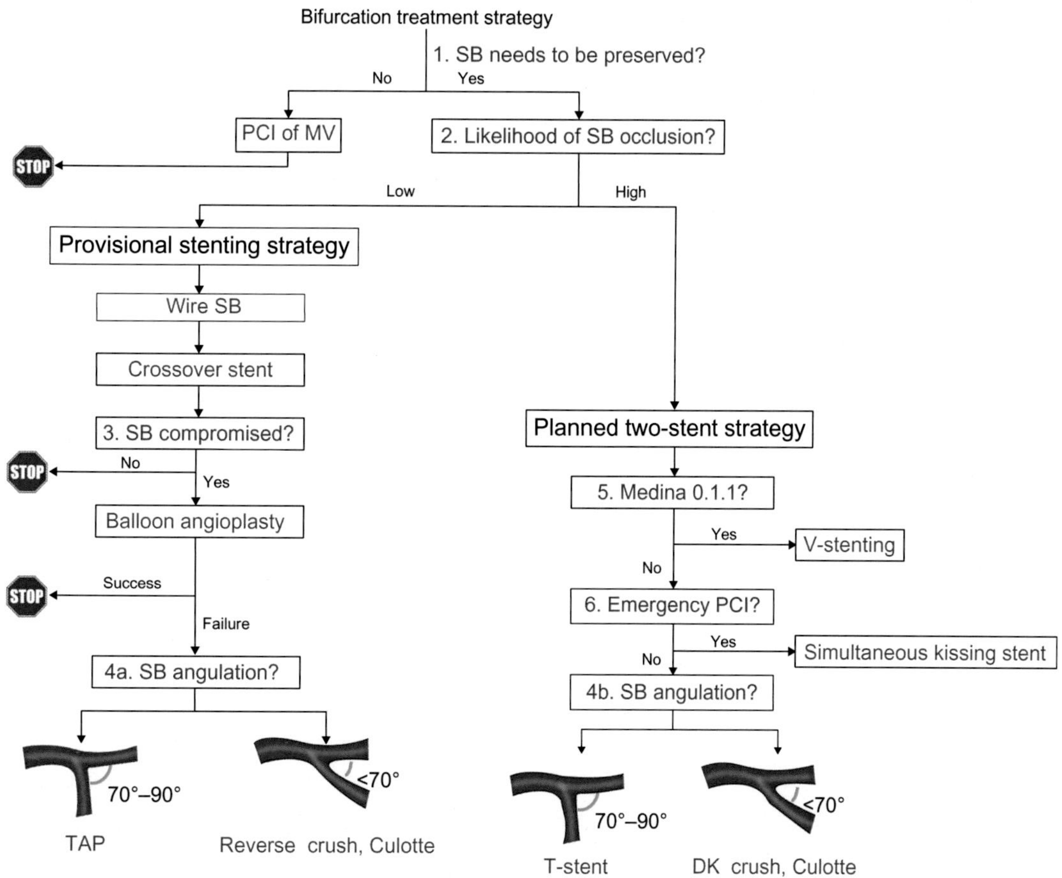

FIGURE 16.1 Algorithm for approaching coronary bifurcation lesions. MV, main vessel; SB, side branch; PCI, percutaneous coronary intervention; TAP, T and protrusion; DK crush, double kissing crush.

Manual of Percutaneous Coronary Interventions. DOI: https://doi.org/10.1016/B978-0-12-819367-9.00016-0

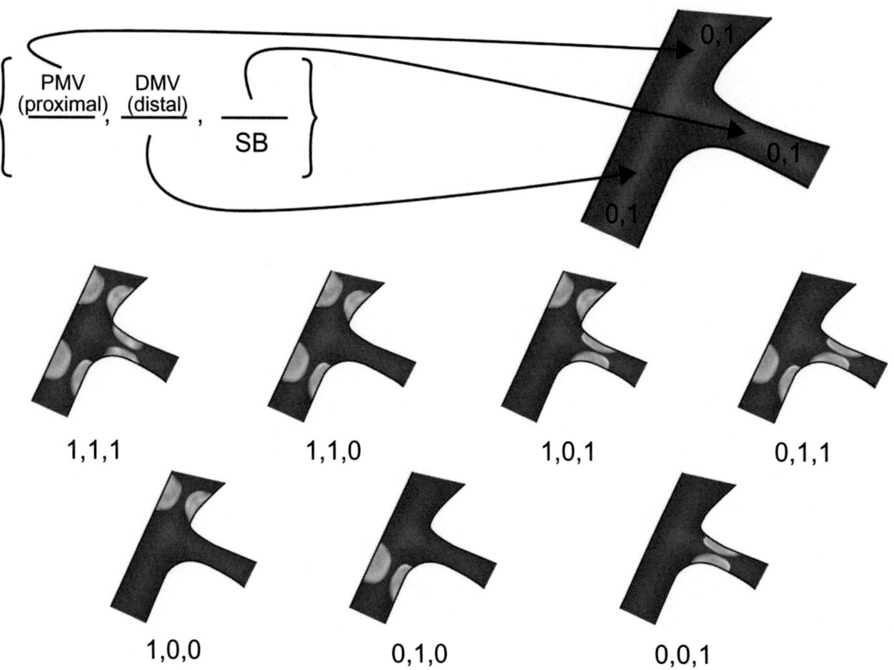

FIGURE 16.2 The Medina classification of bifurcation lesions. The first number describes the proximal main vessel (PMV), the second number the distal main vessel (DMV), and the third number the side branch (SB). 1 signifies ≥ 50% diameter stenosis, whereas 0 signifies <50% diameter stenosis.

16.1.2 What is the likelihood of SB occlusion?

The likelihood of SB occlusion depends on:

1. Location of disease (Medina classification [1], Fig. 16.2).
2. Severity of SB disease.
3. Lesion morphology (calcification, thrombus, length). Use of intravascular imaging can help with lesion morphology assessment.
4. Bifurcation angulation (extreme angles may be challenging to wire).

16.1.3 If a provisional stenting strategy (Section 16.2.10.1) is implemented, is the SB compromised after MV stenting?

If an important SB becomes compromised after MV stenting (TIMI flow < 3, diameter stenosis > 70%, or NHLBI dissection B or more), efforts should be made to restore flow. The risk of SB compromise is reduced by sizing the MV stent according to the diameter of the distal MV. Balloon angioplasty often suffices to restore or improve flow in the SB, but sometimes SB stenting may be necessary.

16.1.4 What is the SB angulation?

The SB angulation is key for selecting the optimal two-stent bifurcation stenting strategy (either primary or in case of SB compromise after MV stenting).

For angulations between 70°–90° T-stenting (Section 16.2.10.2) or T and Protrusion (TAP, Section 16.2.10.3) are usually used.

For angulation <70°, double kissing crush (DK crush) (Section 16.2.10.5), culotte (Section 16.2.10.6), and reverse crush (Section 16.2.10.4) are commonly used, with DK crush being the preferred strategy in most cases.

16.1.5 Medina 0.1.1

V-stenting (Section 16.2.10.7) can be performed in Medina 0.1.1 bifurcations.

16.1.6 Need for emergency stenting

When emergency stenting is needed and access to both MV and SB needs to be maintained, simultaneous kissing stents (SKS, Section 16.2.10.8) can be used. However, subsequent treatment and equipment delivery can be very challenging after SKS, due to formation of a long stent neocarina. SKS should not be performed in non-emergent bifurcation PCI.

16.2 Bifurcation PCI: step-by-step

16.2.1 Preparation

Planning for bifurcation PCI is performed as discussed in Chapter 1: Planning. Some bifurcation stenting techniques, such as V-stenting and SKS require at least 7 French guide catheters (that have large enough lumen to accommodate two stents).

16.2.2 Monitoring

Monitoring for bifurcation PCI is performed as discussed in Chapter 2: Monitoring.

16.2.3 Pharmacology

Pharmacology is as discussed in Chapter 3: Medications.

Periprocedural anticoagulation for bifurcation PCI is similar to standard PCI.

Prolonged dual antiplatelet therapy should be considered in patients who have low risk of bleeding, as bifurcation stenting has been associated with increased risk of stent thrombosis.

16.2.4 Access

Arterial access is described in Chapter 5: Coronary and Graft Engagement.

Radial or femoral access are both acceptable for bifurcation stenting, depending on patient anatomy, lesion complexity, and operator experience. For highly complex cases, femoral access may provide optimal support.

16.2.5 Engagement

Goal: To engage target vessel with a guide catheter that will provide optimal support and facilitate performance of the various steps of bifurcation stenting.

How to?

1. Coronary engagement is covered in detail in Chapter 5: Coronary and Graft Engagement.
2. Guide catheter type: supportive guide catheters are preferred (AL for the right coronary artery and EBU or XB for the left main).
3. Guide catheter size: Most coronary bifurcation techniques can be performed through 6 French guide catheters (except for V-stenting and SKS deployment which need at least 7 French guide catheters). However, it is much easier to pass two pieces of equipment (in particular, previously used angioplasty balloons) through a larger guide catheter. A larger guide catheter also provides more support and facilitates troubleshooting in case of difficulties during the procedure.

16.2.6 Angiography

Performing coronary angiography is described in detail in Chapter 6: Coronary Angiography.

For bifurcation lesions angiography should be performed in angulations that allow optimal separation of the MV and the SB, to facilitate wiring and placement of the stents as well as the stent crush, if applicable.

16.2.7 Selecting target lesion(s)

This is performed as described in detail in Chapter 7: Selecting Target Lesion(s).

16.2.8 Wiring

Performing coronary wiring is described in detail in Chapter 8: Wiring.

Wiring of both MV and SB should be performed in most cases, unless the SB is too small or does not have any significant disease. Standard workhorse wires (Section 30.7.1) are preferred for both the MV and SB.

If SB wiring is challenging, the following techniques can be used (Section 8.6.8):

A. Wire-based options

- Create secondary bend on the wire.
- Use a hydrophilic-coated or polymer-jacketed guidewire (Section 30.7.2).
- Reversed guidewire technique (Chapter 8: Wiring, Section 8.6) [2–4].

B. Microcatheter options

- Use of big (Section 30.6.1) or small (Section 30.6.2) microcatheters improves wire support and allows changing the shape of the wire tip without losing wire position.
- Angulated microcatheters (such as Venture, Swift NINJA, and Supercross, Section 30.6.3)
- Dual lumen microcatheter (such as TwinPass Torque and Sasuke, Section 30.6.4).

C. Modify MV

- With balloon angioplasty (possibly with use of plaque modification balloons) or atherectomy (carries risk of SB occlusion).

The distal position of the MV and SB wires should be constantly monitored, as too distal migration could cause perforation. If polymer-jacketed guidewires were used for wiring they should be replaced for a workhorse guidewire before performing lesion preparation and stenting.

16.2.9 Lesion preparation

Lesion preparation is described in detail in Chapter 9: Lesion Preparation.

- Preparation of the MV lesion should be done in all cases, as it facilitates stent sizing, stent delivery, and stent expansion.
- Preparation of the SB should generally be avoided when provisional stenting is done, as it may lead to SB dissection, necessitating stenting.
- Preparation of the SB (and the MV) should be routinely performed when two-stent techniques are planned.

Intravascular imaging/physiology is performed as described in Chapter 12: Coronary Physiology and Chapter 13: Coronary Intravascular Imaging. Intravascular imaging can be very useful for determining the size and length of the MV and SB lesions and for determining the need for atherectomy or other forms of lesion preparation.

- *Atherectomy*: Atherectomy can be challenging in coronary bifurcation lesions, because only one guidewire (the atherectomy wire) can be in the treated coronary artery during atherectomy. The other guidewire is traditionally removed prior to atherectomy to prevent wire fracture. However, in case of dissection of the non-wired vessel, rewiring can be challenging. If it is highly desirable to not remove the second wire the following techniques could be used:

Solutions for keeping in place both the MV and SB wires during atherectomy:

- Advance atherectomy device (burr or crown) through a guide extension (if the lesion that requires atherectomy is at the proximal SB or the distal portion of the MV). The guide extension is placed over the atherectomy wire only, isolating the burr or crown from the other wire. This requires a 7 French guide catheter (with a 6 French guide catheter extension) which can accommodate a 1.25 mm rotational atherectomy burr or the 1.25 mm orbital atherectomy crown. An 8 French guide catheter with a 7 French guide catheter extension can accommodate a 1.5 mm rotational atherectomy burr.
- Use a microcatheter over the nonatherectomy wire to protect it: a smaller microcatheter, such as the Finecross should be used to minimize the risk of microcatheter damage. If multiple atherectomy runs are performed, the microcatheter should be slowly moved during atherectomy runs to minimize damage of the same portion of the microcatheter (Fig. 16.3, CTO PCI Manual case 136). This requires an 8 French guide catheter. Alternatively, atherectomy can be performed leaving the SB wire in place, accepting a low risk of guidewire damage.
- Use laser atherectomy.

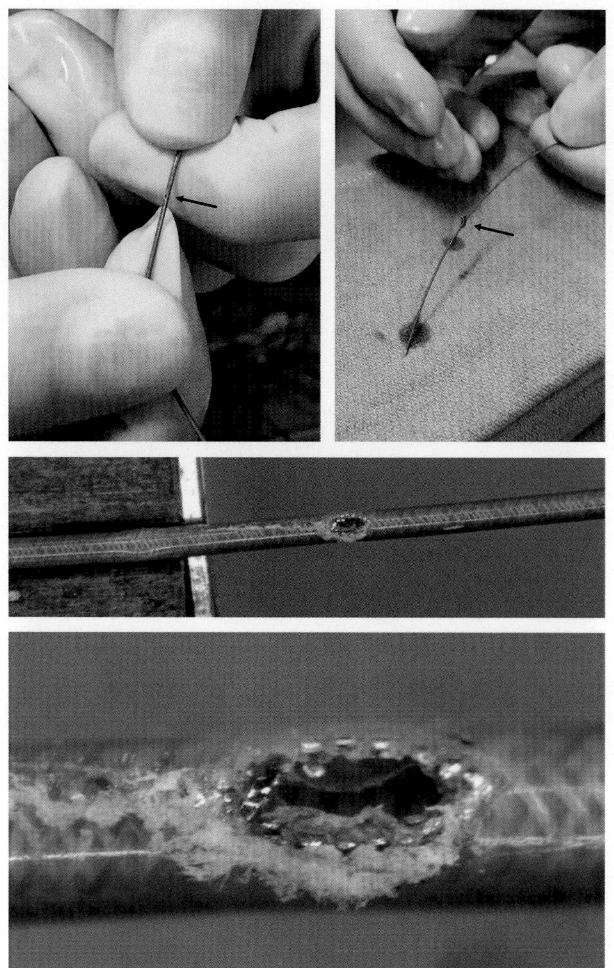

FIGURE 16.3 Perforation of a Finecross microcatheter used to protect the side branch guidewire while performing orbital atherectomy on the main vessel guidewire.

16.2.10 Stenting

16.2.10.1 Provisional stenting

See online video: *"How to do provisional stenting"*

***Background*:** Provisional stenting is the preferred and easiest technique for most bifurcation lesions, if technically feasible without posing significant risk of compromising a significant SB. If the severity of disease at the SB ostium is unclear, hemodynamic assessment or intravascular imaging should be considered.

***Goal*:** Deploy a stent in the MV without compromising the SB.

***How*?**

***Preparation*:**

***Wiring*:** both MV and SB should be wired if the SB is important and there is risk of occlusion (Fig. 16.4).

***Lesion preparation*:** balloon angioplasty is usually done in the MV, but is in most cases avoided in the SB, as it may lead to SB dissection necessitating stenting.

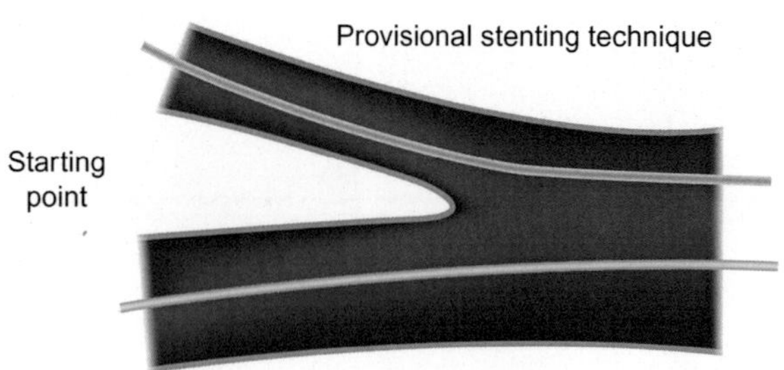

FIGURE 16.4 Provisional bifurcation stenting: wiring both main vessel and side branch.

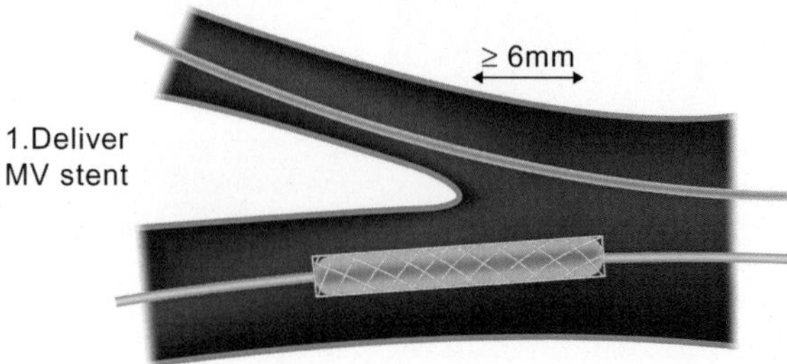

FIGURE 16.5 Provisional bifurcation stenting: a stent is delivered in the main vessel and positioned across the bifurcation.

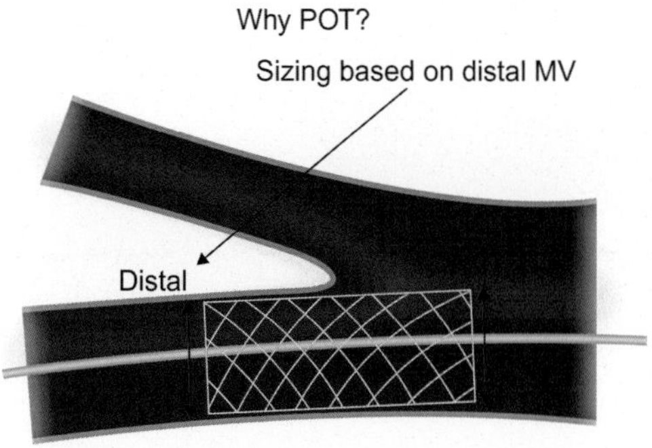

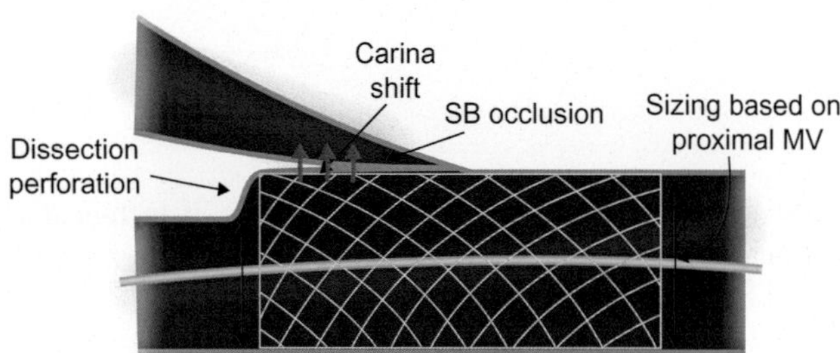

FIGURE 16.6 Why the stent should be sized to match the distal true lumen, followed by POT. If the stent is sized based on the proximal MV diameter, the SB may become occluded and perforation or dissection can occur in the distal MV.

16.2.10.1.1 Provisional stenting steps

Step 1. Deliver MV stent (Fig. 16.5)

Tips:

- The diameter of the MV stent should match the diameter of the distal MV (Fig. 16.6), but should also be able to be expanded to the size of the proximal MV [using the proximal optimization technique (POT)] to achieve full apposition. Knowledge to the individual stent post dilatation limits (Fig. 10.1) is necessary to avoid excessive stent deformation. The MV stent length should be enough to allow use of a balloon for POT (≥6 mm if 6 mm balloons are available).

Step 2. Deploy MV stent, jailing the SB guidewire (Fig. 16.7)

Jailing of the SB wire is recommended in any bifurcation stenting procedure when the SB needs to be preserved for the following reasons [5]:

1. The jailed guidewire helps keep the SB open; also in case of occlusion, it acts as a marker of the SB position, facilitating rewiring attempts.
2. The jailed guidewire facilitates reaccess to the SB by favorably altering the angle of the bifurcation for recrossing.
3. The jailed guidewire provides anchoring that facilitates seating of the guide catheter, and increases support for equipment advancement in subsequent bifurcation PCI steps; and
4. In extreme situations, a low profile balloon (or torquable microcatheter) can be advanced over the jailed guidewire and inflated to urgently rescue the SB.

Several guidewire types (workhorse, with hydrophilic coating, or polymer-jacketed, Section 30.7) can be jailed. There has been controversy about whether polymer-jacketed guidewires should be jailed or not due to risk of stripping the polymer during wire withdrawal, potentially causing distal embolization. In one microscopy-based study of jailed guidewires, only 2 of 115 polymer-coated guidewires showed mild damage while 55% (63 of 120) of the non-polymer-coated guidewires were damaged, with two showing internal fracturing [6]. The length of the jailed guidewire appeared to be a factor contributing to the degree of guidewire damage and wiring the SB was quicker when a polymer-coated guidewire was used [6]. On the other hand, distal embolization of the polymer jacket material was detected downstream of stented coronaries in 10% of autopsied hearts in one study (not in the setting of guidewire jailing) [7].

Tips and tricks:

- Before jailing, partially withdraw the SB guidewire, so that the radiopaque portion of the SB guidewire is closer to the carina making the wire more visible. However, do not jail across the radiopaque portion of the guidewire, as this may increase the risk of guidewire fracture.

Step 3. Perform POT (Fig. 16.8)

Tips and tricks:

- The diameter of the balloon used for POT should match the diameter of the proximal MV.
- The length of the balloon used for POT should be equal or shorter to the MV stent length proximal to the carina. Since the shortest balloon length currently available is 6 mm, the length of the MV stent should be enough to allow at least 6 mm and ideally 8–10 mm of stent proximal to the carina after placement.

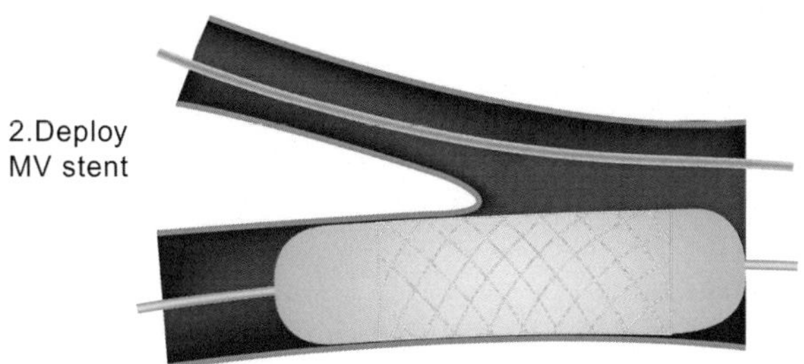

FIGURE 16.7 Provisional bifurcation stenting: main vessel stent deployment.

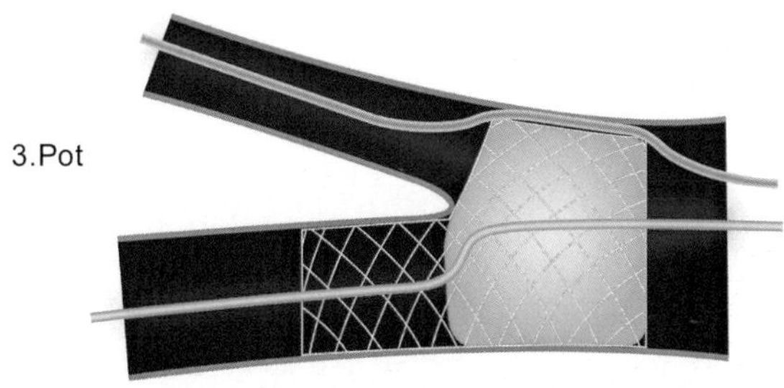

FIGURE 16.8 Illustration of POT (proximal optimization technique). POT is performed by inflating a short balloon in the portion of the stent proximal to the carina. The goal of POT is to optimize deployment of the MV stent proximal to the carina, and is especially important in cases with significant MV diameter discrepancy proximal and distal to the carina. Moreover, POT increases the stent strut diameter, facilitating distal wire crossing.

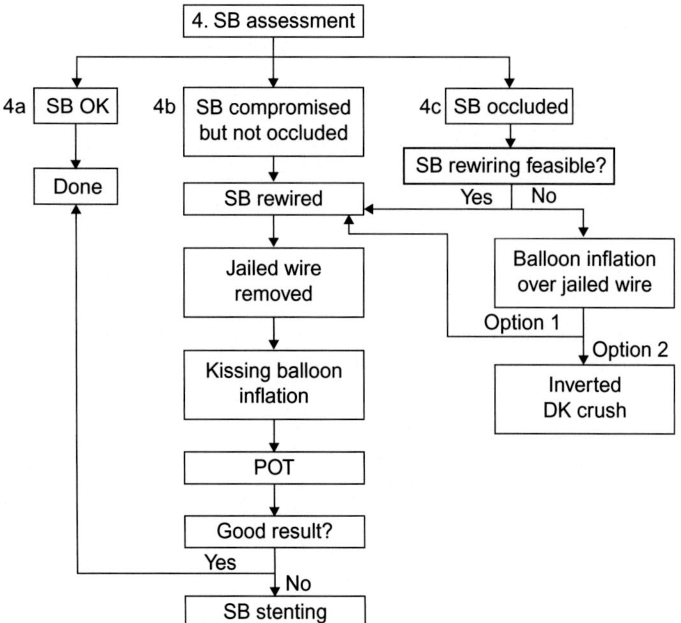

FIGURE 16.9 Provisional bifurcation stenting: assessment of the side branch after MV stenting.

- The distal marker of the balloon used for POT should be placed at the carina. Too distal placement may increase the risk for compromising the SB, whereas too proximal placement will not allow optimal expansion of the MV stent next to the SB ostium.
- If the balloon length is shorter than the MV stent length proximal to the carina, multiple inflations will be needed, changing the position of the balloon. If the balloon is longer than the MV stent length proximal to the carina, it will injure the MV proximal to the stent, increasing the risk of restenosis.

Step 4. Assess SB (Fig. 16.9)
Is the SB compromised?

***How to assess the SB after MV stenting*:**

Antegrade flow: if antegrade flow is decreased or absent, the SB is compromised. Significant disease at the SB ostium or placement of an oversized stent in the MV (or use of an oversized POT balloon) increase the likelihood of SB compromise.

Luminal stenosis: The severity of the SB stenosis can be assessed visually, although visual interpretation frequently overestimates the severity of SB stenosis.

Physiologic assessment: Coronary physiology can be used (in nonculprit lesions for acute coronary syndromes) to determine whether the SB is ischemic post stenting [8].

Intravascular imaging: Intravascular imaging (IVUS or OCT) can also be used to determine the severity of the SB ostial stenosis.

Step 4a. No SB compromise: procedure done!

Step 4b. Yes, SB is compromised but not occluded:

1. ***The SB is rewired through the MV stent (through a distal strut***) (Fig. 16.10)
   The reason for distal strut rewiring is that is provides better scaffolding of the SB ostium (Fig. 16.11) [9]
   Tips and tricks:
   - Some operators use the MV wire for rewiring the SB (so as to avoid substent wiring with a new guidewire), however in our practice we usually use a new wire for SB rewiring.
   - Some operators advance the guidewire that will be used for SB rewiring through a microcatheter. Once the SB is crossed, the microcatheter is advanced into the SB to facilitate subsequent balloon advancement into the SB.

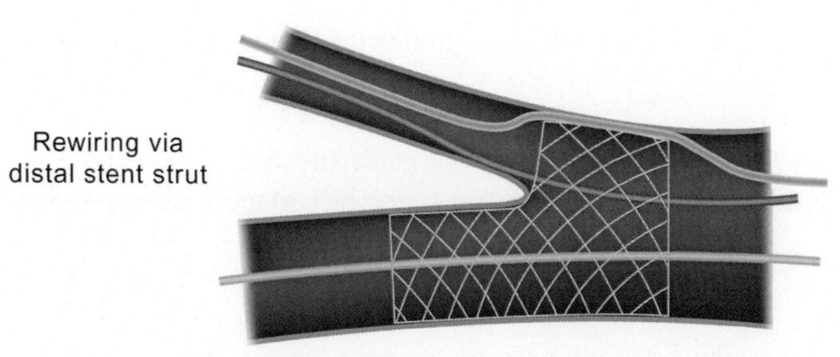

FIGURE 16.10 Provisional bifurcation stenting: side branch rewiring after MV stenting (rewiring is performed through a distal strut, i.e., closer to the carina).

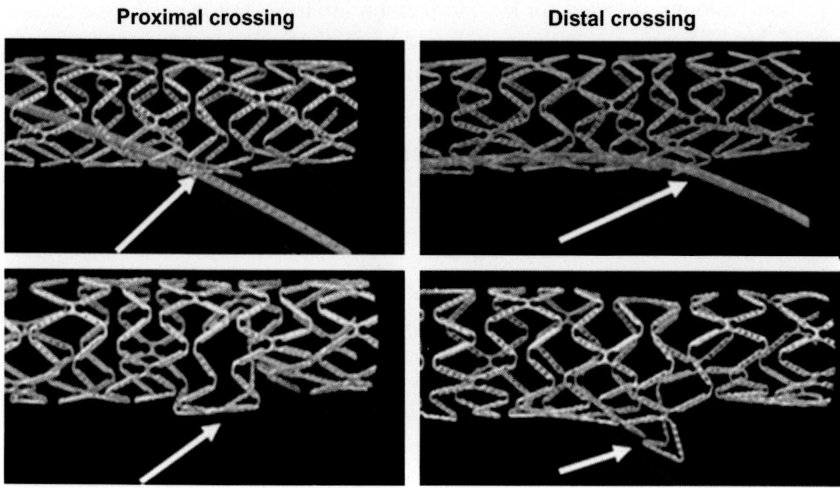

FIGURE 16.11 Proximal versus distal crossing in side branch. This figure shows the importance of distal cell recrossing close to the carina after main vessel stenting. Recrossing wire through the strut closest to the carina showed better scaffolding of the side branch ostium than proximal recrossing that pushes the struts inward toward the main vessel lumen. The *white arrow* indicates location of wire recrossing and its effect on side branch scaffolding. *Reproduced with permission from Sawaya FJ, Lefevre T, Chevalier B, et al. Contemporary approach to coronary bifurcation lesion treatment. JACC Cardiovasc Interv 2016;9:1861–78. Copyright Elsevier.*

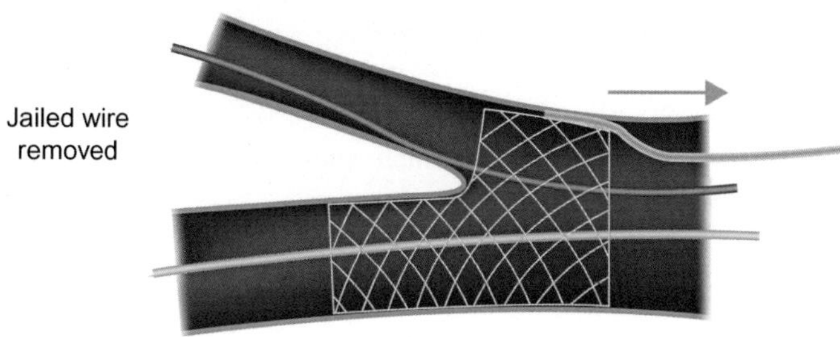

FIGURE 16.12 Provisional bifurcation stenting: removal of the jailed guidewire.

- To ensure a distal strut crossing, pass the wire distal to the bifurcation, then point the wire tip toward the SB. Gently pull the wire back and watch for the wire tip to drop toward the SB, then advance the wire through the stent strut.
- Confirmation of guidewire crossing through a distal strut can be obtained using OCT.

2. ***The jailed wire is removed*** (Fig. 16.12)

Challenge: Difficulty removing the jailed guidewire.

Solutions:

1 Do NOT pull hard! Apply gentle constant traction instead. Pulling hard may lead to guidewire fracture and/or unraveling, possibly requiring emergency cardiac surgery [10]. Also, pulling the guidewire hard may cause deep intubation of the guide catheter that could lead to guide catheter induced dissection.

2 Advance a small balloon or microcatheter all the way to the area of guidewire entrapment, and then gently withdraw.

3 Advance a small balloon all the way distally and inflate to free up the entrapped guidewire. Repeat POT should be performed afterwards.
4 Rotation of the entrapped guidewire using a torquer may help.
5 Before jailing a guidewire, ensure that the tip is straight, as subsequent withdrawal of a guidewire with a loop at the tip may be more challenging as the guidewire can become entangled with the deployed stent (PCI Manual cases 33, 88). Avoid jailing long guidewire segments.

3. ***Kissing balloon inflation*** (Fig. 16.13)

   Although not always necessary, kissing balloon inflation can help optimize the geometry of the stent and the SB ostium coverage and facilitate SB access in the future.

   Balloon type: usually noncompliant, short balloons are used for kissing balloon inflation, but they may be harder to deliver. Compliant balloons can be used, but high-pressure inflations should be minimized.

   Balloon size: According to Finet's law, the effective diameter of simultaneous inflation of two balloons next to each other is equal to:

$$0.678 \times (\text{balloon 1 diameter} + \text{balloon 2 diameter})$$

   The effective diameter of the simultaneously inflated balloons should be similar to the diameter of the proximal MV.

4. ***Final POT of the MV*** (Fig. 16.14)

   Final POT may not be necessary depending on the result achieved with kissing balloon inflation, but should be considered to correct the oval deformation of the MV stent caused by the effect of two balloons inflated in the proximal MV during kissing inflation.

5. ***If the SB result remains suboptimal, a SB stent may need to be placed (using the TAP, reverse crush, or culotte techniques, as described below).***

4c. Yes, SB is occluded:

Rewiring SB feasible?

Yes (next steps are similar to 4b above):

1. The SB is rewired through the MV stent (through a distal strut).
2. The jailed wire is removed after a low profile balloon can be advanced into the SB, which confirms that the guidewire entered through and not underneath the MV stent that is jailing the SB.

FIGURE 16.13 Provisional bifurcation stenting: kissing balloon inflation.

FIGURE 16.14 Provisional bifurcation stenting: final POT.

3. Kissing balloon inflation.
4. Final POT of the MV.
5. If the SB result remains suboptimal, a SB stent may need to be placed (using the T-stenting, TAP, reverse crush, or culotte techniques, as described below).

***No*:**

1. A small (1.0–1.5 mm) balloon is inserted over the jailed guidewire partially into the occluded SB (Fig. 16.15). If a balloon will not cross, consider a torquable microcatheter which often can be spun into the SB, creating a channel for balloon crossing.
2. Balloon angioplasty is performed to restore antegrade flow in the SB (Fig. 16.16)

Option 1.

3. The SB is then rewired through the MV stent (ideally using a dual lumen microcatheter to avoid SB wire advancement behind the MV stent struts that have been "lifted") (Fig. 16.17).
   Step 5. The jailed guidewire is removed (Fig. 16.18)
   Step 6. Kissing balloon inflation (Fig. 16.19)
   Step 7. Final POT of the MV (Fig. 16.20)
4. If the SB result remains suboptimal, a SB stent may need to be placed (using T-stenting, TAP, reverse crush, or culotte, as described below).

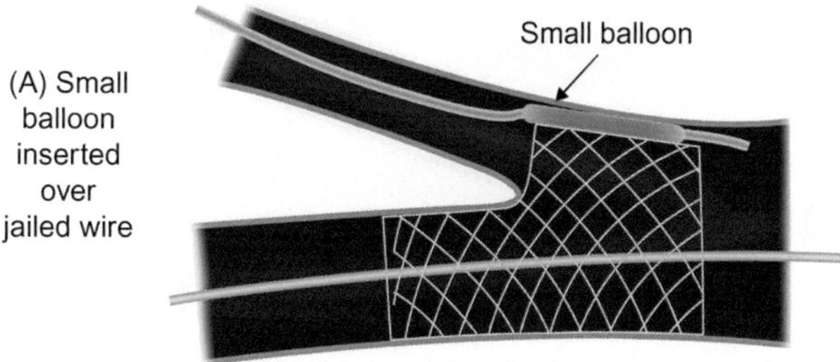

FIGURE 16.15 Provisional bifurcation stenting causing SB occlusion: advancement of a small balloon over the jailed guidewire.

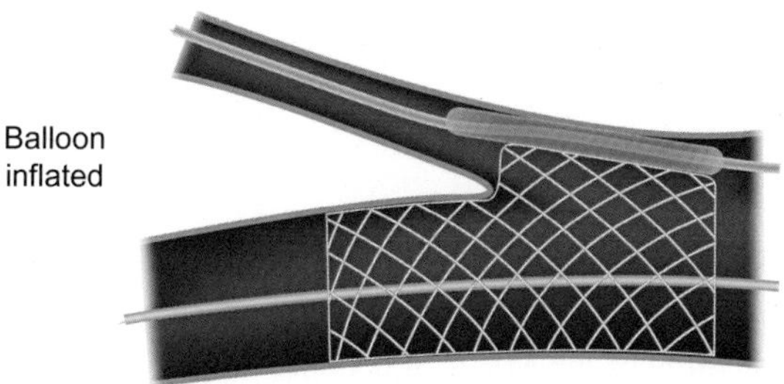

FIGURE 16.16 Provisional bifurcation stenting causing SB occlusion: balloon inflation over jailed guidewire.

FIGURE 16.17 Provisional bifurcation stenting causing SB occlusion: side branch rewiring through the main vessel struts.

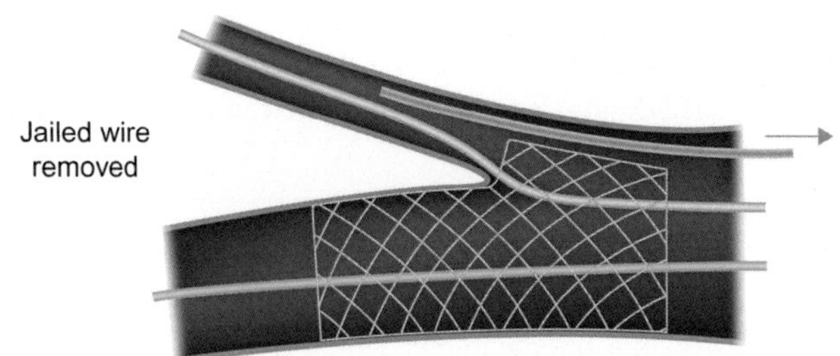

FIGURE 16.18 Provisional bifurcation stenting causing SB occlusion: removal of the jailed guidewire.

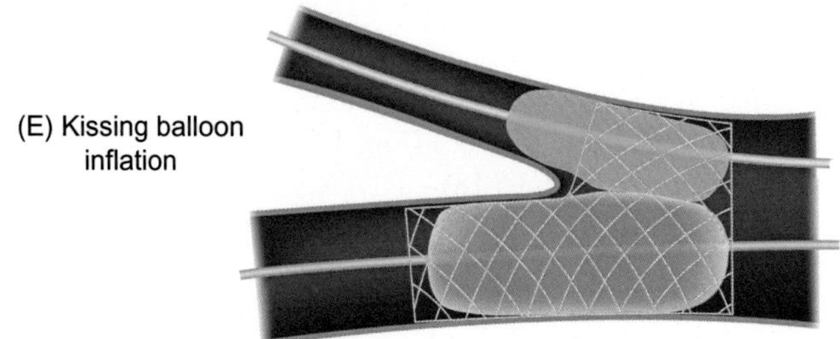

FIGURE 16.19 Provisional bifurcation stenting causing SB occlusion: kissing balloon inflation.

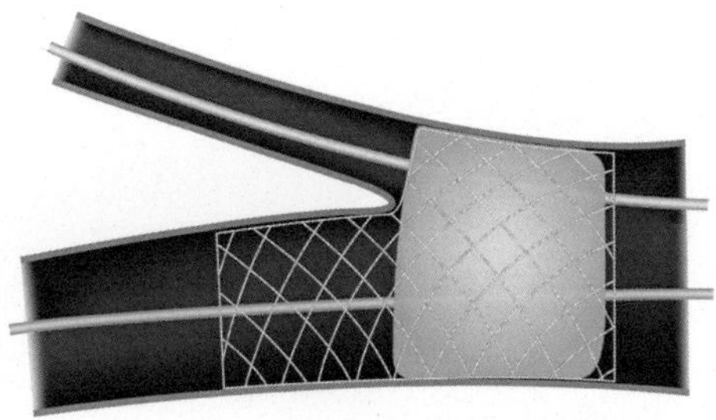

FIGURE 16.20 Provisional bifurcation stenting causing SB occlusion: final POT.

Option 2.

1 Convert to inverted DK crush technique, with the SB acting as MV and the MV acting as SB (see section on DK crush below).

16.2.10.2 T-stenting

***Background*:** T-stenting and TAP are best suited for bifurcations with angulations close to 90°, especially when the MV is large relative to the size of the SB. T stenting and TAP are easier and faster to perform than DK crush or culotte. TAP is performed after provisional stenting of the MV, if it becomes apparent that SB stenting is needed.

***How*?**
T-stenting steps

1. **Insert MV balloon (sized 1:1 with the distal MV) in the MV at or distal to the SB origin.**

2. **Deliver SB stent with the proximal stent marker placed at the origin of the SB, so that the entire ostium of the SB is covered.**

   Challenge: inability to advance stent in SB or MV.
   Solutions:
   - Additional SB or MV preparation (standard balloon angioplasty, additional POT, modified balloon angioplasty, atherectomy)

- Increase support (usually by using the SB anchoring technique with anchoring balloon in the MV or using a guide catheter extension)

Challenge: Poor SB ostium coverage
Solutions:

- The SB stent should be long enough to cover the SB lesion and protrude slightly in the MV. Longer protrusion is OK, but shorter protrusion may lead to suboptimal SB ostium coverage, potentially requiring implantation of another stent.

3. **Deploy SB stent.**
   Tips:
   - The stent is deployed at 10–12 atm (not at higher pressure to minimize the risk of distal edge dissection) and then the stent balloon is pulled slightly back protruding into the MV and inflated at high pressure (20 atm) to optimize stent expansion.
4. **Remove SB balloon and guidewire.**
5. **Deploy MV stent.**
6. **Rewire SB and perform kissing balloon inflation.**

16.2.10.3 T and Protrusion (TAP)

Background: TAP is a technique designed for bailout stenting of the SB if it becomes compromised after deployment of the MV stent. The advantage of the technique is its simplicity (no rewiring of the SB is required). The disadvantage is the creation of a neocarina from the SB stent struts that may impede equipment advancement in the MV.

How?

TAP steps

Preparation: TAP is performed after the MV is stented, the SB is rewired (ideally at a distal stent strut), kissing balloon inflation and repeat POT is performed (Fig. 16.21).

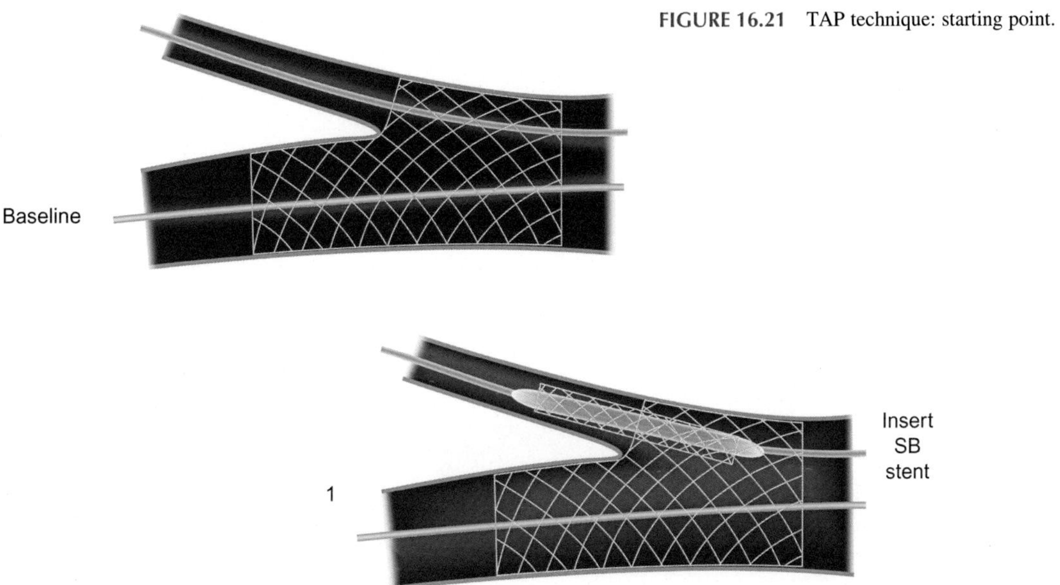

FIGURE 16.21 TAP technique: starting point.

FIGURE 16.22 TAP technique: deliver SB stent across the SB ostium.

Step 1. Deliver SB stent across the SB ostium (Fig. 16.22).

Step 2. Insert MV balloon across the take-off of the SB (Fig. 16.23).

Step 3. Position SB stent with minimal protrusion (1–2 mm) into the MV (Fig. 16.24).

Step 4. Deploy SB stent (Fig. 16.25).

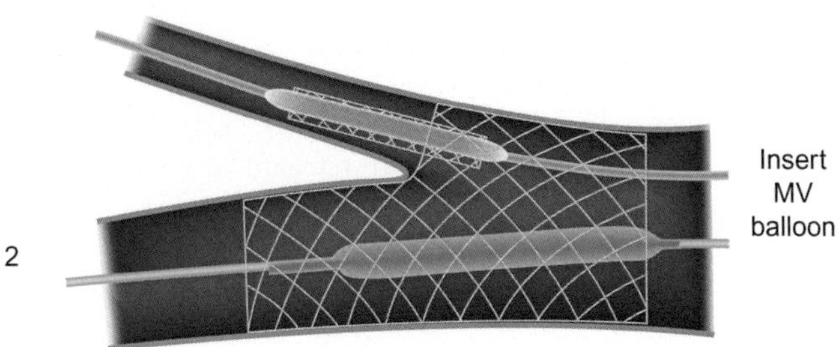

FIGURE 16.23 TAP technique: insertion of the MV balloon across the take-off of the SB.

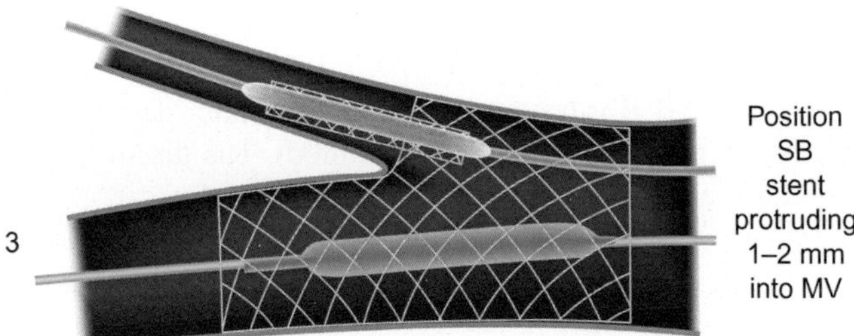

FIGURE 16.24 TAP technique: SB stent positioning.

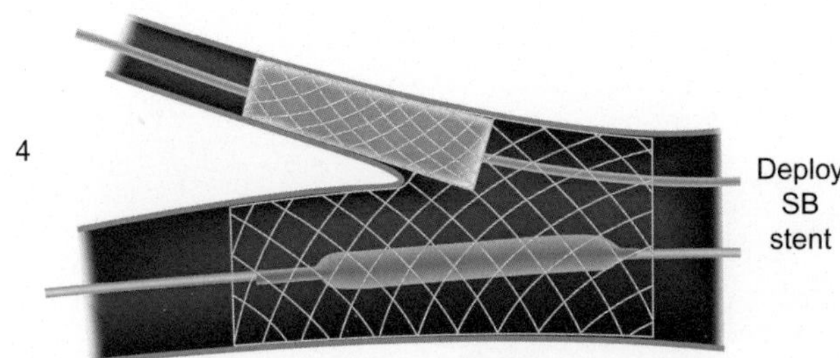

FIGURE 16.25 TAP technique: SB stent deployment.

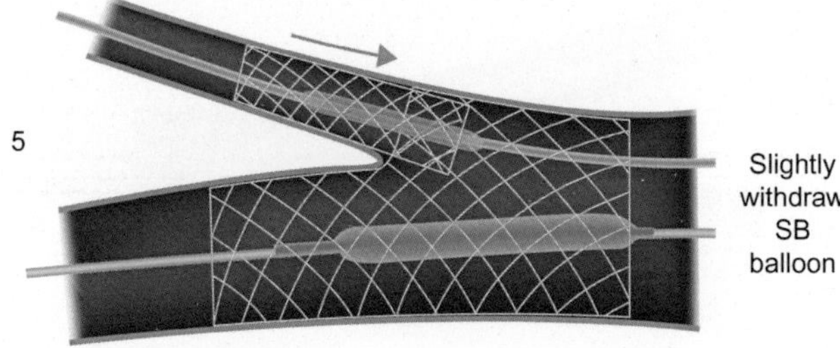

FIGURE 16.26 TAP technique: slight SB balloon withdrawal.

Step 5. Slightly withdraw the SB stent balloon (Fig. 16.26).

Step 6. Kissing balloon inflation of MV and SB stents (Fig. 16.27). Deflate the MV balloon first, to avoid crushing of the protruding stent struts into the SB ostium.

Step 7. Final POT (Fig. 16.28). The distal marker of the POT balloon should be advanced just proximal to the neocarina to avoid crushing of the protruding stent struts.

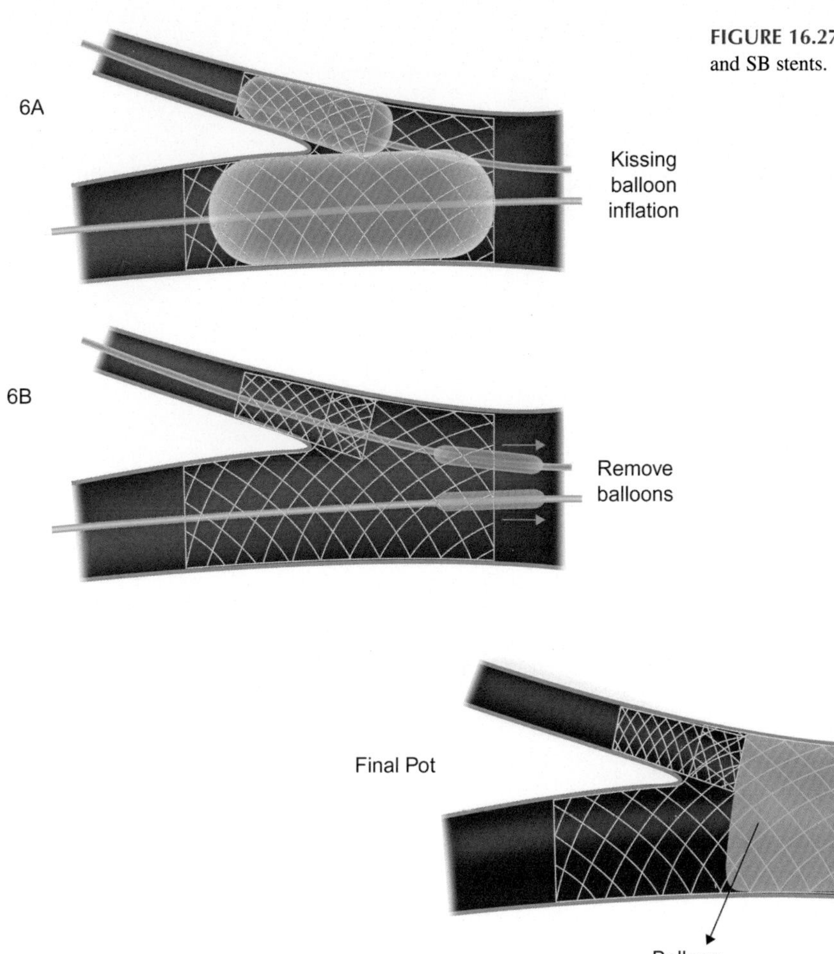

FIGURE 16.27 TAP technique: kissing balloon inflation of MV and SB stents.

FIGURE 16.28 Final POT.

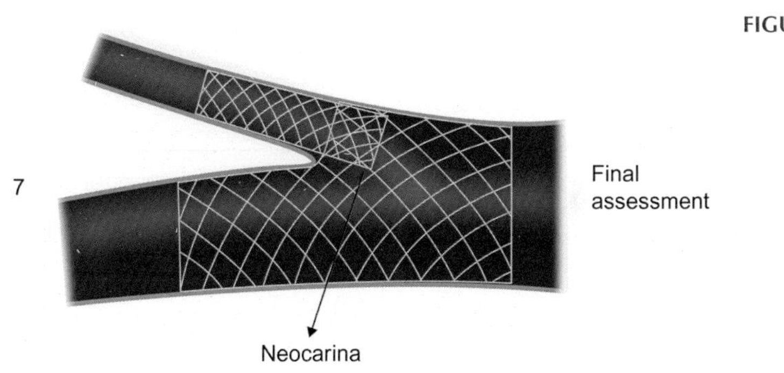

FIGURE 16.29 TAP technique: final assessment.

Step 8. Final assessment (Fig. 16.29).

16.2.10.4 Reverse Crush (also called "internal crush")

See online video: "How to perform the reverse crush technique for bifurcation stenting"

***Background*:** Reverse (also called "internal") crush is a technique for bailout SB stenting in case of SB compromise after MV stenting using the provisional approach.

How?

Reverse crush steps

Step 1. Rewire SB + remove jailed guidewire (Fig. 16.30). A balloon can be advanced into the MV at this time (instead of advancing it in step 3), as after SB ballooning some struts may protrude, making it challenging to advance the MV balloon.

Step 2. Dilate SB (Fig. 16.31).

Step 3. Deliver MV balloon (Fig. 16.32).

Step 4. Deliver SB stent (Fig. 16.33).

Step 5. Deploy SB stent (Fig. 16.34).

Step 6. Remove stent balloon (Fig. 16.35).

Step 7. Crush SB stent (Fig. 16.36).

Step 8. Rewire SB stent + remove jailed guidewire (Fig. 16.37). POT is needed if there is discrepancy in the size of the distal versus proximal MV.

Step 9. Kissing balloon inflation (Fig. 16.38).

Step 10. POT (Fig. 16.39).

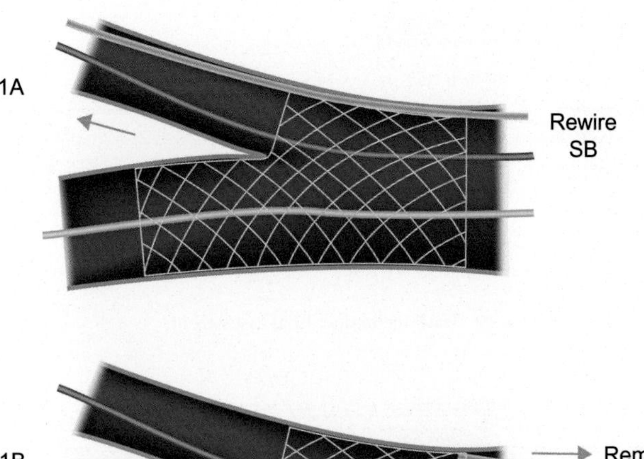

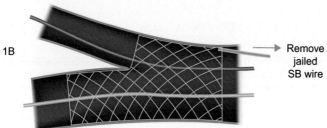

FIGURE 16.30 Reverse crush: SB rewiring and removal of the jailed guidewire.

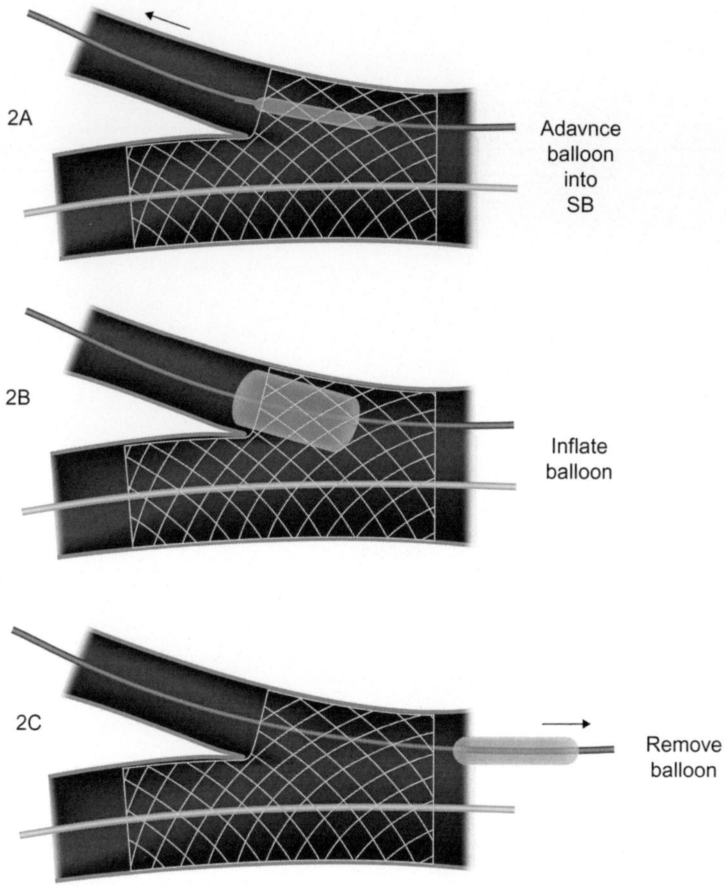

2A
Adavnce balloon into SB

2B
Inflate balloon

2C
Remove balloon

FIGURE 16.31 Reverse crush: SB dilation.

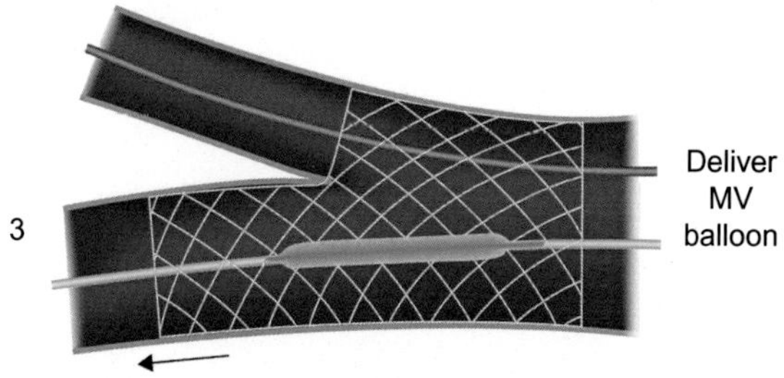

3
Deliver MV balloon

FIGURE 16.32 Deliver MV balloon.

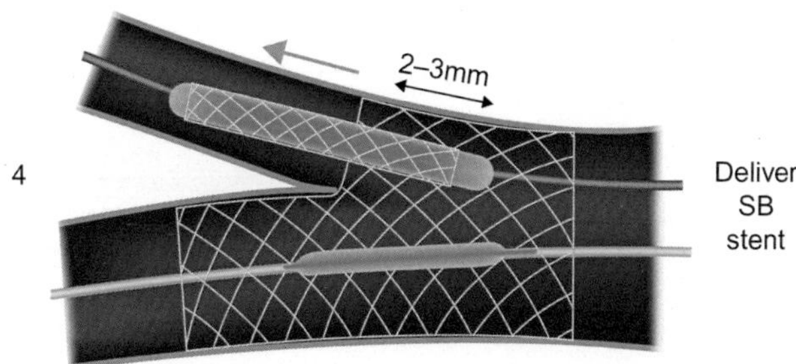

FIGURE 16.33 Reverse crush: delivery of the SB stent.

FIGURE 16.34 Reverse crush: deployment of the SB stent.

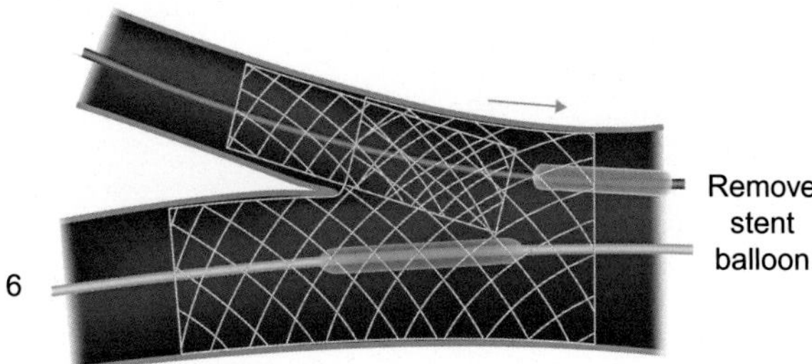

FIGURE 16.35 Reverse crush: removal of the stent balloon.

FIGURE 16.36 Reverse crush: crushing of the SB stent.

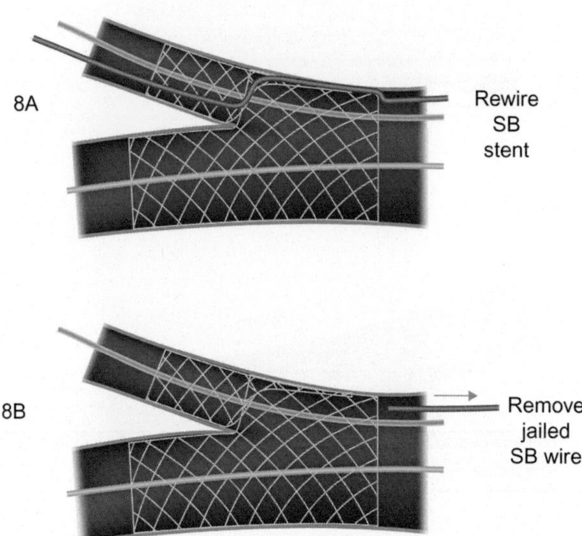

FIGURE 16.37 Reverse crush: rewiring of the SB stent and removal of the jailed guidewire.

FIGURE 16.38 Reverse crush: kissing balloon inflation.

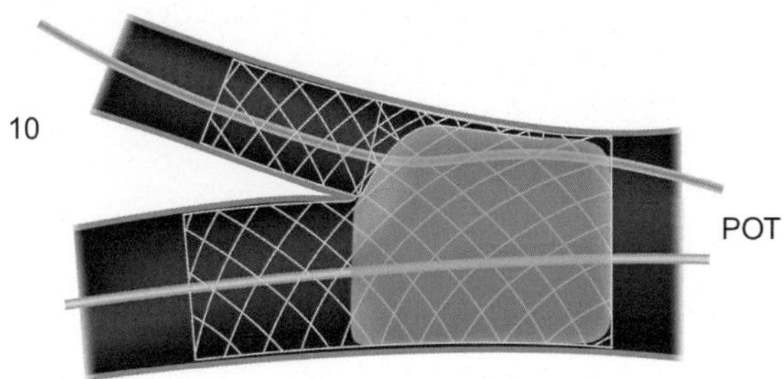

FIGURE 16.39 Reverse crush: proximal optimization technique (POT).

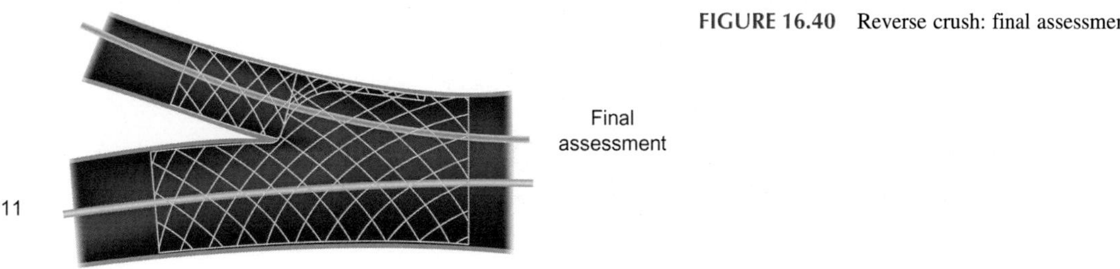

FIGURE 16.40 Reverse crush: final assessment.

Step 11. Final assessment (Fig. 16.40).

16.2.10.5 Double kissing crush

See online videos "bench description of the technique" and "illustrative case presentation".

***Background*:** DK crush has multiple steps and can be cumbersome to perform, with potential difficulties occurring at each step. DK crush has the best supportive clinical trial data, having shown to be superior to culotte in the DK crush III trial [11] and to provisional stenting in the DK Crush V trial (5% target lesion failure at 12 months versus 10% target lesion failure in the provisional group) [12,13] in left main bifurcations. Moreover, DK crush is a great strategy for maintaining wire access to the MV at all times with little to no risk of SB loss, which is particularly important when there are dissections.

***Goal*:** To deliver stents in both SB and MV optimizing the SB coverage and MV stent expansion.

***Preparation*:** Both SB and MV and wired and prepared (with balloon angioplasty and/or atherectomy).

How?

DK crush steps

Step 1. Insert MV balloon (sized 1:1 with the distal MV) in the MV at or distal to the SB origin (Fig. 16.41).

Step 2. Deliver SB stent protruding 2–3 mm in the MV (Fig. 16.42).

Challenge: inability to advance stent in SB or MV.

Solutions:

- Additional SB or MV preparation (standard balloon angioplasty, modified balloon angioplasty, atherectomy).
- Increase support (usually by using the side branch anchoring technique with anchoring balloon in the MV or using a guide catheter extension).

Challenge: Poor SB ostium coverage.

Solutions:

- The SB stent should be long enough to cover the SB lesion and protrude 2–3 mm in the MV. Longer protrusion is OK, but shorter protrusion may lead to suboptimal SB ostium coverage, potentially requiring implantation of another stent.

Step 3. Deploy SB stent (Fig. 16.43).

Tips:

- The stent is deployed at 10–12 atm (not at higher pressure to minimize the risk of distal edge dissection). The stent balloon is then pulled slightly back protruding into the MV and inflated at high pressure (20 atm) to optimize stent expansion.

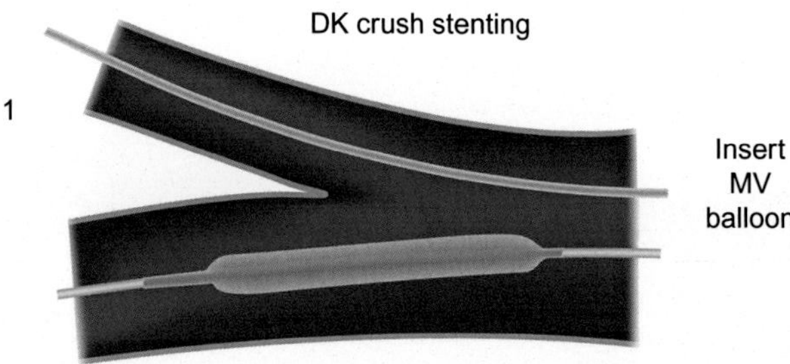

FIGURE 16.41 DK crush: insertion of MV balloon.

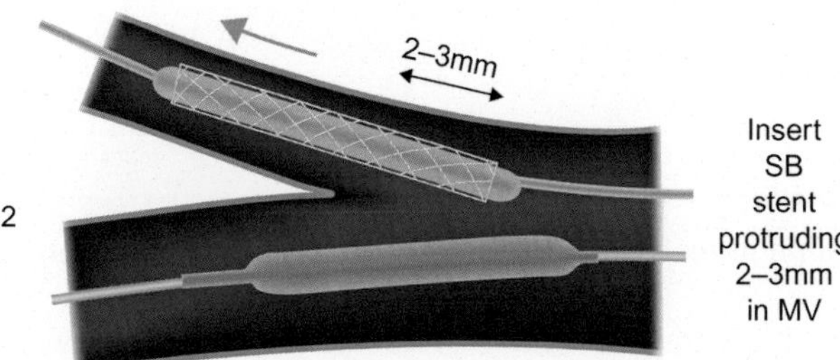

FIGURE 16.42 DK crush: delivery of the SB stent protruding 2–3 mm into the MV.

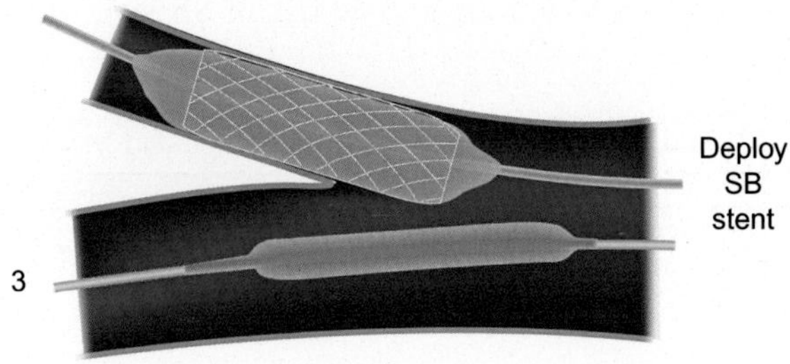

FIGURE 16.43 DK crush: deployment of the SB stent.

Step 4. Remove SB balloon (Fig. 16.44).

Step 5. Perform angiography to assess SB result—additional PCI of the SB may be needed (in case of stent underexpansion or distal dissection); failure to optimize the SB result may hinder rewiring (Fig. 16.45).

Step 6. Crush SB stent by inflating MV balloon (could jail the SB guidewire although most operators remove the SB guidewire prior to crushing the stent) (Fig. 16.46). If the MV balloon is sized according to the proximal main branch (same size for doing a POT as shown in Figure 16.46) it should not be advanced past the carina to minimize the risk of distal MV dissection. Alternatively, the MV balloon is sized according to the distal MV diameter, followed by POT.

Step 7. Rewire SB through crushed stent (through proximal strut), then remove jailed SB wire (Fig. 16.47).

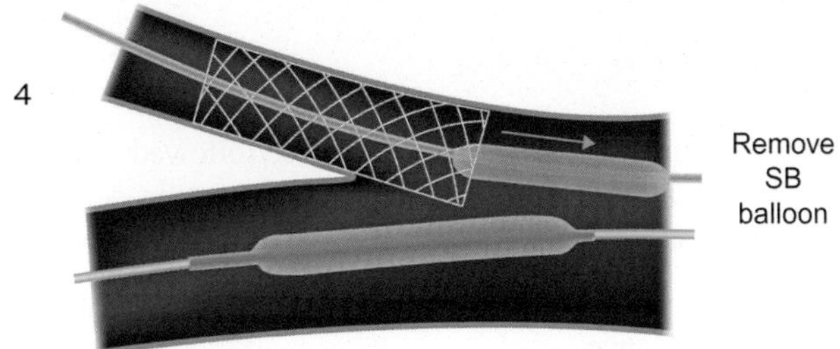

FIGURE 16.44 DK crush: removal of the SB stent balloon.

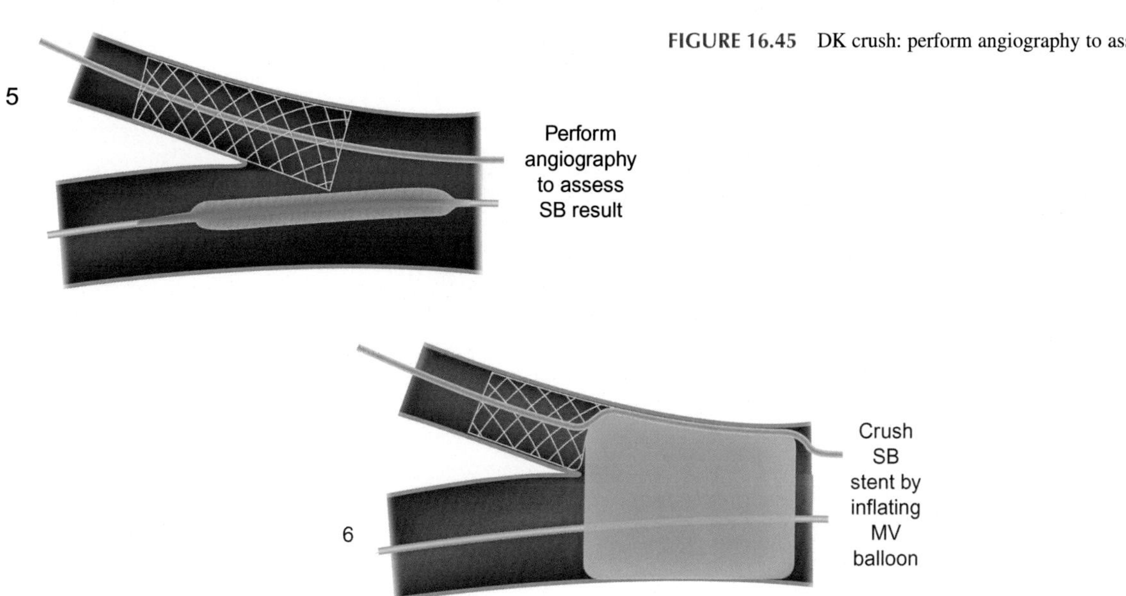

FIGURE 16.45 DK crush: perform angiography to assess SB result.

FIGURE 16.46 DK crush: crushing the SB stent.

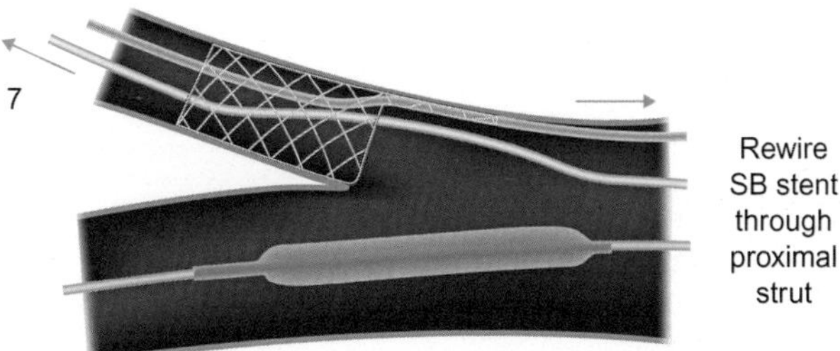

FIGURE 16.47 DK crush: rewiring the SB through the crushed stent (through a proximal strut).

Challenge: inability to rewire SB.
Solutions:

A. Wire-based options

- Insert secondary bend to the guidewire.
- Use hydrophilic or polymer-jacketed guidewire.
- Reversed guidewire (hairpin) technique (Chapter 8: Wiring, Section 8.6).

B. Microcatheter options

- Use standard microcatheter: improves wire support and allows change in guidewire tip without losing wire position.
- Angulated microcatheter.
- Dual lumen microcatheter.

Step 8. Deliver balloon in SB (Fig. 16.48).

Step 9. First kissing balloon inflation (Fig. 16.49).
Tips:

- Two-step kiss: high pressure—20 atm—balloon inflation in SB, followed by inflation of both MV and SB balloons at 10–12 atm) (can use original MV balloon instead of a new one depending on the size of the proximal MV and the size of the SB balloon, using Finet's law).

What to do if the balloon does not advance?

1. Small balloon (i.e., Sapphire Pro 1.0 mm).
2. Anchor in MV.

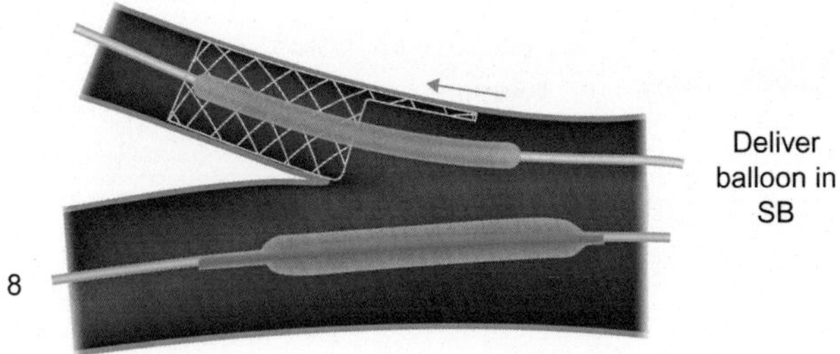

FIGURE 16.48 DK crush: balloon delivery into the SB.

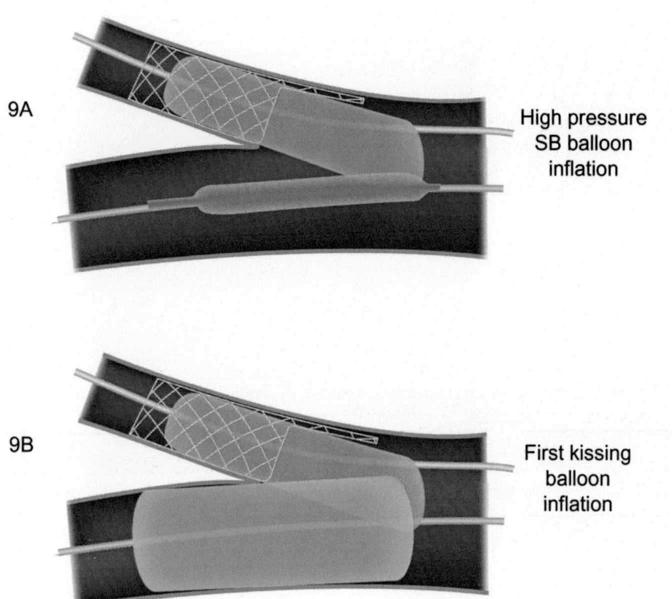

FIGURE 16.49 DK crush: first kissing balloon inflation.

3. Glider balloon.
4. Pass a microcatheter to create a track for balloon advancement.
5. Rewire SB.

Why do two-step kiss (Fig. 16.50) [14]?

Step 10. Remove SB and MV balloon (may or may not leave SB wire in) (Fig. 16.51).

Step 11. Deliver and deploy MV stent (Fig. 16.52).

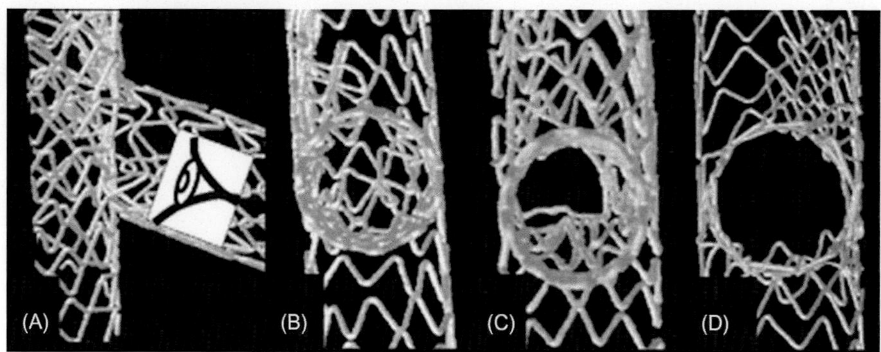

FIGURE 16.50 Two-step kissing balloon postdilation further improves the SB ostium. Stent struts opposite the SB ostium have been removed electronically (A) to allow clear viewing of the SB ostium for these Driver stents. The double layer of stents before postdilation (B) are partially cleared from the SB ostium by one-step kissing postdilation (C) and more fully cleared by two-step kissing postdilation (D). *Reproduced with permission from Ormiston JA, Webster MW, Webber B, Stewart JT, Ruygrok PN, Hatrick RI. The "crush" technique for coronary artery bifurcation stenting: insights from micro-computed tomographic imaging of bench deployments. JACC Cardiovasc Interv 2008;1:351–7. Copyright Elsevier.*

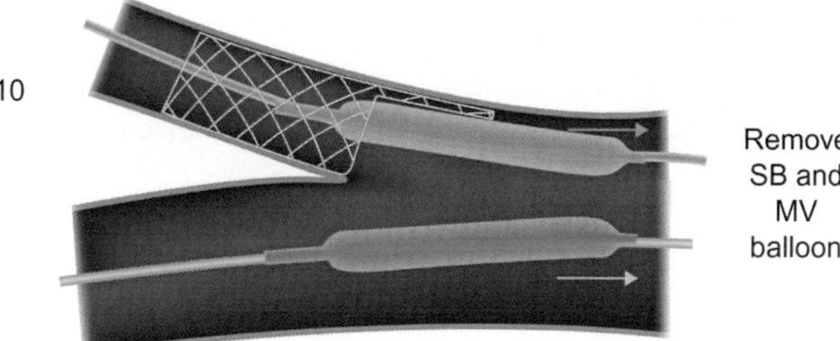

FIGURE 16.51 DK crush: remove SB and MV balloons.

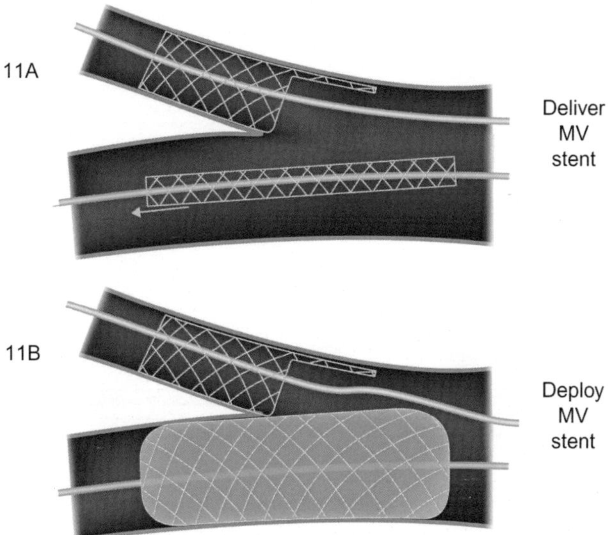

FIGURE 16.52 DK crush: delivery and deployment of the MV stent.

Step 12. Assess MV stent result—additional PCI of the MV distally may be needed (Fig. 16.53).

Step 13. POT (if proximal MV has significant size mismatch from distal MV) (Fig. 16.54).

1. Place POT balloon (sized 1:1 to proximal MV) with its distal edge at the bifurcation carina.
2. Inflate POT balloon.
3. Remove POT balloon.

Why POT? To ensure good stent expansion and apposition, especially when the proximal MV is significantly larger than the distal MV. This will also ensure that the wire going back through into the SB does not advance underneath the MV stent.

Step 14. Rewire SB through MV stent (then remove the jailed SB wire if it had been left in) (Fig. 16.55).

1. Rewire SB though MV stent (through distal strut, in contrast to the first rewiring).

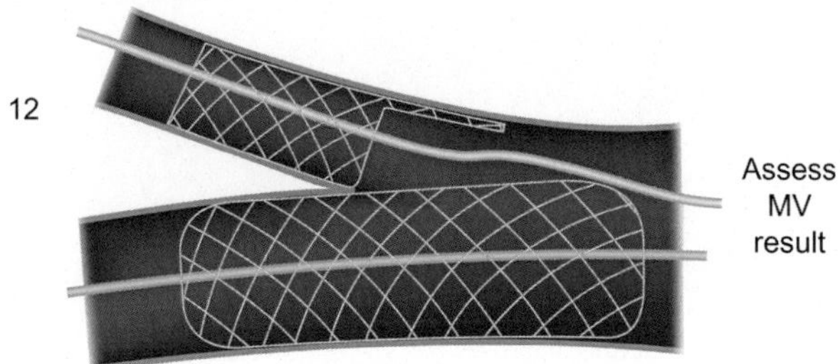

FIGURE 16.53 DK crush: assessment of the MV results and additional MV treatment if needed.

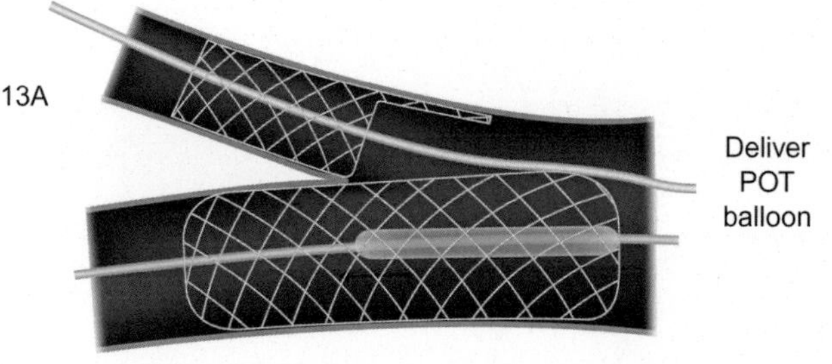

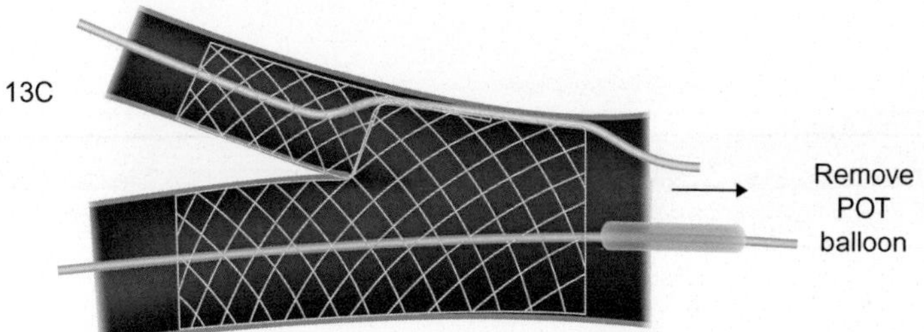

FIGURE 16.54 DK crush: proximal optimization technique (POT).

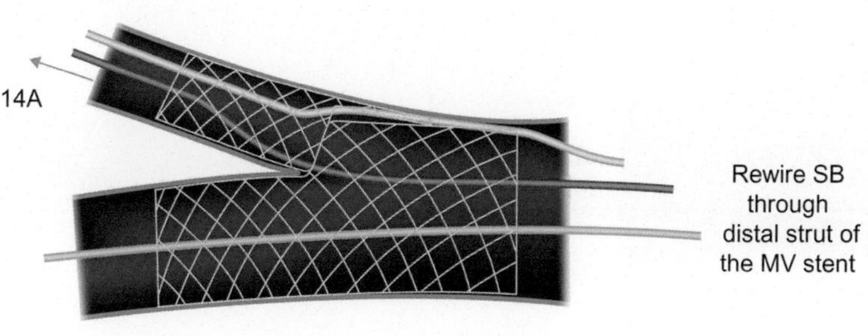

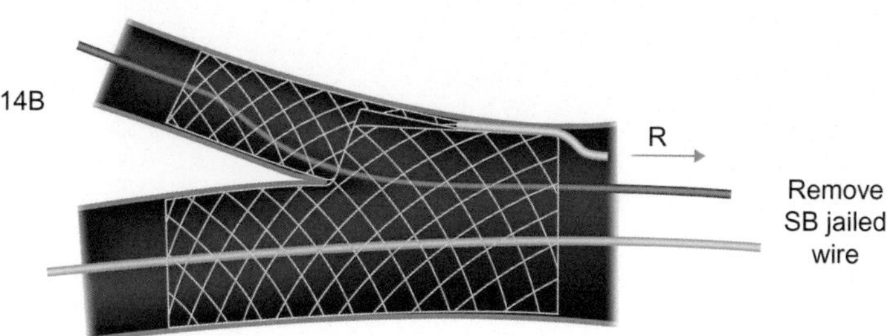

FIGURE 16.55 DK crush: rewiring of the jailed SB through the stented MV.

2. Remove jailed SB wire.
   Challenge: inability to rewire SB
   Solutions: as outlined in step 7 above.

Step 15. Second kissing balloon inflation (two-step kiss) (Fig. 16.56)

a Advance SB balloon and MV balloon (it does not matter which one goes first).
b Inflate SB (or MV) balloon at high pressure.
c Inflate MV (or SB) balloon at high pressure.
d Inflate both MV and SB balloon at 12–14 atm.

Step 16. Final POT (Fig. 16.57)

a Place POT balloon (sized 1:1 to proximal MV) with its distal edge at the carina of the bifurcation.
b Inflate POT balloon.
c Remove POT balloon.

Step 17. Final angiography (and intravascular imaging) to assess final result (Fig. 16.58).

16.2.10.6 Culotte

See online video: "How to do culotte"

***Background*:** Culotte was shown to be inferior to DK Crush in left main bifurcations in the DK crush III trial [11]. Culotte was superior to TAP in the BBK II trial [15]. In the EBC TWO trial outcomes were similar with provisional stenting and culotte in complex coronary bifurcation lesions with large stenosed SBs [16]. Culotte is best suited for lesions that have similarly sized SB and distal MV.

***Goal*:** Deliver stents in both SB and MV while optimizing the SB coverage and MV stent expansion.

How?

Culotte steps

Step 1. Deliver + deploy first stent in most angulated branch (Fig. 16.59).

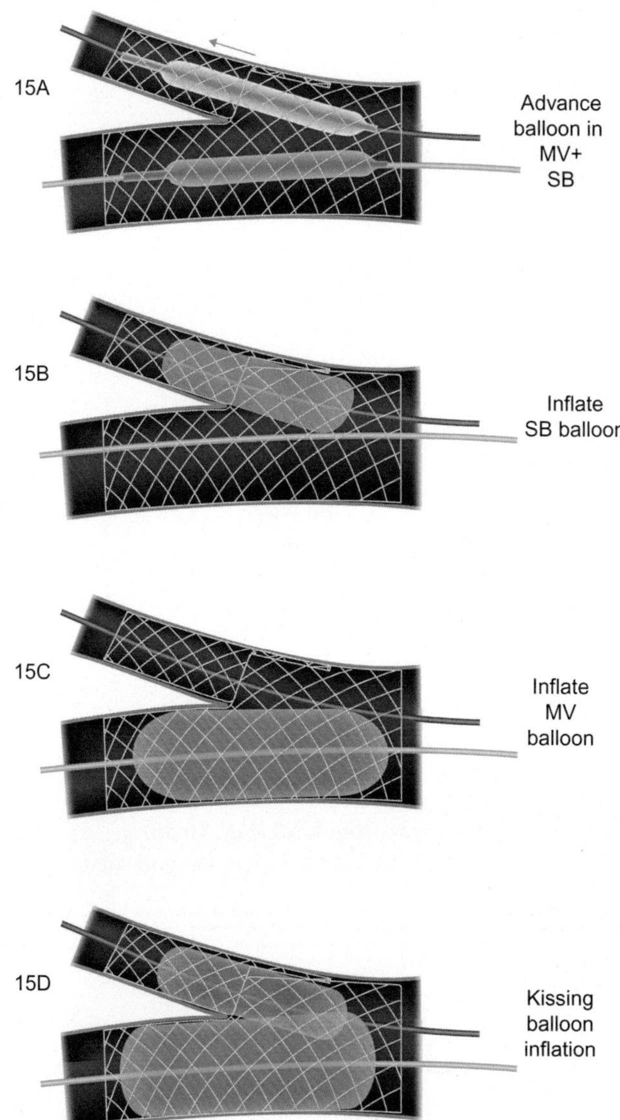

FIGURE 16.56 DK crush: second kissing balloon inflation.

Step 2. Remove stent balloon (Fig. 16.60).

Step 3. Evaluate the stented vessel (is additional PCI needed for optimal result?) (Fig. 16.61)

Step 4. POT (Fig. 16.62).

Step 5. Rewire nonstented vessel through a distal stent cell at the carina + remove jailed guidewire (Fig. 16.63).

Step 6. Dilate struts into nonstented branch (Fig. 16.64).

Step 7. Deliver + deploy second stent (Fig. 16.65).

Step 8. Second POT (Fig. 16.66).

Step 9. Rewire initially stented branch through a distal stent cell + remove jailed guidewire (Fig. 16.67).

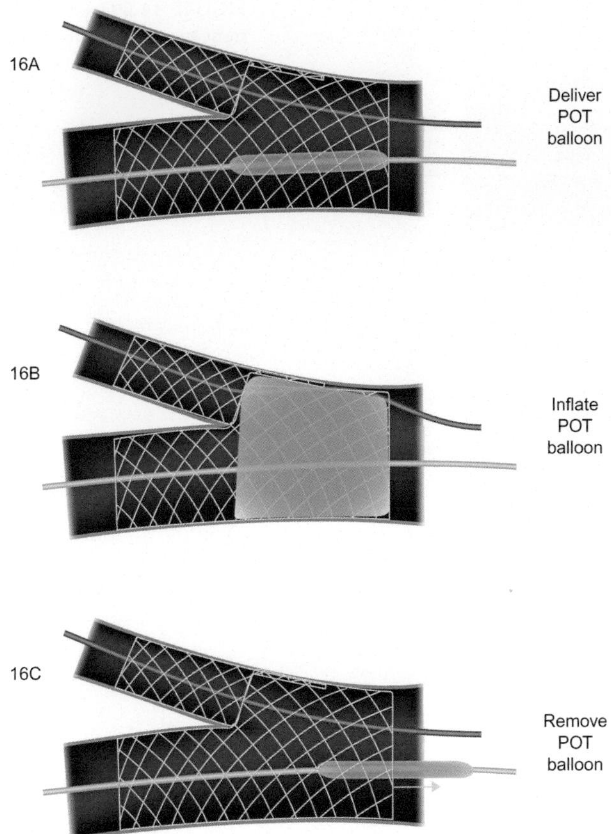

FIGURE 16.57 DK crush: final POT.

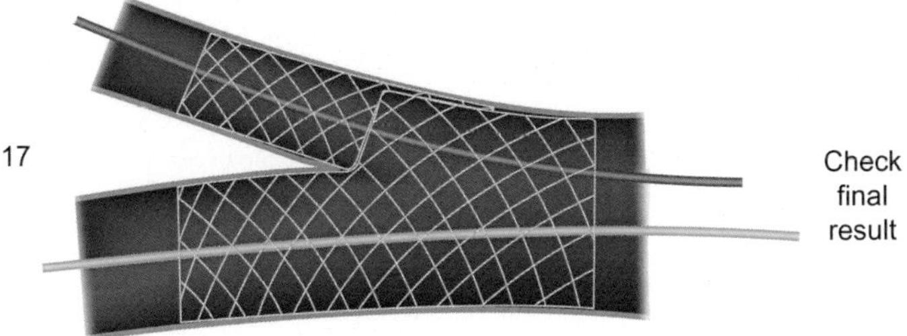

FIGURE 16.58 DK crush: final assessment after DK crush.

Step 10. Kissing balloon inflation (Fig. 16.68).

- Advance SB balloon and MV balloon (which one is first does not matter).
- Inflate SB (or MV) balloon at high pressure.
- Inflate MV (or SB) balloon at high pressure.
- Inflate both MV and SB balloon at 12–14 atm.
- Remove both balloons.
- A final POT is optional.

Step 11. Final vessel assessment (Fig. 16.69).

16.2.10.7 V-stenting

***Background*:** V stenting can be used for Medina 0.1.1. bifurcations (i.e., disease in the distal MV and the SB ostium, without significant disease in the proximal MV). Its advantage is simplicity (no need for rewiring or for kissing balloon

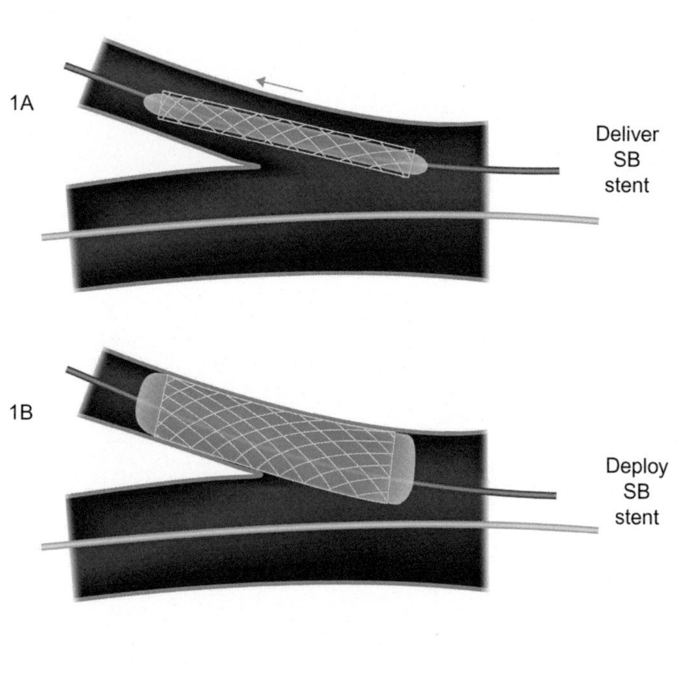

FIGURE 16.59 Culotte: stent delivery and deployment in the most angulated bifurcation branch.

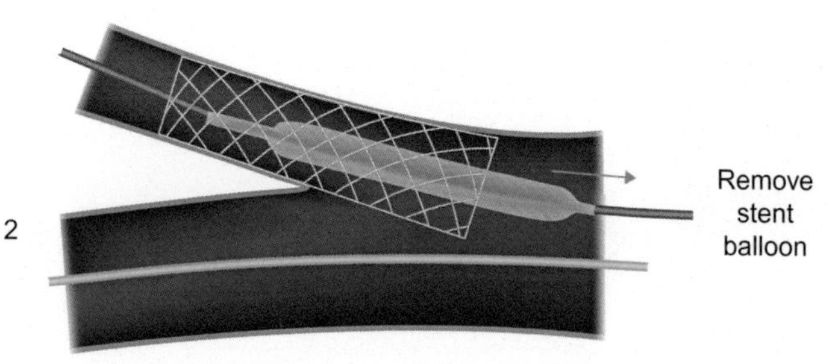

FIGURE 16.60 Culotte: removal of the stent balloon.

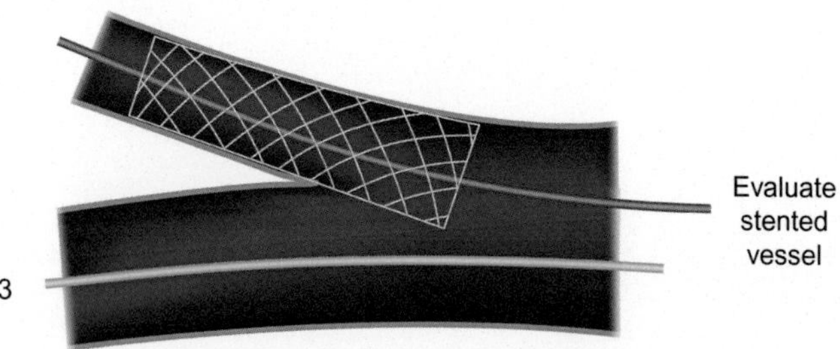

FIGURE 16.61 Culotte: evaluation of the stented vessel.

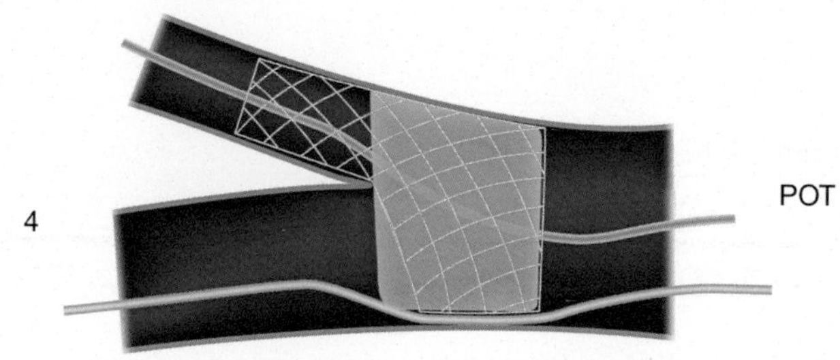

FIGURE 16.62 Culotte: proximal optimization technique (POT).

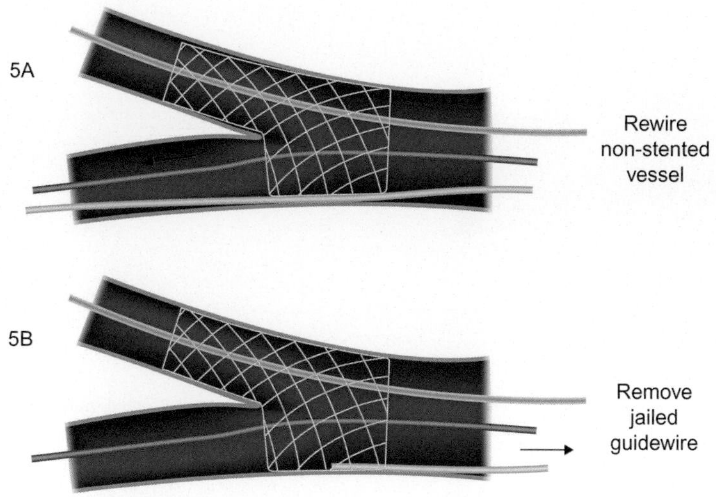

5A

Rewire
non-stented
vessel

5B

Remove
jailed
guidewire

FIGURE 16.63 Culotte: rewiring of the nonstented vessel and removal of the jailed guidewire.

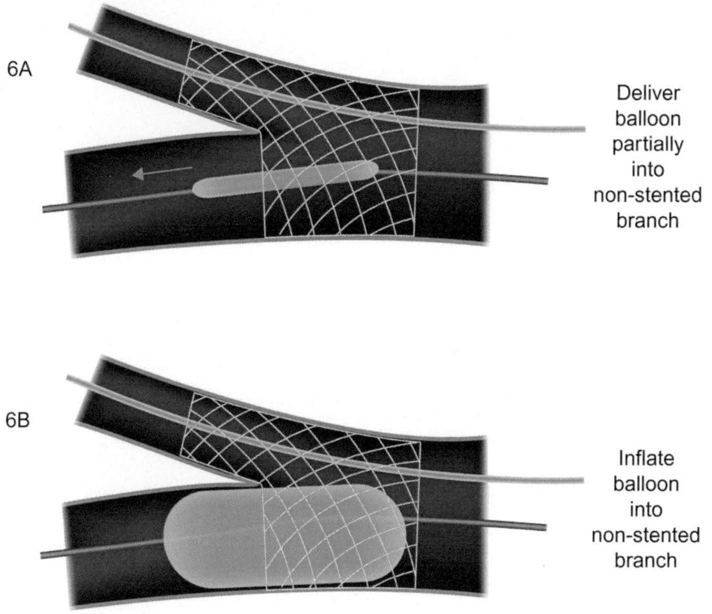

6A

Deliver
balloon
partially
into
non-stented
branch

6B

Inflate
balloon
into
non-stented
branch

FIGURE 16.64 Culotte: dilating the struts jailing the non-stented vessel.

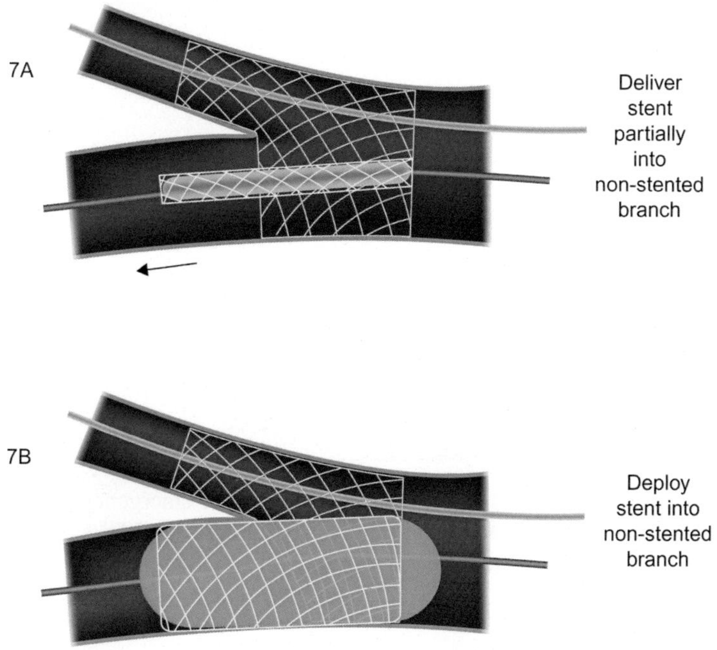

7A

Deliver
stent
partially
into
non-stented
branch

7B

Deploy
stent into
non-stented
branch

FIGURE 16.65 Culotte: delivery and deployment of the second stent.

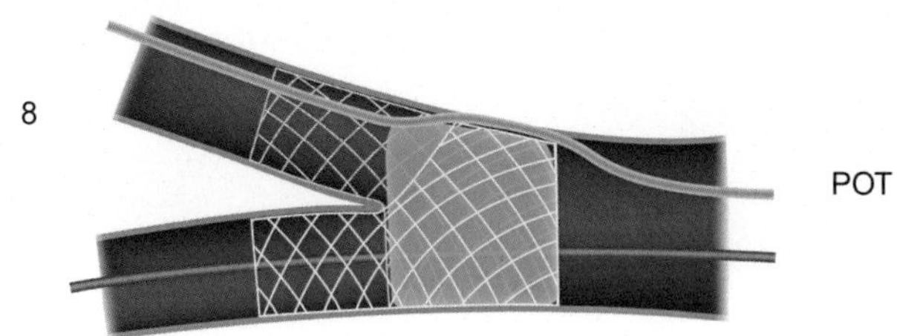

FIGURE 16.66 Second POT.

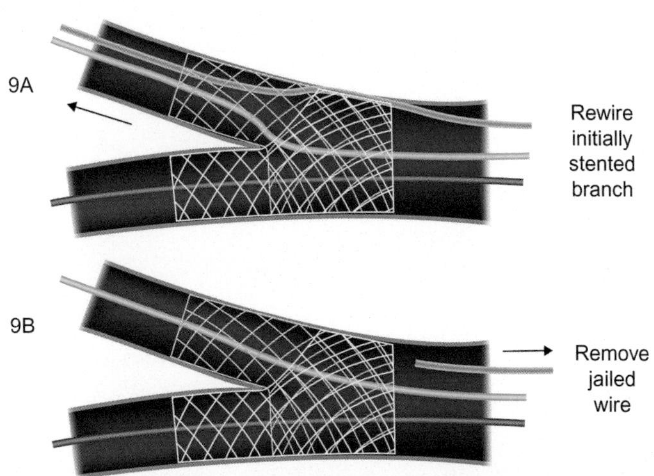

FIGURE 16.67 Culotte: rewiring of the initially stented branch and removal of the jailed guidewire.

inflations), and maintainance of wire access in both MV and SB at all times. Its disadvantage is formation of a neocarina that can hinder subsequent wiring and equipment advancement.

***Preparation*:** At least 7 French guide catheters are required for simultaneously delivering two DES. The distal MV and SB are prepared, as indicated for all bifurcation lesions. The starting point for V-stenting is illustrated in Fig. 16.70.

***How*?**

V-stenting steps

Step 1. Insert stents into both branches, protruding 1–2 mm from the ostium of each branch (Fig. 16.71).

Step 2. Simultaneous stent deployment with kissing balloon inflation (Fig. 16.72).

Step 3. Remove both stent balloons (Fig. 16.73).

Step 4. Final assessment (Fig. 16.74).

16.2.10.8 Simultaneous kissing stents

***Background*:** SKS is a two-stent bifurcation stenting technique that should only be used in case of emergency or for bailout. Its advantage is that it can be performed very quickly, while maintaining wire access in both the MV and SB. Its disadvantage is the creation of a long neocarina, which makes subsequent treatment extremely challenging.

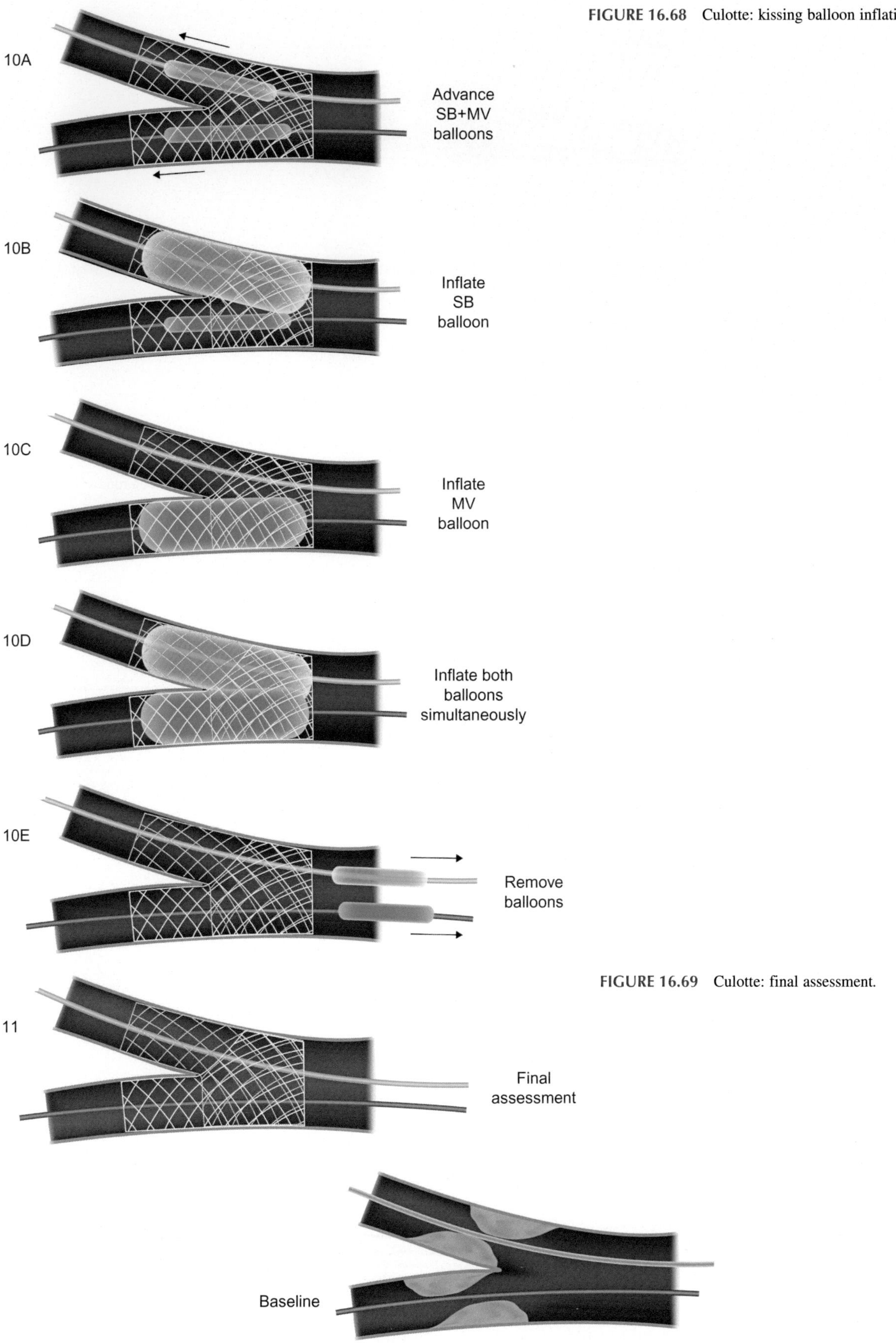

FIGURE 16.68 Culotte: kissing balloon inflation.

FIGURE 16.69 Culotte: final assessment.

FIGURE 16.70 Medina 0.1.1 bifurcation.

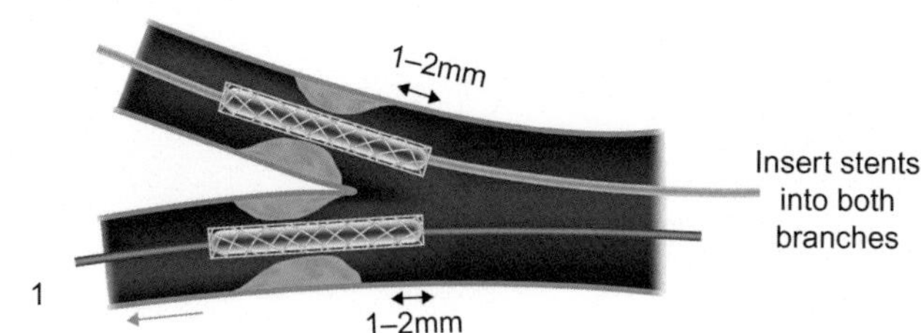

FIGURE 16.71 V-stenting: stents inserted in both branches of the bifurcation.

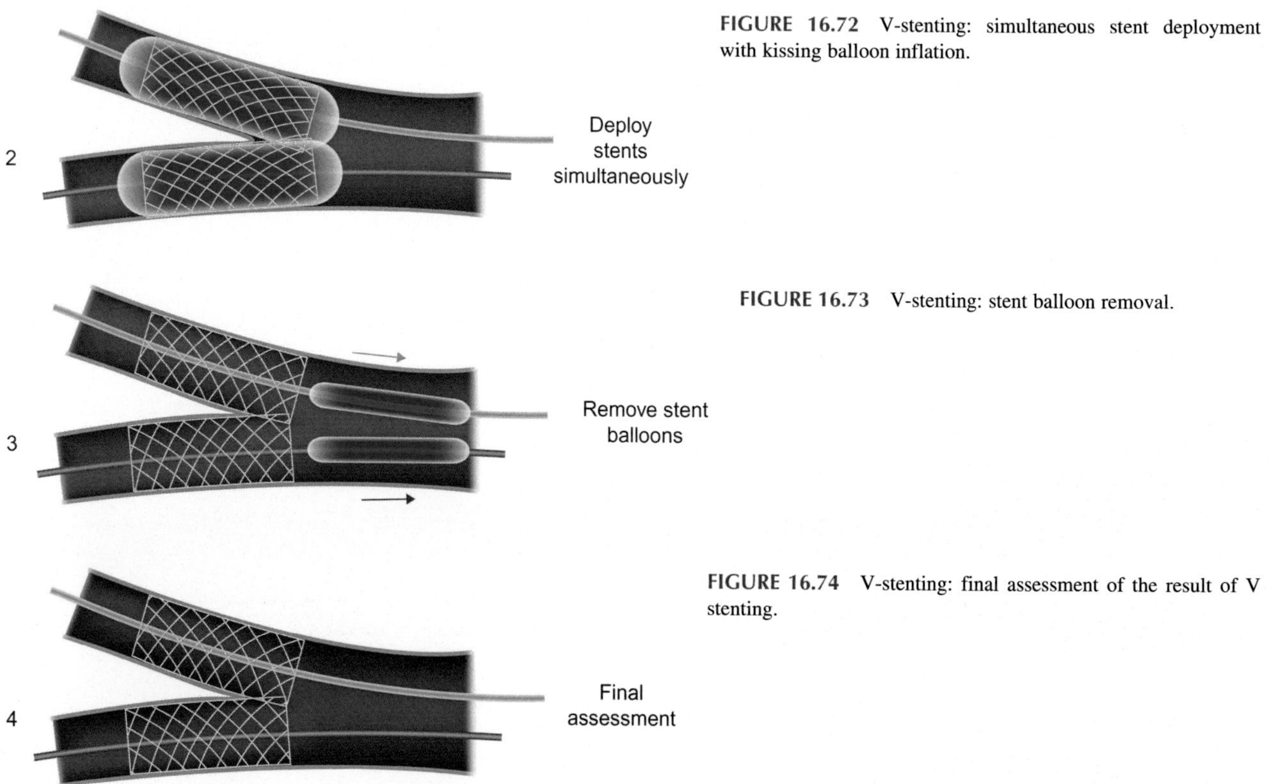

FIGURE 16.72 V-stenting: simultaneous stent deployment with kissing balloon inflation.

FIGURE 16.73 V-stenting: stent balloon removal.

FIGURE 16.74 V-stenting: final assessment of the result of V stenting.

***Preparation*:** At least 7 French guide catheters are required for simultaneously delivering two DES. The distal MV and SB are wired prepared as indicated for all bifurcation lesions.

***How*?**

SKS steps

Step 1. Insert stents into both branches, completely covering the bifurcation (Fig. 16.75).

Step 2. Simultaneous stent deployment with kissing balloon inflation (Fig. 16.76).

Step 3. Remove stent balloons (Fig. 16.77).

Step 4. Final assessment (Fig. 16.78).

16.2.11 Closure

Closure is performed as described in detail in Chapter 11: Access Closure.

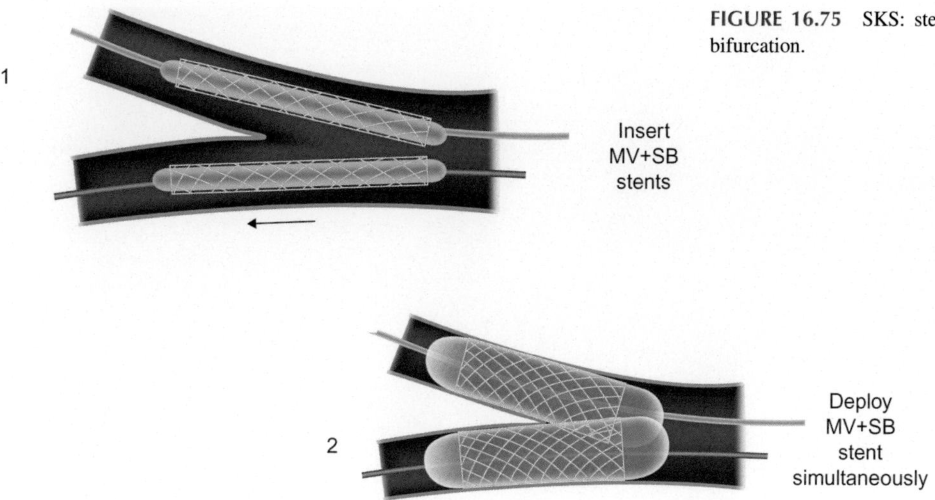

FIGURE 16.75 SKS: stent delivery into both branches of the bifurcation.

FIGURE 16.76 SKS: simultaneous deployment of the two stents.

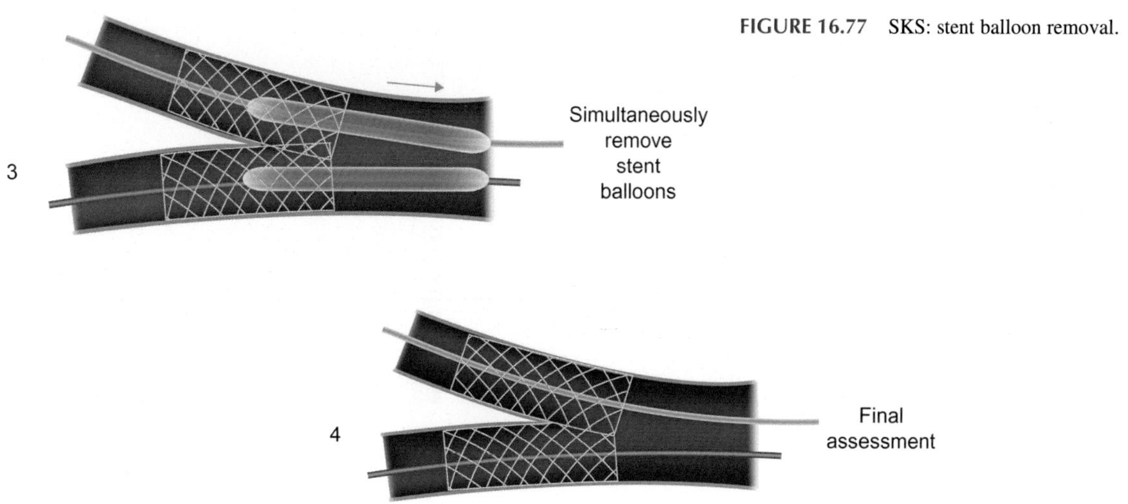

FIGURE 16.77 SKS: stent balloon removal.

FIGURE 16.78 SKS: final assessment.

16.2.12 Physiology

Coronary physiology can assist on multiple aspects of coronary bifurcation stenting:

1. Determining the baseline severity of MV and SB lesions.
2. Determining whether an optimal final result has been achieved.
3. Determining whether a SB has been compromised during provisional stenting (Fig. 16.79).

16.2.13 Intravascular imaging

Intracoronary imaging can assist with multiple aspects of coronary bifurcation stenting:

1. Determine diameter and length of the MV and SB lesions.
2. Detect the presence and severity of calcification.
3. Assess stent expansion. For left main stenting, the risk of restenosis can be reduced if the stent area is larger than the cutoff values show in Fig. 17.2 [17].
4. Determine the cause of complications (such as substent wiring).

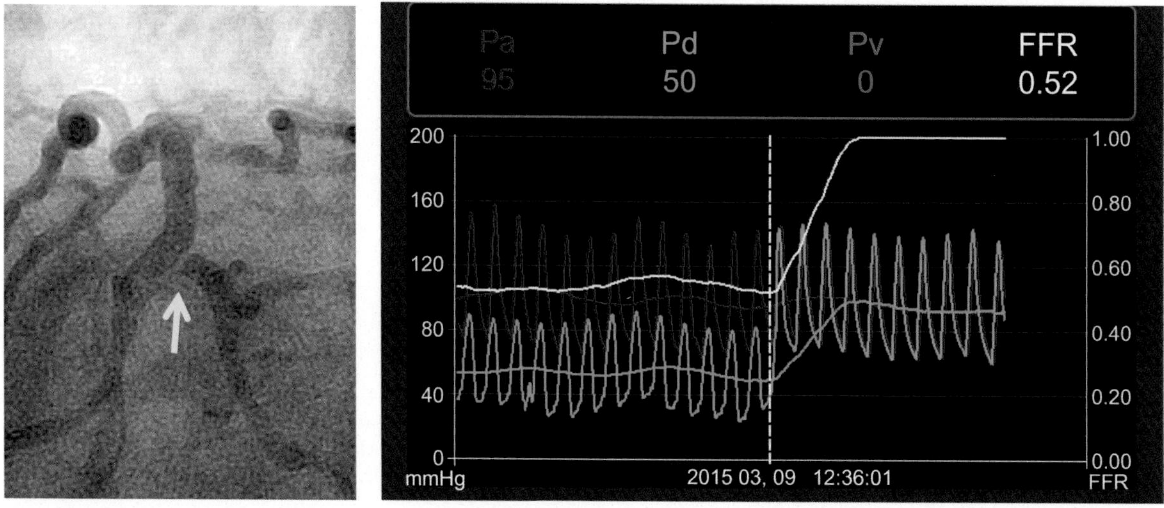

FIGURE 16.79 After left main stenting the ostium of the circumflex appears compromised. Fractional flow reserve measurement confirmed significant focal ostial circumflex stenosis, as illustrated by the pressure step-up at the left main during FFR pullback.

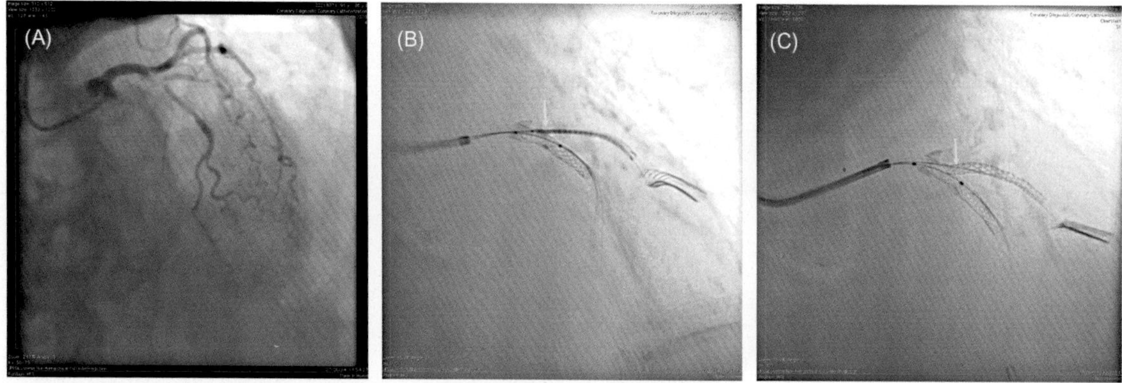

FIGURE 16.80 Use of the CLEARstent stent visualization software (Siemens) to ensure that the ostium of the diagonal side branch is completely covered by a stent (*arrow*, panel B and C).

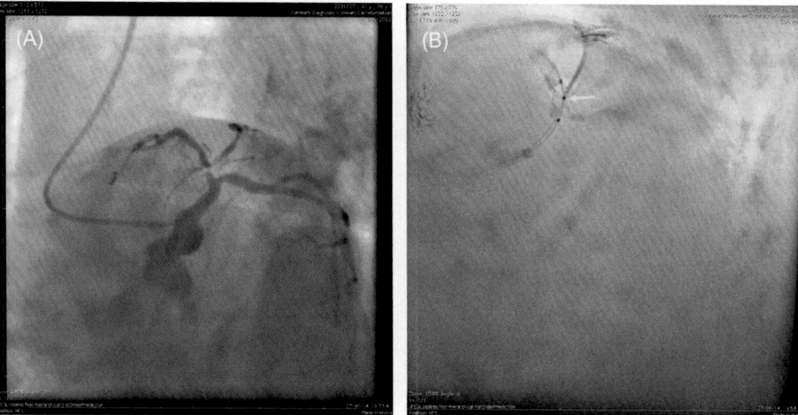

FIGURE 16.81 Use of the CLEARstent stent visualization software (Siemens) to ensure that the ostium of the diagonal side branch is completely covered by a stent (*arrow*, panel B). *Courtesy of Dr. Pedro Cardoso.*

In addition to intravascular imaging, enhanced stent visualization software, such as the StentBoost (Philips) and CLEARstent (Siemens) can significantly facilitate stent placement, for example by ensuring that the SB ostium is completely covered (Figs. 16.80 and 16.81) and helping determine whether stent deformation has occurred.

16.2.14 Hemodynamic support

Use of hemodynamic support depends on patient and procedural factors. Complex left main bifurcation stenting in patients with depressed left ventricular function may be an indication for hemodynamic support, especially in patients with an occluded or severely diseased right coronary artery.

References

[1] Medina A, Suarez de Lezo J, Pan M. A new classification of coronary bifurcation lesions. Rev Esp Cardiol 2006;59:183.

[2] Ide S, Sumitsuji S, Kaneda H, Kassaian SE, Ostovan MA, Nanto S. A case of successful percutaneous coronary intervention for chronic total occlusion using the reversed guidewire technique. Cardiovasc Interv Ther 2013;28:282–6.

[3] Kawasaki T, Koga H, Serikawa T. New bifurcation guidewire technique: a reversed guidewire technique for extremely angulated bifurcation—a case report. Catheter Cardiovasc Interv 2008;71:73–6.

[4] Suzuki G, Nozaki Y, Sakurai M. A novel guidewire approach for handling acute-angle bifurcations: reversed guidewire technique with adjunctive use of a double-lumen microcatheter. J Invasive Cardiol 2013;25:48–54.

[5] Lassen JF, Burzotta F, Banning AP, et al. Percutaneous coronary intervention for the left main stem and other bifurcation lesions: 12th consensus document from the European Bifurcation Club. EuroIntervention 2018;13:1540–53.

[6] Pan M, Ojeda S, Villanueva E, et al. Structural damage of jailed guidewire during the treatment of coronary bifurcation lesions: a microscopic randomized trial. JACC Cardiovasc Interv 2016;9:1917–24.

[7] Grundeken MJ, Li X, Kurpershoek CE, et al. Distal embolization of hydrophilic-coating material from coronary guidewires after percutaneous coronary interventions. Circ Cardiovasc Interv 2015;8:e001816.

[8] Koo BK, Kang HJ, Youn TJ, et al. Physiologic assessment of jailed side branch lesions using fractional flow reserve. J Am Coll Cardiol 2005;46:633–7.

[9] Sawaya FJ, Lefevre T, Chevalier B, et al. Contemporary approach to coronary bifurcation lesion treatment. JACC Cardiovasc Interv 2016;9:1861–78.

[10] Iturbe JM, Abdel-Karim AR, Papayannis A, et al. Frequency, treatment, and consequences of device loss and entrapment in contemporary percutaneous coronary interventions. J Invasive Cardiol 2012;24:215–21.

[11] Chen SL, Xu B, Han YL, et al. Comparison of double kissing crush versus Culotte stenting for unprotected distal left main bifurcation lesions: results from a multicenter, randomized, prospective DKCRUSH-III study. J Am Coll Cardiol 2013;61:1482–8.

[12] Chen SL, Zhang JJ, Han Y, et al. Double kissing crush versus provisional stenting for left main distal bifurcation lesions: DKCRUSH-V randomized trial. J Am Coll Cardiol 2017;70:2605–17.

[13] Chen X, Li X, Zhang JJ, et al. 3-Year outcomes of the DKCRUSH-V trial comparing DK crush with provisional stenting for left main bifurcation lesions. JACC Cardiovasc Interv 2019;12:1927–37.

[14] Ormiston JA, Webster MW, Webber B, Stewart JT, Ruygrok PN, Hatrick RI. The "crush" technique for coronary artery bifurcation stenting: insights from micro-computed tomographic imaging of bench deployments. JACC Cardiovasc Interv 2008;1:351–7.

[15] Ferenc M, Gick M, Comberg T, et al. Culotte stenting vs. TAP stenting for treatment of de-novo coronary bifurcation lesions with the need for side-branch stenting: the Bifurcations Bad Krozingen (BBK) II angiographic trial. Eur Heart J 2016;37:3399–405.

[16] Hildick-Smith D, Behan MW, Lassen JF, et al. The EBC TWO study (European Bifurcation Coronary TWO): a randomized comparison of provisional T-stenting versus a systematic 2 stent culotte strategy in large caliber true bifurcations. Circ Cardiovasc Interv 2016;9.

[17] Kang SJ, Ahn JM, Song H, et al. Comprehensive intravascular ultrasound assessment of stent area and its impact on restenosis and adverse cardiac events in 403 patients with unprotected left main disease. Circ Cardiovasc Interv 2011;4:562–9.

Chapter 17

Left main

CTO Manual Online cases: 11, 14, 48, 67, 105, 127, 135, 136, 148
PCI Manual Online cases: 28, 32, 33, 41, 52, 59, 60, 72, 83, 84, 85, 88, 91, 96, 97, 100

Percutaneous coronary intervention (PCI) in the left main coronary artery can be challenging for many reasons:

- **Severe ischemia and hemodynamic instability**: the left main provides blood flow to most of the myocardium, hence coronary flow compromise during the procedure can result in hypotension, cardiogenic shock, or cardiac arrest.
- **Anatomic complexity:** the left main can have ostial disease, body disease, and/or bifurcation disease (the bifurcation is involved in >80% of left main lesions undergoing PCI [1,2]). Lesions in the left main body are the easiest to treat.

17.1 Planning

It is important to review prior coronary angiograms and operative reports, to answer questions, such as:

- Is there also disease or occlusion of the right coronary artery?
- Are there visible collaterals from the right coronary artery to the left system?
- Is the left main protected by an aorto-coronary bypass graft?
- Is there severe calcification that may require atherectomy?

Additional clinical information, such as left ventricular function and baseline hemodynamic status, clinical presentation and comorbidities, is vital to help assess procedural risk and to determine the need for hemodynamic support. Right heart catheterization can help assess filling pressures and cardiac output, determine the need for hemodynamic support prior to the intervention, and monitor hemodynamics during the procedure.

17.2 Monitoring

Monitoring is performed as described in Chapter 2: Monitoring. Careful monitoring is especially important in patients with left main disease due to risk of rapid hemodynamic deterioration. If hypotension develops, early use of vasopressors (such as norepinephrine) is encouraged to prevent rapid hemodynamic deterioration. Defibrillation pads should ideally be placed on the patient to allow prompt treatment in case of ventricular arrhythmia or fibrillation.

17.3 Medications

Medications are administered as described in Chapter 3: Medications.

17.4 Access

Either femoral or radial artery access can be used for left main PCI. Femoral artery may be preferred in patients with highly complex anatomy. Larger guides (7 or 8 French) provide more support and are preferred for left main bifurcation PCI.

Manual of Percutaneous Coronary Interventions. DOI: https://doi.org/10.1016/B978-0-12-819367-9.00017-2

17.5 Engagement

- Left main engagement can be challenging in patients with ostial left main lesions, as severe pressure dampening can occur causing extensive ischemia. Careful monitoring of the pressure waveform is needed, followed by prompt disengagement in case of dampening before any hemodynamic deterioration or vessel injury occurs.
- Larger guides (7 or 8 French) provide more support and allow some bifurcation stenting techniques, such as V-stenting (Section 16.1.10.7).
- Side hole guides should not be used for engaging the left main (with the exception of left main chronic total occlusions), as they may mask pressure dampening leading to profound ischemia. Side hole catheters are particularly dangerous in patients with short left main because the guide catheter might completely obstruct the ostium of the circumflex with a seemingly normal pressure waveform (Fig. 17.1).
- As described in Chapter 15: Ostial Lesions (Section 15.1.5) in cases of severe left main ostial disease a guidewire (and often a balloon mounted on the wire) should be inserted and advanced to the tip of the guide catheter prior to engagement, to facilitate immediate wiring (and ballooning) upon engagement.
- Intermittently disengage the guide catheter to alleviate ischemia. Use of the independent-hand technique (Section 9.5.8, Fig. 9.5 and Fig. 15.4) or the floating wire technique (Section 15.1.5, Fig. 15.3) can facilitate guide manipulations.
- Avoid contrast injections while pressure is dampened to minimize the risk of left main ostial dissection that can be catastrophic. If a dissection occurs, disengage the guide and perform only cautious contrast injections from a distance to clarify the situation without enlarging the dissection. Also prepare for stent implantation to seal the dissection entry.
- Avoid prolonged contrast injections if the patient has a left dominant coronary circulation or a total occlusion of the right coronary artery, as they can cause hypotension.

17.6 Angiography

- Angiographic evaluation of left main disease has limitations: although it is usually adequate for left main lesions with ≥70% diameter stenosis, there is significant interobserver variability for 30–70% lesions [3]. The high

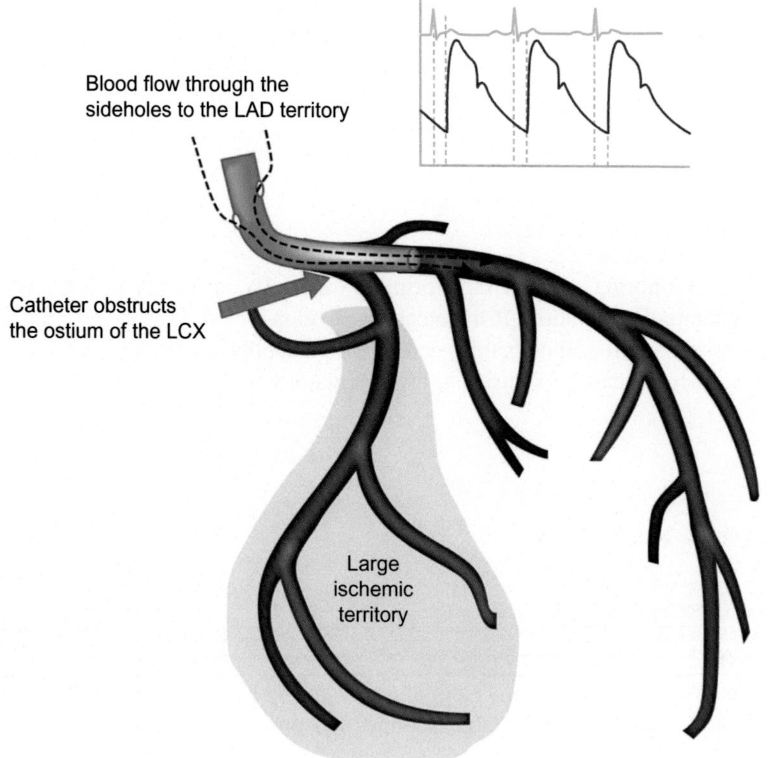

FIGURE 17.1 Deep intubation of the LAD due to short left main causes complete obstruction of the circumflex ostium. The results in a large territory of ischemia despite the pseudonormal pressure waveform due to the side holes of the guide catheter. *Courtesy of Dr. Imre Ungi.*

variability of the visual assessment stems from the lack of reference segment due to the short length and usually diffuse disease of the left main. The most appropriate views to visualize the left main bifurcation are RAO-caudal, AP-caudal, and LAO-caudal ("spider") views. The LAO-cranial or AP-cranial views are the most reliable for accurate aorto-ostial positioning of the proximal stent edge. The ostium of the left main is usually oval, therefore orthogonal views are important for proper assessment. The high anatomical and pathological variability of left main disease makes functional assessment [4] or intravascular imaging [5] essential for assessing intermediate left main lesions.

17.7 Selecting target lesion(s)

As described in Chapter 7: Selecting Target Lesion(s), decision about revascularization in patients with left main disease are challenging and depend on several factors:

1. Clinical presentation (stable vs acute coronary syndrome).
2. Coronary anatomy (extent and complexity) and ability to provide complete revascularization.
3. Left ventricular function.
4. Comorbidities, such as diabetes.
5. Patient preference.

In general the choice of revascularization strategy in patients with left main disease who are good surgical candidates and do not have diabetes, depends on overall CAD complexity. Patients with complex CAD (such as patients with high [$\geq$33] Syntax score) who are good surgical candidates usually benefit more from CABG, whereas those with low (0–22) or intermediate (23–32) Syntax score derive similar benefit from either CABG or PCI [1,2,6,7]. This, however, remains an area of continued controversy with some studies showing better outcomes with CABG. The ultimate decision should be made by a heart team taking into account the patient's preference and comorbidities.

An algorithm about left main revascularization is shown in Fig. 17.2 [6].

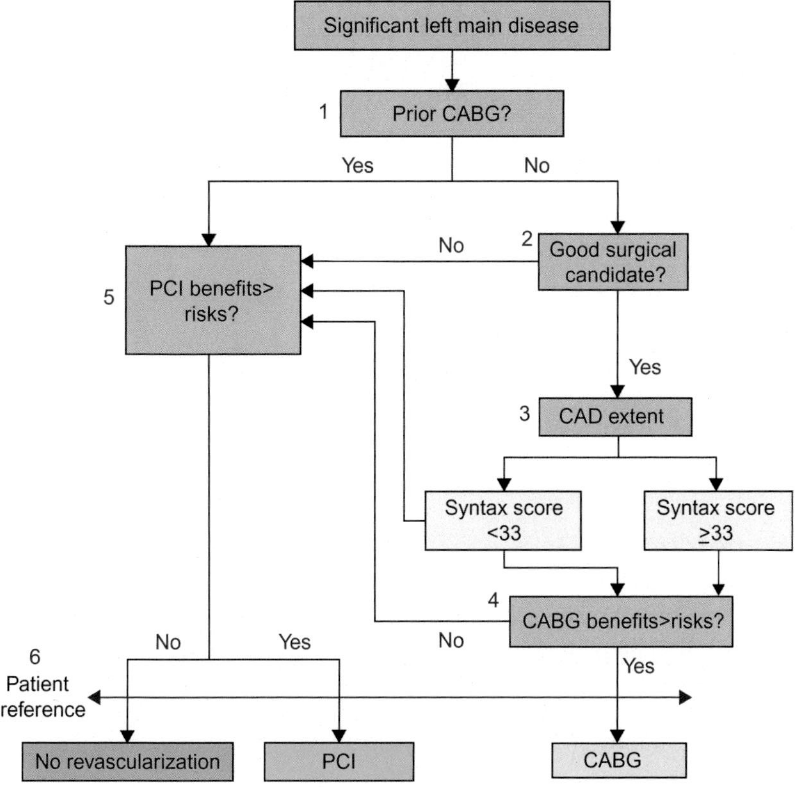

FIGURE 17.2 Algorithm for heart team management of left main coronary artery disease. *CABG*, coronary artery bypass graft surgery; CAD, coronary artery disease; *PCI*, percutaneous coronary intervention. *Reproduced from Ramadan R, Boden WE, Kinlay S. Management of left main coronary artery disease. J Am Heart Assoc 2018;7. Open access from JAHA.*

17.8 Wiring

Challenges with wiring left main lesions, include wiring ostial lesions (as discussed in Chapter 15: Ostial Lesions), bifurcations (as discussed in Chapter 16: Bifurcations), and trifurcations/quadrifications.

17.9 Lesion preparation

- Meticulous lesion preparation (with balloon angioplasty, plaque modification balloons, and possibly atherectomy in heavily calcified lesions) is critical prior to left main stenting.
- Particular attention must be paid to the aortic pressure during left main balloon occlusions. Balloon inflations should be kept as brief as possible to minimize the duration of ischemia, especially when hemodynamic support is not used. Use of a 70% saline/30% contrast mixture in the indeflator can speed up balloon inflation and deflation.

17.10 Stenting

- For ***ostial*** left main lesions, stenting is performed as described in Chapter 15: Ostial Lesions. If the left main is short, the stent may need to cross over into the LAD (provisional stenting, Section 16.1.10.1). Ostial left main stent placement should be confirmed in multiple angiographic views, especially the LAO cranial.
- For ***bifurcation*** left main lesions, stenting is performed as described in Chapter 16: Bifurcations, with provisional stenting preferred for most bifurcations without significant involvement of the side branch (circumflex in most cases) and two-stent bifurcation techniques (especially double kissing crush [DK crush]) preferred for true (Medina 1.1.1. or Medina 0.1.1.) bifurcation lesions [7,8]. The stents deployed from the left main into the LAD or circumflex should be able to be expanded to match the left main diameter (Fig. 10.1).
- If the patient ***also has lesions in the LAD and circumflex*** that require PCI and the left main stenosis is not critical, the LAD and circumflex lesions should ideally be stented first, as equipment delivery may be more challenging after the left main is stented.
- If the patient ***also has significant lesions in the RCA***, consider treating the RCA lesions first to improve the safety of the left main PCI.
- ***Trifurcating*** left mains are encountered rarely. For trifurcating left main lesions, stenting is performed based on involvement of the side branch. Single stent strategy is preferred in cases without significant side branch disease. In case of side branch involvement, any two-stent technique may be performed with consideration to final three kissing balloons or "trissing" to preserve the polygon of confluence. Large (7 or 8 French) guide catheter are needed for such cases.
- There is no clear benefit of routine kissing balloon inflation after provisional left main bifurcation stenting.
- Intravascular imaging should be routinely used for optimizing the outcome of left main PCI, as discussed in Section 17.13.

17.11 Closure

Arterial closure is performed as described in Chapter 11: Access Closure.

17.12 Physiology

17.12.1 Baseline assessment

For ostial left main lesions, the guide catheter should be disengaged to prevent pressure dampening, which is standard practice for all aorto-ostial lesions.

Either FFR or non-hyperemic indices can be used in left main lesions. They should be performed in both the LAD and the circumflex with pullback to demonstrate whether the step-up occurs in the left main. Adenosine FFR ≥ 0.80 (or non-hyperemic index >the cutoff value) in one vessel (such as the LAD), but not the other (such as the circumflex) may suggest that the left main stenosis is not significant. However, downstream stenosis in either the LAD or the circumflex may decrease the maximum achievable blood flow and as a result measure higher distal pressure and underestimate the severity of the left main stenosis.

17.12.2 Final assessment

In case of provisional stenting of the left main to LAD and concern for angiographic stenosis of the jailed ostial circumflex, imaging and physiologic assessment of the circumflex should be performed to guide treatment strategy (Fig. 16.79). If adenosine FFR is >0.80 or the non-hyperemic index is >cutoff value, stenting of the circumflex should be deferred.

In case of FFR ≤ 0.80, a second stent can be placed using T-stenting, TAP, reverse crush, or the culotte technique, as described in Chapter 16: Bifurcations.

17.13 Imaging

17.13.1 Baseline assessment

In contrast to non-left main lesions, IVUS and OCT are useful for determining the hemodynamic significance of left main lesions.

Left main minimum lumen area <6.0 mm^2 by IVUS or <5.4 mm^2 by OCT is considered significant, requiring revascularization. The minimum lumen area cutoff may be different for different populations: a cutoff of 4.5 mm^2 by IVUS has been proposed for Korean patients [9] who have smaller size coronary arteries compared with US patients [10].

17.13.2 Final assessment

IVUS or OCT is mandatory for assessing the final result of left main interventions and should be performed in both branches (LAD and circumflex).

To minimize the risk of restenosis, the minimum IVUS lumen areas shown in Fig. 17.3 should be achieved in each segment of the left main bifurcation [11].

17.14 Hemodynamic support

Hemodynamic support is used as described in Chapter 14: Hemodynamic Support. Given the potential for profound ischemia and hemodynamic instability, hemodynamic support is often used for left main PCI, especially in patients with low ejection fraction, decompensated hemodynamics (high left ventricular end diastolic pressure or high pulmonary capillary wedge pressure), occluded right coronary artery, and patients in whom complex procedures resulting in prolonged ischemia (e.g., atherectomy) are planned.

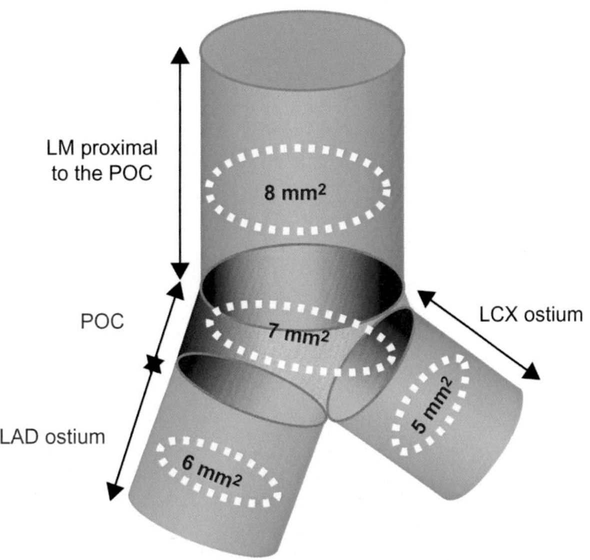

FIGURE 17.3 Minimum lumen areas that need to be achieved to minimize the risk of in-stent restenosis after left main stenting. *LM*, left main artery; *POC*, polygon of confluence; *LAD*, left anterior descending artery; *LCX*, left circumflex artery.

References

[1] Everolimus-eluting stents or bypass surgery for left main coronary artery disease. N Engl J Med 2019;381:1789.

[2] Stone GW, Sabik JF, Serruys PW, et al. Everolimus-eluting stents or bypass surgery for left main coronary artery disease. N Engl J Med 2016;375:2223–35.

[3] Fisher LD, Judkins MP, Lesperance J, et al. Reproducibility of coronary arteriographic reading in the coronary artery surgery study (CASS). Cathet Cardiovasc Diagn 1982;8:565–75.

[4] Hamilos M, Muller O, Cuisset T, et al. Long-term clinical outcome after fractional flow reserve-guided treatment in patients with angiographically equivocal left main coronary artery stenosis. Circulation 2009;120:1505–12.

[5] de la Torre Hernandez JM, Hernandez Hernandez F, Alfonso F, et al. Prospective application of pre-defined intravascular ultrasound criteria for assessment of intermediate left main coronary artery lesions results from the multicenter LITRO study. J Am Coll Cardiol 2011;58:351–8.

[6] Holm NR, Makikallio T, Lindsay MM, et al. Percutaneous coronary angioplasty versus coronary artery bypass grafting in the treatment of unprotected left main stenosis: updated 5-year outcomes from the randomised, non-inferiority NOBLE trial. Lancet 2020;395:191–9.

[7] Makikallio T, Holm NR, Lindsay M, et al. Percutaneous coronary angioplasty versus coronary artery bypass grafting in treatment of unprotected left main stenosis (NOBLE): a prospective, randomised, open-label, non-inferiority trial. Lancet 2016;388:2743–52.

[8] Ramadan R, Boden WE, Kinlay S. Management of left main coronary artery disease. J Am Heart Assoc 2018;7.

[9] Chen SL, Zhang JJ, Han Y, et al. Double kissing crush versus provisional stenting for left main distal bifurcation lesions: DKCRUSH-V randomized trial. J Am Coll Cardiol 2017;70:2605–17.

[10] Chen SL, Xu B, Han YL, et al. Comparison of double kissing crush versus Culotte stenting for unprotected distal left main bifurcation lesions: results from a multicenter, randomized, prospective DKCRUSH-III study. J Am Coll Cardiol 2013;61:1482–8.

[11] Park SJ, Ahn JM, Kang SJ, et al. Intravascular ultrasound-derived minimal lumen area criteria for functionally significant left main coronary artery stenosis. JACC Cardiovasc Interv 2014;7:868–74.

[12] Jasti V, Ivan E, Yalamanchili V, Wongpraparut N, Leesar MA. Correlations between fractional flow reserve and intravascular ultrasound in patients with an ambiguous left main coronary artery stenosis. Circulation 2004;110:2831–6.

[13] Kang SJ, Ahn JM, Song H, et al. Comprehensive intravascular ultrasound assessment of stent area and its impact on restenosis and adverse cardiac events in 403 patients with unprotected left main disease. Circ Cardiovasc Interv 2011;4:562–9.

Chapter 18

Bypass grafts—prior CABG patients

CTO Manual Online cases: 3, 16, 42, 43, 50, 85, 117, 125, 131, 134, 137, 140, 144, 148, 149
PCI Manual Online cases: 22, 27, 34, 39, 41, 46, 52, 56, 57, 58, 60, 62, 64, 73, 82, 87, 92, 96, 100

PCI in patients with prior coronary artery bypass graft surgery (CABG) can be challenging in many ways:

1. In addition to the native coronary arteries, bypass grafts need to be engaged, visualized, and sometimes treated.
2. Native coronary artery lesions are often quite challenging (bypass grafting accelerates native coronary atherosclerosis), and
3. Prior CABG patients have higher baseline risk, as they are often old and have multiple comorbidities [1].

Due to the need to engage and visualize the bypass grafts (and the often high complexity of treated lesions) angiography and PCI in prior CABG patients requires longer procedural and fluoroscopy time, higher radiation dose, and larger volume of contrast [2–4]. As a result (and also because of often decreased baseline renal function), the risk of contrast-induced acute kidney injury and possibly hemodialysis is increased in prior CABG patients.

Even though coronary perforations were previously considered "innocent" complications in prior CABG patients due to pericardial adhesions preventing formation of a pericardial effusion and tamponade, it is now appreciated that they can be lethal events. Coronary perforation in prior CABG patients can lead to loculated hematomas resulting in cardiac chamber compromise and hemodynamic collapse (dry tamponade) [5]. Such loculated effusions may require surgery or computed tomography–guided drainage for treatment. Prompt identification and treatment of coronary or graft perforation is, therefore, critical in prior CABG patients [6,7].

Some of the current recommendations for catheterization and PCI in prior CABG patients are summarized in Fig. 18.1 [1].

18.1 Planning

18.1.1 CABG anatomy

The goal of diagnostic angiography in the post-CABG patient is to visualize the native coronary arteries and all bypass grafts in the safest and most expeditious way. Detailed and meticulous preparation of the angiographer is key and includes:

1. Review of the surgical report;
2. Review of any prior coronary angiograms, including pre- and post-CABG angiograms; and
3. Review of ancillary imaging studies, especially coronary computed tomography angiography [1].

Obtaining the CABG surgical report and reviewing prior coronary angiograms is key for optimal coronary and bypass graft angiography and can save time, contrast, radiation dose, and reduce the number of catheters needed for engaging the grafts [8]. In non-urgent cases it may be best to delay angiography to allow time for obtaining the CABG report. It is important to review the operative report if at all possible instead of relying on what has been thought to be the post-surgical anatomy, as mistakes are frequently made and details can be "lost in translation."

Review of prior angiograms can help determine the location of the bypass graft proximal anastomoses and the catheters that were successful in engaging them. In addition, review of prior chest computed tomograms, even if not geared toward cardiac evaluation, can be very helpful. Thinner slice reconstruction (i.e., ≤ 1 mm) is enough to identify the CABG anatomy, that is, the origin and number of bypass grafts. Although dedicated coronary computed tomography angiography has been shown to be an excellent method to determine the anatomy of the native coronary arteries,

Manual of Percutaneous Coronary Interventions. DOI: https://doi.org/10.1016/B978-0-12-819367-9.00018-4

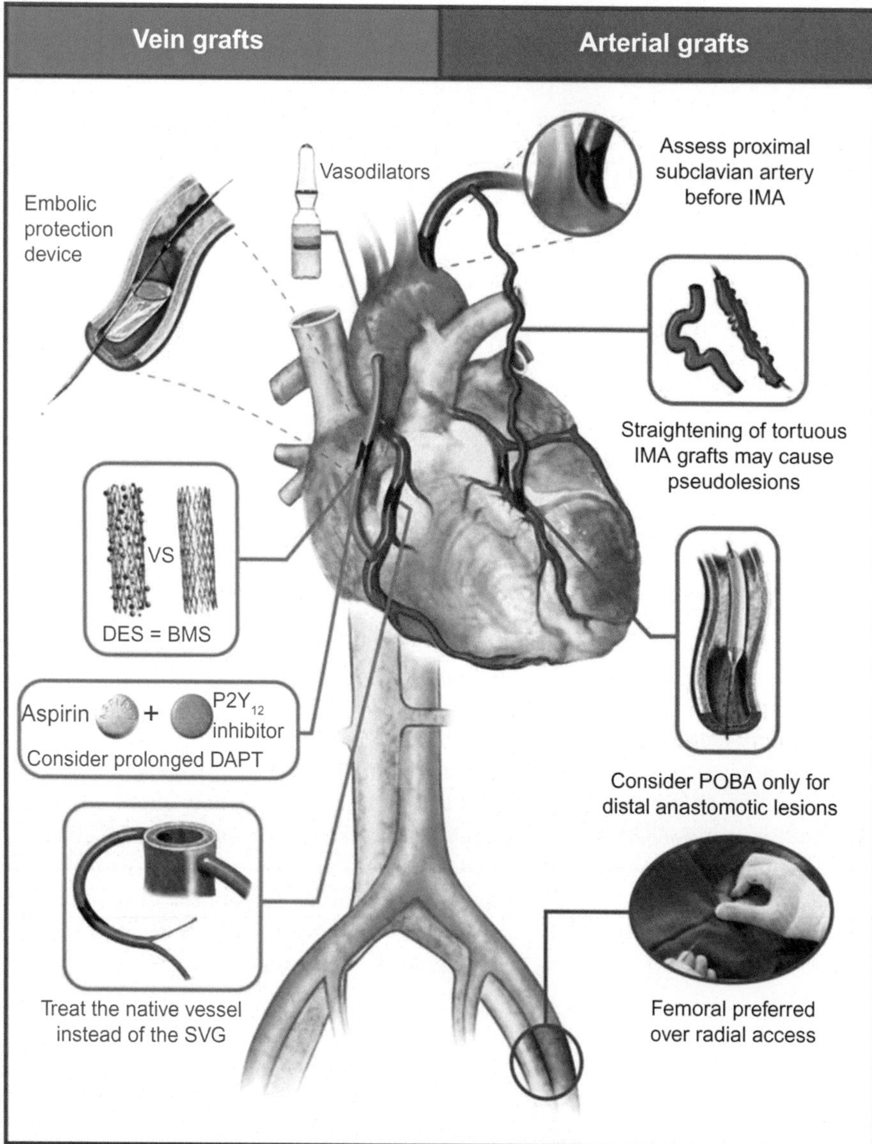

FIGURE 18.1 Overview of recommendations about cardiac catheterization and PCI in prior CABG patients. *Reproduced with permission from Xenogiannis I, Tajti P, Hall AB, et al. Update on cardiac catheterization in patients with prior coronary artery bypass graft surgery. JACC Cardiovasc Interv 2019;12:1635–49. Copyright Elsevier.*

location and patency of bypass grafts, visualization of distal native vessels and graft anastomoses can sometimes be hindered by the presence of calcification and surgical clips [1].

18.1.2 Review of comorbidities and pre-procedural optimization

Prior CABG patients often have multiple noncardiac comorbidities that could impact performance of the procedure, such as chronic kidney disease and peripheral arterial disease. Risk scores can be used to estimate each patient's risk for contrast-induced acute kidney injury [9]: in high-risk patients meticulous attention should be given to pre-procedural hydration, using isoosmolar contrast media, and limiting contrast volume to less than 3.7 × the estimated glomerular filtration rate (GFR) through careful preprocedural planning and contrast saving devices.

Prior CABG patients often experience congestive heart failure; in non-emergent cases, cardiac catheterization may be best deferred in patients who appear to be volume overloaded to allow hemodynamic optimization. This process can often be guided by right heart catheterization. In prior CABG patients with severe peripheral arterial disease access site selection may be difficult and could be facilitated by preprocedure CT angiography.

18.2 Monitoring

Monitoring is performed as described in Chapter 2: Monitoring. Careful monitoring is especially important in prior CABG patients given their higher baseline risk and often complex coronary anatomy.

18.3 Medications

Intragraft vasodilators (such as adenosine [10], nitroprusside [11], nicardipine [12], and verapamil [13]) can be administered both before and after saphenous vein graft (SVG) PCI to prevent or treat no reflow by causing vasodilation and facilitating passage of debris and/or chemical mediators liberated. Nicardipine (100–300 mcg intragraft) is often preferred due to prolonged duration of action and less hypotensive effect and is often administered both before and after PCI, especially if slow flow or no reflow occurs.

In contrast, platelet glycoprotein IIb/IIIa receptor inhibitors can cause harm during SVG intervention and should not be used routinely in SVG PCI [14,15].

18.4 Access

Engagement of arterial grafts and SVGs for angiography and/or PCI (Section 5.7) can be performed using either femoral or radial approach, however femoral access is generally preferred as it is associated with shorter time, fewer catheters and lower contrast and radiation dose [1,16,17].

If radial access is selected, the left radial artery should be used in most cases to facilitate engagement of the left internal mammary artery (LIMA) and the other bypass grafts. If left radial access is not feasible, the right radial artery can be used, but LIMA engagement can be challenging. Use of a Simmons catheter (Fig. 18.2) can facilitate LIMA engagement via right radial access, whereas the Bartorelli-Cozzi catheter (Fig. 18.3) can facilitate LIMA engagement via left radial access. When graft engagement is challenging using radial access, early conversion to femoral access should be considered [18]. In patients with bilateral IMA grafts upfront femoral access is preferred.

18.5 Engagement

18.5.1 Native coronary arteries

Engagement of native coronary arteries is performed as described in Chapter 5: Coronary and Graft Engagement.

18.5.2 Saphenous vein grafts

Knowledge of the coronary and bypass graft anatomy facilitates selective graft engagement. The usual location of bypass grafts is shown in Fig. 18.4.

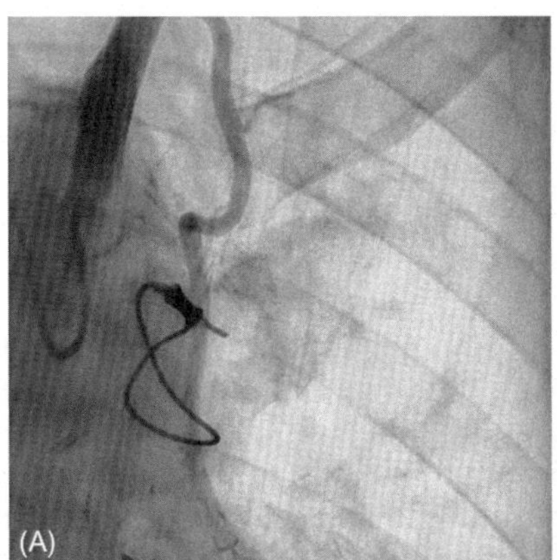

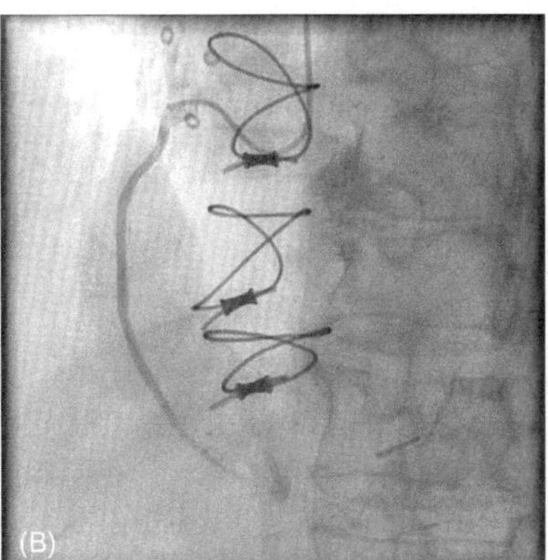

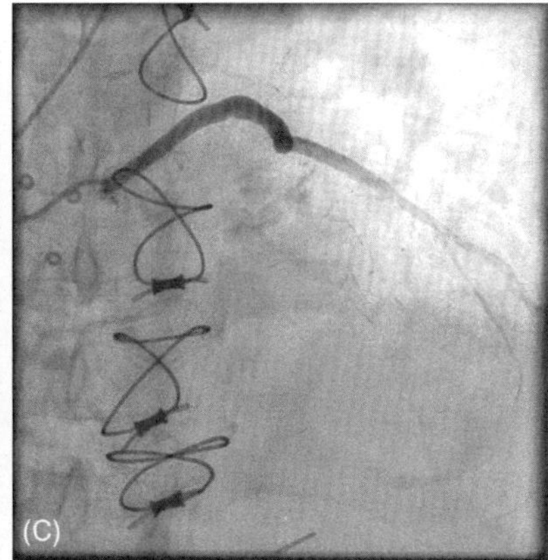

FIGURE 18.2 Use of a Simmons catheter to engage the LIMA (panel A), a saphenous vein graft to the right coronary artery (panel B) and a saphenous vein graft to a diagonal (panel C) via right radial access. *Courtesy of Dr. Abdul Hakeem.*

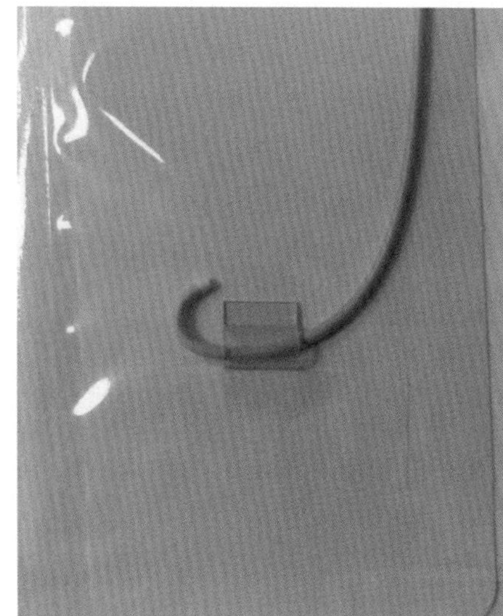

FIGURE 18.3 The Bartorelli-Cozzi catheter (Cordis) can be used to perform selective LIMA angiography via left radial access. *Courtesy of Dr. Jaikirshan Khatri.*

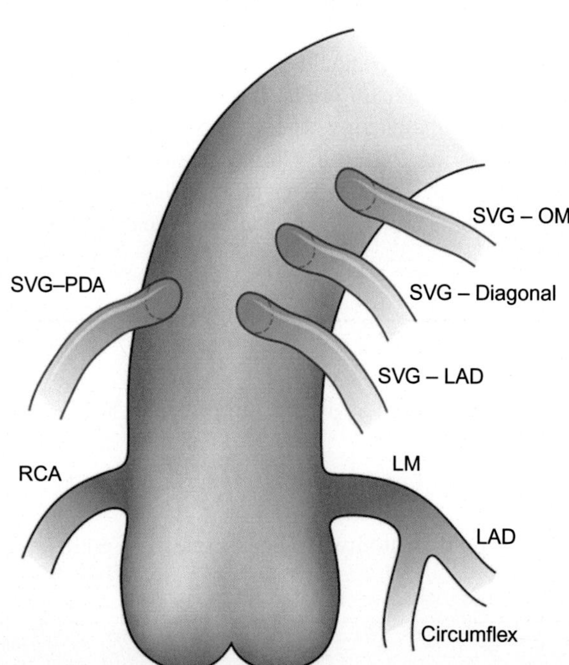

FIGURE 18.4 Usual location of aorto-coronary bypass grafts. *CX*, circumflex; *Dx*, diagonal; *LAD*, left anterior descending artery; *OM*, obtuse marginal; *RCA*, right coronary artery; *SVG*, saphenous vein graft.

- Graft markers can help graft engagement, but are only used in a minority (approximately 10%) of CABG patients [19]. The most commonly used catheter for engaging bypass grafts (Section 5.7) is the multipurpose for right coronary artery grafts, and the Amplatz left (AL) 1 or left coronary bypass (LCB) for left-sided grafts. Often, both right and left grafts can be engaged using a Judkins right (JR) 4 catheter, hence many operators will try this catheter first immediately after right coronary artery angiography in an attempt to save time and equipment cost.
- If the CABG anatomy is unknown, attempts to engage grafts should be continued until sources of coronary flow to all myocardial territories are identified. Bilateral subclavian artery angiography can help determine whether one or both IMAs were utilized as grafts. The presence of surgical clips along the IMA can help determine if it was used by the surgeon or not. In case of challenging graft engagement, aortography can be performed (usually administering 60 mL of contrast at a rate of 20 cc/sec) while performing cineangiography in the left anterior oblique projection.

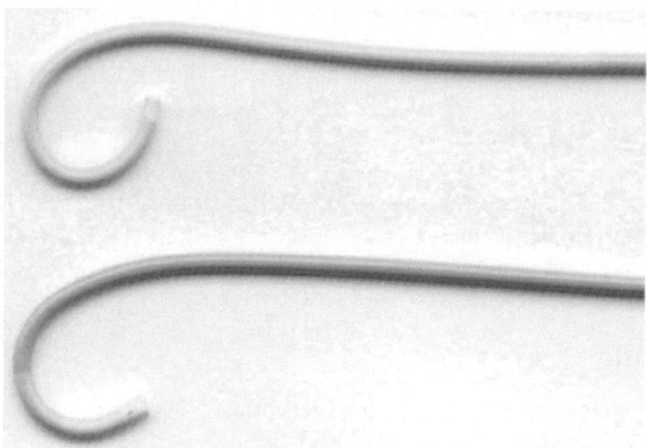

FIGURE 18.5 The IM-VB1 diagnostic (*top*) and guide (*bottom*) catheters. *Reproduced with permission from Warner JJ, Gehrig TR, Behar VS. The VB-1 catheter: an improved catheter for difficult-to-engage internal mammary artery grafts. Catheter Cardiovasc Interv 2003;59:361–5.*

- If graft intervention is needed, obtaining adequate guide catheter support is critical. This can be accomplished by using large guide catheters (such as 7 or 8 French), supportive guide catheter shapes (such as Amplatz left for left-sided grafts), deep guide intubation, or use of a guide catheter extension [20].

18.5.3 Arterial grafts

- Radial grafts are engaged as described for SVGs. Nitroglycerin administration is important, as radial grafts are very prone to spasm.
- Internal mammary artery grafts are usually engaged using the internal mammary (IM) catheter, or the modified internal mammary artery catheter (IM-VB1, named after Victor Behar) [21] that has a shape that is similar to the proximal two-thirds of the curve of a standard pigtail catheter (Fig. 18.5). Alternatively, in the absence of specialized catheters, a pigtail catheter can be used with a coronary wire or by cutting a small part of its tip [22].
- Deep guide intubation and use of guide catheter extensions may lead to IMA dissection and/or perforation.

18.6 Angiography

18.6.1 Native coronary arteries

Angiography of native coronary arteries in prior CABG patients is performed as described in Chapter 6: Coronary Angiography. Given the high prevalence of CTOs in prior CABG patients [23], long cine runs without panning can be useful in determining the presence and course of collateral circulation, as described in Chapter 21: Chronic Total Occlusions.

18.6.2 Saphenous vein grafts

Optimal graft engagement is critical for obtaining excellent quality diagnostic images of the bypass grafts (and the native coronary arteries) in prior CABG patients. Prolonged cineangiography runs can be useful for clarifying distal coronary filling, especially in patients with long grafts supplying a large myocardial territory.

18.6.3 Arterial grafts

In patients with IMA grafts, nonselective angiography of the proximal subclavian artery should be performed before selective angiography of the left or right IMA to identify subclavian lesions that could lead to subclavian steal [24]. If subclavian angiography is not performed, careful attention should be paid to the arterial pressure waveform in the subclavian artery as compared with that of the central aorta. Subclavian steal is a rare cause of recurrent cardiac ischemia and vertebrobasilar insufficiency, such as vertigo, in patients with previous CABG when an IMA graft has been used [25]. It is caused by significant stenosis of the ipsilateral subclavian artery and should be suspected in patients with >20 mmHg difference in systolic pressure between their arms.

18.7 Selecting target lesion(s)

PCI in prior CABG patients may be performed in native coronary arteries or in saphenous vein or arterial bypass grafts.

18.7.1 Native coronary arteries

In general native coronary artery PCI is preferred over SVG PCI in prior CABG patients, given lower periprocedural risk and better long-term outcomes. However, native coronary artery lesions in prior CABG patients can be challenging to treat, for example due to severe calcification or because they are chronic total occlusions.

18.7.2 Saphenous vein grafts

There are several treatment options in patients presenting with saphenous vein graft lesions (Fig. 18.6).

In patients presenting with SVG lesions revascularization is usually performed with PCI because redo CABG carries high risk. In few patients, however, especially those who do not have a LIMA-LAD graft and in whom a LIMA-LAD graft is feasible, redo CABG may be an option.

Otherwise, if the native coronary artery supplied by the SVG can be easily treated with PCI, PCI of the native coronary artery is preferred. If however, PCI of the native coronary artery is challenging (for example the native coronary artery lesion is a complex CTO), and PCI of the SVG appears to be simple, PCI of the SVG is performed instead.

When SVG PCI fails or in patients with recurrent SVG failure, recanalization of the native coronary artery CTO can be performed instead, at experienced centers.

18.7.2.1 Acute SVG occlusions

Acute saphenous vein graft (SVG) thrombosis is challenging to treat due to large thrombus burden, diffuse SVG degeneration, and high recurrent SVG failure rates [26]. Aggressive use of thrombectomy and EPDs is often required to restore luminal patency, but even if acute recanalization is achieved, long-term SVG patency is low [26].

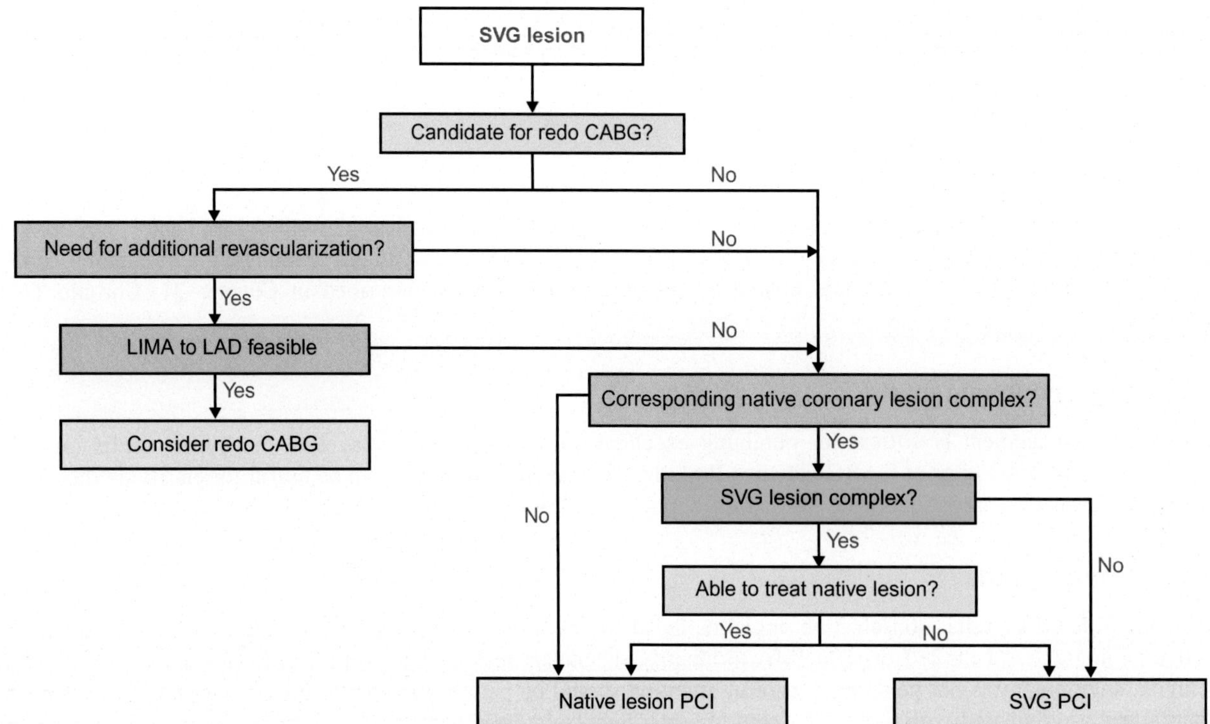

FIGURE 18.6 Revascularization options in prior CABG patients presenting with SVG lesions. *Reproduced with permission from Xenogiannis I, Tajti P, Hall AB, et al. Update on cardiac catheterization in patients with prior coronary artery bypass graft surgery. JACC Cardiovasc Interv 2019;12:1635–49. Copyright Elsevier.*

Alternative revascularization approaches, such as ad hoc [27] or staged [28] PCI of the bypassed native coronary artery [29] may provide better outcomes [30,31], but these can be challenging procedures, requiring dedicated equipment and expertise.

18.7.2.2 Chronic SVG occlusions

Because of high restenosis risk, SVG CTOs should generally not be recanalized (class III indication, level of evidence C) [15], unless no other treatment options exist. Occluded SVGs can be used, however, for retrograde crossing of the corresponding native coronary artery if the occlusion morphology is favorable.

18.7.3 Arterial grafts

- PCI of arterial grafts, especially the IMA grafts, is much less common than SVG PCI. The two main reasons are the higher rates of IMA patency as compared with SVGs and the more frequent performance of redo CABG in cases of IMA failure. Redo CABG is usually avoided in patients with a patent IMA graft to the LAD [32].
- PCI through IMA grafts should be avoided, given high risk of ischemia and complications, such as pseudolesion formation after wiring (see Section 18.8) and dissection.
- Despite the aforementioned limitations, PCI to IMA lesions has been associated with higher rates of restoration of TIMI-flow 3 and lower rates of periprocedural complications compared with SVG PCI. IMA anastomotic lesions sometimes respond to low-pressure balloon dilatation while proximal and mid-segment of IMA graft are stented in most cases.
- In patients with IMA grafts, the proximal subclavian artery should be evaluated, as severe lesions in this location could lead to coronary ischemia, and even acute coronary syndromes [24]. Subclavian artery stenting can be an effective treatment in such cases.

18.8 Wiring

18.8.1 Native coronary arteries

Wiring of native coronary arteries is performed as discussed in Chapter 8: Wiring. If wiring of a native coronary artery lesion is performed via a bypass graft, the distance from the guide tip to the lesion can be very long, sometimes not allowing equipment to reach the target lesion [33].

Potential solutions include:

1. Short guide catheters
2. Deep guide catheter intubation into the target vessel.
3. Use of long (300 cm) guidewires
4. Long-shaft balloon and stents

18.8.2 Saphenous vein grafts

18.8.2.1 Embolic protection devices in saphenous vein grafts

Embolic protection devices should be used in SVG PCI, if technically feasible, to reduce the risk of distal embolization. Two embolic protection devices are currently available in the US, the Filterwire (Boston Scientific) and the Spider (Medtronic) [34,35]. Both require a distal landing zone for deployment, hence they cannot be used in distal anastomotic lesions (unless the filter is deployed in the native coronary artery). The Filterwire is advanced (collapsed within the delivery sheath) through the target lesion, whereas the Spider is advanced over a standard 0.014 in. guidewire that is first advanced through the lesion.

18.8.2.1.1 Filterwire (Section 30.8.1)

Filterwire use: step-by-step.

18.8.2.1.1.1 Step 1: Determine that use of Filterwire is feasible

Criteria:

1. De novo SVG lesion (in-stent restenotic SVG lesions have low risk for embolization and do not require use of an embolic protection device).

2. Reference SVG diameter at the planned filter loop deployment site is between 2.25 and 5.5 mm.
3. Adequate landing zone (25 mm for the Filterwire EZ 2.25–3.5 mm device and 30 mm for the Filterwire EZ 3.5–5.5 mm device)

Challenges

1. Short landing zone
   In patients with short landing zone, the filter loop could potentially be deployed within the native coronary artery supplied by the SVG, although this will offer only partial protection.

18.8.2.1.1.2 Step 2: Preparation of the Filterwire

1. Open the Filterwire pouch and place the packing coils and accessory tool kit on the sterile field.
2. Unclip and carefully remove the yellow protective housing from the filter.
3. Remove the preloaded protection wire from the retaining clip; advance the EZ Delivery sheath until the clear section is exposed; then grasp the clear section of the sheath and remove the preloaded protection wire from the packing coil.
4. Keep the packing coil on the sterile field, as it also contains the EZ retrieval sheath that will be needed for removal of the Filterwire from the SVG.
5. Attach the wire torquer to the Filterwire near the exit port (to facilitate retraction of the filter into the EZ delivery sheath during preparation).
6. Grasp the EZ delivery sheath toward the very distal end of the clear section and submerge the filter and sheath into heparinized saline.
7. While the filter and delivery sheath are submerged in the heparinized saline (Fig. 18.7), sheath the protection wire by slowly retracting it into the EZ delivery sheath until the nose cone is partially retracted into the sheath.

What can go wrong?

1. Inadvertently discarding the packing coils. This is a problem because the coils also contain the EZ retrieval sheath that is needed for removal of the Filterwire. If the packing coils (and retrieval sheath) are accidentally discarded, the bent tip EZ retrieval sheath (sold separately) can be used instead for retrieving the Filterwire.

18.8.2.1.1.3 Step 3: Delivery and deployment of the Filterwire

1. Flush the peel-away introducer.
2. Insert the sheathed Filterwire tip into the peel-away introducer.
3. Create a desired shape and size bend at the tip of the Filterwire.
4. Advance the Filterwire and peel-away introducer assembly through the hemostasis valve of the Y-connector.
5. Remove the peel-away introducer.
6. Advance the Filterwire and EZ delivery sheath to the tip of the guide catheter.

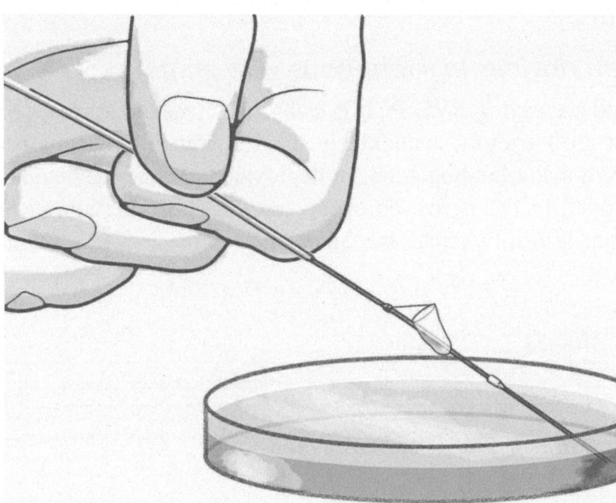

FIGURE 18.7 Preparation of the Filterwire. *Courtesy Boston Scientific © 2019 Boston Scientific Corporation or its affiliates. All rights reserved.*

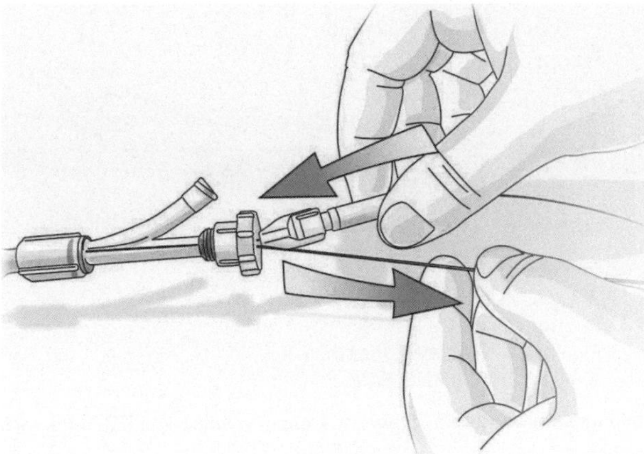

FIGURE 18.8 Deploying the Filterwire. *Courtesy Boston Scientific ©2019 Boston Scientific Corporation or its affiliates. All rights reserved.*

7. Advance the Filterwire and EZ delivery sheath assembly into the target vessel and through the target SVG lesion under fluoroscopic guidance, using one hand to torque the wire and the other hand to advance the EZ delivery sheath.
8. Once the Filterwire loop is advanced to the desired deployment zone the torquer is slid over the Filterwire until it reaches the hemostatic valve and tightened.
9. The filter is deployed by holding the wire in place with the wire torquer pressed against the hemostatic valve, while simultaneously retracting the EZ delivery sheath (Fig. 18.8).
10. Contrast is injected to verify that the protection wire is in proper position and there is adequate flow.

Challenges

1. Inability to cross the lesion. If the Filterwire cannot be delivered through the target SVG lesion, predilation with a small (1.5–2.0 mm balloon), use of a buddy wire or other maneuvers to increase guide catheter support (such as use of larger guide catheters or guide catheter extensions) can facilitate delivery. Buddy wires should be removed immediately after delivery because inadvertent stenting over the buddy wire will result in Filterwire entrapment.
2. No reflow after Filterwire delivery. This is more likely to occur when advancing equipment through diffusely degenerated SVGs. Treatment is with aspiration through the guide and vasodilator administration. Balloon angioplasty and stenting should not be performed until after TIMI 3 flow is restored.

18.8.2.1.1.4 Step 4: PCI is performed

1. Interventional equipment is advanced over the Filterwire, as if it were a standard guidewire.
2. The position of the filter should be constantly monitored to avoid movement. Retraction of the filter is particularly worrisome after stent deployment, as it can lead to filter entanglement with the stent. If repositioning of the Filterwire is needed, it should be performed after collapsing the filter using the Filterwire EZ retrieval sheath.
3. In case of slow flow or no flow during PCI, aspiration should be performed through the guide catheter, followed by filter removal. If additional PCI is needed a new Filterwire should be inserted.

18.8.2.1.1.5 Step 5: Filterwire retrieval

1. The EZ retrieval sheath is removed from the packing coil.
2. The EZ retrieval sheath is flushed with heparinized saline.
3. The EZ retrieval sheath is advanced over the protection wire past any deployed stent(s) until the tip of the EZ retrieval sheath reaches the protection wire's catheter stop.
4. The wire torquer is slid along the protection wire and secured against the hemostatic valve. The protection wire is then retracted back into the EZ retrieval sheath until resistance is felt (Fig. 18.9).
5. The EZ retrieval sheath and protection filter assembly is retracted into the guide catheter and then through the hemostatic valve outside the body.
6. The guide catheter is aspirated.

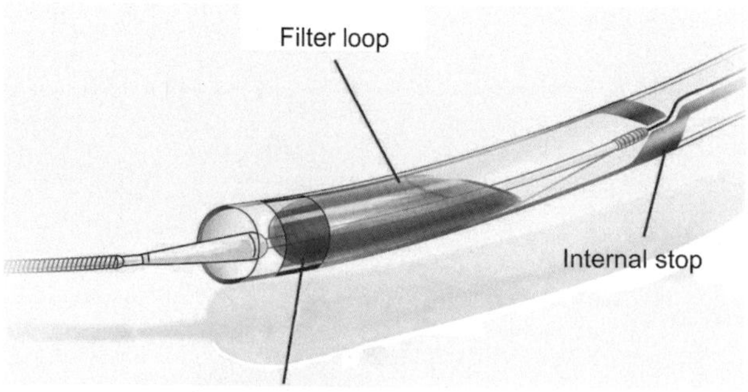

FIGURE 18.9 Retraction of the filter into the retrieval sheath. *Courtesy Boston Scientific ©2019 Boston Scientific Corporation or its affiliates. All rights reserved.*

TABLE 18.1 How to select the size of the Spider Fx embolic protection device.

Target vessel size (mm)	Spider filter size (mm)
3.0	3.0
3.1–4.0	4.0
4.1–5.0	5.0
4.5–6.0	6.0
5.5–6.0	7.0

Challenges

1. Inability to advance the EZ retrieval sheath into the SVG or through deployed stents. One solution is to use the bent tip retrieval sheath or a guide catheter extension.

18.8.2.1.2 SpiderFx (Section 30.8.2)

Spider Fx use: step-by-step.

18.8.2.1.2.1 Step 1: Determine that use of Spider Fx is feasible

***Criteria*:**

1. De novo SVG lesion (in-stent restenotic SVG lesions have low risk of embolization and do not require use of an embolic protection device).
2. Reference SVG diameter 3–6 mm. The size of the Spider is determined by the target vessel size (Table 18.1).
3. Adequate landing zone (4–5 cm distal to the target lesion is recommended in the instructions for use).
4. Minimum guide size: 6 French.

Challenges

1. Short landing zone
   In patients with short landing zone, the Spider filter could potentially be deployed within the native coronary artery supplied by the SVG, although this may offer only partial protection.

18.8.2.1.2.2 Step 2: Preparation of the Spider

1. Open the Spider pouch and place the packing coils and accessory tool kit on the sterile field.
2. Remove the colored portion of the hoop to expose the filter and the delivery end (green) of the SpiderFX™ catheter (Fig. 18.10).
3. Submerge the filter in heparinized saline to wet and remove air (Fig. 18.11).

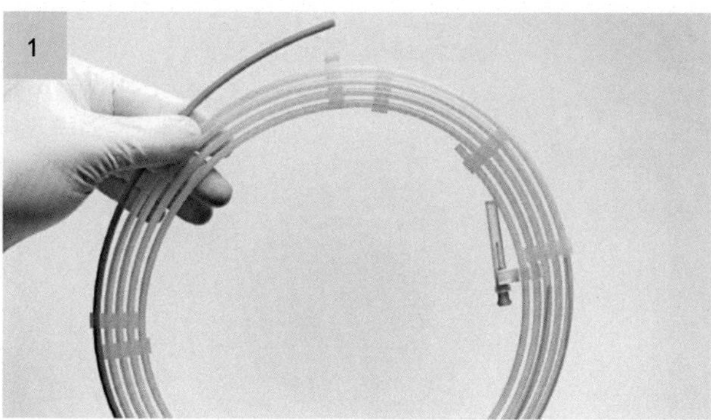

FIGURE 18.10 *©2016 Medtronic. All rights reserved. Used with permission of Medtronic.*

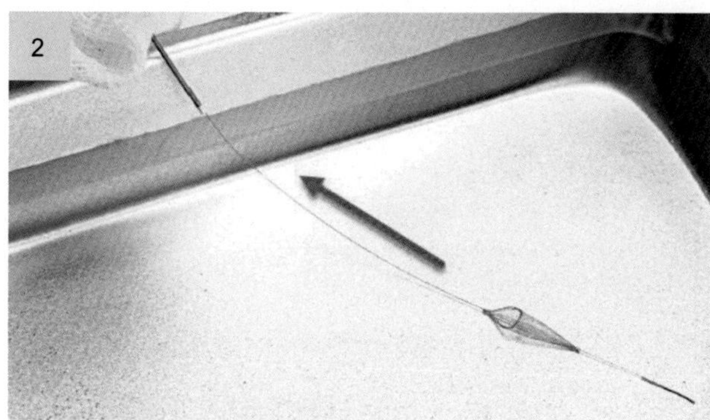

FIGURE 18.11 *©2016 Medtronic. All rights reserved. Used with permission of Medtronic.*

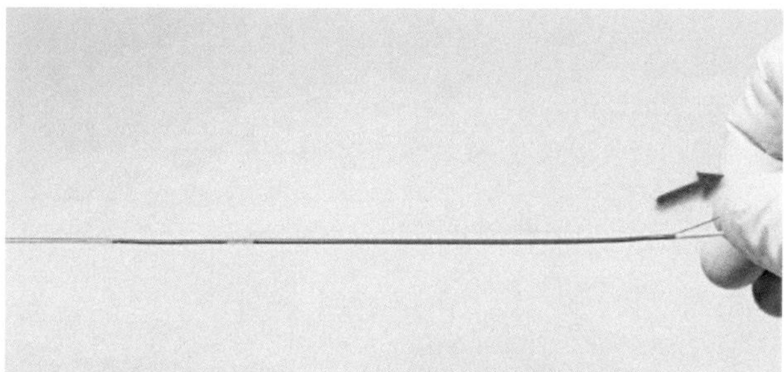

FIGURE 18.12 *©2016 Medtronic. All rights reserved. Used with permission of Medtronic.*

4. Pull the capture wire proximally until the filter and distal floppy tip are fully contained within the distal tip of the SpiderFX catheter (Fig. 18.12).
5. Block the primary guidewire exit port and flush through distal tip of catheter until fluid passes from the capture wire exit port (Fig. 18.13).

18.8.2.1.2.3 Step 3: Delivery and deployment of the Spider

1. Wire the SVG lesion using any 0.014 in. guidewire.
2. Load the delivery end (green) of the catheter over the back end of the primary guidewire and advance until the wire passes through the exit port. Advance delivery end of catheter 4–5 cm beyond the anticipated distal end of interventional devices that will be used (balloons, stents, etc.) (Fig. 18.14).

   Challenges: exiting of the primary guidewire from the wire exit port can be challenging—bending the Spider Fx catheter at the wire exit port can facilitate this step.

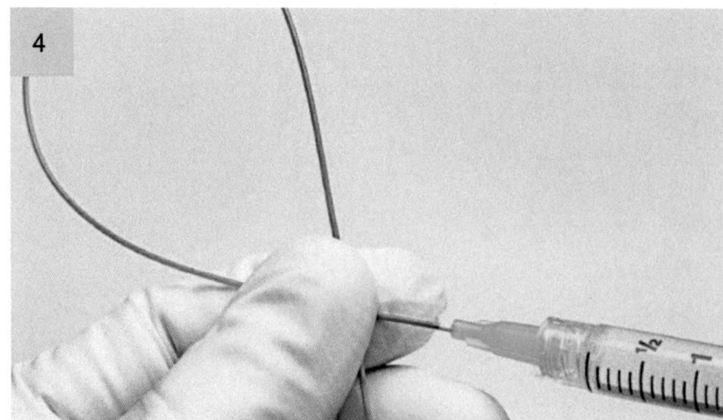

FIGURE 18.13 ©2016 Medtronic. All rights reserved. Used with permission of Medtronic.

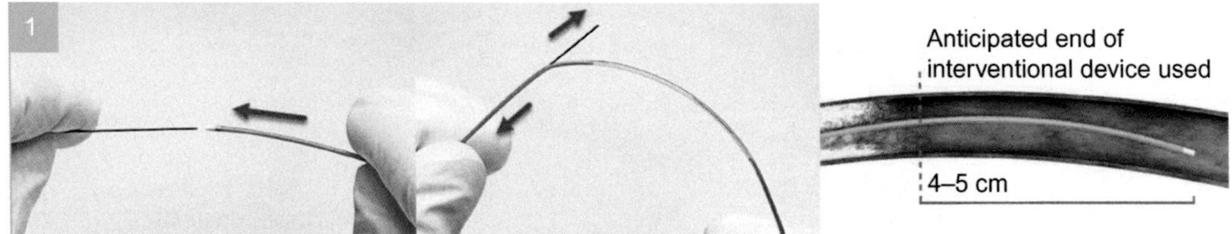

FIGURE 18.14 ©2016 Medtronic. All rights reserved. Used with permission of Medtronic.

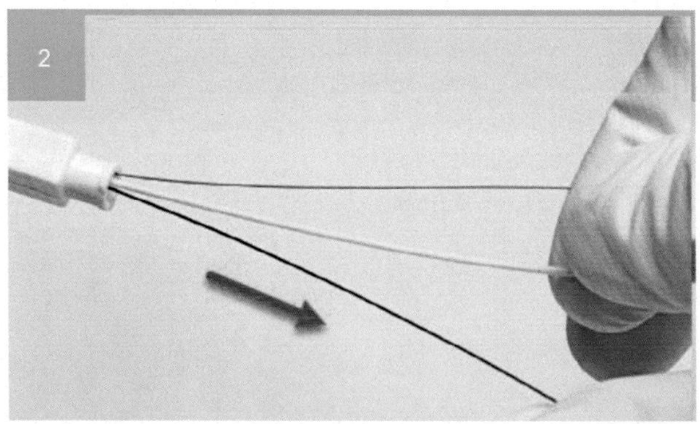

FIGURE 18.15 ©2016 Medtronic. All rights reserved. Used with permission of Medtronic.

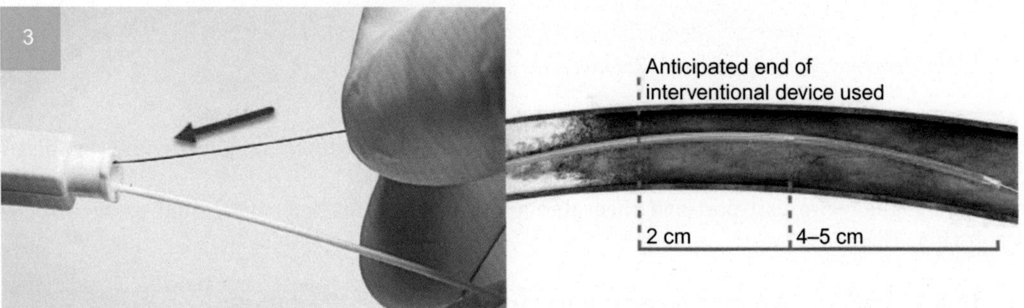

FIGURE 18.16 ©2016 Medtronic. All rights reserved. Used with permission of Medtronic.

3. Fix the delivery catheter and the capture wire and withdraw the initially used 0.014 in. guidewire (Fig. 18.15).
4. Hold catheter stationary and gently advance the capture wire until the distal marker on the filter aligns with the distal marker on the catheter. Ensure that the proximal marker on filter is at least 2 cm from the interventional device used (Fig. 18.16).

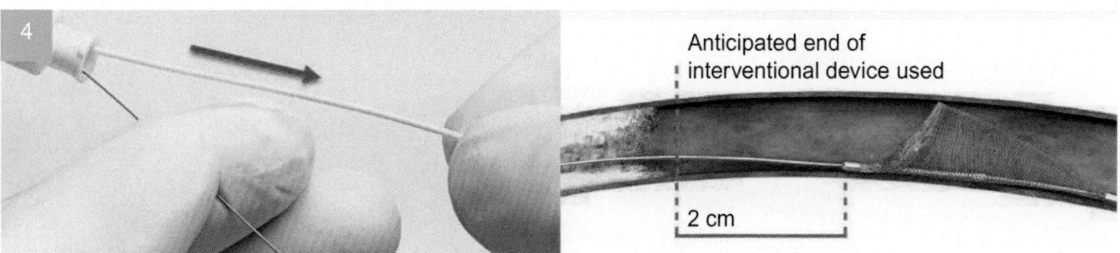

FIGURE 18.17 *©2016 Medtronic. All rights reserved. Used with permission of Medtronic.*

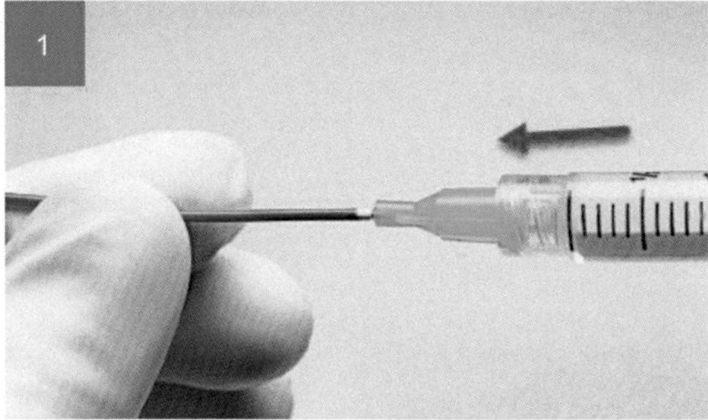

FIGURE 18.18 *©2016 Medtronic. All rights reserved. Used with permission of Medtronic.*

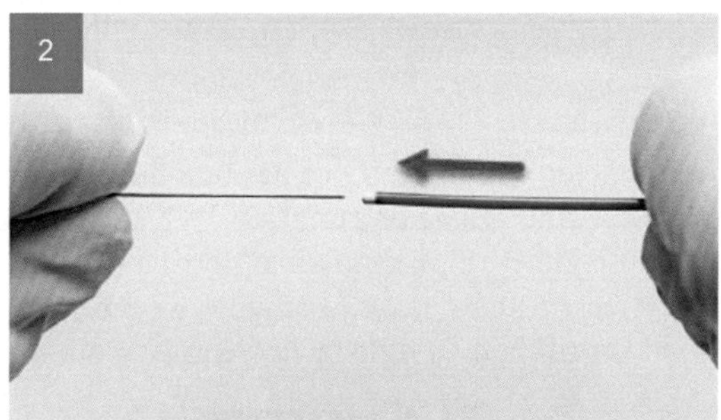

FIGURE 18.19 *©2016 Medtronic. All rights reserved. Used with permission of Medtronic.*

5. Fix the wire and withdraw delivery catheter to deploy filter. Confirm filter mouth is open. Withdraw Spider FX catheter from patient. **DO NOT DISCARD CATHETER** (the blue end of the catheter will be used for capturing the filter at the end of the procedure) (Fig. 18.17).

18.8.2.1.2.4 Step 4: PCI is performed

1. Interventional equipment is advanced over the Spider wire, as if it were a standard guidewire.
2. The position of the filter should be constantly monitored to avoid movement. Retraction of the filter is particularly worrisome after stent deployment, as it can lead to filter entanglement with the stent. If repositioning of the Spider wire is needed it should be performed after collapsing the filter using the retrieval sheath.
3. In case of slow flow or no flow during PCI, aspiration should be performed through the guide catheter, followed by removal of the Spider wire. If additional PCI is needed a new Spider wire should be inserted.

18.8.2.1.2.5 Step 5: Spider wire removal

1. Flush recovery end (blue) of the catheter until fluid passes from distal tip of capture wire exit port (Fig. 18.18).
2. Load recovery end of the SpiderFX catheter on the back end of the Spider wire (Fig. 18.19).
3. Advance the SpiderFx catheter until it reaches the mouth of the filter.

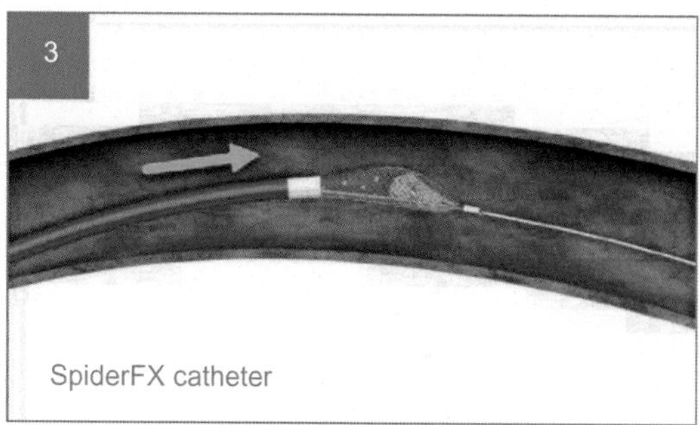

FIGURE 18.20

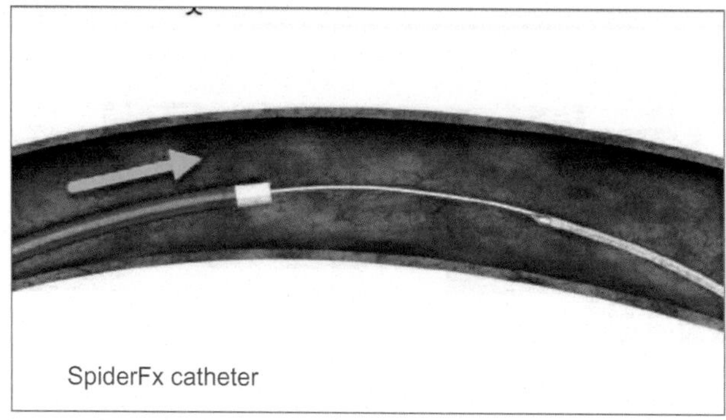

FIGURE 18.21

Challenges

1. Inability to advance the SpiderFx catheter into the SVG or through the deployed stents. One solution is to use the Filterwire bent tip retrieval sheath or a guide catheter extension.

4. Advance the SpiderFx catheter to enclose either (a) the proximal portion of the filter completely covering the proximal mouth indicator (partial recovery, best if there is large amount of debris captured in the filter) (Fig. 18.20) or (b) the entire filter up to the distal marker (if there is little or no debris captured) (Fig. 18.21)

Challenges

1. Aggressive advancement of the SpiderFx catheter over a full filter may result in distal embolization of a portion of the captured debris.

5. Withdraw the SpiderFx catheter and Spider capture wire together.

18.8.3 Arterial grafts

Straightening of a tortuous LIMA during advancement of guidewires and microcatheters may lead to pseudolesions (the so called "accordioning effect"), that can lead to flow compromise and ischemia (Fig. 18.22) [36]. Pseudolesions must be differentiated from vasospam and dissection. Administration of intravenous vasodilators will be ineffective in the presence of the pseudolesions, which should correct with guidewire withdrawal.

18.9 Lesion preparation

18.9.1 Native coronary arteries

Native coronary artery lesions in prior CABG patients are often complex, with high rates of severe calcification, tortuosity, and chronic total occlusions. Hence, aggressive lesion preparation, including plaque modification balloons and atherectomy, is often required.

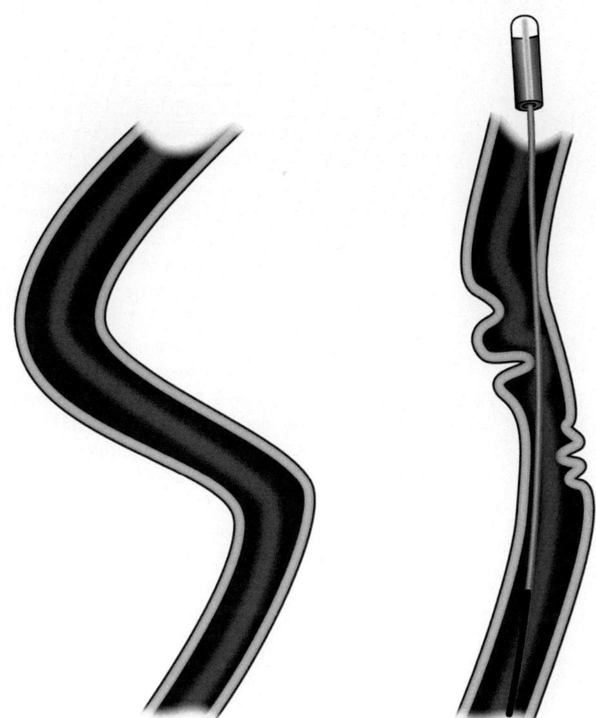

FIGURE 18.22 Illustration of the accordioning effect during PCI of a LIMA graft (or any other tortuous vessel).

18.9.2 Saphenous vein grafts

Primary stenting without predilation has been advocated for SVG PCI to reduce the risk of distal embolization [37].

Atherectomy is generally avoided in SVG lesions due to risk of distal embolization, but may be needed in old grafts with highly-calcified lesions.

Vasodilator administration before and after PCI could reduce the consequences of distal embolization. Nicardipine (100–300 mcg intragraft) is often preferred due to prolonged duration of action and less hypotensive effect and is often administered both before and after PCI, especially if slow flow or no reflow occurs.

18.9.3 Arterial grafts

Balloon dilation alone may suffice in some IMA anastomotic lesions.

18.10 Stenting

18.10.1 Native coronary arteries

- Given multiple levels of coronary lesion complexity (calcification, bifurcation, etc.) stent delivery and expansion can be challenging in prior CABG patients, requiring meticulous lesion preparation and use of various techniques for enhancing guide support.
- Sometimes PCI of a native coronary artery can be performed using two catheters: one catheter to deliver equipment (usually in the native coronary artery) and another catheter for visualization (usually in a bypass graft) (Fig. 18.23) [38].

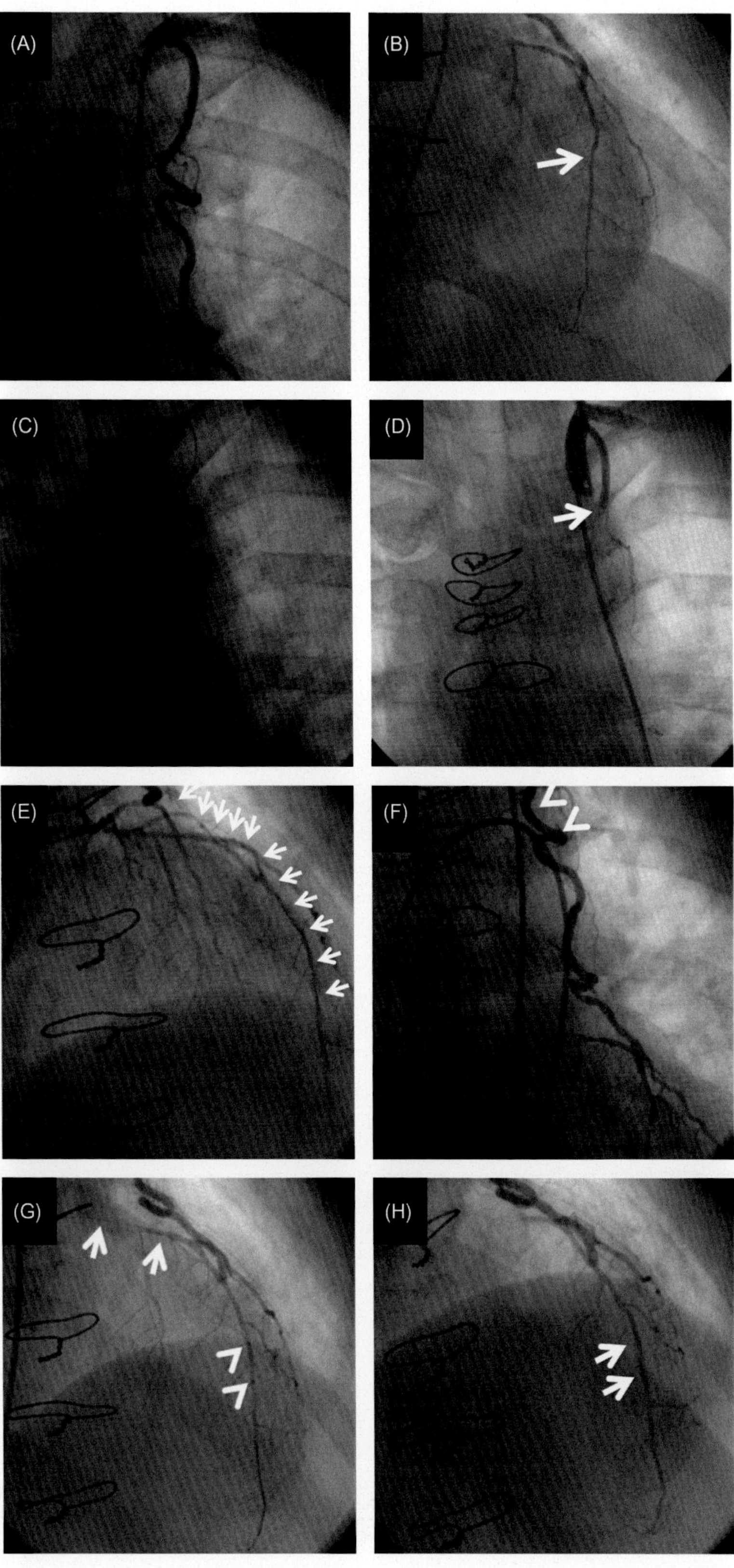

FIGURE 18.23 Left internal mammary artery (LIMA) graft angiography showed marked LIMA tortuosity (panel A) and a left anterior descending artery (LAD) stenosis distal to the LIMA anastomosis (*arrow*, panel B). After a Fielder wire was advanced over a microcatheter to the distal LAD (panel C) antegrade LIMA flow ceased (*arrow*, panel D). Coronary angiography through a second guiding catheter that engaged the left main artery confirmed placement of the wire (*arrows*, panel E) inserted via the LIMA into the distal LAD. After removal of the LIMA guidewire (*arrowheads*, panel F), antegrade flow through the LIMA was restored. A 2.5 × 12 mm stent (*arrowheads*, panel G) was advanced into place over the native LAD guidewire (*arrows*, panel G), while angiography from the LIMA provided lesion visualization (Panel G). Final angiography (via the LIMA) demonstrated successful treatment of the distal LAD lesion (*arrows*, panel H). *Reprinted from Lichtenwalter C, Banerjee S, Brilakis ES. Dual guide catheter technique for treating native coronary artery lesions through tortuous internal mammary grafts: separating equipment delivery from target lesion visualization. J Invasive Cardiol 2010;22:E78–81 with permission.*

18.10.2 Saphenous vein grafts

Saphenous vein grafts is the one coronary lesion subset in which bare metal and drug-eluting stents have similar long-term outcomes [39,40], hence either type can be used.

Undersized stents (stent to vessel diameter ratio <0.89) reduced the risk of distal embolization without significantly increasing the risk of restenosis in one study [41].

18.10.3 Arterial grafts

Balloon dilation alone may suffice in some IMA anastomotic lesions, whereas proximal and mid-segment of IMA graft are stented in most cases.

Using soft guidewires may decrease the risk of IMA kinking, and occasionally use of shortened guide catheters may be required to allow a balloon or stent to reach a distal anastomotic lesion, or a lesion in the native vessel distal to the IMA anastomosis.

18.11 Closure

Arterial closure is performed as described in Chapter 11: Access Closure.

18.12 Physiology

Although fractional flow reserve (FFR) measurement is the standard of care for assessing intermediate native coronary artery lesions, it is of limited value for evaluating intermediate SVG lesions because: (1) FFR also reflects flow from the native coronary artery if patent, hence significant SVG lesions may have normal FFR; and (2) intermediate SVG lesions have high rates of progression, hence a normal FFR does not necessarily mean low rates of subsequent lesion progression and adverse events [42–44].

Similarly, non-invasive physiologic assessment using coronary computed tomography angiography data (FFR-CT), albeit feasible for native coronary artery evaluation, has not been validated in patients with prior coronary stents and/or bypass grafts.

18.13 Imaging

18.13.1 Baseline assessment

The risk of distal embolization may be lower with direct stenting [37] and with slight undersizing [41] of the stent. Moreover, SVGs can sometimes be large and their size difficult to estimate visually. Use of IVUS or OCT can facilitate direct stenting by accurate stent sizing.

18.13.2 Final assessment

IVUS and OCT can confirm good stent expansion and strut apposition and potentially reduce the amount of contrast needed (with IVUS).

18.14 Hemodynamic support

Hemodynamic support is used as per standard practice, as described in Chapter 14: Hemodynamic Support. Prior CABG patients often have low ejection fraction, congestive heart failure, and complex coronary or bypass graft lesions that may benefit from hemodynamic support.

References

[1] Xenogiannis I, Tajti P, Hall AB, et al. Update on cardiac catheterization in patients with prior coronary artery bypass graft surgery. JACC Cardiovasc Interv 2019;12:1635–49.

[2] Michael TT, Karmpaliotis D, Brilakis ES, et al. Impact of prior coronary artery bypass graft surgery on chronic total occlusion revascularisation: insights from a multicentre US registry. Heart 2013;99:1515–18.

[3] Teramoto T, Tsuchikane E, Matsuo H, et al. Initial success rate of percutaneous coronary intervention for chronic total occlusion in a native coronary artery is decreased in patients who underwent previous coronary artery bypass graft surgery. JACC Cardiovasc Interv 2014;7:39–46.
[4] Toma A, Stahli BE, Gick M, et al. Long-term follow-up of patients with previous coronary artery bypass grafting undergoing percutaneous coronary intervention for chronic total occlusion. Am J Cardiol 2016;118:1641–6.
[5] Sapontis J, Salisbury AC, Yeh RW, et al. Early procedural and health status outcomes after chronic total occlusion angioplasty: a report from the OPEN-CTO Registry (outcomes, patient health status, and efficiency in chronic total occlusion hybrid procedures). JACC Cardiovasc Interv 2017;10:1523–34.
[6] Karatasakis A, Akhtar YN, Brilakis ES. Distal coronary perforation in patients with prior coronary artery bypass graft surgery: the importance of early treatment. Cardiovasc Revasc Med 2016;17:412–17.
[7] Wilson WM, Spratt JC, Lombardi WL. Cardiovascular collapse post chronic total occlusion percutaneous coronary intervention due to a compressive left atrial hematoma managed with percutaneous drainage. Catheter Cardiovasc Interv 2015;86:407–11.
[8] Varghese I, Boatman DM, Peters CT, et al. Impact on contrast, fluoroscopy, and catheter utilization from knowing the coronary artery bypass graft anatomy before diagnostic coronary angiography. Am J Cardiol 2008;101:1729–32.
[9] Mehran R, Aymong ED, Nikolsky E, et al. A simple risk score for prediction of contrast-induced nephropathy after percutaneous coronary intervention: development and initial validation. J Am Coll Cardiol 2004;44:1393–9.
[10] Sdringola S, Assali A, Ghani M, et al. Adenosine use during aortocoronary vein graft interventions reverses but does not prevent the slow-no reflow phenomenon. Catheter Cardiovasc Interv 2000;51:394–9.
[11] Zoghbi GJ, Goyal M, Hage F, et al. Pretreatment with nitroprusside for microcirculatory protection in saphenous vein graft interventions. J Invasive Cardiol 2009;21:34–9.
[12] Fischell TA, Subraya RG, Ashraf K, Perry B, Haller S. "Pharmacologic" distal protection using prophylactic, intragraft nicardipine to prevent no-reflow and non-Q-wave myocardial infarction during elective saphenous vein graft intervention. J Invasive Cardiol 2007;19:58–62.
[13] Michaels AD, Appleby M, Otten MH, et al. Pretreatment with intragraft verapamil prior to percutaneous coronary intervention of saphenous vein graft lesions: results of the randomized, controlled vasodilator prevention on no-reflow (VAPOR) trial. J Invasive Cardiol 2002;14:299–302.
[14] Roffi M, Mukherjee D, Chew DP, et al. Lack of benefit from intravenous platelet glycoprotein IIb/IIIa receptor inhibition as adjunctive treatment for percutaneous interventions of aortocoronary bypass grafts: a pooled analysis of five randomized clinical trials. Circulation 2002;106:3063–7.
[15] Levine GN, Bates ER, Blankenship JC, et al. ACCF/AHA/SCAI guideline for percutaneous coronary intervention: a report of the American College of Cardiology Foundation/American Heart Association Task Force on Practice Guidelines and the Society for Cardiovascular Angiography and Interventions. Circulation 2011;124:e574–651 2011.
[16] Rigattieri S, Sciahbasi A, Brilakis ES, et al. Meta-analysis of radial versus femoral artery approach for coronary procedures in patients with previous coronary artery bypass grafting. Am J Cardiol 2016;117:1248–55.
[17] Michael TT, Alomar M, Papayannis A, et al. A randomized comparison of the transradial and transfemoral approaches for coronary artery bypass graft angiography and intervention: the RADIAL-CABG trial (RADIAL versus femoral access for coronary artery bypass graft angiography and intervention). JACC Cardiovasc Interv 2013;6:1138–44.
[18] Cooper L, Banerjee S, Brilakis ES. Crossover from radial to femoral access during a challenging percutaneous coronary intervention can make the difference between success and failure. Cardiovasc Revasc Med 2010;11(266):e5–8.
[19] Varghese I, Samuel J, Banerjee S, Brilakis ES. Comparison of percutaneous coronary intervention in native coronary arteries vs. bypass grafts in patients with prior coronary artery bypass graft surgery. Cardiovasc Revasc Med 2009;10:103–9.
[20] Farooq V, Mamas MA, Fath-Ordoubadi F, Fraser DG. The use of a guide catheter extension system as an aid during transradial percutaneous coronary intervention of coronary artery bypass grafts. Catheter Cardiovasc Interv 2011;78:847–63.
[21] Warner JJ, Gehrig TR, Behar VS. The VB-1 catheter: an improved catheter for difficult-to-engage internal mammary artery grafts. Catheter Cardiovasc Interv 2003;59:361–5.
[22] Lapp H, Haltern G, Kranz T, et al. Use of a pigtail catheter to engage a difficult internal mammary artery. Catheter Cardiovasc Interv 2002;56:489–91.
[23] Jeroudi OM, Alomar ME, Michael TT, et al. Prevalence and management of coronary chronic total occlusions in a tertiary veterans affairs hospital. Catheter Cardiovasc Interv 2014;84:637–43.
[24] Dimas B, Lindsey JB, Banerjee S, Brilakis ES. ST-segment elevation acute myocardial infarction due to severe hypotension and proximal left subclavian artery stenosis in a prior coronary artery bypass graft patient. Cardiovasc Revasc Med 2009;10:191–4.
[25] Philippe F, Folliguet T, Carbogniani D, et al. Coronary subclavian steal syndrome after internal mammary artery bypass grafting. A cause of severe postoperative recurrent myocardial ischemia. Arch Mal Coeur Vaiss 2000;93:1555–9.
[26] Abdel-Karim AR, Banerjee S, Brilakis ES. Percutaneous intervention of acutely occluded saphenous vein grafts: contemporary techniques and outcomes. J Invasive Cardiol 2010;22:253–7.
[27] Brilakis ES, Banerjee S, Lombardi WL. Retrograde recanalization of native coronary artery chronic occlusions via acutely occluded vein grafts. Catheter Cardiovasc Interv 2010;75:109–13.
[28] Xenogiannis I, Tajti P, Burke MN, Brilakis ES. Staged revascularization in patients with acute coronary syndromes due to saphenous vein graft failure and chronic total occlusion of the native vessel: a novel concept. Catheter Cardiovasc Interv 2019;93:440–4.
[29] Nguyen-Trong PK, Alaswad K, Karmpaliotis D, et al. Use of saphenous vein bypass grafts for retrograde recanalization of coronary chronic total occlusions: insights from a multicenter registry. J Invasive Cardiol 2016;28:218–24.

[30] Brilakis ES, Rao SV, Banerjee S, et al. Percutaneous coronary intervention in native arteries versus bypass grafts in prior coronary artery bypass grafting patients a report from the national cardiovascular data registry. JACC Cardiovasc Interv 2011;4:844–50.

[31] Brilakis ES, O'Donnell CI, Penny W, et al. Percutaneous coronary intervention in native coronary arteries versus bypass grafts in patients with prior coronary artery bypass graft surgery: insights from the veterans affairs clinical assessment, reporting, and tracking program. JACC Cardiovasc Interv 2016;9:884–93.

[32] Ibanez B, James S, Agewall S, et al. ESC guidelines for the management of acute myocardial infarction in patients presenting with ST-segment elevation: the task force for the management of acute myocardial infarction in patients presenting with ST-segment elevation of the European Society of Cardiology (ESC). Eur Heart J 2017;2018(39):119–77.

[33] Tajti P, Sandoval Y, Brilakis ES. "Around the world"—how to reach native coronary artery lesions through long and tortuous aortocoronary bypass grafts. Hellenic J Cardiol 2018;59:354–7.

[34] Dixon SR, Mann JT, Lauer MA, et al. A randomized, controlled trial of saphenous vein graft intervention with a filter-based distal embolic protection device: TRAP trial. J Interv Cardiol 2005;18:233–41.

[35] Stone GW, Rogers C, Hermiller J, et al. Randomized comparison of distal protection with a filter-based catheter and a balloon occlusion and aspiration system during percutaneous intervention of diseased saphenous vein aorto-coronary bypass grafts. Circulation 2003;108:548–53.

[36] Muller O, Hamilos M, Ntalianis A, Sarno G, De Bruyne B. Images in cardiovascular medicine. The accordion phenomenon: lesson from a movie. Circulation 2008;118:e677–8.

[37] Leborgne L, Cheneau E, Pichard A, et al. Effect of direct stenting on clinical outcome in patients treated with percutaneous coronary intervention on saphenous vein graft. Am Heart J 2003;146:501–6.

[38] Lichtenwalter C, Banerjee S, Brilakis ES. Dual guide catheter technique for treating native coronary artery lesions through tortuous internal mammary grafts: separating equipment delivery from target lesion visualization. J Invasive Cardiol 2010;22:E78–81.

[39] Brilakis ES, Edson R, Bhatt DL, et al. Drug-eluting stents versus bare-metal stents in saphenous vein grafts: a double-blind, randomised trial. Lancet 2018;391:1997–2007.

[40] Colleran R, Kufner S, Mehilli J, et al. Efficacy Over Time With Drug-Eluting Stents in Saphenous Vein Graft Lesions. J Am Coll Cardiol 2018;71:1973–82.

[41] Hong YJ, Pichard AD, Mintz GS, et al. Outcome of undersized drug-eluting stents for percutaneous coronary intervention of saphenous vein graft lesions. Am J Cardiol 2010;105:179–85.

[42] Almomani A, Pothineni NV, Edupuganti M, et al. Outcomes of fractional flow reserve-based deferral in saphenous vein graft narrowing. Am J Cardiol 2018;122:723–8.

[43] Aqel R, Zoghbi GJ, Hage F, Dell'Italia L, Iskandrian AE. Hemodynamic evaluation of coronary artery bypass graft lesions using fractional flow reserve. Catheter Cardiovasc Interv 2008;72:479–85.

[44] Di Serafino L, De Bruyne B, Mangiacapra F, et al. Long-term clinical outcome after fractional flow reserve- versus angio-guided percutaneous coronary intervention in patients with intermediate stenosis of coronary artery bypass grafts. Am Heart J 2013;166:110–18.

Chapter 19

Calcification

CTO Manual Online cases: 1, 5, 17, 18, 27, 39, 40, 46, 49, 65, 85, 133, 134, 136, 142, 145, 147, 148, 149, 150
PCI Manual Online cases: 3, 6, 8, 17, 20, 21, 22, 24, 32, 45, 46, 47, 60, 62, 64, 71, 85

Severe calcification is usually defined angiographically as radioopacities noted without cardiac motion before contrast injection, involving both sides of the arterial wall. Moderate calcification is defined by densities involving <50% of the reference lesion diameter, often noted only during cardiac movement before contrast injection. However, coronary angiography underestimates the degree of calcification: in a study of 1,155 lesions, calcification was seen in 38% of lesions by angiography and in 73% via IVUS [1]. Imaging with IVUS or OCT can also better quantify the arc of calcification. Greater than 270° arc of calcification is considered severe, 180°–270° arc is considered moderate, and less than 180° arc is considered mild.

Performing PCI of heavily calcified lesions can be challenging due to: (1) difficulty assessing lesion severity; (2) difficulty delivering equipment; (3) difficulty expanding the lesion and stent; and (4) increased risk of complications, both acute (such as dissection, perforation, and equipment loss or entrapment) and late (such as restenosis).

19.1 Planning

Knowing that the target lesion(s) are heavily calcified has important implications for procedural planning. Treatment of heavily calcified lesions can be facilitated by use of large guide catheters, techniques to increase guide catheter support, intravascular imaging, atherectomy, and other plaque modification strategies, such as intracoronary lithotripsy.

19.2 Monitoring

Monitoring is performed as described in Chapter 2: Monitoring.

19.3 Medications

Standard medications (anticoagulation and antiplatelet therapy) are used for PCI of heavily calcified lesions, as described in Chapter 3: Medications. PCI of calcified lesions may have long duration, hence anticoagulation should be carefully monitored.

If atherectomy is performed in the artery that supplies the AV node (usually the right coronary artery), bradycardia may occur. Aminophylline (250–300 mg intravenously over 10 min [2] or 20–40 mg intracoronary injection [3,4]) may be used to prevent bradycardia. Aminophylline is an A1 adenosine receptor antagonist that blocks the effect of adenosine, which is released from red blood cells injured during atherectomy. Alternatively atropine (0.5–1.0 mg) can be administered intravenously prior to performing atherectomy.

Vasodilators, such as nicardipine (100–300 mcg intracoronary) or nitroprusside (100–300 mcg intracoronary) can also be used before atherectomy to facilitate passage of particles released during atherectomy through the microcirculation.

19.4 Access

Most calcified lesions can be successfully treated via radial access (ideally using 7 French low-profile sheaths or sheathless guide catheters), however femoral access can provide superior support that can facilitate treatment of particularly complex calcified lesions.

19.5 Engagement

Strong guide catheter support should be obtained when treating heavily calcified lesions. Consider utilizing large guide catheters (7 or 8 French) with supportive shapes (AL1 or 3D Right for the RCA and XB or EBU for the left main).

Manual of Percutaneous Coronary Interventions. DOI: https://doi.org/10.1016/B978-0-12-819367-9.00019-6

Techniques to improve guide catheter support should be used early and coaxial guide catheter alignment should be maintained during atherectomy.

19.6 Angiography

Multiple views may be needed to assess heavily calcified lesions. Calcification can hinder angiographic interpretation, hence physiologic assessment and intracoronary imaging with IVUS or OCT are encouraged [5].

19.7 Selecting target lesion(s)

Selection of target lesion(s) is performed as discussed in Chapter 7: Selecting Target Lesion(s).

19.8 Wiring

Wiring of calcified lesions is performed as discussed in Chapter 8: Wiring. Use of microcatheters and hydrophilic or polymer-jacketed guidewires may be required for wiring heavily calcified lesions, especially if they are also tortuous. A microcatheter can also be used to exchange the initially used guidewire for an atherectomy wire. Alternatively, a microcatheter can be used to insert a support guidewire (Section 30.7.4) that can facilitate equipment delivery across calcified and tortuous lesions.

19.9 Lesion preparation

Primary stenting should never be performed in heavily calcified lesions, as failure to completely dilate the lesion and expand the stent can result in stent underexpansion that increases the risk of restenosis and stent thrombosis (Fig. 19.1).

Adequate lesion preparation can facilitate both equipment delivery and stent expansion, but carries risks in itself. The following options exist for treating severely calcified lesions (Fig. 19.2):

1. Balloon angioplasty (ideally using noncompliant balloons at high pressure, Chapter 9: Lesion Preparation).
2. Plaque modification balloons (Angiosculpt, Cutting balloon, Chocolate balloon, Section 30.9.3), although delivery can be challenging through heavily calcified lesions.
3. Atherectomy (orbital and rotational; laser is not very effective in severely calcified lesions, but can be used, especially for treating underexpanded stents) (Section 30.10).
4. Novel treatments, not clinically available in the United States as of 2020: intravascular lithoptripsy (Section 30.9.8) and the SIS OPN high-pressure balloon (Section 30.9.7).

19.9.1 Balloon angioplasty versus atherectomy

The key question when performing PCI of severely calcified lesions is whether and when to perform coronary atherectomy in addition to balloon angioplasty.

There are two approaches: (1) upfront or primary atherectomy; and (2) "secondary" atherectomy if the target lesion fails to expand with balloon angioplasty (using standard or plaque modification balloons) (Fig. 19.3).

The advantage of upfront atherectomy is better vessel preparation with lower likelihood of stent loss (which occurred in 2.5% of patients randomized to no rotational atherectomy in the ROTAXUS trial [6]) or stent underexpansion. The disadvantage of upfront atherectomy is the risk of complications and more time and cost. Initial balloon angioplasty may be faster but can lead to false impression of adequate vessel preparation and can cause dissections or perforations that may hinder performing atherectomy. In addition, stent delivery may still be difficult. In the PREPARE-CALC study that randomized severely calcified lesions to a plaque modification balloon or rotational atherectomy, cross-over from the plaque modification balloon group to rotational atherectomy was required in 16% [7]. The ongoing 2,000 patient ECLIPSE trial is comparing both acute and long-term outcomes of balloon angioplasty versus orbital atherectomy.

Coronary atherectomy should be performed when the anticipated benefits exceed the potential risks.

The potential benefits of atherectomy are higher when treating severely calcified and long lesions. The presence of circumferential superficial calcification, obstructive nodular calcium, or thick calcium, as assessed by intravascular imaging, suggests high likelihood of needing atherectomy to modify the plaque and facilitate adequate stent expansion. OCT is particularly useful in determining the arc and depth of calcification (Fig. 13.7).

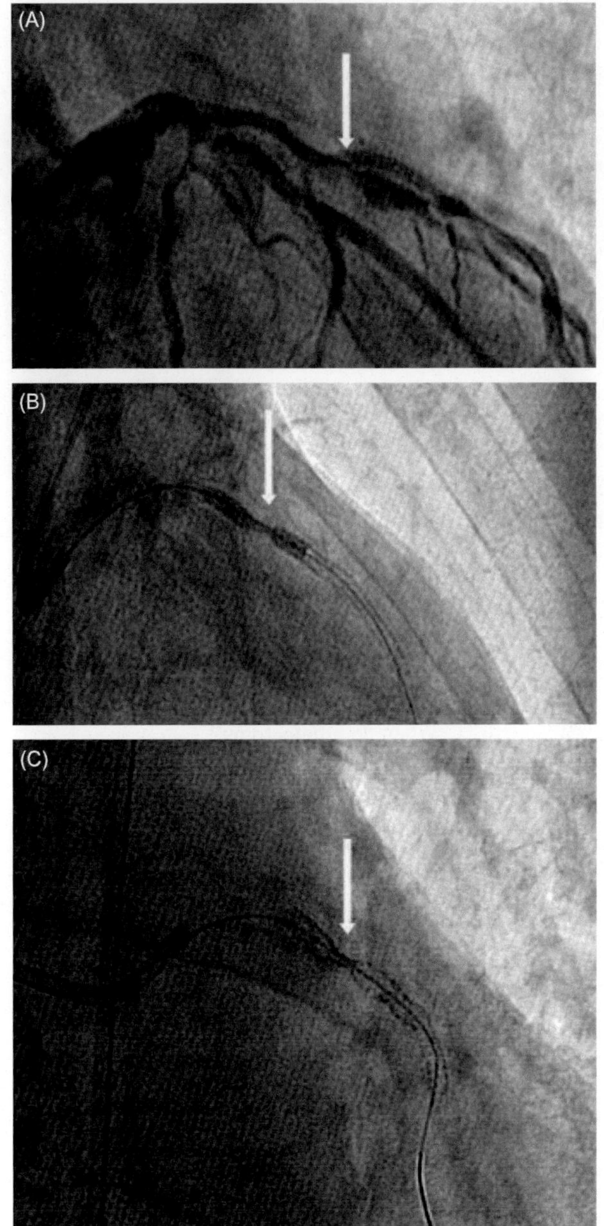

FIGURE 19.1 Stenting of a highly calcified lesion (*arrow*, panel A) resulted in stent underexpansion (*arrows*, panel B and C). *Courtesy of Dr. Jeffrey Chambers.*

The risks of atherectomy (dissection, embolization, perforation, equipment loss, or entrapment) increase with increasing tortuosity (relative contraindication) and in the setting of thrombus, dissections (relative contraindication), and bypass grafts. Atherectomy should generally not be performed in bypass grafts, although rare cases of heavily calcified old saphenous vein graft (SVG) lesions have been successfully treated with atherectomy after failure to expand the lesion with other techniques [8]. Atherectomy is also usually avoided in dissected coronary segments, although it has been performed in this setting to treat chronic total occlusions and balloon undilatable lesions [9].

Intravascular imaging can assist with deciding about whether to perform atherectomy depending on arc of calcium, calcium thickness, and calcium length, as described in Section 13.3.6.2, part C.

19.9.2 Orbital versus rotational atherectomy

(Fig. 19.4)

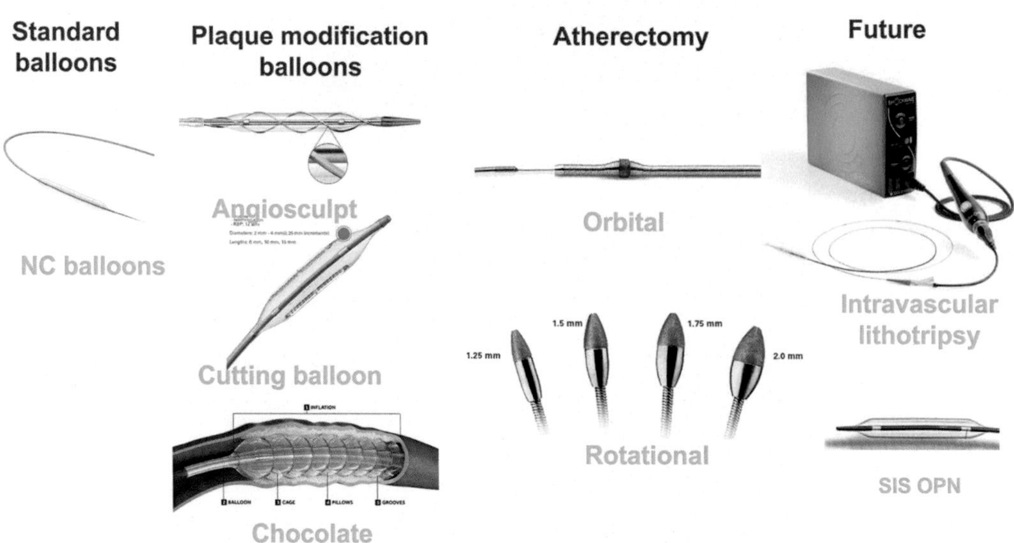

FIGURE 19.2 Lesion preparation options for heavily calcified coronary lesions.

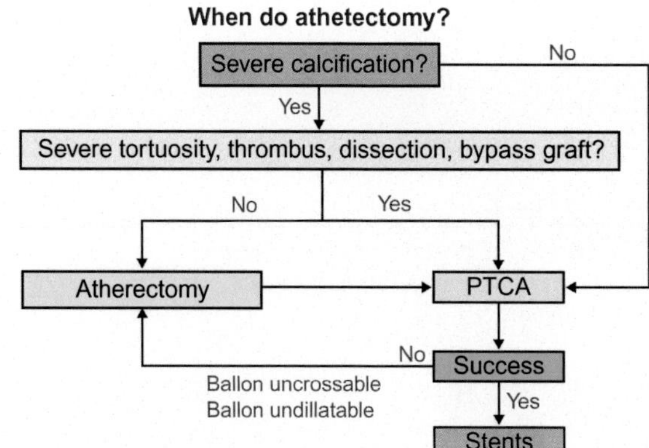

FIGURE 19.3 Algorithm for determining when to do coronary atherectomy.

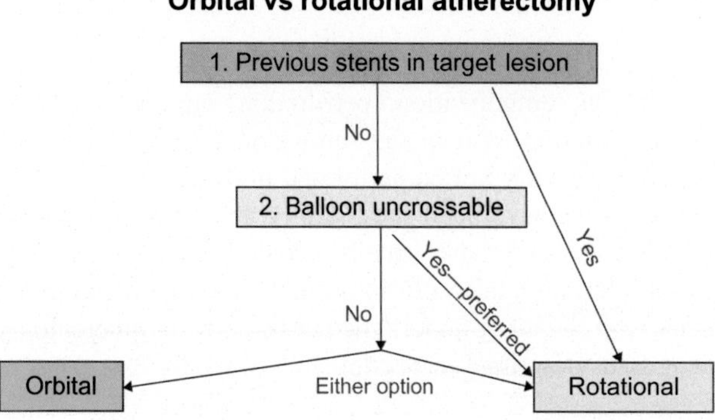

FIGURE 19.4 For most cases either orbital or rotational atherectomy can be successfully used for preparation of calcified lesions. Local availability, experience, and cost determine device selection. Laser is not very effective for treating severe calcification and is infrequently used with the exception of underexpanded stents (Section 23.2).

1. Orbital atherectomy should NOT be done within recently placed stents due to risk of entrapment, but has been successfully performed in stents that have endothelialized. Rotational atherectomy can be done in old stents, but has also been performed within stents placed immediately prior, after they failed to expand. Rotational atherectomy through recently placed stents carries risk of distal embolization, burr entrapment and stent distortion and should be followed in most cases by repeat stent implantation due to disruption of the previously placed stents. When performing rotational atherectomy in recently deployed stents higher speeds should be used to avoid burr and stent entrapment.
2. Rotational atherectomy may be preferred for balloon uncrossable lesions due to its forward ablation (vs side ablation in orbital atherectomy) mechanism of action, although orbital atherectomy has also been successfully used in this setting.
3. Orbital atherectomy may be preferred in larger vessels, as it can provide effective ablation with a single crown, whereas increasingly larger burrs may be needed when performing rotational atherectomy in large diameter vessels.

19.9.3 Balloon angioplasty for severely calcified lesions

Balloon type

Balloon angioplasty is reasonable in deep wall calcification, although superficial and nodular calcification is likely best treated with atherectomy. Balloon angioplasty can be performed with standard noncompliant or plaque modification balloons (Angiosculpt, Chocolate, cutting balloon). Plaque modification balloons facilitate lesion expansion, but are less deliverable than standard balloons. The decision on whether to start with a standard or a plaque modification balloon depends on the severity of calcification (with more severe calcification favoring upfront use of plaque modification balloons) and the likelihood of successful balloon delivery (lower for plaque modification balloons). In most cases, standard balloons are used first, followed by plaque modification balloons if standard balloons fail to expand the lesion. Noncompliant (NC) balloons are preferred over compliant balloons, as higher pressures can be achieved with lower risk of dissection or rupture, but NC balloons have a higher crossing profile and are less flexible which makes them harder to deliver. Balloon rupture leading to vessel dissection is more likely to happen in heavily calcified vessels, particularly with nodular calcification.

The efficacy of balloon inflation in expanding a calcified lesion may be enhanced by the following:

1. High pressure (≥20 atm).
2. Prolonged inflation time (1–2 min).
3. Using one or more buddy wires that can score the vessel.
4. Monitoring the inflation device pressure: decreasing pressure suggests that the lesion is expanding, hence longer balloon inflation is needed [10].

Cutting balloon

In contrast to Angiosculpt and Chocolate, the cutting balloon requires slow inflation and deflation (1 atm every 5 sec) to allow enough time for the blades to exit the balloon folds and rewrap [11].

What to observe?

1. *Balloon expansion*: the balloon (sized 1:1) needs to be fully expanded before stents are implanted to prevent stent underexpansion, which is much harder to treat than balloon undilatable de novo lesions.
2. *Balloon movement*: after inflation the winged balloon can be advanced back and forth through the lesion to determine whether adequate lesion modification has been achieved or whether additional predilation is needed to facilitate subsequent stent delivery.
3. Intravascular imaging: after balloon angioplasty circumferential superficial calcification will ideally have "cracked" in some sectors, which are identifiable on imaging. This provides the operator confidence that the stent will expand adequately.

What can go wrong?

1. *Balloon rupture*. Rapid loss of pressure suggests balloon rupture. The balloon should be immediately deflated and removed, followed by coronary angiography to determine whether balloon rupture caused vessel dissection, perforation, and/or no reflow. If perforation is confirmed, a new balloon is rapidly deployed proximal to the perforation to stop pericardial bleeding and facilitate subsequent treatment steps (Chapter 26: Perforation). Moreover, if the balloon had not been adequately prepared, there is a risk of air embolization (Section 25.2.3.3).
2. *Vessel perforation*. High-pressure balloon inflation can cause perforation, especially in heavily calcified lesions and with use of oversized balloons.

3. *Vessel dissection.* Dissection is part of the mechanism of action of balloon angioplasty, however extensive dissections may compromise antegrade flow or result in side branch occlusion.
4. *Balloon entrapment*, especially in cases of balloon rupture.

19.9.4 Preparation for atherectomy

Two key questions need to be answered before proceeding with atherectomy (orbital or rotational):

1. ***Need for temporary pacemaker*** (Fig. 19.5)

    A temporary pacemaker is not routinely needed for LAD and non-dominant circumflex artery PCI.

    For RCA or dominant circumflex: can use temporary pacemaker, or aminophylline (250–300 mg IV over 10 min or 20–40 mg intracoronary bolus [3,4]), or do brief test runs to determine if the patient develops bradycardia. The advantage of the latter approach (not using pacemaker or aminophylline) is that atherectomy runs can be tailored (atherectomy stops when patient develops bradycardia) and there is no risk of right ventricular perforation from the temporary pacemaker lead. Orbital atherectomy may be less likely to cause bradycardia than rotational atherectomy.

    Bigger burrs and longer runs are more likely to cause bradycardia, than smaller burrs and shorter runs.
2. ***Need for hemodynamic support?***

    Atherectomy may lead to complications, hence prophylactic hemodynamic support should be considered, especially in patients with poor baseline hemodynamics, low ejection fraction, large area of myocardium at risk, severe concomitant valve disease and severe pulmonary hypertension.

19.9.5 Rotational atherectomy

19.9.5.1 Step 1. Select burr size

There are multiple size burrs, ranging from 1.25 mm to 2.5 mm. Most times the 1.5 mm burr is preferred, as it may be less likely to get entrapped as compared with the 1.25 mm burr. Larger burrs (1.75 mm or larger) are infrequently used except for left main lesions, as the goal is to modify the vessel enough to allow balloon expansion rather than full calcium debulking. A burr to artery diameter ratio of 0.5 is usually selected. The minimum guide size required for each burr size is shown in Table 19.1:

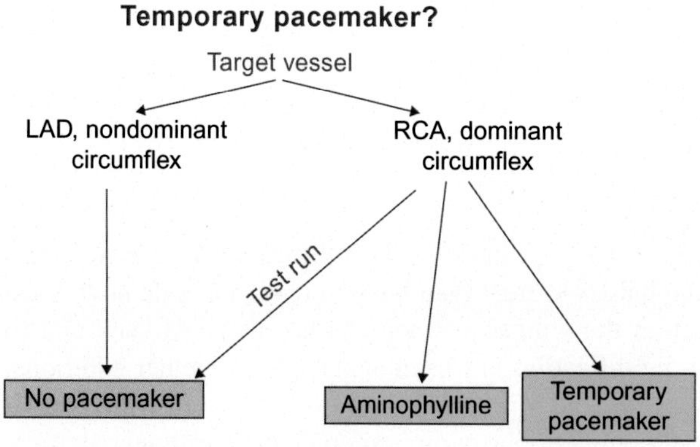

FIGURE 19.5 Is a temporary pacemaker needed during coronary atherectomy?

TABLE 19.1 Compatibility of various size Rotablator burrs with various guide sizes.

Burr size	Guide size
1.25–1.75 mm	≥6 French
2.0 mm	≥7 French
2.15–2.25 mm	≥8 French

19.9.5.2 Step 2. Engage the target vessel

- Engagement is performed as described in Chapter 5: Coronary and Graft Engagement.
- Supportive guide catheters should be used, ensuring coaxial alignment. However, advancement of the Rotablator burr can be challenging through guide catheters with significant bends, such as Amplatz Left.
- Avoid pressure dampening, as it will increase the likelihood of no-reflow or slow flow.
- Ensure coaxial guide catheter orientation with the target vessel.

19.9.5.3 Step 3. Wiring

- Rotational atherectomy should only be performed using a dedicated Rotawire. There are 2 Rotawire types, the Rotawire Floppy and the Rotawire Extra Support (Section 30.10.1.1.3). The Rotawire floppy is used in most cases with the Rotawire extra support used when treating aorto-ostial lesions.
- Although the Rotawire can be used for primary wiring of the target lesion, we recommend wiring the target lesion using a microcatheter and workhorse guidewire and then exchanging for the Rotawire over the microcatheter, as the Rotawire can be more difficult to steer and kinks easily.
- The Rotawire can be torqued using the WireClip torquer (Section 30.10.1.1.4). The WireClip torquer can also prevent the guidewire from spinning during rotational atherectomy if the break is accidentally released by the break defeater.
- If the rotational atherectomy guidewire becomes kinked it should be replaced for a new one. Atherectomy should not be performed over a kinked guidewire due to increased risk of losing wire position during burr movement.
- The Rotawire extra support should be used for preferential cutting on the outer curvature of the vessel and the Rotawire floppy for non-preferential cutting.

19.9.5.4 Step 4. Vasodilator administration

Consider vasodilator administration prior to atherectomy in most patients (unless the patient is hypotensive). Nicardipine is most commonly used, but nitroprusside, verapamil, or adenosine can be used as well. A cocktail of verapamil, nitroglycerin, and heparin is often used during atherectomy along with the Rotaglide solution that contains olive oil, egg yolk, phopholipids, sodium deoxycholate, L-histidine, disodium EDTA, sodium hydroxide, and water.

19.9.5.5 Step 5. Prepare rotational atherectomy device for insertion into the guide catheter

The following checklist (**DRAW**) should be confirmed prior to insertion of the Rotablator burr into the guide catheter:

D: ensure there is sufficient *D*rip of the flush solution from the tip of the burr.

R: Activate the device outside the body (platforming) and confirm that it rotates at the desired speed (usually 140,000–160,000 rpm) (ensure that the burr is not touching a towel or gauze, as this can result in towel wrapping around the burr).

A: *A*dvance and withdraw the burr using the advancer knob to ensure that the burr is moving smoothly.

W: a gentle pull should be done on the *W*ire to ensure that the brake is active.

Before inserting into the guide catheter the advancer knob is locked approximately 2 cm from the distal end.

19.9.5.6 Step 6. Deliver burr proximal to lesion

The Rotablator burr can be advanced under fluoroscopy while fixing the Rotawire, or utilizing the Dynaglide mode (which activates the burr at a reduced speed to decrease friction onto the wire) with the wire and wire clip in the break release. This is traditionally performed by two operators, but can also be done by a single operator [12]. Alternatively, if the guide size is big enough, trapping can be used.

The Rotablator burr is advanced 1–2 cm proximal to the lesion. The advancer knob is then moved back to release any tension in the systems. This is important for preventing the burr from jumping forward into the lesion upon initial activation.

Challenges

1. Inability to advance the burr proximal to the lesion.

Causes:

- Vessel tortuosity and calcification.
- Poor guide catheter support.
- Non coaxial guide catheter engagement.
- Tortuous iliac arteries causing multiple bends on the guide catheter.

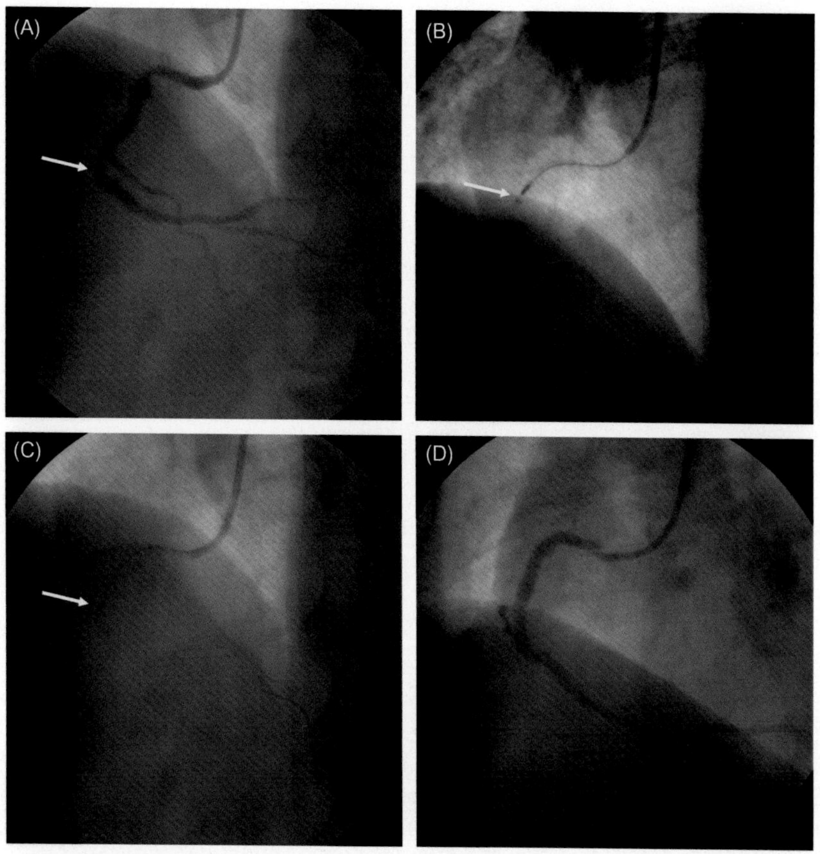

FIGURE 19.6 Guide catheter extension to facilitate rotablation of a lesion (*arrow*, panel B) located distal to a tortuous proximal coronary segment. A guide catheter extension (*arrow*, panel B) was advanced to the mid right coronary artery, allowing delivery of the Rotablator burr (*arrow*, panel C) proximal to the target lesion. An excellent final result (panel D) was achieved after rotational atherectomy. *Courtesy of Dr. Ioannis Tsiafoutis.*

Prevention:

- Use small burrs.
- Balloon angioplasty prior to burr delivery, especially in highly tortuous vessels.
- Ensure coaxial guide catheter engagement.

Solutions:

- Gentle pulling of the Rotawire while simultaneously advancing the burr.
- Activation of the Dynaglide during advancement.
- Use of a guide catheter extension (7 French guide catheter extensions are preferred for 1.25 and 1.5 mm burrs) (Fig. 19.6).
- Perform rotablation in the proximal calcific segments first.
- Brief press of the pedal to facilitate forward motion ("burping").

19.9.5.7 Step 7. Perform rotational atherectomy

1. Confirm that the flush solution drip is on.
2. Confirm that the WireClip torquer is on the back end of the wire.
3. Platform: the burr is activated proximal to the lesion and the desired rotational speed is selected.
4. Advance the burr using a pecking motion, avoiding decelerations >5000 rpm. Visual, tactile and auditory clues can minimize the risk of decelerations. This reduces the risk of burr entrapment, and also reduces the size and volume of particles released during atherectomy, and as a result the risk for slow flow/no-reflow.
5. Duration of runs: aim to keep them short, ideally <15 seconds (Fig. 19.7).
6. Wait for at least 30 sec and ideally longer in-between runs (Fig. 19.8).
7. Monitor ECG and pressure during rotablation to detect bradycardia and ST changes due to coronary flow disturbance.
8. Contrast injections can be performed if complications are suspected.

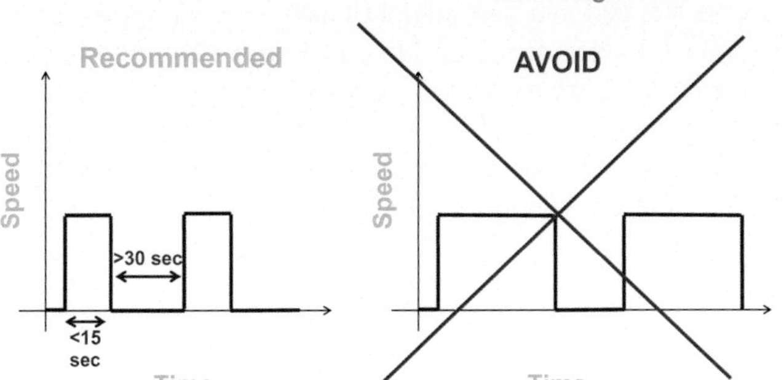

FIGURE 19.7 Short runs are recommended (*left panel*) to minimize the volume of debris released from the target lesion. Long atherectomy runs (*right panel*) result in more particle release and higher likelihood of slow flow/no reflow.

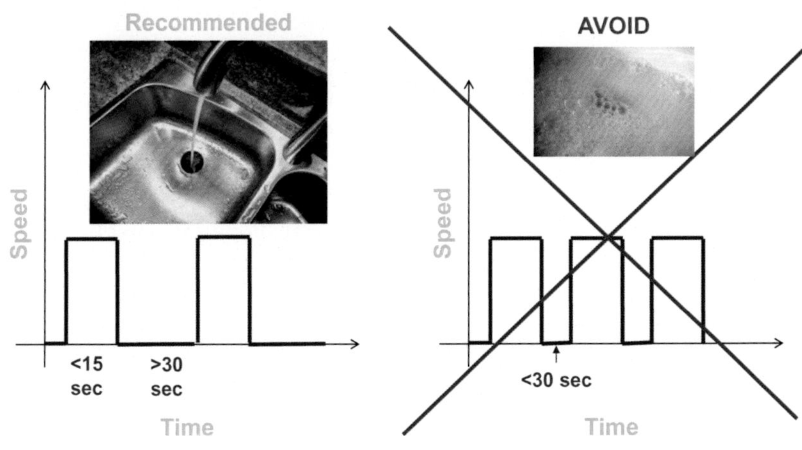

FIGURE 19.8 Visual analog illustrating the importance of pauses in-between atherectomy runs. If flour is continuously thrown into the sink with running water there is not enough time for the flour to go down the drain and the sink plugs up. In contrast, if enough time elapses between throwing flour in, the flour passes through the drain without plugging it.

9. Additional rotablation runs are performed until the burr passes easily through the lesion without resistance or decelerations or a complication occurs (slow flow, ECG changes, etc.) ("polishing" runs).

   What can go wrong?

19.9.5.7.1 Decreased antegrade flow

Causes:
- Dissection.
- Distal embolization.
- Vasospasm.

Prevention:
- Vasodilator administration prior to atherectomy.
- Contrast injection after each atherectomy run.
- Avoid burr oversizing (burr to artery ratio should be <0.5).
- Avoid fast rotation speeds (>180,000 rpm).
- Avoid dampening of the guide pressure.
- Short (<15 sec) rotablation runs, with adequate time between runs (>30 sec).
- Avoid atherectomy if there is a preexisting dissection.

- Ensure guide catheter is coaxial when performing atherectomy of aorto-ostial lesions to minimize the risk of aortic dissection.

Treatment:
- If cause is suspected to be distal embolization: atherectomy is stopped and vasodilators (such as nicardipine, nitroprusside, verapamil, adenosine) are administered. Often a microcatheter or a dual lumen catheter (TwinPass or an aspiration thrombectomy catheter) can be advanced distally over the wire to deliver vasoactive drugs into the distal vascular bed. Intracoronary epinephrine can also be given in refractory no reflow.
- If the cause of slow/no antegrade flow is suspected to be dissection, stenting will likely be required. Distal injection of contrast through a dual lumen catheter or IVUS can be helpful in determining if dissection has occurred.

19.9.5.7.2 Perforation

Causes:
- Small vessel.
- Large burr.
- High rotational atherectomy speed.
- Vessel tortuosity.
- Inappropriate Rotawire position (tip too close to the target lesion).
- Rotawire fracture with subsequent uncontrolled ablation.
- Extreme wire bias.

Prevention:
- Avoid burr oversizing.
- Avoid rotational atherectomy in highly tortuous vessels.

Treatment:
- Perforation is treated as outlined in Chapter 26: Perforation. The first step is to inflate a balloon proximal or at the perforation site to stop bleeding into the pericardium.

19.9.5.7.3 Burr entrapment [13]

Causes:
- Forceful burr advancement with significant decelerations.
- Use of small burrs (especially the 1.25 mm burr—this is often called the "kokesi" effect).
- Rotablation within previously deployed stents.
- Rotablation in very tortuous vessels.

Prevention:
- Gentle burr advancement with short, nonforceful contact with the lesion ("pecking" the lesion).
- Use 1.5 mm or larger burrs (1.25 mm burrs are more likely to advance through the lesion and then fail to come back through the lesion).
- Do not stop the burr distal to the lesion.
- Do not start or stop rotablation within the target lesion.

Treatment:

(Fig. 19.9)

- Straighten out the guide, ensure coaxial alignment, and try again to remove the entrapped burr.
- If rotation is still possible, using the Dynaglide setting may facilitate burr withdrawal.
- A second wire can be advanced next to the entrapped burr, followed by balloon dilation in an effort to dig the burr out [14].
- The Rotablator shaft is cut, and a guide extension is advanced over it (Figs. 19.10, 19.11, and 19.12) [15–17].
- A snare can also be advanced over the Rotablator shaft [15].
- The Rotawire can be withdrawn (gently) possibly helping bring back the burr (the 0.014 in. tip of the Rotawire is larger than the 0.009 lumen of the burr).
- Subintimal plaque modification next to the burr.
- If hemodynamics permit, vasodilators can be administered.
- If all attempts fail, emergency cardiac surgery may be required [18].

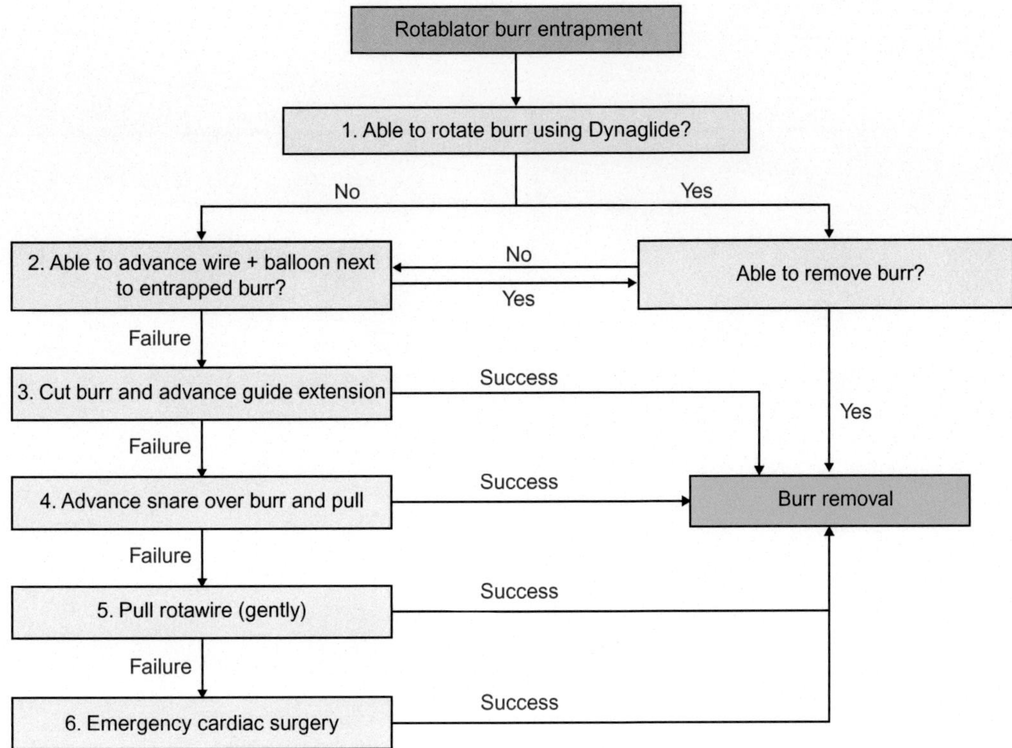

FIGURE 19.9 Approach to entrapped Rotablator catheter.

19.9.5.7.4 Arrhythmias

Causes:

- Atherectomy of a right coronary artery or dominant circumflex.
- Aggressive runs (prolonged duration, forceful advancement).

Prevention:

- Temporary pacemaker, aminophylline, or atropine administration can be used to prevent bradycardia when performing rotational atherectomy of the right coronary artery or a dominant circumflex.
- Avoid aggressive rotational atherectomy runs.
- If a vessel location of particular resistance is identified, the operator should move the Rotawire during the ablation (either gentle push or pull, while briefly releasing the brake), as this will prevent constant damage of the same Rotawire segment, which is compressed by the burr onto the calcified lesion, potentially leading to wire fracture.

Treatment:

- Bradycardia usually resolves after atherectomy is stopped.
- Tachyarrhythmias are usually due to coronary flow disturbances: slow flow/no reflow or dissection are treated accordingly.

19.9.5.7.5 Rotawire fracture

Causes:

- Performing rotational atherectomy close to the 0.014 in. tip of the Rotawire.
- Performing rotational atherectomy over a kinked Rotawire [19].
- Performing rotational atherectomy in highly angulated lesions [20].
- Prolonged lesion contact in tortuous segments.

Prevention:

- Avoid rotablation close to the 0.014 in. tip of the Rotawire.
- In case of Rotawire kinking, replace the wire before performing rotational atherectomy.

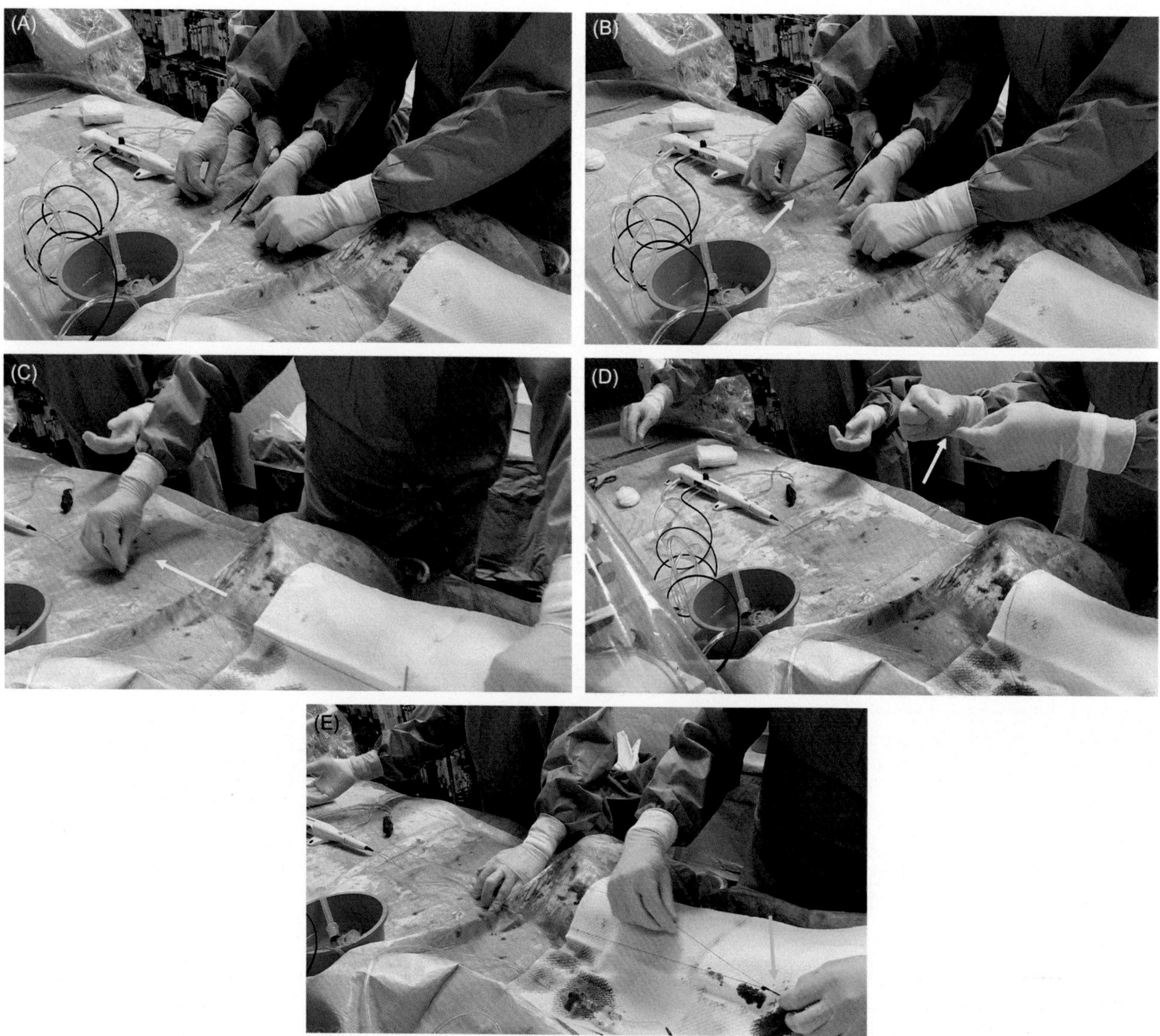

FIGURE 19.10 How to cut the Rotablator shaft and insert a guide catheter extension over it for retrieving an entrapped Rotablator burr. (Panels A and B) The Rotablator shaft is cut (*arrow*). (Panel C) Removal of plastic Rotablator shaft cover. (Panels D and E) Insertion of a guide catheter extension (*arrow*) over the cut Rotablator shaft. *Courtesy of Dr. Roberto Garbo.*

- Ensure smooth movement of the burr. Avoid aggressive persistent burr advancement if resistance is encountered.

Treatment:

- Retrieval could be attempted, using another wire or a snare [21] as described in Section 27.2.
- Alternatively, if the wire fragment embolizes to a distal location it may be left in place to endothelialize over time.

19.9.5.8 Step 8. Remove burr

The burr can be removed either using the Dynaglide function or using the trapping technique (if the guide is big enough to allow insertion of a trapping balloon next to the Rotablator burr).

Option 1: Using Dynaglide

1. Dynaglide is activated by stepping on the Dynaglide button on the foot pedal (for the Rotablator classic) or pressing the Dynaglide button on the Rotapro Advancer (for the Rotapro).
2. It is confirmed that the WireClip torquer is on the back end of the Rotawire.

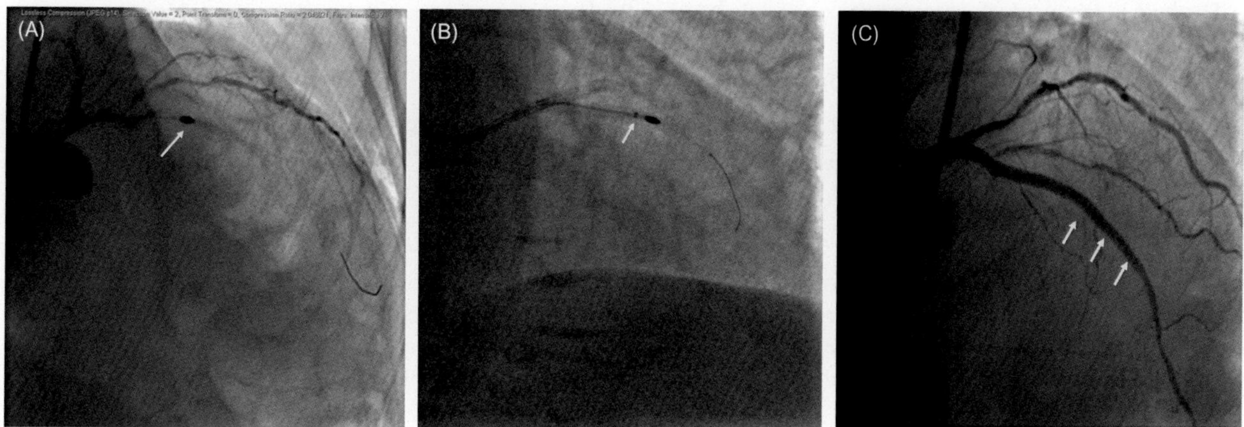

FIGURE 19.11 Example of a Rotablator entrapment case (from the same case presented in Fig. 19.10). (Panel A) After successful rotablation with a 1.25 mm burr, a 1.75 mm burr (*arrow*) was stuck in a proximal calcified LAD stenosis. (Panel B) After cutting of the Rotablator shaft a guide extension (*arrow*) was advanced over it close to the entrapped burr. (Panel C) Final result after withdrawal of entrapped burr, balloon angioplasty and stenting. *Courtesy of Dr. Roberto Garbo.*

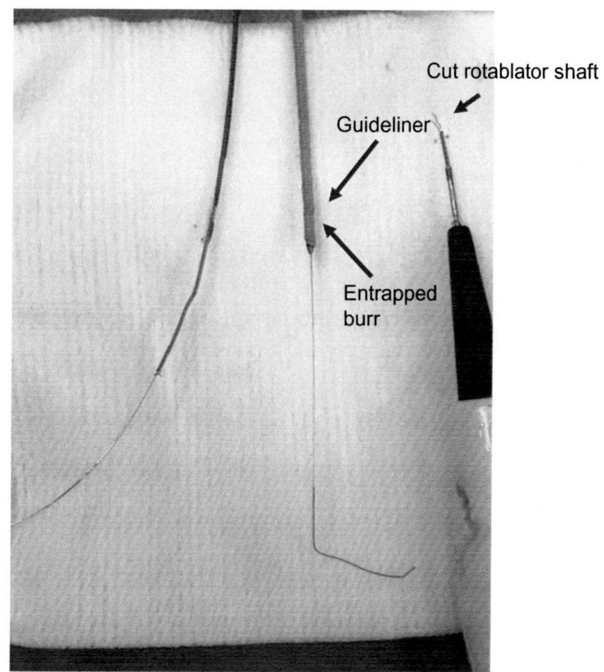

FIGURE 19.12 Entrapped Rotablator burr that was successfully retrieved by cutting the Rotablator shaft and using a guide catheter extension. *Courtesy of Dr. Amir Ravandi.*

3. The brake defeat button is pressed while simultaneously pressing the foot pedal to rotate at 60,000–90,000 rpm (Dynaglide mode).
4. The burr is retracted while constantly monitoring the distal position of the Rotawire. This is typically done by two operators, but can also be done by a single operator (by inserting the WireClip torquer in the docking port of the advancer, which keeps the brake defeat engaged).

Option 2: Using the trapping technique

1. The burr is withdrawn inside the guide catheter.
2. A balloon is advanced inside the guide catheter but distal to the distal end of the burr (not necessary if a Trapliner is used).
3. The balloon is inflated "trapping" the Rotawire.

4. The burr is removed (fluoroscopy is only needed during the beginning of burr withdrawal).

19.9.5.9 Step 9. Check angiographic result

Coronary angiography is performed to identify slow flow/no reflow, dissections, embolization, side branch loss, perforation, or other complications.

19.9.5.10 Step 10. Replace Rotawire with a standard workhorse guidewire

Using the trapping technique, the Rotawire is exchanged for a standard workhorse guidewire over a microcatheter or over-the-wire balloon. PCI can be performed over the Rotawire guidewire (especially if a complication, such as dissection or perforation occurs), but workhorse guidewires are preferred as they are more torquable, less likely to cause distal vessel injury, and easier to work on rather than the thinner and more likely to kink Rotawire.

19.9.5.11 Step 11. Balloon angioplasty

Balloon angioplasty with a 1:1 sized noncompliant balloon should always be performed after atherectomy to ensure that the balloon expands fully. If it does not, repeat atherectomy or other treatments (such as use of plaque modification balloons or intracoronary lithotripsy) may be required.

19.9.6 Orbital atherectomy

19.9.6.1 Step 1. Engage the target vessel

This is done as described in Section 19.9.5.2 for rotational atherectomy.

19.9.6.2 Step 2. Wiring

Orbital atherectomy should only be performed using a dedicated ViperWire Advance guidewire (Section 30.10.2.3). The ViperWire Advance Flex Tip guidewire can be used for primary lesion wiring. The ViperWire Advance is stiffer and more difficult to steer, hence we recommend wiring the target lesion using a microcatheter and workhorse guidewire and then exchanging for the ViperWire Advance over the microcatheter.

19.9.6.3 Step 3. Vasodilator administration

Consider vasodilator administration prior to atherectomy in most patients (unless the patient is hypotensive). Nicardipine is most commonly used, but nitroprusside, verapamil, or adenosine can be used as well.

19.9.6.4 Step 4. Test device outside the body

The orbital atherectomy device is connected to the pump and threaded through the back end of the ViperWire Advance guidewire. Before inserting the device through the Y-connector, the following are checked:

- The advancer knob can move easily back and forth.
- Viperslide is dripping from the tip of the device.
- Test the rotation of the device (brake is pushed down and the on/off button located on top of the crown advancer is pressed and released). During spinning of the device the flow of Viperslide should increase. *Caution*: the crown should be held in the air as it will easily get entangled on towels.
- Lock the crown advancer knob at 1 cm from the fully proximal position by rotating the crown advancer knob to the parallel position.

19.9.6.5 Step 5. Advance crown proximal to lesion

The orbital atherectomy crown can be advanced under fluoroscopy while fixing the ViperWire Advance. Alternatively, if the guide size is big enough (Table 19.2) trapping can be used.

Advancement can be done using trapping when 7 French and 8 French guides are used. Trapping can also be done if an 8 French Trapliner is used.

Keeping the knob in the parallel position and the brake up, advance the crown approximately 3–5 mm proximal to the lesion.

TABLE 19.2 Ability to trap the Diamondback crown with a balloon in various guide catheters.

Guide size	
6 French	No
6 French + Guide extension	No
7 French	Yes
7 French + Guide extension	No
8 French	Yes
8 French + Guide extension	No

Challenges

1. **Inability to advance the crown proximal to the lesion.**

Causes:
- Vessel tortuosity and calcification.
- Multiple bends within guide catheter (for example with Amplatz guides).
- Poor guide catheter support.

Prevention:
- Use larger guide.
- Balloon angioplasty prior to crown delivery.
- Advance the crown through the lesion using the GlideAssist mode, then treat while withdrawing the crown.

Solutions:
- Gentle pulling of the ViperWire Advance while simultaneously advancing the crown.
- Guide extension
- Use the GlideAssist setting that rotates the crown at 5000 rpm. The GlideAssist is enabled by pressing and holding the low speed button. Release the button once the low speed light begins to slowly blink. The slowly blinking light indicates that GlideAssist is enabled. In GlideAssist mode the crown will spin even in the unlocked position. Therefore, the ViperWire should be held either with the fingers or with a guidewire torquer to prevent vessel injury.
- Perform atherectomy of the proximal segment.

19.9.6.6 Step 6. Perform orbital atherectomy [22]

1. **Push brake down.** The crown will not spin if the guidewire brake is not locked (except in the GlideAssist mode, which is why the Viper wire should be held, ideally with a torquer if spinning without the brake on in the GlideAssist mode).
2. **Speed: slow speed (80,000 rpm)** should be the initial treatment speed in all cases, as it carries less risk of dissection, perforation or other complications as compared with the high speed (120,000 rpm). Nearly all procedures can be completed on low speed (80,000 rpm). Consider using high speed if the vessel is >3.5 mm in diameter or if the postdilatation balloon does not fully expand. High speed should be avoided in arteries <3.0 mm in diameter.
3. Unlock the crown advancer knob (turn to the perpendicular position) and retract to the fully proximal position to relieve any tension/torque in the driveshaft (which may cause the device to "jump" after activation if not relieved).
4. Push the "on" button on top of the crown advancer knob to start spinning. Wait for 2 seconds required for the crown to orbit at a stable speed.
5. **Advance SLOWLY** (1 mm per second—even slower if the auditory pitch of the device changes). Do NOT force through lesion. Slowly engage and disengage the lesion to maintain 1:1 motion between the crown and the advancer knob (as viewed via live angiography). The device works both in the antegrade and retrograde direction. The same 1 mm/second 1:1 motion movement should be maintained with both antegrade and retrograde treatments.
6. **Duration: limit runs to <15 seconds** (device will beep at 25 sec). Allow 30–60 seconds in-between runs to reduce the risk of spasm and allow for particulate to advance through the microcirculation. At a very minimum, rest periods should be at least equal to the preceding orbital atherectomy treatment duration.

7. Where to stop: the crown should not be stopped within an obstructive untreated lesion. Instead the crown should be moved slowly distal or proximal to the target lesion [22].
8. **Do not allow the crown to get closer than 10 mm to the radiopaque distal portion of the ViperWire** Advance guidewire to avoid shearing off the guidewire tip.
9. Stop orbital atherectomy when the crown passes easily through the target lesion without changes in the sound generated by the device.
10. Total orbital atherectomy time should be <5 minutes. If >5 minutes are required, a new orbital atherectomy device should be used.
11. Perform angiography intermittently—definitely before removing the device or if the patient develops chest pain, ECG changes, or hypotension.

Challenges

1. ***Aorto-ostial lesions***

Orbital atherectomy can be challenging in aorto-ostial lesions, because the elliptical motion of the crown is not constrained by the vessel wall. The ideal technique for performing orbital atherectomy in such lesions (if the crown can cross the lesion) is to advance the crown through the aorto-ostial lesion to a relatively normal section of the artery, activate the device starting on low speed, and treat the lesion in a retrograde fashion.

If the aorto-ostial lesion is too severe for the non-spinning crown to cross, ensure that the guide catheter is coaxial and place the nose cone of the device into the ostium, keeping the crown 5 mm from the lesion. This will constrain the orbit and allow the lesion to be treated in a more controlled manner. Utilizing low speed, slowly advance the crown through the lesion. After the initial pass treat in a retrograde direction until the plaque is fully modified.

If the nose cone cannot engage the lesion, orbital atherectomy should not be performed.

What can go wrong?

Complications of orbital atherectomy are similar with those of rotational atherectomy discussed in Sections 19.9.5.7.1–19.9.5.7.5.

Atherectomy of vessels that have highly diseased side branches carries risk of side branch dissection and occlusion. Protection of such branches with a guidewire and a microcatheter (such as the Finecross) has been proposed, but carries the risk of microcatheter and guidewire damage (Section 16.1.9, CTO Manual Online case 136) and embolization, requires use of an 8 French guide catheter and should be avoided. Delivery of the atherectomy catheter over a guide catheter extension past the side branch can help protect more proximal branches [23].

19.9.6.7 Step 7. Remove the orbital atherectomy device

The orbital atherectomy crown can be removed either by "walking back" the device under fluoroscopy, by utilizing the GlideAssist mode, or by using the trapping technique (if the guide is big enough to allow insertion of a trapping balloon next to the orbital atherectomy crown—see Table 19.2).

Option 1: Walking device back

1. Lock knob (turn to parallel position).
2. Turn break up.
3. Remove the device while monitoring the position of the ViperWire Advance guidewire distal tip.

Option 2: GlideAssist

1. Release the brake on the end of the device.
2. Hold the low speed button for 2 second until the light blinks.
3. Press the black button on top of the advancer knob to turn GlideAssist on.
4. Remove the device while monitoring the position of the ViperWire Advance guidewire distal tip under fluoroscopy.

Option 3: Using the trapping technique

1. The crown is withdrawn inside the guide catheter.
2. A balloon is advanced inside the guide catheter but distal to the distal end of the crown.
3. The balloon is inflated "trapping" the ViperWire Advance guidewire.
4. The crown is removed (fluoroscopy is only needed at the beginning of crown withdrawal).

19.9.6.8 Step 8. Perform angiography

Coronary angiography is performed to identify slow flow/no reflow, dissection, embolization, side branch loss, perforation, or other complications.

19.9.6.9 Step 9. Replace ViperWire Advance guidewire with a standard workhorse guidewire

Using the trapping technique, the ViperWire Advance is exchanged for a standard workhorse guidewire using a microcatheter or over-the-wire balloon. PCI can be performed over the ViperWire Advance (especially if a complication, such as dissection or perforation occurs), but workhorse guidewires are preferred as they are more torquable and less likely to cause distal vessel injury. The ViperWire Advance Flex Tip guidewire is easier to work on compared with the stiffer ViperWire Advance guidewire.

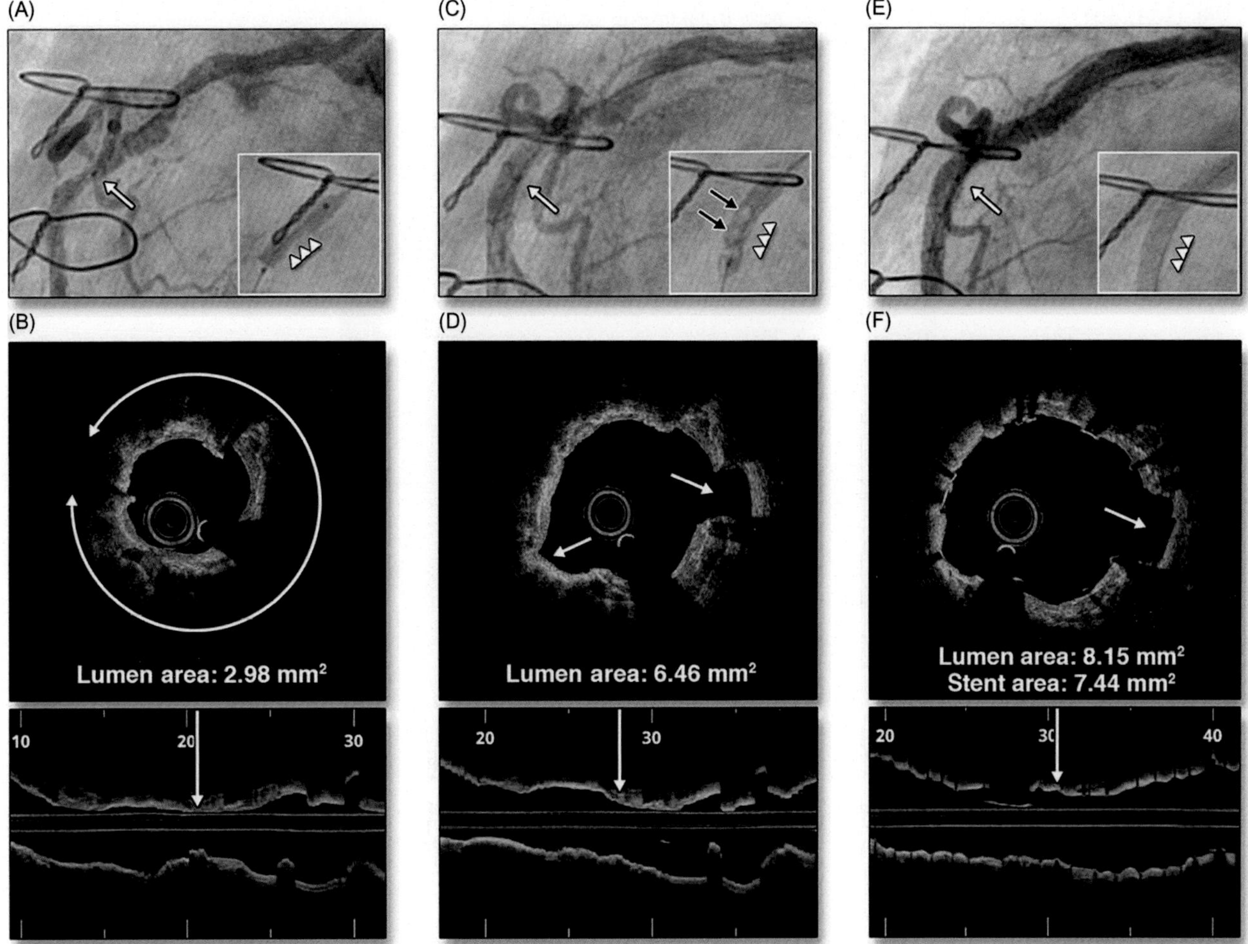

FIGURE 19.13 Plaque modification in severe coronary artery calcification by intracoronary lithotripsy. Angiography demonstrates a stenotic lesion in the mid right coronary artery, undilatable by standard high-pressure balloon angioplasty (*inset*, arrowheads). (B) Optical coherence tomography (OCT) cross-sectional (*top*) and longitudinal (*bottom*) images acquired before lithotripsy and coregistered to the OCT lens (*arrow* in A) demonstrate severe near-circumferential calcification (*double-headed arrow*) in the area of the stenosis. (C) Angiography demonstrates improvement in the area of stenosis after lithotripsy (*inset*; note the cavitation bubbles generated by lithotripsy [*black arrows*]). (D) OCT cross-sectional (*top*) and longitudinal (*bottom*) images acquired post lithotripsy and coregistered to the OCT lens (*white arrow* in C) demonstrate multiple calcium fractures and large acute luminal gain. (E) Angiography demonstrates complete stent expansion with the semicompliant stent balloon (*inset*) without the need for high-pressure noncompliant balloon inflation. (F) OCT cross-sectional (*top*) and longitudinal (*bottom*) images acquired poststenting and coregistered to the OCT lens (*arrow* in E) demonstrate further fracture displacement (*arrow*), with additional increase in the acute area gain (5.17 mm^2), resulting in full stent expansion and minimal malapposition. *Reproduced with permission from Ali ZA, Brinton TJ, Hill JM, Maehara A, Matsumura M, Karimi Galougahi K, et al. Optical coherence tomography characterization of coronary lithoplasty for treatment of calcified lesions: first description. JACC Cardiovasc Imaging 2017;10:897–906. Copyright Elsevier.*

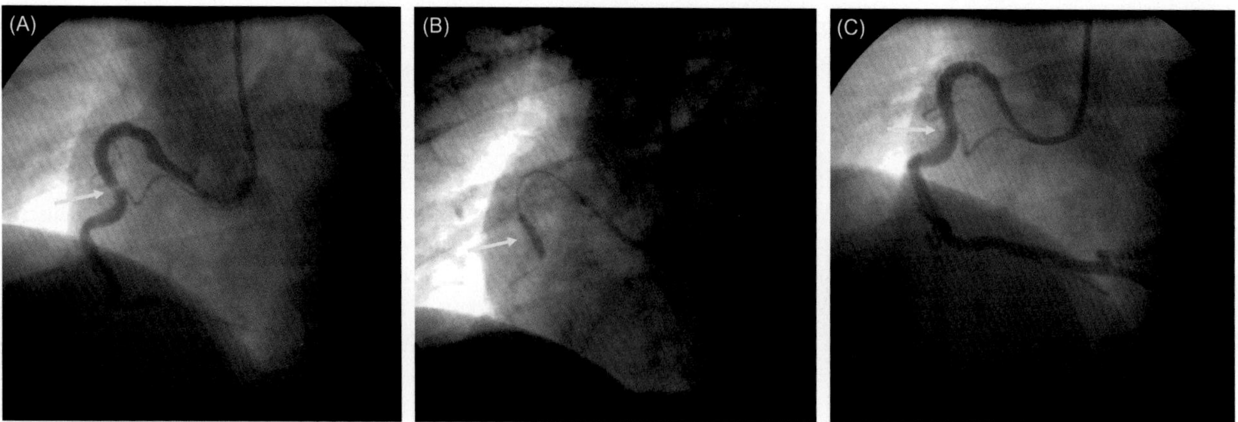

FIGURE 19.14 Use of intracoronary lithotripsy (*arrow*, panel B) for treating a heavily calcified lesion (*arrow*, panel A) in a large vessel with severe proximal tortuosity. An excellent final result was achieved (*arrow*, panel C). *Courtesy of Dr. Ioannis Tsiafoutis.*

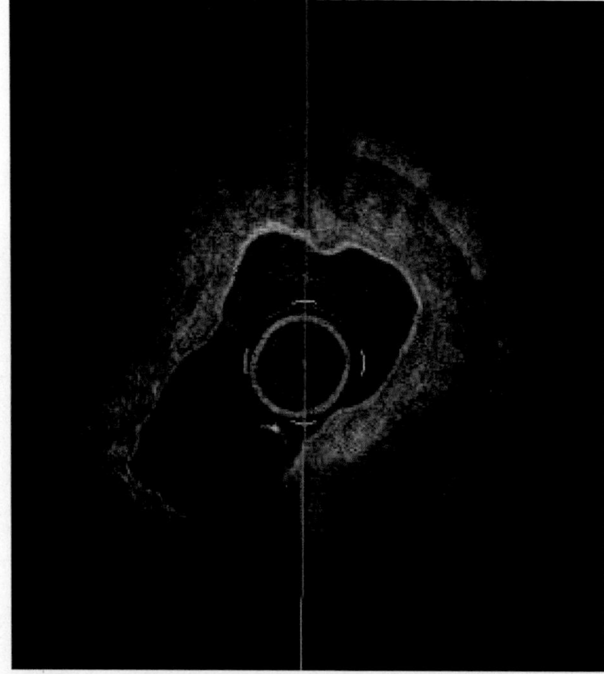

FIGURE 19.15 OCT cross-section demonstrating severe calcification. There is circumferential and superficial calcium. Courtesy of Dr. Evan Shlofmitz.

19.9.6.10 Step 10. Balloon angioplasty

Balloon angioplasty with a 1:1 sized noncompliant balloon should always be performed after atherectomy to ensure that the balloon expands fully. If it does not, repeat atherectomy or other treatments (such as plaque modification balloons) may be required.

19.9.7 Novel calcium treatment modalities

Intravascular lithotripsy (Section 30.9.8, Fig. 19.13) [24] and the SIS OPN high-pressure balloon (Section 30.9.7) are promising modalities for treating highly calcified lesions (especially in larger or highly tortuous vessels—Fig. 19.14), but are not currently available for clinical use in the United States. The peripheral intravascular lithotripsy balloon has been used off label in large coronary arteries with good results (*PCI Manual* Online case 63 and case 100. In a series of

326 balloon undilatable lesions the high-pressure balloon achieved expansion in >90% with 2%–3% risk of vessel rupture [25].

19.10 Stenting

After adequate balloon expansion is confirmed, stents are delivered and deployed at the lesion, followed with postdilation at high pressure (≥20 atm) with a 1:1 sized noncompliant balloon.

(A)

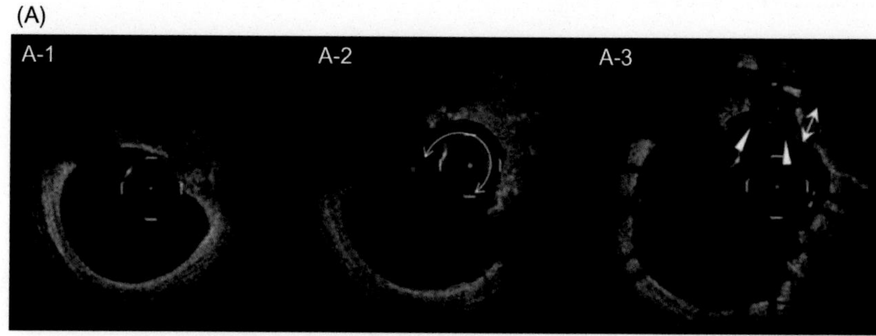

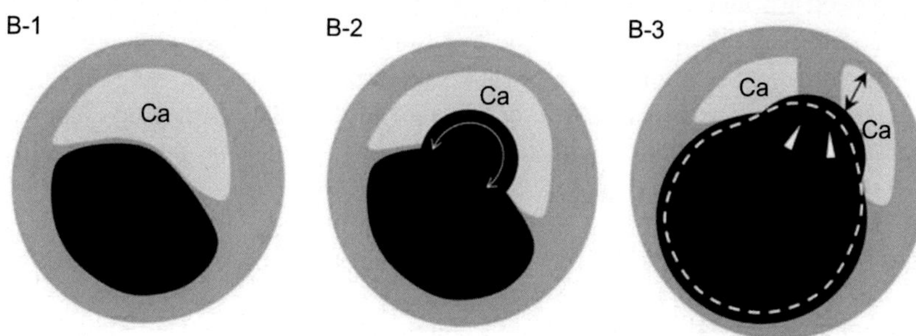

(B)

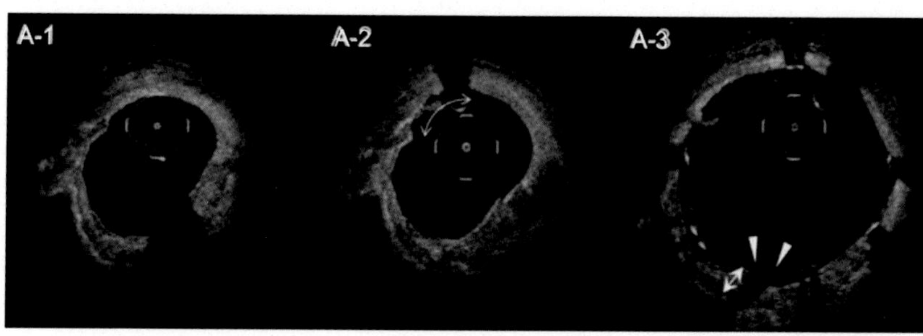

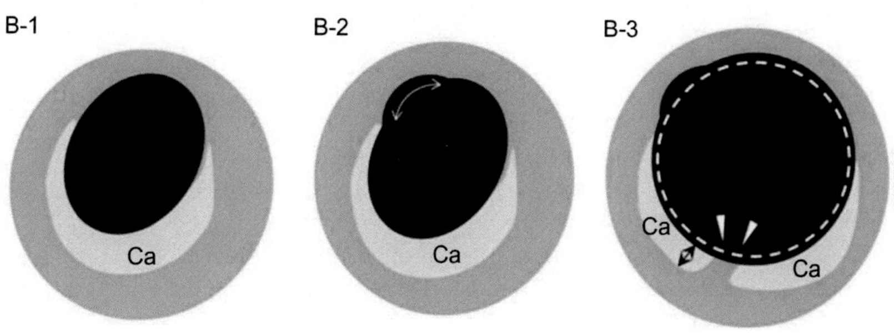

FIGURE 19.16 (A) Representative case of calcium fracture after orbital atherectomy. Cross-sectional preintervention (A-1), postorbital atherectomy (A-2), and poststent (A-3) optical coherence tomography images were matched. Calcium modification was seen postorbital atherectomy (A-2) (*double-headed arrow*). Calcium fracture after stenting (A-3) (*triangles*) was seen at the same site with calcium modification. Maximum fractured calcium thickness throughout a single calcium fracture (*double-headed arrow*) measured 0.60 mm. Corresponding illustrations of each optical coherence tomography image (B-1, B-2, and B-3). (B) Representative case of calcium fracture without orbital atherectomy calcium modification. Cross-sectional preintervention (A-1), postorbital atherectomy (A-2), and poststent (A-3) OCT images were matched. Noncalcified modification was seen postorbital atherectomy (A-2) (*double-headed arrow*). Calcium fracture (A-3) (*triangles*) occurred at the site without calcium modification. Maximum fractured calcium thickness throughout a single calcium fracture (*double-headed arrows*) measured 0.32 mm. Corresponding illustrations of each optical coherence tomography image (B-1, B-2, and B-3). *Reproduced with permission from Yamamoto MH, Maehara A, Kim SS, Koyama K, Kim SY, Ishida M, et al. Effect of orbital atherectomy in calcified coronary artery lesions as assessed by optical coherence tomography. Catheter Cardiovasc Interv 2019;93:1211–8.*

Stent delivery can be challenging requiring use of techniques that enhance guide catheter support and meticulous lesion preparation.

Stent recoil can occur especially in aorto-ostial calcified lesions that can sometimes be treated with implantation of a renal stent (Section 15.1.10.4, Challenge 4 and Fig. 15.13).

Intravascular imaging is recommended both before stenting to ensure sufficient lesion preparation and after stenting to ensure adequate stent expansion. The "StentBoost" and "CLEARstent" X-ray imaging modes can also be very useful.

19.11 Closure

Access closure is performed as described in Chapter 11: Access Closure.

19.12 Physiology

Physiologic assessment is encouraged in heavily calcified lesions, because of poor correlation of angiographic findings with coronary physiology measurements in such lesions [5].

19.13 Imaging

19.13.1 Baseline assessment

Intracoronary imaging is very useful in calcified lesions to confirm the severity and location of the calcium (circumferential vs focal and superficial vs. deep) and the need for atherectomy (Section 13.3.6.2, part C). OCT imaging may be preferred since it can determine the depth of calcium (Fig. 19.15), whereas the IVUS ultrasound images are blocked by the calcium and just show shadowing. Lesions with calcification angle $>180°$, maximum thickness >0.5 mm, and length >5 mm are at increased risk of stent underexpansion [26]. However, delivery of intracoronary imaging catheters can be challenging through calcification, especially when combined with tortuosity. Imaging may be performed after treatment with atherectomy if the catheter will not cross initially.

19.13.2 Final assessment

Intracoronary imaging is very useful after atherectomy (Fig. 19.16). Calcium fracture on OCT has been associated with improved stent expansion [27]. Moreover, the PCI result after stenting should be checked to ensure that adequate stent expansion has been achieved and to detect large edge dissections.

19.14 Hemodynamic support

The need for hemodynamic support depends on the patient's baseline hemodynamics, ejection fraction, size and importance of the target vessel, and the plan for performing atherectomy, as discussed in Chapter 14: Hemodynamic Support. Use of atherectomy is associated with higher risk of hemodynamic compromise, potentially favoring use of hemodynamic support.

References

[1] Mintz GS, Popma JJ, Pichard AD, Kent KM, Satler LF, Chuang YC, et al. Patterns of calcification in coronary artery disease. A statistical analysis of intravascular ultrasound and coronary angiography in 1155 lesions. Circulation 1995;91:1959–65.

[2] Megaly M, Sandoval Y, Lillyblad MP, Brilakis ES. Aminophylline for preventing bradyarrhythmias during orbital or rotational atherectomy of the right coronary artery. J Invasive Cardiol 2018;30:186–9.

[3] Murad B. Intracoronary aminophylline for management of bradyarrhythmias during thrombectomy with the AngioJet catheter. J Invasive Cardiol 2008;20:12A–18AA.

[4] Murad B. Intracoronary aminophylline for heart block with AngioJet thrombectomy. J Invasive Cardiol 2005;17:A4.

[5] Johnson PM, Madamanchi C, Sharalaya ZM, Iqbal Z, Gehi AK, Kaul P, et al. Angiographic severity does not correlate with fractional flow reserve in heavily calcified coronary arteries. Catheter Cardiovasc Interv 2017;89:226–32.

[6] Abdel-Wahab M, Richardt G, Joachim Buttner H, Toelg R, Geist V, Meinertz T, et al. High-speed rotational atherectomy before paclitaxel-eluting stent implantation in complex calcified coronary lesions: the randomized ROTAXUS (Rotational Atherectomy Prior to Taxus Stent Treatment for Complex Native Coronary Artery Disease) trial. JACC Cardiovasc Interv 2013;6:10–19.

[7] Abdel-Wahab M, Toelg R, Byrne RA, Geist V, El-Mawardy M, Allali A, et al. High-speed rotational atherectomy versus modified balloons prior to drug-eluting stent implantation in severely calcified coronary lesions. Circ Cardiovasc Interv 2018;11:e007415.

[8] Don CW, Palacios I, Rosenfield K. Use of rotational atherectomy in the body of a saphenous vein coronary graft. J Invasive Cardiol 2009;21: E168–70.

[9] Hussain F, Golian M. Desperate times, desperate measures: rotablating dissections in acute myocardial infarction. J Invasive Cardiol 2011;23: E226–8.

[10] Saad M, Bavineni M, Uretsky BF, Vallurupalli S. Improved stent expansion with prolonged compared with short balloon inflation: A meta-analysis. Catheter Cardiovasc Interv 2018;92:873–80.

[11] Barbato E, Shlofmitz E, Milkas A, Shlofmitz R, Azzalini L, Colombo A. State of the art: evolving concepts in the treatment of heavily calcified and undilatable coronary stenoses—from debulking to plaque modification, a 40-year-long journey. EuroIntervention 2017;13:696–705.

[12] Lee MS, Wiesner P, Rha SW. Novel technique of advancing the rotational atherectomy device: "single-operator" technique. J Invasive Cardiol 2016;28:183–6.

[13] Fouquet O, Gomez M, Abi-Khalil W, Binuani P. Open heart surgery for a stuck rotablator. Eur J Cardiothorac Surg 2016;49:e149.

[14] Tanaka Y, Saito S. Successful retrieval of a firmly stuck rotablator burr by using a modified STAR technique. Catheter Cardiovasc Interv 2016;87:749–56.

[15] Chiang CH, Liu SC. Successful retrieval of an entrapped rotablator burr by using a guideliner guiding catheter and a snare. Acta Cardiol Sin 2017;33:96–8.

[16] Imamura S, Nishida K, Kawai K, Hamashige N, Kitaoka H. A rare case of Rotablator((R)) driveshaft fracture and successful percutaneous retrieval of a trapped burr using a balloon and GuideLiner((R)). Cardiovasc Interv Ther 2017;32:294–8.

[17] Cunnington M, Egred M. GuideLiner, a child-in-a-mother catheter for successful retrieval of an entrapped rotablator burr. Catheter Cardiovasc Interv 2012;79:271–3.

[18] Sulimov DS, Abdel-Wahab M, Toelg R, Kassner G, Geist V, Richardt G. Stuck rotablator: the nightmare of rotational atherectomy. EuroIntervention 2013;9:251–8.

[19] Foster-Smith K, Garratt KN, Holmes Jr. DR. Guidewire transection during rotational coronary atherectomy due to guide catheter dislodgement and wire kinking. Cathet Cardiovasc Diagn 1995;35:224–7.

[20] Lai CH, Su CS, Wang CY, Lee WL. Heavily calcified plaques in acutely angulated coronary segment: high risk features of rotablation resulting in Rotawire transection and coronary perforation. Int J Cardiol 2015;182:112–14.

[21] Gavlick K, Blankenship JC. Snare retrieval of the distal tip of a fractured rotational atherectomy guidewire: roping the steer by its horns. J Invasive Cardiol 2005;17:E55–8.

[22] Shlofmitz E, Martinsen BJ, Lee M, Rao SV, Genereux P, Higgins J, et al. Orbital atherectomy for the treatment of severely calcified coronary lesions: evidence, technique, and best practices. Expert Rev Med Devices 2017;14:867–79.

[23] Hafiz AM, Smith C, Kakouros N. The GuideLiner-sleeve technique: a novel technique for high-risk PCI using two left main guides for orbital atherectomy of two-vessel calcified ostial coronary artery disease. J Invasive Cardiol 2017;29:E159–60.

[24] Ali ZA, Brinton TJ, Hill JM, Maehara A, Matsumura M, Karimi Galougahi K, et al. Optical coherence tomography characterization of coronary lithoplasty for treatment of calcified lesions: first description. JACC Cardiovasc Imaging 2017;10:897–906.

[25] Secco GG, Buettner A, Parisi R, Pistis G, Vercellino M, Audo A, et al. Clinical experience with very high-pressure dilatation for resistant coronary lesions. Cardiovasc Revasc Med 2019;20:1083–7.

[26] Fujino A, Mintz GS, Matsumura M, Lee T, Kim SY, Hoshino M, et al. A new optical coherence tomography-based calcium scoring system to predict stent underexpansion. EuroIntervention 2018;13:e2182–9.

[27] Yamamoto MH, Maehara A, Kim SS, Koyama K, Kim SY, Ishida M, et al. Effect of orbital atherectomy in calcified coronary artery lesions as assessed by optical coherence tomography. Catheter Cardiovasc Interv 2019;93:1211–18.

Chapter 20

Acute coronary syndromes—thrombus

CTO Manual Online cases: 6, 19, 103

PCI Manual Online cases: 4, 9, 11, 14, 19, 27, 31, 35, 40, 41, 42, 43, 44, 45, 50, 54, 57, 58, 61, 66, 78, 83, 84, 87, 94

Acute coronary syndromes encompass ST-segment elevation acute myocardial infarction (STEMI), non-ST-segment elevation acute myocardial infarction, and unstable angina. STEMI patients require immediate reperfusion to minimize the infarct size and decrease mortality.

20.1 Planning

Given urgency to restore the patency of the culprit vessel in STEMI patients, there is often limited time for planning. Nevertheless, the key components of planning remain essential, such as knowledge of prior revascularization procedures (e.g., STEMI in a patient with prior stenting could be due to stent thrombosis; knowledge of bypass graft anatomy is particularly important in prior CABG patients as it may expedite diagnostic angiography), cardiac function (risk of developing cardiogenic shock is higher in ACS patients who already have decreased ejection fraction) comorbidities (such as renal failure, diabetes, and need for future surgery), and medications that have been administered.

STEMI in postpartum patients may be due to spontaneous coronary dissection (SCAD, Section 22.1). STEMI in women who have recently had a stressful event may be due to stress cardiomyopathy (takotsubo), but coronary angiography often needs to be performed to exclude other possible etiologies.

20.2 Monitoring

Monitoring is of critical importance in ACS patients, as they are more likely to develop hemodynamic instability and arrhythmias. Often accelerated idioventricular rhythm and ST-segment normalization are indications of successful reperfusion in STEMI patients.

20.3 Medications

Potent *antithrombotic medication administration* (both anticoagulation and antiplatelet) is critical for treating thrombotic lesions, to minimize the risk of thrombus expansion or new thrombus formation. Antiplatelet treatment may be best achieved with more potent and rapidly acting oral P2Y12 inhibitors (such as prasugrel and ticagrelor), sometimes in combination with an intravenous antiplatelet agent, such as a glycoprotein IIb/IIIa inhibitor or cangrelor (for cases with large thrombus burden or for bailout in case of thrombotic complications). Anticoagulation is most commonly achieved with unfractionated heparin [1].

If glycoprotein IIb/IIIa (GP IIb/IIIa) receptor antagonists or cangrelor are being used, heparin is given at a dose of 50−70 U/kg IV bolus to achieve an activated clotting time (ACT) of 200−250 seconds (Hemochron device). When GP IIb/IIIa receptor antagonists or cangrelor are not being used, 70−100 U/kg bolus of unfractionated heparin is recommended with an ACT goal of 300−350 seconds (Hemochron device) [2].

20.4 Access

Radial access is generally preferred over femoral access in ACS patients, especially those with STEMI, given multiple prior studies that have shown lower bleeding risk and lower mortality with radial access [3], although the SAFARI-STEMI trial did not show any difference in clinical outcomes with femoral vs. radial access [4]. Femoral access is preferred in prior CABG patients, in many of whom graft engagement can be challenging using radial access. Moreover, timing is important

Manual of Percutaneous Coronary Interventions. DOI: https://doi.org/10.1016/B978-0-12-819367-9.00020-2

in STEMI patients: if vessel engagement is challenging through radial access causing delays in reperfusion (such as in patients with arteria lusoria) [5], prompt conversion to femoral access is indicated, since "time = muscle."

20.5 Engagement

Vessel engagement is performed as outlined in Chapter 5: Coronary and Graft Engagement.

In STEMI patients it is often debated whether angiography of both the suspected culprit and the nonculprit vessels should be performed before performing PCI or whether the target vessel should be immediately engaged with a guide catheter with immediate PCI of the culprit lesion.

Immediate engagement of the suspected culprit vessel can shorten door-to-balloon time by 8–13 minutes [6,7]. However, if the culprit lesion is in the RCA and PCI is done without knowledge of the left coronary anatomy, outcomes may be suboptimal, as some patients may have significant left main or multivessel CAD and may: (1) require hemodynamic support; or (2) be best served with CABG to improve their long-term outcomes, hence balloon angioplasty only of the culprit lesion may be preferable without stent implantation.

A recommended approach to coronary vessel engagement in STEMI patients is shown in Fig. 20.1.

20.6 Angiography

The goal of diagnostic angiography in ACS patients is to identify the culprit lesion(s), and define the overall coronary anatomy, so that the optimal revascularization strategy can be selected.

20.7 Determine target lesion(s)

Identification of the ACS culprit lesion(s) is often easy on diagnostic angiography (vessel occlusion with thrombus), but can sometimes be challenging. The following parameters can help identify the culprit lesion (*PCI Manual* Online cases 9, 11, 42, 50, 56, 78, 87):

1. Electrocardiogram.
2. Slow antegrade flow.
3. Luminal haziness, suggestive of thrombus.
4. Collateral flow.
5. Occlusion of distal branches, suggestive of prior coronary embolism.

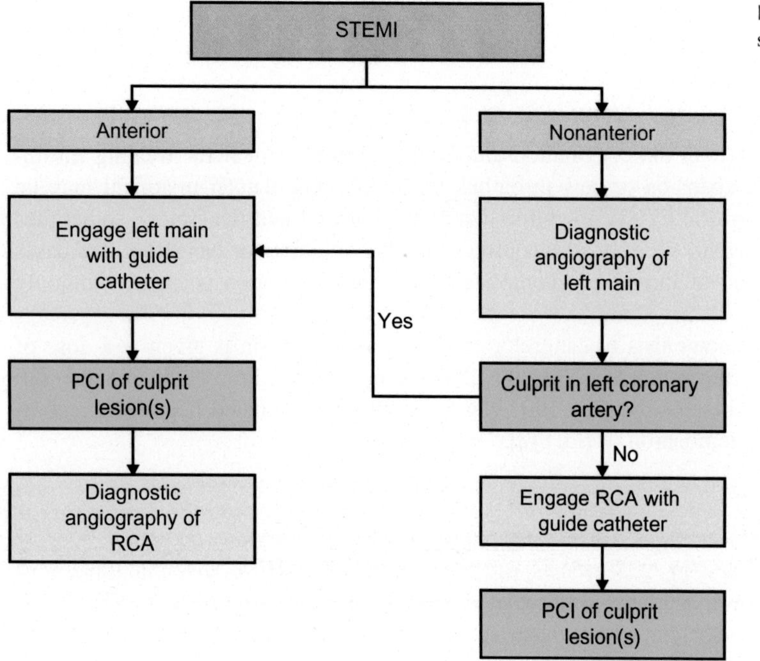

FIGURE 20.1 Recommended vessel engagement sequence in STEMI patients.

6. Left ventriculography (can reveal area of hypokinesis or suggest alternative diagnoses, such as takotsubo cardiomyopathy).
7. Intravascular imaging with optical coherence tomography (OCT), revealing plaque rupture or erosion and thrombus.
8. Transthoracic echocardiography that can reveal areas of hypokinesis/akinesis and can also help with diagnosis of mechanical complications, such as ventricular rupture, ventricular septal defect, and papillary muscle rupture.

In some cases a culprit lesion cannot be identified, because the presenting symptoms are not due to ischemia (such as in takotsubo cardiomyopathy).

In other cases, the culprit lesion may not be angiographically apparent. Intravascular imaging with OCT can help determine if a culprit lesion is present.

In cases where it remains unclear what the culprit lesion(s) are, subsequent magnetic resonance imaging can help detect the area of myocardial injury and infer what the culprit lesion(s) and vessel(s) are [8].

Complete revascularization (either at the time of emergent PCI or staged) is recommended for ST-segment elevation myocardial infarction patients, given the results of the COMPLETE trial [9]. In patients with cardiogenic shock culprit-only PCI is recommended in the acute setting based on the results of the CULPRIT-SHOCK trial [10,11].

20.8 Wiring

Wiring in ACS patients is performed as described in Chapter 8: Wiring. A specific challenge for wiring (and also lesion preparation and stenting) is the presence of intracoronary thrombus, in part because the distal vessel may be poorly or not at all visible, since collaterals may not have had enough time to form [12]. Moreover, it may be impossible to determine whether there is angulation within the occluded segment, or whether the guidewire has entered into side branches. In case of guidewire exit from the vessel subsequent advancement of a balloon can result in *perforation* (*PCI Manual* Online case 45), further complicating an already high-risk clinical presentation.

20.8.1 Guidewire choice

1. Workhorse guidewire

   To minimize the risk of perforation, initial wiring through an acute complete thrombotic occlusion is usually attempted with soft workhorse guidewires without hydrophilic tip coating. Occasionally advancement of the guidewire alone will result in restoration of some antegrade flow, that can facilitate subsequent wiring attempts and management of intracoronary thrombus.
2. Hydrophilic coated guidewires

   If a soft workhorse guidewire fails to advance through the occlusion, then a soft guidewire with hydrophilic tip coating can be used (such as the Sion or BMW Universal, see Section 30.7.1).
3. Polymer-jacketed guidewires.
4. Filterwire or Spider (Section 30.8). Although these devices are currently only approved for use in SVGs in the United States, they may capture thrombi liberated during thrombectomy, balloon angioplasty, and stenting) (*PCI Manual* Online case 61, 66).

If crossing with workhorse (with or without hydrophilic coating, Section 30.7.1) guidewires fails, soft polymer-jacketed guidewires, such as the Fielder FC or Sion black (Asahi Intecc, Nagoya, Japan, Section 30.7.2) are often used next. Escalation to stiffer polymer-jacketed guidewires (such as the Pilot 200) should be avoided, if possible, as it increases the risk of perforation, but may be needed in some challenging lesions [12].

20.8.2 Microcatheter and knuckling

Guidewire advancement over a microcatheter or a small balloon (1.0–1.5 mm in diameter) can also be useful for crossing by increasing support and allowing guidewire exchanges without losing wire position.

Occasionally advancing a knuckled guidewire (*PCI Manual* Online case 54) may help advance through areas of tortuosity with low (although not zero) risk of perforation, but may require strong guide catheter support, for example, by using a guide catheter extension.

20.8.3 Confirmation of guidewire position

If the location of the guidewire tip is uncertain after advancing it through an occlusion, it may be best to not advance a balloon but attempt to clarify the guidewire position first. If there is absolutely no antegrade flow after wiring, obtaining a second arterial access and performing contralateral injection may help clarify guidewire position through collateral filling of the occluded vessel. Injecting contrast through an over-the-wire balloon or microcatheter should be avoided, as it carries risk of extending a subintimal dissection if the microcatheter or balloon tip is in the subintimal space. Use of intravascular ultrasound can often aid in confirming intraluminal placement of the guidewire. Careful "dottering" of the thrombotic lesion with a low-profile short balloon may also be performed, however it is not a reliable strategy to confirm true lumen guidewire position; sometimes, it may restore enough flow to confirm guidewire location.

20.8.4 Wiring the wrong vessel?

Sometimes, wiring may fail because the lesion being crossed is not the culprit lesion for the acute event (it may be a chronic total occlusion). Alternatively some patients may have more than one culprit lesion. Repeat review of the angiogram (and of prior angiograms if available) and correlation of the electrocardiographic and angiographic findings may be helpful in such cases. Intravascular imaging (especially with OCT, which is the best imaging modality for detecting intracoronary thrombus) can be especially helpful in cases with unclear or multiple potential culprit lesions, as presence of thrombus or ulceration within a lesion strongly suggests it is a culprit. Thrombus aspiration can clarify the anatomy distal to the occlusion, and help clarify if the guidewire is in a side branch or in the main vessel.

20.9 Lesion preparation

A unique characteristic of ACS lesions is intracoronary *thrombus* that can be managed as outlined in Fig. 20.2.

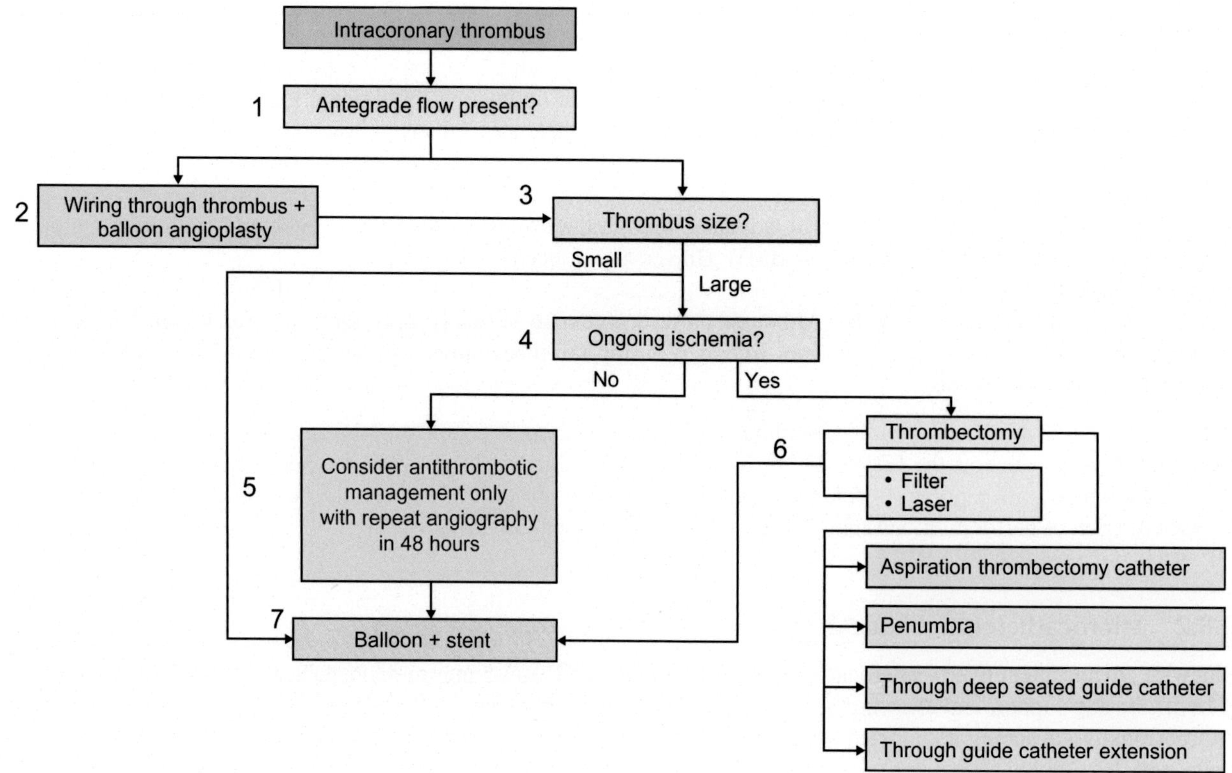

FIGURE 20.2 Management of intracoronary thrombus.

20.9.1 Antegrade flow present?

Sometimes guidewire crossing alone will restore some antegrade flow. If not, inflation of a small balloon (usually 2 mm in diameter) at low pressure (6–8 atm) may restore some flow and allow planning of subsequent treatment steps.

Failing to restore antegrade flow after ballooning could be due to large residual thrombus, balloon inflation proximal to the culprit stenosis in vessels with TIMI 0 flow, severe distal vessel disease or failure to be in the distal segment with the guidewire. If large thrombus burden is suspected, aspiration thrombectomy or laser may be beneficial. If wiring into a small branch is suspected, it may be best to leave the original wire in place and attempt wiring with another wire, followed by balloon angioplasty. Wire perforation may have occurred if wiring is very challenging.

Administration of anticoagulant and antiplatelet medications is crucial to prevent thrombus expansion, as described in Chapter 3: Medications.

20.9.2 Wiring through thrombus and balloon angioplasty

Balloon angioplasty can restore antegrade flow and also assist with assessing the lesion length and vessel diameter.

20.9.3 Thrombus size

In patients with large thrombus size balloon angioplasty and stenting may cause severe distal embolization that could be difficulty to treat.

20.9.4 Ongoing ischemia

The presence of ongoing ischemia is critical for making decisions about optimal management of patients with large thrombus burden.

20.9.5 Antithrombotic management only with repeat angiography

If antegrade flow is restored yet large thrombus remains, one treatment option is to *defer further intervention* by allowing 24–48 hours of intense antithrombotic therapy (including aspirin, oral P2Y12 inhibitor, intravenous platelet inhibitor and/or intravenous antithrombotic therapy). Subsequent angiography frequently shows reduced thrombus burden, allowing balloon angioplasty and stenting with lower risk of distal embolization and no reflow [13,14].

20.9.6 Thrombectomy, filter, or laser

In lesions with large thrombus, if a decision is made to proceed with PCI due to ongoing symptoms or high-risk plaque features, consideration should be given to *thrombectomy* before further balloon angioplasty and stenting. Although thrombectomy in this setting is logical in selected cases, routine use is not supported by randomized controlled trial data.

Use of a filter (Sections 30.8 and 18.8.1) may capture thrombus and prevent distal embolization during treatment of lesions with large thrombus burden.

Laser (Section 30.11) may "burn" through the thrombus and also inhibits platelet aggregation [15]. The 0.9 mm laser catheter (Philips) is most commonly used in native coronary arteries with the larger size catheters reserved for treating saphenous vein grafts or very large native coronary arteries. However, a 5-minute warm up period is required for the laser to become ready for use.

20.9.6.1 Thrombectomy

There are multiple ways to perform thrombectomy:

1. **Aspiration thrombectomy** using a dedicated catheter (Section 30.12.1). Larger (such as 7 French) aspiration thrombectomy catheters are more effective than smaller ones (such as 6 French). Delivery of the aspiration catheter can be difficult. Some aspiration thrombectomy catheters have stylets that can facilitate delivery. Alternatively, insertion of a 0.035 in. guidewire into the aspiration lumen can make it stiffer and more deliverable ("armored aspiration catheter" technique) [16].
2. **Aspiration using the Penumbra system** (Section 30.12.2), which is similar to aspiration thrombectomy but has a pump creating stronger and continuous suction.

3. Through a deep-seated guide catheter [that can sometimes be accomplished using the "balloon-assisted deep intubation" ("BADI") technique] and aspirating through the manifold [17].
4. Through a distally inserted guide catheter extension. However, deep guide seating and use of guide catheter extensions, carries risk of complications, such as distal dissection, ischemia leading to hemodynamic compromise, and stent deformation during delivery attempts.

To minimize the risk of embolization (both distally and in the aorta) suction should be maintained until the thrombectomy catheter is removed from the guide catheter, followed by aspiration of the guide catheter. This minimizes the risk of thrombi remaining within the guide catheter and re-injected into the coronary artery or into the aorta. Occasionally there is limited blood return from the aspiration catheter, which may be due to a large thrombus obstructing the tip of the catheter. Sometimes distal embolic protection devices (filters) (Sections 30.8 and 18.8.1) may be used to capture any liberated thrombi.

Multiple runs of aspiration thrombectomy or laser may be needed in cases with large thrombus burden. The over-the-wire lumen of the aspiration catheter could then be used to administer vasodilators or antiplatelet agents directly within the thrombotic coronary artery.

In cases of large thrombus adherent to the coronary wall, thrombus aspiration may fail, resulting in downstream thrombus embolization after balloon inflation or stent deployment. The embolized thrombus is often easier to aspirate.

20.9.6.2 How to do aspiration (or Penumbra) thrombectomy

20.9.6.2.1 Step 1. Select aspiration thrombectomy catheter

Goals:

1. Select the catheter that is likely to provide the best thrombectomy result (Section 30.12).

   How?
   1. Consider guide catheter size (7 French thrombectomy catheters provide more effective aspiration but cannot be used through 6 French guide catheters).
   2. Consider coronary artery size, tortuosity, and calcification (larger thrombectomy catheters may be challenging to deliver through tortuous and calcified coronary segments—thrombectomy catheters with a stylet are more deliverable).

   Challenges
   1. Local availability (you cannot use thrombectomy catheters you don't have on the shelf!).

What can go wrong?

1. Selection of a catheter that cannot be delivered due to size mismatch (7 French thrombectomy catheter cannot fit into a 6 French guide; a 6 French aspiration thrombectomy catheter cannot be inserted into the guide if there is a buddy wire) or due to poor deliverability to the thrombotic coronary segment.

20.9.6.2.2 Step 2. Prepare the aspiration thrombectomy catheter

Goals:

1. Promptly and meticulously prepare the aspiration thrombectomy catheter for use.

   How?
   1. The aspiration or Penumbra thrombectomy catheter is flushed.
   2. For aspiration thrombectomy: the vacuum syringe is attached to the back end of the aspiration catheter, the three-way stopcock is closed, the syringe plunger is pulled back and locked.
   3. For Penumbra: the catheter is flushed and closed off to the patient.

What can go wrong?

1. Incorrect connection of the aspiration syringe (connecting to aspiration catheter without the connecting segment with the on/off switch).

   Causes:
   - Misunderstanding of thrombectomy catheter design.
   - Being in a hurry to complete the procedure.

   Prevention:
   - Attention to catheter assembly.

   Treatment:
   - Disconnect thrombectomy catheter and reconnect with aspiration syringe in the proper fashion.

20.9.6.2.3 Step 3. Load the thrombectomy catheter on the guidewire

***Goals*:**

1. Load the thrombectomy catheter on the guidewire before delivering to the target thrombotic lesion.
   ***How*?**
   1. The thrombectomy catheter monorail lumen is inserted over the coronary guidewire.

   What can go wrong?

1. **Insertion of the guidewire through the aspiration lumen of the thrombectomy catheter.** This will be promptly identified if short (180 cm long) guidewires are being used.
   Causes:
   - Misunderstanding of the thrombectomy catheter design.
   - Urgency and time pressure to recanalize vessel.

   Prevention:
   - Closely observe that the guidewire is inserted through the monorail segment of the thrombectomy catheter.

   Treatment:
   - Remove thrombectomy catheter and reinsert the wire in the monorail segment.

20.9.6.2.4 Step 4. Deliver the thrombectomy catheter to the thrombotic lesion

***Goals*:**

1. Deliver the thrombectomy catheter to the target thrombotic lesion(s).
   ***How*?**
   1. Advance the thrombectomy catheter over the guidewire to the target thrombotic lesion.

   Challenges

1. **Failure to advance the thrombectomy catheter.**
   Causes:
   - Tortuosity.
   - Calcification.
   - Prior stents.
   - Poor guide catheter support.

   Prevention:
   - Balloon angioplasty of vessel proximal to the thrombus.
   - Obtain strong guide catheter support (as outlined in Sections 9.8.1.1-9.8.1.6).

   Treatment:
   - Modify the proximal vessel, for example, with balloon predilation.
   - Improve guide catheter support, for example, by deep intubation, using a guide extension (however, the thrombectomy catheter may not fit within the extension), or side branch anchor.

   What can go wrong?

1. **Loss of guide catheter and wire position.**
   Causes:
   - Tortuosity.
   - Calcification.
   - Prior stents.
   - Poor guide catheter support.

   Prevention:
   - Avoid excessive force while attempting to advance the thrombectomy catheter.
   - Obtain strong guide catheter support (Sections 9.8.1.1-9.8.1.6).
   - Continually monitor the position of the guide catheter and the guidewire.

   Treatment:
   - Modify the proximal vessel, for example, with balloon predilation.
   - Improve guide catheter support. This may be a good time for exchanging guide catheter for a more supportive one.

20.9.6.2.5 Step 5. Aspirate thrombus

***Goals*:**

- Remove ideally all thrombus from the target thrombotic lesion(s).
  ***How*?**
- The connecting switch is opened transmitting the vacuum to the aspiration lumen of the thrombectomy catheter.
- The catheter is advanced and withdrawn slowly within the area of thrombus.

Challenges

1. **Unable to aspirate blood or thrombus.**
   Causes:
   - Large thrombus size.
   - Poor preparation of the thrombectomy catheter.
   - Positioning in a small vessel or distal to the thrombus.

   Prevention:
   - Use the largest possible thrombectomy catheter.
   - Excellent preparation of the thrombectomy catheter.
   - Initiate suction prior to reaching the occlusion.

   Treatment:
   - Remove thrombectomy catheter while maintaining suction, (to prevent a thrombus from embolizing into the coronary artery, the aorta, or the guide catheter). Aspiration thrombectomy has been associated with increased risk of stroke [18], hence every effort should be made to minimize the risk of systemic embolization.

***What can go wrong*?**

1. **Distal embolization.**
   Prevention:
   - Do not stop aspirating until the thrombectomy catheter is removed from the guide catheter.
   - Back bleed and aspirate the guide catheter after removing the thrombectomy catheter.

   Treatment:
   - Repeat thrombectomy of the embolized thrombus.
   - In case of embolization in small branches, administer intracoronary vasodilators, such as nicardipine and adenosine.

20.9.6.2.6 Step 6. Remove thrombectomy catheter

***Goals*:**

- Remove thrombectomy catheter and all aspirated thrombus, while avoiding embolization.
  ***How*?**
  - Withdraw thrombectomy catheter while maintaining suction.

***What can go wrong*?**

1. **Thrombus embolization**.
   Causes:
   - Large thrombus size.
   - Losing suction while removing catheter.

   Prevention:
   - Keep suction while withdrawing catheter.
   - Remove while enough volume remains in the aspiration syringe.

   Treatment:
   - Aspirate guide catheter.
   - Change aspiration syringe and resume suction.
   - Repeat thrombectomy of the embolized thrombus.

20.9.6.2.7 Step 7. Aspirate guide catheter

Goals:

Remove thrombus that may be dislodged inside the guide catheter.

How?

Aspirate guide catheter and discard the aspirated blood.

What can go wrong?

1. **Thrombus embolization.**

 Causes:
 - Incomplete clearance of the guide catheter.
 - Reinjection of the aspirated thrombus.

 Prevention:
 - Discard aspirated blood.

 Treatment:
 - Repeat thrombectomy of the embolized thrombus.

20.9.6.2.8 Step 8. Examine aspirated blood

Goals:

- Determine if thrombus has been removed.

 How?
 - Inject the aspirated blood through the provided capture basket or over a clean gauze and examine what is retrieved.

20.9.6.2.9 Step 9. Perform repeat angiography

Goals:

- Determine if thrombus has been removed and antegrade coronary flow has been restored.

20.10 Stenting

Stenting is usually delayed until after thrombectomy in cases with large thrombus burden, but in some cases in which thrombectomy fails, *stenting* over the thrombus may effectively "trap" the thrombus behind the stent struts and restore antegrade flow. However, stenting can still cause distal embolization, in part due to a "cheese grater" effect of the stent struts. Moreover, thrombus can lead to stent undersizing and malapposition.

20.10.1 No stenting for ACS culprit lesions?

In some cases an excellent angiographic result may be obtained after thrombus aspiration alone; in such cases *observation without stent implantation* has been advocated, especially when the culprit vessel is small. Escaned et al reported outcomes on 28 STEMI patients in whom thrombus aspiration resulted in: (1) restoration of TIMI flow grade 3; (2) residual TIMI thrombus grade ≤ 3; (3) absence of a residual significant stenosis; and (4) absence of significant distal thrombus embolization. Despite not performing stenting, the culprit vessel was patent when repeat angiography was performed after 6 ± 2 days [19].

20.11 Closure

Radial access is advantageous for treatment of ACS patients, especially STEMI patients, because potent anticoagulation and antiplatelet regimens are often used, increasing the risk of bleeding.

When femoral access is used, the femoral sheath can sometimes be left in place for a few hours after the procedure in patients with borderline hemodynamics that may subsequently need a hemodynamic support device, however, this carries increased risk of bleeding.

20.12 Physiology

20.12.1 Culprit lesions

Coronary physiology should not be performed in the suspected culprit lesion(s) for the acute coronary syndrome, due to possible microcirculation injury from the ACS.

20.12.2 Nonculprit lesions

Nonculprit lesions in ACS patients can undergo physiologic assessment, as measurements do not significantly change over time [20].

20.13 Imaging

Imaging can be of critical importance during PCI for ACS.

20.13.1 Before stenting

Intravascular imaging with OCT remains the gold standard for identifying the ACS culprit lesion, and is of particular importance in patients with unclear culprit lesion(s). Ulceration and thrombus are the hallmarks of the culprit lesion, although some lesions only have erosion.

20.13.2 After stenting

Intravascular imaging can help optimize the stenting result (stent expansion and apposition and potential geographic miss) and detect residual thrombus or dissection.

20.14 Hemodynamic support

The likelihood of requiring hemodynamic support (Chapter 14: Hemodynamic Support) is higher in ACS as compared with stable angina patients, given myocardial injury and possible hemodynamic compromise.

References

[1] Erlinge D, Omerovic E, Frobert O, et al. Bivalirudin versus heparin monotherapy in myocardial infarction. N Engl J Med 2017;377:1132–42.

[2] O'Gara PT, Kushner FG, Ascheim DD, et al. ACCF/AHA guideline for the management of ST-elevation myocardial infarction: a report of the American College of Cardiology Foundation/American Heart Association Task Force on Practice Guidelines. Circulation 2013;127:e362–425.

[3] Ferrante G, Rao SV, Juni P, et al. Radial versus femoral access for coronary interventions across the entire spectrum of patients with coronary artery disease: a meta-analysis of randomized trials. JACC Cardiovasc Interv 2016;9:1419–34.

[4] Le May M, Wells G, So D, et al. Safety and efficacy of femoral access vs radial access in st-segment elevation myocardial infarction: the safari-stemi randomized clinical trial. JAMA Cardiol 2020;5:126–34.

[5] Khalili H, Banerjee S, Brilakis ES. Arteria lusoria in a patient with ST-segment elevation acute myocardial infarction: implications for primary PCI. J Invasive Cardiol 2015;27:E106.

[6] Applegate RJ, Graham SH, Gandhi SK, et al. Culprit vessel PCI versus traditional cath and PCI for STEMI. J Invasive Cardiol 2008;20:224–8.

[7] Plourde G, Abdelaal E, Bataille Y, et al. Effect on door-to-balloon time of immediate transradial percutaneous coronary intervention on culprit lesion in ST-elevation myocardial infarction compared to diagnostic angiography followed by primary percutaneous coronary intervention. Am J Cardiol 2013;111:836–40.

[8] Xenogiannis I, Lin D, Lesser JR, et al. Finding the culprit: combining cardiac magnetic resonance imaging with optical coherence tomography. JACC Cardiovasc Interv 2019;12:2106–9.

[9] Mehta SR, Wood DA, Meeks B, et al. Design and rationale of the COMPLETE trial: a randomized, comparative effectiveness study of complete versus culprit-only percutaneous coronary intervention to treat multivessel coronary artery disease in patients presenting with ST-segment elevation myocardial infarction. Am Heart J 2019;215:157–66.

[10] Thiele H, Akin I, Sandri M, et al. One-year outcomes after PCI strategies in cardiogenic shock. N Engl J Med 2018;379:1699–710.

[11] Thiele H, Akin I, Sandri M, et al. PCI strategies in patients with acute myocardial infarction and cardiogenic shock. N Engl J Med 2017;377:2419–32.

[12] Karacsonyi J, Henry T, Ungi I, Banerjee S, Brilakis ES. The impact of thrombus as a cause and as a result of complicated percutaneous coronary intervention. In: Topaz O, editor. Cardiovascular thrombus: from pathology and clinical presentations to imaging, pharmacotherapy, and interventions. Academic Press; 2018.

[13] Echavarria-Pinto M, Lopes R, Gorgadze T, et al. Safety and efficacy of intense antithrombotic treatment and percutaneous coronary intervention deferral in patients with large intracoronary thrombus. Am J Cardiol 2013;111:1745–50.

[14] De Maria GL, Alkhalil M, Oikonomou EK, Wolfrum M, Choudhury RP, Banning AP. Role of deferred stenting in patients with ST elevation myocardial infarction treated with primary percutaneous coronary intervention: a systematic review and meta-analysis. J Interv Cardiol 2017;30:264–73.

[15] Topaz O, Shah R, Mohanty PK, McQueen RA, Janin Y, Bernardo NL. Application of excimer laser angioplasty in acute myocardial infarction. Lasers Surg Med 2001;29:185–92.

[16] Brilakis E, Lichtenwalter C, Banerjee S. "Armored" aspiration catheter technique to enhance aspiration catheter delivery in challenging thrombus-containing lesions. Catheter Cardiovasc Interv 2009;74:846–9.

[17] Stys AT, Stys TP, Rajpurohit N, Khan MA. A novel application of GuideLiner catheter for thrombectomy in acute myocardial infarction: a case series. J Invasive Cardiol 2013;25:620–4.

[18] Jolly SS, Cairns JA, Yusuf S, et al. Randomized trial of primary PCI with or without routine manual thrombectomy. N Engl J Med 2015;372:1389–98.

[19] Escaned J, Echavarria-Pinto M, Gorgadze T, et al. Safety of lone thrombus aspiration without concomitant coronary stenting in selected patients with acute myocardial infarction. EuroIntervention 2013;8:1149–56.

[20] Ntalianis A, Sels J-W, Davidavicius G, et al. Fractional flow reserve for the assessment of nonculprit coronary artery stenoses in patients with acute myocardial infarction. J Am Coll Cardiol Intv 2010;3:1274–81.

Chapter 21

Chronic total occlusions

Chronic total occlusions (CTOs) are defined as completely occluded coronary arteries with Thrombolysis In Myocardial Infarction (TIMI) 0 flow with an estimated duration of at least 3 months [1]. CTOs can be challenging to recanalize, requiring specialized equipment and techniques [2], which is covered in detail in the *Manual of Chronic Total Occlusion Interventions* [3]. A global consensus document on the key principles underlying CTO PCI was recently published and will be briefly discussed in this chapter [1].

21.1 Planning

Planning is key for the success and safety of CTO PCI. The key indication for CTO PCI is symptom improvement [1]. To allow adequate time for procedural planning and preparation and for proper counseling of patients, *ad hoc CTO PCI is discouraged* in most cases [1,4]. CTO PCI preplanning can also help minimize contrast and radiation dose, reduce patient and operator fatigue, allow additional evaluation (such as myocardial viability) to be performed, and enable detailed discussion with the patient about all aspects of the CTO PCI procedure.

Coronary computed tomography angiography (CCTA) can be useful for evaluating the CTO segment and planning PCI. Several CCTA-based scores have been developed, such as the CT-RECTOR multicenter registry (Computed Tomography Registry of CTO revascularization) score [5] and the Korean Multicenter CTO CT Registry score [6]. CCTA can also help identify the optimal fluoroscopic projection angles with the least foreshortening to use during CTO crossing attempts (Fig. 21.1).

In selected cases, however, ad hoc PCI may be the best option, such as in patients who present with an acute coronary syndrome due to failure of a highly diseased saphenous vein graft. In such patients treatment of the native coronary artery CTO is preferred, if feasible (Section 18.7) [7]. Another possible scenario is patients in whom obtaining arterial access is very challenging.

21.2 Monitoring

CTO PCI can be technically challenging and time consuming, and may require high radiation and contrast doses. Careful attention to the patient's ECG and hemodynamics, as well as radiation and contrast use, is needed throughout the case. Moreover, the operators should be ready to manage any complication that may occur, such as perforation.

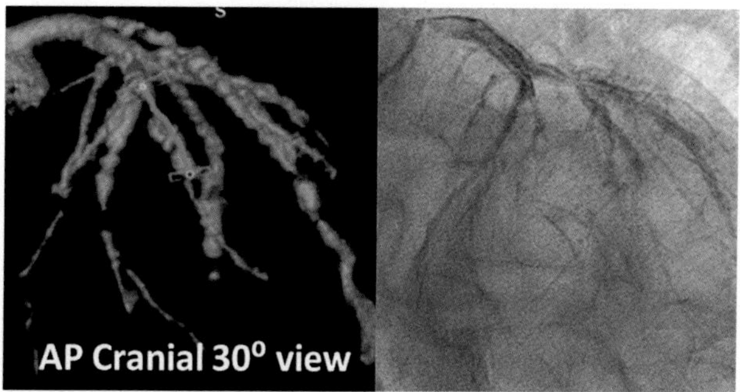

FIGURE 21.1 Use of CCTA for determining the optimal fluoroscopic projection for CTO crossing. *Courtesy of Dr. Yangsoo Jang.*

Manual of Percutaneous Coronary Interventions. DOI: https://doi.org/10.1016/B978-0-12-819367-9.00021-4

21.3 Medications

- Unfractionated heparin is the preferred agent, because it can be reversed in case of severe perforation. The recommended activated clotting times (ACT) are:
  - > 300 seconds (Hemochron) for antegrade CTO PCI (some operators use >250 s).
  - > 350 seconds (Hemochron) for retrograde CTO PCI (some operators use >300 s but check ACT very frequently if it is in the low 300s range).
- The ACT should be checked every 20–30 minutes once therapeutic.
- Bivalirudin is best avoided because its anticoagulant effect cannot be reversed. Moreover, there are unpublished cases in which guide thrombosis occurred during long procedures.
- Glycoprotein IIb/IIIa inhibitors and cangrelor should NOT be given, *even after successful crossing and stenting* of the CTO, because minor wire perforations could reopen and cause delayed pericardial effusion and tamponade.

21.4 Arterial access

CTO PCI is unique in that dual (or sometimes triple) arterial access is commonly required to allow dual angiography (see angiography section below). Both femoral and wrist access (proximal radial, distal radial [8], and ulnar) can be used, with use of radial access, including biradial access [9], significantly increasing in recent years [10,11]. Use of distal radial access (Section 4.3.2) [8,12] or use of long radial artery sheaths may facilitate use of the left radial artery for CTO PCI. However, many operators still utilize at least one femoral access (often with 45 cm long sheaths) to increase guide catheter support.

21.5 Engagement

Vessel engagement is performed as outlined in Chapter 5: Coronary and Graft Engagement. Use of large (7 or 8 French) guide catheters with supportive shapes (AL for the right coronary artery and EBU or XB for the left main) are particularly important in CTO PCI, since strong support is often required to cross the CTO. Using 6 French guide catheters may limit the option of using IVUS-guided cap penetration or management of subsequent complications.

21.6 Angiography

A key difference of CTO PCI compared with non-CTO PCI is that in most cases it requires dual injection for better understanding of the CTO anatomy and guiding CTO crossing attempts. Dual injection angiography is the simplest and most effective technique for increasing CTO PCI success rates and should be performed in all patients with contralateral collaterals [13]. It also improves procedural safety by elucidating the guidewire location during crossing attempts and facilitating management of periprocedural complications, such as perforation [1].

The following four characteristics need to be assessed for every CTO: (1) proximal cap morphology, (2) occlusion length, course, and composition (e.g., calcium), (3) quality of the distal vessel, and (4) characteristics of the collateral circulation (Fig. 21.2).

21.6.1 Proximal cap morphology

Understanding the location and morphology of the proximal cap is critical for selecting an optimal CTO crossing strategy, because trying to cross an ambiguous proximal cap may lead to perforation. Several techniques can be used to clarify the location of the proximal cap and allow safe antegrade crossing, such as:

- Additional angiographic projections.
- Selective contrast injection through a microcatheter located near the proximal cap.
- Use of intravascular ultrasound (IVUS) [14].
- Preprocedural CCTA or real-time CCTA coregistration [15].
- Use of dissection/reentry techniques (move the cap techniques) [16,17].

If the location of the proximal cap cannot be resolved, a primary retrograde approach or move the cap techniques are often recommended, if technically feasible.

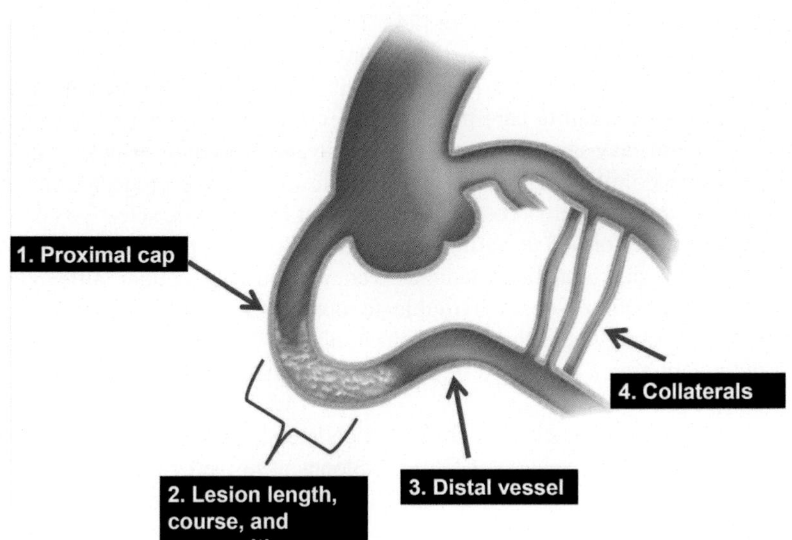

FIGURE 21.2 The four key angiographic characteristics of chronic total occlusions. *Reproduced with permission from Brilakis ES. Manual of coronary chronic total occlusion interventions. A step-by-step approach. 2nd ed. Cambridge, MA: Elsevier; 2017 (Figure 3.9).*

21.6.2 Lesion length, course, and composition

Lesion length is often overestimated with antegrade only injections due to underfilling and poor opacification of the distal vessel from competing antegrade and retrograde coronary flow, leaving uncertainty about the location and morphology of the distal cap. Dual injection or preprocedural CCTA [18] allows more accurate estimation of the CTO proximal cap location, length, and distal cap anatomy.

Severe calcification and tortuosity of the occluded segment can adversely affect CTO crossing and increase the likelihood of subintimal (also called extraplaque) guidewire entry. Advancing a knuckled (J-shaped) guidewire or changing to the retrograde approach is often preferred when the vessel course is unclear or highly tortuous [19], since a knuckled guidewire can facilitate advancement within the vessel architecture with low risk of perforation [20].

21.6.3 Distal vessel

The quality of the distal vessel can significantly impact the likelihood of CTO crossing: distal vessels of large caliber (>2.0 mm) that fill well, do not have significant disease and are free from major branches may facilitate CTO recanalization [1]. Conversely, small, diffusely diseased distal vessels are more challenging to recanalize, especially following subintimal guidewire entry. In some cases, however, distal vessels are small due to hypoperfusion, leading to negative remodeling and will increase in size after recanalization [21]. Distal CTO caps in native coronary artery CTOs are more likely to be calcified and resistant to guidewire penetration in vessels previously bypassed distal to the CTO [22]. Moreover, distal vessel calcification may hinder wire reentry in case of subintimal guidewire entry. The presence of a bifurcation at the distal cap (as well as at the proximal cap or within the occluded segment) may hinder antegrade wiring of the main branch and also increases the likelihood of side branch loss. The retrograde approach is favored in cases of CTOs with a bifurcation at the distal cap because antegrade techniques often lead to occlusion of one of the two branches [16].

21.6.4 Collateral circulation

Evaluation of the collateral circulation is critical for determining the feasibility of the retrograde approach [23]. High-quality angiography (ideally obtained on low magnification during breath hold and without panning) allowing complete opacification of collateral vessels and obtained in optimal angiographic projections, should, therefore, be encouraged as part of the routine diagnostic studies when a CTO is discovered.

Retrograde access to the distal vessel can be achieved via septal collaterals, epicardial collaterals, or (patent or occluded) coronary bypass grafts. When assessing collateral channels it is important to consider size, tortuosity, bifurcations, angle of entry to and exit from the collateral, jailing of entry or exit by a previously deployed stent, which may hinder guidewire crossing, and distance between the collateral exit and the distal cap. The most important predictor of successful guidewire and device crossing is lack of tortuosity, followed by size [24]. The size of the collaterals is often assessed using the Werner classification (CC0: no continuous connection; CC1: threadlike connection; CC2: side branch-like connection) [25]. Crossing invisible septal collateral channels is often possible with the surfing technique, letting the wire find the path of least resistance [26]. It is important to carefully study previous angiograms for multiple potential collateral pathways, as the predominant collateral may change over time prior to the procedure or during the course of PCI ("shifting collaterals"). Previously visualized collaterals that disappear at the time of the procedure may still be crossable. Whenever required, and after ensuring adequate blood backflow to prevent barotrauma, selective contrast tip injections through the microcatheter can be safely performed to outline collateral anatomy. Patent bypass grafts represent an ideal retrograde conduit due to the absence of side branches, predictable course and large caliber, although the angle of the distal bypass graft anastomosis can sometimes be unfavorable to obtain retrograde access. Even occluded grafts can be used as retrograde pathways. In cases where the collateral circulation originates from the left anterior descending artery that is supplied by a mammary artery graft, access through the IMA graft increases the risk of global ischemia and should be avoided whenever possible [27].

Septal collaterals are usually safer and easier to navigate compared with epicardial collaterals [28,29]. In contrast to epicardial collaterals, septal collaterals can be safely dilated with small (≤1.5 mm) balloons at low pressure (2–4 atm) to facilitate microcatheter or device crossing, if required. The donor vessel proximal to the collateral origin, as well as collateral dominance (i.e., presence of a single large visible collateral), should also be assessed during retrograde procedures to determine the risk of ischemia during retrograde crossing attempts. Careful review of collaterals prior to the procedure can reduce contrast and radiation dose as well as the duration of the procedure. In cases where the collateral anatomy is unclear or ambiguous, it can be helpful to perform selective contrast injection into the collateral through the central lumen of a microcatheter placed into the collateral using a 2–3 cc luer lock syringe, such as the Medallion syringe (Merit Medical). In cases where unfavorable or non-interventional epicardial collaterals provide the dominant blood flow to the CTO, gentle balloon occlusion of the epicardial collateral for 2–4 minutes may allow recruitment of more favorable interventional collaterals that can be used for retrograde crossing [18].

21.7 Determine target lesion(s)

This step is performed as discussed in Chapter 7: Selecting Target Lesion(s).

21.8 Wiring

Advancing a guidewire through the CTO is the most challenging part of CTO PCI. There are four CTO crossing strategies, classified according to wiring direction (antegrade and retrograde) and whether or not the subintimal space is utilized (wiring vs dissection and reentry) (Fig. 21.3) [1].

21.8.1 Antegrade wiring

Antegrade wiring (also called antegrade wire escalation) is the most widely used CTO crossing technique [14,30–32]. Various guidewires are advanced in the antegrade direction (original direction of blood flow). The choice of guidewire depends on CTO characteristics. If there is a tapered proximal cap or a functional occlusion with a visible channel, a polymer jacketed, low penetration force, tapered guidewire is used initially with subsequent escalation to intermediate and high penetration force guidewires, as required. If there is a blunt proximal cap, antegrade wiring is usually started with an intermediate penetration force polymer-jacketed guidewire or a composite core guidewire. Stiff, high penetration force guidewires may be required in highly resistant proximal caps or when areas of resistance are encountered within the body of the occlusion. After proximal cap crossing of 1–2 mm, however, deescalation to less penetrating guidewires should follow to safely navigate through the CTO segment ("step-up/step-down" technique).

Contralateral injection and orthogonal angiographic projections help determine the guidewire position during crossing attempts. If the guidewire enters into the distal true lumen, the microcatheter is advanced into the distal true lumen and the dedicated CTO guidewire is exchanged for a workhorse guidewire through the microcatheter to minimize the risk of distal vessel injury and perforation during subsequent balloon angioplasty and stenting. If the guidewire exits the vessel structure it should be withdrawn and redirected *without* advancing microcatheters, balloons, or stents over it to prevent enlarging (or creating) a perforation. If the guidewire enters the subintimal space it can be redirected, but if this maneuver fails, the wire can be left in place to aid directing a second guidewire into the distal true lumen (parallel wire technique), which can be assisted by a dual lumen microcatheter, or facilitated by the use of IVUS [14]. Alternatively,

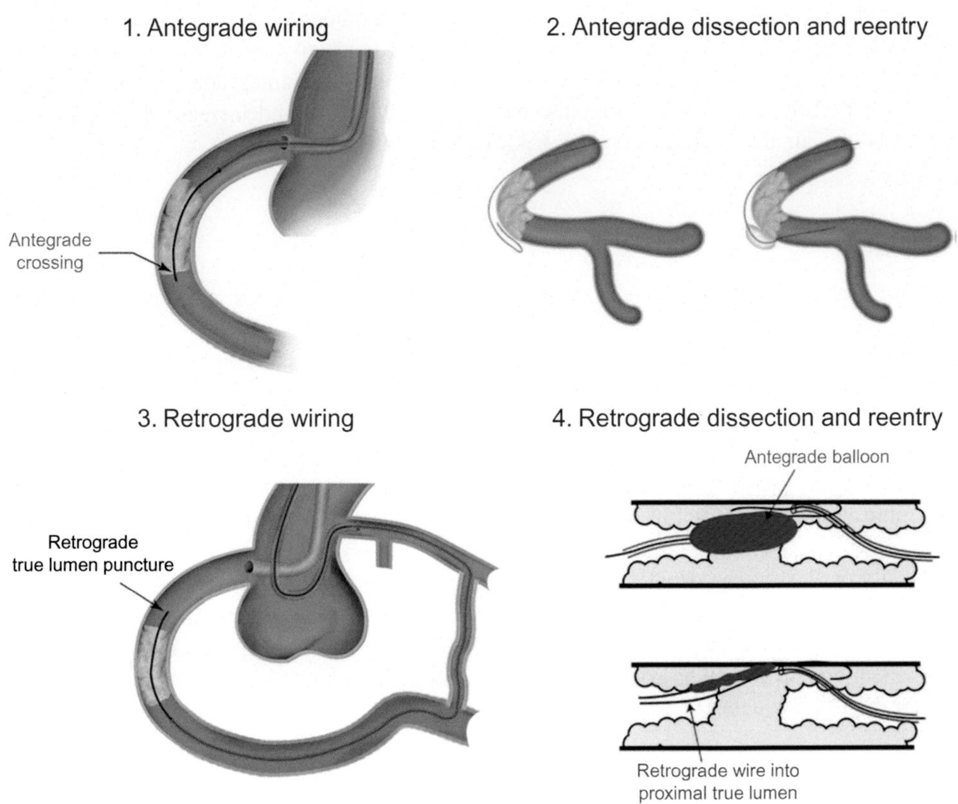

FIGURE 21.3 CTO crossing strategies. *Reproduced with permission from Brilakis ES. Manual of coronary chronic total occlusion interventions. A step-by-step approach. 2nd ed. Cambridge, MA: Elsevier; 2017 (Figure 3.9).*

antegrade dissection/reentry techniques can be used to reenter into the distal true lumen, as described below. Subintimal guidewire advancement significantly distal to the distal cap should be avoided, as it can lead to hematoma formation, causing luminal compression and reducing the likelihood of success. Antegrade vessel reentry can be guided by IVUS, although this approach requires 8 French guide catheters and expertise in IVUS interpretation and may be hindered by limited wire maneuverability in the presence of the subintimal IVUS catheter.

21.8.2 Antegrade dissection and reentry

Antegrade dissection and reentry involves entering the subintimal space, followed by subintimal crossing of the CTO with subsequent reentry into the distal true lumen. Antegrade dissection may be intentional or unintentional during antegrade wiring attempts. The initially developed dissection reentry technique was named STAR (subintimal tracking and reentry) and used uncontrollable reentry into the distal lumen [33]. This frequently necessitated stenting long coronary segments with occlusion of numerous side branches, with extensive vascular injury and high rates of in-stent restenosis and reocclusion [33–35]. As such, the STAR technique has evolved to a bailout strategy without stent implantation after ballooning, in preparation for a repeat CTO PCI attempt (subintimal plaque modification, also called "investment" procedure) after 2–3 months [36–38]. The development of limited dissection/reentry techniques (using dedicated reentry systems [39,40] or wire-based strategies [41,42]) was an important advancement, as they minimize vascular injury, limit the length of dissection and subsequent stent length, and increase the likelihood of side branch preservation [20,39,43]. Such approaches have been associated with favorable clinical outcomes [43–47].

21.8.3 The retrograde approach

The retrograde technique differs from the antegrade approach in that the occlusion is approached from the distal vessel with guidewire advancement against the original direction of blood flow [48]. A guidewire is advanced into the artery distal to the occlusion through a collateral channel or through a bypass graft, followed by placement of a microcatheter at the distal CTO cap. Retrograde CTO crossing is then attempted either with retrograde wiring (usually for short occlusions, especially when the distal cap is tapered [22]) or using retrograde dissection/reentry techniques.

The most commonly used retrograde crossing technique is the reverse controlled antegrade and retrograde tracking (reverse CART), in which a balloon is inflated over the antegrade guidewire which is usually located in the subintimal space, followed by retrograde guidewire advancement into the space created by the deflated antegrade balloon. In challenging reverse CART cases, intravascular ultrasound can clarify the mechanism of failure and increase the likelihood of success [49]. Guide catheter extensions can also facilitate reverse CART [50].

21.8.4 Crossing strategy selection

Selecting the initial and subsequent crossing strategies depends on the CTO lesion characteristics and local equipment availability and expertise.

Several algorithms have been developed to facilitate crossing strategy selection, such as the hybrid [51] and Asia Pacific [19] algorithm. Antegrade crossing is generally preferred over retrograde crossing as the initial crossing strategy, given the higher risk of complications, longer procedure time and more radiation with the retrograde approach [30–32] and need for antegrade lesion preparation even when the retrograde approach is used. Some retrograde CTO PCI complications, however, are caused by antegrade crossing attempts. The retrograde approach remains critical for achieving high success rates, especially in more complex CTOs [30,32] and has been associated with favorable long-term outcomes [52].

CTOs with proximal cap ambiguity or flush aorto-ostial CTOs are often approached with a primary retrograde strategy. Alternatively, CTOs with ambiguous proximal caps can be approached in the antegrade direction, especially when no collateral or graft is available by using: (1) intravascular ultrasound or preprocedural CCTA for determining the location of the proximal cap and vessel course [16,19,53] or (2) techniques to facilitate entry into the subintimal space proximal to the occlusion (move the cap techniques) [17].

21.8.5 Change of crossing strategy

If the initial or subsequent crossing strategy fails to achieve progress, small changes (such as modifying the guidewire tip angulation or changing guidewire) or more significant changes (such as converting from an antegrade to a retrograde approach) should be made, based on pre-procedural planning and the case progression [1,19,51].

Similar to selection of the initial crossing strategy, the timing and choice of subsequent crossing strategies depends on lesion characteristics, challenges encountered with the original technique, and equipment availability and expertise. Strategy selection can be guided by various crossing algorithms, such as the hybrid algorithm (Fig. 21.4) [19,51,54].

Reasons to stop a CTO PCI attempt include occurrence of a complication, high radiation dose (usually >5 Gray air kerma dose in the absence of lesion crossing or substantial progress), large contrast volume administration (>3.7 × the estimated creatinine clearance [55]), exhaustion of crossing options, or patient or physician fatigue. As with all interventions, careful assessment of individual risk versus benefit should guide decision-making and choice of strategy during different stages of the procedure. On many occasions, it may be best to perform STAR or fail rather than pursue highly aggressive strategies that may lead to serious complications [1,30].

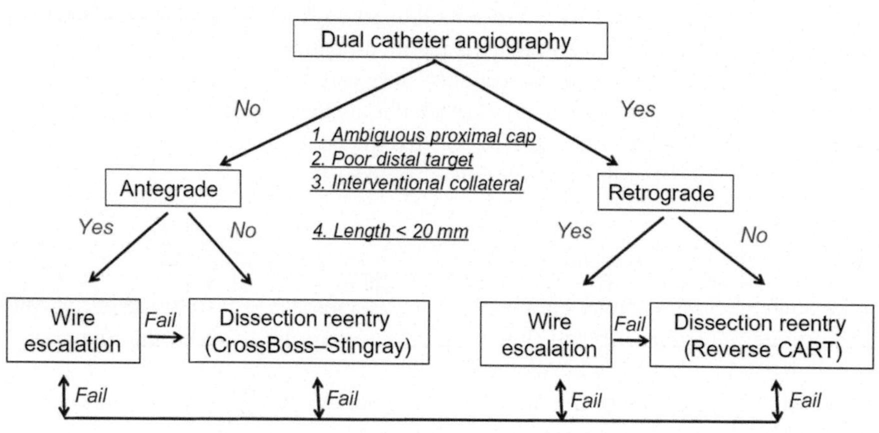

FIGURE 21.4 The hybrid crossing algorithm. *Reproduced with permission from Brilakis ES. Manual of coronary chronic total occlusion interventions. A step-by-step approach. 2nd ed. Cambridge, MA: Elsevier; 2017 (Figure 7.1).*

21.9 Lesion preparation

Lesion preparation is performed as described in Chapter 9: Lesion Preparation.

21.10 Stenting

Stenting is performed as described in Chapter 10: Stenting. Given long lesion length, multiple long stents are often required, especially in the right coronary artery. Stent optimization using intravascular imaging is critical to prevent subsequent events [1].

Donor artery lesions located in the path of a retrograde route should be treated to avoid donor artery thrombosis during retrograde CTO PCI.

21.11 Closure

Closure is performed as described in Chapter 11: Access Closure.

21.12 Physiology

21.12.1 CTO vessel

Coronary physiology is not typically used in the CTO target vessel, as prior studies have shown that the CTO supplied territory is always ischemic, even in the presence of good collateral circulation [56,57]. Post-stenting coronary physiology can help assess the PCI result: presence of a step-up suggests that additional treatment is needed in the coronary segment where the step-up was observed.

21.12.2 Non-CTO vessel

The hemodynamic significance of a lesion in the donor vessel that collateralizes the CTO vessel may change after successful CTO recanalization, with FFR increasing by 0.03 [58] to 0.10 [59]. This should be taken into account when performing physiologic assessment of donor vessels supplying collaterals to a CTO territory [3].

21.13 Imaging

Imaging can be of critical importance of CTO PCI, both for CTO crossing and for stent optimization.

21.13.1 CTO crossing

IVUS can facilitate CTO crossing by resolving proximal cap ambiguity and facilitating and confirming true lumen wire reentry in case of initial subintimal wire crossing in both the antegrade and retrograde direction (for example by selecting the appropriate balloon size during the reverse CART technique) [53,60].

21.13.2 Stenting optimization

Intravascular imaging can help optimize the stenting result by achieving good stent expansion and stent strut apposition, which is of particular importance in prior CTO patients given long stent length, and negative remodeling of the distal vessel [61–64].

21.14 Hemodynamic support

Hemodynamic support is used in approximately 4% of CTO PCI cases in the United States [65].

Use of hemodynamic support is based on the principles described in Chapter 14: Hemodynamic Support.

References

[1] Brilakis ES, Mashayekhi K, Tsuchikane E, et al. Guiding principles for chronic total occlusion percutaneous coronary intervention. Circulation 2019;140:420–33.

[2] Tajti P, Burke MN, Karmpaliotis D, et al. Update in the percutaneous management of coronary chronic total occlusions. JACC Cardiovasc Interv 2018;11:615–25.

[3] Brilakis ES. Manual of coronary chronic total occlusion interventions. A step-by-step approach. 2nd ed. Cambridge, MA: Elsevier; 2017.

[4] Sandoval Y, Tajti P, Karatasakis A, et al. Frequency and outcomes of ad hoc versus planned chronic total occlusion percutaneous coronary intervention: multicenter experience. J Invasive Cardiol 2019;31:133–9.

[5] Opolski MP, Achenbach S, Schuhback A, et al. Coronary computed tomographic prediction rule for time-efficient guidewire crossing through chronic total occlusion: insights from the CT-RECTOR multicenter registry (Computed Tomography Registry of Chronic Total Occlusion Revascularization). JACC Cardiovasc Interv 2015;8:257–67.

[6] Yu CW, Lee HJ, Suh J, et al. Coronary computed tomography angiography predicts guidewire crossing and success of percutaneous intervention for chronic total occlusion: Korean multicenter CTO CT registry score as a tool for assessing difficulty in chronic total occlusion percutaneous coronary intervention. Circ Cardiovasc Imaging 2017;10.

[7] Brilakis ES, Banerjee S, Lombardi WL. Retrograde recanalization of native coronary artery chronic occlusions via acutely occluded vein grafts. Catheter Cardiovasc Interv 2010;75:109–13.

[8] Gasparini GL, Garbo R, Gagnor A, Oreglia J, Mazzarotto P. First prospective multicentre experience with left distal transradial approach for coronary chronic total occlusion interventions using a 7 Fr Glidesheath Slender. EuroIntervention 2019;15:126–8.

[9] Bakker EJ, Maeremans J, Zivelonghi C, et al. Fully transradial versus transfemoral approach for percutaneous intervention of coronary chronic total occlusions applying the hybrid algorithm: insights from RECHARGE registry. Circ Cardiovasc Interv 2017;10.

[10] Tajti P, Alaswad K, Karmpaliotis D, et al. Procedural outcomes of percutaneous coronary interventions for chronic total occlusions via the radial approach: insights from an international chronic total occlusion registry. JACC Cardiovasc Interv 2019;12:346–58.

[11] Megaly M, Karatasakis A, Abraham B, et al. Radial versus femoral access in chronic total occlusion percutaneous coronary intervention. Circ Cardiovasc Interv 2019;12:e007778.

[12] Gasparini GL, Garbo R, Gagnor A, Oreglia J, Mazzarotto P. Feasibility and safety of distal radial access for percutaneous coronary intervention using a 7 FR sheath. Catheter Cardiovasc Interv 2019;94:902.

[13] Singh M, Bell MR, Berger PB, Holmes Jr. DR. Utility of bilateral coronary injections during complex coronary angioplasty. J Invasive Cardiol 1999;11:70–4.

[14] Suzuki Y, Tsuchikane E, Katoh O, et al. Outcomes of percutaneous coronary interventions for chronic total occlusion performed by highly experienced Japanese specialists: the first report from the Japanese CTO-PCI expert registry. JACC Cardiovasc Interv 2017;10:2144–54.

[15] Karatasakis A, Danek BA, Karmpaliotis D, et al. Impact of proximal cap ambiguity on outcomes of chronic total occlusion percutaneous coronary intervention: insights from a multicenter US registry. J Invasive Cardiol 2016;28:391–6.

[16] Ghoshhajra BB, Takx RAP, Stone LL, et al. Real-time fusion of coronary CT angiography with x-ray fluoroscopy during chronic total occlusion PCI. Eur Radiol 2017;27:2464–73.

[17] Vo MN, Karmpaliotis D, Brilakis ES. "Move the cap" technique for ambiguous or impenetrable proximal cap of coronary total occlusion. Catheter Cardiovasc Interv 2016;87:742–8.

[18] Fujino A, Otsuji S, Hasegawa K, et al. Accuracy of J-CTO score derived from computed tomography versus angiography to predict successful percutaneous coronary intervention. JACC Cardiovasc Imaging 2018;11:209–17.

[19] Harding SA, Wu EB, Lo S, et al. A new algorithm for crossing chronic total occlusions from the Asia Pacific Chronic Total Occlusion Club. JACC Cardiovasc Interv 2017;10:2135–43.

[20] Azzalini L, Carlino M, Brilakis ES, et al. Subadventitial techniques for chronic total occlusion percutaneous coronary intervention: the concept of "vessel architecture". Catheter Cardiovasc Interv 2018;91:725–34.

[21] Galassi AR, Tomasello SD, Crea F, et al. Transient impairment of vasomotion function after successful chronic total occlusion recanalization. J Am Coll Cardiol 2012;59:711–18.

[22] Sakakura K, Nakano M, Otsuka F, et al. Comparison of pathology of chronic total occlusion with and without coronary artery bypass graft. Eur Heart J 2014;35:1683–93.

[23] McEntegart MB, Badar AA, Ahmad FA, et al. The collateral circulation of coronary chronic total occlusions. EuroIntervention 2016;11: e1596–603.

[24] Huang CC, Lee CK, Meng SW, et al. Collateral channel size and tortuosity predict retrograde percutaneous coronary intervention success for chronic total occlusion. Circ Cardiovasc Interv 2018;11:e005124.

[25] Werner GS, Ferrari M, Heinke S, et al. Angiographic assessment of collateral connections in comparison with invasively determined collateral function in chronic coronary occlusions. Circulation 2003;107:1972–7.

[26] Dautov R, Urena M, Nguyen CM, Gibrat C, Rinfret S. Safety and effectiveness of the surfing technique to cross septal collateral channels during retrograde chronic total occlusion percutaneous coronary intervention. EuroIntervention 2017;12:e1859–67.

[27] Tajti P, Karatasakis A, Karmpaliotis D, et al. Retrograde CTO-PCI of native coronary arteries via left internal mammary artery grafts: insights from a multicenter U.S. registry. J Invasive Cardiol 2018;30:89–96.

[28] Rathore S, Katoh O, Matsuo H, et al. Retrograde percutaneous recanalization of chronic total occlusion of the coronary arteries: procedural outcomes and predictors of success in contemporary practice. Circ Cardiovasc Interv 2009;2:124–32.

[29] Benincasa S, Azzalini L, Carlino M, et al. Outcomes of the retrograde approach through epicardial versus non-epicardial collaterals in chronic total occlusion percutaneous coronary intervention. Cardiovasc Revasc Med 2017;18:393–8.

[30] Tajti P, Karmpaliotis D, Alaswad K, et al. The hybrid approach to chronic total occlusion percutaneous coronary intervention: update from the PROGRESS CTO registry. JACC Cardiovasc Interv 2018;11:1325–35.

[31] Maeremans J, Walsh S, Knaapen P, et al. The hybrid algorithm for treating chronic total occlusions in Europe: the RECHARGE registry. J Am Coll Cardiol 2016;68:1958–70.

[32] Wilson WM, Walsh SJ, Yan AT, et al. Hybrid approach improves success of chronic total occlusion angioplasty. Heart 2016;102:1486–93.

[33] Colombo A, Mikhail GW, Michev I, et al. Treating chronic total occlusions using subintimal tracking and reentry: the STAR technique. Catheter Cardiovasc Interv 2005;64:407–11.

[34] Valenti R, Vergara R, Migliorini A, et al. Predictors of reocclusion after successful drug-eluting stent-supported percutaneous coronary intervention of chronic total occlusion. J Am Coll Cardiol 2013;61:545–50.

[35] Godino C, Latib A, Economou FI, et al. Coronary chronic total occlusions: mid-term comparison of clinical outcome following the use of the guided-STAR technique and conventional anterograde approaches. Catheter Cardiovasc Interv 2012;79:20–7.

[36] Hirai T, Grantham JA, Sapontis J, et al. Impact of subintimal plaque modification procedures on health status after unsuccessful chronic total occlusion angioplasty. Catheter Cardiovasc Interv 2018;91:1035–42.

[37] Wilson WM, Bagnall AJ, Spratt JC. In case of procedure failure: facilitating future success. Interventional Cardiol 2013;5:521–31.

[38] Goleski PJ, Nakamura K, Liebeskind E, et al. Revascularization of coronary chronic total occlusions with subintimal tracking and reentry followed by deferred stenting: experience from a high-volume referral center. Catheter Cardiovasc Interv 2019;93:191–8.

[39] Whitlow PL, Burke MN, Lombardi WL, et al. Use of a novel crossing and re-entry system in coronary chronic total occlusions that have failed standard crossing techniques: results of the FAST-CTOs (Facilitated Antegrade Steering Technique in Chronic Total Occlusions) trial. JACC Cardiovasc Interv 2012;5:393–401.

[40] Karacsonyi J, Tajti P, Rangan BV, et al. Randomized comparison of a CrossBoss first versus standard wire escalation strategy for crossing coronary chronic total occlusions: the CrossBoss first trial. JACC Cardiovasc Interv 2018;11:225–33.

[41] Galassi AR, Boukhris M, Tomasello SD, et al. Long-term clinical and angiographic outcomes of the mini-STAR technique as a bailout strategy for percutaneous coronary intervention of chronic total occlusion. Can J Cardiol 2014;30:1400–6.

[42] Galassi AR, Tomasello SD, Costanzo L, et al. Mini-STAR as bail-out strategy for percutaneous coronary intervention of chronic total occlusion. Catheter Cardiovasc Interv 2012;79:30–40.

[43] Karatasakis A, Danek BA, Karacsonyi J, et al. Mid-term outcomes of chronic total occlusion percutaneous coronary intervention with subadventitial vs. intraplaque crossing: a systematic review and meta-analysis. Int J Cardiol 2018;253:29–34.

[44] Wilson WM, Walsh SJ, Bagnall A, et al. One-year outcomes after successful chronic total occlusion percutaneous coronary intervention: the impact of dissection re-entry techniques. Catheter Cardiovasc Interv 2017;90:703–12.

[45] Maeremans J, Dens J, Spratt JC, et al. Antegrade dissection and reentry as part of the hybrid chronic total occlusion revascularization strategy: a subanalysis of the RECHARGE registry (Registry of CrossBoss and Hybrid Procedures in France, the Netherlands, Belgium and United Kingdom). Circ Cardiovasc Interv 2017;10.

[46] Azzalini L, Dautov R, Brilakis ES, et al. Procedural and longer-term outcomes of wire- versus device-based antegrade dissection and re-entry techniques for the percutaneous revascularization of coronary chronic total occlusions. Int J Cardiol 2017;231:78–83.

[47] Danek BA, Karatasakis A, Karmpaliotis D, et al. Use of antegrade dissection re-entry in coronary chronic total occlusion percutaneous coronary intervention in a contemporary multicenter registry. Int J Cardiol 2016;214:428–37.

[48] Joyal D, Thompson CA, Grantham JA, Buller CEH, Rinfret S. The retrograde technique for recanalization of chronic total occlusions: a step-by-step approach. JACC Cardiovasc Interv 2012;5:1–11.

[49] Wu EB, Tsuchikane E, Lo S, et al. Retrograde algorithm for chronic total occlusion from the Asia Pacific Chronic Total Occlusion club. AsiaIntervention 2018;4:98–107.

[50] Huang Z, Zhang B, Chai W, et al. Usefulness and safety of a novel modification of the retrograde approach for the long tortuous chronic total occlusion of coronary arteries. Int Heart J 2017;58:351–6.

[51] Brilakis ES, Grantham JA, Rinfret S, et al. A percutaneous treatment algorithm for crossing coronary chronic total occlusions. JACC Cardiovasc Interv 2012;5:367–79.

[52] Galassi AR, Sianos G, Werner GS, et al. Retrograde recanalization of chronic total occlusions in Europe: procedural, in-hospital, and long-term outcomes from the multicenter ERCTO registry. J Am Coll Cardiol 2015;65:2388–400.

[53] Galassi AR, Sumitsuji S, Boukhris M, et al. Utility of intravascular ultrasound in percutaneous revascularization of chronic total occlusion: an overview. JACC Cardiovasc Interv 2016;9:1979–91.

[54] Galassi AR, Werner GS, Boukhris M, et al. Percutaneous recanalisation of chronic total occlusions: 2019 consensus document from the EuroCTO Club. EuroIntervention 2019;15:198–208.

[55] Laskey WK, Jenkins C, Selzer F, et al. Volume-to-creatinine clearance ratio: a pharmacokinetically based risk factor for prediction of early creatinine increase after percutaneous coronary intervention. J Am Coll Cardiol 2007;50:584–90.

[56] Werner GS, Surber R, Ferrari M, Fritzenwanger M, Figulla HR. The functional reserve of collaterals supplying long-term chronic total coronary occlusions in patients without prior myocardial infarction. Eur Heart J 2006;27:2406–12.

[57] Sachdeva R, Agrawal M, Flynn SE, Werner GS, Uretsky BF. The myocardium supplied by a chronic total occlusion is a persistently ischemic zone. Catheter Cardiovasc Interv 2014;83:9–16.

[58] Ladwiniec A, Cunnington MS, Rossington J, et al. Collateral donor artery physiology and the influence of a chronic total occlusion on fractional flow reserve. Circ Cardiovasc Interv 2015;8.

[59] Sachdeva R, Agrawal M, Flynn SE, Werner GS, Uretsky BF. Reversal of ischemia of donor artery myocardium after recanalization of a chronic total occlusion. Catheter Cardiovasc Interv 2013;82:E453–8.

[60] Xenogiannis I, Tajti P, Karmpaliotis D, et al. Intravascular imaging for chronic total occlusion intervention. Curr Cardiovasc Imaging Rep 2018;11:31.

[61] Kim BK, Shin DH, Hong MK, et al. Clinical impact of intravascular ultrasound-guided chronic total occlusion intervention with zotarolimus-eluting versus biolimus-eluting stent implantation: randomized study. Circ Cardiovasc Interv 2015;8:e002592.

[62] Tian NL, Gami SK, Ye F, et al. Angiographic and clinical comparisons of intravascular ultrasound- versus angiography-guided drug-eluting stent implantation for patients with chronic total occlusion lesions: two-year results from a randomised AIR-CTO study. EuroIntervention 2015;10:1409–17.

[63] Karacsonyi J, Alaswad K, Jaffer FA, et al. Use of intravascular imaging during chronic total occlusion percutaneous coronary intervention: insights from a contemporary multicenter registry. J Am Heart Assoc 2016;5.

[64] Zhang J, Gao X, Kan J, et al. Intravascular ultrasound versus angiography-guided drug-eluting stent implantation: the ULTIMATE trial. J Am Coll Cardiol 2018;72:3126–37.

[65] Danek BA, Basir MB, O'Neill WW, et al. Mechanical circulatory support in chronic total occlusion percutaneous coronary intervention: insights from a multicenter U.S. registry. J Invasive Cardiol 2018;30:81–7.

Chapter 22

Other complex lesions

In addition to calcification, thrombus, and CTOs, other complex lesion types include:

- Spontaneous coronary artery dissection (SCAD) (PCI Manual Online case 72),
- Stent failure (in-stent restenosis and stent thrombosis),
  CTO Manual Online cases: 19, 52, 54, 58, 72, 84, 91, 101; *PCI Manual* Online cases: 63, 89, 94
- Small and large vessels, and
- Long lesions.

22.1 Spontaneous coronary artery dissection

SCAD is defined as an epicardial coronary artery dissection that is not associated with atherosclerosis or trauma and is not iatrogenic [1]. The predominant mechanism of myocardial injury occurring as a result of SCAD is coronary artery obstruction caused by formation of an intramural hematoma (IMH) or intimal disruption rather than atherosclerotic plaque rupture or intraluminal thrombus [1].

22.1.1 Planning

SCAD should be suspected in young women, especially in the peripartum setting (Fig. 22.1) [1].

22.1.2 Monitoring

Given high risk of complications and hemodynamic deterioration, careful monitoring is needed throughout the procedure.

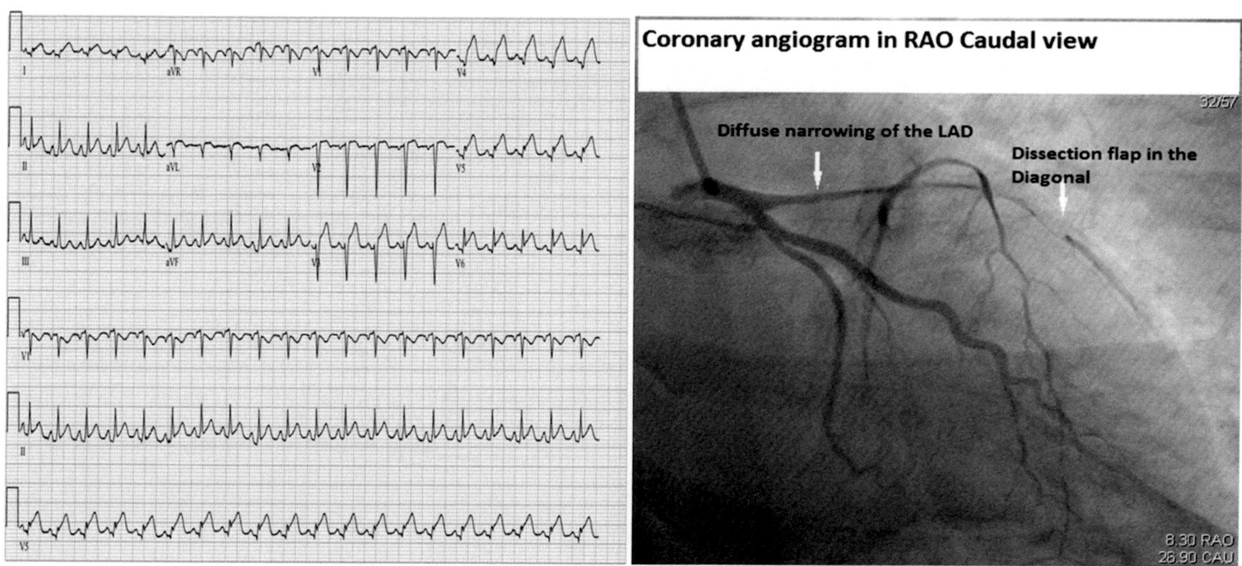

FIGURE 22.1 Acute anterior myocardial infarction in a term pregnant woman. *Courtesy of Dr. Abdul Hakeem.*

Manual of Percutaneous Coronary Interventions. DOI: https://doi.org/10.1016/B978-0-12-819367-9.00022-6

22.1.3 Medications

Anticoagulation and oral antiplatelet therapy is administered as in non-SCAD cases. Glycoprotein IIb/IIIa inhibitors and cangrelor should generally be avoided given risk of extending an IMH.

22.1.4 Access

Femoral access is preferred for coronary angiography of patients with suspected SCAD, given a threefold higher risk for catheter-induced iatrogenic dissection with radial access observed in one series [2].

22.1.5 Engagement

Coronary engagement should be performed with extreme care aiming for coaxial positioning, and avoiding deep coronary artery intubation [1].

22.1.6 Angiography

Strong contrast injections should be avoided to reduce the risk of iatrogenic dissection [1].

Identification of SCAD can be challenging and requires awareness and careful angiographic interpretation. There are three angiographic types of SCAD (Fig. 22.2) [3]:

Type 1: Classic appearance of multiple radiolucent lumens or arterial wall contrast staining.

Type 2: Diffuse stenosis that can be of varying severity and length (usually >20 mm). Variant 2A is diffuse arterial narrowing bordered by normal segments proximal and distal to the IMH, and variant 2B is diffuse narrowing that extends to the distal tip of the artery.

Type 3: Focal or tubular stenosis, usually <20 mm in length. Type 3 mimics atherosclerosis and requires intracoronary imaging (if safe) to diagnose SCAD.

22.1.7 Selecting target lesion(s)

PCI of SCAD lesions carries high risk of complications and is often quoted as "a temptation to be avoided." Conservative management is preferred (if feasible), however in patients with active ischemia or hemodynamic instability coronary revascularization may be required, as follows [1,3]:

- Stable patients without left main or proximal two-vessel coronary dissection: in-hospital observation for 3–5 days without revascularization.

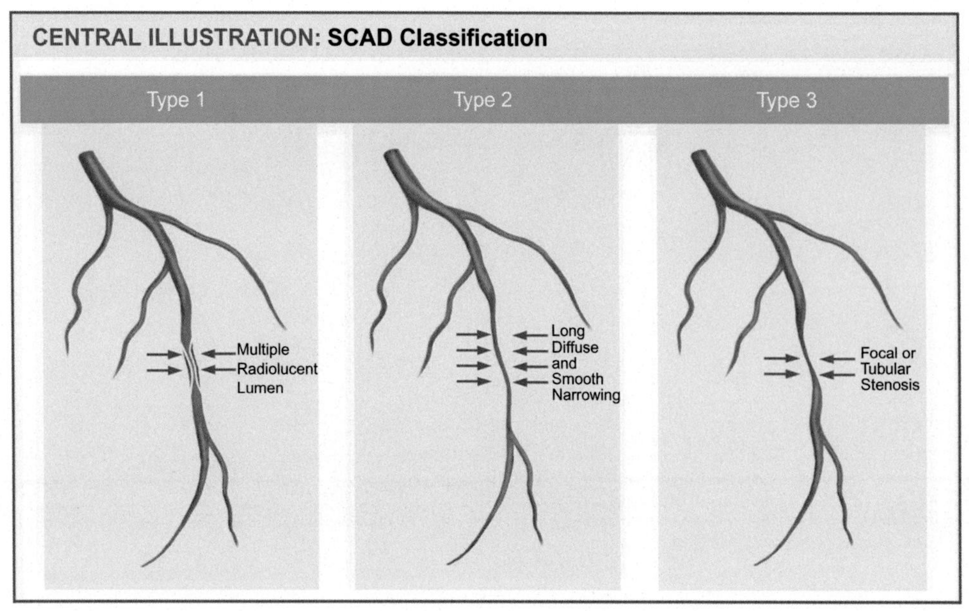

FIGURE 22.2 The Saw SCAD angiographic classification. *Reproduced with permission from Saw J, Humphries K, Aymong E, et al. Spontaneous coronary artery dissection: clinical outcomes and risk of recurrence. J Am Coll Cardiol 2017;70:1148–58. Copyright Elsevier.*

- Stable patients with left main or proximal two-vessel coronary dissection: CABG versus continued in-hospital observation.
- Unstable patients or active ischemia: PCI or CABG depending on anatomy.

22.1.8 Wiring

Given the risk of subintimal guidewire entry, polymer-jacketed guidewires should be avoided in patients with suspected SCAD.

22.1.9 Lesion preparation

Direct stenting has been proposed for SCAD lesions to minimize the risk of IMH expansion. A cutting balloon can sometimes be used to decompress the hematoma, but it carries a risk of perforation.

The cutting balloon has been used for fenestration of an IMH with luminal compression at multiple sites prior to stenting [4].

22.1.10 Stenting

Implanting long stents that extend 5–10 mm on both sides of the IMH has been proposed in SCAD patients to minimize the risk of subintimal hematoma expansion [1]. The stent diameter should generally be smaller than the size used in similar atherosclerotic lesions to avoid hematoma expansion.

22.1.11 Closure

Access closure is performed as described in Chapter 11: Access Closure.

22.1.12 Coronary physiology

Coronary physiology has limited role in SCAD cases.

22.1.13 Intravascular imaging

Intravascular imaging, especially OCT, can confirm the diagnosis of SCAD, but may also cause extension of the dissection and should, therefore, be used cautiously.

22.1.14 Hemodynamic support

Because of high risk of acute vessel closure during PCI of SCAD lesions, access to hemodynamic support devices, such as VA-ECMO is recommended (PCI Manual Online case 72).

22.2 Stent failure

Stent failure can manifest as stent thrombosis or in-stent restenosis [5]. Stent thrombosis is a life-threatening complication and usually presents as ST-segment elevation acute myocardial infarction. In-stent restenosis usually presents with stable angina but can sometimes present as an acute coronary syndrome.

Stent thrombosis is often caused by stent underexpansion or insufficient antithrombotic therapy. Stent thrombosis is rarely caused by stent fracture.

22.2.1 Planning

Special emphasis should be given to a detailed history of prior percutaneous coronary interventions, including stent types, stent sizes, prior stent failure (stent thrombosis or in-stent restenosis), and prior brachytherapy [that requires lifelong dual antiplatelet therapy due to continued risk of very late (>1 year) stent thrombosis].

Moreover, it is important to know the current antiplatelet therapy and compliance with the therapy, any recent antiplatelet therapy discontinuation and any recent noncardiac surgery.

22.2.2 Monitoring

Given high risk of complications and hemodynamic deterioration of stent thrombosis patients, careful monitoring is needed throughout the procedure.

22.2.3 Medications

Lifelong dual antiplatelet therapy is recommended after development of stent thrombosis unless the patient is at high risk of bleeding.

Cilostazol may decrease the risk of in-stent restenosis, but is contraindicated in patients with heart failure.

22.2.4 Access

Radial or femoral access can be used. Radial access is generally preferred in stent thrombosis patients presenting with a STEMI.

22.2.5 Engagement

Performed as described in Chapter 5: Coronary and Graft Engagement.

22.2.6 Angiography

Performed as described in Chapter 6: Coronary Angiography.

Using X-ray image enhancing technologies, such as the StentBoost (Philips) or CLEARstent (Siemens) (by placing a balloon within the occluded stent) can help identify stent fracture or stent gap.

22.2.7 Selecting target lesion(s)

Stent thrombosis requires immediate PCI, as is standard for STEMI patients.

The type of revascularization of in-stent restenotic lesions depends on the type of restenosis, prior restenotic events, overall coronary anatomy and feasibility of coronary bypass graft surgery.

22.2.8 Wiring

Stent thrombosis: wiring can be challenging due to thrombus and is performed as described in Chapter 8: Wiring and Chapter 20: Acute Coronary Syndromes—Thrombus, Section 20.9. Use of polymer-jacketed guidewires can facilitate crossing.

A unique potential challenge with wiring through prior stents (whether in the setting of stent thrombosis or in-stent restenosis) is guidewire passage under the struts of the previously implanted stent(s) (CTO PCI Manual case 143). This can manifest as difficulty advancing equipment and can often be prevented by advancing a knuckled guidewire through the prior stent.

Sometimes stent failure is due to stent fracture, which can be very challenging to wire through.

22.2.9 Lesion preparation

Meticulous lesion preparation (guided by intravascular imaging as described in Section 22.2.13) is critical prior to additional stent implantation. Stent underexpansion can lead to both restenosis and stent thrombosis and can be treated as described in Chapter 23: Balloon Uncrossable and Balloon Undilatable Lesions, Section 23.2 on balloon undilatable lesions.

Coronary brachytherapy (Section 30) is an additional treatment option for recurrent in-stent restenosis. Beta-radiation is administered within the restenosed coronary segment to reduce the risk of restenosis. Indefinite dual antiplatelet therapy is required afterwards due to increased risk of stent thrombosis.

22.2.10 Stenting

Stent thrombosis: repeat stenting may be needed if a suboptimal result is obtained after PCI of the stent thrombosis lesion.

In-stent restenosis: the first episode of in-stent restenosis is typically treated with repeat DES implantation, after confirming that the lesion is well dilated. Recurrent in-stent restenosis is best treated with brachytherapy or drug-coated balloons (which are not currently available in the United States) (Section 30.9.6).

If the mechanism of stent occlusion is stent undersizing or underexpansion (Section 23.2), intravascular imaging is critical for stent sizing and assessment of expansion of the new stent. High-pressure postdilatation with a noncompliant balloon is recommended.

If the mechanism of stent occlusion is stent fracture, a new stent is usually implanted.

If the mechanism of the stent occlusion is a gap, intravascular imaging and stent imaging enhancing technologies can help fully cover the gap and the old stent.

22.2.11 Closure

Access closure is performed as described in Chapter 11: Access Closure.

22.2.12 Coronary physiology

Coronary physiology can help assess the significance of stent failure lesions as described in Chapter 12: Coronary Physiology, although intracoronary imaging is preferred in most cases.

22.2.13 Intravascular imaging

Intravascular imaging, is essential for stent failure cases (both stent thrombosis and in-stent restenosis). It allows determination of the mechanism of stent failure, such as neointima formation, neoatherosclerosis, stent underexpansion or stent fracture, and helps guide treatment, as described in Chapter 13: Coronary Intravascular Imaging.

22.2.14 Hemodynamic support

Hemodynamic support may be needed, especially in cases of stent thrombosis, as described in Chapter 14: Hemodynamic Support.

22.3 Small and large vessels

Small (<2.5 mm) or large (≥ 4.0 mm) vessels pose additional challenges to PCI.

22.3.1 Planning

Planning is performed as described in Chapter 1: Planning.

22.3.2 Monitoring

Monitoring is performed as described in Chapter 2: Monitoring.

22.3.3 Medications

Medication administration is similar to standard PCI, as described in Chapter 3: Medications.

22.3.4 Access

Radial or femoral access can be used, as described in Chapter 4: Access.

22.3.5 Engagement

Performed as described in Chapter 5: Coronary and Graft Engagement.

22.3.6 Angiography

Angiography is performed as described in Chapter 6: Coronary Angiography.
Small vessels: administration of nitroglycerin is critical in patients with small coronary arteries, to ensure that the target arteries do not appear to be small due to spasm.
Large vessels: forceful contrast injection and possibly use of a guide catheter are important for completely filling large coronary vessels.

Sometimes, large vessel size is due to a focal (coronary artery aneurysm) or diffuse (coronary artery ectasia) dilation of coronary segments to a size at least 1.5 times bigger than the adjacent normal segment [6].

22.3.7 Selecting target lesion(s)

Small vessels: Both PCI and CABG are more challenging in patients with small coronary arteries. Stenting of small coronary arteries carries increased risk of restenosis, especially in patients with diabetes mellitus.

22.3.8 Wiring

Small vessels: aggressive wire manipulation should be avoided to minimize the risk of vessel injury and dissection.
Large vessels: wiring side branches through large vessels, can be challenging due to guidewire prolapse. Use of polymer jacketed guidewires and angulated microcatheters can assist wiring side branches in large vessels.

22.3.9 Lesion preparation

Small vessels: Atherectomy is generally avoided in small coronary vessels, especially if they are highly tortuous. Oversized balloons should also be avoided as they can cause dissections requiring longer stent length. Drug-coated balloons can also be used in this setting [7].
Large vessels: Use of intravascular imaging can be very helpful for optimal selection of balloon and stent size.

22.3.10 Stenting

Small vessels: the smallest currently available stent is the 2.0 mm Resolute Onyx stent. Minimizing stent length (while covering the target lesion) may help reduce the risk of restenosis.
Large vessels: the largest diameter DES currently available is the Resolute Onyx that is available in diameters up to 5.0 mm (and can be expanded up to 5.75 mm). The 4.0 mm Synergy stent can also be expanded up to 5.75 mm and the 3.5 and 4.0 mm Xience Sierra stents can be expanded up to 5.5 mm (Chapter 10: Stenting, Section 10.2.3 and Fig. 10.1). The stent diameter should be selected based on the distal reference diameter, followed by postdilation of the more proximal segment of the stent with a larger and shorter noncompliant balloon sized according to the proximal reference segment.

22.3.11 Closure

Access closure is performed as described in Chapter 11: Access Closure.

22.3.12 Coronary physiology

Small vessels: coronary physiology can be very helpful in determining the physiologic significance of lesions, as functional assessment depends both on the severity of the stenosis and on the size of the territory supplied by this vessel.
Large vessels: assessment of lesion severity can also be challenging in large vessels, potentially benefiting from physiologic assessment.

22.3.13 Intravascular imaging

Small vessels: Intravascular imaging can help determine the true size of the target vessel and determine whether there is positive remodeling, which may enable placement of larger stents.
Large vessels: Intravascular imaging can help accurately determine the true size of the vessel. IVUS is preferred over OCT, as complete vessel filling with contrast can be challenging in large vessels.

22.3.14 Hemodynamic support

Hemodynamic support may be needed, as described in Chapter 14: Hemodynamic Support.

22.4 Long lesions

22.4.1 Planning

Planning is performed as discussed in Chapter 1: Planning.

22.4.2 Monitoring

Monitoring is performed as described in Chapter 2: Monitoring.

22.4.3 Medications

Medication administration is similar to standard PCI, as described in Chapter 3: Medications.

22.4.4 Access

Radial or femoral access can be used, as described in Chapter 4: Access.

22.4.5 Engagement

Performed as described in Chapter 5: Coronary and Graft Engagement.

22.4.6 Angiography

Performed as descried in Chapter 6: Coronary Angiography.

22.4.7 Selecting target lesion(s)

The presence of long coronary lesions is associated with increased risk of restenosis after stenting, which should be taken into account before deciding on the type (PCI vs CABG) of coronary revascularization.

22.4.8 Wiring

Performed as described in Chapter 8: Wiring. Aggressive wire manipulation should be avoided to reduce the risk of dissection or vessel injury.

22.4.9 Lesion preparation

Performed as described in Chapter 9: Lesion Preparation.

22.4.10 Stenting

The longest stent available in the United States is currently 40 mm (Orsiro, Biotronik). Hence, multiple stents may be needed for stenting long coronary lesions, such as chronic total occlusions.

22.4.11 Closure

Access closure is performed as described in Chapter 11: Access Closure.

22.4.12 Coronary physiology

Coronary physiology can help determine the location of the stenosis by doing a pullback: the presence of a step-up may suggest that the lesion is more focal than apparent by angiography.

22.4.13 Intravascular imaging

Intravascular imaging can facilitate treatment of long coronary lesions that have higher risk of restenosis, by accurately measuring the proximal and distal reference vessel diameter and the lesion length. IVUS-guided PCI of long lesions (≥28 mm in length) was associated with lower incidence of subsequent major adverse cardiac events compared with angiography-only guided PCI in the IVUS-XPL trial [8].

22.4.14 Hemodynamic support

Hemodynamic support may be needed, as discussed in Chapter 14: Hemodynamic Support.

References

[1] Hayes SN, Kim ESH, Saw J, et al. Spontaneous coronary artery dissection: current state of the science: a scientific statement from the American Heart Association. Circulation 2018;137:e523–57.

[2] Saw J, Aymong E, Sedlak T, et al. Spontaneous coronary artery dissection: association with predisposing arteriopathies and precipitating stressors and cardiovascular outcomes. Circ Cardiovasc Interv 2014;7:645–55.

[3] Saw J, Humphries K, Aymong E, et al. Spontaneous coronary artery dissection: clinical outcomes and risk of recurrence. J Am Coll Cardiol 2017;70:1148–58.

[4] Bresson D, Calcaianu M, Lawson B, Jacquemin L. Coronary artery fenestration as rescue management of intramural haematoma with luminal compression. Catheter Cardiovasc Interv 2019;94:E17–19.

[5] Torrado J, Buckley L, Duran A, et al. Restenosis, stent thrombosis, and bleeding complications: navigating between Scylla and Charybdis. J Am Coll Cardiol 2018;71:1676–95.

[6] Kawsara A, Nunez Gil IJ, Alqahtani F, Moreland J, Rihal CS, Alkhouli M. Management of coronary artery aneurysms. JACC Cardiovasc Interv 2018;11:1211–23.

[7] Megaly M, Rofael M, Saad M, et al. Outcomes with drug-coated balloons in small-vessel coronary artery disease. Catheter Cardiovasc Interv 2019;93:E277–86.

[8] Hong SJ, Kim BK, Shin DH, et al. Effect of intravascular ultrasound-guided vs angiography-guided everolimus-eluting stent implantation: the IVUS-XPL randomized clinical trial. JAMA 2015;314:2155–63.

Chapter 23

Balloon uncrossable and balloon undilatable lesions

23.1 Balloon uncrossable lesions

CTO Manual Online cases: 1, 5, 15, 18, 27, 30, 31, 47, 49, 52, 53, 57, 73, 124, 126
PCI Manual Online case: 17, 64, 99
Balloon uncrossable lesions are lesions that cannot be crossed with a balloon after successful guidewire crossing. Fig. 23.1 outlines a step-by-step algorithm for approaching such lesions.

23.1.1 Small balloon

- The first step when trying to cross a balloon uncrossable lesion is to use a small balloon (Section 30.9.2).
- Use single marker, rapid exchange compliant balloons with a low crossing profile (1.0, 1.20, 1.25, and 1.5 mm in diameter) and long length (20–30 mm). The balloon profile is highest at the marker segment, hence longer balloons may allow for deeper lesion penetration with the lower profile segment of the balloon before the balloon marker reaches the uncrossable segment of the lesion. The Blimp Balloon (IMDS, not available in the United States) may also help crossing such lesions.
- The Sapphire Pro 1.0 mm (OrbusNeich), the Ryurei 1.0 mm (Terumo, not available in the United States), the Ikazuchi Zero (Kaneka, not available in the United States) and the NIC Nano 0.85 mm balloon (SIS, not available in the United States) are the lowest-profile balloons currently available and are, thus, the first choice for balloon uncrossable lesions.
- Other options are the Threader (1.2 mm, Boston Scientific) and the Takeru (Terumo) balloon that have low crossing profiles and stiff shafts to facilitate advancement though challenging lesions.
- The Threader (Boston Scientific) has a hydrophilic coating and 0.017 in. lesion entry profile. The Threader is available in both rapid exchange and over-the-wire versions. The rapid exchange Threader is preferred to the over-the-

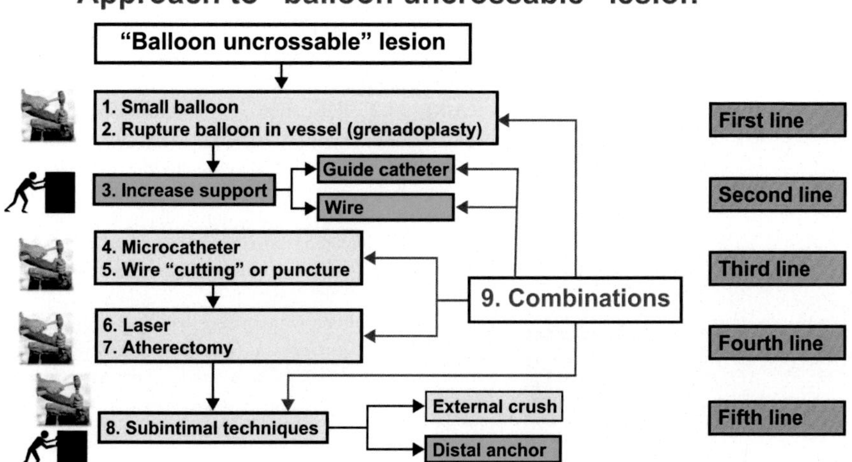

FIGURE 23.1 Algorithm for crossing a balloon uncrossable lesion. *Yellow*, plaque modification; *green*, increase support.

Manual of Percutaneous Coronary Interventions. DOI: https://doi.org/10.1016/B978-0-12-819367-9.00023-8

wire version for balloon uncrossable lesions, as it has more penetrating capacity (likely due to stiffer shaft). However, the over-the-wire Threader allows guidewire changes and contrast injection.

- The Glider balloon (Teleflex) has a beveled tip and was developed to cross through the struts of stents during bifurcation stenting, but can also be useful in difficult to cross lesions, as it can be torqued to present alternative tip configurations to the lesion.
- The Blimp balloon (Section 30.9.3.4) may also help cross some of the balloon uncrossable lesions.
- If the balloon stops advancing, it can be inflated while maintaining forward pressure. This may dilate the entry to the lesion and allow lesion crossing, sometimes even with the same balloon ("balloon wedge" technique).
- If the balloon fails to advance, consider balloon rotation. Avoid rotating more than two to three times in each direction so as to prevent the balloon from getting "stuck" on the guidewire.
- If the balloon fails to advance after inflation, the operator can reshape it using his or her fingers while applying vacuum, or try a new small balloon (balloons lose their original profile after inflation), or one manufactured by another company, as different crossing profile and tip characteristics may assist in crossing. Rapid exchange balloon catheters allow more pushability into the lesion.
- Alternatively, one can attempt crossing with a larger 2.5–3.0 mm diameter rapid exchange balloon. Sometimes inflation with a larger diameter balloon just proximal to the lesion will disrupt the architecture of the lesion entry enough to allow subsequent passage of a small profile balloon or microcatheter.

What can go wrong?

1. Guide catheter and guidewire position can be lost during attempts to advance the balloon or microcatheter. Carefully monitor the guide catheter position and stop advancing if the guide catheter starts backing out of the coronary ostium or if the distal guidewire position is being compromised. Forceful back and forth balloon manipulation can cause ostial dissection, especially if there is a preexisting ostial lesion.
2. Injury of the distal target vessel can occur (dissection or perforation) due to significant distal guidewire movement ("see-saw" action of wire with forward push and retraction of the balloon), especially when stiff (such as Confianza Pro 12, Asahi) or polymer-jacketed (such as the Pilot 200, Abbott Vascular) guidewires are used.
3. Balloon entrapment can occur within the lesion, although this is highly unlikely.

23.1.2 Grenadoplasty (intentional balloon rupture; also called "balloon assisted microdissection" or "BAM")

This is a simple, safe, and often effective technique, that is increasingly being used in the treatment algorithm for balloon uncrossable lesions [1].

How?

A small (usually 1.0–1.5 mm) balloon is advanced as far as possible into the lesion and inflated at high pressure until it ruptures (Fig. 23.2) [2]. When the balloon ruptures, suction should immediately be applied through the inflating device to avoid unnecessary barotrauma to the vessel. The balloon rupture can often sufficiently modify the plaque, resulting in subsequent successful crossing with a new balloon.

What can go wrong?

1. Proximal vessel dissection and perforation. This is extremely unlikely when small (1.0–1.5 mm) balloons are used. Larger balloons (≥ 2.0 mm) should NOT be used for grenadoplasty.

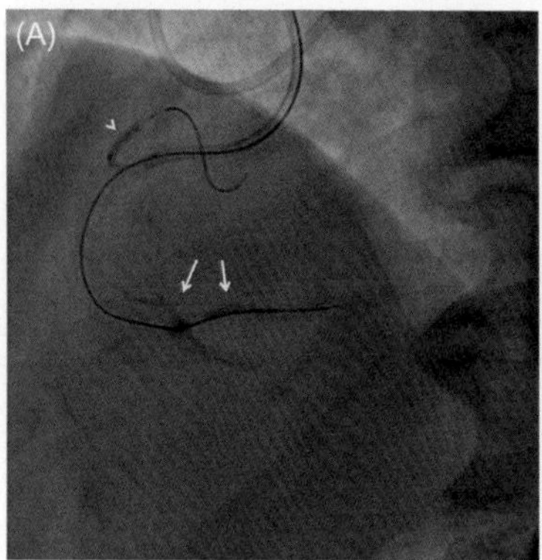

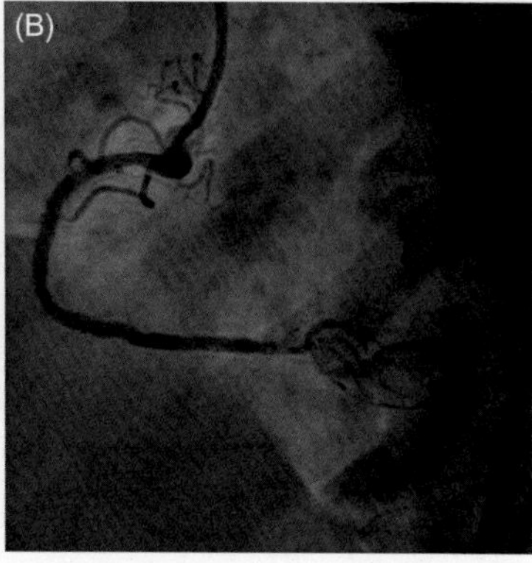

FIGURE 23.2 Illustration of the grenadoplasty technique to cross a distal right coronary artery balloon uncrossable CTO. The lesion could not be crossed despite using an 8 Fr Amplatz guide and an anchor balloon (*arrowhead*, panel A). A 1.2 mm balloon was ruptured with contrast spreading proximally and distally (*arrows*, panel A). A Finecross microcatheter could then be advanced through the lesion followed by wire exchange and a 2.0 mm balloon with an excellent final result (panel B). *Courtesy of Dr. Gabriele Gasparini. Reproduced with permission from Brilakis ES. Manual of coronary chronic total occlusion interventions. A step-by-step approach. 2nd ed. Cambridge, MA: Elsevier; 2017 (Figure 8.2).*

2. The balloon should be meticulously prepared to empty all air and hence minimize the risk of air embolism.
3. Watching the indeflator rather than the screen allows more rapid deflation of the balloon immediately upon rupture. This will reduce the chance of pinhole contrast-induced vessel injury from the rupture site of the balloon.
4. One may encounter difficulty removing the ruptured balloon. In some cases, the ruptured balloon becomes entangled with the guidewire, requiring removal of both, hence losing guidewire position.

23.1.3 Increase support

Increasing support is critical for crossing a balloon uncrossable lesion and is discussed in detail in Section 9.5.8. Key elements for increasing support are the guide catheter and the guidewire.

23.1.3.1 Guide catheter

Better guide catheter support increases the likelihood of successful balloon or microcatheter crossing. Guide support can be increased by using larger guide catheters with supportive shapes, using femoral access, forward push or deep intubation, guide catheter extensions [3] and various anchoring techniques [4], as described in detail in Sections 9.5.8.1–9.5.8.7. Changing guide catheter after guidewire crossing can be challenging and may lead to guidewire position loss, which may not be acceptable in some cases (such as chronic total occlusions).

23.1.3.1.1 Guide catheter extension

How?

A Guideliner (Teleflex), Guidezilla (Boston Scientific), Guidion (IMDS), or Telescope (Medtronic) guide catheter extension (Section 30.3) is advanced into the vessel, enhancing guide catheter support and the pushability of balloons/ microcatheters. In a randomized trial, use of a 5-in-6 guide catheter extension was more effective and efficient in facilitating the success of transradial PCI for complex coronary lesions, as compared with buddy-wire or balloon-anchoring [5].

What can go wrong?

1. Guidewire and guide catheter position loss or distal vessel injury during attempts to advance the guide catheter extension.
2. Guide catheter extension advancement can cause ostial or mid target vessel dissection [3]. When a guide extension catheter is advanced distally into a wedged position, a dampened pressure tracing is observed, and guide catheter injections are avoided due to high risk of hydraulic dissection and damage to the proximal vessel.
3. Dislodgement of the guide catheter extension distal marker can also occur [6].
4. When equipment is advanced through the guide catheter extension, another potential complication is deformation or separation (stripping) of the stent from the stent balloon during attempts to advance it through the guide catheter extension proximal "collar." This may occur around the subclavian curve with radial access, hence consider using guide catheter extensions with a long (40 cm instead of 25 cm) cylinder length in radial access cases.

23.1.3.1.2 Anchor strategies

How?

1. ***Side branch anchor technique*** (Fig. 23.3). A workhorse guidewire is advanced into a side branch (usually a conus or acute marginal branch for the right coronary artery or a diagonal for the left anterior descending artery), followed by a small balloon (usually 1.5–2.0 mm in diameter depending on the side branch vessel size). The balloon is inflated usually at 6–8 atm "anchoring" the guide into the vessel and enhancing advancement of balloons, stents or microcatheters. Sometimes, patients may develop chest pain during inflation of the balloon in the side branch [2,4,7].
2. **Buddy wire stent anchor** (Section 9.5.8.1.7). If the proximal vessel requires stenting, a buddy wire can be inserted and a stent deployed over it, effectively "trapping" the buddy wire, which then provides strong guide catheter support.
3. **Both antegrade and retrograde anchoring** can be performed in challenging retrograde CTO PCI cases [7].

What can go wrong?

1. Guidewire and guide catheter position loss or distal vessel injury can occur during attempts to advance an anchor balloon.

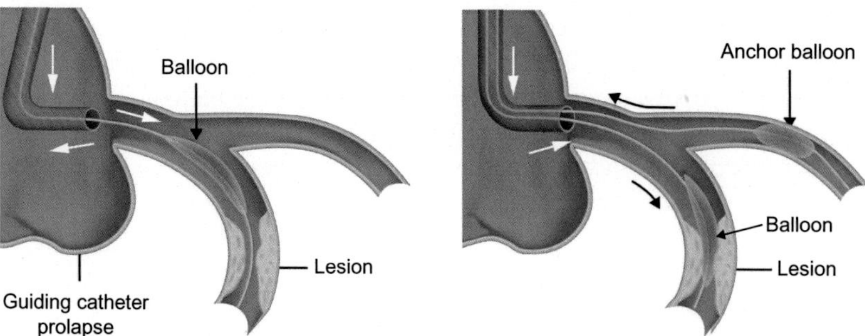

FIGURE 23.3 Illustration of the side branch anchor technique for treating a balloon uncrossable lesion. *Modified with permission from Brilakis ES. Manual of coronary chronic total occlusion interventions. A step-by-step approach. 2nd ed. Cambridge, MA: Elsevier; 2017 (Figure 8.4)*

2. A side branch anchor can cause injury or dissection of the side branch, however this is infrequent and usually does not lead to significant adverse consequences.
3. Perforation of the side branch may rarely occur. Oversizing of the anchor balloon should be avoided to minimize the risk of both side branch perforation and dissection and a workhorse wire should be used to minimize wire-related vessel injuries.
4. Potential risks of the "buddy wire stent anchor" technique include: inability to remove the buddy wire or failure to advance equipment through the proximal stent.

23.1.3.2 Guidewire

Guidewire exchange may not be feasible or desirable in difficult to cross lesions, such as chronic total occlusions. However, leaving the original guidewire in place and crossing the lesion with a second guidewire may succeed and can sometimes be performed through a second guide catheter [8].

Using a support guidewire, a Wiggle guidewire, or the deep wiring technique can significantly facilitate crossing of balloon uncrossable lesions, as described in Section 9.5.8.2.

23.1.4 Microcatheter advancement

How?

1. The concept behind use of a microcatheter is that advancement of a microcatheter through the lesion can modify the occlusion, enabling subsequent crossing with a balloon.
2. There are several microcatheters that can be utilized as described in Section 30.6.
3. The following microcatheters are especially designed for balloon uncrossable lesions (Section 30.6.5):
   - The Tornus catheter (Asahi Intecc) was designed for advancing through calcified and difficult to penetrate lesions and should be advanced using counterclockwise rotation and withdrawn using clockwise rotation [9].
   - The Turnpike Spiral and Turnpike Gold catheters (Teleflex) were also designed with threads to "screw into the lesion" and modify it. In contrast to the Tornus catheter, they are advanced by turning clockwise and withdrawn by turning counterclockwise.
4. Standard microcatheters can also be used:
   - The Corsair Pro and Corsair XS microcatheters (Asahi Intecc) (Section 30.6.1) can be advanced by rotating in either direction (in contrast to the Tornus catheter).
   - The Turnpike and Turnpike LP catheters (Teleflex) (Sections 30.6.1 and 30.6.2) can also be rotated in either direction.
   - Similarly, the Finecross (Terumo) or MicroCross 14 (Boston Scientific) (Section 30.6.2) can be rotated in either direction, although rotation may be challenging and there is a risk of tip dislodgement if aggressively torqued.
   - The Caravel (Asahi Intecc) is a low-profile microcatheter, but is not designed for aggressive torquing as is often done with the Corsair and Turnpike family of microcatheters and should generally not be used in balloon uncrossable lesions to minimize the risk of tip fracture and separation.
5. If successful advancement of a microcatheter is achieved, a balloon can often subsequently cross the lesion. Alternatively, the guidewire can be exchanged for a more supportive guidewire or an atherectomy wire (*PCI Manual* Online case: 64), if the latter is planned as the next lesion preparation step.

What can go wrong?

1. Guide catheter and guidewire position may be lost with aggressive pushing of the microcatheters.
2. Distal vessel injury can occur from uncontrolled guidewire movement during microcatheter advancement attempts.
3. The microcatheter can get damaged if overtorqued, leading to catheter tip entrapment or tip/shaft fracture. If the tip of the microcatheter breaks off it can become entrapped in the lesion. Rotation should not exceed 10 turns before allowing the catheter to "unwind." A guidewire should always be kept within the microcatheter lumen to prevent kinking and possible entrapment. If the tip of the microcatheter breaks off it can become entrapped in the lesion.
4. Rarely, excessive manipulation of the microcatheter can disrupt the device and/or the guidewire and lock both devices together, requiring withdrawal of both. A polymer-jacketed guidewire can sometimes be advanced through the track that has been established, allowing the crossing attempts to restart.

23.1.5.1 Wire-cutting

See Fig. 23.4 [10].

How?

1. A second guidewire is advanced through the lesion (which may be challenging to accomplish).
2. A balloon is advanced over the first guidewire, as far as possible into the proximal cap and inflated.
3. The second guidewire is withdrawn while the balloon is inflated, effectively "cutting" the proximal cap and modifying it.
4. After deflation and removal of the original balloon, a new balloon is advanced over the first guidewire, often successfully crossing the modified lesion.
5. A modified version of this technique is the "*seesaw wire-cutting technique*." In this technique two balloons are advanced over the two guidewires (one balloon over each wire). One of the balloons is first advanced as distally as possible and inflated, pressing the other wire against the proximal cap. The first balloon is then pulled back and the other balloon is advanced distally and inflated producing a similar cutting effect to modify the cap on the other side. This process is repeated multiple times until one of the balloons crosses the lesion. A retrospective study of 80

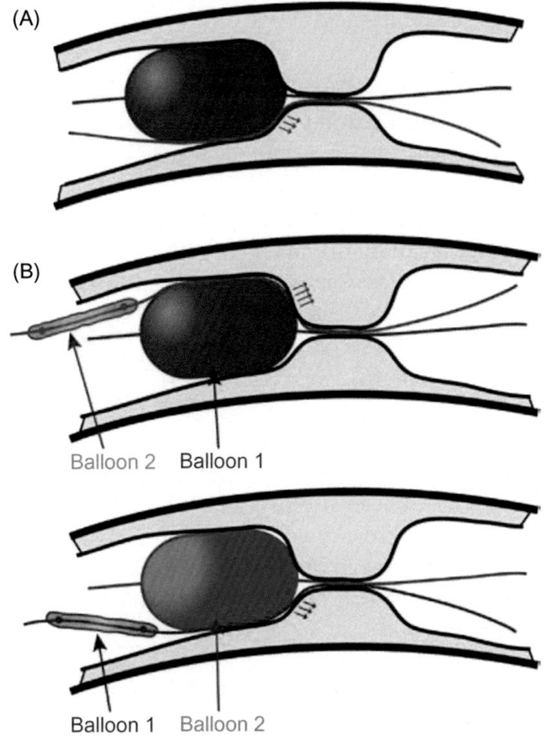

FIGURE 23.4 Illustration of the wire-cutting (A) and seesaw wire-cutting (B) techniques. *Reproduced with permission from Brilakis ES. Manual of coronary chronic total occlusion interventions. A step-by-step approach. 2nd ed. Cambridge, MA: Elsevier; 2017 (Figure 8.3).*

patients found this technique to be associated with higher device and procedural success rates and shorter procedure time as compared with use of the Tornus catheter [11,12].

What can go wrong?

1. Withdrawal of the second guidewire may result in deep intubation of the guide catheter, potentially leading to ostial or proximal vessel dissection. This complication can be prevented by carefully watching the guide catheter during withdrawal of the second guidewire, and promptly backing it out of the vessel as needed. If, however, a dissection occurs it could be used for subintimal wire advancement and subintimal lesion modification (or subintimal distal anchor) as described in Section 23.1.8, provided that the operator has experience with these techniques.

23.1.5.2 Wire puncture

How?

1. A stiff guidewire (such as Hornet 14 [Boston Scientific] or Confianza Pro 12 [Asahi Intecc]) is advanced to the lesion over a microcatheter or over-the-wire balloon.
2. The lesion entry is "punctured" or "jabbed" several times with the stiff guidewire tip, which may modify the lesion and facilitate equipment crossing.
3. Advancing a supportive second guidewire through the balloon uncrossable lesion may in itself facilitate lesion crossing with a balloon or microcatheter.

What can go wrong?

1. Perforation, although this is unlikely even if the guidewire exits the vessel architecture (unless it is followed by a balloon or microcatheter).

23.1.6 Laser

- Online video: "Impact of contrast on laser activation"
- *PCI Manual* Online cases: 23, 64
- *CTO Manual* Online cases: 5, 18, 27, 47, 52, 73

Laser atherectomy (Section 30.11) is increasingly utilized for balloon uncrossable and balloon undilatable lesions, in part because laser atherectomy can be performed over any standard 0.014 in. guidewire in contrast to rotational and orbital atherectomy, which require dedicated guidewires [13].

How?

1. The *0.9 mm excimer laser atherectomy catheter* should be used, usually at maximum repetition and fluence levels (repetition rate 80 Hz and fluence of 80 mJ/mm^2) [14]. The coronary laser catheter is programmed to minimize the risk of coronary ischemia, with laser activation designed to work with a slow, continuous saline infusion [15], however the peripheral "turbo elite" catheter (Philips) is generally preferred over the coronary catheter since it does not automatically stop after 10 seconds of treatment. The laser can modify the uncrossable segment and facilitate balloon entry into the lesion [16].
2. The laser console requires a *5-minute warming period* as well as calibration prior to use. Anticipating its potential use and setting up the system early can minimize delays and improve the efficiency of the procedure.
3. *A variation of the above technique* is to perform laser while injecting contrast through the guide catheter or via the over-the-wire side port of the laser catheter itself (if an over-the-wire laser is used, which is more commonly done in peripheral arteries and rarely in coronary arteries). Laser activation with simultaneous contrast injection can cause profound plaque modification through the acousticomechanical effect of the rapidly exploding bubbles. Laser with contrast, however, increases the risk of vessel injury and/or perforation and should generally only be done within previously deployed stents.

What can go wrong?

1. Vessel perforation, especially when performing laser in tortuous coronary vessels (*PCI Manual* Online case: 55). The risk may be higher when the high laser settings are used (repetition rate 80 Hz and fluence of 80 mJ/mm^2) and when contrast is injected; the risk is lower when laser is activated within previously placed stents.

2. Laser activation over polymer-jacketed guidewires can result in polymer damage, making the wire "sticky" or "gummy."

23.1.7 Atherectomy

How?

1. Rotational or orbital atherectomy (Section 30.10 and Chapter 19: Calcification) can greatly facilitate lesion crossing with a balloon. These atherectomy modalities, however, require wire exchange for a dedicated guidewire (Rotawire floppy or Rotawire extra support for rotational atherectomy and ViperWire Advance or ViperWire Advance Flex Tip for orbital atherectomy), which may not always be feasible through a balloon uncrossable lesion [17].
2. If no other maneuver is successful in crossing the balloon uncrossable lesion, it is sometimes possible to "bury" a microcatheter as far as possible into the lesion, pull the original guidewire, and attempt to rewire with a dedicated atherectomy guidewire. If rewiring is successful, then atherectomy can be performed.
3. Rotational atherectomy may be advantageous over the current version of orbital atherectomy in balloon uncrossable lesions, as it provides forward "end-on" cutting in contrast to orbital atherectomy that provides "sideways" cutting.
4. Atherectomy will differentially cut calcific tissue but will not cut through elastic tissue, such as the adventitia.

What can go wrong?

1. Loss of guidewire position across the lesion (if the guidewire has to be removed and replaced with a dedicated atherectomy guidewire) that may fail to recross the lesion.
2. Vessel perforation, which is why a small diameter rotational atherectomy burr (1.25–1.50 mm) is preferred. The risk of perforation may be higher with atherectomy in the subintimal space.
3. Burr entrapment upon forceful forward advancement. The burr should be advanced in a repetitive and gentle manner avoiding forceful "wedging" into the occlusion with burr deceleration. Prevention and management of this complication is discussed in Chapter 19, Calcification, Section 19.9.5.7.3.

23.1.8 Subintimal techniques

CTO Manual Online case: 1

Subintimal techniques (Fig. 23.5) can significantly facilitate treatment of "balloon uncrossable" lesions, but the operators need to be experienced in the use of these CTO PCI techniques.

23.1.8.1 External crush

How?

CTO Manual Online case: 15

A second guidewire (antegrade or retrograde) is advanced subintimally around the balloon uncrossable segment of the lesion. Alternatively the CrossBoss catheter can be used to achieve subintimal position and then be exchanged for a guidewire (Fig. 23.5, panels A, B, and C). A balloon is advanced over the subintimal guidewire next to the lesion and inflated (usually at 8–10 atm), "crushing" the plaque from the outside (Fig. 23.5, panel D1). This can often sufficiently modify the plaque to allow passage of a balloon over the guidewire that had previously entered the distal true lumen [18,19].

23.1.8.2 Subintimal distal anchor

In a variation of the subintimal external crush technique entitled "subintimal distal anchor" a second guidewire is advanced subintimally distal to the uncrossable lesion. A balloon is advanced over the subintimal guidewire distal to the CTO. The balloon is inflated distally, "anchoring" the true lumen guidewire, and enabling antegrade delivery of a balloon over the true lumen guidewire (Fig. 23.5, panel D2) [20].

What can go wrong?

1. The subintimal techniques require subintimal wire crossing, which may not always be feasible (e.g., the guidewire may track side branches).

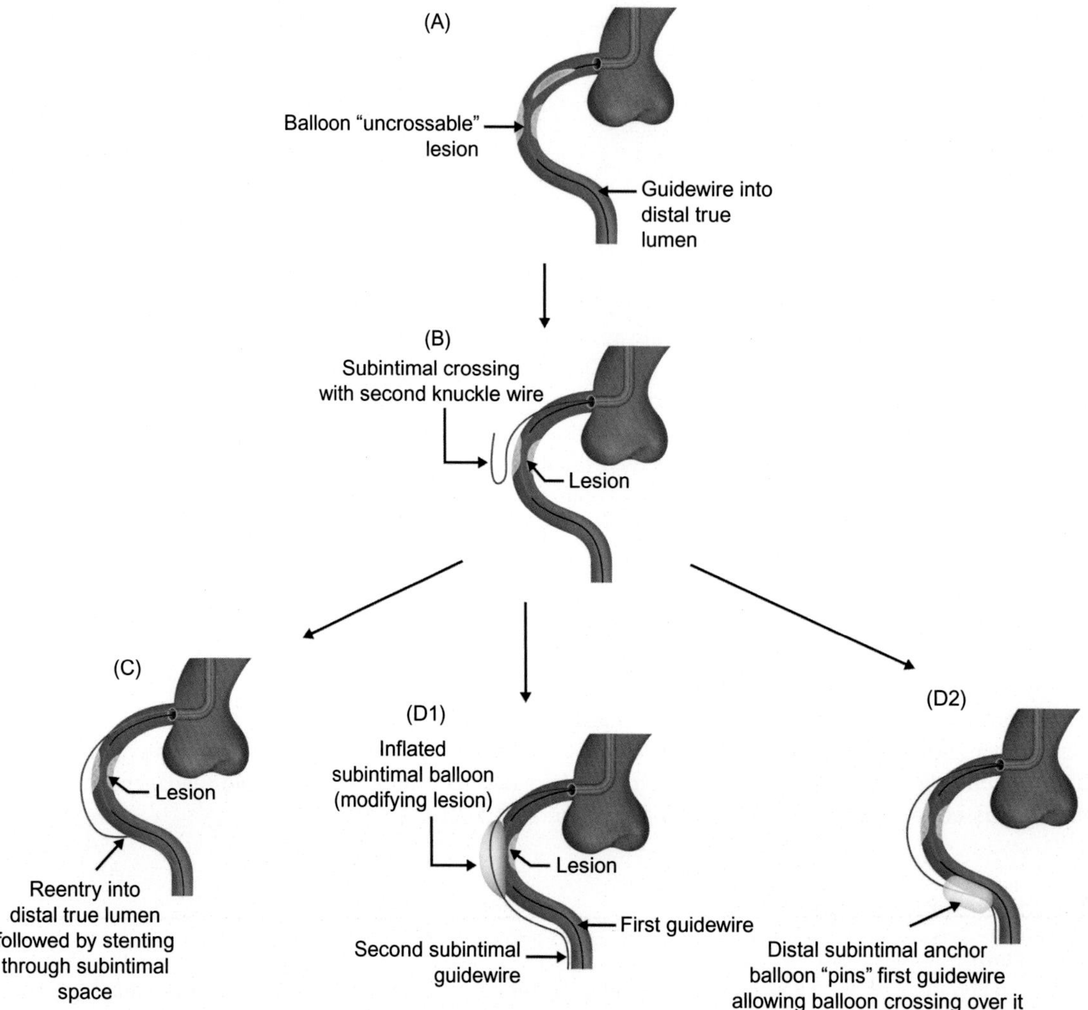

FIGURE 23.5 Illustration of subintimal crossing techniques for crossing a "balloon uncrossable" lesion. If a balloon cannot cross the lesion after successful guidewire crossing (panel A), a second guidewire is used to subintimally cross the lesion (panel B) and reenter into the distal true lumen (panel C). A balloon can be inflated over the subintimal guidewire next to the balloon uncrossable lesion to "crush" and modify it (panel D1), or distal to the lesion to "anchor" the true lumen guidewire and allow balloon crossing over the true lumen guidewire (panel D2). *Modified with permission from Brilakis ES. Manual of coronary chronic total occlusion interventions. A step-by-step approach. 2nd ed. Cambridge, MA: Elsevier; 2017 (Figure 8.6).*

2. Positioning of the subintimal guidewire within the vessel "architecture" should always be confirmed to prevent perforation, for example, if the guidewire enters side branches or if it exits the vessel and enters the pericardium. Confirmation of guidewire position is usually done using contralateral injection in two orthogonal projections.
3. Subintimal crossing and balloon inflation in the subintimal space can cause a large subintimal hematoma that may compress the distal true lumen.

23.1.9 Simultaneous versus sequential use of strategies

Simultaneous application of lesion-modification strategies and techniques that increase guide catheter support can enhance the likelihood of successful balloon crossing. For example, the "Anchor-Tornus" [21], "Proxis-Tornus" [22], and "Anchor-Laser" (Fig. 23.6) [14] techniques have been described for crossing balloon uncrossable lesions. In addition to simultaneous application, sequential application of various techniques is used until a final successful outcome is achieved [20,23]. An operator must be creative in using the above strategies in various combinations to achieve the desired goal. This process can be difficult and laborious, but also very rewarding.

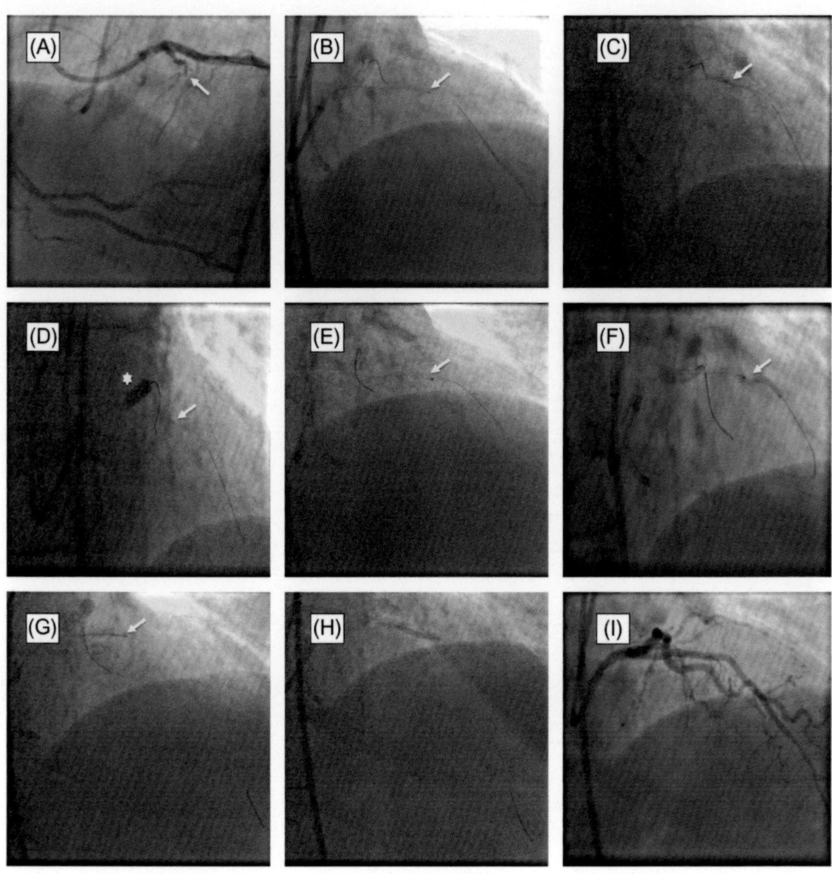

FIGURE 23.6 (Panel A) Dual injection demonstrates the mid LAD CTO (*arrow*) involving a semiambiguous cap, short length (<20 mm), significant calcification and a distal vessel filling mainly via septal collaterals from the right coronary artery. (Panel B) Successful antegrade wire escalation, but the microcatheter (*arrow*) could not be advanced through the lesion due to severe calcification. (Panels C and D) Use of Threader catheter alone (Panel C) and in conjunction with a 4.0 mm anchor balloon (*yellow star*) in the circumflex stent (Panel D). (Panel E) Laser (*arrow*). (Panel F) Severe balloon underexpansion. (Panel G) Orbital atherectomy (*arrow*). (Panel H) Adequate balloon expansion. (Panel I) Final angiography after stenting. *Reproduced with permission from Sandoval Y, Lobo AS, Tajti P, Brilakis ES. Laser-assisted orbital or rotational atherectomy: a hybrid treatment strategy for balloon-uncrossable lesions. Hellenic J Cardiol 2018 (Figure 2). Copyright Elsevier.*

23.2 Balloon undilatable lesions

CTO Manual Online cases: 26, 86, 124

PCI Manual Online cases: 8, 15, 17, 21, 23, 67, 99, 100

Balloon undilatable lesions are lesions that cannot be expanded despite multiple high-pressure balloon inflations. Balloon undilatable lesions often have severe calcification and are also often balloon uncrossable (Section 23.1).

Prevention:

Balloon undilatable lesions are more frequent in CTOs than non-CTO lesions [24]. It is important to avoid implanting a stent in a balloon undilatable lesion. Adequate predilation with a balloon sized according to the vessel reference diameter is critical before stenting a coronary lesion to ensure proper stent expansion, especially when the lesion is severely calcified. Intravascular imaging can be useful in determining the appropriate predilation balloon size and plaque characteristics that may benefit from atherectomy [25,26]. If the lesion is resistant, additional predilation and/or atherectomy should be pursued prior to stent deployment.

Treatment:

Fig. 23.7 outlines an algorithm for approaching balloon undilatable lesions.

De novo versus ISR

A key determinant of treatment options for balloon undilatable lesions is whether the undilatable lesion is de novo or is within a stent (implanted during the same procedure or during a prior procedure). Orbital atherectomy should not be done in recently implanted stents [27] and rotational atherectomy carries increased risk of burr entrapment and stent distortion when performed within recently implanted stents [28]. Laser with simultaneous contrast injection can often expand in-stent undilatable lesions (ESLAP = Extra-Stent Laser-Assisted Plaque modification) [13], but is infrequently used in de novo lesions because it can lead to dissections or perforations.

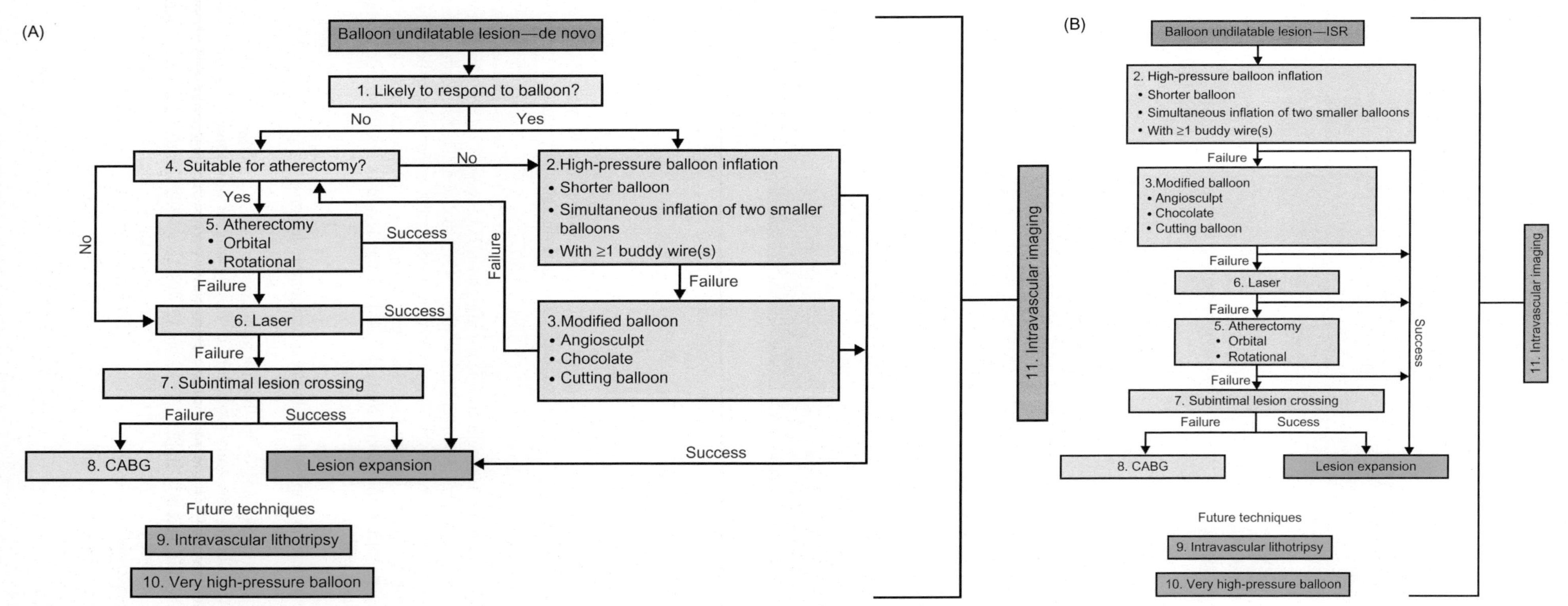

FIGURE 23.7 Algorithm for dilating a de novo (panel A) or in-stent (panel B) balloon undilatable lesion.

23.2.1 Likely to respond to balloon?

Heavily calcified lesions may be less likely to respond to balloon dilation. Intravascular imaging can help determine whether balloon angioplasty is unlikely to be successful. The combination of the following three OCT findings have been associated with stent underexpansion: (1) maximum calcium angle >180°; (2) maximum calcium thickness >0.5 mm; and (3) calcium length >5.0 mm (Fig. 13.7) (Section 13.3.6.2.3) [29].

23.2.2 High-pressure balloon inflation

How?

1. Using noncompliant balloons inflated at high pressures (often up to 26–28 atm) [30]. It is important to be familiar with the rated burst pressure and percentage diameter growth of the various noncompliant balloons.
2. Performing prolonged inflations (30–60 seconds or more). Alternatively, once at high pressure, reducing and increasing the pressure quickly multiple times may help "break" the resistant lesion.
3. Using slightly undersized balloons (usually by 0.5 mm).
4. Avoiding balloon injury of non-stented vessel segments. All ballooned segments should subsequently be covered with stent(s) to minimize the risk of restenosis (geographic miss).
5. Using shorter balloons might help to concentrate the force of the dilatation and achieve the desired balloon expansion inside the lesion.
6. Simultaneous inflation of two undersized balloons positioned side by side within the lesion creates asymmetric pressure, sometimes resulting in lesion expansion.
7. Advancing one [31–33] or more [34] buddy wires through the undilatable lesion, followed by high-pressure balloon inflation: the wires can modify the balloon forces exerted on the vessel wall, leading to plaque modification and expansion (this is a makeshift version of a cutting balloon or scoring balloon).

What can go wrong?

1. Vessel perforation is best avoided by conservative balloon sizing (in nearly all cases 1:1 or lower vessel to balloon ratio should be used) and careful balloon preparation that removes all air from the balloon.
2. Balloon rupture leading to vessel dissection and/or perforation. If balloon rupture occurs, immediate angiography should be performed to determine whether perforation, air embolization, or flow-limiting dissection occurred.
3. After high-pressure inflation many balloons can become "sticky" and challenging to remove, sometimes leading to loss of wire position. Fluoroscopy can be used in such cases during balloon removal to minimize the risk of wire position loss.

23.2.3 Plaque modification balloons

How?

Plaque modification balloons (Angiosculpt, cutting balloon, Chocolate, Section 30.9.3) facilitate lesion expansion through application of focused pressure [35], but are less deliverable and more costly than standard balloons as described in Chapter 19, Calcification, Section 19.9.3.

What can go wrong?

1. Loss of guidewire and guide catheter position, as the modified balloons may not deliver easily through the balloon undilatable lesion: use of enhanced guide catheter support techniques (such as anchor techniques and guide catheter extensions) may be needed to deliver the plaque modification balloon through the lesion.
2. Vessel rupture or perforation.
3. Balloon entrapment (especially if the balloon ruptures). Excessive inflation forces or balloon rupture within the lesion may make equipment withdrawal challenging, causing fracture of the cutting balloon blade [36], stent strut avulsion [37,38], and balloon entrapment [39–42]. The cutting balloon should be inflated slowly and >14 atm inflation pressures should be avoided.

23.2.4 Suitable for atherectomy?

As described in Chapter 19, Calcification, Section 19.9.1 the risks of atherectomy (dissection, embolization, perforation, equipment loss, or entrapment) increase with increasing tortuosity (relative contraindication) and in the setting of thrombus, dissections, and bypass grafts.

23.2.5 Atherectomy

Rotational (Chapter 19: Calcification, Section 19.9.5) and orbital (Chapter 19: Calcification, Section 19.9.6) atherectomy can be used to expand a balloon undilatable lesion. The combination of laser and rotational atherectomy (RASER) has been described for highly resistant lesions [43] and can offer a solution to lesions resistant to either approach applied independently.

Discovering that a lesion is "balloon undilatable" *after* stenting poses significant challenges (PCI Manual online case 67), as stent underexpansion can predispose to stent thrombosis and restenosis. Prevention is key and can be achieved by use of intravascular imaging and/or performing careful balloon predilation with a 1:1 sized balloon and ensuring complete balloon expansion before stent implantation. If the balloon fails to fully expand, additional vessel preparation (often with atherectomy) is needed before stent implantation.

Atherectomy ("stentablation") [44–48] can be performed for in-stent undilatable lesions, but is infrequently done due to the risk of stent material or plaque embolization, burr entrapment and stent damage necessitating implantation of additional stents (*PCI Manual* Online case: 23) [47]. The risk of complications is higher when performing atherectomy within recently placed stents. To avoid burr entrapment, larger sized burrs (≥1.75 mm) should be selectively used for undilatable stents. Moreover, in-stent atherectomy carries higher risk of restenosis as compared with use of laser [44,49]. Orbital atherectomy can be performed within previously implanted stents but should be avoided within recently implanted stents.

In general, atherectomy should be reserved for cases where other strategies, such as high-pressure balloon inflations (with and without buddy wires), laser (Fig. 23.8) [49], or novel plaque modification techniques where available (intravascular lithotripsy or very high-pressure balloon) fail to expand the stent.

23.2.6 Laser

How?

1. The *0.9 mm excimer laser atherectomy catheter* should be used, usually at maximum repetition and fluence levels (repetition rate 80 Hz and fluence of 80 mJ/mm^2) [14,50–54].
2. Laser is activated during saline flushing (or with simultaneous contrast injection if performed within an underexpanded stent) [14,50,53,55].
3. High-pressure balloon inflation is subsequently performed to determine whether sufficient lesion modification has occurred, to allow expansion of the underexpanded lesion.

What can go wrong?

1. Vessel dissection, which is the reason why laser activation with contrast is typically reserved for use within an underexpanded stent.
2. Perforation.

23.2.7 Subintimal lesion crossing

This may also allow expansion of the lesion, as discussed in the treatment of balloon uncrossable lesions (Sections 23.1.8 and 23.1.10 and *PCI Manual* online case 67).

23.2.8 CABG

In rare cases when all treatment options fail, CABG may be required. Leaving an undilatable lesion without treatment is associated with increased risk of stent thrombosis and in-stent restenosis.

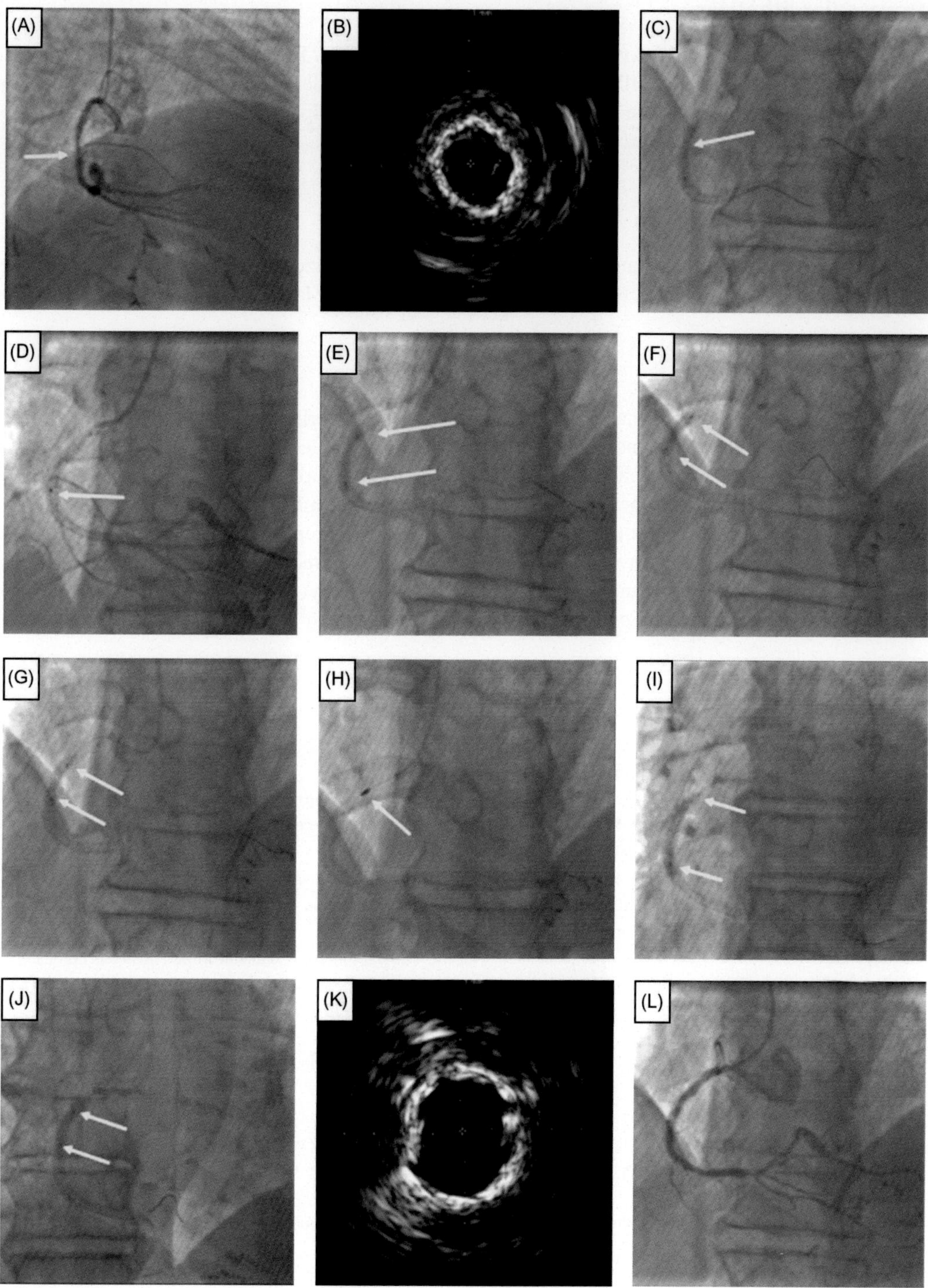

FIGURE 23.8 Example of balloon undilatable lesion treated by laser and rotational atherectomy. A patient presented with recurrent angina after right coronary artery (RCA) stenting and was found to have in-stent restenosis in the mid RCA (panel A). Intravascular ultrasound (IVUS) showed stent underexpansion with minimum lumen area of 3.3 mm^2 and severe calcification (panel B). The lesion was predilated with 3.5 × 20 mm NC balloon, but it did not expand (panel C). Laser atherectomy was performed with the Rx 0.9 mm laser catheter with multiple runs at maximum 80 mJ/mm^2 fluence and 80 Hz frequency (panel D). Further dilations were attempted with 3.5 × 15 mm, 3.5 × 20 mm NC balloons and a 3.5 × 15 mm Angiosculpt balloon without success as confirmed by IVUS (panels E, F, and G). Rotational atherectomy was performed with a 1.5 mm burr (panel H) followed by dilations with a 3.5 × 15 mm Angiosculpt balloon and a 4.0 × 15 mm noncompliant balloon (panels I and J) achieving stent expansion as confirmed by IVUS (panel K) and angiography (panel L). *Reproduced with permission from Brilakis ES. Manual of coronary chronic total occlusion interventions. A step-by-step approach. 2nd ed. Cambridge, MA: Elsevier; 2017 (Figure 8.10).*

23.2.9 Intravascular lithotripsy

The intravascular lithotripsy system (Shockwave Medical, Inc., Section 30.9.8) uses ultrasound shockwaves to fracture calcific plaque [56] and has been successfully used for expanding undilatable lesions with circumferential calcification [57]. It is not yet clinically available in the United States, however the peripheral catheter can sometimes be used off label in large coronary arteries (*PCI Manual* Online cases: 63, 100).

23.2.10 Very high-pressure balloon

The OPN balloon (SIS Medical, Switzerland, Section 30.9.7), is a specialized noncompliant balloon that can deliver very high ($>$40 atm) pressures, but is not currently available in the United States [30].

23.2.11 Intravascular imaging

Intravascular imaging (Chapter 13: Coronary intravascular imaging) is critical for all treatment stages of balloon undilatable lesions, which may sometimes not be apparent on coronary angiography.

Intravascular imaging can detect possible challenges with lesion expansion before treatment and also ensure that a good result has been achieved after treatment. It is not uncommon, especially for heavily calcified lesions, to appear well expanded during balloon inflations but have a small minimum lumen area when evaluated with intravascular imaging. Heavy circumferential calcification favors early use of coronary atherectomy, whereas less calcified lesions are usually initially approached with balloon angioplasty (Chapter 13: Coronary intravascular imaging, Section 13.3.6.2.2, point 3 and Chapter 19: Calcification, Section 19.13) [24].

References

[1] Vo MN, Christopoulos G, Karmpaliotis D, Lombardi WL, Grantham JA, Brilakis ES. Balloon-assisted microdissection "BAM" technique for balloon-uncrossable chronic total occlusions. J Invasive Cardiol 2016;28:E37–41.

[2] Brilakis ES. Manual of coronary chronic total occlusion interventions. A step-by-step approach. 2nd ed Cambridge, MA: Elsevier; 2017.

[3] Luna M, Papayannis A, Holper EM, Banerjee S, Brilakis ES. Transfemoral use of the GuideLiner catheter in complex coronary and bypass graft interventions. Catheter Cardiovasc Interv 2012;80:437–46.

[4] Di Mario C, Ramasami N. Techniques to enhance guide catheter support. Catheter Cardiovasc Interv 2008;72:505–12.

[5] Zhang Q, Zhang RY, Kirtane AJ, et al. The utility of a 5-in-6 double catheter technique in treating complex coronary lesions via transradial approach: the DOCA-TRI study. EuroIntervention 2012;8:848–54.

[6] Papayannis AC, Michael TT, Brilakis ES. Challenges associated with use of the GuideLiner catheter in percutaneous coronary interventions. J Invasive Cardiol 2012;24:370–1.

[7] Synetos A, Toutouzas K, Latsios G, et al. Proximal anchoring distal trapping technique in a chronic total occlusion unable to cross. Cardiovasc Revasc Med 2018;19:887–9.

[8] Koutouzis M, Avdikos G, Nikitas G, et al. "Ping-pong" technique for treating a balloon uncrossable chronic total occlusion. Cardiovasc Revasc Med 2018;19:117–19.

[9] Fang HY, Lee CH, Fang CY, et al. Application of penetration device (Tornus) for percutaneous coronary intervention in balloon uncrossable chronic total occlusion-procedure outcomes, complications, and predictors of device success. Catheter Cardiovasc Interv 2011;78:356–62.

[10] Hu XQ, Tang L, Zhou SH, Fang ZF, Shen XQ. A novel approach to facilitating balloon crossing chronic total occlusions: the "wire-cutting" technique. J Interv Cardiol 2012;25:297–303.

[11] Xue J, Li J, Wang H, et al. "Seesaw balloon-wire cutting" technique is superior to Tornus catheter in balloon uncrossable chronic total occlusions. Int J Cardiol 2017;228:523–7.

[12] Li Y, Li J, Sheng L, et al. "Seesaw balloon-wire cutting" technique as a novel approach to "balloon-uncrossable" chronic total occlusions. J Invasive Cardiol 2014;26:167–70.

[13] Karacsonyi J, Armstrong EJ, Truong HTD, et al. Contemporary use of laser during percutaneous coronary interventions: insights from the laser veterans affairs (LAVA) multicenter registry. J Invasive Cardiol 2018;30:195–201.

[14] Ben-Dor I, Maluenda G, Pichard AD, et al. The use of excimer laser for complex coronary artery lesions. Cardiovasc Revasc Med 2011;12: e1–8.

[15] Shen ZJ, Garcia-Garcia HM, Schultz C, van der Ent M, Serruys PW. Crossing of a calcified "balloon uncrossable" coronary chronic total occlusion facilitated by a laser catheter: a case report and review recent four years' experience at the Thoraxcenter. Int J Cardiol 2010;145:251–4.

[16] Niccoli G, Giubilato S, Conte M, et al. Laser for complex coronary lesions: impact of excimer lasers and technical advancements. Int J Cardiol 2011;146:296–9.

[17] Pagnotta P, Briguori C, Mango R, et al. Rotational atherectomy in resistant chronic total occlusions. Catheter Cardiovasc Interv 2010;76:366–71.

[18] Vo MN, Ravandi A, Grantham JA. Subintimal space plaque modification for "balloon-uncrossable" chronic total occlusions. J Invasive Cardiol 2014;26:E133–6.

[19] Christopoulos G, Kotsia AP, Rangan BV, et al. "Subintimal external crush" technique for a "balloon uncrossable" chronic total occlusion. Cardiovasc Revasc Med 2017;18:63–5.

[20] Michael TT, Banerjee S, Brilakis ES. Subintimal distal anchor technique for "balloon-uncrossable" chronic total occlusions. J Invasive Cardiol 2013;25:552–4.

[21] Kirtane AJ, Stone GW. The Anchor-Tornus technique: a novel approach to "uncrossable" chronic total occlusions. Catheter Cardiovasc Interv 2007;70:554–7.

[22] Brilakis ES, Banerjee S. The "Proxis-Tornus" technique for a difficult-to-cross calcified saphenous vein graft lesion. J Invasive Cardiol 2008;20:E258–61.

[23] Sandoval Y, Lobo AS, Tajti P, Brilakis ES. Laser-assisted orbital or rotational atherectomy: a hybrid treatment strategy for balloon-uncrossable lesions. Hellenic J Cardiol 2020;61:57–9.

[24] Tajti P, Karmpaliotis D, Alaswad K, et al. Prevalence, presentation and treatment of 'balloon undilatable' chronic total occlusions: insights from a multicenter US registry. Catheter Cardiovasc Interv 2018;91:657–66.

[25] Kim BK, Shin DH, Hong MK, et al. Clinical impact of intravascular ultrasound-guided chronic total occlusion intervention with zotarolimus-eluting versus biolimus-eluting stent implantation: randomized study. Circ Cardiovasc Interv 2015;8:e002592.

[26] Tian NL, Gami SK, Ye F, et al. Angiographic and clinical comparisons of intravascular ultrasound- versus angiography-guided drug-eluting stent implantation for patients with chronic total occlusion lesions: two-year results from a randomised AIR-CTO study. EuroIntervention 2015;10:1409–17.

[27] Shaikh K, Kelly S, Gedela M, Kumar V, Stys A, Stys T. Novel use of an orbital atherectomy device for in-stent restenosis: lessons learned. Case Reports in Cardiol 2016;2016(4).

[28] Si D, Liu G, Tong Y, He Y. Rotational atherectomy ablation for an unexpandable stent under the guide of IVUS: a case report. Medicine (Baltimore) 2018;97:e9978.

[29] Fujino A, Mintz GS, Matsumura M, et al. A new optical coherence tomography-based calcium scoring system to predict stent underexpansion. EuroIntervention 2018;13:e2182–9.

[30] Raja Y, Routledge HC, Doshi SN. A noncompliant, high pressure balloon to manage undilatable coronary lesions. Catheter Cardiovasc Interv 2010;75:1067–73.

[31] Yazdanfar S, Ledley GS, Alfieri A, Strauss C, Kotler MN. Parallel angioplasty dilatation catheter and guide wire: a new technique for the dilatation of calcified coronary arteries. Cathet Cardiovasc Diagn 1993;28:72–5.

[32] Stillabower ME. Longitudinal force focused coronary angioplasty: a technique for resistant lesions. Cathet Cardiovasc Diagn 1994;32:196–8.

[33] Meerkin D. My buddy, my friend: focused force angioplasty using the buddy wire technique in an inadequately expanded stent. Catheter Cardiovasc Interv 2005;65:513–15.

[34] Lindsey JB, Banerjee S, Brilakis ES. Two "buddies" may be better than one: use of two buddy wires to expand an underexpanded left main coronary stent. J Invasive Cardiol 2007;19:E355–8.

[35] Wilson A, Ardehali R, Brinton TJ, Yeung AC, Lee DP. Cutting balloon inflation for drug-eluting stent underexpansion due to unrecognized coronary arterial calcification. Cardiovasc Revasc Med 2006;7:185–8.

[36] Haridas KK, Vijayakumar M, Viveka K, Rajesh T, Mahesh NK. Fracture of cutting balloon microsurgical blade inside coronary artery during angioplasty of tough restenotic lesion: a case report. Catheter Cardiovasc Interv 2003;58:199–201.

[37] Harb TS, Ling FS. Inadvertent stent extraction six months after implantation by an entrapped cutting balloon. Catheter Cardiovasc Interv 2001;53:415–19.

[38] Wang HJ, Kao HL, Liau CS, Lee YT. Coronary stent strut avulsion in aorto-ostial in-stent restenosis: potential complication after cutting balloon angioplasty. Catheter Cardiovasc Interv 2002;56:215–19.

[39] Kawamura A, Asakura Y, Ishikawa S, et al. Extraction of previously deployed stent by an entrapped cutting balloon due to the blade fracture. Catheter Cardiovasc Interv 2002;57:239–43.

[40] Sanchez-Recalde A, Galeote G, Martin-Reyes R, Moreno R. AngioSculpt PTCA balloon entrapment during dilatation of a heavily calcified lesion. Rev Esp Cardiol 2008;61:1361–3.

[41] Giugliano GR, Cox N, Popma J. Cutting balloon entrapment during treatment of in-stent restenosis: an unusual complication and its management. J Invasive Cardiol 2005;17:168–70.

[42] Pappy R, Gautam A, Abu-Fadel MS. AngioSculpt PTCA Balloon Catheter entrapment and detachment managed with stent jailing. J Invasive Cardiol 2010;22:E208–10.

[43] Egred M. RASER angioplasty. Catheter Cardiovasc Interv 2012;79:1009–12.

[44] Ferri LA, Jabbour RJ, Giannini F, et al. Safety and efficacy of rotational atherectomy for the treatment of undilatable underexpanded stents implanted in calcific lesions. Catheter Cardiovasc Interv 2017;90:E19–24.

[45] Vales L, Coppola J, Kwan T. Successful expansion of an underexpanded stent by rotational atherectomy. Int J Angiol 2013;22:63–8.

[46] Hernandez J, Galeote G, Moreno R. Rotational atherectomy: if you do not do it before, you can do it after stenting. J Invasive Cardiol 2014;26:E122–3.

[47] Medina A, de Lezo JS, Melian F, Hernandez E, Pan M, Romero M. Successful stent ablation with rotational atherectomy. Catheter Cardiovasc Interv 2003;60:501–4.

[48] Kobayashi Y, Teirstein P, Linnemeier T, Stone G, Leon M, Moses J. Rotational atherectomy (stentablation) in a lesion with stent underexpansion due to heavily calcified plaque. Catheter Cardiovasc Interv 2001;52:208–11.

[49] Latib A, Takagi K, Chizzola G, et al. Excimer Laser LEsion modification to expand non-dilatable stents: the ELLEMENT registry. Cardiovasc Revasc Med 2014;15:8–12.

[50] Karacsonyi J, Danek BA, Karatasakis A, Ungi I, Banerjee S, Brilakis ES. Laser coronary atherectomy during contrast injection for treating an underexpanded stent. JACC Cardiovasc Interv 2016;9:e147–8.

[51] Fernandez JP, Hobson AR, McKenzie D, et al. Beyond the balloon: excimer coronary laser atherectomy used alone or in combination with rotational atherectomy in the treatment of chronic total occlusions, non-crossable and non-expansible coronary lesions. EuroIntervention 2013;9:243–50.

[52] Badr S, Ben-Dor I, Dvir D, et al. The state of the excimer laser for coronary intervention in the drug-eluting stent era. Cardiovasc Revasc Med 2013;14:93–8.

[53] Sunew J, Chandwaney RH, Stein DW, Meyers S, Davidson CJ. Excimer laser facilitated percutaneous coronary intervention of a nondilatable coronary stent. Catheter Cardiovasc Interv 2001;53:513–17.

[54] Veerasamy M, Gamal AS, Jabbar A, Ahmed JM, Egred M. Excimer laser with and without contrast for the management of under-expanded stents. J Invasive Cardiol 2017;29:364–9.

[55] Egred M. A novel approach for under-expanded stent: excimer laser in contrast medium. J Invasive Cardiol 2012;24:E161–3.

[56] Brinton TJ, Ali ZA, Hill JM, et al. Feasibility of shockwave coronary intravascular lithotripsy for the treatment of calcified coronary stenoses. Circulation 2019;139:834–6.

[57] Watkins S, Good R, Hill J, Brinton TJ, Oldroyd KG. Intravascular lithotripsy to treat a severely underexpanded coronary stent. EuroIntervention 2019;15:124–5.

Chapter 24

Complex patient subgroups

Complex patient subgroups that will be discussed in this chapter include:

- TAVR patients
- Cardiogenic shock patients

24.1 TAVR patients

CTO Manual Online case: 113
PCI Manual Online case: 62

Transcatheter aortic valve replacement (TAVR) is increasingly being performed for symptomatic aortic stenosis patients, even those at low surgical risk. Coronary angiography and percutaneous coronary intervention (PCI) is often required in TAVR patients and carries unique challenges, such as coronary artery engagement, especially in patients with self-expanding valves, which extend above the coronary ostia [1].

24.1.1 Planning

Given frequent difficulty in engaging the coronary arteries in TAVR patients, preprocedural CT angiography can be performed to help understand the three-dimensional geometric interaction among the valve prosthesis, the aortic root, and the coronary ostia to help predict and prepare for potential challenges of coronary reaccess (Fig. 24.1) [1].

However, use of CT angiography in prior TAVR patients has several limitations [1]:

1. CT cannot be performed in urgent or emergent situations, such as in patients with an acute coronary syndrome requiring urgent coronary angiography and PCI.
2. CT angiography requires contrast administration that increases the risk of contrast-induced acute kidney injury after additional contrast administration during cardiac catheterization.
3. Motion artifact and image quality may limit the ability to visualize leaflet orientation of the transcatheter valve relative to the coronary ostia, making it difficult to determine whether the commissural post may impede the ability to reaccess the coronary arteries.

24.1.2 Monitoring

Monitoring is performed as described in Chapter 2: Monitoring.

24.1.3 Medications

Medication administration is performed as described in Chapter 3: Medications.

24.1.4 Access

Given frequent challenges associated with engagement in prior TAVR patients, femoral access may be preferred.

Manual of Percutaneous Coronary Interventions. DOI: https://doi.org/10.1016/B978-0-12-819367-9.00024-X

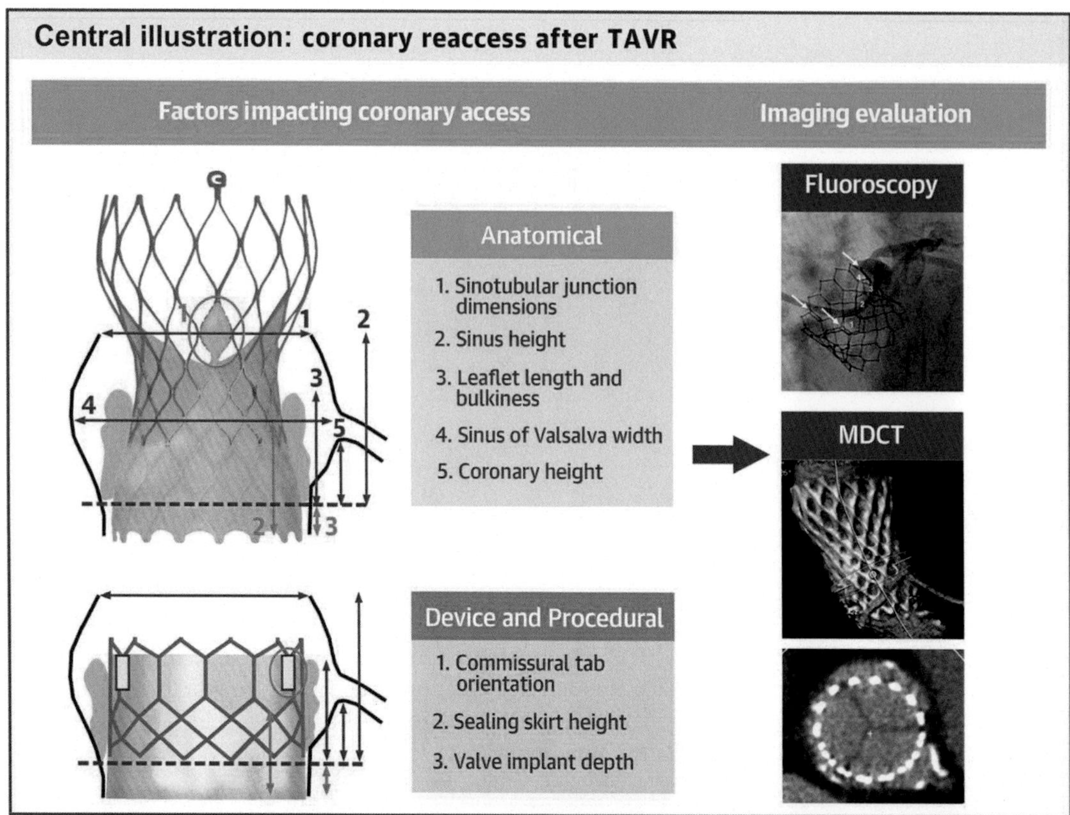

FIGURE 24.1 Coronary reaccess after TAVR. Summary of factors impacting coronary access and imaging evaluation after TAVR. MDCT, multidetector computed tomography; TAVR, transcatheter aortic valve replacement. *Reproduced with permission from Yudi MB, Sharma SK, Tang GHL, Kini A. Coronary angiography and percutaneous coronary intervention after transcatheter aortic valve replacement. J Am Coll Cardiol 2018;71:1360–78 (Central Illustration). Copyright Elsevier.*

24.1.5 Engagement

Coronary engagement should be performed with extreme care aiming for coaxial positioning, and avoiding deep coronary artery intubation [2].

24.1.5.1 Evolut-PRO CoreValve

Engagement depending on depth of valve implantation

If the Evolut-PRO CoreValve is positioned optimally (skirt below coronary ostia), it is feasible to engage the coronary artery in a coaxial manner, assuming the native aortic valve leaflets will not interfere with the path to the coronary ostium (Fig. 24.2, panel A).

If the valve is deployed high (Fig. 24.2, panel B) coronary obstruction would not occur due to the narrow waist of the valve and sufficient sinus of Valsalva width [1]. However, selective coronary angiography would be difficult in this scenario and would have to occur from a diamond above the ostium, given that the supra-annular valve and its covered segment (e.g., sealing skirt) would be above the level of the ostium. A straighter catheter with a short tip, such as a Judkins right (JR) 4, could be used in this scenario, even for left main artery engagement [1].

Engagement depending on the position of the transcatheter valve commissures in relation to those of the native aortic valve

The circumferential sealing skirt of the Evolut-PRO CoreValve is 13 mm in height (14 mm in the 34-mm Evolut-R), however it rises up to 26 mm at the commissural insertion point (Fig. 24.3). If a commissure ends up being positioned directly in front of the coronary ostium coaxial engagement of the coronary ostia would be challenging, if not impossible [1]. Engagement also depends on the width of the sinus of Valsalva that determines the space between the valve frame and the coronary ostia; the wider the sinus the more room there is to manipulate a catheter toward the coronary ostia. A narrow sinus would require a very acute angle for the catheter to be pointing toward the ostia for a nonselective

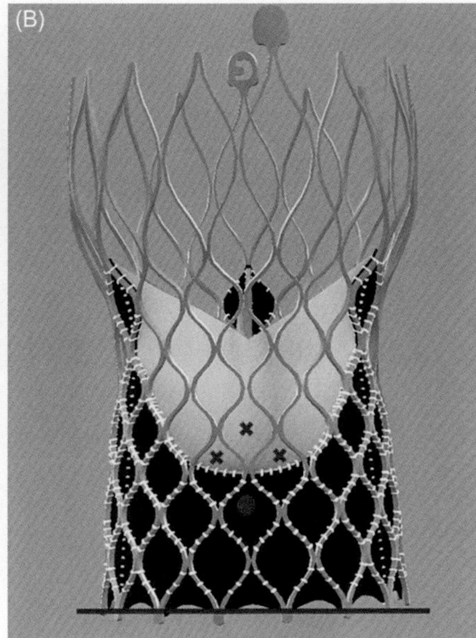

FIGURE 24.2 Self-expanding valve and coronary access depending on level of implantation across the annulus. *Red dot* represents the location of the coronary ostium in relation to the valve frame, and the *red line* represents the annular plane. The *red x's* depict the closest diamonds that can be used to access the coronary ostium. An optimally positioned Evolut-R (Medtronic, Galway, Ireland) (A) would make coronary access potentially easier than one with a higher implant (B). *Reproduced with permission from Yudi MB, Sharma SK, Tang GHL, Kini A. Coronary angiography and percutaneous coronary intervention after transcatheter aortic valve replacement. J Am Coll Cardiol 2018;71:1360–78 (Figure 2). Copyright Elsevier.*

FIGURE 24.3 Self-expanding valve and coronary access if ostia line up with commissural post. *Red line* represents the annular plane. The three *red dots* depict coronary ostia heights of approximately 10, 14, and 18 mm above the annular plane, respectively. The *red x's* depict the closest diamonds that can be used to access the coronaries. The commissural post of an Evolut-R (Medtronic, Galway, Ireland) is 26 mm in height (panel A). Depending on the height of coronary ostia, a different catheter and approach is necessary for coronary reaccess, when the ostium faces the side of the commissural post (panel B). *Reproduced with permission from Yudi MB, Sharma SK, Tang GHL, Kini A. Coronary angiography and percutaneous coronary intervention after transcatheter aortic valve replacement. J Am Coll Cardiol 2018;71:1360–78 (Figure 3). Copyright Elsevier.*

coronary angiogram. If selective engagement is required, a coronary wire would have to be manipulated into the coronary artery, and the guide, or a guide catheter extension, would then have to be railed into the ostium. This represents the most difficult scenario: a valve commissure overlying a low coronary ostium in a patient with a narrow sinus of Valsalva. The aforementioned description does not account for the native aortic leaflet height and severity of calcification facing the left and right sinuses. A tall and bulky leaflet may extend beyond the 13- or 14-mm sealing skirt of the repositionable Evolut-PRO self-expanding valve and would likely further add to the challenge of coronary reaccess [1].

Catheter selection

***Left coronary artery*:** Smaller catheters, such as a JL3.5 or JL3, are frequently used to engage the left main. On the contrary, engagement of the RCA can usually be managed with a JR4 catheter. There are reports of XB guide catheter

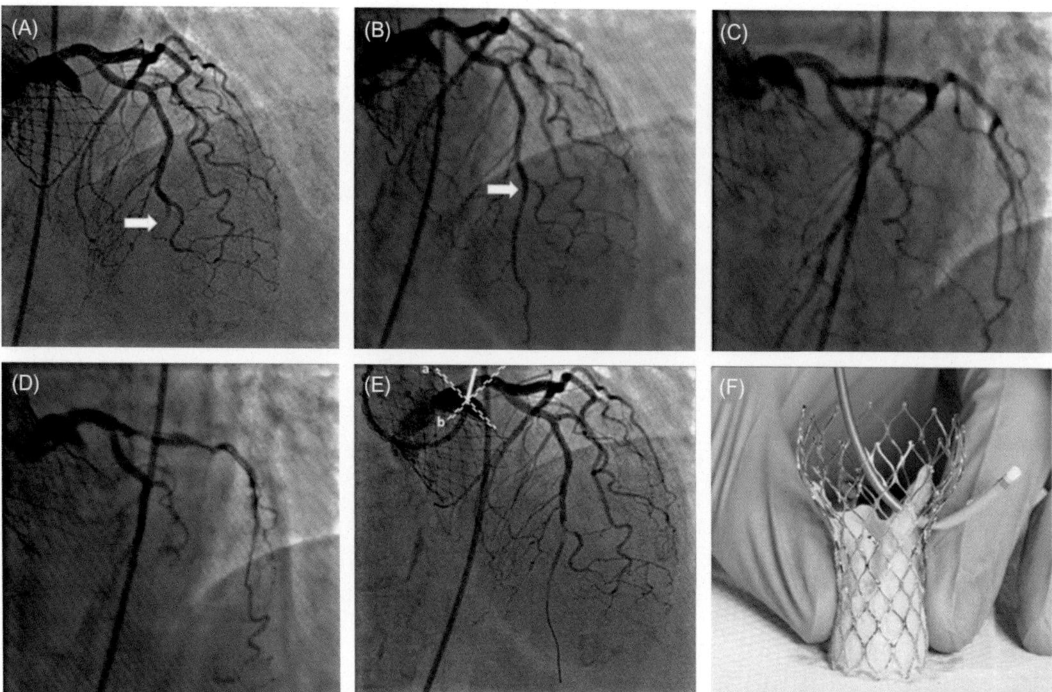

FIGURE 24.4 Guide catheter entrapment. (A) Left coronary angiography showing total occlusion of the distal left anterior descending artery (*arrow*). (B) Left coronary angiography postpercutaneous coronary intervention and drug-eluting stent deployment (*arrow*). (C and D) Dissection of the left main and left anterior descending coronary arteries. (E) (*a*) *Interrupted line* showing guide catheter engagement into the left coronary ostium at an acute angle with the vertical axis of the stent frame. (*b*) *Interrupted line* showing crossing of the stent frame at a perpendicular angle. (F) Ex vivo simulation using an extra back-up 3.5 guide catheter, showing catheter entrapment within the stent frame when crossed at an acute angle. *Reproduced with permission from Harhash A, Ansari J, Mandel L, Kipperman R. STEMI after TAVR: procedural challenge and catastrophic outcome. JACC Cardiovasc Interv 2016;9:1412–3 (Figure 1). Copyright Elsevier.*

kinking and entrapment through the valve diamonds (Fig. 24.4) [3]. The following measures may minimize the risk of catheter entrapment:

1. Crossing the stent frame perpendicularly through a diamond at the same level with the coronary ostium.
2. Using catheters with favorable geometry (e.g., left Judkins for left and right Amplatz for the right coronary artery).
3. Using a balloon and/or a guidewire to back the catheter out of the coronary ostium.

***Right coronary artery*:** JR4 or Ikari Right catheters are usually used. A JR4.5, JR5, or Amplatz right (AR) 2 catheter may be preferable if the sinus width is large, creating a larger distance from the valve frame to the ostium [1].

24.1.5.2 Sapien 3

Coronary engagement is often easier in patients who have a Sapien valve compared with a self-expanding valve.

Engagement depending on depth of valve implantation (Fig. 24.5)

24.1.6 Angiography

Coronary angiography is performed as described in Chapter 6: Coronary Angiography.

24.1.7 Selecting target lesion(s)

Performing PCI in patients with severe aortic stenosis who require TAVR remains controversial, as it carries increased risk, but could potentially decrease the risk of complications during TAVR. PCI performed at the same time as TAVR is logistically convenient, however it increases complexity and contrast use. Deferring PCI until after TAVR allows better assessment of symptoms caused by coronary artery disease (as compared with symptoms caused by aortic stenosis), but coronary artery engagement can be challenging, as described in Section 24.1.5.

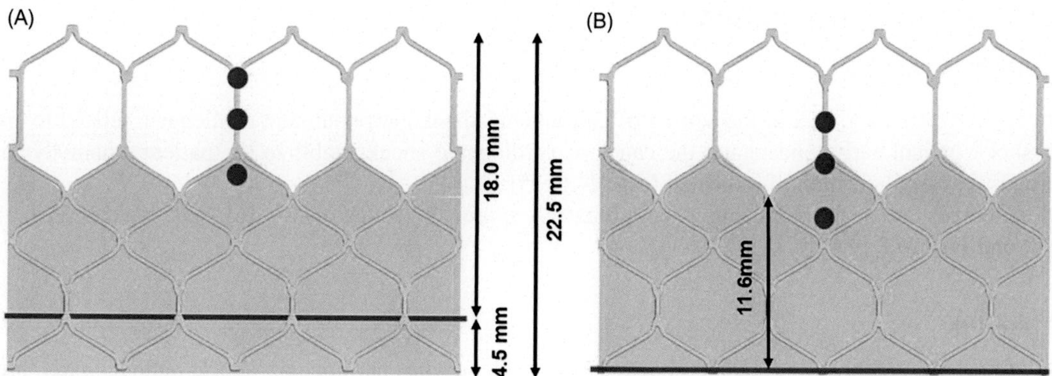

FIGURE 24.5 Balloon-expandable valve and coronary ostia based on depth of implant. *Red dots* represent the different locations of the coronary ostium in relation to the valve frame of a 29 mm Sapien 3 valve (Edwards Lifesciences, Irvine, California), and the *red line* represents the annular plane. An optimally positioned Sapien 3 valve (Edwards Lifesciences, Irvine, California) (panel A) would make coronary access potentially easier than one with a higher implant (panel B), where the coronary ostium will be located below the seal skirt. Tall native leaflet or bulky calcium at the leaflet tip may further increase difficulty of coronary access in a high valve implant. *Reproduced with permission from Yudi MB, Sharma SK, Tang GHL, Kini A. Coronary angiography and percutaneous coronary intervention after transcatheter aortic valve replacement. J Am Coll Cardiol 2018;71:1360–78 (Figure 6). Copyright Elsevier.*

24.1.8 Wiring

Wiring of the coronary artery may be required prior to full catheter engagement to facilitate engagement.

24.1.9 Lesion preparation

Lesion preparation is performed as described in Chapter 9: Lesion Preparation. Prior TAVR patients often have complex coronary anatomy, such as severe calcification and chronic total occlusions, requiring meticulous lesion preparation prior to stenting.

24.1.10 Stenting

Stenting is performed as described in Chapter 10: Stenting.

24.1.11 Closure

Vascular closure is performed as described in Chapter 11: Access Closure.

24.1.12 Coronary physiology

Fractional flow reserve decreases significantly after TAVR due to a significant increase in hyperemic flow [4]. In contrast, iFR does not significantly change post-TAVR [4]. Hence in patients who have severe aortic stenosis FFR should be used with caution as it may underestimate the severity of coronary stenoses.

24.1.13 Intravascular imaging

Intravascular imaging is performed as described in Chapter 13: Coronary Intravascular Imaging.

24.1.14 Hemodynamic support

Hemodynamic support is used as described in Chapter 14: Hemodynamic Support.

24.2 Cardiogenic shock patients

PCI Manual Online case: 27

Cardiac catheterization and PCI in the setting of cardiogenic shock carries unique challenges, related to the need to identify the shock patient early, understand the cause of cardiogenic shock, stabilize the patient's hemodynamics, and perform coronary revascularization, if indicated (Fig. 24.6) [5].

Optimal outcomes are achieved using a collaborative approach (shock team) [6] and standardized algorithms (Fig. 24.7 [8] and Fig. 24.8 [9,10]).

24.2.1 Planning

Planning is critical for treating cardiogenic shock patients. Identifying cardiogenic shock patients early prior to the development of refractory shock and metabolic shock can improve the likelihood of survival. Developing local shock algorithms and a dedicated shock team (Figs. 24.7 and 24.8) can optimize and standardize delivery of complex and advanced care to these critically ill patients.

24.2.2 Monitoring

Monitoring is performed as described in Chapter 2: Monitoring. Cardiogenic shock patients can have rapid and significant changes in clinical status that need to be promptly identified and treated. Serial measures of perfusion such as lactate levels and invasive hemodynamics from right heart catheterization can provide important diagnostic and monitoring information. Regularly performing assessment of RV failure (RA Pressure, PAPI) and maintaining a cardiac power output > 0.6 W while limiting escalating dose of vasopressors and inotropes are key for improving survival.

24.2.3 Medications

Use of vasopressors and inotropes is often required in shock (Section 3.6), but can adversely affect the likelihood of mycardial recovery by increasing myocardial oxygen consumption and may cause arrhythmias. In a study of 1,679 shock patients (19% of whom had cardiogenic shock) use of dopamine was associated with more arrhythmias compared with norepinephrine [11]. Similarly, alpha agonist should be avoided as they do not provide inotropic

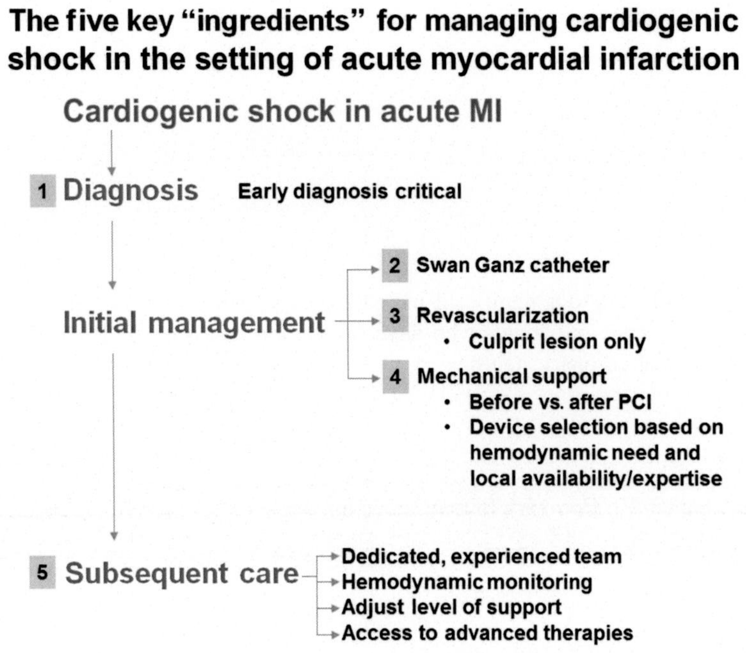

FIGURE 24.6 The five key "ingredients" for managing cardiogenic shock. *Reproduced with permission from Brilakis ES, Eckman P. The five key "ingredients" for improving outcomes in cardiogenic shock complicating acute myocardial infarction. Catheter Cardiovasc Interv 2018;91:462–3.*

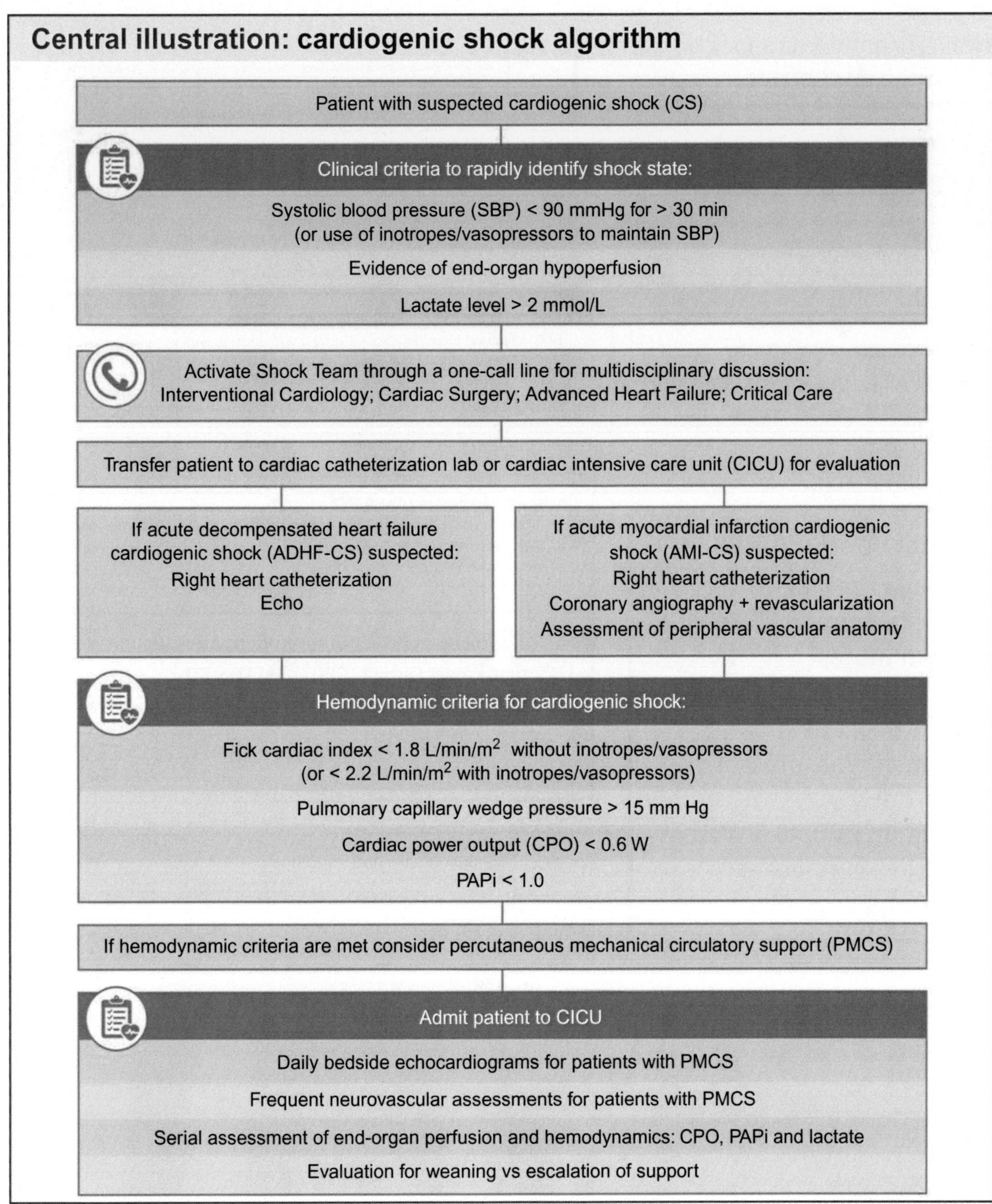

FIGURE 24.7 Cardiogenic shock algorithm. Schematic representation of the care pathways in the upstream and critical care management of patients with acute myocardial infarction (AMI) and acute decompensated heart failure (ADHF) cardiogenic shock at the INOVA Heart and Vascular Institute. CPO = [mean arterial pressure × cardiac output]/451; PAPi = [systolic pulmonary arterial pressure − diastolic pulmonary arterial pressure]/[right atrial pressure]. *Reproduced with permission from Tehrani BN, Truesdell AG, Sherwood MW, et al. Standardized team-based care for cardiogenic shock. J Am Coll Cardiol 2019;73:1659–69 (Central Figure). Copyright Elsevier.*

support and increase afterload. Early use of mechanical circulatory support can help reduce the need for such medications and their adverse effects.

24.2.4 Access

Given the frequent need for mechanical circulatory support and rapid delineation of coronary anatomy, femoral access may be preferred in shock patients.

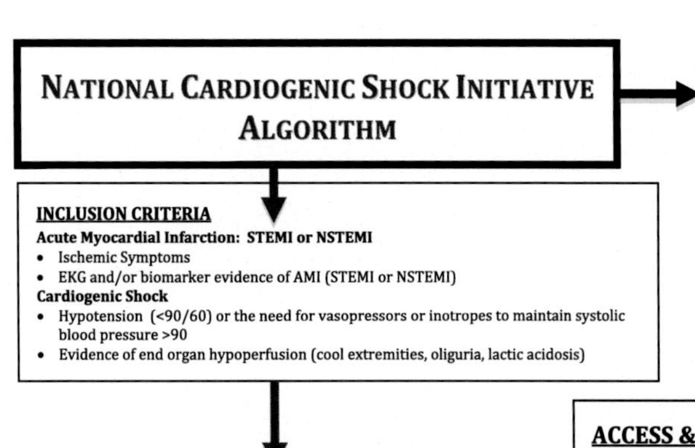

NATIONAL CARDIOGENIC SHOCK INITIATIVE ALGORITHM
INCLUSION CRITERIA
Acute Myocardial Infarction: STEMI or NSTEMI
• Ischemic Symptoms
• EKG and/or biomarker evidence of AMI (STEMI or NSTEMI)
Cardiogenic Shock
• Hypotension (<90/60) or the need for vasopressors or inotropes to maintain systolic blood pressure >90
• Evidence of end organ hypoperfusion (cool extremities, oliguria, lactic acidosis)

EXCLUSION CRITERIA
- Evidence of Anoxic Brain Injury
- Unwitnessed out of hospital cardiac arrest or any cardiac arrest in which ROSC is not achieved in 30 minutes
- IABP placed prior to Impella
- Septic, anaphylactic, hemorrhagic, and neurologic causes of shock
- Non-ischemic causes of shock/hypotension *(Pulmonary Embolism, Pneumothorax, Myocarditis, Tamponade, etc.)*
- Active Bleeding
- Recent major surgery
- Mechanical Complications of AMI
- Known left ventricular thrombus
- Patient who did not receive revascularization
- Contraindication to intravenous systemic anticoagulation
- Mechanical aortic valve

ACTIVATE CATH LAB

ACCESS & HEMODYNAMIC SUPPORT
- Obtain femoral arterial access (via direct visualization with use of ultrasound and fluoro)
- Obtain venous access (Femoral or Internal Jugular)
- Obtain either Fick calculated cardiac index or LVEDP

IF LVEDP >15 *or* Cardiac Index < 2.2 AND anatomy suitable, place IMPELLA

Coronary Angiography & PCI
- Attempt to provide TIMI III flow in all major epicardial vessels other than CTO
- If unable to obtain TIMI III flow, consider administration of intra-coronary vasodilators

Perform Post-PCI Hemodynamic Calculations

1. Cardiac Power Output (CPO): $\frac{MAP \times CO}{451}$
2. Pulmonary Artery Pulsatility Index (PAPI): $\frac{sPAP - dPAP}{RA}$

Wean OFF Vasopressors and Inotropes
If CPO is >0.6 and PAPI >0.9, operators should wean vasopressors and inotropes and determine if Impella can be weaned and removed in the Cath Lab or left in place with transfer to ICU.

Escalation of Support
If CPO remains <0.6 operators should consider the following options:
- PAPI is <0.9 consider right sided hemodynamic support
- PAPI >0.9 consideration for additional hemodynamic support

Local practice patterns should dictate the next steps:
- Placement of more robust MCS device(s)
- Transfer to LVAD/Transplant center

If CPO is >0.6 and PAPI <0.9 consider providing right sided hemodynamic support if clinical suspicion for RV dysfunction/failure

Vascular Assessment
- Prior to discharge from the Cath Lab, a detailed vascular exam should be performed including femoral angiogram and Doppler assessment of the affected limb.
- If indicated, external bypass should be performed.

ICU Care
- Daily hemodynamic assessments should be performed, including detailed vascular assessment
- Monitor for signs of hemolysis and adjust Impella position as indicated

Device Weaning
Impella should only be considered for explantation once the following criteria are met:
- Weaning off from all inotropes and vasopressors
- CPO >0.6, and PAPI > 0.9

Bridge to Decision
Patients who do not regain myocardial recovery within 3-5 days, as clinically indicated, should be transferred to an LVAD/Transplant center. If patients are not candidates, palliative care options should be considered.

**** QUALITY MEASURES ****
- Impella Pre-PCI
- Door to Support Time < 90 minutes
- Establish TIMI III Flow
- Right Heart Cath
- Wean off Vasopressors & Inotropes
- Maintain CPO >0.6 Watts
- Improve survival to discharge to >80%

NATIONAL CARDIOGENIC SHOCK INITIATIVE

NationalCSI@hfhs.org

www.henryford.com/cardiogenicshock

NationalCSI - Algorithm - v1.5 - 11/2017

FIGURE 24.8 National cardiogenic shock initiative algorithm. *Reproduced with permission from Basir MB, Kapur NK, Patel K, et al. Improved outcomes associated with the use of shock protocols: updates from the National Cardiogenic Shock Initiative. Catheter Cardiovasc Interv 2019;93:1173–83.*

24.2.5 Engagement

Coronary engagement is performed as described in Chapter 5: Coronary and Graft Engagement.

24.2.6 Angiography

Coronary angiography is performed as described in Chapter 6: Coronary Angiography.

24.2.7 Selecting target lesion(s)

In patients who present with cardiogenic shock due to acute myocardial infarction revascularization should be provided. Early revascularization improves mortality in cardiogenic shock patients [12]. Approximately two thirds of acute myocardial infarction patients with cardiogenic shock have multivessel coronary artery disease. In such patients culprit-only PCI is recommended at the time of presentation based on the results of the CULPRIT-SHOCK trial [13,14]. Staged revascularization can be performed in the remaining coronary lesions.

Identification of the culprit lesion(s) may be challenging and can be facilitated using electrocardiographic, angiographic, intravascular imaging, and myocardial imaging data (Section 20.7). Few cardiogenic shock patients have more than one culprit lesions which should be treated.

24.2.8 Wiring

Wiring is performed as described in Chapter 8: Wiring.

24.2.9 Lesion preparation

Lesion preparation is performed as described in Chapter 9: Lesion Preparation.

24.2.10 Stenting

Stenting is performed as described in Chapter 10: Stenting.

24.2.11 Closure

Vascular closure is performed as described in Chapter 11: Access Closure.

24.2.12 Coronary physiology

Fractional flow reserve is affected by coronary afterload, hence it may be inaccurate in the setting of cardiogenic shock. In patients with cardiogenic shock coronary physiology should be used with caution as it may not accurately depict the severity of coronary stenoses.

24.2.13 Intravascular imaging

Intravascular imaging is performed as described in Chapter 13: Coronary Intravascular Imaging.

24.2.14 Hemodynamic support

Evaluating the need for hemodynamic support is crucial in the treatment of cardiogenic shock patients and is discussed in Chapter 14: Hemodynamic Support. The decision about initiating hemodynamic support can be optimized by using right heart catheterization (to determine the mechanism and severity of cardiogenic shock) and a team approach with input from advanced heart failure specialists, cardiac surgeons, and cardiac intensivists [6,7]. There is, however, limited data that use of hemodynamic support improves the outcomes of cardiogenic shock patients [15–19].

References

[1] Yudi MB, Sharma SK, Tang GHL, Kini A. Coronary angiography and percutaneous coronary intervention after transcatheter aortic valve replacement. J Am Coll Cardiol 2018;71:1360–78.

[2] Hayes SN, Kim ESH, Saw J, et al. Spontaneous coronary artery dissection: current state of the science: a scientific statement from the American Heart Association. Circulation 2018;137:e523–57.

[3] Harhash A, Ansari J, Mandel L, Kipperman R. STEMI after TAVR: procedural challenge and catastrophic outcome. JACC Cardiovasc Interv 2016;9:1412–13.

[4] Ahmad Y, Gotberg M, Cook C, et al. Coronary hemodynamics in patients with severe aortic stenosis and coronary artery disease undergoing transcatheter aortic valve replacement: implications for clinical indices of coronary stenosis severity. JACC Cardiovasc Interv 2018;11:2019–31.

[5] Brilakis ES, Eckman P. The five key "ingredients" for improving outcomes in cardiogenic shock complicating acute myocardial infarction. Catheter Cardiovasc Interv 2018;91:462–3.

[6] Taleb I, Koliopoulou AG, Tandar A, et al. Shock team approach in refractory cardiogenic shock requiring short-term mechanical circulatory support. Circulation 2019;140:98–100.

[7] Tehrani BN, Truesdell AG, Sherwood MW, et al. Standardized team-based care for cardiogenic shock. J Am Coll Cardiol 2019;73:1659–69.

[8] Doll JA, Ohman EM, Patel MR, et al. A team-based approach to patients in cardiogenic shock. Catheter Cardiovasc Interv 2016;88:424–33.

[9] Basir MB, Kapur NK, Patel K, et al. Improved outcomes associated with the use of shock protocols: updates from the National Cardiogenic Shock Initiative. Catheter Cardiovasc Interv 2019;93:1173–83.

[10] Basir MB, Schreiber T, Dixon S, et al. Feasibility of early mechanical circulatory support in acute myocardial infarction complicated by cardiogenic shock: the Detroit cardiogenic shock initiative. Catheter Cardiovasc Interv 2018;91:454–61.

[11] De Backer D, Biston P, Devriendt J, et al. Comparison of dopamine and norepinephrine in the treatment of shock. N Engl J Med 2010;362:779–89.

[12] Hochman JS, Sleeper LA, Webb JG, et al. Early revascularization and long-term survival in cardiogenic shock complicating acute myocardial infarction. JAMA 2006;295:2511–15.

[13] Thiele H, Akin I, Sandri M, et al. One-year outcomes after PCI strategies in cardiogenic shock. N Engl J Med 2018;379:1699–710.

[14] Thiele H, Akin I, Sandri M, et al. PCI strategies in patients with acute myocardial infarction and cardiogenic shock. N Engl J Med 2017;377:2419–32.

[15] Rao P, Khalpey Z, Smith R, Burkhoff D, Kociol RD. Venoarterial extracorporeal membrane oxygenation for cardiogenic shock and cardiac arrest. Circ Heart Fail 2018;11:e004905.

[16] Ouweneel DM, Eriksen E, Sjauw KD, et al. Percutaneous mechanical circulatory support versus intra-aortic balloon pump in cardiogenic shock after acute myocardial infarction. J Am Coll Cardiol 2017;69:278–87.

[17] Thiele H, Zeymer U, Neumann FJ, et al. Intraaortic balloon support for myocardial infarction with cardiogenic shock. N Engl J Med 2012;367:1287–96.

[18] Amin AP, Spertus JA, Curtis JP, et al. The evolving landscape of Impella use in the United States among patients undergoing percutaneous coronary intervention with mechanical circulatory support. Circulation 2020;141:273–84.

[19] Dhruva SS, Ross JS, Mortazavi BJ et al. Association of use of an intravascular microaxial left ventricular assist device vs intra-aortic balloon pump with in-hospital mortality and major bleeding among patients with acute myocardial infarction complicated by cardiogenic shock. Jama 2020;323:734–745.

Part C

Complications

Complications of PCI can be classified according to timing (as acute and long-term) and according to location (cardiac coronary, cardiac noncoronary, and noncardiac). The acute complications of PCI are summarized in Fig. C.1.

Another potential classification is by complication severity (Fig. C.2). Preventing complications is the goal, but if a complication occurs every effort should be made to minimize its impact, that is, shifting toward the left in Fig. C.2.

"An ounce of prevention is worth a pound of cure." Here are some suggestions about how to prevent complications during coronary angiography and percutaneous coronary intervention:

1. Be aware of what can go wrong, that is, the spectrum of possible complications.
2. Be present (no joking, no multitasking, do not take things lightly, explain every unexpected finding).
3. Pay close attention to the basics, such as always looking at the pressure waveform before and after each injection.

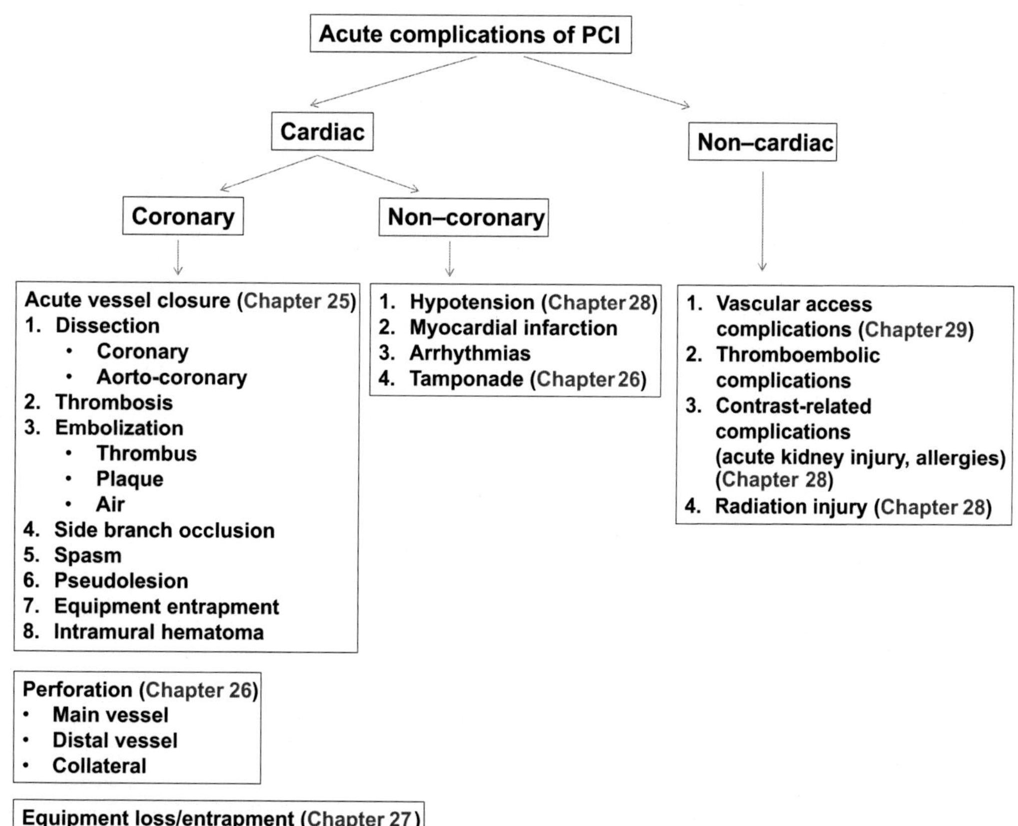

FIGURE C.1 Classification of the acute complications of percutaneous coronary intervention. *PCI*, Percutaneous coronary intervention.

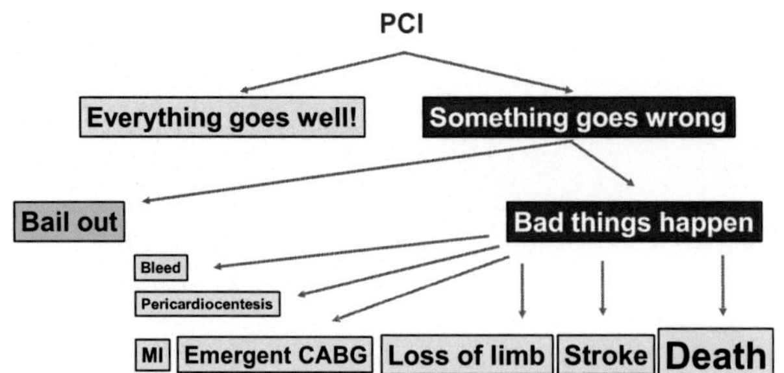

FIGURE C.2 Classification of the acute complications of percutaneous coronary intervention by severity. *PCI*, Percutaneous coronary intervention.

Chapter 25

Acute vessel closure

CTO PCI Manual Online cases: 38, 92, 98, 112, 125
PCI Manual Online cases: 18, 21, 40, 44, 57, 68, 72, 81

Acute vessel closure is defined as (partial or complete) decrease in antegrade coronary flow that occurs during or immediately after percutaneous coronary intervention.

Fig. 25.1 outlines a step-by-step algorithm for approaching such lesions.

25.1 Maintain guidewire position

Acute vessel closure can lead to significant patient and operator stress, which may in turn lead to inadvertent removal of the previously delivered guidewire. ***Losing guidewire position can have catastrophic consequences, especially if acute vessel closure is due to a dissection, and should be avoided at all cost.*** Losing guidewire position can be the result of guide catheter manipulation, hence continuous attention to the guide catheter position is recommended throughout the procedure.

Remaining calm and maintaining true lumen guidewire position (if the guidewire was placed in the true lumen prior to the complication) is key for successfully managing acute vessel closure.

25.2 Determine the cause of acute vessel closure and treat accordingly

Treatment of acute vessel closure depends on the cause (Table 25.1). The key differentiation is between dissection (that requires stenting) and distal embolization (that may actually worsen from stenting and requires physical removal of thrombus or debris and/or vasodilator administration).

25.2.1 Dissection

In coronary dissection there is separation of the various layers of the coronary arterial wall. Angiographically, a dissection appears as a linear or spiral filling defect within the vessel lumen, although in severe cases complete vessel occlusion may occur (Section 6.8.6).

The consequences of dissection depend on:

- the severity of coronary flow obstruction (none, partial, or complete cessation of coronary flow), and
- the location of the dissection (more proximal dissections, such as in the left main, can have more profound consequences as they cause a larger area of ischemia, potentially leading to arrhythmias and hemodynamic compromise).

Causes:

- Injection of contrast despite dampened pressure waveform (Chapter 5: Coronary and graft engagement, Section 5.6).
- Non-coaxial guide catheter position.
- Guidewire crossing attempts (Chapter 8: Wiring, Sections 8.6, 8.7, and 8.8).
- Lesion preparation (balloon angioplasty, atherectomy, laser, especially use of oversized balloons).
- Balloon rupture (Fig. 25.2).
- Heavily calcified and tortuous lesions.
- Stenting (Chapter 10: Stenting, Sections 10.2.3 and 10.2.13).
- Spontaneous coronary artery dissection (Chapter 22: Other complex lesion types, Section 22.1).
- Aortic dissection: aortic dissection may involve the coronary ostia causing partial or complete occlusion.

Manual of Percutaneous Coronary Interventions. DOI: https://doi.org/10.1016/B978-0-12-819367-9.00025-1

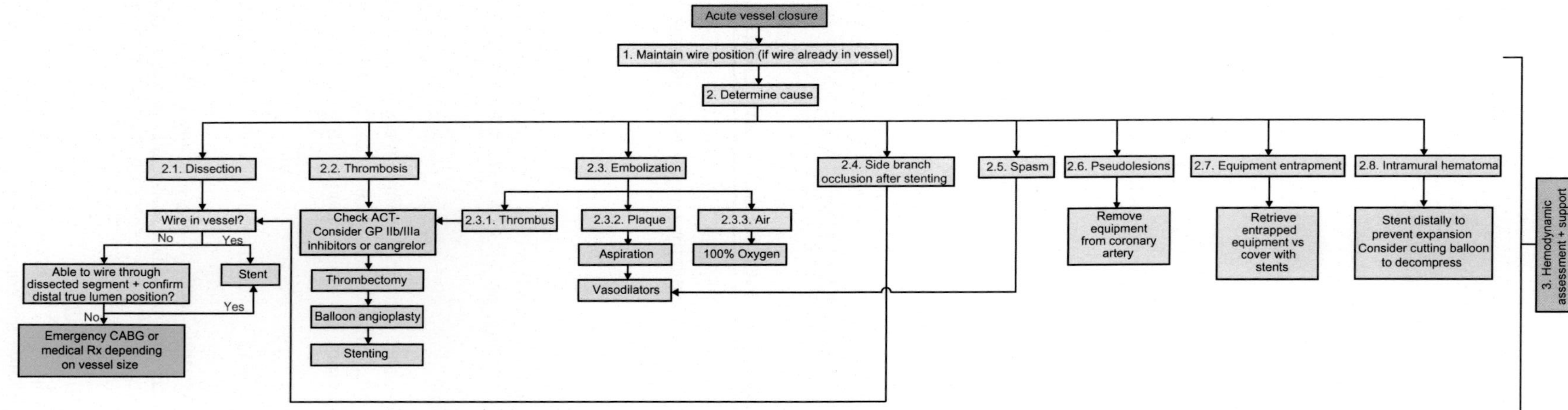

FIGURE 25.1 Approach to acute vessel closure. *ACT*, activated clotting time; *CABG*, coronary artery bypass graft surgery; *GP IIa/IIIa*, glycoprotein IIb/IIIa; *Rx*, treatment.

TABLE 25.1 Causes of acute vessel closure.

Mechanism	Risk factors	Diagnosis	Treatment
1. Dissection	• SCAD • Iatrogenic from guide catheter or guide catheter extension, guidewire, lesion preparation, stent • Severe tortuosity • Aortic dissection	Angiography OCT, IVUS	• Stenting • STAR or Stingray-based reentry as bailout (Section 21.8.2)
2. Thrombosis	• Suboptimal anticoagulation and antiplatelet Rx • Suboptimal stent implantation	Angiography IVUS OCT	• Optimize anticoagulation/ antiplatelet Rx • Thrombectomy • Balloon angioplasty • Stenting
3.1. Embolization—thrombus	• Suboptimal catheter preparation • Suboptimal balloon/ stent/indeflator preparation • Thrombus formation in sheath	Angiography IVUS OCT	• Optimize anticoagulation/ antiplatelet Rx • Thrombectomy • Balloon angioplasty • Stenting
3.2. Embolization—plaque	• Lipid-rich lesions • Atherectomy	Angiography IVUS OCT	• Aspiration • Vasodilators • Balloon angioplasty (if epicardial embolization)
3.3. Embolization—air	• Inadequate catheter preparation	Angiography ECG changes	• 100% oxygen • Aspiration • Vasodilators • IC epinephrine
4. Side branch occlusion after stenting	• Ostial disease • Dissection • Large plaque burden	Angiography ECG changes	• Rewire branch • Balloon angioplasty, including kissing balloon inflation • Stenting
5. Spasm	• Catheter manipulation	Angiography + response to vasodilators	• Vasodilators
6. Pseudolesions	• Severe tortuosity	Angiography	• Remove equipment from tortuous vessel
7. Equipment entrapment	• Calcification • Tortuosity • Poor lesion preparation • Rotational atherectomy • Equipment advancement through stents	Angiography	• Equipment retrieval (Chapter 27: Equipment loss and entrapment) • Stenting over entrapped equipment
8. Intramural hematoma	• SCAD, large lipid-rich plaque, use of cutting balloons	IVUS, OCT	• Stent distally first, then proximally • Cutting balloon • Medical therapy

Note: *ECG*, electrocardiographic; *IC*, intracoronary; *IVUS*, intravascular ultrasound; *OCT*, optical coherence tomography; *Rx*, treatment; *SCAD*, spontaneous coronary artery dissection; *STAR*, subintimal tracking and re-entry

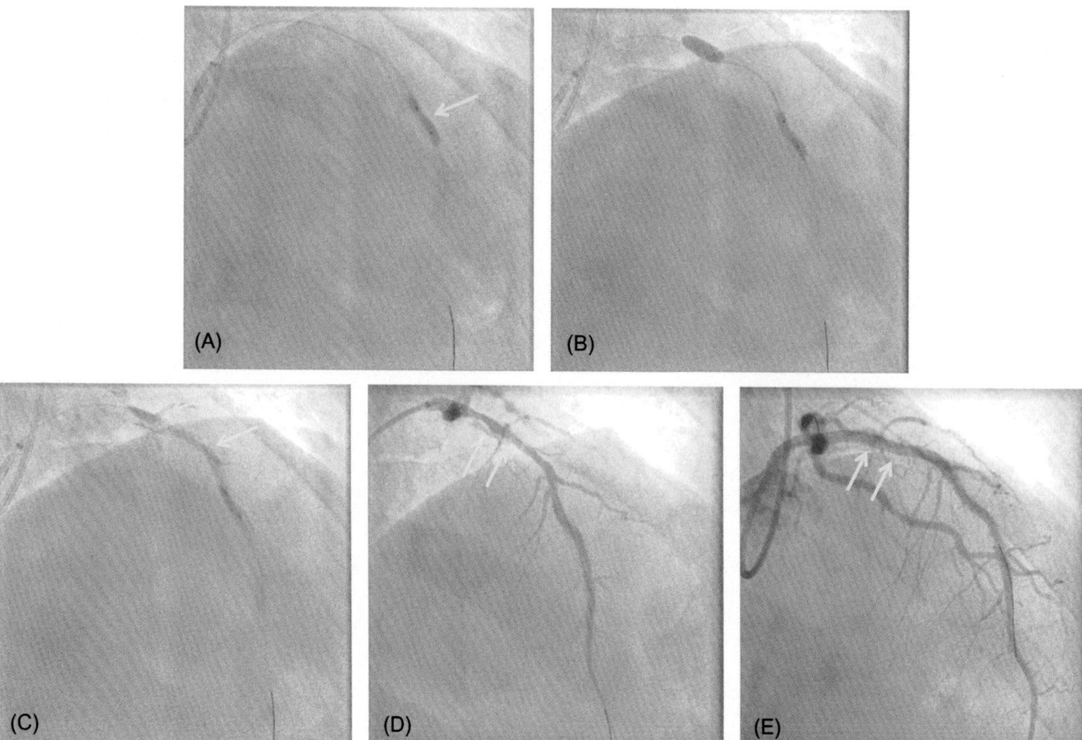

FIGURE 25.2 Dissection due to pinhole balloon rupture. An undilatable mid LAD lesion was ballooned with a 3.0 × 8 mm noncompliant balloon at 30 atm (*arrow*, panel A). A pinhole balloon shaft rupture created a subintimal dissection (*arrows*, panels C and D), that was treated with stenting and an excellent final result (*arrows*, panel E). *Courtesy of Dr. Abdul Hakeem.*

Prevention:

- Do not inject contrast if there is dampened pressure waveform.
- Achieve coaxial catheter position.
- Avoid aggressive wiring strategies; avoid using polymer-jacketed guidewires as workhorse.
- Avoid use of oversized balloons and very high-pressure balloon inflations.
- Preparation of calcified lesions.

Treatment:

- Small, non-flow limiting dissections do not require treatment. OCT (Chapter 13: Coronary intravascular imaging) is very sensitive in identifying dissections (as compared with IVUS) [1]. Proximal edge dissections after stenting most often do not require treatment, as the stent prevents forward propagation of the dissection [1].
- If wire position is maintained into the distal true lumen, stenting is performed.
- If there is no wire into the distal true lumen, various guidewires can be used in an attempt to enter into the distal true lumen (parallel wiring).
- IVUS can help confirm distal true lumen position. Injection of a collateralizing vessel, if available, can also help opacify the distal true lumen. Alternatively a microcatheter can be advanced distally and contrast injected, however, if the wire is subintimal, contrast injection will enlarge the subintimal space and make reentry into the distal true lumen more challenging. A safer strategy is to advance the microcatheter distally and then attempt to advance a workhorse guidewire. If the guidewire is in the true lumen it will reach the distal vessel without resistance. If the guidewire is in the subintimal space, significance resistance will be encountered when attempting to advance it.
- If wiring into the true lumen fails, CTO techniques (such as use of the Stingray balloon [2,3], retrograde crossing [4], or STAR [subintimal tracking and reentry]) can be used to achieve wire crossing into the distal true lumen.
- Avoid contrast injections; if they are absolutely needed they should be performed nonselectively with the catheter disengaged from the coronary ostium to avoid propagation of the dissection.
- Rapid stent delivery and deployment is needed, especially for proximal dissections, such as left main dissections [5,6].

- If a stent cannot be delivered due to significant tortuosity, calcification, or other reasons, balloon angioplasty or using a guide catheter extension may be needed first. Sometimes prolonged balloon angioplasty alone may be sufficient (although stenting is preferable, if feasible).
- If attempts to advance a guidewire into the distal true lumen fail, emergent surgery may be needed (e.g., left main dissections). Medical therapy may suffice in patients with small dissected vessels.
- If the cause of an ostial lesion is aortic dissection, emergency cardiac surgery is required with replacement of the aortic root.

25.2.1.1 Aorto-coronary dissection

CTO Manual Online case: 10, 112, 118 PCI Manual Online cases: 45, 79

Aorto-coronary dissection is a rare complication which can occur with any PCI, but is more common with CTO PCI (especially retrograde procedures) (frequency was 0.8% to 1.8% in two contemporary series [7,8]) and most commonly occurs in the right coronary artery [9] (Fig. 25.3) [10]. Dissection may be limited to the coronary sinus, but may extend to the proximal ascending aorta or even beyond the ascending aorta [11].

Causes (Fig. 25.4) [8]:

- Deep coronary engagement and utilization of aggressive guide catheters, such as 8 French Amplatz left catheters.
- Guide catheter pressure dampening.
- Forceful contrast injection, especially through "wedged" guide catheters with dampened pressure waveform.
- Predilation of the coronary ostium.
- Balloon rupture.
- Retrograde wire advancement into the subintimal and subaortic space during the reverse CART procedure.
- Antegrade attempts to recanalize a true aorto-ostial CTO.
- Aortic dissection extending into the coronary ostia.

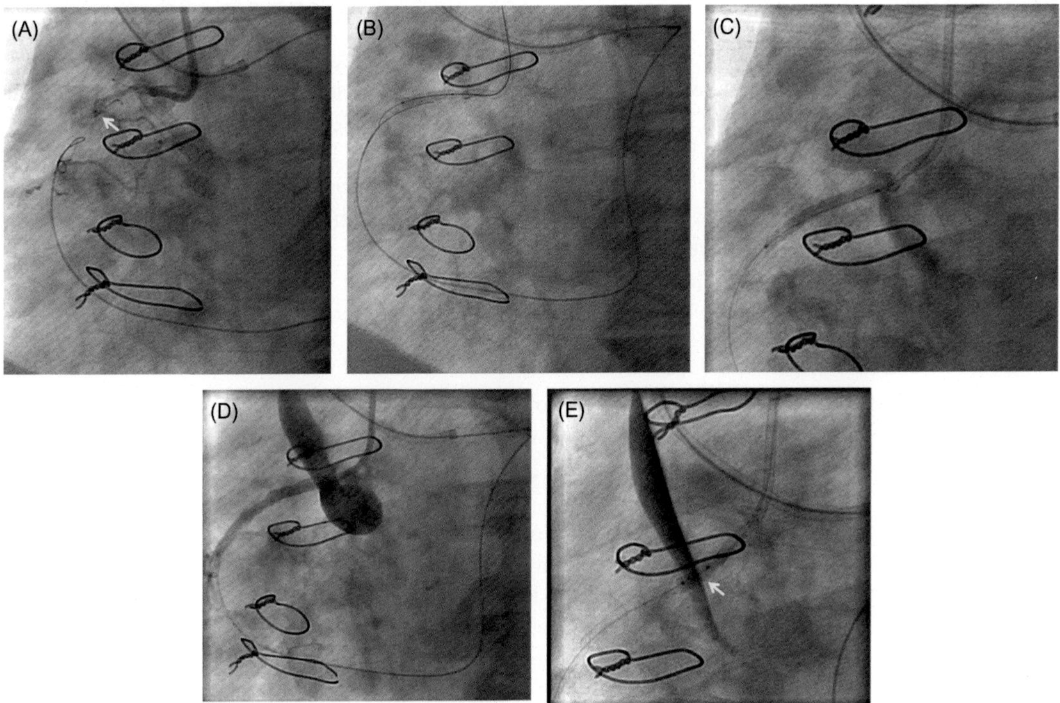

FIGURE 25.3 Illustration of an aorto-coronary dissection during retrograde CTO intervention. Retrograde CTO intervention was performed to recanalize a proximal right coronary artery CTO (*arrow*, panel A), using the reverse controlled antegrade and retrograde tracking (reverse CART) technique (panel B). Staining of the aorto-coronary junction was observed with test injections during stent placement (panel C), that expanded when cine angiography was performed (panel D). Stenting of the right coronary artery ostium was performed (*arrow*, panel E) without further antegrade contrast injections. The patient had an uneventful recovery. This case illustrates the importance of stopping antegrade contrast injections and stenting the vessel ostium if aorto-coronary dissection occurs, in order to seal the dissection flap at the entry point of the dissection. *Courtesy of Dr. Parag Doshi. Reproduced with permission from Brilakis ES. Manual of coronary chronic total occlusion interventions. A step-by-step approach. 2nd ed. Cambridge, MA: Elsevier; 2017 (Figure 12.4).*

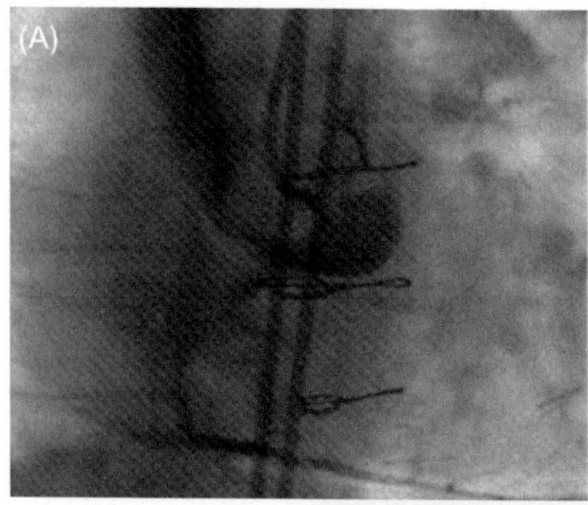

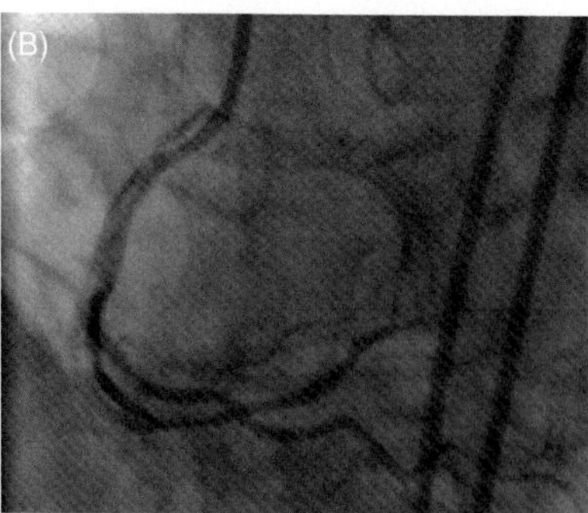

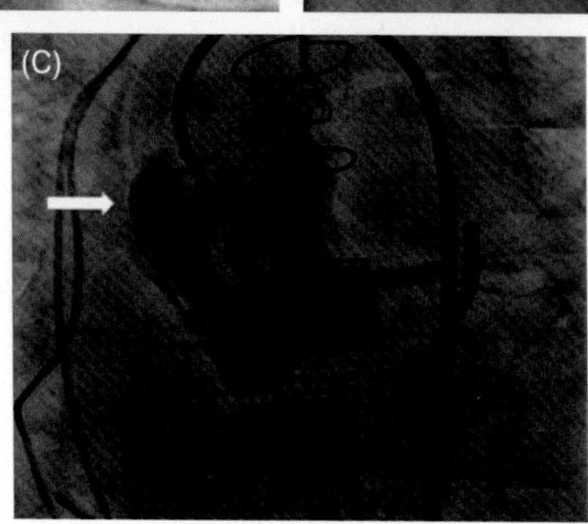

FIGURE 25.4 Examples of aorto-coronary dissection. (Panel A) Anteroposterior cranial view showing a class 2 aorto-coronary dissection caused by a retrograde approach to a proximal CTO lesion of the RCA. Guide catheter: left Amplatz 1 (Cordis). Presumed mechanism: contrast injection with a wedged catheter. (Panel B) Left anterior oblique view showing class 1 (limited to right sinus of Valsalva) aorto-coronary dissection caused by an antegrade approach of the ostial CTO lesion of the RCA. Guide catheter: Judkins right 4 (Cordis). Presumed mechanism: catheter trauma. (Panel C) Anteroposterior view showing class 2 aorto-coronary dissection with a parietal hematoma (*arrow*) caused by a retrograde approach to the proximal CTO lesion of the RCA. Guide catheter: right Amplatz 1. Presumed mechanism: catheter trauma. *CTO*, Chronic total occlusion; *RCA*, right coronary artery. *Reproduced with permission from Boukhris M, Tomasello SD, Marza F, Azzarelli S, Galassi AR. Iatrogenic aortic dissection complicating percutaneous coronary intervention for chronic total occlusion. Can J Cardiol 2015;31:320–7. Copyright Elsevier.*

Prevention:

- Consider using anchor techniques (Section 3.6.5) as an alternative to aggressive guide catheter intubation to enhance guide catheter support.
- If a guide catheter extension is being used, avoid delivery into the coronary artery directly over the guidewire. Delivering the guide catheter extension over a balloon, for example, using the "inchworming" technique (Section 30.3.3 and Fig. 30.28) is less likely to injure the vessel. Also use extreme caution when injecting through a guide catheter extension, as there is increased risk of hydraulic vessel dissection.
- Use of guide catheters with side holes in occluded right coronary arteries may decrease the risk of barotrauma, but may also provide a false sense of security, as the pressure waveform may appear normal, but ischemia and aorto-coronary dissection can still occur.
- Power injectors should also be avoided or used with caution after the proximal segment of a CTO vessel has been dilated; manual injections and contralateral injections from the donor vessel are preferred.

Treatment:

- Stop injecting contrast into the coronary a (as injections can expand the dissection plane).
- Stent the ostium of the dissected coronary artery with a stent that can expand to a diameter that will seal the dissection. The stent should be protruding 1–2 mm into the aorta to cover the ostium of the dissected vessel.
- Use intravascular ultrasonography to guide stent placement and ensure complete ostial coverage [12].
- If contrast injection is considered absolutely essential to check the status of the distal vessel, it is best performed through a dual lumen microcatheter or an aspiration thrombectomy catheter (that allows better vessel filling, but is bulkier and harder to deliver) advanced into the distal vessel [13].
- If the aorto-coronary dissection is large, perform serial non-invasive imaging (with computed tomography or transesophageal echocardiography) to ensure that the dissection has stabilized and/or resolved (Fig. 25.5) [8]. This is of particular importance if the dissection involves the ascending aorta. The blood pressure should be carefully controlled after the procedure.
- Emergency surgery is rarely needed except in patients who develop aortic regurgitation, tamponade due to rupture into the pericardium, or extension of the dissection (> 40 mm from the coronary ostia has been proposed as a cutoff [14]) (Fig. 25.6) [7–9,15].
- Emergency replacement of the aortic root is needed in patients who present with aortic dissection involving the coronary ostia (in contrast to PCI-induced aorto-coronary dissections that rarely require surgery).

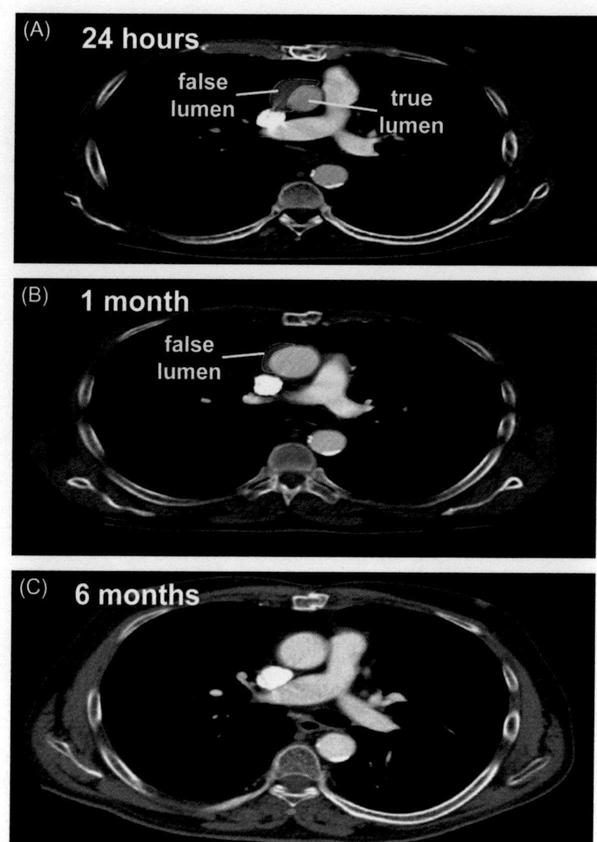

FIGURE 25.5 Computed tomography (CT) follow-up of the aorto-coronary dissection shown in panel A of Fig. 25.4. (Panel A) Twenty-four hours after the procedure, CT angiogram examination showing dislocation of intimal calcification with an eccentric double lumen. (Panel B) One-month CT control examination demonstrating almost total resolution of the dissected thrombosed lumen. (Panel C) Six-month CT control examination demonstrating total resolution of the dissection. *Reproduced with permission from Boukhris M, Tomasello SD, Marza F, Azzarelli S, Galassi AR. Iatrogenic aortic dissection complicating percutaneous coronary intervention for chronic total occlusion. Can J Cardiol 2015;31:320–7. Copyright Elsevier.*

25.2.2 Thrombosis

Causes:

- Inadequate anticoagulant and antiplatelet treatment.
- Suboptimal lesion treatment (such as stent underexpansion).
- Hypercoagulable state, including heparin-induced thrombocytopenia.
- Venous access failure resulting in inadequate intravenous administration of the anticoagulation agent.

Prevention:

- Maintain therapeutic ACT (> 250 [Hemotec] or >300 [Hemochron] seconds depending on lesion complexity and concomitant use of intravenous antiplatelet agents, such as glycoprotein IIb/IIIa inhibitors and cangrelor) throughout the case. ACTs measured using the Hemochron devices are higher than ACTs measured with the Hemotec device [16].
- Administer a P2Y12 inhibitor (pretreatment is ideal for planned PCI).
- Meticulous lesion preparation to achieve excellent stent expansion.
- Use of intravascular imaging to confirm than an excellent PCI result has been achieved.

Treatment:

- Treatment of intracoronary thrombus is performed as described in Chapter 20, Acute coronary syndromes—thrombus.

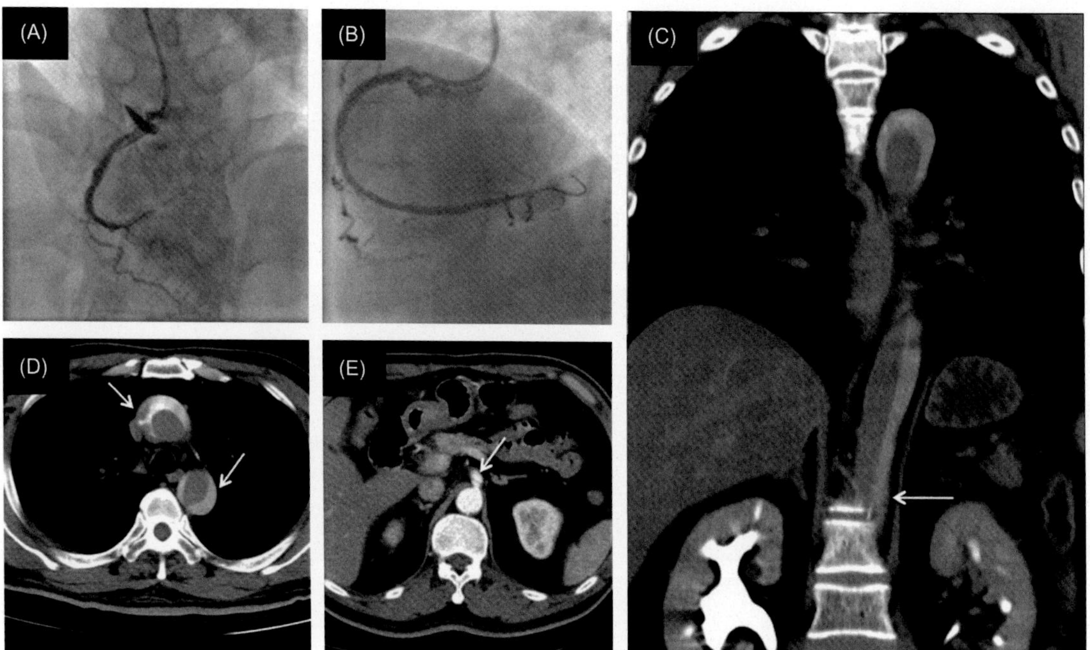

FIGURE 25.6 Example of aorto-coronary dissection extending into the descending aorta after CTO intervention. Angiography demonstrating proximal long segment dissection of the right coronary artery, extending to sinus of Valsalva (panels A and B). After stenting with a 3.5 × 24 mm bare metal stent, the final angiogram revealed limited dissection to the sinus of Valsalva (panel B). Computed tomography imaging demonstrated a type A aortic dissection extending from the ascending aorta to the suprarenal abdominal level (panels C and D) with involvement of the aortic arch and celiac trunk (panel E). *Reproduced with permission from Liao MT, Liu SC, Lee JK, Chiang FT, Wu CK. Aortocoronary dissection with extension to the suprarenal abdominal aorta: a rare complication after percutaneous coronary intervention. JACC Cardiovasc Interv 2012;5:1292–3. Copyright Elsevier.*

- Ensure optimal anticoagulation and antiplatelet therapy is achieved.
- Thrombectomy may be needed in cases of large thrombus burden.
- Occasionally stenting may trap thrombus and restore antegrade flow.

25.2.3 Embolization

Embolization, if severe, can cause profound hemodynamic compromise and require prompt treatment. Embolization may be due to thrombus, plaque/debris, and air.

25.2.3.1 Thrombus embolization

Causes:

Causes are the same as described in Section 25.2.2.

Embolization can also occur if thrombus develops in the arterial sheath and is then picked up by catheters advanced through the sheath.

Prevention:

Prevention is as described in Section 25.2.2.

Routine aspiration, followed by flushing (using a different syringe) of the arterial sheath prior to inserting a catheter.

Treatment:

Treatment is as described in Section 25.2.2. Thrombectomy may be needed for removal of the embolized thrombus.

25.2.3.2 Plaque embolization

Causes:

- Atherectomy.
- Lipid-rich plaque (Fig. 25.7) [17].
- Treatment of saphenous vein graft lesions (Chapter 18: Bypass grafts—prior CABG patients).

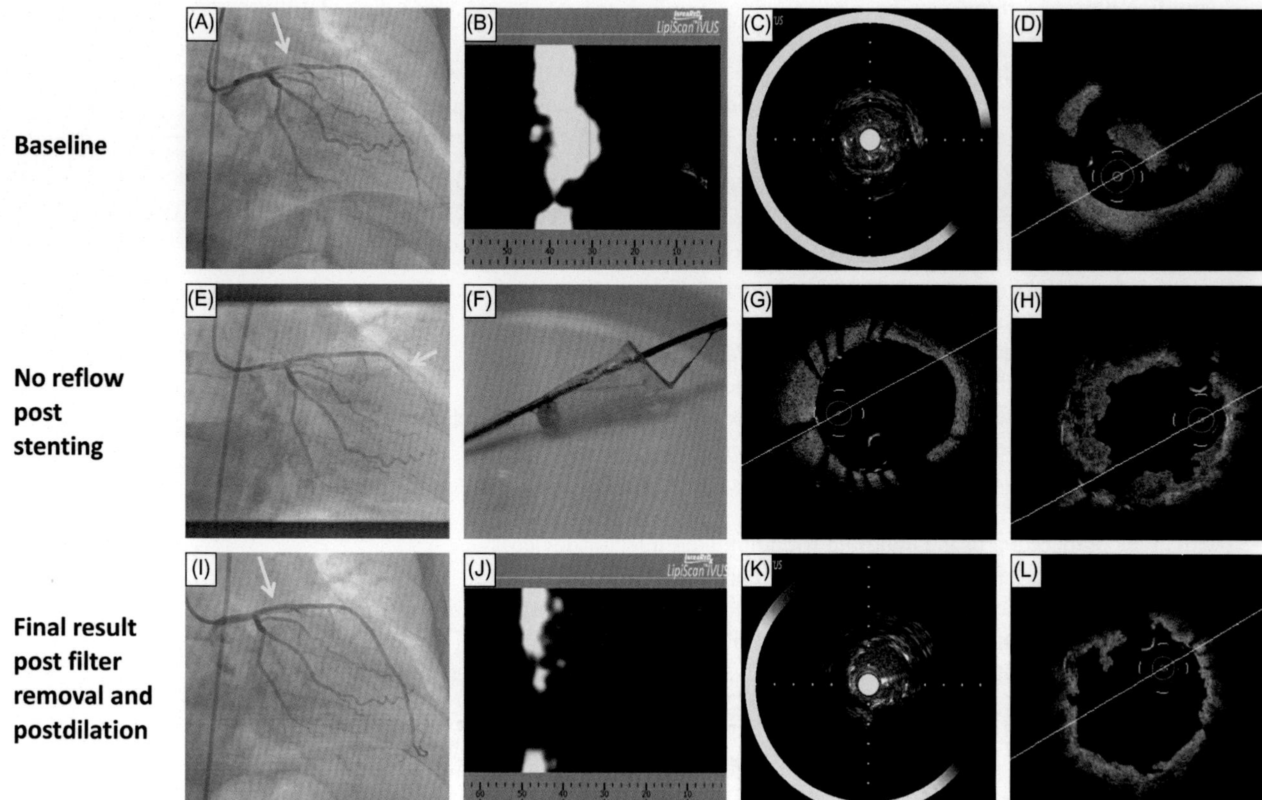

FIGURE 25.7 Stenting of a lipid core plaque rich lesion complicated by distal embolization and intrastent thrombus formation, detected by OCT. Diagnostic coronary angiography revealed a 90% lesion in the proximal left anterior descending artery (panel A). Baseline near-infrared spectroscopy with intravascular ultrasound demonstrated a large circumferential lipid core plaque (panel B) causing significant luminal stenosis (panel C). Optical coherence tomography revealed an intraluminal mass suggestive of thrombus (panel D). A Filterwire (Boston Scientific, Natick, Massachusetts) was deployed in the distal left anterior descending artery. Poststenting, antegrade flow ceased (panel E), but was restored after retrieval of the Filterwire. Examination of the Filterwire revealed a yellow tissue fragment (panel F). Optical coherence tomography immediately poststenting did not show any in-stent thrombus (panel G), however after a few minutes a filling defect appeared within the stent and optical coherence tomography revealed thrombus formation within the deployed stent (panel H). Thrombus decreased significantly after eptifibatide administration and postdilation (panel L), with an excellent angiographic result. Repeat near-infrared spectroscopy with intravascular ultrasonography showed a decrease in the size of the lipid core plaque (panels J and K). *Reproduced with permission from Papayannis AC, Abdel-Karim AR, Mahmood A, et al. Association of coronary lipid core plaque with intrastent thrombus formation: a near-infrared spectroscopy and optical coherence tomography study. Catheter Cardiovasc Interv 2013;81:488–93.*

- Inadequate cleaning of the catheter after advancement through the aorta, resulting in aortic or iliac plaque embolization.
- Aspiration thrombectomy or initial balloon predilation of a culprit lesion in STEMI.

Prevention:

- Avoidance of atherectomy; if atherectomy is used consider using lower speed (140,000–150,000 rpm for rotational atherectomy and 80,000 rpm for orbital atherectomy) which may be associated with lower risk of distal embolization and no reflow.
- Use of embolic protection devices for saphenous vein graft lesions (Chapter 18: Bypass grafts—prior CABG patients, Section 18.8.1).
- Thoroughly clear the catheter after advancement through the aorta before performing contrast injection.

Treatment:

- Stenting should be avoided in cases of plaque embolization, as it can result in release of more plaque or thrombus into the coronary circulation and worsening of the no reflow phenomenon. Covered stents may prevent distal embolization but carry high risk of restenosis and stent thrombosis.

- Aspiration thrombectomy can be performed to remove thrombus and remaining debris. (*CTO Manual* Online case: 19).
- If aspiration fails, laser could be used.
- Vasodilators (to facilitate passage or remaining debris through the microcirculation).

25.2.3.3 Air embolization

Air embolization can be a dramatic event. Large air embolization can potentially lead to cardiac arrest [18] or death.

Causes:

- Suboptimal manifold (or automated injector device) preparation.
- Poor connections of the manifold.
- Pressure dampening due to ostial disease or because the tip of the catheter is against the aortic wall.
- Use of large devices that may entrain air into the guide catheter.
- Use of the trapping technique without back bleeding the catheter afterwards.
- Rupture of a poorly prepared balloon that contains air.

Prevention:

- Meticulous preparation of the manifold (or automated injector device).
- Meticulous clearing of the manifold after advancement through the aorta.
- Always back bleed the hemostatic connector after using the trapping technique.
- Meticulous preparation of balloons to remove all air.
- Meticulous preparation of other large devices inserted through guide catheter.

Treatment:

- Administer 100% oxygen using a non-rebreather mask (helps with resorption of the air).
- Advance a coronary workhorse wire to the distal vessel and into different side branches. The tip of the wire can rupture any remaining bubbles in the distal circulation.
- Aspirate through the guide catheter [19] or through an aspiration thrombectomy catheter [20,21] (if a large amount of air has embolized).
- Intracoronary epinephrine if the patient develops cardiac arrest [18].
- Hemodynamic deterioration can be very rapid, requiring rapid intervention.
- Large systemic air embolization could be treated by hyperbaric chamber, whereas placing the patient in Trendelenburg position is controversial as it may increase cerebral edema [22,23].

25.2.4 Side branch occlusion after stenting

This is discussed in Chapter 16, Bifurcations, Section 16.2.10.1.

25.2.5 Spasm

Spasm is infrequently the cause of acute complete flow cessation, but can contribute in multiple other settings of acute vessel occlusion, such as in the setting of dissection and distal embolization.

Causes:

- Deep guide catheter engagement.
- Aggressive guide catheter manipulations.
- Coronary artery instrumentation (wiring, balloon, stenting).

Prevention:

- Avoid deep engagement and aggressive catheter manipulations.
- Administer nitroglycerin after coronary engagement, before performing coronary angiography.

Treatment:

- Intracoronary nitroglycerin (usual dose is 100–300 mcg; may need to repeat administration).
- Intracoronary calcium channel blockers.

25.2.6 Pseudolesions

Causes:

- Catheter and wire insertion through highly tortuous vessels, such as left internal mammary artery grafts (see Section 18.8.3)

Prevention:

- Avoid instrumentation of highly tortuous vessels.
- Use soft body guidewires and microcatheters when treating highly tortuous vessels.

Treatment:

- Remove equipment from the tortuous vessel. Prior to removal, ensure that no dissection has occurred (e.g., by using intravascular ultrasound).

25.2.7 Equipment entrapment

Equipment loss and entrapment is discussed in detail in Chapter 27, Equipment loss and entrapment. Rotablator entrapment is discussed in Chapter 19, Section 19.9.5.7.3.

25.2.8 Intramural hematoma (Fig. 13.16)

Post-PCI intramural hematomas can cause compression of the true lumen of the vessel leading to acute closure [24].

Causes:

- Spontaneous coronary artery dissection (Section 12.1).
- Complication of balloon angioplasty or stenting.

Prevention:

- Proper balloon and stent sizing.
- Avoid very high stent deployment pressures.
- Avoid protrusion of the postdilation balloon from the proximal and distal stent edge.

Treatment:

- Stent distally first (to prevent distal propagation of the hematoma) and then more proximally.
- Alternatively, a cutting a balloon can be used to decompress the hematoma [25,26].
- Small hematomas may be treated conservatively, especially if the inciting tear has already been sealed/stented.

25.3 Hemodynamic support

Acute closure of large coronary vessels can rapidly lead to arrhythmias and hemodynamic compromise necessitating use of hemodynamic support, as described in Chapter 14, Hemodynamic support.

References

[1] Prati F, Romagnoli E, Burzotta F, et al. Clinical impact of OCT findings during PCI: the CLI-OPCI II study. JACC Cardiovasc Imaging 2015;8:1297–305.

[2] Martinez-Rumayor AA, Banerjee S, Brilakis ES. Knuckle wire and stingray balloon for recrossing a coronary dissection after loss of guidewire position. JACC Cardiovasc Interv 2012;5:e31–2.

[3] Shaukat A, Mooney M, Burke MN, Brilakis ES. Use of chronic total occlusion percutaneous coronary intervention techniques for treating acute vessel closure. Catheter Cardiovasc Interv 2018;92:1297–300.

[4] Kotsia A, Banerjee S, Brilakis ES. Acute vessel closure salvaged by use of the retrograde approach. Interventional Cardiology 2014;6:145–7.

[5] Abdel-Karim AR, Gadiparthi C, Banerjee S, Brilakis ES. Catastrophic left main coronary artery occlusion following diagnostic coronary angiography: salvage by emergency left main coronary artery stenting. Acute Card Care 2011;13:170–3.

[6] Koza Y, Tas H, Sarac I. Successful management of an iatrogenic left main coronary artery occlusion during coronary angiography: a case report and brief review. Cardiovasc Revasc Med 2019;20:432–5.

[7] Shorrock D, Michael TT, Patel V, et al. Frequency and outcomes of aortocoronary dissection during percutaneous coronary intervention of chronic total occlusions: a case series and systematic review of the literature. Catheter Cardiovasc Interv 2014;84:670–5.

[8] Boukhris M, Tomasello SD, Marza F, Azzarelli S, Galassi AR. Iatrogenic aortic dissection complicating percutaneous coronary intervention for chronic total occlusion. Can J Cardiol 2015;31:320–7.

[9] Carstensen S, Ward MR. Iatrogenic aortocoronary dissection: the case for immediate aortoostial stenting. Heart Lung Circ 2008;17:325–9.

[10] Brilakis ES. Manual of coronary chronic total occlusion interventions. A step-by-step approach. 2nd ed. Cambridge, MA: Elsevier; 2017.

[11] Gomez-Moreno S, Sabate M, Jimenez-Quevedo P, et al. Iatrogenic dissection of the ascending aorta following heart catheterisation: incidence, management and outcome. EuroIntervention 2006;2:197–202.

[12] Abdou SM, Wu CJ. Treatment of aortocoronary dissection complicating anomalous origin right coronary artery and chronic total intervention with intravascular ultrasound guided stenting. Catheter Cardiovasc Interv 2011;78:914–19.

[13] Al Salti Al Krad H, Kaminsky B, Brilakis ES. Use of a thrombectomy catheter for contrast injection: a novel technique for preventing extension of an aortocoronary dissection during the retrograde approach to a chronic total occlusion. J Invasive Cardiol 2014;26:E54–5.

[14] Dunning DW, Kahn JK, Hawkins ET, O'Neill WW. Iatrogenic coronary artery dissections extending into and involving the aortic root. Catheter Cardiovasc Interv 2000;51:387–93.

[15] Liao MT, Liu SC, Lee JK, Chiang FT, Wu CK. Aortocoronary dissection with extension to the suprarenal abdominal aorta: a rare complication after percutaneous coronary intervention. JACC Cardiovasc Interv 2012;5:1292–3.

[16] Avendano A, Ferguson JJ. Comparison of Hemochron and HemoTec activated coagulation time target values during percutaneous transluminal coronary angioplasty. J Am Coll Cardiol 1994;23:907–10.

[17] Papayannis AC, Abdel-Karim AR, Mahmood A, et al. Association of coronary lipid core plaque with intrastent thrombus formation: a near-infrared spectroscopy and optical coherence tomography study. Catheter Cardiovasc Interv 2013;81:488–93.

[18] Prasad A, Banerjee S, Brilakis ES. Images in cardiovascular medicine. Hemodynamic consequences of massive coronary air embolism. Circulation 2007;115:e51–3.

[19] Sinha SK, Madaan A, Thakur R, Pandey U, Bhagat K, Punia S. Massive coronary air embolism treated successfully by simple aspiration by guiding catheter. Cardiol Res 2015;6:236–8.

[20] Patterson MS, Kiemeneij F. Coronary air embolism treated with aspiration catheter. Heart 2005;91:e36.

[21] Yew KL, Razali F. Massive coronary air embolism successfully treated with intracoronary catheter aspiration and intracoronary adenosine. Int J Cardiol 2015;188:56–7.

[22] Shaikh N, Ummunisa F. Acute management of vascular air embolism. J Emerg Trauma Shock 2009;2:180–5.

[23] McCarthy CJ, Behravesh S, Naidu SG, Oklu R. Air embolism: practical tips for prevention and treatment. J Clin Med 2016;5.

[24] Maehara A, Mintz GS, Bui AB, et al. Incidence, morphology, angiographic findings, and outcomes of intramural hematomas after percutaneous coronary interventions: an intravascular ultrasound study. Circulation 2002;105:2037–42.

[25] Vo MN, Brilakis ES, Grantham JA. Novel use of cutting balloon to treat subintimal hematomas during chronic total occlusion interventions. Catheter Cardiovasc Interv 2018;91:53–6.

[26] Alsanjari O, Myat A, Cockburn J, Karamasis GV, Hildick-Smith D, Kalogeropoulos AS. A case of an obstructive intramural haematoma during percutaneous coronary intervention successfully treated with intima microfenestrations utilising a cutting balloon inflation technique. Case Rep Cardiol 2018;2018:4875041.

Chapter 26

Perforation

CTO PCI Manual Online cases: 3, 13, 17, 26, 39, 40, 41, 42, 63, 89, 90, 112, 118, 119, 125, 131, 134, 139, 140, 146

PCI Manual Online cases: 6, 16, 20, 45, 55

Coronary perforation is one of the most feared complications of PCI, as it can lead to pericardial effusion and tamponade. Because effusions from coronary perforations accumulate rapidly, they often cause hypotension, sometimes necessitating emergency pericardiocentesis (and rarely cardiac surgery). Sometimes perforation may not lead to classic tamponade, but instead create a loculated effusion (especially in prior coronary artery bypass graft surgery patients) [1–5], causing compression of cardiac chambers, intramyocardial hematoma [6], or intracavitary bleeding [7].

Coronary perforations occur more frequently in complex and chronic total occlusion (CTO) PCIs. Although coronary perforations are common in CTO PCI [8], most perforations do not have serious consequences, and the risk of tamponade is low, approximately 0.3% [9]. However, the risk is higher with retrograde CTO PCI (approximately 1.3%) [10,11]. In contrast to PCI of non-CTO vessels, occlusion of a perforated target vessel in CTO PCI usually does not cause myocardial ischemia, allowing for testing sequential strategies, preparing hardware, etc.

26.1 Perforation classification, causes, and prevention

Coronary perforations are best classified according to location, as location has important implications regarding management [12]. There are three main perforation locations: (1) large vessel perforation, (2) distal vessel perforation, and (3) collateral vessel perforation, in either a septal or an epicardial collateral (Figs. 26.1 and 26.2) [13–15]. Most coronary perforations (75% in one series) were large vessel perforations, followed by distal vessel perforation (25%) [16].

The severity of coronary perforations has traditionally been graded using the Ellis classification (Fig. 26.3) [17]:

1. Class I: A crater extending outside the lumen only, in the absence of linear staining angiographically suggestive of dissection.
2. Class II: Pericardial or myocardial blush without a ≥ 1 mm exit hole.
3. Class III: Frank streaming of contrast through a ≥ 1 mm exit hole.
4. Class III-cavity spilling: Perforation into an anatomic cavity chamber, such as the coronary sinus (Fig. 26.4), the right ventricle, the left ventricle, etc.

The above classification has to be adapted to various scenarios discussed below that were not contemplated at the time the Ellis classification was developed (i.e., perforation of epicardial and septal collateral channels).

26.2 General treatment of perforations

Treatments specific to each perforation location are described in the following section. The following general measures are useful for managing coronary perforations (Fig. 26.5) [12]:

1. *Balloon inflation* proximal to or at the site of the perforation to stop the bleeding. This should be performed *immediately* to prevent accelerated accumulation of blood in the pericardial space and cardiac tamponade. The balloon should be same size as the vessel and must be semicompliant and inflated to no more than 8–10 atm to ensure occlusion of antegrade flow, without over stretching the vessel. Balloon inflation should be prolonged, lasting at least 10–15 minutes. In some cases, this may be sufficient to achieve sealing of the perforation (particularly if the perforation is less severe, such as Ellis class I or II).
2. In large vessel perforations, securing a second arterial access should be strongly considered if the guide catheter used is not 8 French, as additional bulky equipment is likely to be needed. The second access can be used to introduce a second guide catheter ("ping pong technique") with specific hardware to treat the perforation if needed, while

Manual of Percutaneous Coronary Interventions. DOI: https://doi.org/10.1016/B978-0-12-819367-9.00026-3

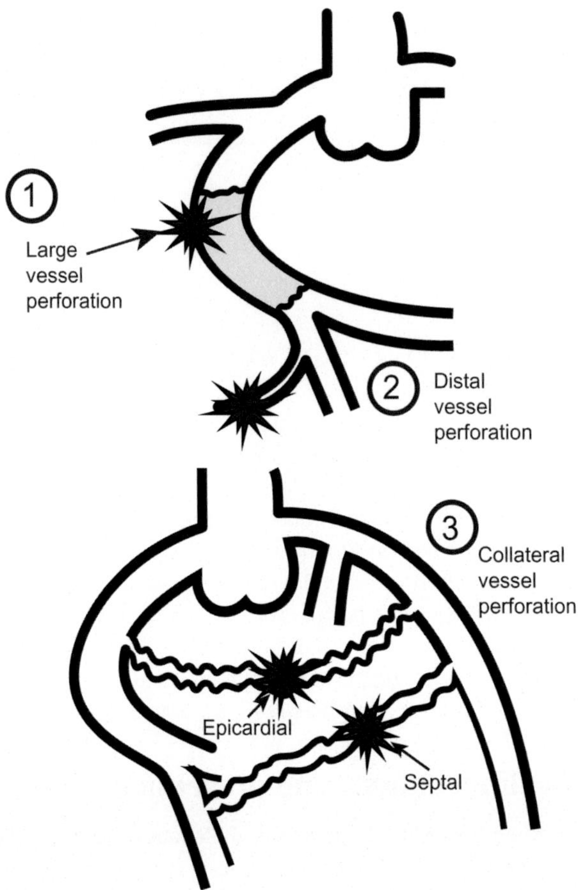

FIGURE 26.1 Types of coronary perforation. *Reproduced with permission from Brilakis ES. Manual of coronary chronic total occlusion interventions. A step-by-step approach. 2nd ed. Cambridge, MA: Elsevier; 2017 (Figure 12.15).*

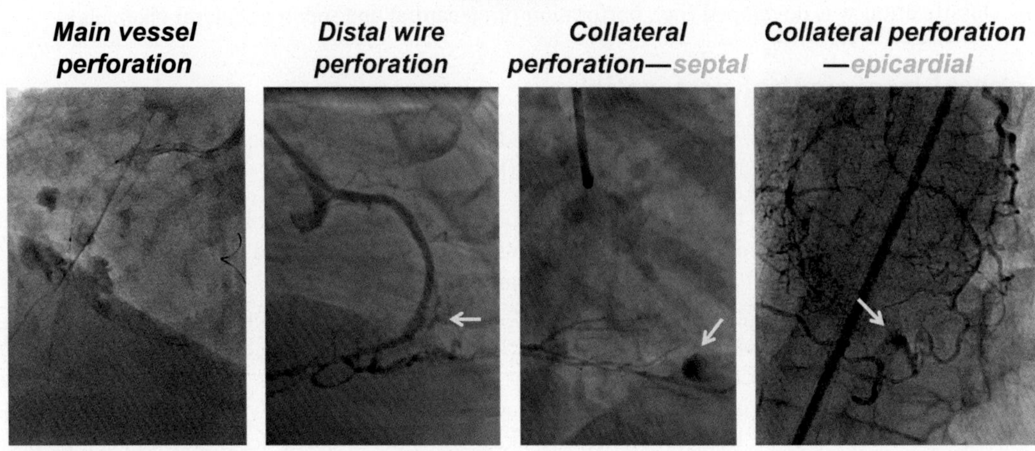

FIGURE 26.2 Examples of the different types of coronary perforation.

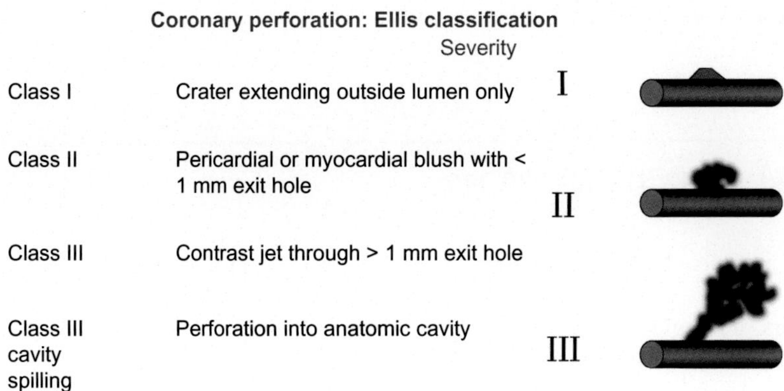

Coronary perforation: Ellis classification
Severity
Class I
Crater extending outside lumen only
I
Class II
Pericardial or myocardial blush with < 1 mm exit hole
II
Class III
Contrast jet through > 1 mm exit hole
Class III cavity spilling
Perforation into anatomic cavity
III
Ellis et al. Circulation 1994;90:2725

FIGURE 26.3 The Ellis classification of coronary perforation severity.

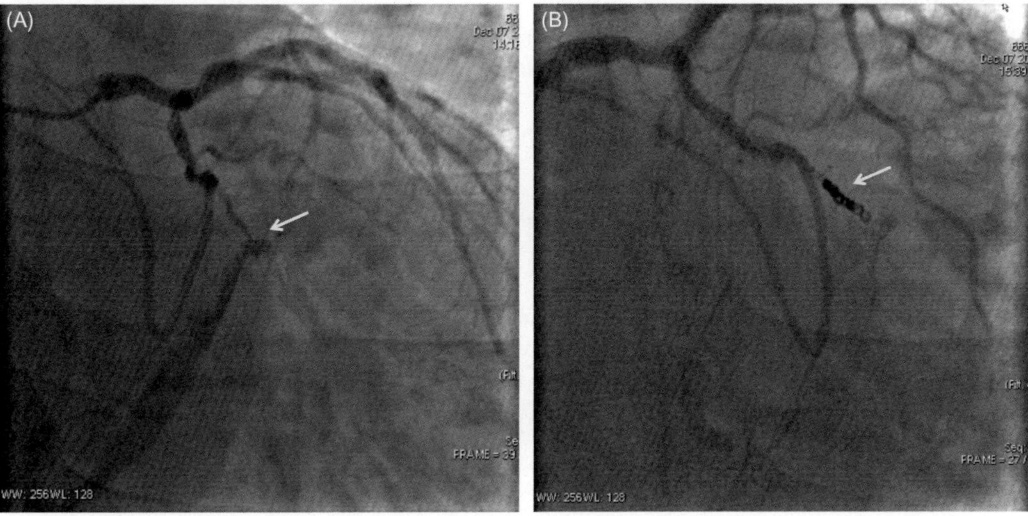

FIGURE 26.4 Perforation into the coronary sinus. Wire and microcatheter-induced perforation into the coronary sinus (*arrow*, panel A). The patient became hypotensive, requiring coiling of the perforated vessel (*arrow*, panel B). *Courtesy of Dr. Artis Kalnins.*

maintaining hemostasis at the site of perforation with an inflated balloon through the first guide catheter. Hemostasis should be confirmed using contrast injection.

3. Administration of *intravenous fluids and vasopressors* (and atropine if the patient develops bradycardia due to a vasovagal reaction).
4. Appropriate timing for performing *pericardiocentesis*: hemodynamic instability requires immediate pericardiocentesis, yet smaller size pericardial effusions may be best managed conservatively, as the elevated pericardial pressure due to the entrance of blood into the pericardial space may help "tamponade the perforation site" and minimize the risk of further bleeding. *Pericardiocentesis* can frequently be performed using X-ray guidance due to contrast exit into the pericardial space. Echocardiography remains important for assessing the size of pericardial effusion, guiding pericardiocentesis (if time allows) and evaluating the result of pericardiocentesis. Use of an echocardiographic contrast agent can be useful for detecting ongoing bleeding into the pericardial space [18].
5. *Cardiac surgery notification*: notifying cardiac surgery early may facilitate subsequent treatment, if pericardial bleeding continues despite percutaneous management attempts [18].

Reversal of anticoagulation should in most cases NOT be performed until after removal of all interventional equipment and complete evacuation of blood from the pericardial space, because reversing the effect of heparin carries the risk of guide and/or target vessel thrombosis, as well as thrombosis of the pericardial blood making it inaccessible to percutaneous drainage. Protamine dose for heparin reversal is 1 mg per 100 units of heparin (maximum dose 50 mg), administered at a rate not to exceed 5 mg per minute. Protamine administration may cause anaphylactic reactions in patients treated with NPH insulin in the past or with a history of fish allergy [19].

If glycoprotein IIb/IIIa inhibitors or cangrelor were used they should be discontinued if a perforation occurs.

26.3 Large vessel perforation

PCI Manual Online cases: 45, 55

CTO Manual Online cases: 1, 3, 39, 40, 63, 89, 90, 112, 118, 134, 140, 146

Causes

1. Implantation of oversized stents or high-pressure balloon inflations (especially in heavily calcified vessels (Fig. 26.6) or saphenous vein grafts).
2. Balloon rupture (when balloon rupture occurs, an angiogram should be obtained immediately after removal of the ruptured balloon to determine whether vessel perforation has occurred).
3. Atherectomy.
4. Guidewire exit from the vessel during lesion crossing attempts, followed by inadvertent advancement of equipment (such as balloons or microcatheters) into the pericardial space. Whereas guidewire perforation alone seldom causes blood extravasation and pericardial effusion (because it creates a very small, self-sealing hole), catheter/balloon advancement over that guidewire enlarges the hole, increasing the risk of blood extravasation. Occasionally the contrast extravasation may not occur until after a stent is placed over the perforated area.

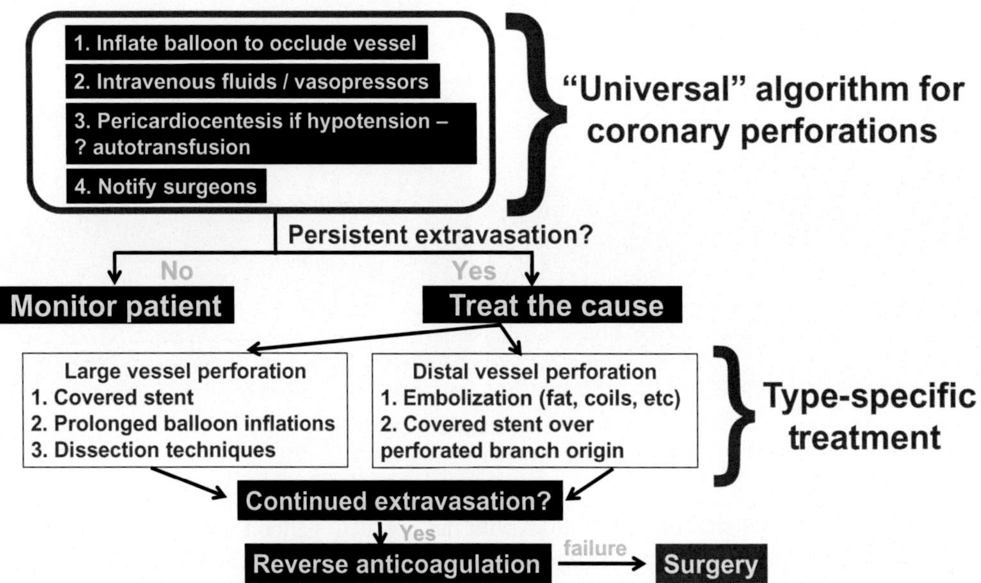

FIGURE 26.5 Overview of the management of coronary perforations.

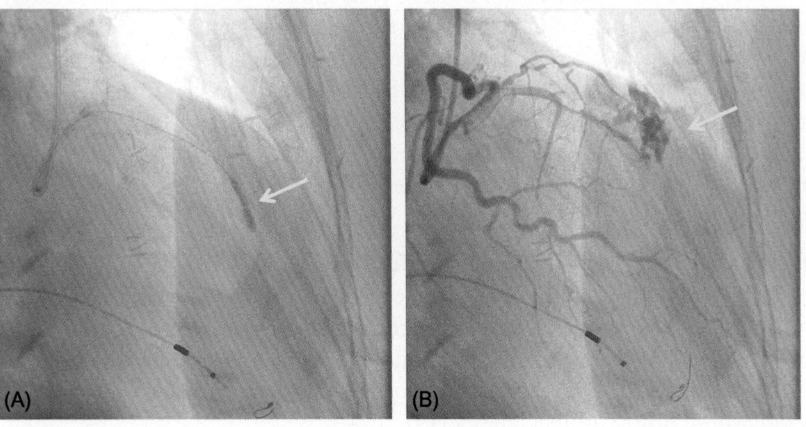

FIGURE 26.6 Vessel rupture due to high-pressure balloon inflation. A noncompliant balloon was inflated at 25 atm to treat a balloon undilatable lesion (*arrow*, panel A), resulting in vessel rupture (*arrow*, panel B). *Courtesy of Dr. Abdul Hakeem.*

Prevention

1. Avoid use of oversized stents and balloons (intravascular imaging can help guide balloon and stent size selection as outlined in Chapter 13: Coronary Intravascular Imaging).
2. Avoid very high-pressure balloon inflations.
3. Always confirm guidewire position within the vessel "architecture" (true lumen or subintimal space) before advancing other equipment.

Treatment

1. Inflate a balloon proximal to the perforation to stop the bleeding, as described in section 26.2 General Treatment of Perforations.
2. If extravasation persists despite prolonged balloon inflations, place a **covered stent**, such as the Graftmaster Rx (Section 30.18.1.1) [20] or PK Papyrus (Section 30.18.1.2) in the United States; the 5 French guide catheter compatible Begraft covered stent (Section 30.18.1.3) is available outside the United States] (Table 26.1) [20,21].
3. Depending on the size of the guide catheter being used and the size of the covered stent, delivery of the covered stent could be achieved using (1) a single guide catheter (also called "**block and deliver**" technique [22]) (Fig. 26.7) or (2) two guide catheters (dual guide catheter, also called "ping pong" guide catheter, or "dueling" guide catheter technique) (Figs. 26.8 and 26.9) [23]. The goals of both techniques is to minimize bleeding into the pericardium while preparing for covered stent delivery and deployment. If the balloon used for hemostasis and the covered stent can fit through a single (usually 8 Fr for Graftmaster covered stents) guide catheter, then the single guide catheter technique is used, otherwise two guide catheters are required. Covered stents are generally bulky and require excellent guide catheter support and possibly other maneuvers such as distal anchor balloon for delivery. After deployment, a covered stent should be postdilated aggressively to achieve good expansion and reduce the long-term risk of thrombosis. If there are residual dissections beside the covered stents they should be sealed by additional stenting, as a residual dissection can be a reentry point for bleeding.
4. ***Prolonged balloon inflations***

   If a covered stent cannot be delivered to the perforation site, prolonged balloon inflation could lead to hemostasis. Usually heparin is not reversed, until after hemostasis is achieved and all equipment is removed from the coronary artery. If pericardial bleeding continues despite prolonged balloon inflations, emergency cardiac surgery may be required (*CTO Manual* Online cases: 17 and 40).
5. ***Dissection techniques for treating a large vessel perforation.***

An alternative treatment strategy for a large vessel perforation is to create a subintimal dissection plane (proximal or distal to the perforation) that can help seal the perforation [24]. Application of this technique requires extensive experience with dissection/reentry techniques used in CTO PCI.

26.4 Distal vessel perforation

PCI Manual Online cases: 6, 20

CTO Manual Online cases: 26, 41, 42, 43, 125, 131, 134, 139

Distal vessel perforations can occasionally be difficult to diagnose, especially when collimation is used to minimize radiation exposure. Because blood flow into the pericardium may be slow, tamponade may not occur until several hours after the procedure has ended [25]. Patients with distal vessel perforation should be

TABLE 26.1 Comparison of the Graftmaster and the PK Papyrus covered stents.

	Graftmaster	PK Papyrus
Design	Two stents (sandwich)	Single stent
Material	ePTFE	Polyurethane
Guide needed	6 Fr (7 Fr for 4.5 and 4.8 mm stents)	5 Fr (6 Fr for 4.5 and 5.0 mm stents)
Available diameters (mm)	2.8, 3.5, 4.0,4.5, 4.8	2.5, 3.0, 3.5, 4.0,4.5, 5.0
Available lengths (mm)	16, 19, 26	15, 20, 26

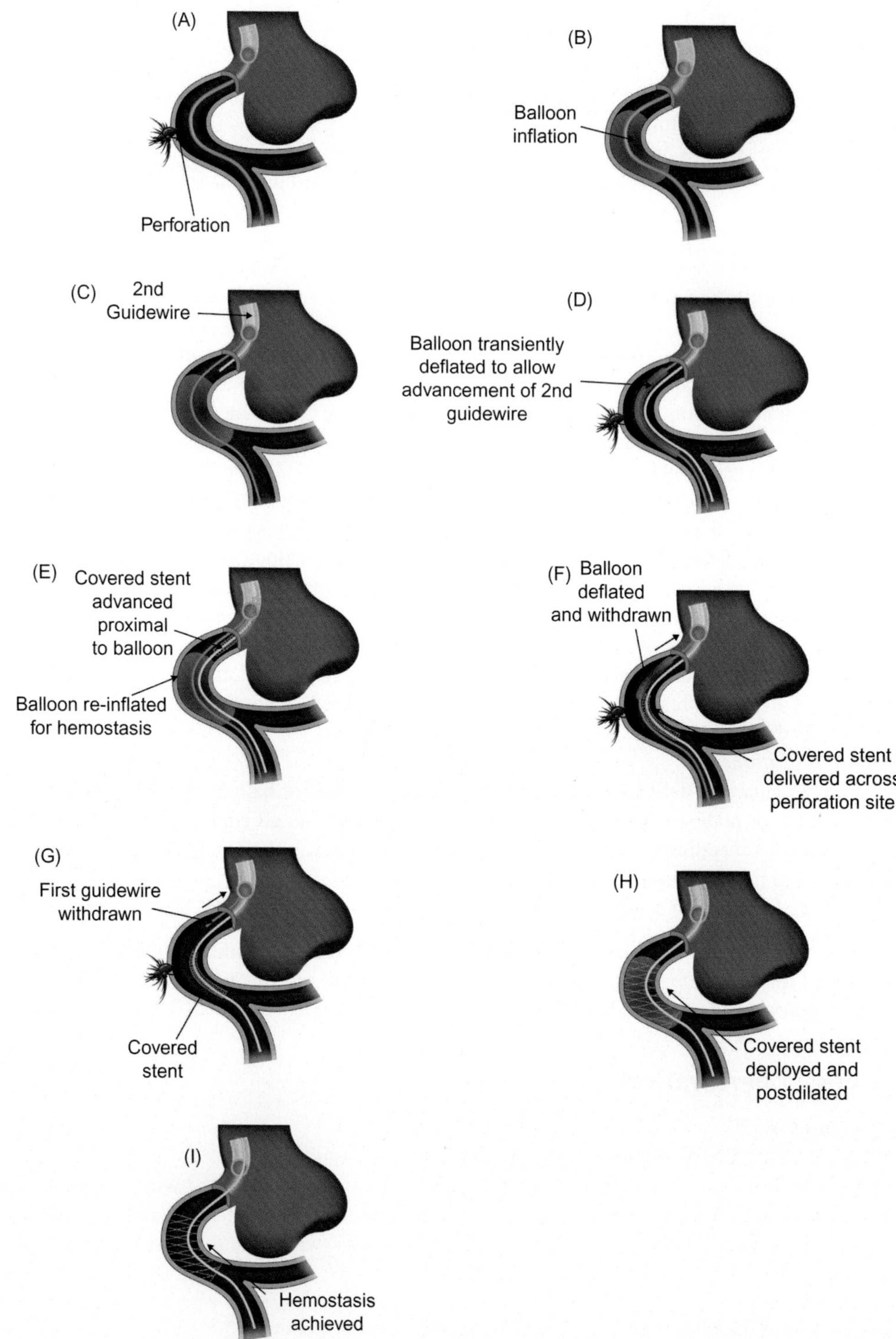

FIGURE 26.7 **Single guide catheter technique for delivering a covered stent.** (Panel A) Large vessel perforation has occurred. (Panel B) A balloon is inflated at the perforation site, stopping pericardial bleeding. (Panel C) A second guidewire is advanced to the tip of the guide catheter. (Panel D) The balloon achieving hemostasis is transiently deflated to allow advancement of the second guidewire. (Panel E) The balloon that is over the first guidewire is re-inflated and a covered stent is advanced to the tip of the guide catheter. The covered stent is advanced toward the inflated balloon that acts as a distal anchor, facilitating delivery of the covered stent to the perforation site. (Panel F) The balloon that is over the first guidewire is deflated and the covered stent is delivered across the perforation site. If the covered stent cannot be delivered to the perforation site, the blocking balloon is reinflated to prevent pericardial bleeding until other techniques to facilitate delivery of the covered stent are employed. The inflated balloon may also provide strong support facilitating delivery of the covered stent. (Panel G) The first guidewire and the hemostasis balloon are removed. (Panel H) The covered stent is deployed and postdilated (the Graftmaster requires high-pressure postdilation to achieve hemostasis). (Panel I) The perforation is sealed. Sometimes implantation of another stent proximal or distal may be necessary to seal any residual dissection that can serve as reentry point for bleeding. Reproduced with permission from Brilakis ES. Manual of coronary chronic total occlusion interventions. A step-by-step approach. 2nd ed. Cambridge, MA: Elsevier; 2017 (Figure 12.19).

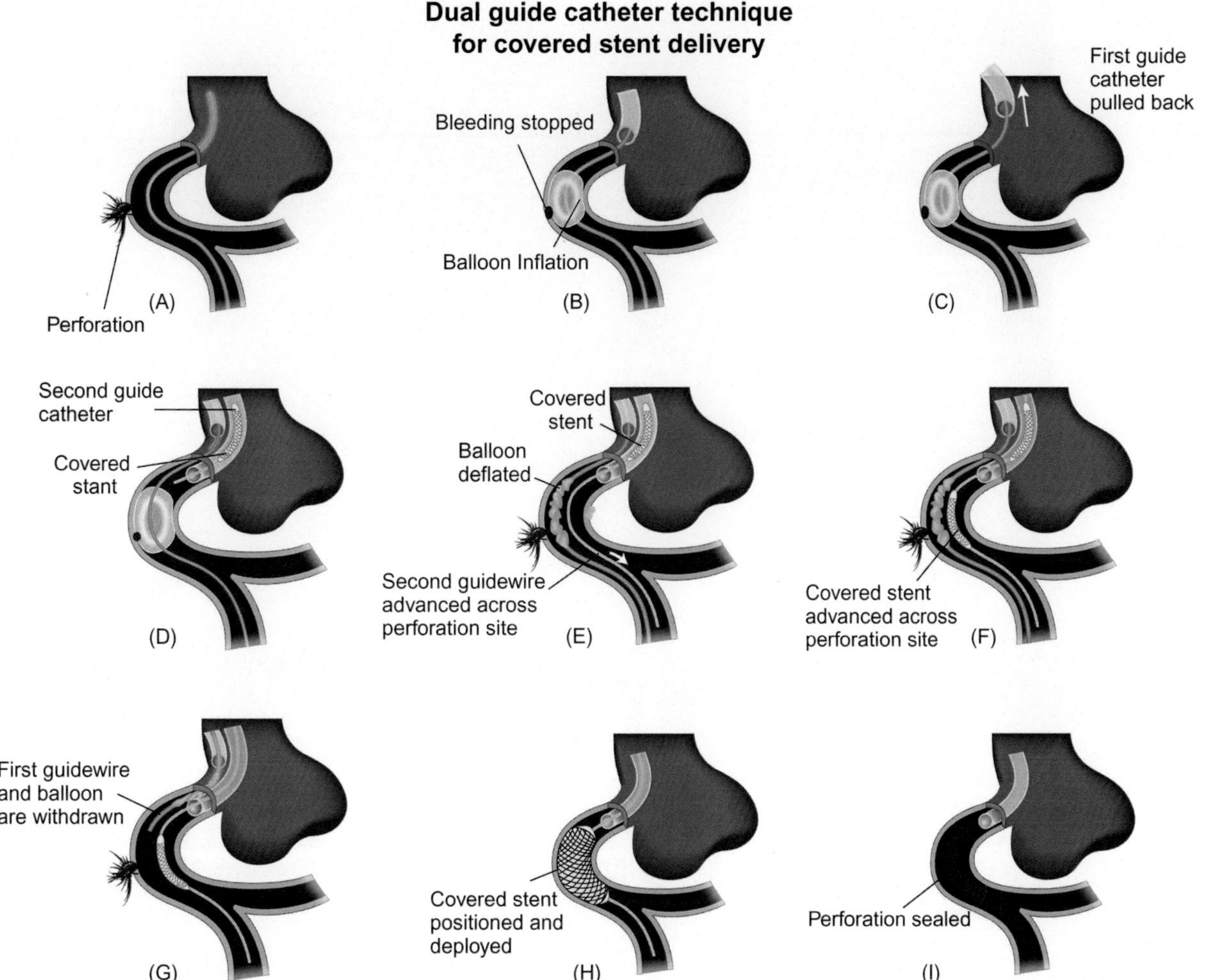

FIGURE 26.8 **Illustration of the dual guide catheter technique for delivering a covered stent.** (Panel A) A large vessel perforation has occurred. (Panel B) A balloon is inflated at the perforation site, stopping bleeding into the pericardium. (Panel C) The guide catheter is pulled back into the aorta. (Panel D) A second guide catheter (ideally 6 Fr or larger for the PK Papyrus stents and 8 Fr for the Graftmaster stent) is advanced to the perforated vessel ostium. A second guidewire is inserted into the second guide catheter along with a covered stent. (Panel E) The balloon that is over the first guidewire is deflated to allow advancement of the second guidewire across the perforation site. (Panel F) The covered stent is delivered to the perforation site. If the covered stent cannot be delivered to the perforation site, the blocking balloon is reinflated to prevent pericardial bleeding until other techniques to facilitate delivery are employed. (Panel G) The first guidewire and the balloon achieving hemostasis are removed. (Panel H) The covered stent is deployed and postdilated (the Graftmaster requires high-pressure postdilation to achieve hemostasis). (Panel I) The perforation is sealed. *Reproduced with permission from Brilakis ES. Manual of coronary chronic total occlusion interventions. A step-by-step approach. 2nd ed. Cambridge, MA: Elsevier; 2017 (Figure 12.20).*

monitored closely and should not receive a glycoprotein IIb/IIIa inhibitor or cangrelor, as tamponade may not develop until hours after the end of PCI.

Causes

1. Inadvertent advancement of a guidewire, balloon and/or microcatheter into a distal small branch. Stiff, tapered, and polymer-jacketed guidewires are more likely to cause such perforations. Knuckled guidewires are not safe, as they can still exit the vessel and cause a perforation.
2. During over-the-wire device exchanges, uncontrolled advancement of a hydrophilic-coated, polymer-jacketed, or stiff guidewire to a distal small branch may cause distal vessel perforation. Use of balloon trapping (first choice) or wire extensions (second choice) are preferred whereas Nanto's maneuver (saline injection through the microcatheter—"hydraulic exchange," Section 3.7.2) should be avoided.

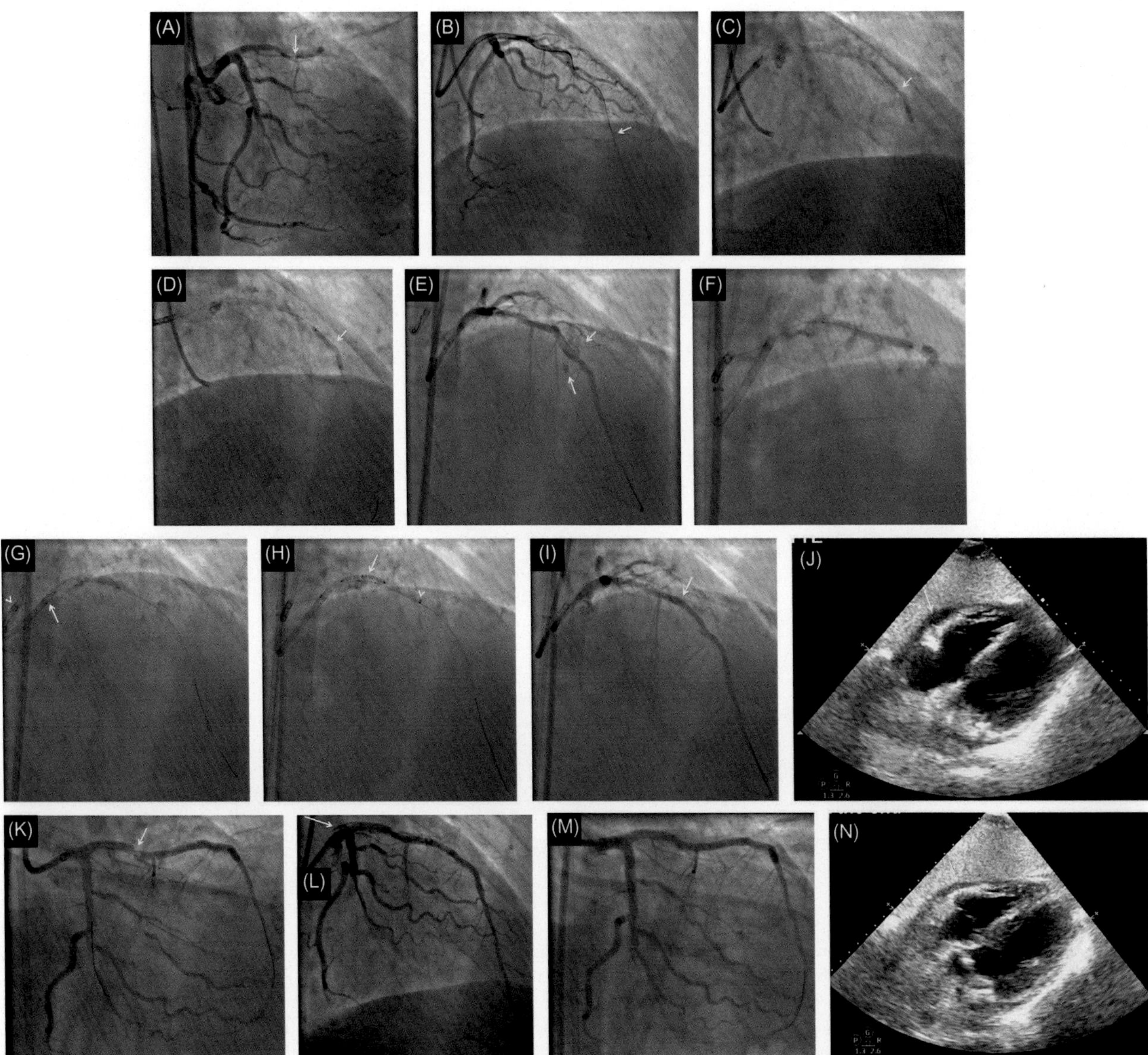

FIGURE 26.9 Illustrative case of large vessel perforation during CTO PCI. (Panel A) Chronic total occlusion of a calcified mid left anterior descending artery. (Panel B) Successful crossing of the CTO with a Fielder XT guidewire. (Panel C) Balloon undilatable lesion. (Panel D) After orbital atherectomy the balloon undilatable lesion is expanding. (Panel E) Perforation of the mid left anterior descending artery after balloon rupture. (Panel F) Balloon inflated proximal to the perforation, preventing pericardial bleeding. (Panel G) In addition to the first guide catheter (*arrowhead*) through which the balloon covering the perforation site is inserted, a second guide catheter (*arrow*) is inserted through which a second guidewire is advanced distal to the perforation site. (Panel H) A 2.8 × 19 mm Graftmaster covered stent (*arrow*) is advanced over an 8 Fr Guideliner (*arrowhead*) to the perforation site. (Panel I) After deployment of the covered stent the perforation is sealed (*arrow*). (Panel J) Echocardiogram showing a small pericardial effusion. (Panel K) Thrombus formation (*arrow*) within the stents placed in the left anterior descending artery. (Panel L) Dissection of the proximal left anterior descending artery extending into the left main (*arrow*). (Panel M) Excellent final angiographic result after thrombus aspiration and stenting of the left main. (Panel N) Echocardiogram at the end of the procedure demonstrating a small pericardial effusion. *Reproduced with permission from Brilakis ES. Manual of coronary chronic total occlusion interventions. A step-by-step approach. 2nd ed. Cambridge, MA: Elsevier; 2017 (Figure 12.21).*

Prevention

Distal wire perforation can be prevented by:

1. Paying meticulous attention to distal guidewire position during attempts to deliver equipment or when multiple guidewires are being used simultaneously, especially when stiff and polymer-jacketed guidewires are used, as those are more likely to perforate compared with workhorse guidewires.
2. Using the trapping technique to minimize guidewire movement during equipment exchanges.
3. Exchanging a stiff or polymer-jacketed guidewire for a workhorse guidewire immediately after confirmation of successful crossing.

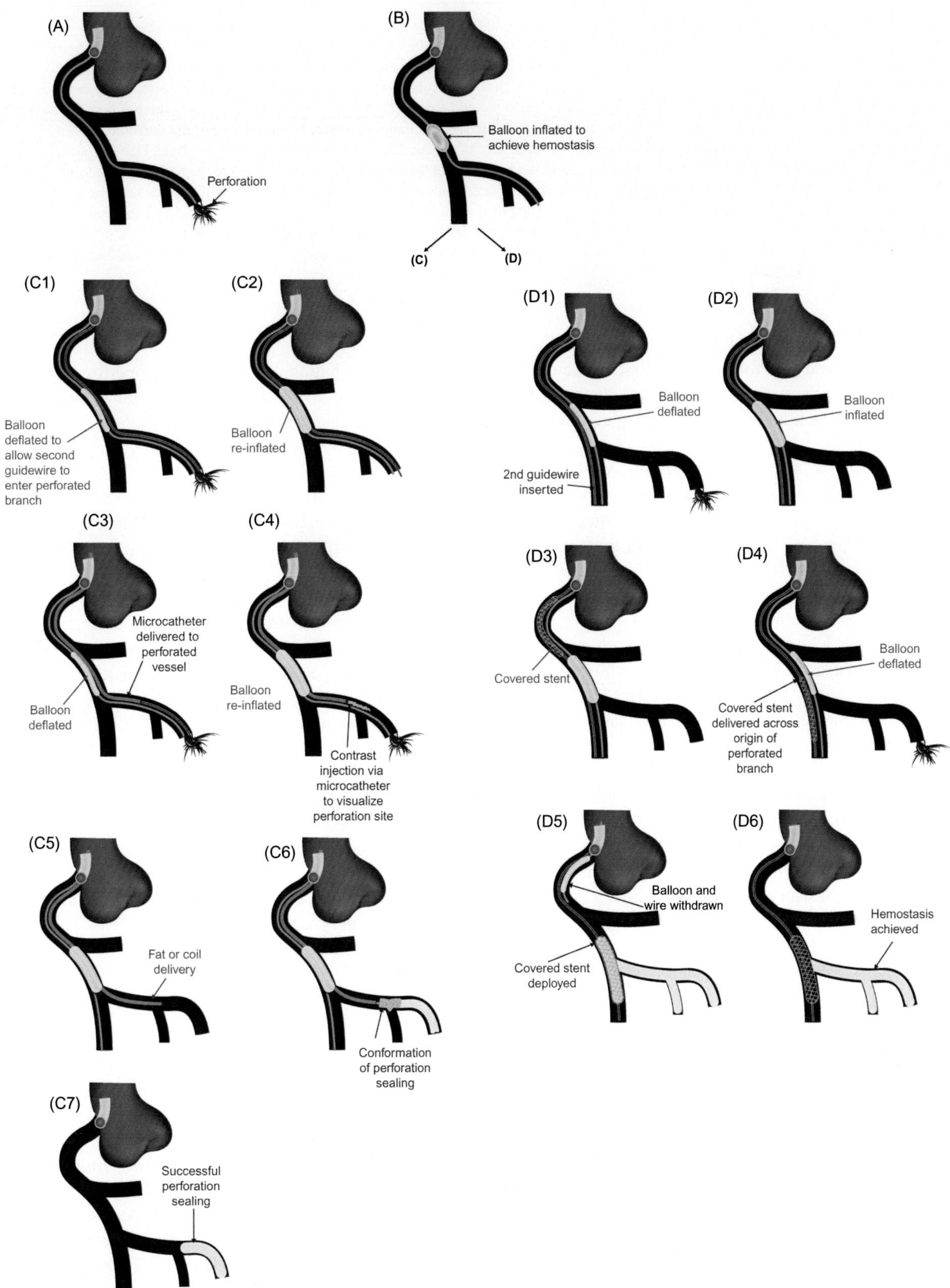

FIGURE 26.10 Treatment of distal vessel perforations. (Panel A) Distal vessel perforation with active bleeding into the pericardium. (Panel B) A balloon (blocking balloon) is inflated proximal to the perforation site to stop pericardial bleeding. (Panel C1) The blocking balloon is

(Continued)

temporarily deflated to allow advancement of a second guidewire into the perforated branch. (Panel C2) The blocking balloon is reinflated to stop pericardial bleeding. (Panel C3) The blocking balloon is transiently deflated to allow delivery of a microcatheter into the perforated vessel. (Panel C4) The blocking balloon is reinflated. Injection of contrast is performed through the microcatheter to clarify the location of the perforation. (Panel C5) Fat or a coil (or other material) is delivered through the microcatheter proximal to the perforation site. If the coil position is considered satisfactory the coil is released. (Panel C6) Contrast is injected through the microcatheter to determine whether the perforation has been sealed (sometimes sealing is delayed for a few minutes after coil delivery). (Panel C7) Successful sealing of the perforation.
(Panel D1) The blocking balloon is temporarily deflated to allow advancement of a second guidewire into the main vessel. (Panel D2) The blocking balloon is reinflated to stop pericardial bleeding. (Panel D3) A covered stent is advanced over the second guidewire proximal to the blocking balloon (which acts as distal anchor). (Panel D4) The blocking balloon is deflated and the covered stent is advanced across the ostium of the perforated vessel. (Panel D5) The covered stent is deployed. High pressure postdilation with a noncompliant balloon should be performed in nearly all cases. (Panel D6) Successful sealing of the perforated branch. This needs to be confirmed with contralateral injection to rule out retrograde filling of the perforated branch.
Reproduced with permission from Brilakis ES. Manual of coronary chronic total occlusion interventions. A step-by-step approach. 2nd ed. Cambridge, MA: Elsevier; 2017 (Figure 12.22).

Treatment (Fig. 26.10)
STEP 1: Inflate a balloon proximal to perforation
As in all coronary perforations the first management step is to inflate a balloon proximal to the perforation site to stop bleeding into the pericardium (Section 26.2 and Fig. 26.10, panels A and B). Pericardiocentesis may be required if the patient develops hypotension. Notifying cardiac surgery could expedite management in case percutaneous treatment fails.
STEP 2: Assess for continued pericardial bleeding
If balloon inflation seals the perforation, observation and heparin reversal (after removal of equipment from the coronary artery) may be all that is needed. Sometimes, suction applied through a microcatheter may collapse the vessel and achieve hemostasis [26]. However, in most cases definitive treatment with embolization or a covered stent is preferred to minimize the risk of late reopening and late tamponade.
STEP 3: Decide about embolization or covered stent implantation
Embolization is the most common treatment for distal vessel perforation and can usually be achieved using fat or coils (Fig. 26.8, panels C1–C7). Other material, such as thrombin, thrombus [27], gelfoam, microparticles, and portion of a guidewire [28] have been used for embolization. Embolization can in most cases be achieved through a single guide catheter using the "block and deliver" technique [22,29]. The "block and deliver" technique consists in simultaneous delivery of a balloon and a microcatheter through a single ≥6-French guide catheter, obviating the need for a second guide catheter. An important benefit of this technique is the ability to assess sealing of the perforation before and after release of microcoils by tip injections from the microcatheter, without the need to deflate the proximal occluding balloon [29]. The minimum guide catheter size required for various microcatheters to allow use of the "block and deliver" technique is shown in Table 26.2.

Embolization may not be feasible in some cases, for example when the perforated branch is too small or too angulated to allow wiring and delivery of a microcatheter. In such cases an alternative treatment strategy is implantation of a *covered stent* across the ostium of the perforated branch.

In rare cases in which neither embolization nor covered stent delivery are feasible, *prolonged balloon inflations* may lead to hemostasis, otherwise cardiac surgery may be required. Cardiac surgery was required in only 3% of 1,762 coronary perforations reported by the British Cardiovascular Society between 2006 and 2013 [30].

26.4.1 Embolization for treating distal vessel perforation

(Fig. 26.8, panels C1–C7)

Step 1 (Panel C1): The blocking balloon is temporarily deflated to allow advancement of a second guidewire into the perforated branch.

Step 2 (Panel C2): The blocking balloon is reinflated to stop bleeding into the pericardium.

Step 3 (Panel C3): The blocking balloon is transiently deflated to allow delivery of a microcatheter into the perforated vessel.

Fat can be delivered through any microcatheter, but if the plan is to deliver a coil the choice of microcatheter is critical. Most commercially available coils, such as the Azur (Terumo), Interlock (Boston Scientific) and Cook Micronester, are compatible with 0.018 in. microcatheters, such as the Progreat (Terumo) or the Renegade (Boston Scientific) and *cannot* be delivered through the standard 0.014 in. microcatheters with the exception of the Finecross (Table 26.2). There are neurovascular coils compatible with all 0.014 in. microcatheters, such as the Axium coils (Medtronic) (Section 30.18.2).

Step 4 (Panel C4): The blocking balloon is reinflated. Injection of contrast is performed through the microcatheter to clarify the location of the perforation.

If the microcatheter is too proximal it can be repositioned so that the length of the proximal vessel occluded with the coil is minimized.

Step 5. Embolization

Embolization is most commonly done using fat or coils. Fat is preferred in most cases (except for very large perforations) because of universal availability, low cost and biologic compatibility. However delivery is not as controlled as when a coil is used (Table 26.3).

Fat embolization

1. Fat can be harvested by advancing a hemostat in the femoral arteriotomy site (Fig. 26.11). Larger pieces can be cut into smaller ones using a scalpel.
2. Fat is then dipped into contrast for about a minute to absorb contrast and become visible under X-ray.
3. Loading the fat into the microcatheter can be challenging because fat has low density and floats on water. Turning the microcatheter hub upside down can facilitate this step (Fig. 26.12).

TABLE 26.2 **Microcatheters and their compatibility with the "Block and Deliver" technique using guiding catheters of different diameters.**

Manufacturer	Microcatheter	Length (cm)	Coil diameter compatibility	Block and deliver compatibility (Fr)
Stryker Neurovascular	EXCELSIOR SL10	150	≤0.014 in.	≥6
	EXCELSIOR XT17	150	≤0.014 in.	≥6
	EXCELSIOR 1018	150	≤0.018 in.	≥6
Cordis	RAPID TRANSIT	70/150/170	≤0.018 in.	≥7
Terumo	PROGREAT	110/130/150	≤0.018 in.	8
	FINECROSS	130/150	≤0.014 in.	≥6
Asahi	CORSAIR PRO	135/150	≤0.014 in.	≥7
	CARAVEL	135/150	≤0.014 in.	≥6
	STRIDESMOOTH	125/150	≤0.018 in.	≥7
IMDS	NHANCER PRO X	135/155	≤0.014 in.	≥6
Teleflex	TURNPIKE	135/150	≤0.014 in.	≥7
	TURNPIKE LP	135/150	≤0.014 in.	≥6
	TURNPIKE SPIRAL	135/150	≤0.014 in.	≥7
Boston Scientific	MAMBA/MAMBA FLEX	135/150	≤0.014 in.	≥7
Orbus Neich	TELEPORT	135/150	≤0.014 in.	≥6
	TELEPORT CONTROL	135/150	≤0.014 in.	≥7
Acrostak	M-CATH	135	≤0.014 in.	8

Courtesy of Dr. Gabriele Gasparini.

TABLE 26.3 **Advantages and disadvantages of fat versus coil embolization for treating distal coronary perforations.**

	Fat	Coil
Visibility	0/+	+
Controlled delivery	0	+
Catheter needed for delivery	Any microcatheter	Bigger microcatheter
Availability	Universal	Often limited
Cost	0	High

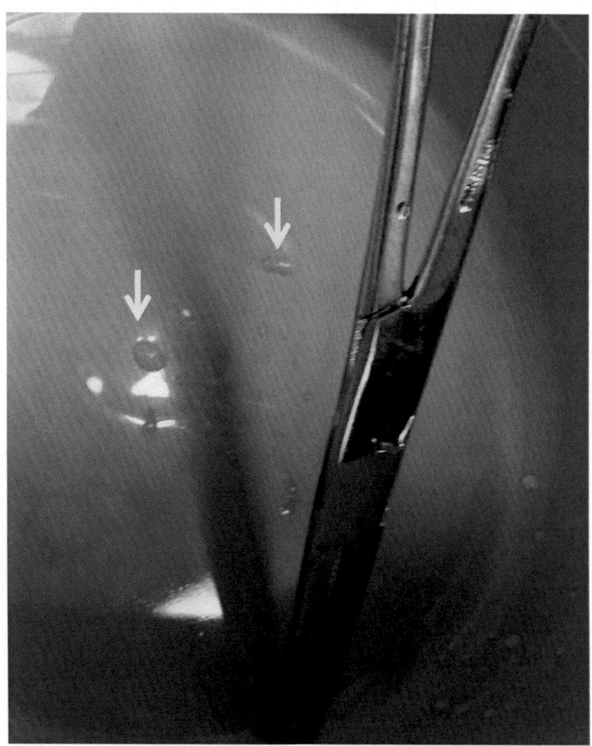

FIGURE 26.11 Harvested fat pieces (*arrows*) from the femoral arteriotomy site using a forceps. *Reproduced with permission from Brilakis ES. Manual of coronary chronic total occlusion interventions. A step-by-step approach. 2nd ed. Cambridge, MA: Elsevier; 2017 (Figure 12.24).*

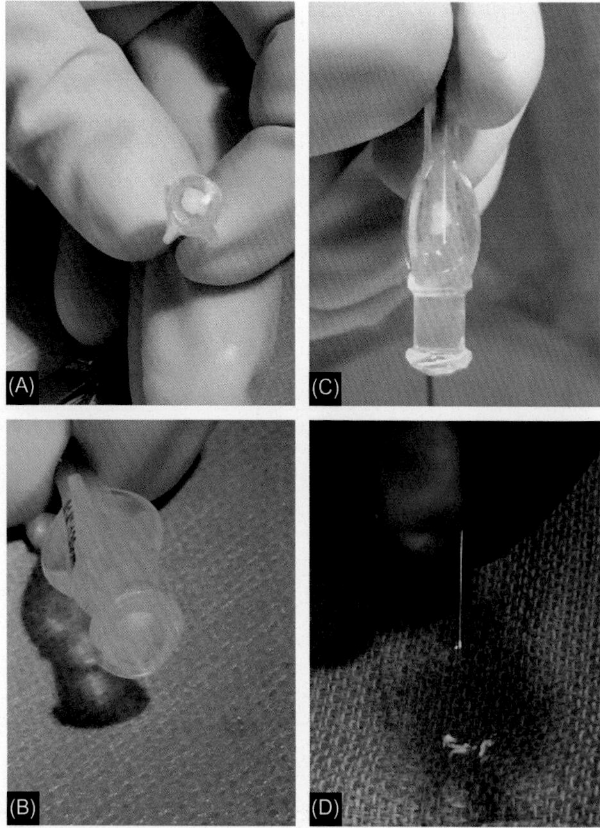

FIGURE 26.12 How to deliver fat for sealing a distal vessel perforation. Upon insertion into the hub of the microcatheter, fat floats (panel A). Upon turning the microcatheter hub upside down (panels B and C) the fat fragments advance into the microcatheter lumen (panel C). After injection with a syringe the fat particles are delivered through the tip of the microcatheter (panel D). *Reproduced with permission from Shemisa K, Karatasakis A, Brilakis ES. Management of guidewire-induced distal coronary perforation using autologous fat particles versus coil embolization. Catheter Cardiovasc Interv 2017;89:253–8 [31].*

4. The fat is injected through the microcatheter by flushing the microcatheter with saline.
5. Several fat pieces may need to be delivered to seal the distal vessel perforation.

Coil embolization

1. Since coiling is very infrequent in the cardiac catheterization laboratory, achieving familiarity with how to deliver and deploy a coil before a complication occurs can significantly facilitate management. Alternatively, obtaining help from an interventional radiologist (radiologists have significant experience with coiling and embolization) can be very helpful.
2. Having only one or two types of coils is sufficient.
3. Coil essentials: as described in Section 30.18.2 the most important characteristics of the coils are: (1) microcatheter compatibility (0.018 vs 0.014 in.) and (2) mechanism of release (pushable vs detachable). It is ideal to have 0.014 compatible coils (which can be deployed through standard 0.014 inch microcatheters) that are detachable, which allows for accurate delivery to the desired location. Table 26.4 describes various 0.014 microcatheter compatible coils currently available in the US. In our laboratory we currently use the Axium coils (Fig. 30.112). The smaller sizes are usually used for coronary perforations.
4. The coil is inserted into the delivery microcatheter and is advanced into the target vessel.
5. Pushable coils cannot be retrieved after delivery, whereas detachable coils can, if the location and configuration of the detachable coil is satisfactory. The coil is connected to the deployment system and released.

Delivery and deployment of the Axium coils is described in the following video "How to deliver an Axium coil".

Step 6 (Panel C6): Contrast is injected through the microcatheter to determine whether the perforation has been sealed (sometimes sealing is delayed for a few minutes after coil delivery).

If bleeding through the perforation site continues additional fat pieces or coils are delivered.

Step 7 (Panel C7): Angiography through the guide is performed to confirm complete sealing of the perforation.

An example of coil embolization is shown in Fig. 26.13.

26.4.2 Covered stent for treating distal vessel perforation

(Fig. 26.10, panels D1–D6)

When the perforated vessel is too small or too tortuous, advancing a guidewire into it may not be feasible. Such cases could be treated with coiling of a more proximal larger branch, but if the perforated vessel is originating from a large vessel, occlusion of that vessel can be undesirable. An alternative solution is implantation of a covered stent over the origin of the perforated branch [32].

Step 1 (Panel D1). The blocking balloon is temporarily deflated to allow advancement of a second guidewire into the main vessel.

Step 2 (Panel D2): The blocking balloon is reinflated to stop pericardial bleeding.

TABLE 26.4 Commercially available neurovascular coils in the US (compatible with 0.014 in. microcatheters). All neurovascular coils are detachable.

Coil name	Manufacturer	Description	Detachment system
Axium	Medtronic	Bare platinum coil with or without PGLA or nylon microfilaments enlaced through the coil	Axium I.D. (mechanical)
Hydrocoil (HES) MicroPlex (MCS)	Microvention	*HES*: Bare platinum coil combined with an expanding hydrogel polymer. *MCS*: Bare platinum coil with various shapes and softness profiles	V-Grip (thermomechanical)
Orbit/Cerecyte	Codman	*Orbit:* Bare platinum coil with various shapes and softness profiles. *Cerecyte*: Bare platinum coil with PGA member within coil core	EnPower (thermomechanical)
Target	Stryker	Bare platinum coil with various shapes and softness profiles	InZone (electrolytic)

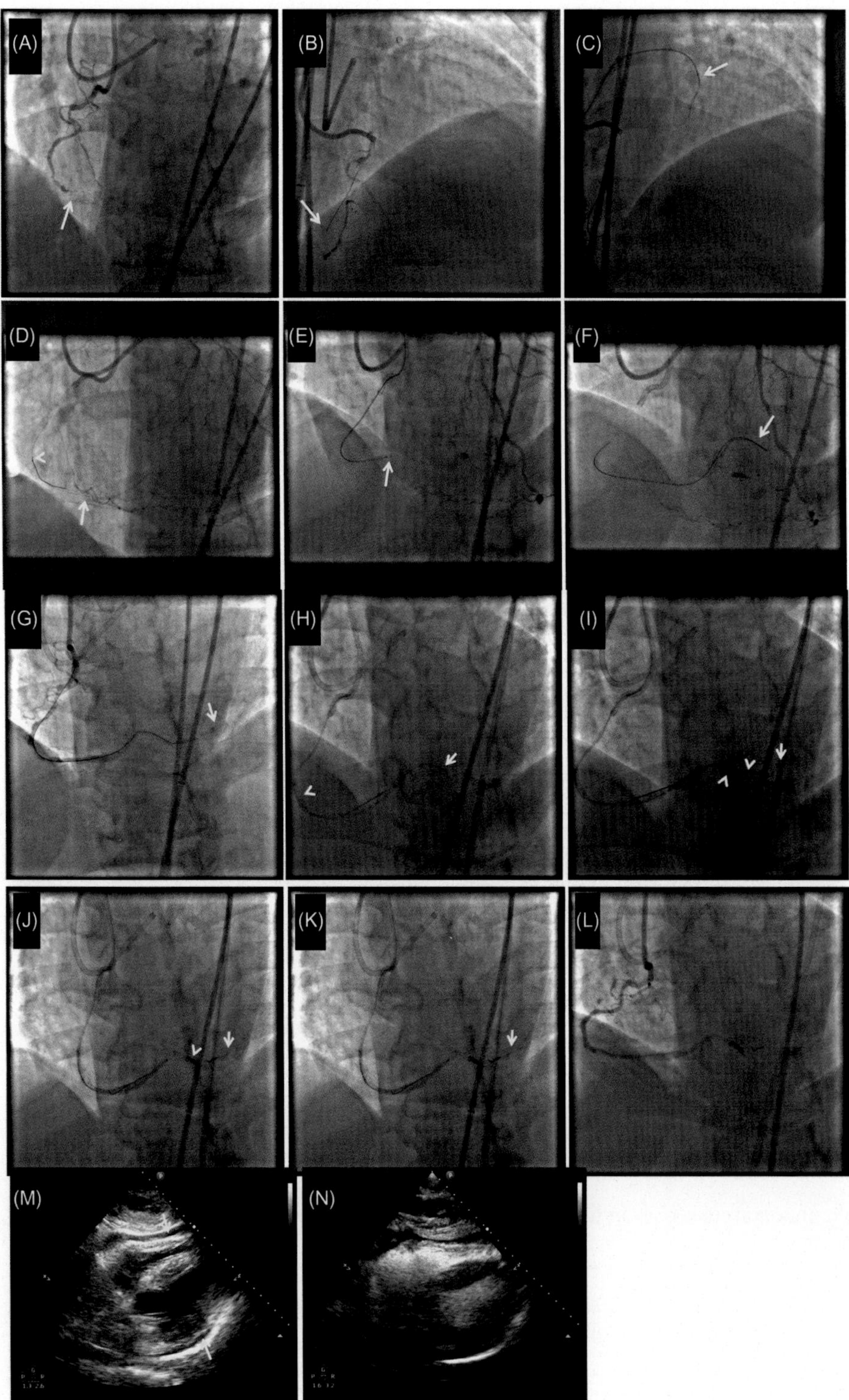

FIGURE 26.13 Example of distal vessel perforation treated with coil embolization. Coronary angiography using dual injection demonstrating a CTO of the right coronary artery (*arrow*, panel A). Antegrade crossing attempts with a guidewire failed (*arrow*, panel B). Retrograde crossing attempts through a septal collateral (*arrow*, panel C) also failed. Repeat antegrade crossing with a CrossBoss catheter (*arrowhead*, panel D) and a knuckled Fielder XT guidewire (*arrow*, panel D) achieved subintimal crossing. Reentry attempts with a Stingray balloon and guidewire (*arrow*, panel E) failed. but distal true lumen entry was achieved advancing the knuckled guidewire (*arrow*, panel F) (Subintimal Tracking And Reentry—STAR technique). Coronary angiography after balloon predilation demonstrated distal vessel perforation (*arrow*, panel G). A balloon (*arrowhead*, panel H) was inflated stopping pericardial bleeding and a Progreat microcatheter (*arrow*, panel H) was delivered to the perforation site. Bleeding (*arrow*, panel I) continued after deployment of two 2 × 5 mm coils (*arrowheads*, panel I). Bleeding (*arrow*, panel J) slowed after deployment of a 2 mm × 2 cm detachable helical hydrocoil (*arrowhead*, panel J) and stopped 15 minutes later (*arrow*, panel K). Final angiography (panel L) revealed complete occlusion of the perforation site. Transthoracic echocardiography demonstrating a small pericardial effusion (*arrows*, panel M). No bleeding into the pericardium could be seen with administration of echocardiography contrast (panel N). CTO Manual Online case: 41. *Reproduced with permission from Tarar MN, Christakopoulos GE, Brilakis ES. Successful management of a distal vessel perforation through a single 8-French guide catheter: combining balloon inflation for bleeding control with coil embolization. Catheter Cardiovasc Interv 2015;86:412–6.*

The goal of the blocking balloon is to stop bleeding from the perforation site into the pericardium, reducing the risk of tamponade. If a blocking balloon is used promptly, only a small pericardial effusion may develop that may not require pericardiocentesis.

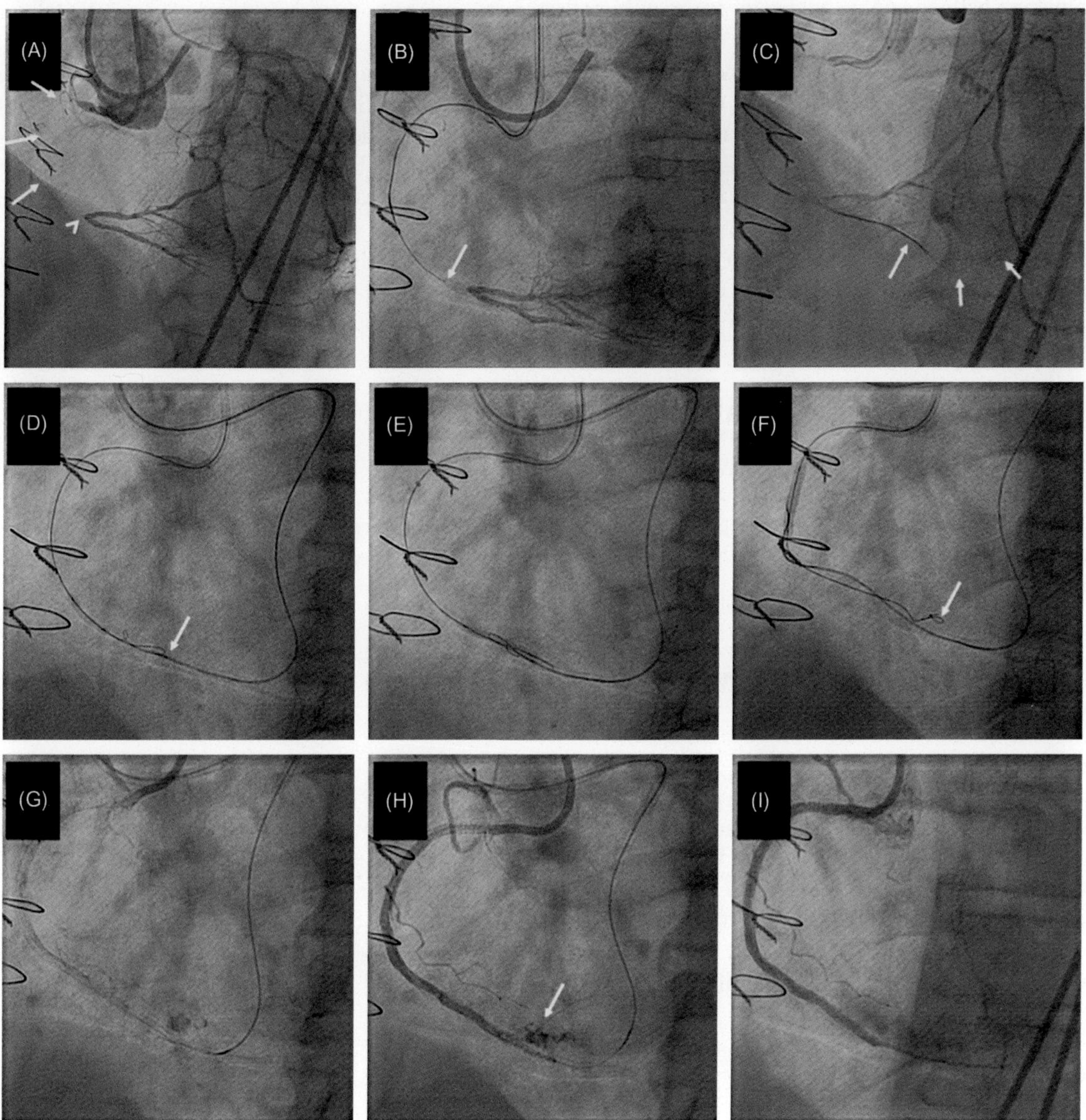

FIGURE 26.14 Example of distal vessel perforation treated with covered stent implantation. Coronary angiography demonstrating patent left anterior descending and circumflex arteries with a chronic total occlusion of the proximal right coronary artery (*arrows*, panel A) with a calcified distal cap at the bifurcation of the right posterior descending and posterolateral arteries (*arrowhead*, panel A). Antegrade wire escalation with multiple guidewires (Pilot 200, Gaia 2nd and 3rd and Confianza Pro 12, Abbott Vascular and Asahi Intecc) failed to penetrate the distal cap (*arrow*, panel B). After multiple unsuccessful attempts retrograde crossing was performed with a Sion guidewire (Asahi Intecc, *arrow*, panel C) through a septal collateral. After delivery of the Corsair catheter to the distal cap (*arrow*, panel D) a retrograde knuckle was advanced through the distal cap (*arrow*, panel E). Guide catheter extension-assisted reverse controlled antegrade and retrograde tracking and dissection was performed in the mid right coronary artery (panel F) leading to externalization of a RG3 guidewire (Asahi Intecc) (panel G). After stenting, perforation of a small branch of the right posterior descending artery (*arrow*, panel H) was seen and was successfully sealed with implantation of a 2.8 × 19 mm Graftmaster Rx covered stent (Abbott Vascular) (panel I). *Reproduced with permission from Karatasakis A, Akhtar YN, Brilakis ES. Distal coronary perforation in patients with prior coronary artery bypass graft surgery: the importance of early treatment. Cardiovasc Revasc Med 2016;17:412–7. Elsevier.*

Step 3 (Panel D3): A covered stent is advanced over the second guidewire proximal to the blocking balloon (which acts as distal anchor).

The second guidewire allows for delivery of a covered stent while maintaining hemostasis by the blocking balloon. The blocking balloon provides extra support of the second guidewire (distal anchor), facilitating delivery of the bulky covered stent.

Step 4 (Panel D4): The blocking balloon is deflated and the covered stent is advanced across the ostium of the perforated vessel.

The blocking balloon and its guidewire are usually not removed until after the covered stent has reached the perforation site.

Panel D5: The covered stent is deployed. High pressure postdilation should be performed in nearly all cases.

High pressure postdilation is important for covered stents (especially the Graftmaster stent) because they are hard to expand (the Graftmaster consists of two bare metal stents with a PTFE sandwiched in-between).

Panel D6: Successful sealing of the perforated branch. If the perforated branch was filling via contralateral collaterals prior to the perforation, contralateral injection should be performed to rule out retrograde filling of the perforated branch.

Contralateral injection is required to confirm there is no continued bleeding into the perforation site via collaterals from the contralateral coronary artery.

Fig. 26.14 demonstrates a case of distal vessel perforation treated with covered stent implantation.

26.5 Collateral vessel perforation

Perforation of an epicardial collateral branch is a serious complication of retrograde CTO PCI, as it can rapidly lead to tamponade and may be difficult to control [33,34]. In contrast, perforation of septal collaterals is unlikely to have adverse consequences [35,36], although septal hematomas [37,38], right ventricular wall hematomas [39], and even dry tamponade [27] have been reported following septal wire perforation. Collateral perforations are discussed in detail in the *Manual of CTO Interventions* [12].

26.6 Perforation in patients with prior coronary artery bypass graft surgery carries very high risk

CTO Manual Online cases: 42, 43

Although in the past prior coronary bypass graft surgery was considered protective from tamponade in patients in whom perforation occurs, we currently know that loculated effusions can develop in these patients that can compress various cardiac structures [1] (such as the left atrium [2–4] or the right ventricle [5]). Such loculated effusions can be lethal, as they can be impossible to reach and drain percutaneously. Computed tomography-guided drainage or emergency surgery may be needed to drain the loculated hematoma.

Therefore, perforations in prior CABG patients should be immediately treated (for example with covered stents or coils) to minimize the risk of loculated effusion development [1].

References

[1] Karatasakis A, Akhtar YN, Brilakis ES. Distal coronary perforation in patients with prior coronary artery bypass graft surgery: the importance of early treatment. Cardiovasc Revasc Med 2016;17:412–17.

[2] Aggarwal C, Varghese J, Uretsky BF. Left atrial inflow and outflow obstruction as a complication of retrograde approach for chronic total occlusion: report of a case and literature review of left atrial hematoma after percutaneous coronary intervention. Catheter Cardiovasc Interv 2013;82:770–5.

[3] Wilson WM, Spratt JC, Lombardi WL. Cardiovascular collapse post chronic total occlusion percutaneous coronary intervention due to a compressive left atrial hematoma managed with percutaneous drainage. Catheter Cardiovasc Interv 2015;86:407–11.

[4] Franks RJ, de Souza A, Di Mario C. Left atrial intramural hematoma after percutaneous coronary intervention. Catheter Cardiovasc Interv 2015;86:E150–2.

[5] Adusumalli S, Morris M, Pershad A. Pseudo-pericardial tamponade from right ventricular hematoma after chronic total occlusion percutaneous coronary intervention of the right coronary artery: successfully managed percutaneously with computerized tomographic guided drainage. Catheter Cardiovasc Interv 2016;88:86–8.

[6] Kawana M, Lee AM, Liang DH, Yeung AC. Acute right ventricular failure after successful opening of chronic total occlusion in right coronary artery caused by a large intramural hematoma. Circ Cardiovasc Interv 2017;10.

[7] Mertens A, Dalal P, Ashbrook M, Hanson I. Coil embolization of coronary-cameral fistula complicating revascularization of chronic total occlusion. Case Rep Cardiol 2018;2018:4.

[8] Rathore S, Matsuo H, Terashima M, et al. Procedural and in-hospital outcomes after percutaneous coronary intervention for chronic total occlusions of coronary arteries 2002 to 2008: impact of novel guidewire techniques. JACC Cardiovasc Interv 2009;2:489–97.

[9] Patel VG, Brayton KM, Tamayo A, et al. Angiographic success and procedural complications in patients undergoing percutaneous coronary chronic total occlusion interventions: a weighted meta-analysis of 18,061 patients from 65 studies. JACC Cardiovasc Interv 2013;6:128–36.

[10] Karmpaliotis D, Karatasakis A, Alaswad K, et al. Outcomes with the use of the retrograde approach for coronary chronic total occlusion interventions in a contemporary multicenter US registry. Circ Cardiovasc Interv 2016;9.

[11] Danek BA, Karatasakis A, Karmpaliotis D, et al. Development and validation of a scoring system for predicting periprocedural complications during percutaneous coronary interventions of chronic total occlusions: the prospective global registry for the study of chronic total occlusion intervention (PROGRESS CTO) complications score. J Am Heart Assoc 2016;5.

[12] Brilakis ES. Manual of coronary chronic total occlusion interventions. A step-by-step approach. 2nd ed. Cambridge, MA: Elsevier; 2017.

[13] Brilakis ES, Karmpaliotis D, Patel V, Banerjee S. Complications of chronic total occlusion angioplasty. Interv Cardiol Clin 2012;1:373–89.

[14] Xenogiannis I, Brilakis ES. Advances in the treatment of coronary perforations. Catheter Cardiovasc Interv 2019;93:921–2.

[15] Tajti P, Xenogiannis I, Chavez I, et al. Expecting the unexpected: preventing and managing the consequences of coronary perforations. Expert Rev Cardiovasc Ther 2018;16:805–14.

[16] Shaukat A, Tajti P, Sandoval Y, et al. Incidence, predictors, management and outcomes of coronary perforations. Catheter Cardiovasc Interv 2019;93:48–56.

[17] Ellis SG, Ajluni S, Arnold AZ, et al. Increased coronary perforation in the new device era. Incidence, classification, management, and outcome. Circulation 1994;90:2725–30.

[18] Bagur R, Bernier M, Kandzari DE, Karmpaliotis D, Lembo NJ, Rinfret S. A novel application of contrast echocardiography to exclude active coronary perforation bleeding in patients with pericardial effusion. Catheter Cardiovasc Interv 2013;82:221–9.

[19] Stewart WJ, McSweeney SM, Kellett MA, Faxon DP, Ryan TJ. Increased risk of severe protamine reactions in NPH insulin-dependent diabetics undergoing cardiac catheterization. Circulation 1984;70:788–92.

[20] Briguori C, Nishida T, Anzuini A, Di Mario C, Grube E, Colombo A. Emergency polytetrafluoroethylene-covered stent implantation to treat coronary ruptures. Circulation 2000;102:3028–31.

[21] Romaguera R, Waksman R. Covered stents for coronary perforations: is there enough evidence? Catheter Cardiovasc Interv 2011;78:246–53.

[22] Tarar MN, Christakopoulos GE, Brilakis ES. Successful management of a distal vessel perforation through a single 8-French guide catheter: combining balloon inflation for bleeding control with coil embolization. Catheter Cardiovasc Interv 2015;86:412–16.

[23] Ben-Gal Y, Weisz G, Collins MB, et al. Dual catheter technique for the treatment of severe coronary artery perforations. Catheter Cardiovasc Interv 2010;75:708–12.

[24] Xenogiannis I, Tajti P, Nicholas Burke M, Brilakis ES. An alternative treatment strategy for large vessel coronary perforations. Catheter Cardiovasc Interv 2019;93:635–8.

[25] Stathopoulos IA, Kossidas K, Garratt KN. Delayed perforation after percutaneous coronary intervention: rare and potentially lethal. Catheter Cardiovasc Interv 2014;83:E45–50.

[26] Yasuoka Y, Sasaki T. Successful collapse vessel treatment with a syringe for thrombus-aspiration after the guidewire-induced coronary artery perforation. Cardiovasc Revasc Med 2010;11:263e1–e3.

[27] Matsumi J, Adachi K, Saito S. A unique complication of the retrograde approach in angioplasty for chronic total occlusion of the coronary artery. Catheter Cardiovasc Interv 2008;72:371–8.

[28] Hartono B, Widito S, Munawar M. Sealing of a dual feeding coronary artery perforation with homemade spring guidewire. Cardiovasc Interv Ther 2015;30:347–50.

[29] Garbo R, Oreglia JA, Gasparini GL. The balloon-microcatheter technique for treatment of coronary artery perforations. Catheter Cardiovasc Interv 2017;89:E75–83.

[30] Kinnaird T, Kwok CS, Kontopantelis E, et al. Incidence, determinants, and outcomes of coronary perforation during percutaneous coronary intervention in the United Kingdom between 2006 and 2013: an analysis of 527 121 cases from the British Cardiovascular Intervention Society database. Circ Cardiovasc Interv 2016;9.

[31] Shemisa K, Karatasakis A, Brilakis ES. Management of guidewire-induced distal coronary perforation using autologous fat particles versus coil embolization. Catheter Cardiovasc Interv 2017;89:253–8.

[32] Sandoval Y, Lobo AS, Brilakis ES. Covered stent implantation through a single 8-french guide catheter for the management of a distal coronary perforation. Catheter Cardiovasc Interv 2017;90:584–8.

[33] Boukhris M, Tomasello SD, Azzarelli S, Elhadj ZI, Marza F, Galassi AR. Coronary perforation with tamponade successfully managed by retrograde and antegrade coil embolization. J Saudi Heart Assoc 2015;27:216–21.

[34] Ngo C, Christopoulos G, Brilakis ES. Conservative management of an epicardial collateral perforation during retrograde chronic total occlusion percutaneous coronary intervention. J Invasive Cardiol 2016;28:E11–12.

[35] Lee NH, Seo HS, Choi JH, Suh J, Cho YH. Recanalization strategy of retrograde angioplasty in patients with coronary chronic total occlusion—analysis of 24 cases, focusing on technical aspects and complications. Int J Cardiol 2010;144:219–29.

[36] Araki M, Murai T, Kanaji Y, et al. Interventricular septal hematoma after retrograde intervention for a chronic total occlusion of a right coronary artery: echocardiographic and magnetic resonance imaging-diagnosis and follow-up. Case Rep Med 2016;2016:8514068.

[37] Lin TH, Wu DK, Su HM, et al. Septum hematoma: a complication of retrograde wiring in chronic total occlusion. Int J Cardiol 2006;113:e64–6.

[38] Abdel-Karim AR, Vo M, Main ML, Grantham JA. Interventricular septal hematoma and coronary-ventricular fistula: a complication of retrograde chronic total occlusion intervention. Case Rep Cardiol 2016;2016:8750603.

[39] Ghobrial MSA, Egred M. Right ventricular wall hematoma following angioplasty to right coronary artery occlusion. J Invasive Cardiol 2019;31:E66.

Chapter 27

Equipment loss and entrapment

Various types or equipment, such as stents, guidewires, and various catheters can be lost or entrapped either within or outside the coronary artery [1]. Such equipment can lead to occlusion or perforation of the coronary vessel. It could also lead to systemic embolization, such as embolization to an intracranial artery causing a stroke. Device entrapment is a more grave complication than device loss, and may require emergency surgery for removal [1]. In this chapter we discuss how to prevent, diagnose and treat such complications. Entrapment of the Rotablator burr is discussed in Section 19.9.5.7.3.

27.1 Stent loss or entrapment

CTO PCI Manual Online cases 74, 122, 128
PCI Manual Online cases 10, 93

27.1.1 Causes

- Coronary tortuosity and calcification [2,3].
- Poor vessel preparation prior to attempting stent delivery. In the ROTAXUS trial stent loss occurred in 2% of lesions in the no atherectomy arm of the study versus 0.5% of lesions in the rotational atherectomy arm of the study [4]. Poor vessel preparation may result in stent deformation during attempts to deliver the stent, followed by stent loss when attempting to withdraw the stent into the guide catheter (Fig. 27.1).
- Direct stenting.
- Use of small (such as 5 French guide catheters).
- Forceful withdrawal of the stent inside the guide catheter (or inside a guide catheter extension, as described below) when resistance is felt [2].
- Use of guide catheter extensions. Stent loss can occur both during advancement of the stent if the stent catches the proximal collar or during stent withdrawal, especially when the stent is deformed.
- Stent advancement through a previously deployed stent [5].
- Attempting to deliver equipment via a collateral during the retrograde approach to CTO interventions (which can predispose to both stent loss [6] and wire entrapment [7]).

27.1.2 Prevention

- Avoid direct stenting, especially in tortuous and calcified vessels.
- Meticulous vessel preparation, often using atherectomy in calcified lesions and intracoronary imaging to confirm adequate lesion expansion before attempting stent delivery.
- Stent from distal to proximal. Sometimes, however, the need for distal stenting does not arise until after a proximal stent is deployed, for example, in cases of distal edge dissection [1].
- Use a guide catheter extension for delivering stents in tortuous and calcified vessels.
- Avoid forceful advancement attempts.
- When using a guide catheter extension, always place the external push rod under a towel at the side of the Y-connector to reduce the risk of the guidewires "wrapping around" the guide catheter extension delivery rod.
- Do not apply force if resistance is felt while advancing a stent through a guide catheter extension. Instead, remove the guide catheter extension and reinsert it, paying particular attention to avoiding wrapping up of the guidewire and

Manual of Percutaneous Coronary Interventions. DOI: https://doi.org/10.1016/B978-0-12-819367-9.00027-5

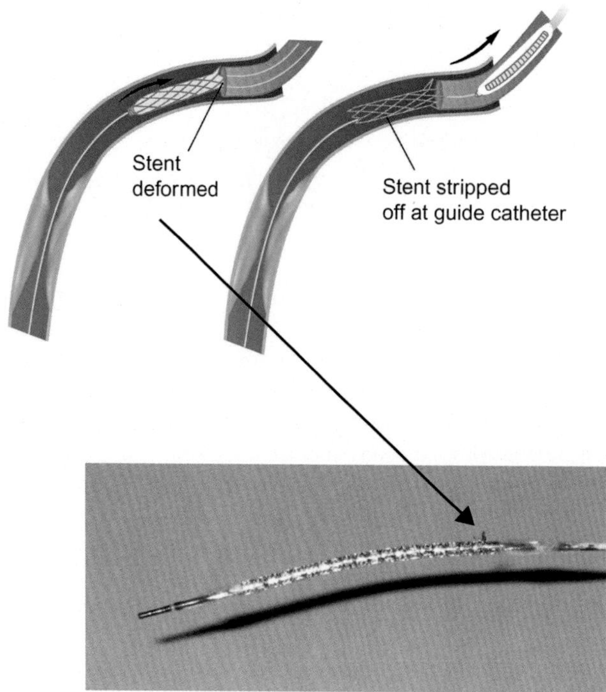

FIGURE 27.1 Mechanism of stent loss: stent is deformed during delivery attempts, followed by "catching" on the guide catheter upon withdrawal and being stripped off the stent delivery balloon.

the guide extension delivery rod. Inserting the stent into the guide catheter extension outside the guide catheter and advancing both into the guide catheter at the same time can help avoid stent dislodgement or loss.

- Avoid using small (such as 5 French) guide catheters when treating complex lesions, as smaller guide catheters have less room to allow for withdrawal of deformed stents.

27.1.3 Treatment (Fig. 27.2)

27.1.3.1 Location of the lost stent

The location where the stent is lost determines the subsequent steps. If a stent is lost in a coronary artery it should be either retrieved or deployed/crushed. If a stent is lost in a noncritical location in the peripheral circulation (such as the lower extremity or pelvic vessels), it can often be left in place without attempting retrieval [3].

Occasionally, the lost stents may be difficult to visualize, especially thin strut stents in obese patients with calcified or previously stented coronary arteries. In such cases, intravascular ultrasonography may facilitate localization of the stent [1].

27.1.3.2 Retrieval needed/desired?

A lost stent does NOT always need to be retrieved. If the stent is lost in a noncritical location in the coronary artery, stent deployment (if wire position is maintained through the stent) or stent crushing (if wire position through the stent has been lost) can be a faster and safer approach. Similarly, if the stent is lost in a noncritical location of the peripheral circulation (such as the lower extremities), the risks of retrieval attempts often outweigh any potential benefit [3]. Conversely, if the lost stent is located within a critical coronary artery (such as left main or major bifurcation) or peripheral artery (such as cerebral or renal arteries) location, retrieval is required.

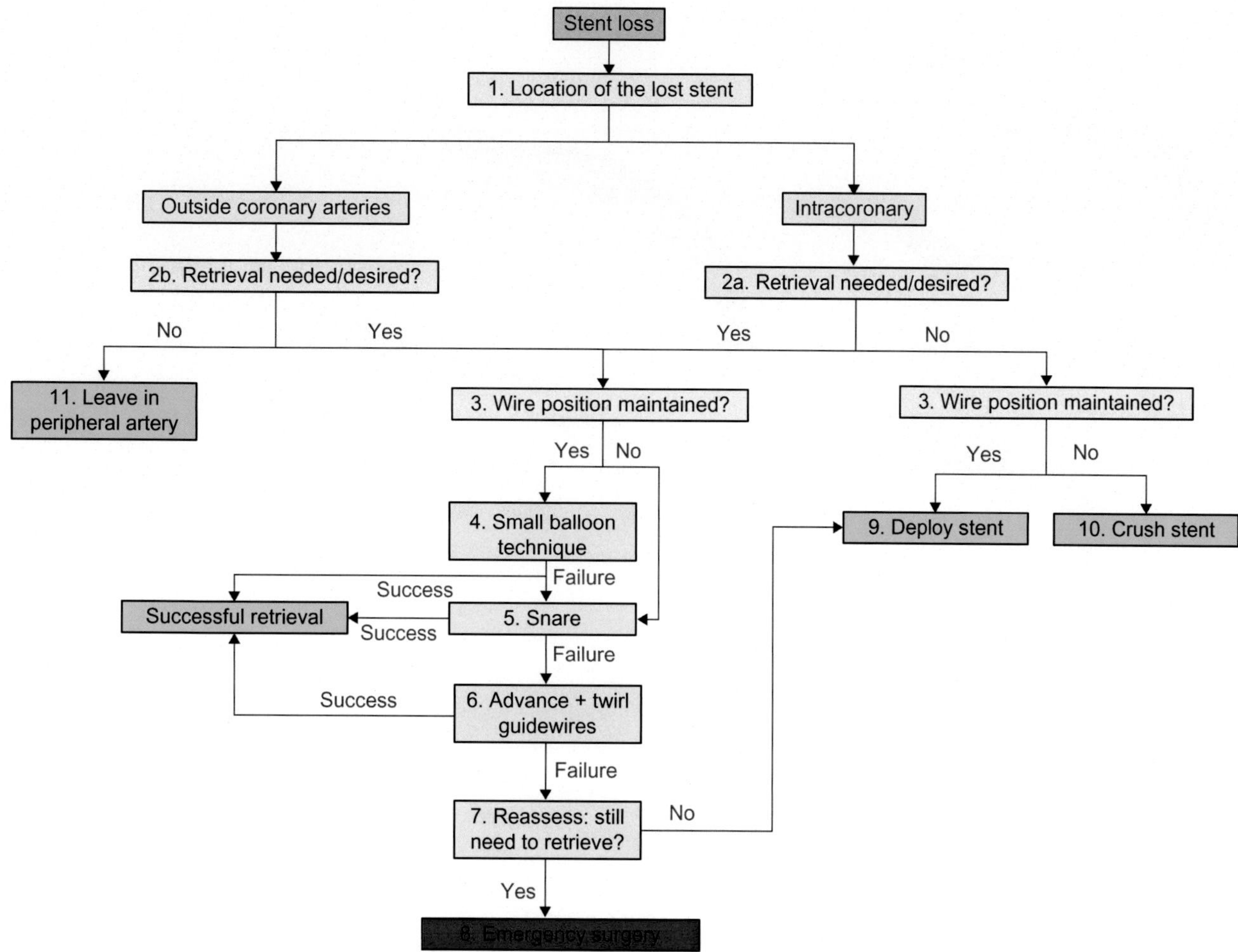

FIGURE 27.2 Algorithm for approaching stent loss.

27.1.3.3 Wire position maintained?

Maintaining wire position through the lost stent can greatly facilitate retrieval attempts (by using the small-balloon technique), and also allows for easier deployment if selected. If wire position is lost it is often impossible to rewire the lost stent, limiting retrieval options to snares or the "guidewire twirling" technique. Similarly loss of wire position does not allow deployment, making crushing the only option if retrieval is not needed or desired.

27.1.3.4 Small-balloon technique (Fig. 27.3)

The small-balloon technique [3,8] can be used when a stent is dislodged from its delivery balloon but guidewire access is maintained through the lost stent. A small balloon is advanced through the stent, inflated distal to the stent, and then withdrawn together with the lost stent.

If the stent is not significantly deformed and the guide catheter is large, the balloon and the stent can be retrieved inside the guide catheter and removed. Often, however, the stent is too deformed to be withdrawn inside the guide catheter: in such cases the inflated balloon, lost stent, and the guide catheter are all removed together from the body (*PCI Manual* Online case 10). It may be difficulty to advance the balloon through the stent, possibly pushing the stent more distally in the vessel. If the balloon is partially advanced through the stent, it can sometimes be inflated in the proximal-mid part of the lost stent, followed by removal of the entire system.

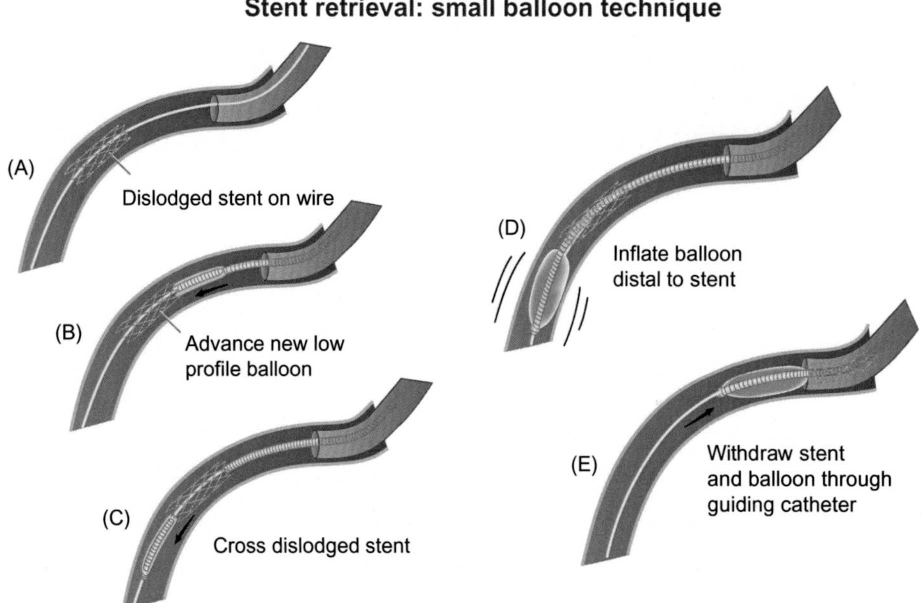

FIGURE 27.3 Illustration of the small-balloon technique.

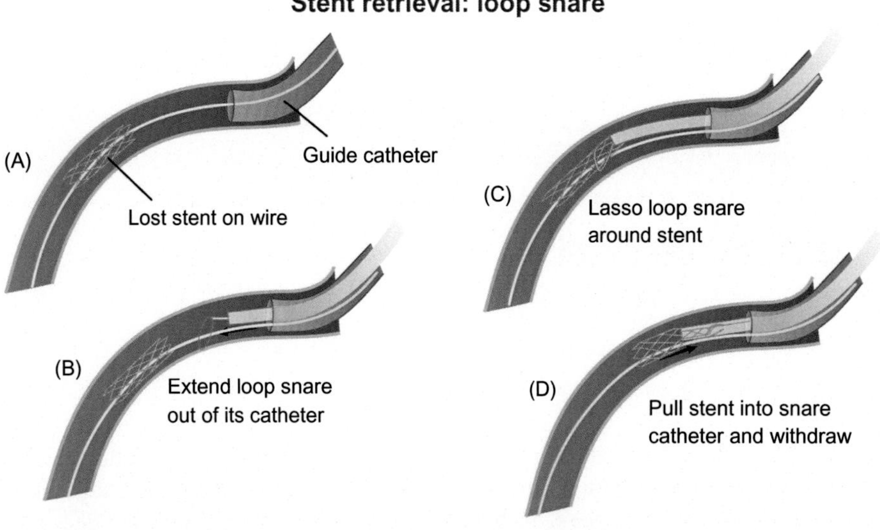

FIGURE 27.4 Retrieval of a lost stent in the coronary circulation using a snare.

27.1.3.5 *Snares*

Snaring is an effective technique for retrieving lost or entrapped equipment. Available snares are discussed in Section 30.18.5.

27.1.3.5.1 Stents lost in the coronary circulation (Fig. 27.4)

Once a loop snare is positioned around the stent to be snared, the snare wire is held still and the snare catheter advanced forward until the lasso loop secures the lost stent, followed by withdrawal of the assembly into the guide catheter.

For retrieval of stents lost in the coronary circulation the Micro-Elite snare (Vascular Solutions) (which is 0.014 in. in diameter, has loop sizes of 2, 4, or 7 mm, is 180 cm in length and does not require a delivery catheter) or the 2 mm or 4 mm En Snare (Merit Medical) are commonly utilized. An example of lost stent snaring from the coronary circulation is shown in Fig. 27.5.

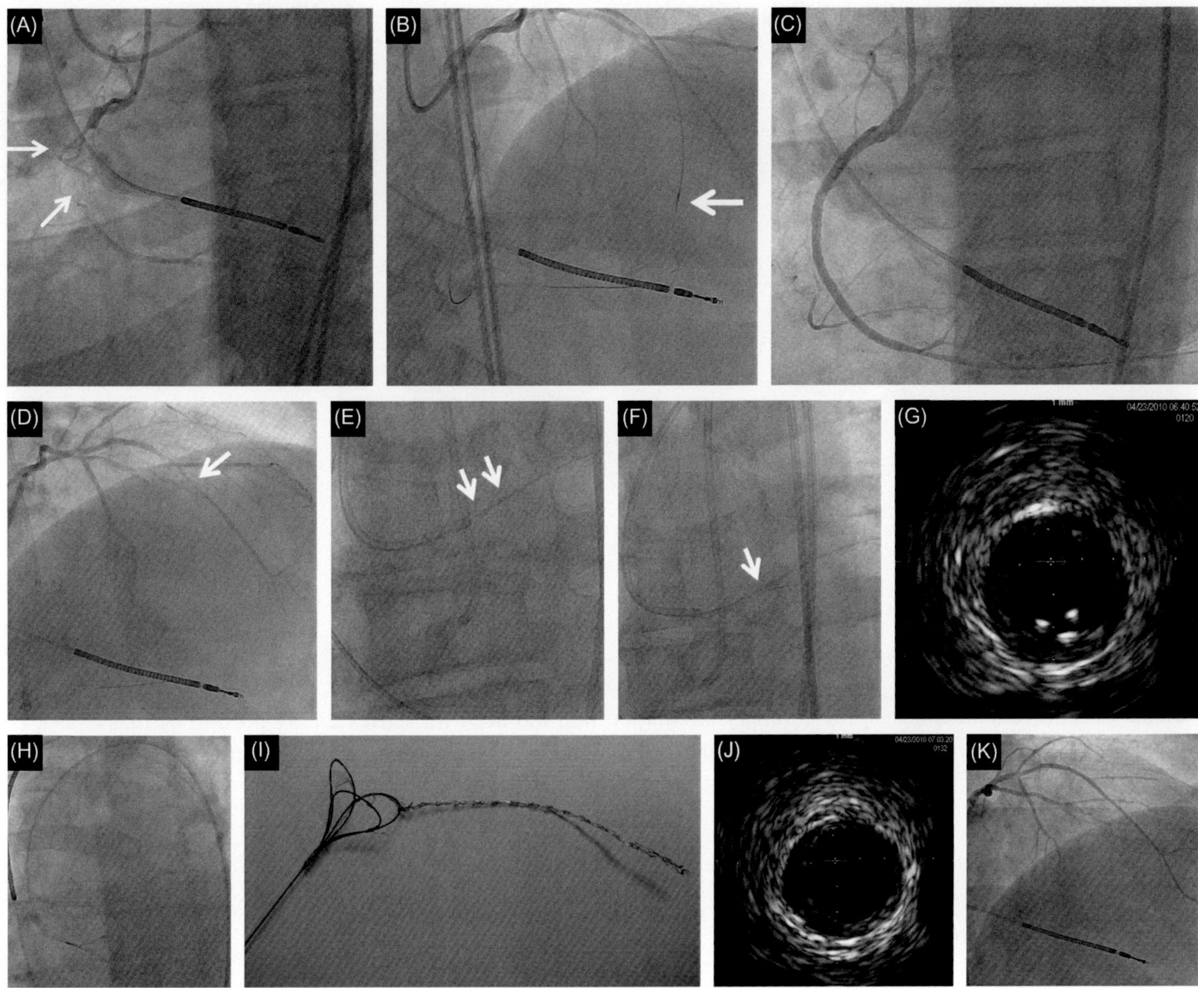

FIGURE 27.5 (CTO PCI Manual online case 74) Illustration of stent loss during retrograde chronic total occlusion (CTO) PCI. Antegrade attempts for crossing a mid right coronary artery CTO (*arrows*, panel A) failed due to subintimal wire passage. A Fielder FC (Abbott Vascular) guidewire was advanced retrogradely via a septal collateral over a Corsair microcatheter (Asahi, *arrow*, panel B). The Corsair catheter was advanced distal to the CTO, followed by retrograde true lumen puncture, as confirmed by intravascular ultrasonography. Following retrograde balloon dilatation antegrade wiring was successful and after drug-eluting stent implantation in the right coronary artery TIMI 3 flow was restored (panel C). Imaging of the left anterior descending artery lesion revealed a lesion (*arrow*, panel D) at the site of the crossed collateral. During attempts to treat this lesion, a 2.5 × 28 mm stent was lost in the left main artery (*arrows*, panel E) and was snared by a Micro Snare Elite (Teleflex) (*arrow*, panel F), but remained partially in the aorta and partially in the left main, as confirmed by intravascular ultrasound (*arrows*, panel G). After snaring with an En Snare (Merit Medical) (panel H) the stent was successfully retrieved (panel I), as confirmed by intravascular ultrasound (panel J). The left anterior descending artery patency was restored after stenting (panel K). *Reproduced/Reprinted from Iturbe JM, Abdel-Karim AR, Papayannis A, et al. Frequency, treatment, and consequences of device loss and entrapment in contemporary percutaneous coronary interventions. J Invasive Cardiol 2012;24:215–21 (Figure 1) with permission.*

27.1.3.5.2 Stents lost in the peripheral circulation

For retrieving stents lost in the peripheral circulation an 18–30 mm En Snare through a JR4 or multipurpose diagnostic or guide catheter is often the snare of choice, because of the three-loop design that facilitates object retrieval. However, it should be used with caution as the snare wires may still cause vessel injury [1].

27.1.3.6 Guidewire "twirling" (Fig. 27.6)

In this technique, two or more 0.014 in. guidewires are advanced through or around the lost stent, and rotated several times to entangle their distal ends. The guidewires are then withdrawn often bringing the lost stent with them.

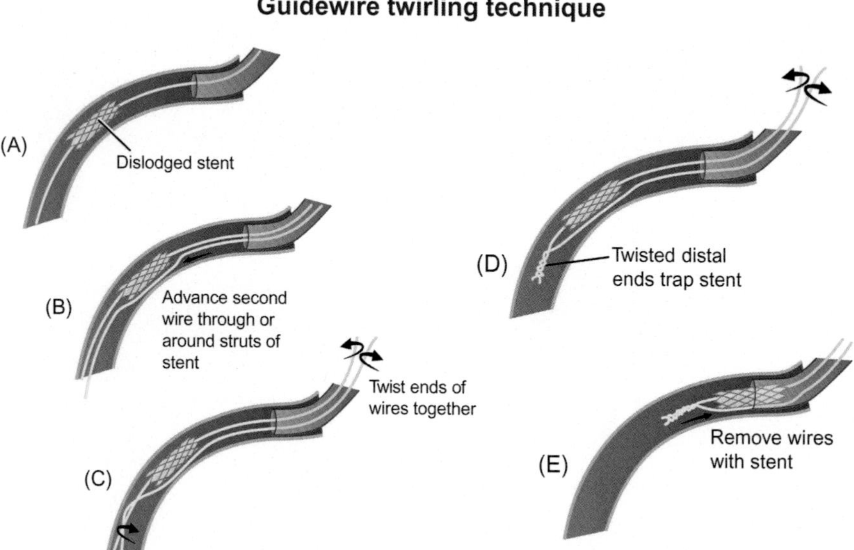

FIGURE 27.6 Illustration of the guidewire "twirling" technique.

27.1.3.7 *Reassess need for retrieval*

If attempts to retrieve a stent lost in the coronary circulation fail, the need for retrieval may need to be reassessed, considering the risks and benefits associated with further retrieval attempts, including surgery versus deploying or crushing the stent. In most cases, deploying or crushing the stent is the fastest and safest approach.

27.1.3.8 *Emergency surgery*

Surgical retrieval of lost stents can be challenging and carries high risk of complications. It should, therefore, only be done under extenuating circumstances.

27.1.3.9 *Stent deployment*

If a stent is lost inside the coronary circulation and wire position is maintained, deployment may be the safest and fastest treatment strategy. Crossing the lost stent with a balloon can sometimes be challenging: using a small balloon can allow advancement within the lost stent, followed by increasingly larger balloons until deployment is optimized. Sometimes, attempts to advance a balloon through a lost stent may lead to more distal displacement of the stent. If a balloon cannot advance through the lost stent (usually due to stent deformation), crushing of the stent could be performed, as described below.

27.1.3.10 *Stent crushing (Fig. 27.7)*

If wire position is lost within the lost stent and retrieval is not feasible or desired, crushing the stent with another stent can be the strategy of choice, unless the stent is located in a critical location, such as the left main. A coronary guidewire is advanced around the lost stent, a balloon is used to crush the stent against the coronary artery wall, and another stent is placed, "excluding" the lost stent from the coronary circulation. It is important to avoid inadvertently passing through one or more stent struts with the second wire, especially if the lost stent has been partially deployed. This should be suspected if balloon delivery is challenging around the lost stent and rewiring should be performed, ideally using a knuckled polymer-jacketed guidewire (Fig. 27.8). If balloon delivery remains challenging after rewiring, consider rewiring again or using a Glider balloon or a Wiggle guidewire.

Meticulous attention should be given to completely appose the stent struts to the vessel wall to avoid limitations of blood flow through the coronary artery, ideally using intracoronary imaging [2,9]. Both stent crushing and stent deployment carry risk of restenosis, which is, however, much lower than the potential risk of aggressive stent retrieval

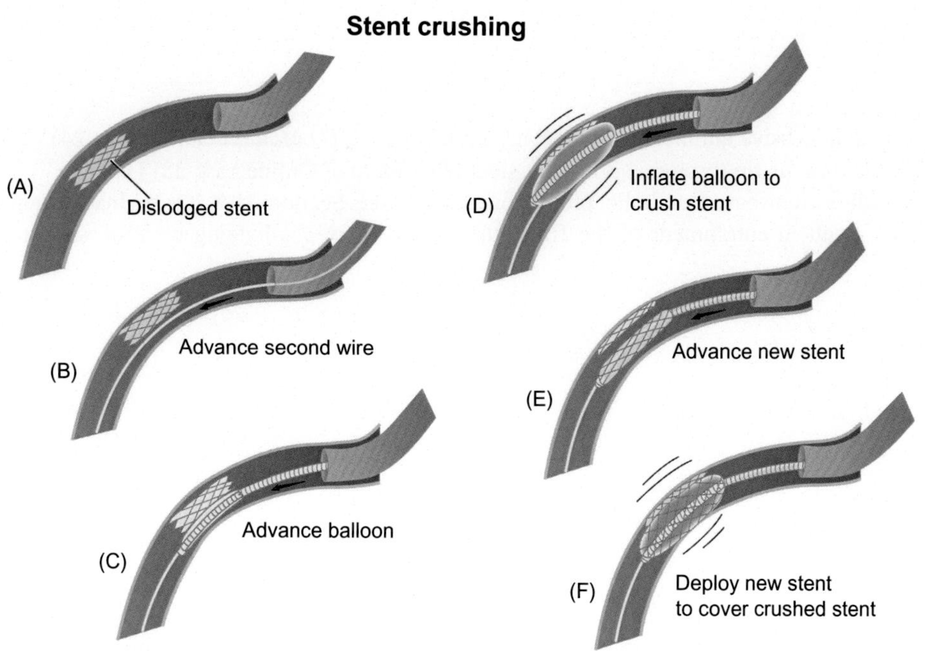

FIGURE 27.7 Crushing of a dislodged stent.

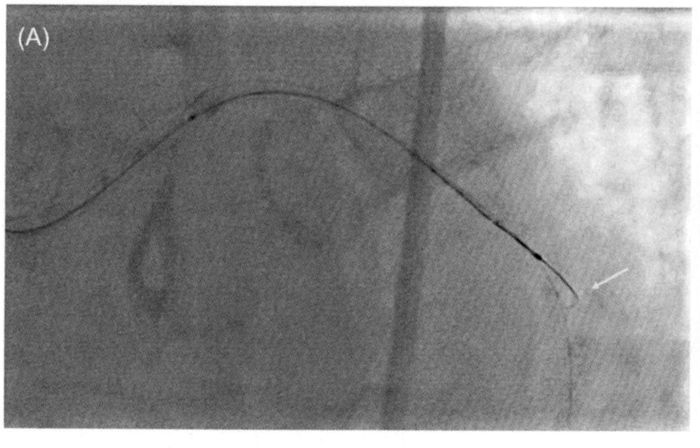

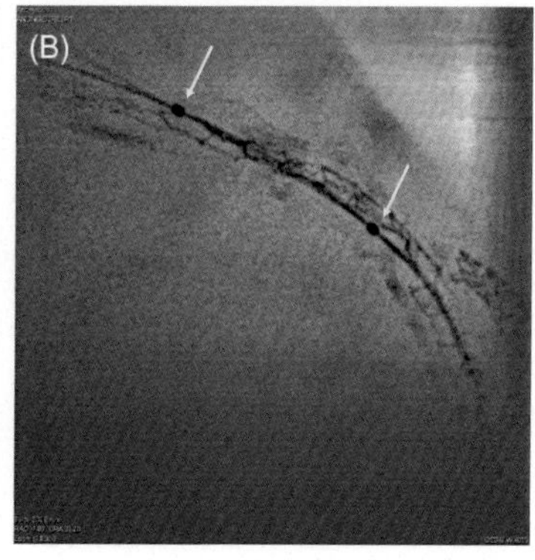

FIGURE 27.8 A patient had a deformed stent in the mid LAD; despite successful crossing with a guidewire no balloon or stent could be advanced through the deformed stent. After using a knuckled Fielder XT guidewire (*arrow*, panel A) to cross through the deformed stent, balloon (*arrow*, panel B) and stent delivery succeeded. *Courtesy of Dr. Goran Olivecrona.*

attempts. Intentional stent crushing is also being performed during PCI of in-stent coronary chronic total occlusions with encouraging results [10,11].

27.1.3.11 Leave in peripheral artery

If the lost stent is removed from the coronary artery into the iliac or femoral vessels but cannot subsequently be removed through the sheath or the vessel wall, the lost stent could be crushed against the iliac or femoral artery wall using a peripheral stent (Fig. 5.8). Distal stent embolization into a small arterial branch (such as a femoral artery branch) may be left untreated, since the distally embolized stents appear to have a benign clinical course. None of the 12 patients with distal stent embolization in three series [3,8,12] had clinical sequela. Hence, the risk of extraction usually exceeds the risk of local complications [2,3].

27.2 Guidewire entrapment and fracture

CTO Manual Online case 24

PCI Manual Online cases 32, 33, 88, 90

Guidewire entrapment and fracture is an infrequent and often preventable complication, but can potentially have devastating consequences, especially if the guidewire unravels [1].

27.2.1 Causes

27.2.1.1 Guidewire entrapment

- Jailing of the guidewire during bifurcation stenting.
- Guidewire deformation (e.g., during aggressive attempts for guidewire knuckling) (*CTO Manual* Online case 24).
- Withdrawal of a guidewire with a loop at its tip through a proximal stent (*PCI Manual* Online case 33) [1].
- Use of a buddy wire together with a filter-based embolic protection device (see Section 18.8.1.1.3). Inadvertent stenting over the buddy wire can result in entrapment of the filter, that can be very challenging or impossible to remove [1].
- Severe lesion calcification.

27.2.1.2 Guidewire fracture

- Aggressive pulling of an entrapped guidewire.
- Guidewire over-rotation.
- Atherectomy over a kinked guidewire.

27.2.2 Prevention

27.2.2.1 Guidewire entrapment

- Some advocate to avoid jailing the side branch guidewire during bifurcation stenting; the risk of side branch occlusion in most cases, however, exceeds the risk of guidewire entrapment.
- Avoid aggressive guidewire manipulation, especially in small branches.
- Straighten guidewire tip before removing through a stent (*PCI Manual* Online case 33, Fig. 27.9).

27.2.2.2 Guidewire fracture

- Do NOT pull hard if a guidewire is difficult to retrieve!
- Do not perform atherectomy over a kinked guidewire. Replace the guidewire with a new one before proceeding with atherectomy.

27.2.3 Treatment (Fig. 27.10)

27.2.3.1 Guidewire entrapment

27.2.3.1.1 Do NOT pull hard!

This is critically important, because if the structural integrity of the entrapped guidewire becomes compromised, it can lead to fracture or unraveling that can be very challenging to treat and may require cardiac surgery.

27.2.3.1.2 Balloon or microcatheter

Advance a microcatheter or balloon over the entrapped guidewire, as far as possible, then pull the guidewire gently. The balloon or microcatheter allows more focused application of the withdrawal force, facilitating retrieval. A guide catheter extension could also be used.

27.2.3.1.3 Inflate balloon

If step 2 fails, inflate a balloon advanced as far distally as possible over the entrapped guidewire and attempt to withdraw again. In most cases this will allow release of the entrapped guidewire.

27.2.3.1.4 Second guidewire + balloon

If step 3 fails, advance a second guidewire next to entrapped guidewire and inflate a balloon to help “free” the entrapped guidewire (Fig. 27.11) [13].

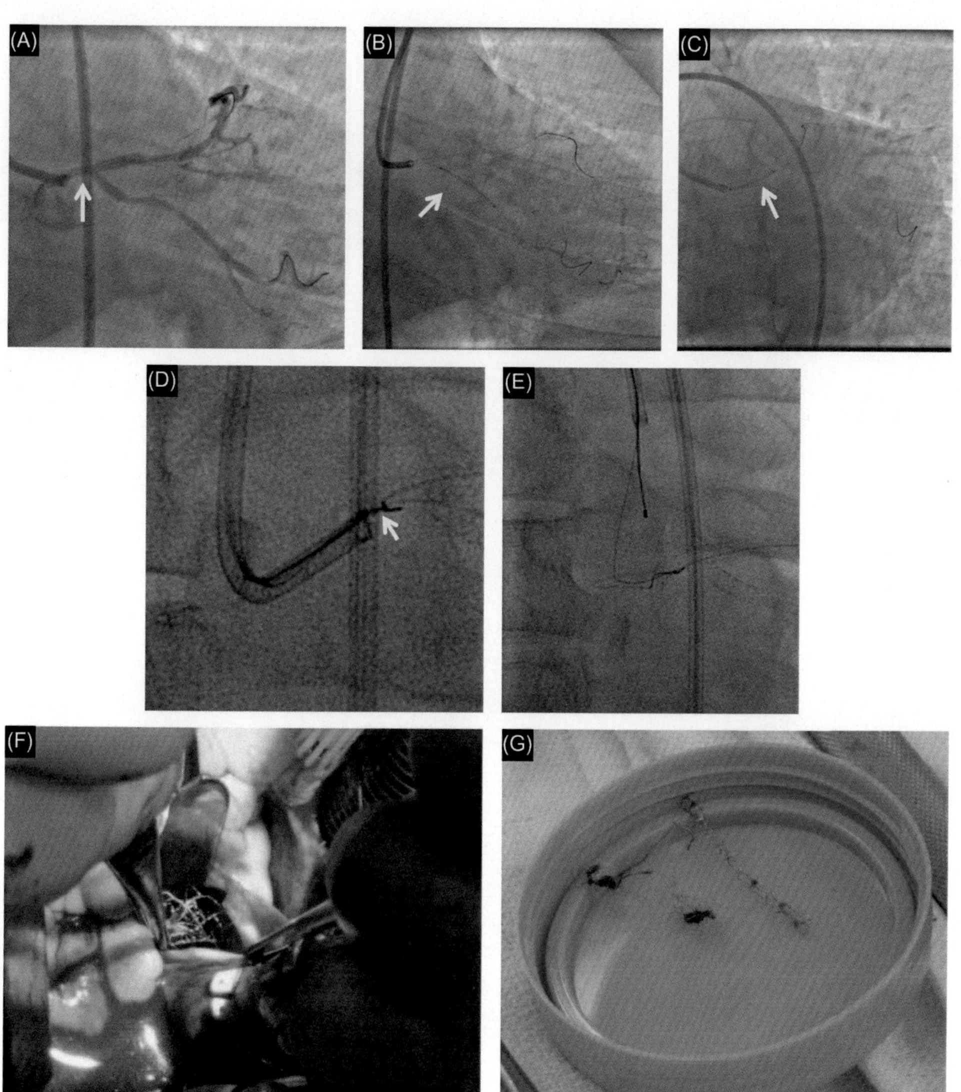

FIGURE 27.9 Illustration of an obtuse marginal guidewire entrapment following proximal left main artery stent deployment. Diagnostic coronary angiography demonstrated severe left main disease (*arrow*, panel A). Both the left anterior descending and the first obtuse marginal arteries were wired. After successful stenting of the first obtuse marginal branch lesion (*arrow*, panel B), the left main lesion was stented with a 3.0 × 18 mm stent over the LAD guidewire (*arrow*, panel C). The obtuse marginal artery wire was left in place to allow stabilization during deployment of the left main stent. Attempts to withdraw the obtuse marginal artery wire caused wire entanglement in the stent (*arrow*, panel D). Several attempts to remove the wire were unsuccessful (panel E), requiring surgical intervention (panels F and G). *Reprinted from Iturbe JM, Abdel-Karim AR, Papayannis A, et al. Frequency, treatment, and consequences of device loss and entrapment in contemporary percutaneous coronary interventions. J Invasive Cardiol 2012;24:215–21 (Figure 4) with permission.*

27.2.3.1.5 Emergency surgery

If all retrieval attempts fail, surgery may be required. A "last straw" option would be to pull "hard" on the guidewire (having a balloon or microcatheter in place) hoping that the guidewire will fracture without unraveling [14].

27.2.3.2 Guidewire fracture

27.2.3.2.1 Wire unraveling

The key consideration in managing a fractured guidewire is whether guidewire unraveling has occurred. Guidewire unraveling is much more challenging to treat than a "clean" fracture, as the guidewire distal spring coil may unravel creating a metal "bird's nest" that could predispose to thrombosis (Fig. 27.9, panels F and G and Fig. 27.12) [1]. In cases of guidewire fracture it is important to perform intravascular ultrasonography to confirm that no wire coil unraveling has occurred [7].

27.2.3.2.2 Able to cover wire fragments with a stent?

If guidewire unraveling has occurred, subsequent treatment depends on the location of the guidewire unraveled fragments. If they are extending into the aorta, or if they are located in a critical coronary location (such as the left main or other major bifurcation), emergency surgery is needed to remove those fragments and prevent future thromboembolic complications (both coronary and systemic). Attempts to retrieve unraveled guidewire fragments should generally be

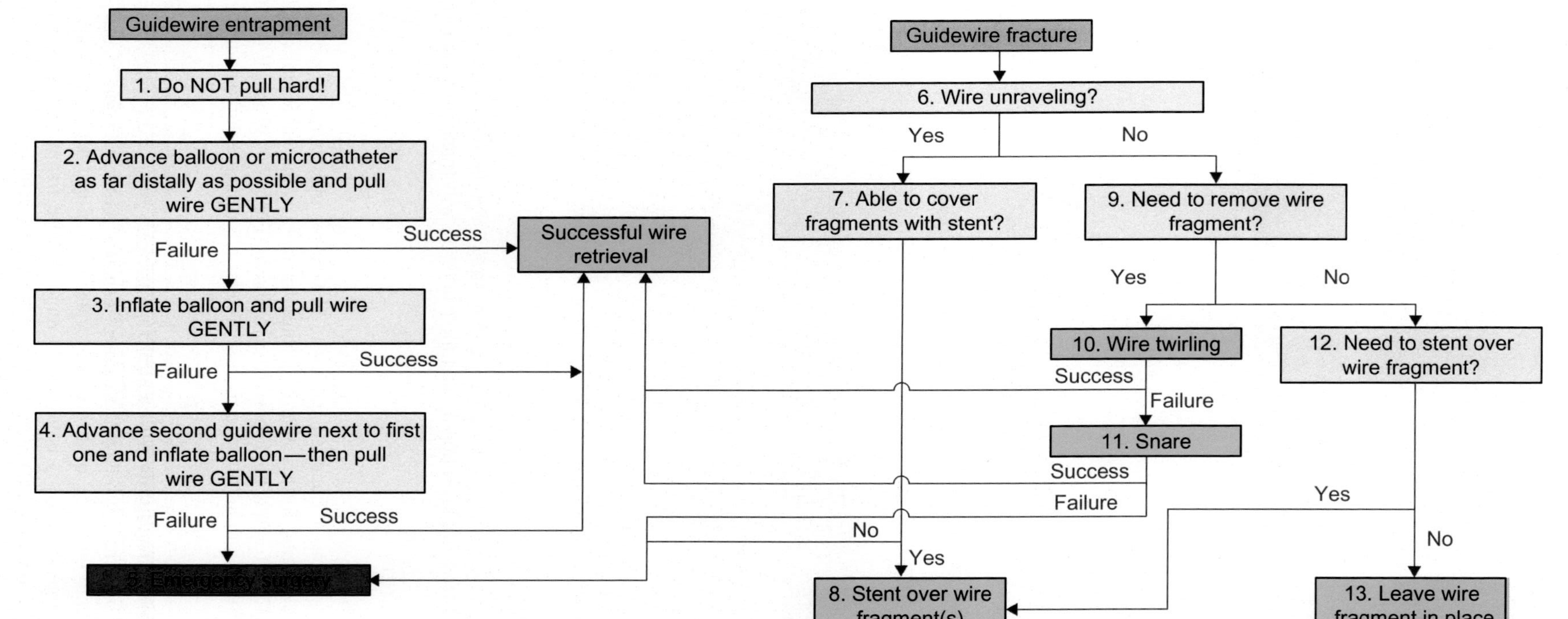

FIGURE 27.10 Algorithmic approach to guidewire entrapment and fracture.

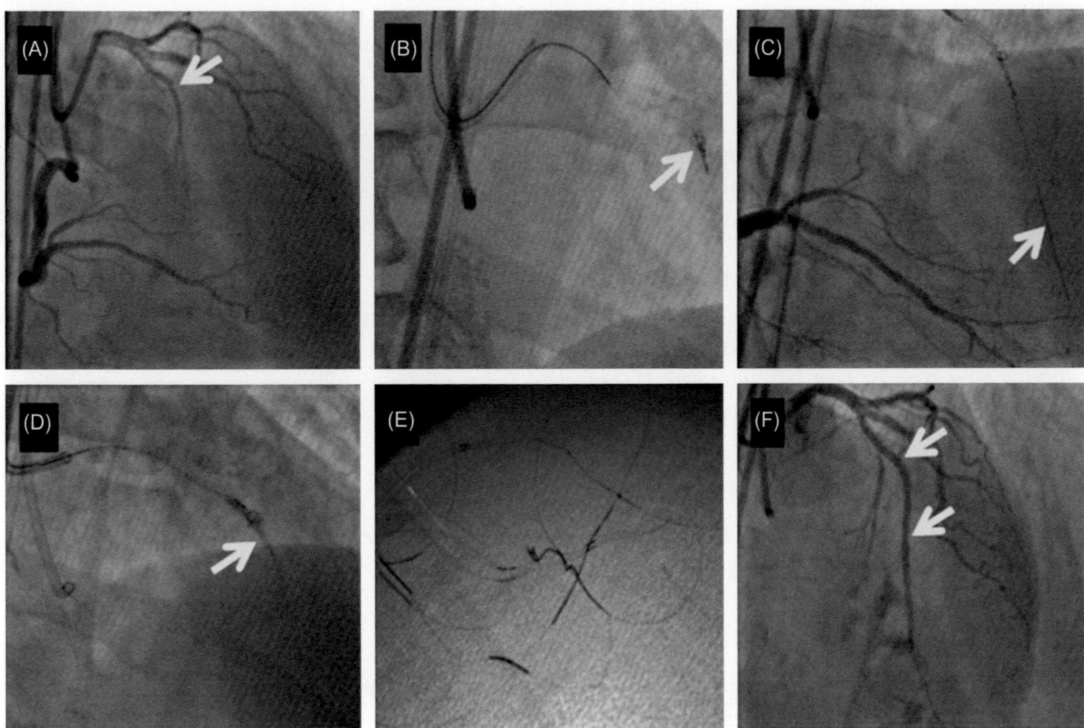

FIGURE 27.11 Bilateral coronary angiography demonstrating a chronic total occlusion of the mid left anterior descending artery (*arrow*, panel A). Entrapment of a knuckled Fielder XT guidewire (*arrow*, panel B). The chronic total occlusion was successfully crossed with a Pilot 200 guidewire (*arrow*, panel C) advanced parallel to the entrapped guidewire. After balloon angioplasty was performed around the entrapped guidewire (*arrow*, panel D) it was successfully retrieved (panel E) with an excellent final angiographic result (*arrow*, panel F). *Reproduced from Danek BA, Karatasakis A, Brilakis ES. Consequences and treatment of guidewire entrapment and fracture during percutaneous coronary intervention. Cardiovasc Revasc Med 2016;17:129–33 with permission. Copyright Elsevier.*

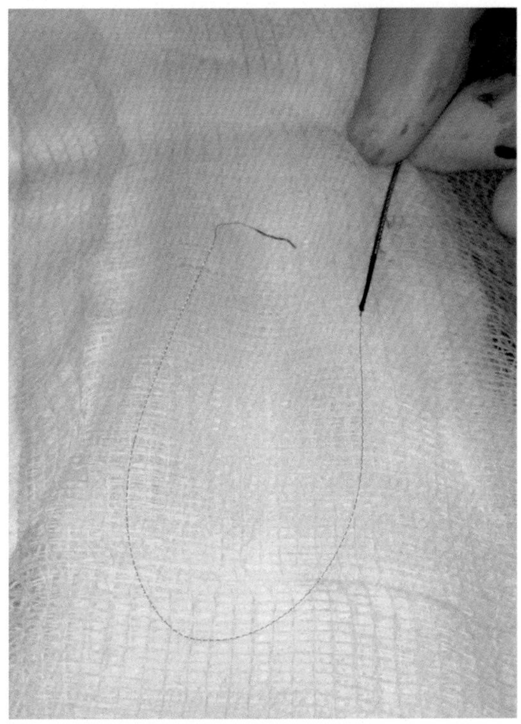

FIGURE 27.12 Guidewire fracture resulting in distal spring coil unraveling. Guidewire fracture was suspected after aggressive guidewire torquing resulted in unresponsiveness of the guidewire tip on fluoroscopy. The fractured guidewire was removed in its entirety with a microcatheter. *Courtesy of Dr. Jaikirshan Khatri.*

avoided (because they can cause even more unraveling), although successful percutaneous guidewire fragment removal has been described [7]. A snare can be used to break the unraveled guidewire as close as possible to the coronary ostium (or even within the coronary artery) to allow subsequent stenting over the wire fragment.

27.2.3.2.3 Stenting over guidewire fragments

If the unraveled fragments can be "trapped" behind a stent, stenting is performed (Fig. 27.13).

27.2.3.2.4 Need to remove guidewire fragments?

If there is no guidewire unraveling after guidewire fracture (confirmed by intravascular ultrasonography), the key question is whether the guidewire fragment needs to be removed. Small guidewire fragments located in small distal branches or in collateral vessels may be best left untreated. If, however, there are long guidewire fragments, especially if located within a large coronary artery, retrieval is preferred.

27.2.3.2.5 "Wire twirling" technique (Fig. 27.14)

In this technique one or more coronary guidewires are advanced next to the guidewire fragment and twisted several times. The wires may intermingle, allowing subsequent retrieval.

27.2.3.2.6 Snaring

Use of small coronary snares (as described in Section 27.1.3.5) may allow retrieval of the guidewire fragments (Fig. 27.15). If it fails emergency surgery may be required for removal [13].

27.2.3.2.7 Need to stent over guidewire fragment?

Similar to Section 27.2.3.2.2 if the guidewire fragment is in a distal location in a small branch, it is usually left in place. If, however, it is located more centrally, stenting can "trap" the guidewire against the arterial wall, preventing future migration and reducing the likelihood of intracoronary thrombus formation [7].

27.2.3.2.8 Leave guidewire fragment "in situ"

This can be done for small guidewire fragments located distally, as described in Section 27.2.3.2.7 [13].

27.3 Balloon entrapment and fracture

CTO Manual Online case 84

PCI Manual Online case 8

Balloons can get entrapped and fractured. Subsequent treatment depends on the type and location of the entrapped and/or fractured balloon.

27.3.1 Causes

27.3.1.1 Balloon entrapment

- Balloon rupture (resulting in the balloon getting "stuck" in the lesion).
- Balloon failure to deflate (discussed in Sections 9.11 and 23.3.11–23.3.14)
- Balloon interaction with previously placed stent (especially true when plaque modification balloons are used). Occasionally previously placed stents have been extracted while retrieving cutting balloons (see Section 9.11) [15–17].

27.3.1.2 Balloon fracture

- Kinking of the (balloon or stent) catheter shaft (Fig. 27.16).

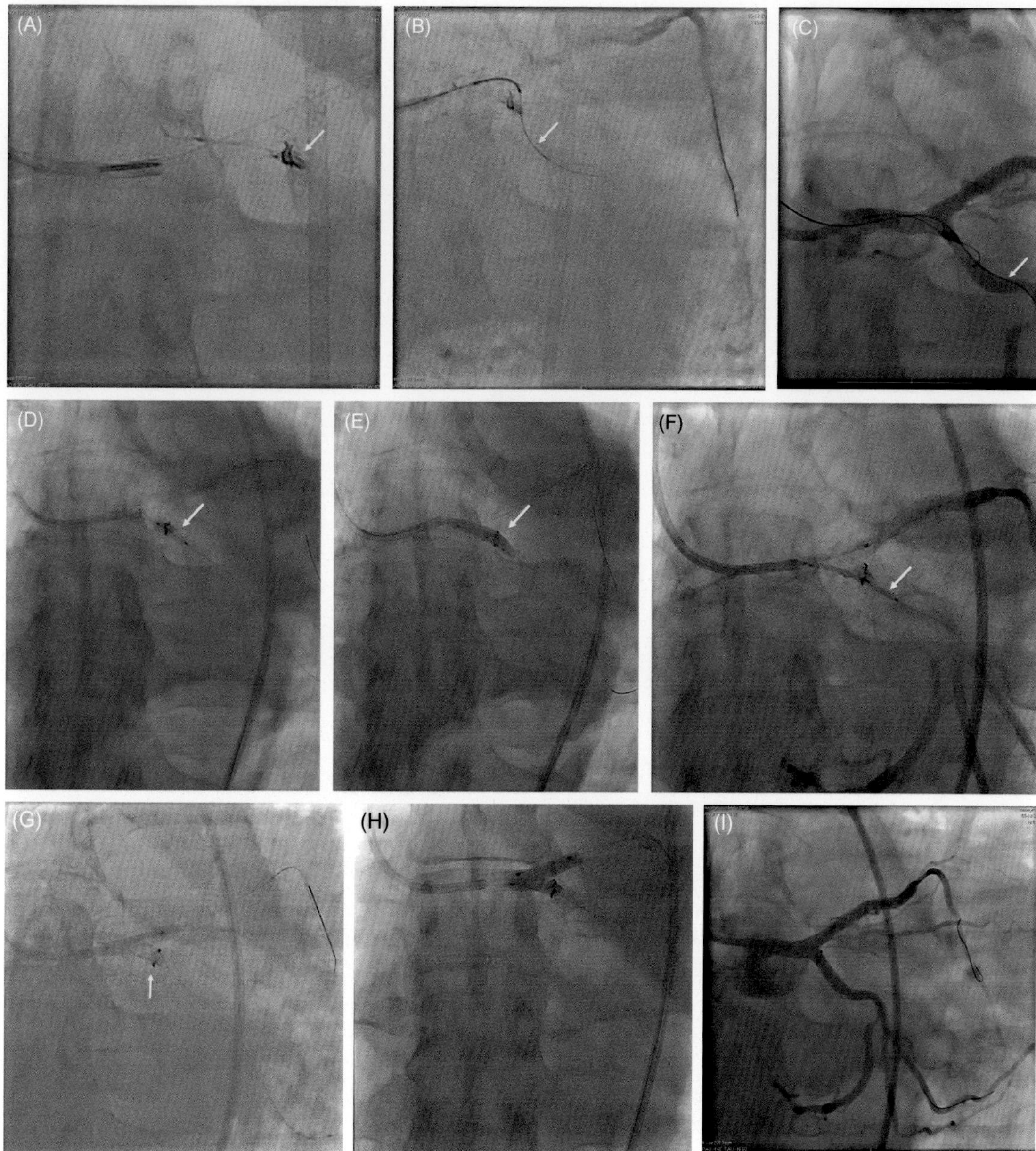

FIGURE 27.13 Stent loss that was further complicated with guidewire entrapment and fracture (*arrow*, panel A) in the proximal circumflex. Another guidewire was advanced past the entrapped equipment (*arrow*, panels B and C), followed by balloon angioplasty (*arrow*, panels D and E) and stenting (*arrow*, panels F and G) across the entrapped equipment. After stent postdilation (panel H) an excellent final result was achieved (panel I). *Courtesy of Dr. Goran Olivecrona.*

27.3.2 Prevention

27.3.2.1 Balloon entrapment

- Consider primary atherectomy for very heavily calcified coronary lesions.
- Avoid very high-pressure balloon inflations in highly calcified lesions that may lead to balloon rupture and entrapment.

27.3.2.2 Balloon fracture

- If the shaft of a balloon (or stent) catheter kinks, discard the catheter and use a new one (see Section 9.5.7).

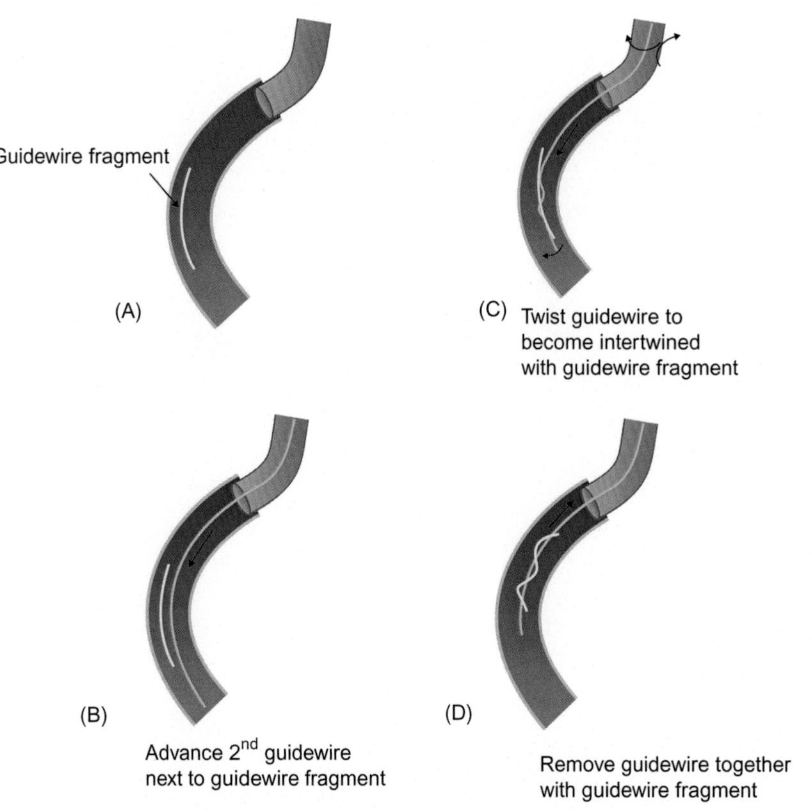

FIGURE 27.14 Illustration of the "guidewire twirling" technique for removing a guidewire fragment.

27.3.3 Treatment (Fig. 27.17)

In addition to attempting to remove the entrapped balloon, balloon entrapment may compromise coronary flow and result in hemodynamic instability, potentially requiring hemodynamic support.

Balloon entrapment

27.3.3.1 Failure to deflate?

If the reason for entrapment is failure to deflate, specifics steps (Sections 27.3.3.3.1–27.3.3.3.4) are followed.

27.3.3.2 Gentle rotation and back and forth movement

Sometimes balloon rotation may help rewrapping the balloon and facilitate retrieval. Similarly gentle back and forth movements may help free the entrapped balloon.

27.3.3.3 Inflate second balloon inside guide catheter and pull

Another balloon is inserted inside the guide catheter as close to the guide catheter tip as possible, but without exiting the guide catheter. The balloon is inflated at high pressure ($\geq$20 atm) and the guide catheter with both balloons (the entrapped one and the second one that is inflated) are retracted as a unit.

27.3.3.4 Guide catheter extension

Cut the balloon shaft and advance a guide catheter extension as close as possible to the entrapped balloon, followed by pulling the entrapped balloon. Using a Trapliner may be advantageous, as it allows simultaneous trapping of the entrapped balloon inside the guide catheter.

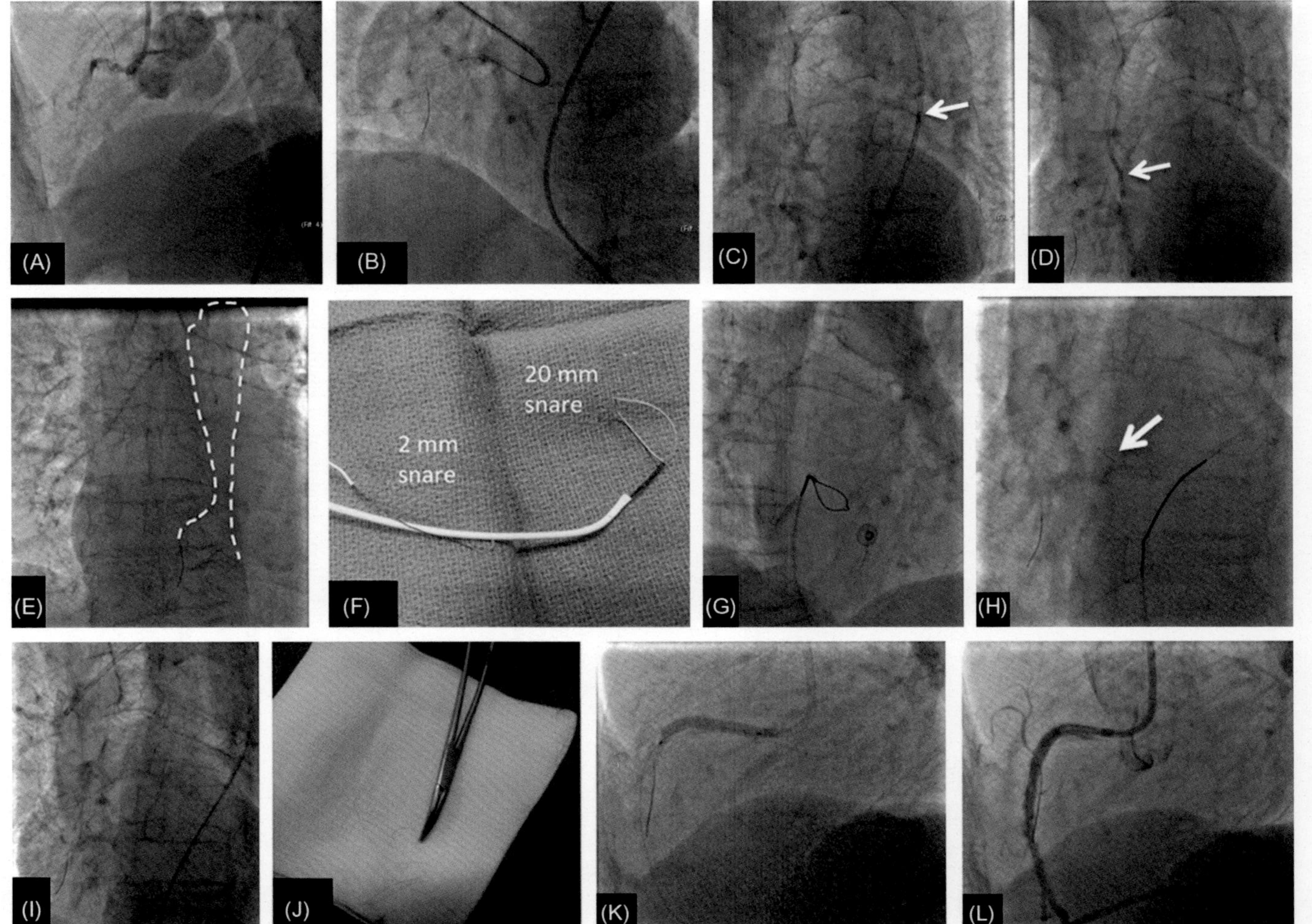

FIGURE 27.15 Retrieval of an entrapped and unraveled guidewire. Right coronary artery calcific acute occlusion (panel A); guidewire tip entrapment and unraveling (panel B); failed attempt to advance a 2 mm snare (*arrow*, panel C); failed attempt to retrieve the guidewire with a trapping balloon (*arrow*, panel D); guidewire unraveled (*dotted line*) with proximal end in the abdominal aorta (panel E); double gooseneck system: a 20 mm gooseneck is mounted inside a 2 mm gooseneck (panel F); snaring of the proximal end of the unraveled guidewire with the 20 mm snare (panel G); advancement of the 2 mm gooseneck (*arrow*) on the unraveled guidewire (panel H); snaring of the guidewire at the ostium and pullback with guidewire fracture and removal (panel I); unraveled guidewire removed (panel J); stenting over the entrapped guidewire (panel K); final result (panel L). *Courtesy of Dr. Jacopo Oreglia.*

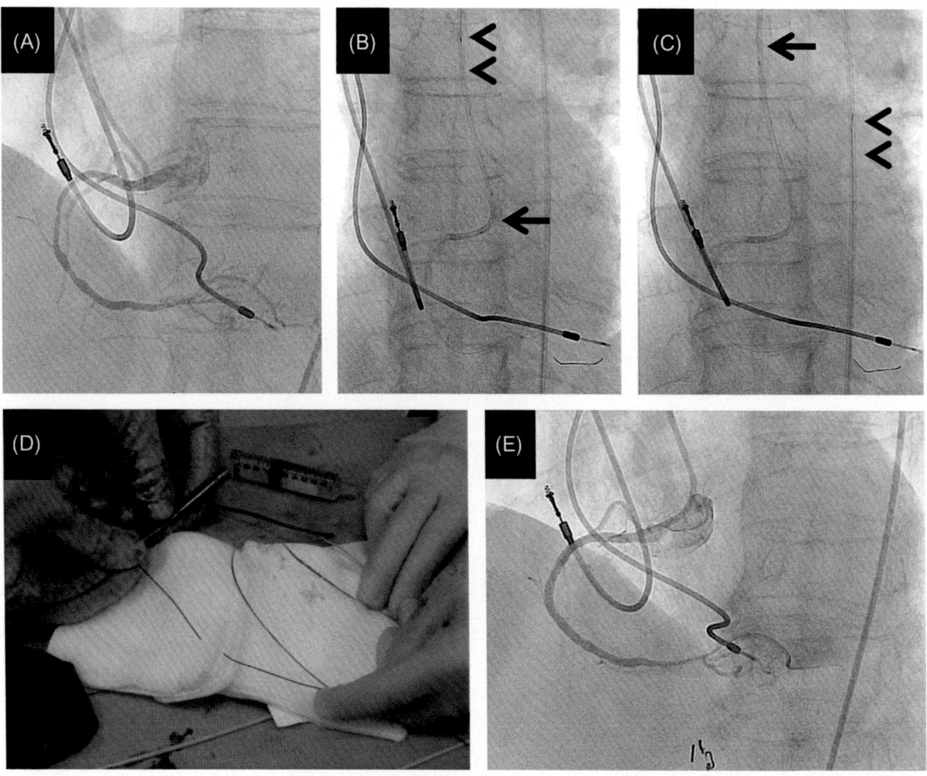

FIGURE 27.16 Illustration of balloon shaft fracture with embolization followed by successful retrieval. During PCI of a 90% mid right coronary artery lesion (panel A), a 2.75 mm × 28 mm Maverick balloon (Boston Scientific) shaft became kinked during insertion. The balloon could not be advanced through the lesion and its shaft fractured in two. The distal balloon shaft portion remained in the right coronary artery. A Gooseneck Micro Snare (Medtronic) (*arrowheads*, panels B and C) was used to successfully retrieve the distal fragment (*arrow*, panels B and C and panel D). Postprocedure angiography confirmed TIMI 3 blood flow in the right coronary artery (panel E). *Reprinted from Iturbe JM, Abdel-Karim AR, Papayannis A, et al. Frequency, treatment, and consequences of device loss and entrapment in contemporary percutaneous coronary interventions. J Invasive Cardiol 2012;24:215–21 (Figure 2) with permission.*

27.3.3.5 Snare

Advance a snare as close to the balloon as possible, tighten it, and withdraw.

27.3.3.6 Surgery

Emergency surgery may be needed if all efforts to retrieve the balloon fail [18].

Balloon fracture

27.3.3.7 Balloon fragment inside guide?

If the balloon shaft breaks, the key question is where the fracture occurred and whether a portion of the retained balloon fragment is inside the guide catheter. One way to tell is by comparing the retrieved balloon catheter fragment (Fig. 27.16) with an intact balloon catheter. If the retained balloon fragment is inside the guide catheter, it can potentially be trapped with another balloon as described in step 8, followed by removal of both the guide catheter and the balloon fragment. If not, then the balloon fragment can be snared.

27.3.3.8 Inflate second balloon inside guide catheter and retract

The second balloon is inflated at high pressure (≥20 atm), trapping the balloon fragment and the entire assembly (guide catheter, balloon fragment and second inflated balloon) is withdrawn under fluoroscopy maintaining the balloon inflation.

27.3.3.9 Snares

Snaring of the retained balloon catheter fragment is performed, followed by withdrawal.

27.3.3.10 Guidewire and balloon around the entrapped balloon

A guidewire is advanced around the entrapped balloon (usually through a second guide catheter), followed by balloon angioplasty around the entrapped balloon fragment, aiming to free the balloon fragment from the coronary artery wall.

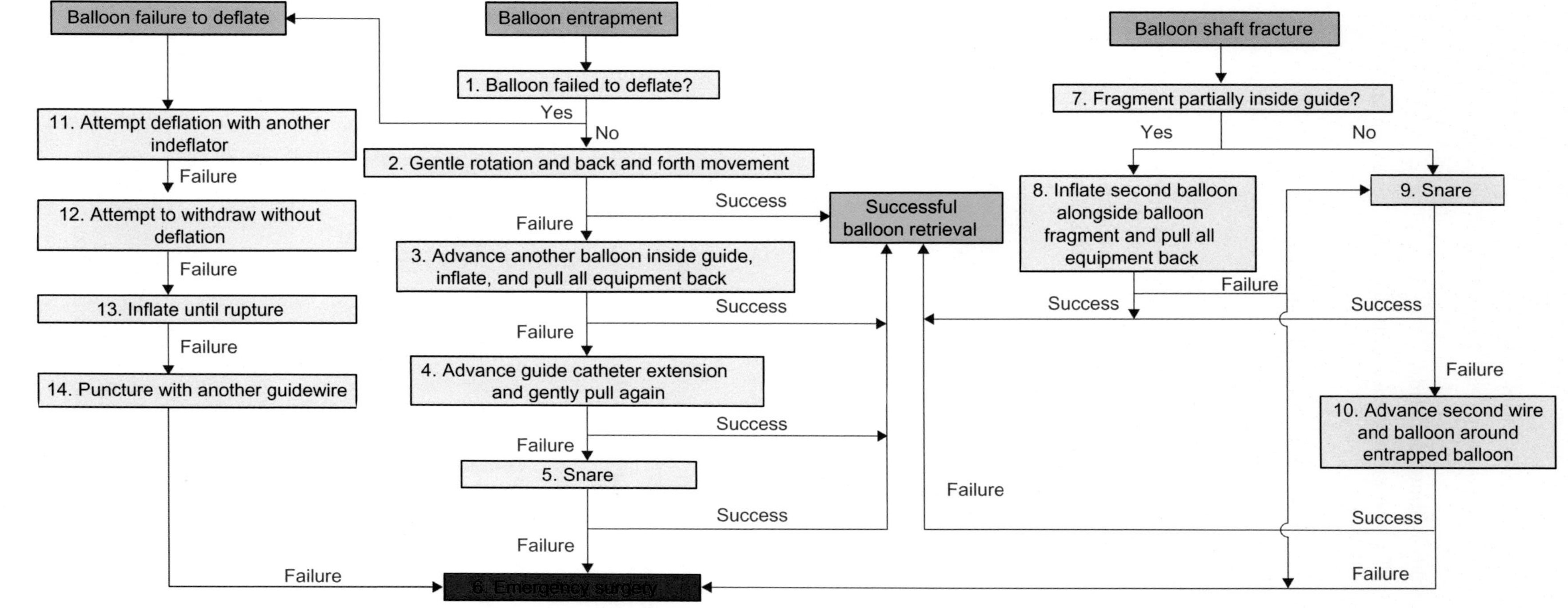

FIGURE 27.17 Algorithmic approach to balloon entrapment, fracture, or failure to deflate.

Balloon failure to deflate

27.3.3.11 Attempt deflation with another indeflator

Attempt deflation with another indeflator (ideally filled with saline) or with a luer-lock syringe (also filled with saline).

27.3.3.12 Attempt to withdraw without deflation

If the size of the balloon is smaller than the reference diameter of the vessel, an attempt could be made to withdraw it into the guide catheter without deflation.

27.3.3.13 Inflate at high pressure to rupture the balloon

If the balloon diameter is smaller than the reference vessel, it can be inflated at high pressure until it ruptures.

27.3.3.14 Puncture the inflated balloon with a guidewire

An over-the-wire balloon is advanced over a second guidewire next to the "failing to deflate" balloon and inflated at low pressure (2–4 atm). The wire is then removed and a stiff guidewire (such as Astato 20, Confianza Pro 12, or Hornet 14) (or alternatively the back end of a guidewire) is advanced through the OTW balloon aiming to puncture the "failing to deflate" balloon. Use of an inflated OTW balloon minimizes the risk of vessel injury by centering the wire and helping direct it towards the "failing to deflate" balloon.

Alternatively, the balloon shaft can be cut and a guide catheter extension advanced and gently pushed against the "failing to deflate" balloon. A stiff guidewire or the back end of a guidewire can then be advanced through the guide catheter extension to puncture the balloon, minimizing the risk of proximal vessel injury.

If all attempts to remove the non-deflating balloon fail, emergency surgery may be required.

27.4 Microcatheter entrapment and fracture

CTO Manual Online case 87
PCI Manual online case 29

27.4.1 Causes

27.4.1.1 Microcatheter entrapment

- Aggressive microcatheter advancement through complex (tortuous and calcified) lesions.
- Advancement of two microcatheters over the externalized guidewire with both ends of the microcatheters meeting and "interlocking."
- Microcatheter manipulations without having a guidewire inside the microcatheter lumen.

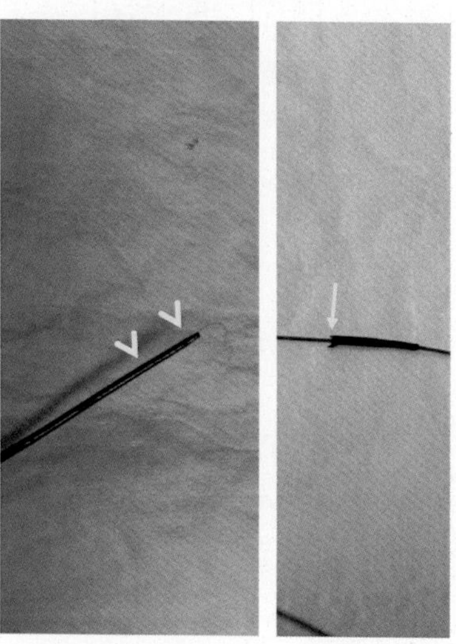

FIGURE 27.18 Fracture of the tip of a Corsair microcatheter after advancement attempts through a heavily calcified lesion. *Courtesy of Dr. Gabriele Gasparini.*

27.4.1.2 Microcatheter fracture

- Aggressive manipulation, especially of softer tip microcatheters (such as the Caravel) through complex lesions (Fig. 27.18).
- Overrotating a microcatheter: this may compromise the structural integrity of the microcatheter and hinder retrieval (and advancement) attempts.

27.4.2 Prevention

27.4.2.1 Microcatheter entrapment

- Avoid aggressive manipulation of microcatheters through highly complex and calcified lesions.
- When performing retrograde CTO PCI over an externalized guidewire, the tip of the antegrade equipment should not come in contact with the tip of the retrograde microcatheter.
- Avoid microcatheter overrotation.
- Always keep a guidewire within the microcatheter lumen during microcatheter manipulations.

27.4.2.2 Microcatheter fracture

- Avoid aggressive manipulation of microcatheters (especially soft-tip, such as the Caravel) through highly complex and calcified lesions or across stent struts.
- Microcatheters with stronger tip construction (such as the Mamba or Teleport) may have lower risk of tip fracture [19].

27.4.3 Treatment

Retrieving the entrapped and/or fractured microcatheter is important for preventing acute and chronic complications. Retrieval is performed using techniques similar to those described for retrieving a fractured balloon catheter, such as snares, use of guide catheter extensions and the "trapping" technique inside the guide, or wiring through the area of entrapment with another guidewire and performing balloon angioplasty. If those attempts fail, emergency surgery may be required.

In cases of microcatheter tip fracture, retrieval can be very challenging or impossible. In such cases stenting over the tip fragment may be the fastest and safest solution.

References

[1] Iturbe JM, Abdel-Karim AR, Papayannis A, et al. Frequency, treatment, and consequences of device loss and entrapment in contemporary percutaneous coronary interventions. J Invasive Cardiol 2012;24:215–21.

[2] Brilakis ES, Garrat KN. Device loss during percutaneous coronary intervention: incidence, complications, and retrieval methods. In: Ellis SG, Holmes DRJ, editors. Strategic approaches in coronary intervention. Lippincott, Williams and Wilkins; 2005. p. 325–31.

[3] Brilakis ES, Best PJ, Elesber AA, et al. Incidence, retrieval methods, and outcomes of stent loss during percutaneous coronary intervention: a large single-center experience. Catheter Cardiovasc Interv 2005;66:333–40.

[4] Abdel-Wahab M, Richardt G, Joachim Buttner H, et al. High-speed rotational atherectomy before paclitaxel-eluting stent implantation in complex calcified coronary lesions: the randomized ROTAXUS (Rotational Atherectomy Prior to Taxus Stent Treatment for Complex Native Coronary Artery Disease) trial. JACC Cardiovasc Interv 2013;6:10–19.

[5] Kozman H, Wiseman AH, Cook JR. Long-term outcome following coronary stent embolization or misdeployment. Am J Cardiol 2001;88:630–4.

[6] Utsunomiya M, Kobayashi T, Nakamura S. Case of dislodged stent lost in septal channel during stent delivery in complex chronic total occlusion of right coronary artery. J Invasive Cardiol 2009;21:E229–33.

[7] Sianos G, Papafaklis MI. Septal wire entrapment during recanalisation of a chronic total occlusion with the retrograde approach. Hellenic J Cardiol 2011;52:79–83.

[8] Eggebrecht H, Haude M, von Birgelen C, et al. Nonsurgical retrieval of embolized coronary stents. Catheter Cardiovasc Interv 2000;51:432–40.

[9] Candilio L, Mitomo S, Carlino M, Colombo A, Azzalini L. Stent loss during chronic total occlusion percutaneous coronary intervention: optical coherence tomography-guided stent 'crushing and trapping'. Cardiovasc Revasc Med 2017;18:531–4.

[10] Capretti G, Mitomo S, Giglio M, Carlino M, Colombo A, Azzalini L. Subintimal crush of an occluded stent to recanalize a chronic total occlusion due to in-stent restenosis: insights from a multimodality imaging approach. JACC Cardiovasc Interv 2017;10:e81–3.

[11] Azzalini L, Karatasakis A, Spratt JC, et al. Subadventitial stenting around occluded stents: A bailout technique to recanalize in-stent chronic total occlusions. Catheter Cardiovasc Interv 2018;92:466–74.

[12] Alfonso F, Martinez D, Hernandez R, et al. Stent embolization during intracoronary stenting. Am J Cardiol 1996;78:833–5.

[13] Danek BA, Karatasakis A, Brilakis ES. Consequences and treatment of guidewire entrapment and fracture during percutaneous coronary intervention. Cardiovasc Revasc Med 2016;17:129–33.

[14] Karacsonyi J, Martinez-Parachini JR, Danek BA, et al. Management of guidewire entrapment with laser atherectomy. J Invasive Cardiol 2017;29:E61–2.

[15] Pappy R, Gautam A, Abu-Fadel MS. AngioSculpt PTCA balloon catheter entrapment and detachment managed with stent jailing. J Invasive Cardiol 2010;22:E208–10.

[16] Sanchez-Recalde A, Galeote G, Martin-Reyes R, Moreno R. AngioSculpt PTCA balloon entrapment during dilatation of a heavily calcified lesion. Rev Esp Cardiol 2008;61:1361–3.

[17] Giugliano GR, Cox N, Popma J. Cutting balloon entrapment during treatment of in-stent restenosis: an unusual complication and its management. J Invasive Cardiol 2005;17:168–70.

[18] Lorusso R, De Cicco G, Ettori F, Curello S, Gelsomino S, Fucci C. Emergency surgery after saphenous vein graft perforation complicated by catheter balloon entrapment and hemorrhagic shock. Ann Thorac Surg 2008;86:1002–4.

[19] Vemmou E, Nikolakopoulos I, Xenogiannis I, et al. Recent advances in microcatheter technology for the treatment of chronic total occlusions. Expert Rev Med Devices 2019;16:267–73.

Chapter 28

Other complications: hypotension, radiation skin injury, contrast-induced acute kidney injury

In this chapter we discuss three non-coronary complications: hypotension, radiation skin injury, and contrast-induced acute kidney injury (CI-AKI).

28.1 Hypotension

CTO Manual Online cases 69, 85, 92, 129, 146
PCI Manual Online cases 23, 27, 40, 45, 72

Continuous careful monitoring of the pressure and electrocardiographic tracing is critical for enhancing the safety of PCI (Section 2.2). Awareness of the differential diagnostic algorithm if hypotension occurs during angiography or PCI can facilitate rapid decision making and initiation of corrective actions.

28.1.1 Causes

The appearance of low blood pressure on hemodynamic monitoring does not necessarily mean that the patient's systemic blood pressure is low, as "hypotension" could be due to technical issues (Fig. 28.1). Therefore, when hypotension occurs, it should be immediately assessed to determine whether the patient is truly hypotensive or not. If the patient's systemic blood pressure is indeed low, the differential diagnosis should be immediately considered to allow prompt diagnosis and treatment.

28.1.1.1 False hypotension

(Systemic pressure is normal, but appears low on the arterial pressure tracing)

1. Hemostatic valve of the Y-connector is open.
2. Connection between catheter and pressure transducer is open or has air in the line.
3. Pressure dampening. This is common, especially in patients with ostial coronary lesions and with use of large guide catheters, such as 8 French. Dampening can be masked when side hole guide catheters are used, hence the latter should never be used to engage the left main coronary artery (with the exception of ostial left main CTOs), as they can mask ischemia and lead to patient hemodynamic collapse.
4. Catheter obstruction by air, thrombus, or contrast (contrast can cause pressure dampening, especially in 4 and 5 French catheters). The catheter should be aspirated. If aspiration fails, the catheter should be removed without performing any injection.
5. Bulky equipment within the catheter.

28.1.1.2 True hypotension

There are three major causes of true hypotension: hypovolemia, cardiac failure, and peripheral vasodilation (Fig. 28.1). If the cause of hypotension is not immediately apparent, right heart catheterization can be very useful for determining the cause of hypotension.

Manual of Percutaneous Coronary Interventions. DOI: https://doi.org/10.1016/B978-0-12-819367-9.00028-7

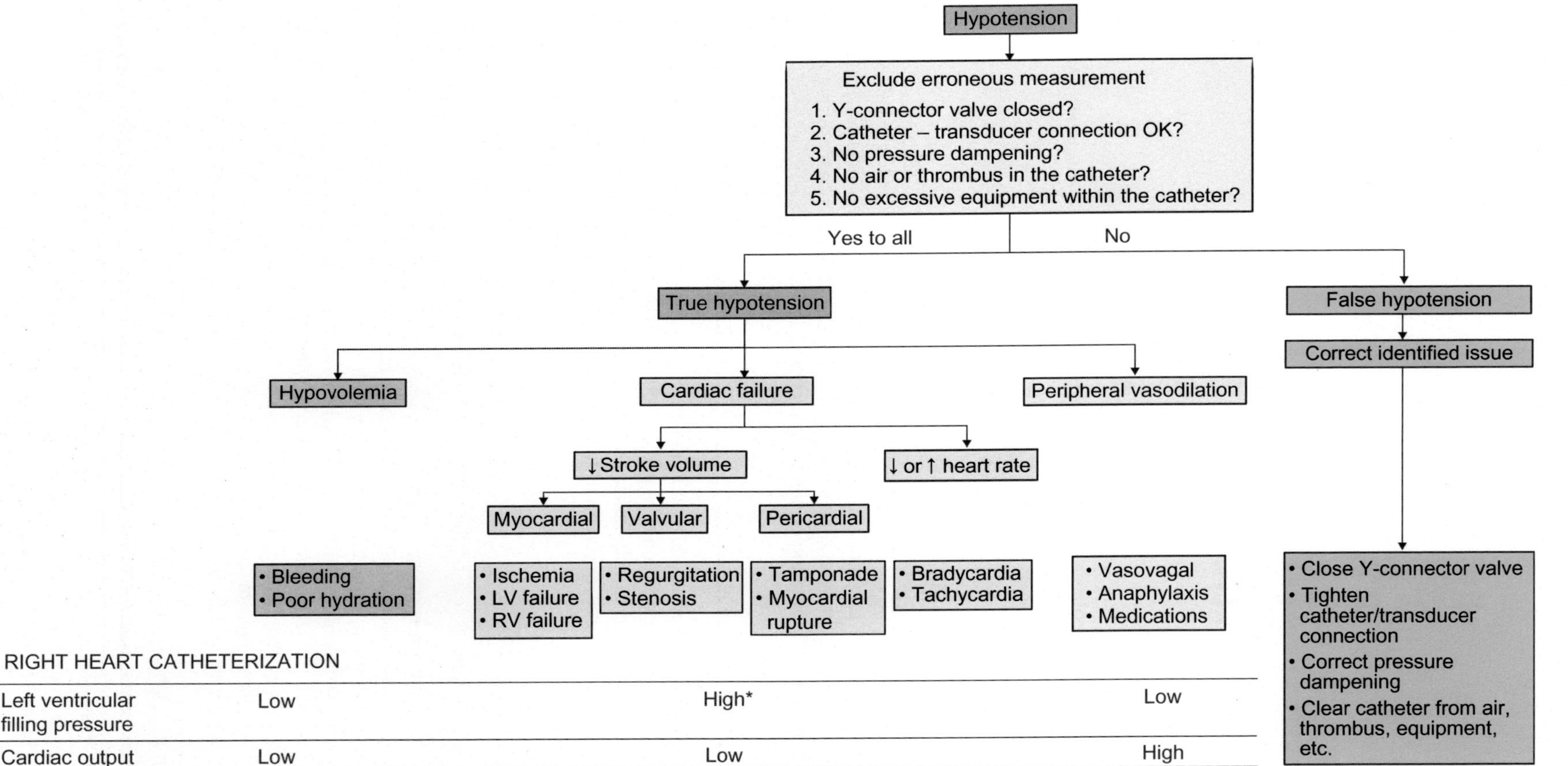

FIGURE 28.1 Differential diagnosis of hypotension. * left ventricular filling pressure can be low in cases of right ventricular failure.

28.1.1.2.1 Hypovolemia

Hypovolemia is most commonly caused by bleeding. Inspection of the access sites may reveal a hematoma. Fluoroscopy of the bladder may demonstrate the "dented bladder" sign (Fig. 29.3) that suggests retroperitoneal hematoma.

28.1.1.2.2 Cardiac failure

Cardiac output is the product of (stroke volume) × (heart rate). Decreased stroke volume may be due to: (1) left or right ventricular dysfunction, which is most commonly caused by ischemia; (2) valvular abnormalities, such as acute valvular regurgitation; and (3) pericardial tamponade.

Amplatz catheters (and deeply curved EBU catheters) may push the aortic cusp open and cause acute aortic regurgitation (Fig. 28.2); simple guide catheter repositioning can immediately correct the hypotension.

Both tachyarrhythmias and bradyarrhythmias can also reduce cardiac output.

28.1.1.2.3 Peripheral vasodilation

Peripheral vasodilation may be due to a vasovagal reaction, a systemic anaphylactic reaction, or medication administration, such as nitroglycerin (especially if the patient had been recently exposed to phosphodiesterase 5 inhibitors) or verapamil (intracoronary or intraarterial for preventing radial spasm).

28.1.2 Prevention

Preventing *false hypotension* can be accomplished using meticulous techniques:

1. Careful assessment and verification of all connections.
2. Meticulous care in clearing the catheter after insertion.
3. Avoiding deep catheter engagement causing dampening.

Preventing true hypotension can be achieved via careful planning of the procedure and attention to each step, as follows:

1. Careful arterial access technique and use of radial access, as well as avoiding excessive anticoagulation can decrease the risk of bleeding.
2. Adequate pre-procedural hydration.
3. Considering prophylactic hemodynamic support in high-risk patients undergoing PCI, as described in Chapter 14: Hemodynamic Support.
4. Preventing prolonged coronary artery occlusion will minimize ischemia.
5. Avoiding perforations and the resultant tamponade.

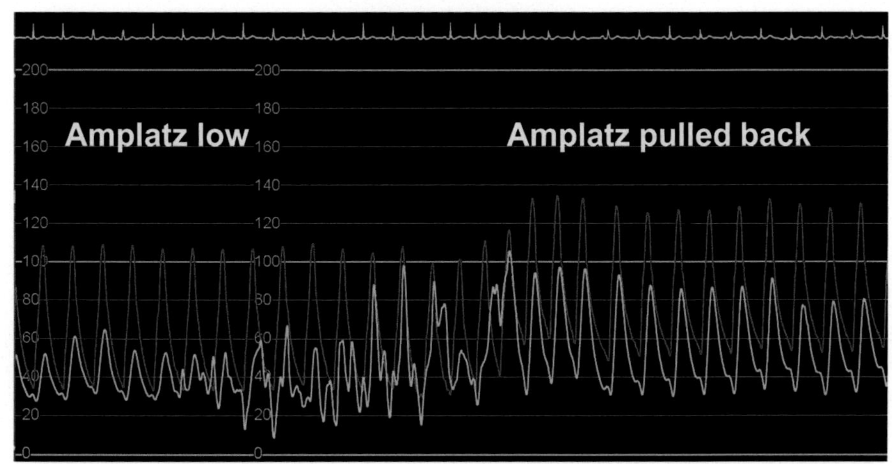

FIGURE 28.2 Blood pressure is lower in the in the left half of the screen due to deep advancement of an Amplatz catheter on the aortic valve. Blood pressure increased after withdrawal of the Amplatz catheter (*right half of the screen*). The *green* pressure tracing has dampening.

6. Adequate premedication for patients with contrast allergy will decrease the risk of anaphylactic shock (Section 3.3).
7. Early diagnosis and correction of tachyarrhythmias and bradyarrhythmias.
8. Adequate sedation and hydration may minimize the risk of a vasovagal reaction.
9. Avoiding excessive doses of nitroglycerin and other vasodilators.

28.1.3 Treatment

Immediate and complete treatment of the underlying cause is best for correcting hypotension, but may not always be feasible, for example, in patients who develop acute vessel closure that cannot be recanalized. Therefore, concomitant implementation of measures that can increase the systemic blood pressure and maintain systemic perfusion is often needed.

28.1.3.1 Maintain systemic perfusion

While treating the underlying cause of hypotension, administration of vasopressors and inotropes and in some patients initiation of mechanical circulatory support (Chapter 14: Hemodynamic Support) may be necessary. If the patient develops cardiac arrest, cardiopulmonary resuscitation is performed (ideally using an automated system, such as the Lucas system), followed by veno-arterial extracorporeal membrane oxygenation (VA-ECMO) initiation if the patient does not promptly recover.

28.1.3.2 Treat underlying cause

***Hypovolemia*:** Normal saline administration and blood transfusion in case of bleeding. If bleeding is due to an arterial access complication, a balloon should be inflated to stop active bleeding, followed by endovascular or surgical repair, as described in Chapter 29 - Vascular Access Complications.

***Myocardial dysfunction*:** This is usually caused by ischemia and can be improved with coronary revascularization and in some cases with mechanical circulatory support.

***Valvular disease*:** Although acute mitral or aortic regurgitation may benefit from vasodilator administration, they are often poorly tolerated requiring urgent or emergent surgery. Acute mechanical circulatory support may also improve hemodynamics, such as an intraaortic balloon pump for acute mitral regurgitation.

***Pericardial disease*:** Tamponade is treated with pericardiocentesis. If the patient has myocardial rupture (usually in the setting of acute myocardial infarction) emergency surgery is needed.

***Arrhythmias*:** Arrhythmias occurring in the cardiac catheterization laboratory are usually due to ischemia, hence treatment of ischemia will improve or correct them. Arrhythmias often occur after reperfusion (post-reperfusion arrhythmias). Bradycardia and atrioventricular block is often caused by medication administration, such as adenosine (Section 3.2.4), however the duration of action of adenosine is very short.

***Peripheral vasodilation*:** Epinephrine for anaphylactic shock and fluid administration.

28.2 Radiation skin injury

Excessive radiation dose can result in deterministic complications (e.g., radiation skin injury), but can also lead to stochastic complications (e.g., cancer or birth defects) [1]. Significant technological developments can help reduce radiation dose [2].

28.2.1 Causes

Deterministic radiation effects, such as skin injury and cataracts, correlate directly with the *air kerma dose* to a particular skin area (Fig. 28.3) [3].

The following AK dose thresholds are important to remember [4]:

<5 Gray: Below this threshold skin injury is unlikely to occur.

5–10 Gray: Skin injury is possible.

10–15 Gray: Skin injury is likely, requiring physicist assessment of the case.

>15 Gray: This is considered a sentinel event by the Joint Commission for Hospital Accreditation and requires reporting to the regulatory authorities in the U.S.

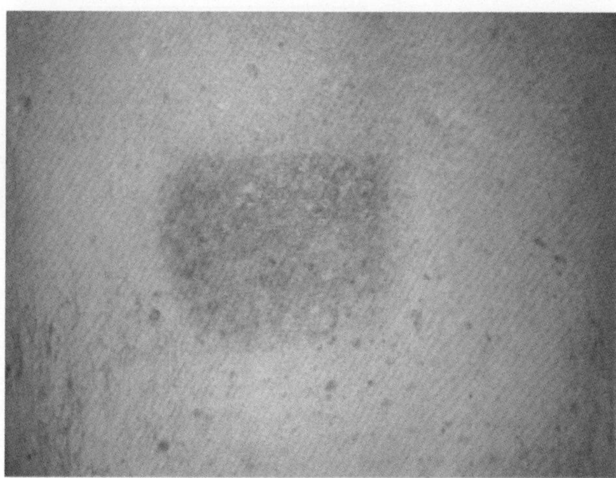

FIGURE 28.3 Example of radiation-induced skin injury after CTO PCI. Erythema and epilation developed on a patient's back 1 month after CTO PCI, during which he received 11.8 Gray air kerma dose. *Reproduced with permission from Chambers CE. Radiation dose in percutaneous coronary intervention OUCH did that hurt? JACC Cardiovasc Interv 2011;4:344–6. Copyright Elsevier.*

28.2.2 Prevention

Preventing radiation skin injury can be achieved by limiting the overall radiation dose and by rotating the image intensifier to distribute the radiation dose to various skin areas (Fig. 28.4).

Reducing patient (and operator) radiation dose can be achieved both before and after the procedure:

Before the procedure

28.2.2.1 *Planning*

Careful planning of the procedure can prevent unnecessary steps, facilitate procedural success and minimize contrast and radiation dose. Careful procedural planning applies not only to complex procedures, such as chronic total occlusion interventions [5], but every procedure. For example, carefully studying the location of aorto-coronary bypass grafts on previous angiograms can expedite bypass graft engagement. Catheterization in very obese patients should be performed with newer X-ray systems.

28.2.2.2 *Equipment*

28.2.2.2.1 X-ray machines

Newer X-ray machines achieve satisfactory image quality with lower radiation dose [6,7].

28.2.2.2.2 Control Rad

This is an add-on device to the X-ray system that allows the operator to identify a part of the screen to view in full resolution, while reducing radiation dose rate to the other parts of the screen, resulting in lower (but still adequate) image resolution in those areas and $\approx 75\%$ radiation dose reduction.

28.2.2.2.3 Zero Gravity (reduces operator dose only)

The Zero Gravity ceiling-suspended lead (Biotronik) not only provides radiation protection to the operator, but also obviates the need for wearing lead and the associated orthopedic injuries.

28.2.2.2.4 Robotic PCI (reduces operator dose only)

Robotic PCI (CorPath, Corindus) allows near elimination of operator radiation dose [8], but is currently available only at a few centers.

28.2.2.2.5 EggNest

The EggNest-XR System (Egg Medical) includes a series of shields that reduce scatter radiation.

Before the procedure

1. Planning
   1. Development of plan prior to the procedure based on prior studies and all clinical information

2. Equipment
   1. New X-ray machines
   2. Control Rad
   3. Zero Gravity (operator only)
   4. Robotic PCI (operator only)
   5. EGGNEST (operator only)

During the procedure

3. Do not use radiation
   1. No pedal "lag time"
   2. Use balloon and wire markers
   3. Use trapping technique for equipment exchanges
   4. Fluoro Store
   5. Use of intravascular imaging

4. Minimize radiation dose
   1. Fluoroscopy rate: 3.75–7.5 fps
   2. Low magnification
   3. Collimation, especially when increasing field of view
   4. Avoid extreme angles
   5. Position patient close to image intensifier
   6. Reposition image intensifier often
   7. Real time dose rate monitoring—make adjustments if >20 mGy/min
   8. Real time patient radiation dose monitoring
   9. Real time personnel dose monitoring

5. Shielding and distance (operator only)
   1. Take a step back
   2. Use manifold tubing extensions
   3. Place fixed radiation shields close to the patient and not too high
   4. Disposable shields

FIGURE 28.4 How to prevent radiation injury during cardiac catheterization.

During the procedure

28.2.2.3 Do not use radiation unless absolutely necessary

28.2.2.3.1 No pedal "lag time"

The "heavy foot" syndrome is defined as using X-ray when it is not needed, for example, when the operator is not looking at the screen! A "lag time" in releasing the pedal is common, especially in early stages of training, and should be a major focus for improvement.

28.2.2.3.2 Use balloon and wire markers

When advancing equipment through the guide catheter, use the balloon, wire, and stent shaft markers to determine if the device is close to the tip of the guide catheter at which time fluoroscopy is needed. Knowing the length of the guide catheter is critical, since using the more proximal marker in 90 cm guide catheters may result in the device exiting the guide catheter before the marker reaches the Y-connector.

28.2.2.3.3 Use trapping technique for equipment exchanges

The trapping technique (Section 8.9.1) allows secure equipment exchanges while minimizing use of X-ray.

28.2.2.3.4 Fluoro store

Cineangiography exposes the patient to $\approx 10\times$ higher dose compared to fluoroscopy and is not reflected in the fluoroscopy time. The "image store" or "fluoro save" function, is available in most modern X-ray equipment (Fig. 28.5) and should be used instead of cine to document balloon and stent inflations.

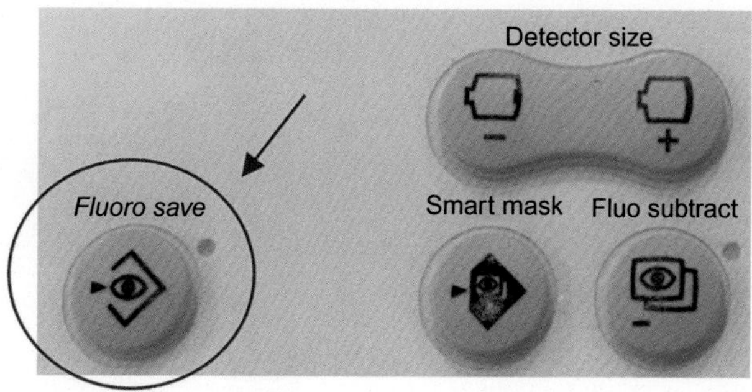

FIGURE 28.5 The fluoro-store button.

28.2.2.3.5 Use of intravascular imaging

Intravascular imaging can limit the need for both contrast and radiation administered during PCI.

28.2.2.4 Minimize radiation dose

28.2.2.4.1 Fluoroscopy frame rate: 6–7.5 fps

Fluoroscopy at 6–7.5 frames per second (fps) (or even 3.75 fps in thin patients) provides in nearly all cases adequate image quality, while reducing radiation dose by half compared with 15 fps fluoroscopy and should be routinely used [9].

28.2.2.4.2 Low magnification

Lower magnification requires less radiation exposure. Similar to using 7.5 fps, using lower magnification requires a "learning curve" to adjust to the change in image size, although with contemporary large screen monitors the images are still quite large.

28.2.2.4.3 Collimation, especially when increasing field of view

Collimation reduces the size of the skin area exposed to radiation and reduces the overall dose area product dose received by the patient, even though the total air kerma is not changed. Collimation allows a smaller skin exposure in any one projection, lessening potential skin injury from overlapping exposures when imaging angles are changed [10].

A caveat of collimation is that some equipment (e.g., the tips of guidewires or guide catheters) may not be included in the field of view, requiring intermittent monitoring to ensure that no significant changes have occurred (e.g., excessive distal migration of a guidewire that can lead to distal vessel perforation or deep engagement of the guide catheter that can lead to aorto-coronary dissection) [10].

28.2.2.4.4 Avoid extreme angles

When performing PCI the working angle should be minimized. Steep angles, such as greater than 30° from anteroposterior (AP), are associated with significantly higher radiation exposure due to the penetration through more tissue, hence less steep angles are preferred [10]. This increased dose is often not recognized by the operator and reflects the automatic dose increase by the equipment to maintain image quality. *The AP (anteroposterior) projection may not be the optimal view, because the spine is included in the field*, hence slightly angulated views are preferable [11].

The right anterior oblique projection can result in less operator radiation exposure, but can be challenging for right coronary artery wiring, although it is excellent for working in the mid left anterior descending artery.

28.2.2.4.5 Position patient close to image intensifier

The table should be placed as high as possible and the image intensifier as close to the patient, as possible (Fig. 28.6) [1].

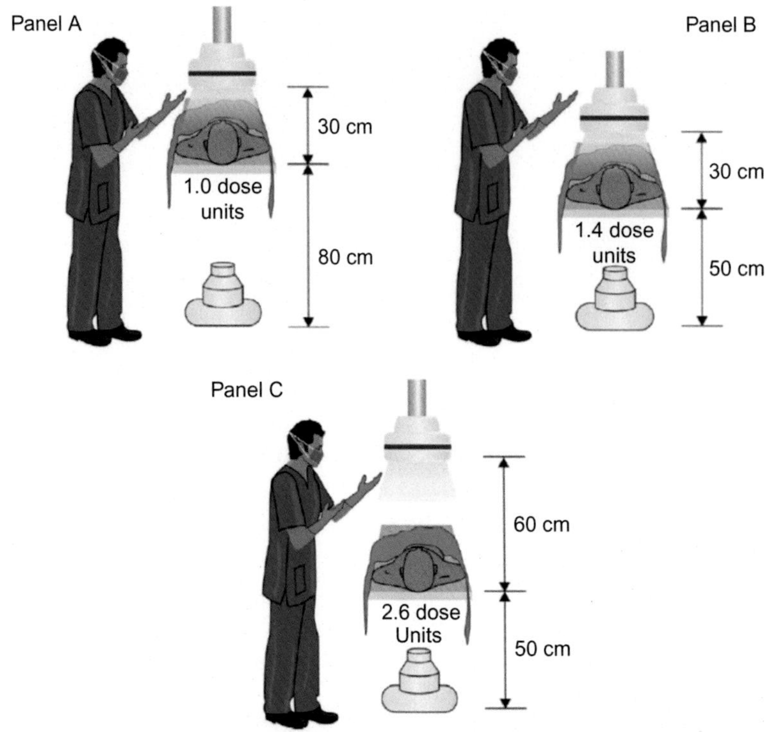

FIGURE 28.6 Example of optimal table positioning to minimize patient (and operator) radiation exposure. Panel A: the physician performs the procedure with the patient table elevated and the image intensifier close to the patient (total distance from the X-ray tube to the detector = 110 cm). Panel B: the physician employs a lower table setting but maintains the image intensifier close to the patient's chest (total distance from the X-ray tube to the detector = 80 cm). Because of the closer proximity to the X-ray tube, the dose rate to the patient at the beam entrance port will be about 40% higher. Panel C: the physician employs a low table height but has elevated the image intensifier (total distance from the X-ray tube to the detector = 110 cm). The skin dose to the patient on the *right* is 260% that of the patient in panel A. (The image generated by the configuration in panel C is 40% to 50% larger owing to geometric magnification caused by the elevated image intensifier). If the procedure in Panel A required a 3 Gray skin dose, the same procedure employing the Panel B configuration would result in a 4.2 Gray dose, whereas the one performed employing the configuration in Panel C would result in 7.8 Gray. *Reproduced with permission from Writing Committee M, Hirshfeld JW, Jr., Ferrari VA, et al. 2018 ACC/HRS/NASCI/SCAI/SCCT expert consensus document on optimal use of ionizing radiation in cardiovascular imaging: best practices for safety and effectiveness. Catheter Cardiovasc Interv 2018;92:E35–97. Copyright Elsevier.*

28.2.2.4.6 Reposition image intensifier often

Using multiple angles during fluoroscopy and cineangiography is critical during long procedures to minimize radiation exposure to the same skin entry point. High radiation dose procedures may not be as deleterious if the radiation is applied to multiple areas of skin, because the dose to each particular skin area is reduced [10].

28.2.2.4.7 Real-time dose rate monitoring

Many operators do not typically look at the radiation dose rate, even though it can provide real-time feedback on the radiation dose being administered, prompting changes if the dose rate is high (> 20 mGy/min).

28.2.2.4.8 Real-time patient dose monitoring

The operator should continually monitor the *cumulative air kerma dose*: consider stopping the procedure if 7–8 Gray air kerma dose is administered and consider staging the procedure for a later time. It is recommended that each catheterization laboratory have a protocol for alerting the operator on radiation dose (e.g., announcing the dose used every 1 Gray and/or every 30 min).

28.2.2.4.9 Real-time personnel dose rate monitoring

Real time operator radiation dose monitors (such as Dose Aware, Philips) are available and have been shown to reduce operator dose by approximately 30% [12].

28.2.2.5 Shielding and distance

These measures only reduce operator dose, which is still very important given daily exposure of operators to ionizing radiation over many years.

28.2.2.5.1 Distance—take a step back

Operator and staff must maximize their distance from the X-ray tube, as radiation dose decreases exponentially with distance (inverse square law) [13]. All appendages—operators' and patients'—should be out of the imaging field!

28.2.2.5.2 Use manifold tubing extensions (Fig. 28.7)

28.2.2.5.3 Place fixed radiation shields close to the patient and not too high (Fig. 28.8)

28.2.2.5.4 Disposable shields

Additional disposable radiation absorbing pads can be placed in the sterile field to reduce scatter radiation, such as the RadPad shields (Worldwide Innovations & Technologies) (Section 29.19) (that should be placed over the patient's abdomen) [14,15] and the Steradian vertical radiation shield (RADUX Devices, LLC, Maple Grove, Minnesota). Disposable shields should not be placed in the field of view, as this will increase radiation emitted from the X-ray tube, due to less radiation received by the detector.

28.2.3 Treatment

All patients who receive >5 Gray air kerma dose should be followed-up within a month and their skin examination documented in the medical record. During physical examination the back of the patients should be inspected to detect any radiation injury; if such an injury is diagnosed, they should be referred to specialists (dermatologist, plastic surgeons) for further evaluation and treatment. This is critical because if the patient is not aware of this risk and develops erythema and local discomfort (Fig. 28.3), he/she may see a dermatologist who may biopsy the lesion, potentially leading to a non-healing ulcer.

For >10 Gray air kerma dose, a qualified physicist should promptly calculate peak skin dose and the patient's skin should be examined in 2–4 weeks.

The Joint Commission identifies peak skin doses >15 Gray as a sentinel event; hospital risk management and regulatory agencies need to be contacted within 24 hours.

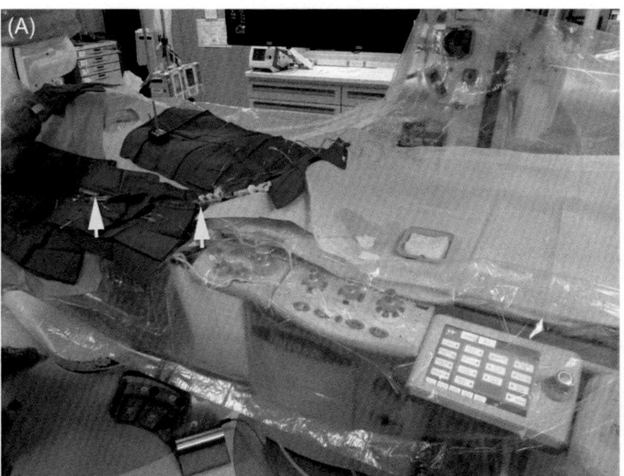

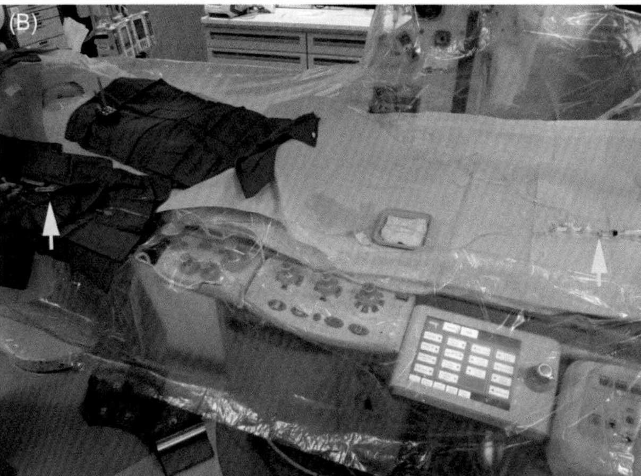

FIGURE 28.7 Manifold extension used to protect the operator from radiation scatter during transradial intervention. Without the manifold extension (panel A) the operator needs to be closer to the radiation source and the patient and receives higher dose as compared with use of a manifold extension (panel B). *Reproduced with permission from Christopoulos G, Makke L, Christakopoulos G, et al. Optimizing radiation safety in the cardiac catheterization laboratory: a practical approach. Catheter Cardiovasc Interv 2016;87:291–301 (Figure 2).*

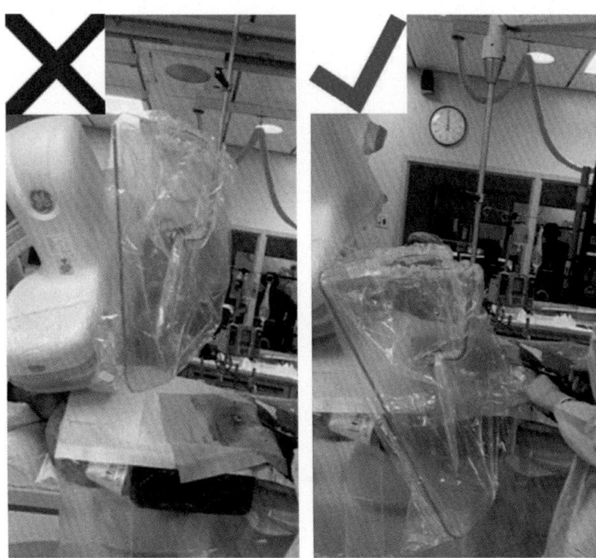

FIGURE 28.8 Suboptimal (*left panel*) and optimal (*right panel*) shield positioning.

28.3 Contrast-induced acute kidney injury

PCI Manual Online case 31, 36

28.3.1 Causes

There are several risk factor for CI-AKI [16]:

- Chronic kidney disease. In patients with advanced chronic kidney disease (CKD), defined as an estimated glomerular filtration rate (eGFR) <30 mL/min/1.73 m^2, the incidence of CI-AKI can be as high as 27% [16].
- Contrast volume. A contrast volume to creatinine clearance (CV/CrCl) ratio >2 has been identified as an independent predictor of CI-AKI in patients with an eGFR <30 mL/min/1.73 m^2 [17].
- Hypotension in various settings: cardiogenic shock, acute heart failure, acute coronary syndrome.
- Advanced age (> 75 years).
- Diabetes.
- Anemia.
- Low ejection fraction.
- Volume depletion.

There are several scores that can be used to identify patients at high risk for CI-AKI, such as the Mehran risk score [18], the Blue Cross Blue Shield of Michigan Cardiovascular Collaborative (BMC2) model [19], or the National Cardiovascular Data Registry Cath-PCI registry AKI prediction model [20].

28.3.2 Prevention

Several strategies can be used to minimize the risk for CI-AKI before, during, and after the procedure (Fig. 28.9).

Before procedure

- Hydration. This is the single most important pre-procedural (and post-procedural) measure to reduce the occurrence of CI-AKI. Hydration should be done by intravenous administration of normal saline, because no other solutions (e.g., bicarbonate, half-normal saline) have be shown to be advantageous [16].
- High dose statins.
- Discontinue nephrotoxic medications, such as NSAIDS, if possible.

During procedure

- Limiting contrast volume. In advanced CKD patients, administering contrast volume less than the creatinine clearance is ideal to minimize the risk of CI-AKI. There are several contrast saving options during PCI (Table 28.1) [16].

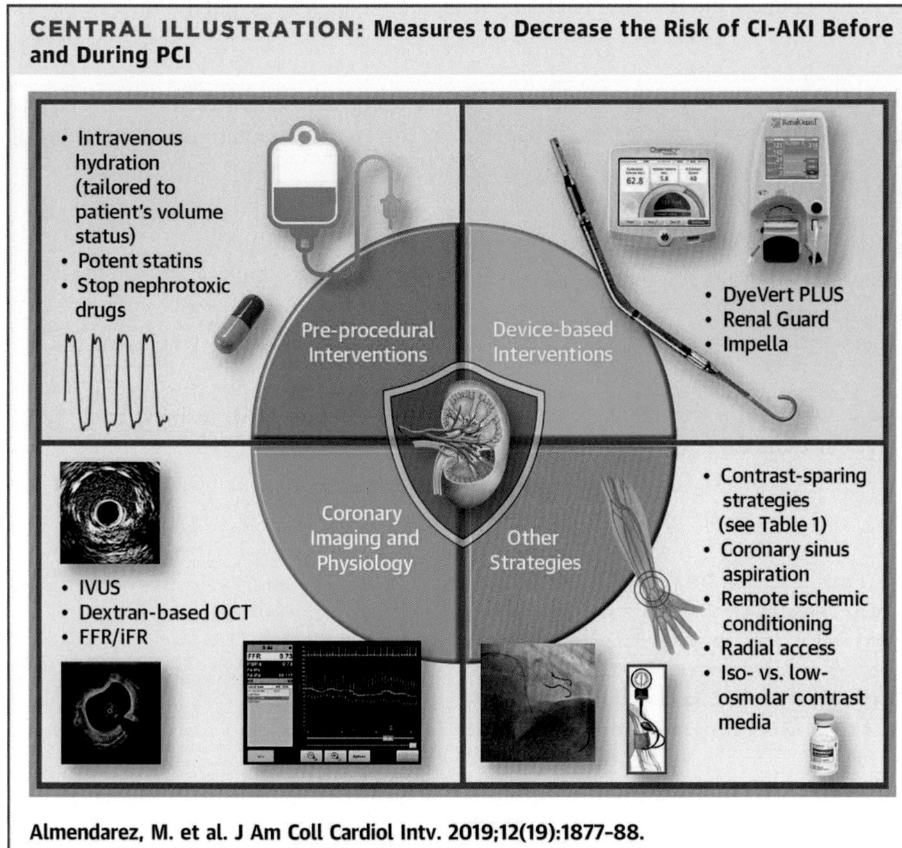
CENTRAL ILLUSTRATION: Measures to Decrease the Risk of CI-AKI Before and During PCI
• Intravenous hydration (tailored to patient's volume status)
• Potent statins
• Stop nephrotoxic drugs
Pre-procedural Interventions
Device-based Interventions
• DyeVert PLUS
• Renal Guard
• Impella
Coronary Imaging and Physiology
Other Strategies
• Contrast-sparing strategies (see Table 1)
• Coronary sinus aspiration
• Remote ischemic conditioning
• Radial access
• Iso- vs. low-osmolar contrast media
• IVUS
• Dextran-based OCT
• FFR/iFR
Almendarez, M. et al. J Am Coll Cardiol Intv. 2019;12(19):1877-88.

FIGURE 28.9 Measures to decrease the risk of CI-AKI before and during PCI. *Reproduced with permission from Almendarez M, Gurm HS, Mariani J, Jr., et al. Procedural strategies to reduce the incidence of contrast-induced acute kidney injury during percutaneous coronary intervention. JACC Cardiovasc Interv 2019;12:1877–88 (Central Illustration). Copyright Elsevier.*

TABLE 28.1 Contrast sparing strategies for PCI.

- Use 5 French catheters with no side holes for coronary angiography.
- Display previous coronary angiograms on cath lab monitors (if available) to avoid acquiring new diagnostic images.
- Use biplane or rotational angiography.
- Limit the volume of contrast per injection (ideally, 2–3 mL/injection).
- Use diluted contrast media.
- Test injections:
  - Enter coronary artery ostium with guidewire to confirm guide catheter engagement.
  - If unable to wire side branches, use intravascular ultrasound in live-view mode.
- Use stent enhancement techniques, e.g., StentBoost (Philips), CLEARstent (Siemens Healthcare).
- Use high fps acquisition rates (15 or 25 fps) to improve image quality during diagnostic angiogram and to evaluate the final result.
- Allow for elimination of contrast from the guiding catheter by back bleeding or aspirating before inserting and advancing new equipment.
- Use additional guidewires to create a roadmap of the target vessel and its side branches, or use dedicated software, e.g., Dynamic 3D Roadmap (Philips).
- Extensive use of intravascular ultrasound, dextran-based optical coherence tomography, and coronary physiology testing.
- If zero-contrast PCI is performed, a transthoracic echocardiogram should be done before and after the procedure to exclude development of a pericardial effusion that could be due to a perforation.

Reproduced with permission from Almendarez M, Gurm HS, Mariani J, Jr., et al. Procedural strategies to reduce the incidence of contrast-induced acute kidney injury during percutaneous coronary intervention. JACC Cardiovasc Interv 2019;12:1877–88 (Table 1). Copyright Elsevier.

- If needed, hemodynamic support may reduce the risk of CI-AKI by reducing the incidence and duration of hemodynamic instability.
- Use of iso-osmolar contrast media.
- When using contrast injection systems (such as the ACIST device) avoid using the system to flush with saline proximal to the Y-connector, other than prior to starting the procedure, as this may add additional contrast to fill the tubing.

After procedure

- Hydration is the key post-procedure action to reduce the incidence of CI-AKI.
- Serum creatinine should be monitored for at least 48 hours post-procedure to detect CI-AKI.

28.3.3 Treatment

- Once CI-AKI is established, there is no specific treatment, hence the goal is prevention [16]. In most cases renal function returns to baseline, however in some cases CI-AKI may require dialysis.

References

[1] Writing Committee M, Hirshfeld Jr. JW, Ferrari VA, et al. 2018 ACC/HRS/NASCI/SCAI/SCCT expert consensus document on optimal use of ionizing radiation in cardiovascular imaging: best practices for safety and effectiveness. Catheter Cardiovasc Interv 2018;92:E35–97.

[2] Brilakis ES. Innovations in radiation safety during cardiovascular catheterization. Circulation 2018;137:1317–19.

[3] Chambers CE. Radiation dose in percutaneous coronary intervention OUCH did that hurt? JACC Cardiovasc Interv 2011;4:344–6.

[4] Brilakis ES, Patel VG. What you can't see can hurt you!. J Invasive Cardiol 2012;24:421.

[5] Brilakis ES, Mashayekhi K, Tsuchikane E, et al. Guiding principles for chronic total occlusion percutaneous coronary intervention. Circulation 2019;140:420–33.

[6] Christopoulos G, Christakopoulos GE, Rangan BV, et al. Comparison of radiation dose between different fluoroscopy systems in the modern catheterization laboratory: results from bench testing using an anthropomorphic phantom. Catheter Cardiovasc Interv 2015;86:927–32.

[7] McNeice AH, Brooks M, Hanratty CG, Stevenson M, Spratt JC, Walsh SJ. A retrospective study of radiation dose measurements comparing different cath lab X-ray systems in a sample population of patients undergoing percutaneous coronary intervention for chronic total occlusions. Catheter Cardiovasc Interv 2018;92:E254–61.

[8] Madder RD, VanOosterhout S, Mulder A, et al. Impact of robotics and a suspended lead suit on physician radiation exposure during percutaneous coronary intervention. Cardiovasc Revasc Med 2017;18:190–6.

[9] Abdelaal E, Plourde G, MacHaalany J, et al. Effectiveness of low rate fluoroscopy at reducing operator and patient radiation dose during transradial coronary angiography and interventions. JACC Cardiovasc Interv 2014;7:567–74.

[10] Brilakis ES. Manual of coronary chronic total occlusion interventions. A step-by-step approach. 2nd ed Cambridge, MA: Elsevier; 2017.

[11] Agarwal S, Parashar A, Bajaj NS, et al. Relationship of beam angulation and radiation exposure in the cardiac catheterization laboratory. JACC Cardiovasc Interv 2014;7:558–66.

[12] Christopoulos G, Papayannis AC, Alomar M, et al. Effect of a real-time radiation monitoring device on operator radiation exposure during cardiac catheterization: the radiation reduction during cardiac catheterization using real-time monitoring study. Circ Cardiovasc Interv 2014;7:744–50.

[13] Christopoulos G, Makke L, Christakopoulos G, et al. Optimizing radiation safety in the cardiac catheterization laboratory: a practical approach. Catheter Cardiovasc Interv 2016;87:291–301.

[14] Shorrock D, Christopoulos G, Wosik J, et al. Impact of a disposable sterile radiation shield on operator radiation exposure during percutaneous coronary intervention of chronic total occlusions. J Invasive Cardiol 2015;27:313–16.

[15] Panetta CJ, Galbraith EM, Yanavitsk M, et al. Reduced radiation exposure in the cardiac catheterization laboratory with a novel vertical radiation shield. Catheter Cardiovasc Interv 2020;95:7–12.

[16] Almendarez M, Gurm HS, Mariani Jr. J, et al. Procedural strategies to reduce the incidence of contrast-induced acute kidney injury during percutaneous coronary intervention. JACC Cardiovasc Interv 2019;12:1877–88.

[17] Gurm HS, Dixon SR, Smith DE, et al. Renal function-based contrast dosing to define safe limits of radiographic contrast media in patients undergoing percutaneous coronary interventions. J Am Coll Cardiol 2011;58:907–14.

[18] Mehran R, Aymong ED, Nikolsky E, et al. A simple risk score for prediction of contrast-induced nephropathy after percutaneous coronary intervention: development and initial validation. J Am Coll Cardiol 2004;44:1393–9.

[19] Gurm HS, Seth M, Kooiman J, Share D. A novel tool for reliable and accurate prediction of renal complications in patients undergoing percutaneous coronary intervention. J Am Coll Cardiol 2013;61:2242–8.

[20] Tsai TT, Patel UD, Chang TI, et al. Contemporary incidence, predictors, and outcomes of acute kidney injury in patients undergoing percutaneous coronary interventions: insights from the NCDR Cath-PCI registry. JACC Cardiovasc Interv 2014;7:1–9.

Chapter 29

Vascular access complications

Both femoral and radial access may lead to complications that are discussed in this chapter.

29.1 Femoral access complications

Femoral access complications include lower extremity ischemia (due to dissection, embolization, or thrombosis), bleeding (groin hematoma and retroperitoneal hematoma), arteriovenous fistula, pseudoaneurysm, infection, and nerve injury.

29.1.1 Lower extremity ischemia

29.1.1.1 Diagnosis

Lower extremity ischemia has various manifestations:

- Acute limb ischemia: manifests as the 6Ps (pain, pallor, paralysis, pulse deficit, paresthesias, poikilothermia) and is usually caused by thrombosis, dissection, peripheral embolization of large thrombi (in contrast to the cholesterol embolization syndrome described below), or vascular closure device complications.
- Cholesterol embolization syndrome: this is usually caused by showers of microemboli and presents with the classic triad of leg/foot pain, livedo reticularis, and intact peripheral pulses [1,2]. Patients often develop hypereosinophilia and sometimes acute renal failure.

If the patient is in the catheterization laboratory, angiography of the affected limb can confirm the diagnosis and plan treatment. If the patient has left the cardiac catheterization laboratory usually ultrasound or computed tomography is used to confirm the diagnosis (Fig. 29.1).

29.1.1.2 Causes

- Arterial dissection.
- Thrombosis.
- Distal embolization (of thrombus, closure device components, cholesterol, etc.).
- Vascular closure device complications.
- Occlusive sheath (diseased and/or small size femoral artery, large diameter sheath).
- Large sheath size.
- Prolonged sheath and catheter dwelling time.
- Poor sheath management technique (sheath should always be aspirated first discarding the aspirated blood and then flushed before inserting equipment or administering medications).
- Prolonged or overzealous common femoral artery compression after sheath removal (which may also cause deep venous thrombosis).
- Suboptimal anticoagulation.

29.1.1.3 Prevention

- Meticulous access technique (Chapter 4: Access).
- Fluoroscopic guidance of femoral access.
- Ultrasound-guided access—ultrasound can help avoid areas with significant atherosclerosis or calcification.
- Use soft deflectable guidewires.
- Use the smallest sheath size possible.
- Meticulous sheath management technique (always aspirate first and discard the blood, as described above).

Manual of Percutaneous Coronary Interventions. DOI: https://doi.org/10.1016/B978-0-12-819367-9.00029-9

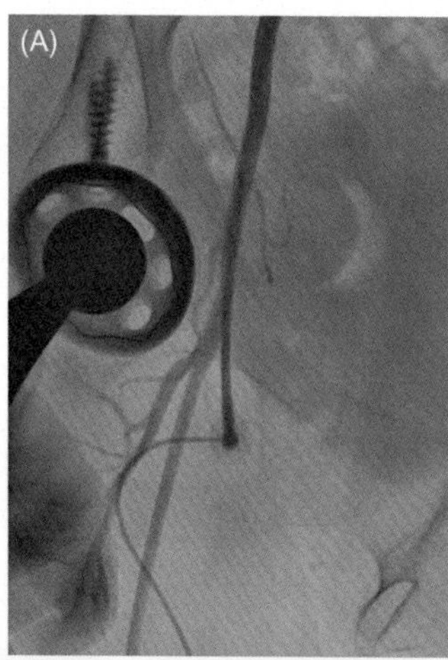

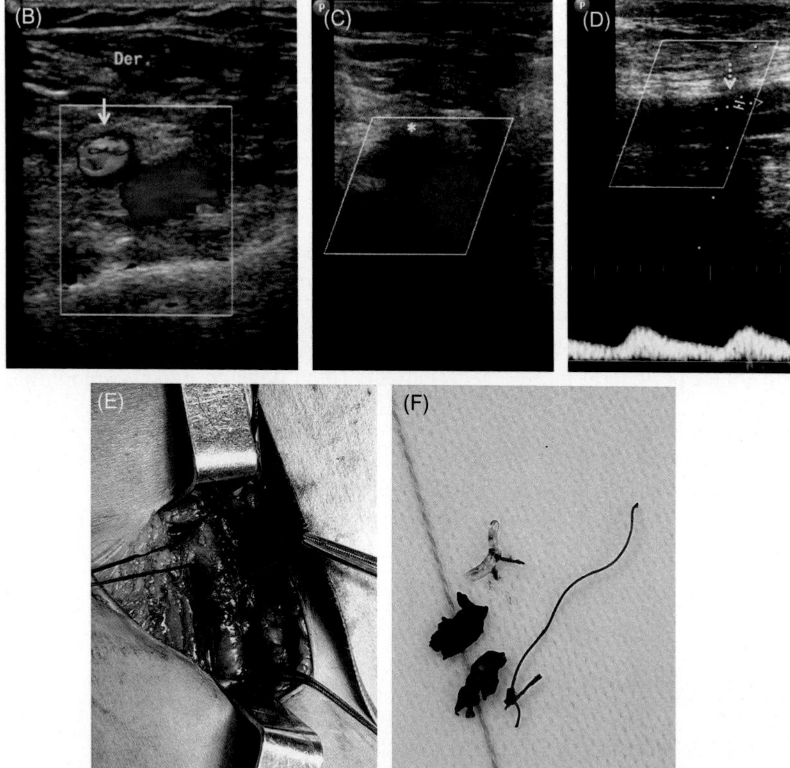

FIGURE 29.1 Acute limb ischemia after vascular closure device insertion. An Angioseal (Terumo) was used to seal the common femoral artery after cardiac catheterization (panel A). Ultrasound assessment of femoral artery before obtaining access showed a patent right common femoral artery (*yellow arrow*, panel B). Groin ultrasound after Angioseal deployment showed common femoral artery occlusion (*asterisk*, panel C) with monophasic flow in the superficial femoral artery (panel D). Emergent surgery was performed (panel E) with removal of the Angioseal (panel F). *Courtesy of Dr. Lucio Padilla.*

- Avoid pushing catheters against resistance.
- Periodic assessment of anticoagulation to ensure optimal ACT levels.
- Radial access

29.1.1.4 Treatment

- Acute limb ischemia: contralateral femoral access is obtained, followed by angiography to determine the cause of acute limb ischemia and often provide treatment (balloon angioplasty, endovascular thrombectomy or thrombolysis). Surgical thrombectomy, endarterectomy or bypass surgery may be needed in some cases (Fig. 29.1).
- Cholesterol embolization syndrome: there is no specific treatment—prevention is key. Statins may help with acute renal injury.

29.1.2 Bleeding related to femoral access

Bleeding related to femoral access may manifest as groin hematoma (can also affect the abdominal wall and testicles) or retroperitoneal hematoma.

29.1.2.1 Diagnosis

***Groin hematoma*:** this is usually evident as a groin mass that resolves after manual pressure, however diagnosis can be challenging and/or delayed in morbidly obese patients. Ultrasound may be needed to exclude the presence of a pseudoaneurysm or arteriovenous fistula. CT scan can provide accurate assessment of the hematoma location and size (Fig. 29.2).

***Retroperitoneal hematoma*:** retroperitoneal hematomas can be challenging to diagnose. The most common presenting signs are hypotension, diaphoresis, and lower abdominal or back pain. Patients often develop bradycardia and have tenderness on abdominal or suprapubic palpation. Retroperitoneal hematomas usually develop during the first 3 hours after the procedure. Use of vascular closure devices should not exclude the possibility of a retroperitoneal hematoma.

Operators should have a high level of suspicion for retroperitoneal hematoma, especially in obese patients, patients with uncontrolled hypertension prior to puncture and/or during the procedure, patients with high femoral puncture or femoral puncture that was not guided by ultrasound or fluoroscopy and when multiple femoral arterial puncture attempts were made.

If contrast has been given, fluoroscopy of the bladder may reveal displacement ("dented bladder") sign, which is highly suggestive of retroperitoneal hematoma (Fig. 29.3).

If at the time of diagnosis the patient is still in the cardiac catheterization laboratory, contralateral femoral access is obtained and angiography is performed to determine the source of bleeding and provide treatment (with prolonged balloon inflation and covered stents or coils, as described in Section 29.1.2.4).

29.1.2.2 Causes

- High/noncompressible puncture (above the inferior border of the inferior epigastric artery). Use of the Angioseal closure device should be avoided in high punctures, as the collagen plug may become entangled within the abdominal wall muscle layers.

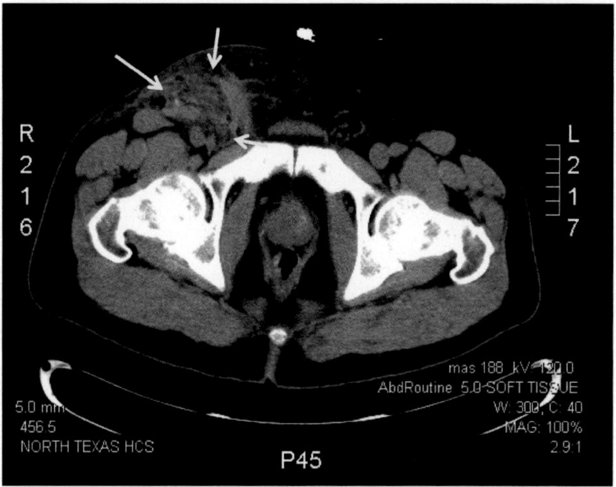

FIGURE 29.2 CT scan of the abdomen and pelvis demonstrating a right groin hematoma.

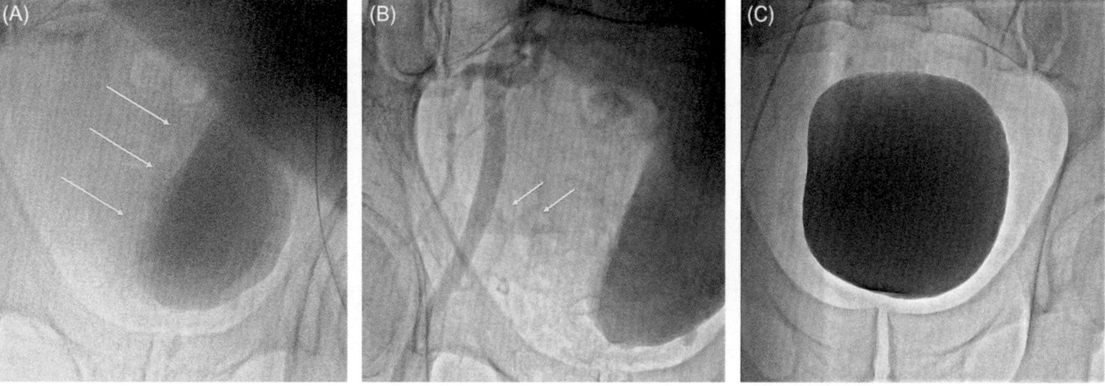

FIGURE 29.3 "Dented bladder sign" (*arrows*, panel A), suggestive of retroperitoneal hematoma. Iliac angiography revealed a perforation of the right external iliac artery (*arrows*, panel B). A normal bladder after cardiac catheterization (panel C).

- Puncture of the back wall of the femoral artery.
- Supratherapeutic or prolonged anticoagulation and use of glycoprotein IIb/IIIa inhibitors or cangrelor.
- Hypertension.
- Inadvertent perforation of renal or other peripheral arteries, usually by a polymer-jacketed guidewire, such as the Glidewire (Terumo).

29.1.2.3 Prevention

- Meticulous arterial access technique with fluoroscopic and ultrasound guidance and immediate femoral angiography, as described in Chapter 4: Access.
- Avoid supratherapeutic anticoagulation
- Control hypertension.
- Advance guidewires in the iliac arteries and the aorta under fluoroscopic guidance. NEVER push wires hard if resistance to advancement is felt.

29.1.2.4 Treatment

Groin hematoma: manual compression proximal to the skin entry site. This will often be painful for the patient, requiring use of analgesics. After bleeding is stopped, a compressive bandage should be applied and the patient placed in bed rest for 6–8 hours. Duplex ultrasound can help ensure there is no ongoing bleeding or pseudoaneurysm formation. In case of failure to control the hematoma, emergency endovascular or surgical treatment may be needed as described for retroperitoneal hematoma below.

Suspected retroperitoneal hematoma (Fig. 29.4):

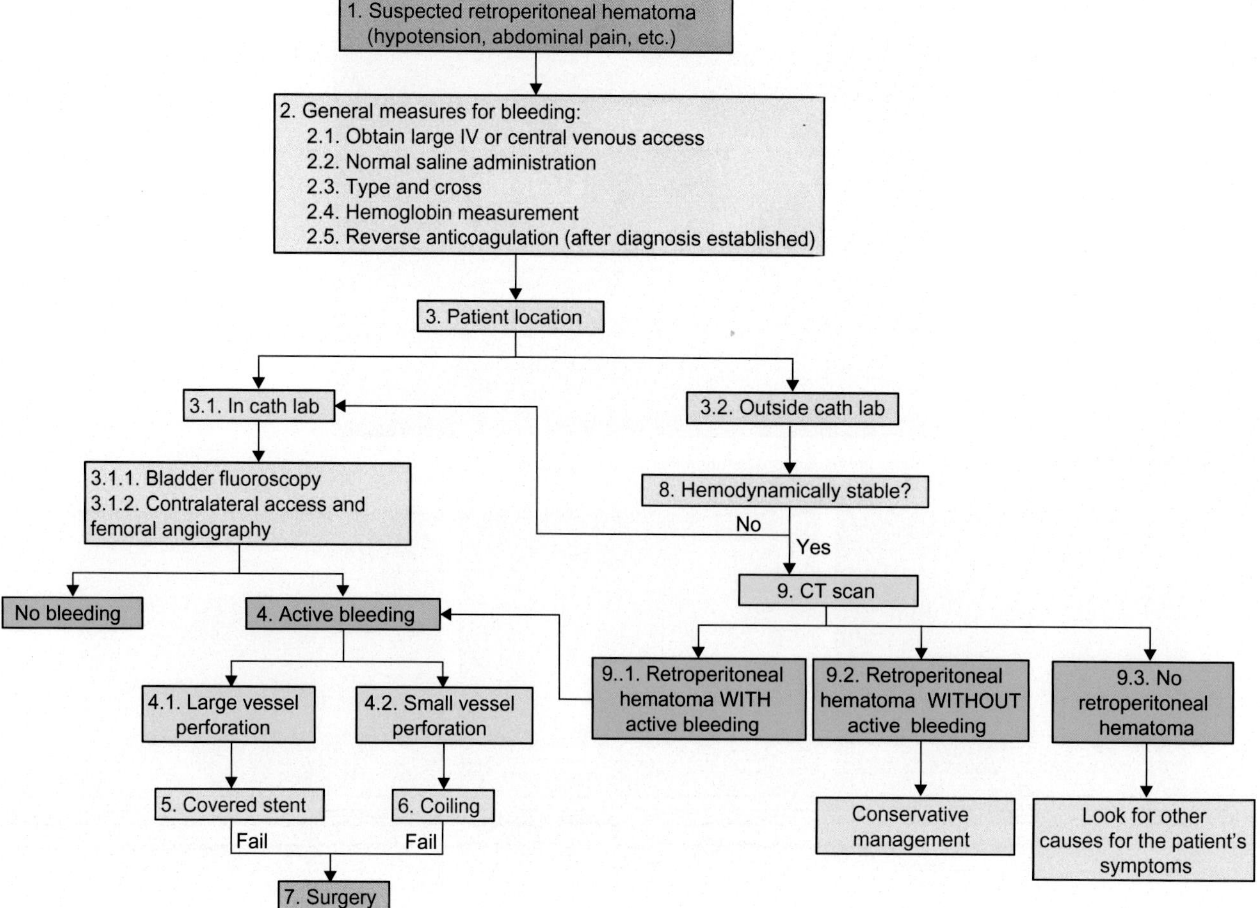

FIGURE 29.4 Algorithm for approaching patients with suspected retroperitoneal hematoma.

1. **Retroperitoneal hematoma should be suspected in every patient who develops hypotension during or after cardiac catheterization using femoral access.**
2. **General measures** for bleeding include:
   - Large intravenous access—sometimes a central line may be needed to administer vasopressors and large volume of crystalloids or blood.
   - Administration of normal saline (several liters may be required).
   - Blood should be sent for type and cross, followed by transfusion depending on the magnitude of bleeding and the patient's hemodynamics.
   - Hemoglobin measurement may be useful, although often acute bleeding does not result in immediate decrease in hemoglobin.
   - Anticoagulation should be reversed (protamine for heparin, discontinue glycoprotein IIb/IIIa inhibitors and cangrelor).
   - Manual pressure should be held over the access site, especially if there is also a groin hematoma.
3. **Patient location.**

   If the patient is in the cath lab, fluoroscopy of the abdomen may show bladder displacement (dented bladder sign, Fig. 29.3, panel A). Femoral angiography can demonstrate the presence and site of bleeding (Fig. 29.3, panel B).
4. **Active bleeding**

   Similar to coronary perforations, the mechanism of bleeding is important for implementing subsequent treatment. There are two main types of iliofemoral perforations: large vessel perforation or small vessel perforation (such as perforation of the inferior epigastric or lateral circumflex iliac artery). Prior to pursuing definitive treatment a balloon is inflated (at low pressure) over the site of perforation to stop ongoing bleeding.
5. **Covered stent**

   If prolonged balloon inflation fails to achieve hemostasis, placement of a covered stent across the site of perforation can seal large vessel perforations (Fig. 29.5) [3], but may be challenging to deliver. Moreover, stents deployed in the common femoral artery are prone to fracture.

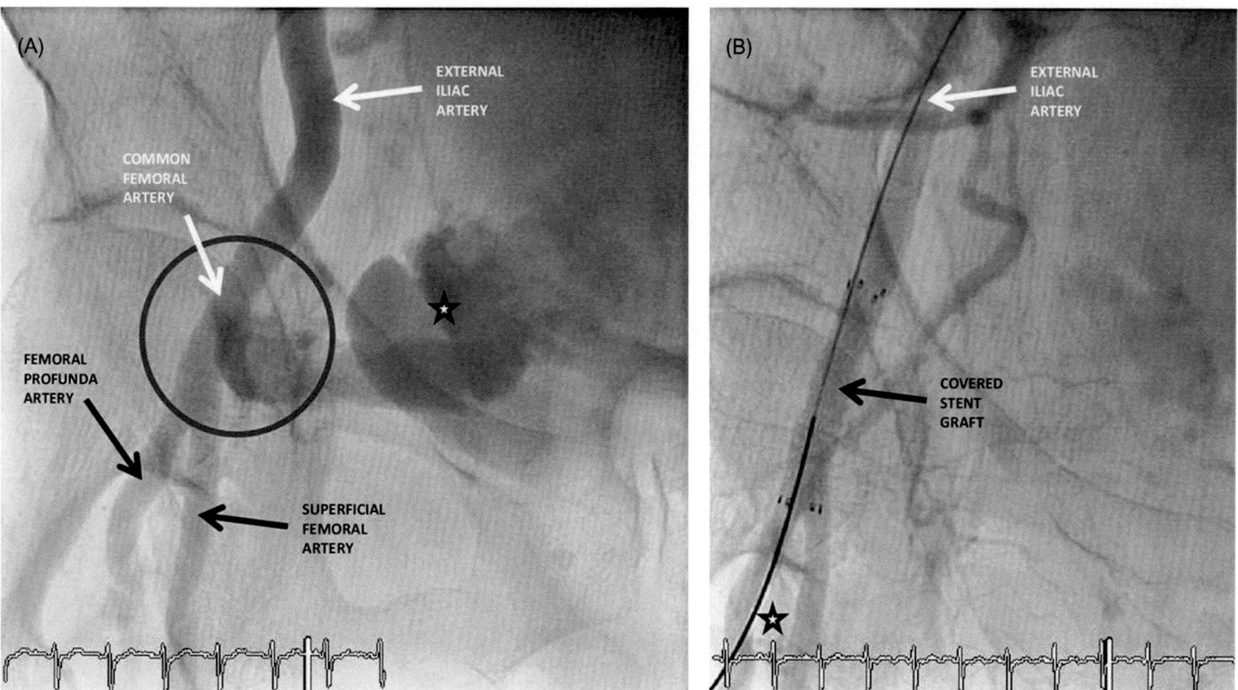

FIGURE 29.5 (A) Arteriotomy closure failure resulting in brisk bleeding from the arterial access site. The *star* indicates extravasation. (B) Secondary vessel closure after implantation of a Fluency Plus (Bard Peripheral Vascular) covered stent graft. The *star* indicates Magic Torque Wire (Boston Scientific) placed in the superficial femoral artery. *Reproduced with permission from Stortecky S, Wenaweser P, Diehm N, et al. Percutaneous management of vascular complications in patients undergoing transcatheter aortic valve implantation. JACC Cardiovasc Interv 2012;5:515–24. Copyright Elsevier.*

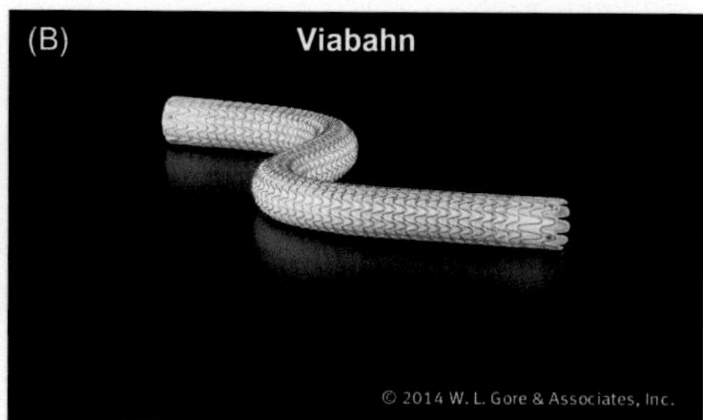

FIGURE 29.6 The Viabahn VBX balloon expandable covered stent (panel A) and Viabahn self-expanding covered stent (panel B). *(A) GORE VIABAHN VBX Balloon Expandable Endoprosthesis. ©2019. (B) GORE VIABAHN Endoprosthesis. ©2019.*

There are several endovascular peripheral covered stents that can be used, both balloon expandable and self-expanding. Balloon expandable covered stents, such as the Viabahn VBX (W.L. Gore, Fig. 29.6, panel A) and iCast (Atrium) allow more precise positioning than self-expanding covered stents, such as the Viabahn (W.L. Gore, Fig. 29.6, panel B) and Fluency Plus (Bard). Self-expanding stents are preferred for the common femoral artery, as they are more resistant to deformation during flexion.

6. **Coil or microsphere embolization**

   Coiling embolization can be used to stop bleeding in case of small branch perforation. If coiling is used, retrograde bleeding through the perforation should be excluded.

   Another option is microsphere, "cyanoacrylate glue," or thrombin embolization, especially in cases of very distal vessel perforation, where delivering coils is not always possible. Particular attention should be paid in those cases to avoiding leakage of the sealant outside the perforated vessel.

   Alternatively prolonged balloon inflation or implantation of a covered stent across the origin of the perforated vessel may achieve hemostasis [4].

7. **Surgery**

   If placement of a covered stent across the perforation site fails or is not desirable (e.g., for perforations at the superficial and profunda femoral artery bifurcation), surgical treatment of the perforation may be required.

8. **Hemodynamic stability**

   Patients who have left the catheterization laboratory and are hemodynamically stable are usually evaluated with computed tomography to confirm the diagnosis of retroperitoneal hematoma and determine whether there is ongoing active bleeding or not.

   Patients who are not hemodynamically stable need to go emergently either to the cardiac catheterization laboratory or to the operating room for surgical correction of the bleeding site.

   Patients who have a retroperitoneal hematoma but do not have ongoing bleeding are treated conservatively, as described in Section 29.1.2.4, point 2.

9. **CT scan**

CT scan of the abdomen and pelvis can help establish the diagnosis of retroperitoneal hematoma (Fig. 29.7) and determine whether there is ongoing bleeding (active contrast extravasation) or not.

29.1.3 Pseudoaneurysm

Pseudoaneurysm (also known as false aneurysm) is a collection of blood between the two outer layers of the artery, the media and the adventitia. In contrast, true aneurysms involve all 3 layers of the artery. Pseudoaneurysm formation carries is a risk of infection, rupture, compression neuropathy, or deep venous thrombosis.

29.1.3.1 Diagnosis

Pseudoaneurysms usually manifest as a pulsatile tender groin mass, often with a bruit. Diagnosis is usually made with Duplex ultrasound that demonstrates blood flow in and out of the pseudoaneurysm cavity.

29.1.3.2 Causes

- Low femoral artery puncture.

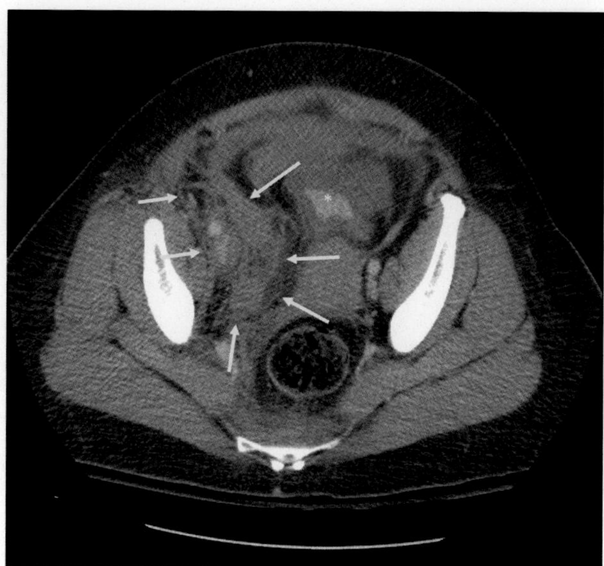

FIGURE 29.7 Right retroperitoneal hematoma (*arrows*) on CT scan of the abdomen and pelvis with bladder (*star*) displacement to the left.

- Suboptimal compression of the puncture site at the end of the procedure.
- Challenges with obtaining access, such as kinking of the guidewire.
- Access of both the femoral artery and vein.
- Intensive anticoagulation.
- Vascular closure device failure.

29.1.3.3 Prevention

- Use optimal femoral access technique, as described in Chapter 4: Access.
- Good hemostasis technique.

29.1.3.4 Treatment

- Small pseudoaneurysms (<2 cm in diameter): conservative management is followed in most patients (unless they have severe groin pain), as many pseudoaneurysms close spontaneously.
- Large pseudoaneurysms (≥2 cm): ultrasound-guided thrombin injection is used in most patients, with follow-up ultrasonography in 2–3 days to confirm no recurrence (Fig. 29.8). This, however, requires that the pseudoaneurysm has a narrow communication "neck." In pseudoaneurysms with wide neck, there is risk of distal thrombin embolization causing acute limb ischemia. In such cases, surgical repair may be preferable. Additionally, surgery may be needed in patients with very large (>6 cm) pseudoaneurysms, synthetic grafts, infections, limb ischemia, or skin necrosis. In some cases, prolonged compression (occasionally >30–45 min) using the ultrasound probe as a compression arm may be used to achieve closure of the pseudoaneurysm. During the compression process, the operator continuously verifies that pressing the probe over the "neck" of the pseudoaneurysm, sufficiently stops the blood flow though the "neck". Eventually, complete thrombosis of the pseudoaneurysm is achieved. Pseudoaneurysm compression can cause significant patient discomfort.

29.1.4 Arteriovenous fistula

Arteriovenous fistula is a communication between the femoral artery and femoral vein.

29.1.4.1 Diagnosis

Patients who develop arteriovenous fistulas after transfemoral cardiac catheterization are usually asymptomatic, although they can rarely develop lower extremity edema or high-output cardiac failure if there is large shunting from

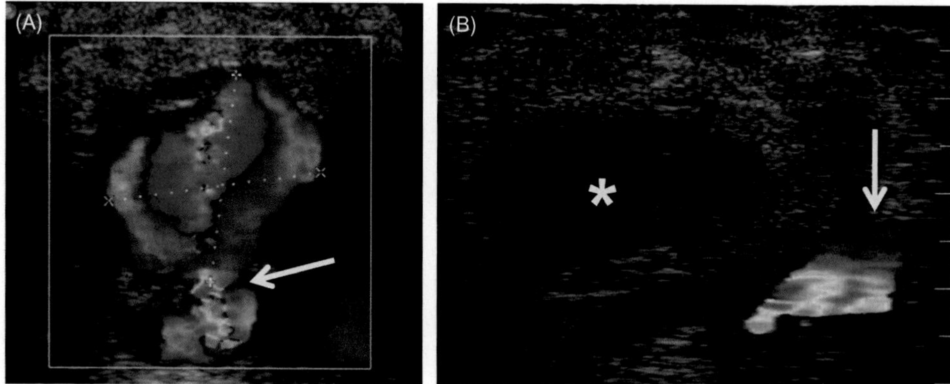

FIGURE 29.8 Example of thrombin injection for sealing a femoral pseudoaneurysm. Color Doppler ultrasound image showing a transverse section of the pseudoaneurysm communicating with the femoral artery with a narrow neck (*arrow*, panel A). After thrombin injection and compression, flow within the cavity (*asterisk*, panel B) stopped corresponding to the thrombosed pseudoaneurysm, whereas flow continued in the slightly compressed femoral artery (*arrow*, panel B). *Courtesy of Dr. Lucio Padilla.*

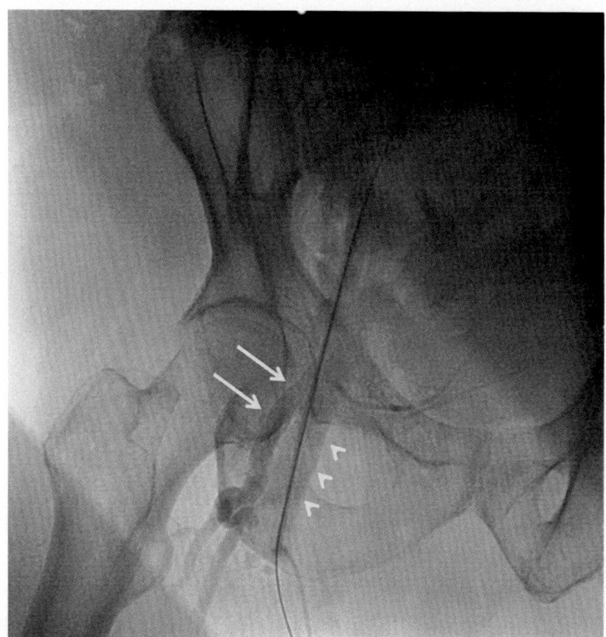

FIGURE 29.9 Simultaneous filling of the right common femoral vein (*arrowheads*) during femoral angiography (*arrows* point to the right common femoral artery) suggesting the presence of an arteriovenous fistula.

the femoral artery to the femoral vein. Patients may have a bruit and palpable thrill over the femoral artery. Diagnosis is usually made by Doppler ultrasound or angiography (Fig. 29.9).

29.1.4.2 Causes

- Low puncture: the femoral vein is often located behind the common femoral artery (or the superficial femoral artery) lower in the groin. Use of ultrasound can reveal the relative location of the femoral artery and vein and prevent access of the femoral vein through the femoral artery.
- Multiple arterial punctures.
- Access of both the femoral artery and femoral vein during the same procedure, especially if done at the same level.

29.1.4.3 Prevention

- Optimal access technique, as described in Chapter 4: Access.

- If both arterial and venous access are obtained, the arterial sheath should be removed first, followed by removal of the venous sheath a few minutes later, instead of simultaneous removal.

29.1.4.4 Treatment

In most cases no treatment is needed. Many arteriovenous fistulae close spontaneously within a year. Rarely, specific treatment may be needed for large arteriovenous fistulas, with covered stent placement in the femoral artery (if the common femoral artery bifurcation into the profunda and the superficial femoral artery is not involved) or surgery.

29.1.5 Infection

Access site infection can be a devastating complication, often requiring surgery and prolonged healing.

29.1.5.1 Diagnosis

Pain, erythema, and discharge from the access site are suggestive of infection, but patients may also have systemic signs of infection, such as fever and leukocytosis. Blood cultures are often performed as well as duplex ultrasound to exclude the presence of a pseudoaneurysm.

29.1.5.2 Causes

- Groin hematoma or pseudoaneurysm.
- Prolonged dwelling of the femoral sheath (e.g., overnight).
- Immunocompromised status (e.g., in transplant patients).
- Poor hygiene of the access site.
- Use of arterial closure devices.

29.1.5.3 Prevention

- Meticulous attention to sterile technique.
- Repeat preparation of the skin with an antiseptic solution before deploying a vascular closure device, especially after long procedures.
- Avoiding use of arterial closure devices in patients at increased risk of infection (such as immunocompromised patients or patients with skin infection at the access site).
- State-of-the-art access technique as described in Chapter 4: Access.
- Sheath removal as soon as possible after the procedure.
- Prophylactic antibiotic administration in patients at high risk of infection.
- Hematomas can act as "culture medium" for microbial growth and infection, therefore every measure should be taken to prevent hematoma formation.

29.1.5.4 Treatment

- Intravenous antibiotics.
- Surgery in case of vascular closure device infection, abscess formation, prior synthetic bypass graft, or persistent bacteremia despite antibiotic treatment.

29.1.6 Nerve injury

29.1.6.1 Diagnosis

Patients rarely develop injury of the lateral cutaneous or femoral nerves, that manifests as leg pain.

29.1.6.2 Causes

- Direct injury from the access needle.
- Nerve compression from a hematoma or pseudoaneurysm.

29.1.6.3 Prevention

- State-of-the-art access technique as described in Chapter 4: Access.

29.1.6.4 Treatment

- Conservative treatment is used. Symptoms resolve in most cases, although it may take several months for complete resolution.

29.2 Radial access complications

Radial access complications include bleeding (althrough bleeding occurs less often with radial as compared with femoral access, it is very important to detect and treat it early as it may be lead to compartment syndrome), radial artery occlusion, spasm, pseudoaneurysm, sterile granuloma, and nerve injury [5].

29.2.1 Bleeding

Bleeding can occur: (a) at the access site due to challenges with hemostasis; or (b) higher in the radial artery.

29.2.1.1 Diagnosis

Bleeding is infrequent with radial access, as the artery can be compressed against the wrist using a compression band as described in Chapter 11: Access Closure. Bleeding can manifest as local hematoma, but if it continues uncontrolled in the forearm it may lead to compartment syndrome that requires emergent surgery (fasciotomy).

29.2.1.2 Causes

- Suboptimal placement of the compression device (the arterial entry point is often more proximal than the skin entry, especially if a shallow angle is used when advancing the needle). Too distal placement of the compression device can result in bleeding into the forearm, whereas too proximal placement can result in a wrist hematoma.
- Intensive anticoagulation.
- Guidewire entry into a side branch of the radial artery.
- Avulsion of side branches of the radial artery from catheter advancement.
- Rupture of the radial artery during aggressive manipulation of the radial sheath and/or catheters in the setting of radial spasm.
- Aggressive manipulation of the catheter, especially through radial loops.

29.2.1.3 Prevention

- Meticulous access technique (Chapter 4: Access).
- Close monitoring of the arterial access site.
- Careful positioning of the compression band (slightly proximal to the puncture site)
- Never force the guidewire, sheath, or (diagnostic or guide) catheter in case of resistance.
- Confirm correct position of the guidewire under X-ray before inserting the sheath.
- Avoid inserting large sheaths relative to the size of the radial artery; the latter can be assessed using ultrasound.
- Use balloon assisted tracking technique when crossing radial loops to avoid the razor effect of the catheter.

29.2.1.4 Treatment

(Fig. 29.10)

- **Access site bleeding:** reposition the compression device. Alternatively, a second compression device can be placed, or manual pressure can be held and the upper extremity can be held elevated. A blood pressure cuff can also be inflated above systolic pressure to assist with hemostasis.
- **Proximal bleeding:** if the radial artery can be safely crossed and a catheter advanced through the area of perforation, hemostasis will be achieved in most cases. In cases of radial artery rupture associated with severe bleeding, it is crucial to maintain wire position. If the catheter, either diagnostic or guide, does not achieve hemostasis, a long peripheral or coronary balloon, of adequate size and type inflated at low pressure for a prolonged period of time,

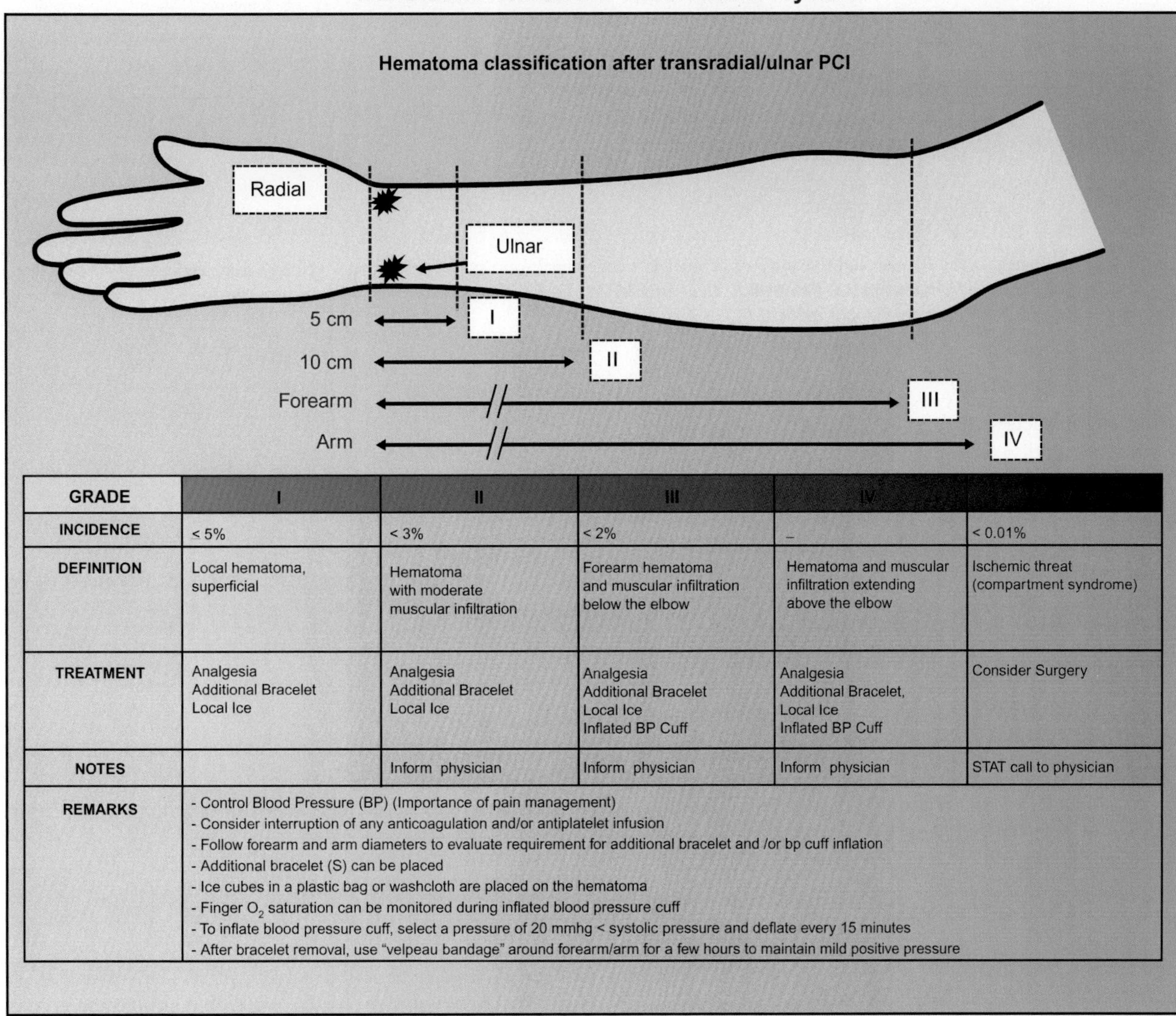

GRADE	I	II	III	IV	V
INCIDENCE	< 5%	< 3%	< 2%	–	< 0.01%
DEFINITION	Local hematoma, superficial	Hematoma with moderate muscular infiltration	Forearm hematoma and muscular infiltration below the elbow	Hematoma and muscular infiltration extending above the elbow	Ischemic threat (compartment syndrome)
TREATMENT	Analgesia Additional Bracelet Local Ice	Analgesia Additional Bracelet Local Ice	Analgesia Additional Bracelet Local Ice Inflated BP Cuff	Analgesia Additional Bracelet, Local Ice Inflated BP Cuff	Consider Surgery
NOTES		Inform physician	Inform physician	Inform physician	STAT call to physician
REMARKS	- Control Blood Pressure (BP) (Importance of pain management) - Consider interruption of any anticoagulation and/or antiplatelet infusion - Follow forearm and arm diameters to evaluate requirement for additional bracelet and /or bp cuff inflation - Additional bracelet (S) can be placed - Ice cubes in a plastic bag or washcloth are placed on the hematoma - Finger O_2 saturation can be monitored during inflated blood pressure cuff - To inflate blood pressure cuff, select a pressure of 20 mmhg < systolic pressure and deflate every 15 minutes - After bracelet removal, use "velpeau bandage" around forearm/arm for a few hours to maintain mild positive pressure				

FIGURE 29.10 Treatment of radial access bleeding.

often seals the perforation. Wrapping the arm with a compression dressing and ice can also help. In case of compartment syndrome, emergent fasciotomy is needed (Fig. 29.11) [6].

29.2.2 Radial artery occlusion

29.2.2.1 Diagnosis

Most patients who develop radial artery occlusion do not have any symptoms. Rarely, patients may develop pain due to acute hand ischemia. Duplex ultrasound can help confirm the diagnosis.

29.2.2.2 Causes

- Small size radial artery or use of large sheaths relative to the radial artery diameter.
- Nonpatent hemostasis after sheath removal.
- Absence of procedural anticoagulation or low intensity anticoagulation.
- Repeat access of the radial artery.

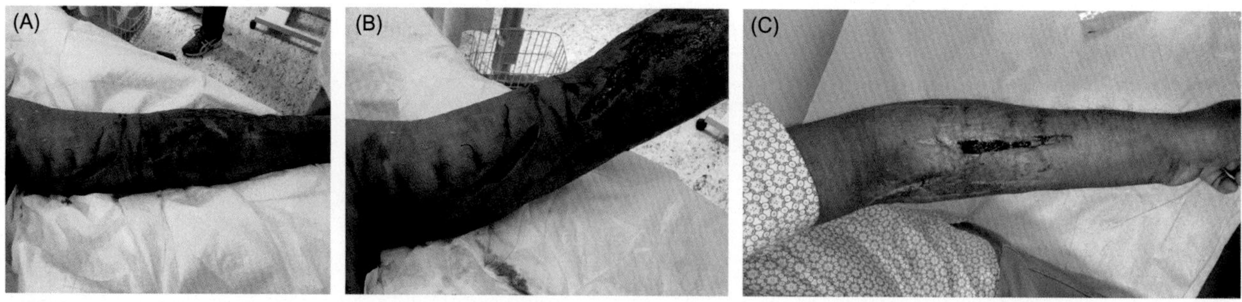

FIGURE 29.11 (Panels A and B) Arm and forearm fasciotomies to treat compartment syndrome. (Panel C) Nearly complete healing 40 days later. *Reproduced with permission from Tsiafoutis I, Katsanou K, Koutouzis M, Zografos T. Compartment syndrome: a rare and frightening complication of transradial catheterization. J Invasive Cardiol 2018;30:E111–2.*

29.2.2.3 Prevention

- Meticulous access technique (Chapter 4: Access).
- Anticoagulation with unfractionated heparin.
- Limit duration of compression (2 h or less).
- Use of the patent hemostasis technique (Section 11.2.5) [7,8].
- Use of the distal instead of the proximal radial artery.
- The Allen's or Barbeau's test is not useful in predicting the risk of radial artery occlusion and does not need to be routinely performed.

29.2.2.4 Treatment

- Most patients are asymptomatic and do not require treatment.
- Anticoagulation may be considered, as well as nonsteroidal antiinflammatory medications in case of significant local pain.
- Percutaneous recanalization of the occluded radial may be attempted in selected, highly symptomatic patients using brachial or distal radial access.

29.2.3 Radial artery spasm

29.2.3.1 Diagnosis

Radial spasm can cause hand discomfort and also hinder advancement or withdrawal of interventional equipment.

29.2.3.2 Causes

- Small size radial artery or use of large sheaths relative to the radial artery diameter.
- Excessive catheter manipulations and multiple catheter exchanges.
- Spasm is more common in women and current smokers.
- Anxiety (i.e., inadequate sedation).
- No administration of vasodilators after obtaining access.
- Use of intravenous vasopressors.

29.2.3.3 Prevention

- Meticulous access technique (Chapter 4: Access). Ultrasound guidance has been shown to reduce the number of attempts required to access the radial artery, which can reduce the likelihood of radial spasm [9].
- Vasodilator administration.
- Use a thin-walled sheath of the smallest diameter possible.
- Do not use very small radial arteries—use ulnar or femoral access instead.
- Avoid excessive catheter manipulations.
- Adequate moderate sedation.
- Adequate local anesthesia before puncture.

FIGURE 29.12 Applying excessive force to remove catheter or sheath in the setting of radial artery spasm can result in avulsion of the entire radial artery. *Courtesy of Dr. Jaikirshan Khatri.*

- Use of long (24 cm) radial sheaths that reach the brachial artery and hence exclude the radial artery from catheter manipulation and minimize the risk of radial artery endothelial trauma and spasm.

29.2.3.4 Treatment

- Vasodilator administration (nitroglycerin, verapamil). Nitroglycerin can be injected around the radial artery.
- Ensure adequate sedation is achieved. Rarely, administration of propofol or even general anesthesia may need to be used in case of equipment entrapment due to severe radial artery spasm.
- Warming the affected arm. Application of warm blankets or gauze pads, usually wet with warm sterile saline, over the affected forearm can create vasodilation and reduce the intensity or even completely resolve spasm.
- Administration of a lubricious solution, such as Rotaglide or Viperglide [10].
- Ischemia-mediated vasodilation: a blood pressure cuff is inflated in the upper arm above systolic pressure for 5 minutes, after which deflation can facilitate relaxation of the radial artery and relief of the spasm.
- Forceful withdrawal of the sheath should be avoided to minimize the risk of radial artery avulsion (Fig. 29.12).

29.2.4 Radial artery pseudoaneurysm

29.2.4.1 Diagnosis

Radial artery pseudoaneurysm presents as pulsatile mass over the radial artery access site. The diagnosis is confirmed by duplex ultrasound.

29.2.4.2 Causes

- Multiple puncture of the radial artery.
- Large sheath size relative to the size of the radial artery.
- Aggressive anticoagulation.
- Suboptimal hemostasis.

29.2.4.3 Prevention

- Meticulous access technique (Chapter 4: Access).
- Do not use very small radial arteries—use ulnar of femoral access instead.
- Use the smallest possible sheath size.

29.2.4.4 *Treatment*

- Most radial artery pseudoaneurysms thrombose spontaneously without specific treatment.
- If the pseudoaneurysm does not spontaneously resolve, ultrasound-guided compression, thrombin injection, or surgery can be used for treatment.
- Endovascular repair with a covered stent has also been reported [11].

29.2.5 Radial artery sterile granuloma

29.2.5.1 *Diagnosis*

Radial artery sterile granulomas present as erythematous nodules sometimes with sterile discharge over the radial access site. Diagnosis is made clinically after excluding infection. Sterile granulomas can sometimes lead to delayed pseudoaneurysm formation [12].

29.2.5.2 *Causes*

- Sterile granuloma is a foreign body reaction to the polymer of the radial sheath.
- Sterile granulomas were more common with earlier generation radial sheaths (Cook) [13], but can still occur with contemporary devices.

29.2.5.3 *Prevention*

- Avoid using Cook hydrophilic sheaths.

29.2.5.4 *Treatment*

- Usually symptoms resolve over time. A steroid cream may help.
- Surgical excision may be needed if the erythematous nodule fails to resolve.

References

[1] Johnson LW, Esente P, Giambartolomei A, et al. Peripheral vascular complications of coronary angioplasty by the femoral and brachial techniques. Cathet Cardiovasc Diagn 1994;31:165–72.

[2] Kronzon I, Saric M. Cholesterol embolization syndrome. Circulation 2010;122:631–41.

[3] Stortecky S, Wenaweser P, Diehm N, et al. Percutaneous management of vascular complications in patients undergoing transcatheter aortic valve implantation. JACC Cardiovasc Interv 2012;5:515–24.

[4] Sanchez CE, Helmy T. Percutaneous management of inferior epigastric artery injury after cardiac catheterization. Catheter Cardiovasc Interv 2012;79:633–7.

[5] Sandoval Y, Bell MR, Gulati R. Transradial artery access complications. Circ Cardiovasc Interv 2019;12:e007386.

[6] Tsiafoutis I, Katsanou K, Koutouzis M, Zografos T. Compartment syndrome: a rare and frightening complication of transradial catheterization. J Invasive Cardiol 2018;30:E111–12.

[7] Koutouzis MJ, Maniotis CD, Avdikos G, Tsoumeleas A, Andreou C, Kyriakides ZS. Ulnar artery transient compression facilitating radial artery patent hemostasis (ULTRA): a novel technique to reduce radial artery occlusion after transradial coronary catheterization. J Invasive Cardiol 2016;28:451–4.

[8] Pancholy S, Coppola J, Patel T, Roke-Thomas M. Prevention of radial artery occlusion-patent hemostasis evaluation trial (PROPHET study): a randomized comparison of traditional versus patency documented hemostasis after transradial catheterization. Catheter Cardiovasc Interv 2008;72:335–40.

[9] Seto AH, Roberts JS, Abu-Fadel MS, et al. Real-time ultrasound guidance facilitates transradial access: RAUST (Radial Artery access with Ultrasound Trial). JACC Cardiovasc Interv 2015;8:283–91.

[10] Repanas T, Christopoulos G, Brilakis ES. Administration of ViperSlide for treating severe radial artery spasm: case report and systematic review of the literature. Cardiovasc Revasc Med 2015;16:243–5.

[11] Tsiafoutis I, Zografos T, Koutouzis M, Katsivas A. Percutaneous endovascular repair of a radial artery pseudoaneurysm using a covered stent. JACC Cardiovasc Interv 2018;11:e91–2.

[12] Korabathina R, Eckstein D, Coppola JT. Delayed occurrence of radial artery pseudoaneurysm following transradial percutaneous coronary intervention. J Cardiol Cases 2015;11:117–19.

[13] Zellner C, Ports TA, Yeghiazarians Y, Boyle AJ. Sterile radial artery granuloma after transradial procedures: a unique and avoidable complication. Catheter Cardiovasc Interv 2010;76:673–6.

Part D

Equipment

Chapter 30

Equipment

Introduction

Equipment is necessary for performing PCI. Although many operators would like to have everything available, the reality is that equipment cost and space limitations require prioritization. Here are some criteria to use when deciding the "must have" equipment for PCI:

1. At least one item that fulfills each of the requisite steps in PCI (e.g., obtaining access, engaging the coronary artery, assessing the coronary artery, wiring, preparing the lesion, stenting the lesion, hemodynamic support, etc.) should be available, and
2. The operator should be familiar with each piece of equipment, understand its strengths and limitations and be willing to actually use it when required (otherwise it will expire on the shelf). In some cases, such as covered stents and coils, equipment expiration is to some extent expected given the low frequency of complications requiring their use.

Table 30.1 PCI equipment classified into 22 categories [1–4].

TABLE 30.1 Equipment needed for PCI.

Category no.	Equipment	
1.	Sheaths	a. Femoral access b. Radial access
2.	Catheters	a. Diagnostic b. Guide
3.	Guide catheter extensions	a. Guideliner V3 b. Trapliner c. Guidezilla II d. Telescope e. Guidion f. Heartrail 5 in 6 system
4.	Support catheters	a. MultiCross b. CenterCross c. Novacross d. Prodigy
5.	Y connectors	a. Y-connector with hemostatic valve
6.	Microcatheters	a. Large b. Small c. Angulated d. Dual lumen e. Plaque modification
7.	Guidewires	0.014 in. a. Workhorse b. Polymer-jacketed c. Stiff

(Continued)

Manual of Percutaneous Coronary Interventions. DOI: https://doi.org/10.1016/B978-0-12-819367-9.00030-5

TABLE 30.1 (Continued)

Category no.	Equipment	
		d. Support e. Atherectomy f. Externalization g. Pressure wire 0.035 in. a. Standard J-tip b. Soft tip c. Polymer-jacketed d. Stiff body
8.	Embolic protection devices	a. Filterwire b. Spider
9.	Balloons	a. Standard balloons b. Small balloons c. Plaque modification balloons d. Trapping balloons e. Ostial Flash f. Drug-coated balloons (DCB) g. Very-high pressure balloon h. Coronary lithotripsy balloon
10.	Atherectomy	a. Rotational b. Orbital
11.	Laser	
12.	Thrombectomy	a. Aspiration thrombectomy b. Penumbra c. Rheolytic thrombectomy (infrequently used)
13.	Ostial lesion equipment	a. Ostial Pro b. Ostial Flash
14.	Stents	a. Bare metal stents b. Drug-eluting stents c. Covered stents (discussed in complication management equipment)
15.	Arterial closure	a. Angioseal b. Perclose c. Radial compression bands d. Other
16.	CTO PCI Dissection/reentry equipment	a. CrossBoss catheter b. Stingray LP balloon and wire
17.	Intravascular imaging	a. IVUS b. OCT
18.	Complication management	a. Covered stents b. Coils c. Pericardiocentesis kit d. Snares (gooseneck, 3-loop)
19.	Radiation protection	a. Radiation scatter shields b. Zero Gravity system
20.	Hemodynamic support	a. Intraaortic balloon pump b. Impella CP c. Tandem Heart d. VA-ECMO e. External compression devices
21.	Contrast management	a. Dyevert system
22.	Brachytherapy	a. Novoste beta radiation system

30.1 Sheaths

Femoral and radial are the most commonly used access sites for cardiac catheterization and PCI. Other sites (such as ulnar and brachial) are rarely used.

30.1.1 Femoral access

30.1.1.1 Sheath diameter

The sheath diameter is measured in French, which refers to the size of the catheter that the sheath can accommodate: 1 French = 0.33 mm diameter. Therefore a 6 French sheath can accommodate a 2.0 mm outer diameter catheter, a 7 French sheath can accommodate a 2.33 mm outer diameter catheter, and an 8 French sheath can accommodate a 2.67 mm outer diameter catheter.

To accommodate a catheter, the internal diameter of the sheath has to be larger than the catheter. For example, the internal diameter of a 6 French Terumo Pinnacle sheath is 2.22 mm tapering to 2.06 mm at its tip, which is larger than the 2.0 mm external diameter of a 6 French catheter (Fig. 30.1).

The larger the sheath the more space it occupies within the artery, hence the higher the likelihood of complications, such as limb ischemia and arterial thrombosis (Fig. 30.2).

The outer diameter of a sheath is approximately the same as the outer diameter of a guide catheter two sizes bigger: a 6 French sheath outer diameter (2.62 mm) is similar to that of an 8 French guide (2.67 mm), highlighting the potential advantages of sheathless guide catheter insertion.

Four to eight French sheaths are most commonly used, depending on the type of the procedure planned. For diagnostic angiography, 4–6 French sheaths suffice, although for 4 Fr sheaths an automated injector, such as the ACIST CVi (ACIST Medical), is often needed. For PCI, 5–8 French sheaths and guide catheters are used, with 7–8 French sheaths and guide catheters preferred for more complex procedures. For very tortuous iliac vessels, using a long sheath that is 1 French larger than the diagnostic or guide catheter may facilitate catheter manipulations.

Larger sheaths are needed for hemodynamic support devices: 7–9 French for intraaortic balloon pump (although IABP can also be inserted without a sheath), 13 French for Impella 2.5, 14 French for Impella CP, and 15–19 French for the arterial cannula and 25 French for the venous cannula of VA-ECMO.

30.1.1.2 Sheath length

The usual sheath length is 10 cm. However, in some cases longer sheaths (such as 25 cm or 45 cm long) are preferred, for example, in obese patients, patients with significant iliac or aortic tortuosity, or when strong guide catheter support is needed. Longer sheaths provide better guide catheter support and torque response compared with shorter sheaths.

PINNACLE Introducer Sheath (Standard)

SIZE (Fr)	SHEATH OD (mm / in)	TIP ID (mm / in)	SHEATH ID (mm / in)
4	1.96 / 0.077	1.40 / 0.055	1.56 / 0.061
5	2.29 / 0.090	1.73 / 0.068	1.89 / 0.074
6	2.62 / 0.103	2.06 / 0.081	2.22 / 0.087
7	2.95 / 0.116	2.40 / 0.094	2.55 / 0.100
8	3.32 / 0.131	2.73 / 0.107	2.92 / 0.115
9	3.73 / 0.147	3.09 / 0.122	3.33 / 0.131
10	4.39 / 0.173	3.46 / 0.136	3.69 / 0.145
11	4.83 / 0.190	3.82 / 0.150	4.13 / 0.163

Sheath French Size Color Indicator

4Fr	5Fr	6Fr	7Fr
8Fr	9 Fr	10Fr	11Fr

FIGURE 30.1 Inner and outer diameter of various sheath sizes (Pinnacle sheaths, Terumo). *©2020 Terumo Medical Corporation. All rights reserved.*

The tip of the 10 cm sheath usually reaches the origin of the internal iliac artery, the 25 cm sheath usually reaches the distal aortic bifurcation, and the tip of the 45 cm long sheath usually reaches the level of the diaphragm (Figs. 30.3 and 30.4).

30.1.2 Radial access

30.1.2.1 Sheath diameter

For radial access 5 or 6 French are the most commonly used sheath sizes, although 7 or 8 French can often be used in larger radial arteries.

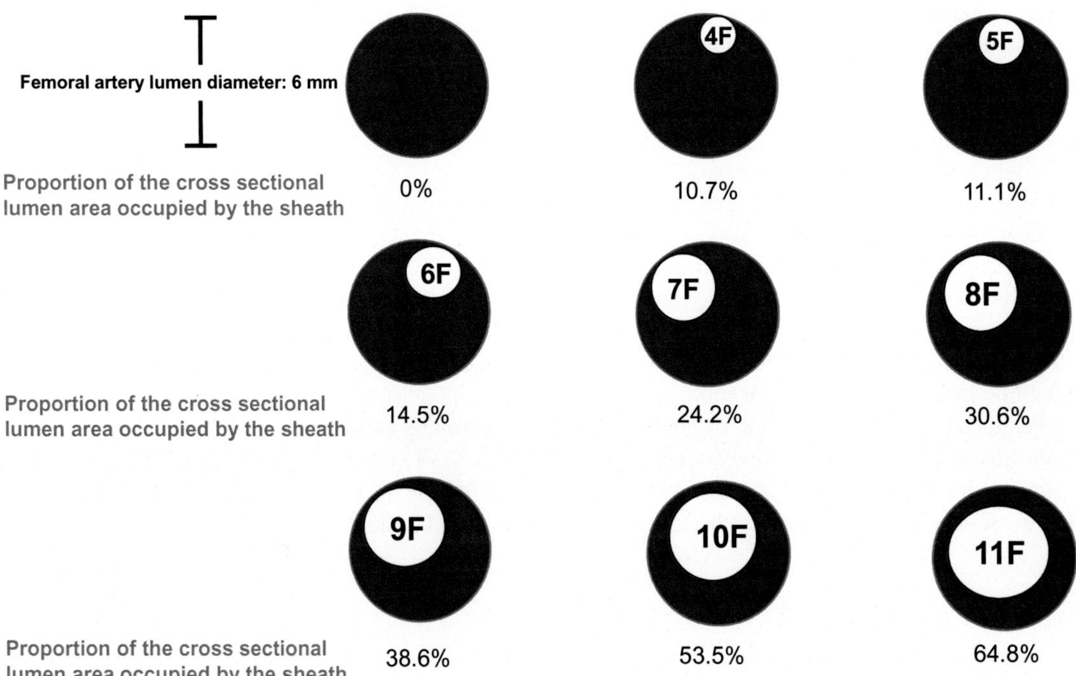

FIGURE 30.2 Outer dimensions of various sheath sizes and how much of the femoral artery (which is assumed to have 6 mm diameter) they occupy.

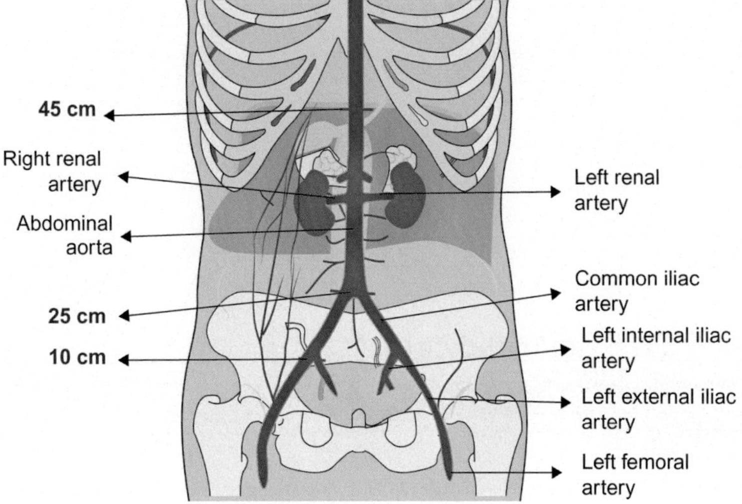

FIGURE 30.3 Location of the sheath tip depending on the sheath length.

Thin wall sheaths, such as the Glidesheath Slender (Terumo) and the Prelude Ideal (Merit Medical) have thinner walls than the standard sheaths, increasing the feasibility of sheath insertion and reducing the likelihood of spasm (Table 30.2), but may be more prone to kinking.

A sheathless guide catheter system (Eaucath, Asahi Intecc) allows CTO PCI with 7.5 French guide catheters through an arterial puncture equivalent to that created by a 5 French sheath. An alternative approach is to use standard 7 or 8 French guide catheter delivered through a short 7 or 8 French sheath.

The Railway system (Cordis) allows the sheathless insertion of any guide catheter. The Railway is a dilator that is inserted through the guide catheter facilitating entry into the radial artery (Fig. 30.5).

In a similar fashion, a 125 cm long diagnostic catheter can be used as a dilator to insert a 90 or 100 cm guide catheter that is 2 French sizes larger in diameter (Fig. 30.6).

Alternatively the balloon-assisted tracking (BAT) technique can be used for sheathless placement of guide catheters [5].

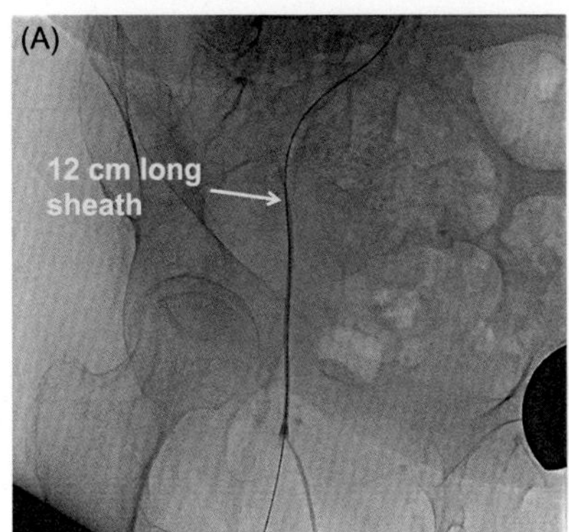

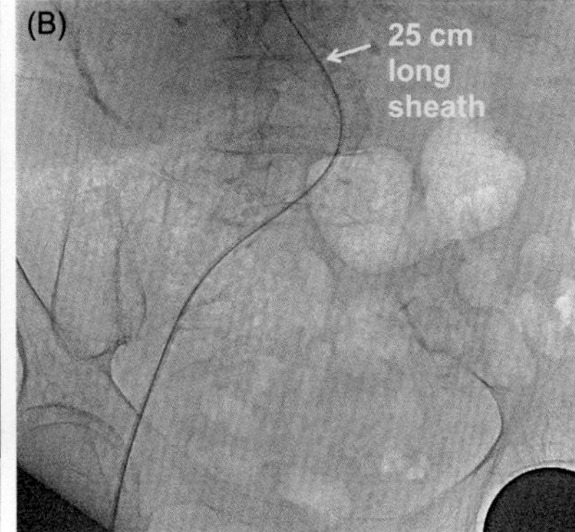

FIGURE 30.4 Example of the location of the tip of a 12 cm sheath (panel A) and a 25 cm sheath (panel B) in the same patient.

TABLE 30.2 Comparison of the outer diameter of standard vs. thin-walled sheaths.

	Sheath outer diameter (mm)	
	Terumo Pinnacle	Terumo Glidesheath Slender
5 Fr	2.29	2.14
6 Fr	2.62	2.44
7 Fr	2.95	2.79

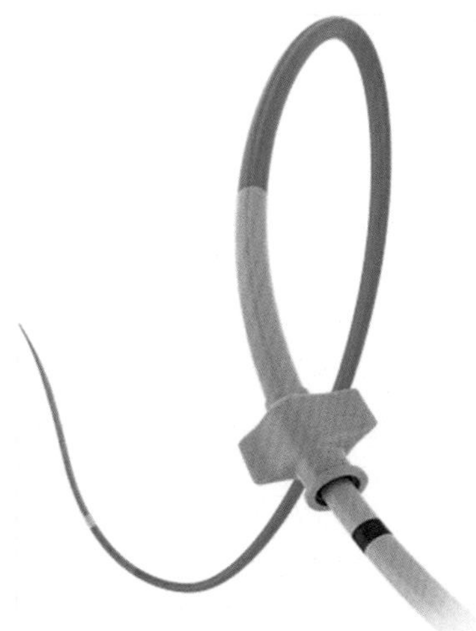

FIGURE 30.5 The Railway system for sheathless insertion of guide catheters.

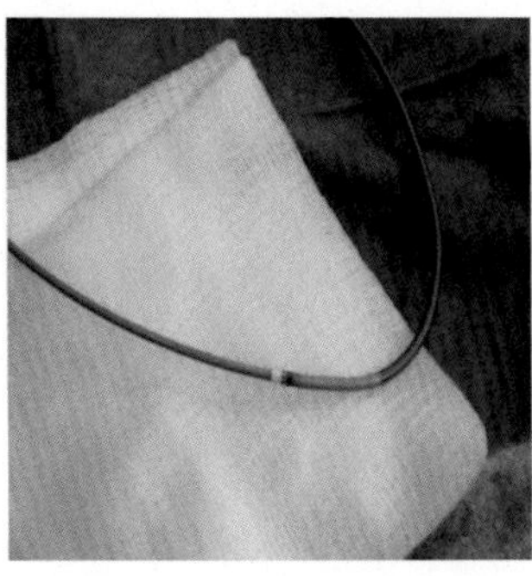
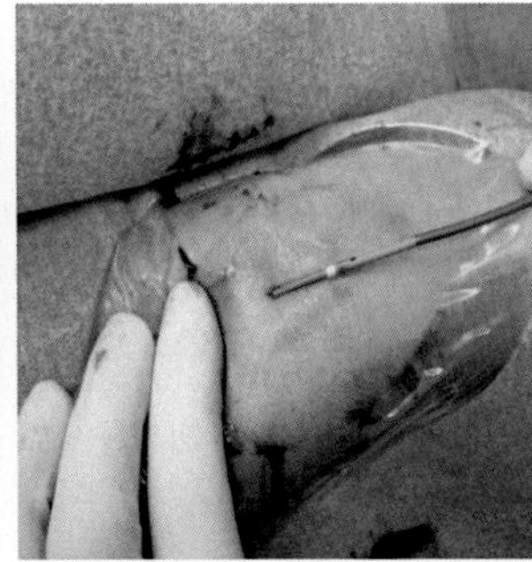
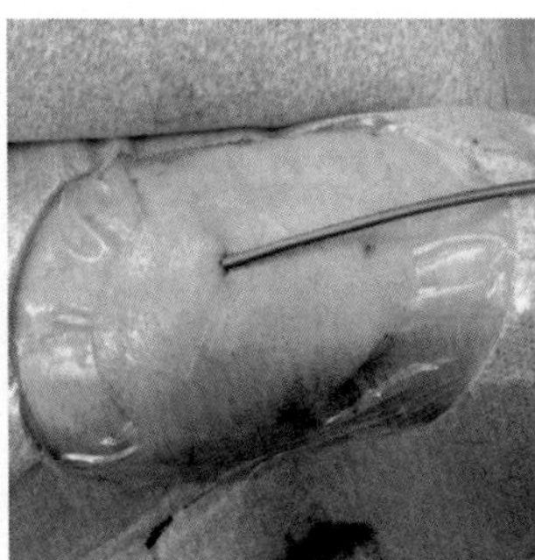

FIGURE 30.6 Use of a 5 French, 125 cm long, Cook Select catheter to insert a 7 French 100 cm long EBU 3.5 guide. *Courtesy of Dr. Jaikirshan Khatri.*

30.1.3 Sheath length

The usual sheath length is 10 cm, however longer sheaths (such as 16 or 25 cm long) can be used, especially when using left radial access (the sheath may be left partially outside the radial artery to facilitate equipment insertion) (Fig. 4.32).

30.2 Catheters

The etymology of the term "catheter" is from Greek: *kathe-*, to send down + *enai*, to send. Catheters used for coronary angiography and PCI have multiple roles:

1. Reach and engage the coronary artery for contrast injection and coronary visualization.
2. Monitor pressure.
3. Deliver equipment (such as wires, microcatheters, imaging catheters, balloons, and stents) into the coronary artery.

The first selective coronary angiography was performed by accident on October 30, 1958 by Mason Sones at the Cleveland Clinic. During an aortic root injection in a 26-year-old patient with rheumatic heart disease the catheter whiplashed into the ostium of the right coronary artery [6]. At the time, selective injection was considered "lethal", but the patient survived and coronary angiography was born! Several improvements in catheter shapes were subsequently made by pioneers, such as Melvin Judkins and Kurt Amplatz, whose catheter designs are still used to this day.

30.2.1 Guide versus diagnostic catheter

A guide catheter differs from a diagnostic catheter: it has thinner walls and as a result larger lumen that can accommodate various equipment (Table 30.3). The guide walls are made thinner through the use of a layer of metal mesh that prevents the guide catheter walls from collapsing. Given their thicker walls, diagnostic catheters are more torquable and less likely to kink. In some cases of challenging guide catheter engagement, a diagnostic catheter can be used to engage and wire the target coronary artery, followed by exchange for a guide catheter over a supportive 300 cm long 0.014 in. guidewire, such as the Iron Man and the Grand Slam (Section 30.8.4), as described in Chapter 5: Coronary and graft engagement, step 7). Guide catheters have softer tips compared with diagnostic catheters to allow for longer coronary engagement and deep seating, while limiting the risk of coronary dissection.

30.2.2 Guide outer and inner diameter

Similar to sheaths, the guide outer diameter is measured using French: 1 French is equivalent to 0.33 mm, i.e., the outside diameter of a 6 French guide catheter is 2.0 mm. Most PCIs can be performed through 6 French guide catheters. However, larger guide catheters (7 or 8 French) provide better support and improve vessel visualization and may be required for delivery of some equipment (e.g., the 2.0 mm Rotablator burrs require 7 French guide catheters, Section 19.9.5) or in cases where multiple simultaneous equipment might be necessary (e.g., certain two-stent strategies, such as V-stenting and simultaneous kissing stent that also require 7 French guide catheters [Chapter 16: Bifurcations]).

The lumen size of each guide catheter depends on the wall thickness and varies slightly by manufacturers (Table 30.4). For example, for 6 French (2.0 mm or 0.079 in. outer diameter) guide catheters, the Cordis Adroit guide catheters inner diameter is 1.83 mm or 0.072 in., while the Launcher and Ikari guide catheters inner diameter is 1.80 mm or 0.071 in., and the Cordis Vista Brite and Boston Scientific Mach 1 guide catheters inner diameter is 1.78 mm or 0.070 in.

TABLE 30.3 Comparison of a 6 French diagnostic and a 6 French guide catheter.

	Diagnostic	Guide
Inner diameter (6Fr)	0.057 in. (1.45 mm)	0.069–0.072 in. (1.78 mm)
Outer diameter	0.079 in. (2 mm)	0.079 in. (2 mm)
Layers	2	3
Price[a]	$10	$50

[a]*Price can vary depending on manufacturer and local agreements.*

TABLE 30.4 Outer and inner diameters of commonly used guide catheters.

	5Fr	6Fr	7Fr	8Fr
Outer diameter	0.066 in. (1.67 mm)	0.079 in. (2.0 mm)	0.092 in. (2.33 mm)	0.105 in. (2.67 mm)
Inner diameter				
Cordis Adroit		0.072 (1.83 mm)		
Cordis Vista Brite		0.070 (1.78 mm)		
Medtronic Launcher	0.058 (1.47 mm)	0.071 (1.80 mm)	0.081 (2.06 mm)	0.090 (2.29 mm)
Boston Scientific, Mach1	NA	0.070 (1.78 mm)	0.081 (2.06 mm)	0.091 (2.31 mm)
Terumo Ikari	0.059 (1.50 mm)	0.071 (1.80 mm)	0.081 (2.06 mm)	NA

Choosing the right guide catheter size is a critical step in complex PCI. The "Complex PCI Solutions" app (only available for mobile devices) can be used to choose guide size based on planned equipment utilization (Fig. 30.7):

https://itunes.apple.com/us/app/complex-pci-solutions/id1355728598?mt = 8

30.2.3 Catheter length

Most catheters are 100 cm long, however, *shorter* guide catheters (usually 90 cm long, which are commercially available from multiple manufacturers) may be needed for retrograde CTO PCI, or when attempting to deliver equipment to very distal target lesions (e.g., when delivering equipment to native coronary artery lesions through bypass grafts) [7]. Most IM guide catheters that are designed for PCI in LIMA grafts are 90 cm long for this reason. If short guide catheters are not available, any guide catheter can be shortened, as described in Section 30.2.4.

Longer guide catheters (125 long) are also available and may be needed to reach the coronary arteries in very tall patients or patients with very tortuous aortas (see Chapter 5: Coronary and graft engagement, Section 5.3.4. *Failure to reach the aortic root*).

30.2.4 Shortening the guide catheter

With the availability of long externalization guidewires, such as the RG3 and R350, and short (90 cm) guide catheters, guide catheter shortening is rarely performed. Guide catheter shortening may, however, still be needed for some retrograde cases via bypass grafts or apical collaterals, as in such cases retrograde microcatheters might not be long enough to reach the antegrade guide catheter.

If premanufactured short guide catheters are not available, any 100 cm long guide catheter can be shortened using the following technique (Fig. 30.8) [9] (video: How to shorten a coronary guide catheter):

1. The guide catheter is inserted into the body to engage the target coronary artery and the length of the guide that is outside the femoral sheath is marked.

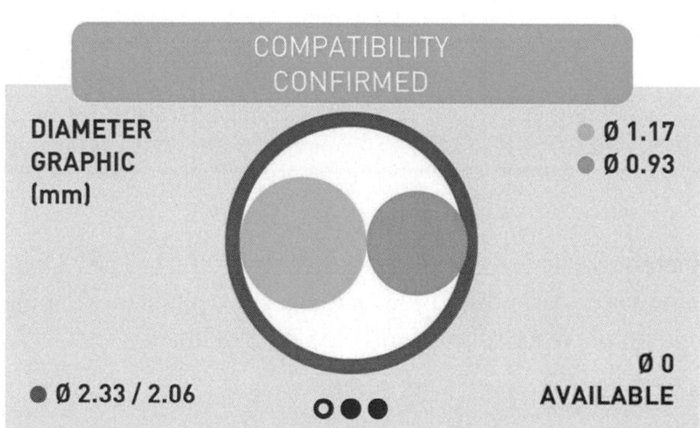

FIGURE 30.7 Example of equipment compatibility using the PCI Solutions app: an Eagle Eye IVUS and a Corsair Pro microcatheter can fit within a 7 French Launcher guide catheter.

2. The guide is removed from the body and the marked segment is cut using sterile scissors and removed (Fig. 30.8, panel A).
3. A sheath (1 Fr size smaller than the guide catheter, i.e., 6 Fr sheath for a 7 Fr guide catheter, 7 Fr sheath for a 8 Fr guide catheter, etc.) is cut to create a 3–4 cm connecting segment for the two guide catheter pieces (Fig. 30.8, panels B and C). Both ends of this connecting segment are flared with a dilator (of equal size to the guide catheter) to facilitate insertion (Fig. 30.8, panel D).
4. This connecting sheath segment is used to reconnect the proximal and distal guide catheter pieces minus the portion that was removed to shorten the guide catheter (Fig. 30.8, panels E and F—final result in panel G). Placing a Tegaderm (3M) over the connection site may help prevent accidental disconnection.

A limitation of shortened guide catheters is that they have poor torque transmission during vessel engagement and guide manipulations (especially during long procedures) and that the two guide catheter pieces may become disconnected.

30.2.5 Side holes

Guide catheters with side-holes (Fig. 30.9) can be used for ostial right coronary artery or ostial bypass graft lesions, because they can prevent pressure dampening, may allow antegrade flow into the vessel, and may decrease the risk of hydraulic dissection during antegrade contrast injection. Side-hole guides may, however, provide a false sense of security,

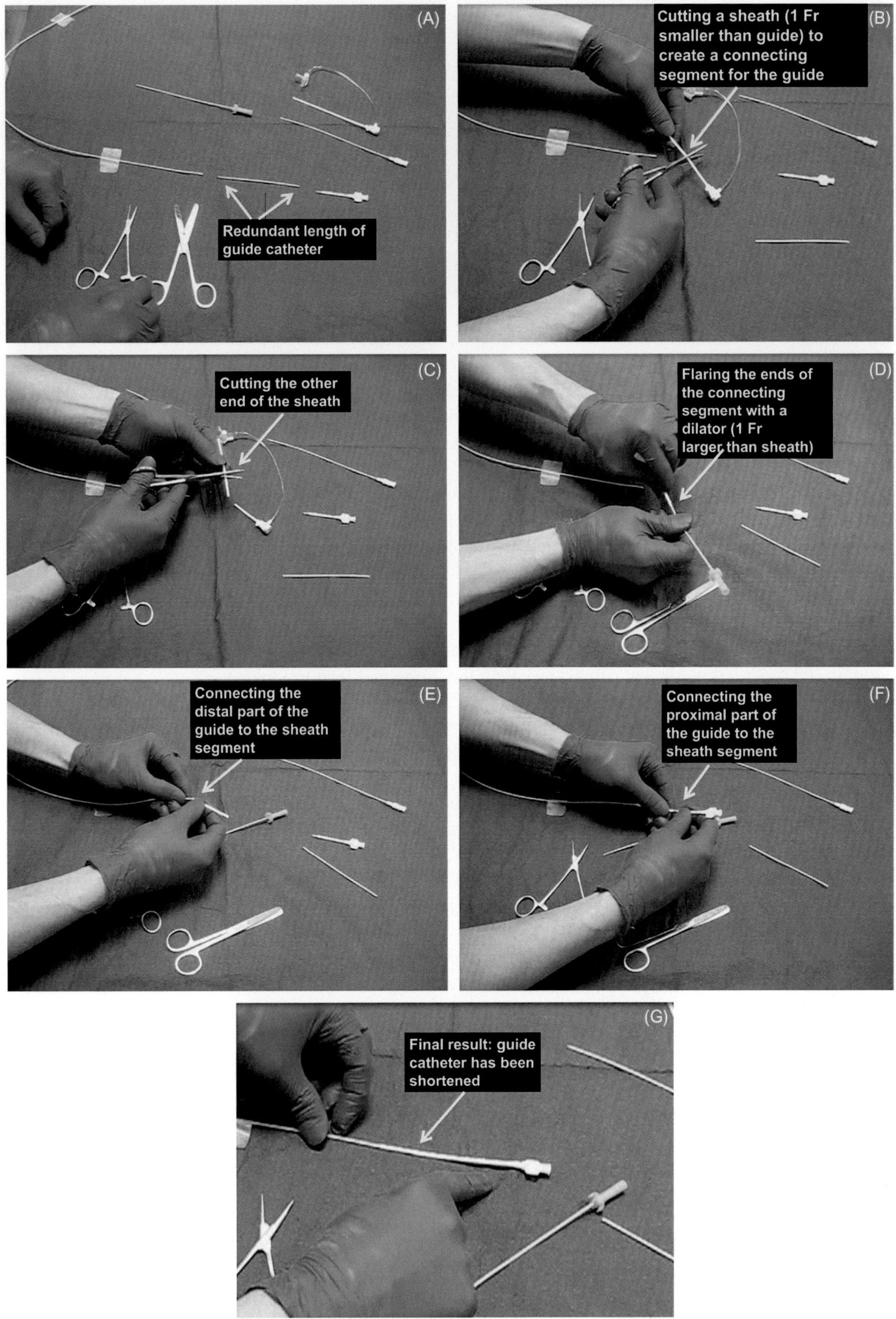

FIGURE 30.8 Overview of the guide shortening technique. *Reproduced with permission from Brilakis ES. Manual of coronary chronic total occlusion interventions. A step-by-step approach. 2nd ed. Cambridge, MA: Elsevier; 2017 (Figure 2.2).*

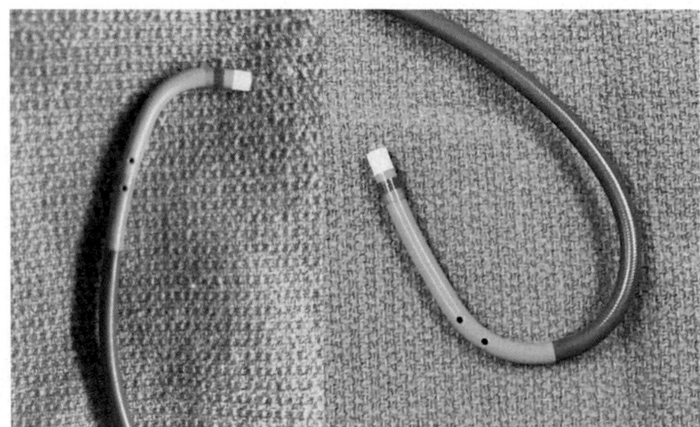

FIGURE 30.9 Example of guide catheters with side holes. *Courtesy of Dr. William Nicholson. Reproduced with permission from Brilakis ES. Manual of coronary chronic total occlusion interventions. A step-by-step approach. 2nd ed. Cambridge, MA: Elsevier; 2017 (Figure 2.3).*

as hydraulic dissections can still occur upon injection. Dampening of the pressure waveform is desired when antegrade dissection-reentry techniques are used for CTO crossing to minimize antegrade flow and subintimal hematoma expansion. A strategy of thoughtful active guide catheter manipulation is often chosen over use of side hole guide catheters, depending on the preference and comfort of the operator.

In contrast, engagement of an unprotected left main coronary artery with side-hole guide catheters should be avoided (with the exception of ostial left main CTOs), as suboptimal guide catheter position may not be recognized, leading to decreased antegrade left main flow, global ischemia and hemodynamic collapse.

Another disadvantage of side-hole guide catheters is that they lead to higher contrast use and image quality degradation due to escape of some contrast into the aorta through the side holes during contrast injection [3]. If no side-hole guide catheter are available, an 18–23 gauge needle or a scalpel can be used to create side holes in the guide catheter, followed by flushing with saline before use. Side holes made by hand may prevent advancement of a guide catheter extension within the modified guide and can also weaken the guide and lead to kinking.

30.2.6 Guide catheter shapes

Most guide catheter shapes are specific for the right or left coronary artery or for bypass grafts, although some guide catheters can be used to engage various vessels. The guide catheters are described by the shape name and the length of their distal segment (Fig. 30.10).

For example, for the commonly used Judkins Right (JR) catheters, the catheter size represents the length of the secondary curve and is available in 3.5, 4, and 5 cm sizes (Fig. 30.10). For Judkinks left (JL) catheters, the catheter size represents the length between the primary and secondary curve and is available in 3.5, 4, 5, and 6 cm sizes (Fig. 30.11).

Selecting diagnostic and guide catheter size depends on the size of the aortic root, with larger catheters needed for larger aortic root sizes. For example, in patients with ascending aorta dilation, using the diameter of the ascending aorta and adding 1 would give the needed JL catheter size (e.g., is the ascending aorta is 4 cm in diameter, a JL5 catheter should be used). Moreover, with aortic root dilation, the root dilates and rotates counterclockwise, with the right coronary artery becoming an anterior structure, hence an AR-1 or 2 or AL 1 or 2 catheters may facilitate engagement.

30.2.6.1 Right coronary artery

The most commonly used guide catheters for engaging the RCA are the following:

Femoral access (Fig. 30.12)

Diagnostic angiography: JR4, Williams, Hockeystick or IM for shepherd's crook RCA.

PCI: JR4, AL1 for complex RCA PCI, 3D Right.

Radial access (Fig. 30.13)

Diagnostic angiography: catheters that can engage both the RCA and left main (such as TIG, Kimny, TRA series, etc.), JR4.

PCI: AL1, Ikari Right, and 3D Right (Fig. 30.14). Judkins Right catheters should generally be avoided due to poor support.

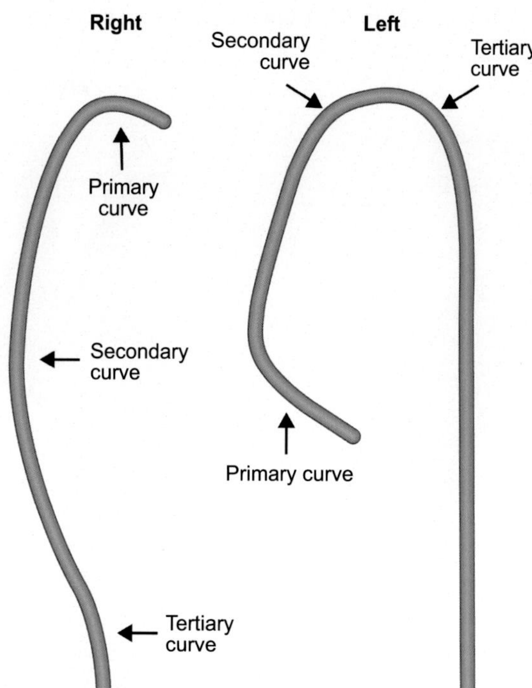

FIGURE 30.10 Illustration of the Judkins Right and Judkins Left catheters.

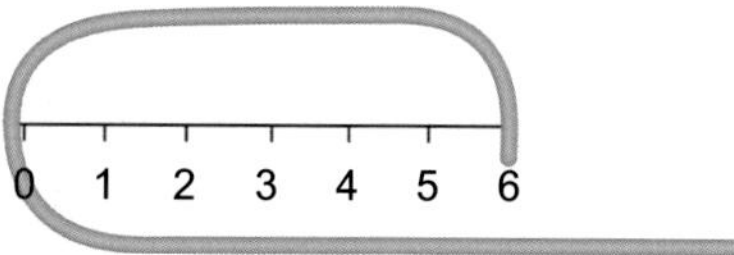

FIGURE 30.11 Sizing of the Judkins Left (JL) catheters.

30.2.6.2 Left main

The most commonly used catheters for engaging the left main are the following:

Femoral access (Fig. 30.15)

Diagnostic angiography: JL4

PCI: EBU, XB, and Voda lines of catheters. The EBU 3.75 and XB 3.5 are used in most cases. Larger catheters, for example, EBU 4.0, XB 4.0, or Voda 4.0 may facilitate PCI of the circumflex, whereas smaller catheters such as EBU 3.5 may facilitate PCI of the LAD. JL guide catheters are generally avoided for PCI, as they provide poor support.

***Radial access*:**

Diagnostic angiography: catheters that can engage both the RCA and left main (such as TIG, Kimny, TRA series, etc.).

PCI: EBU and XB lines of catheters, Ikari left.

30.2.6.3 Internal mammary grafts (LIMA and RIMA) (Fig. 30.16)

The JR4 catheter can engage most IM grafts, although the IM catheter has a bigger distal bend and facilitates engagement. The modified IM catheter (VB1, Cordis, Fig. 18.5) and the Bartorelli-Cozzi catheter (BC, Cordis, designed to engage the LIMA via left radial access, Fig. 18.3) have an even larger distal bend.

30.2.6.4 Left-sided bypass grafts (Fig. 30.17)

For diagnostic angiography a JR4 catheter can engage most left-sided bypass grafts. For PCI or when using radial access an AL guide catheter may facilitate graft engagement and provide stronger support. Multipurpose and LCB catheters can also be used, but often provide less support than the AL catheters.

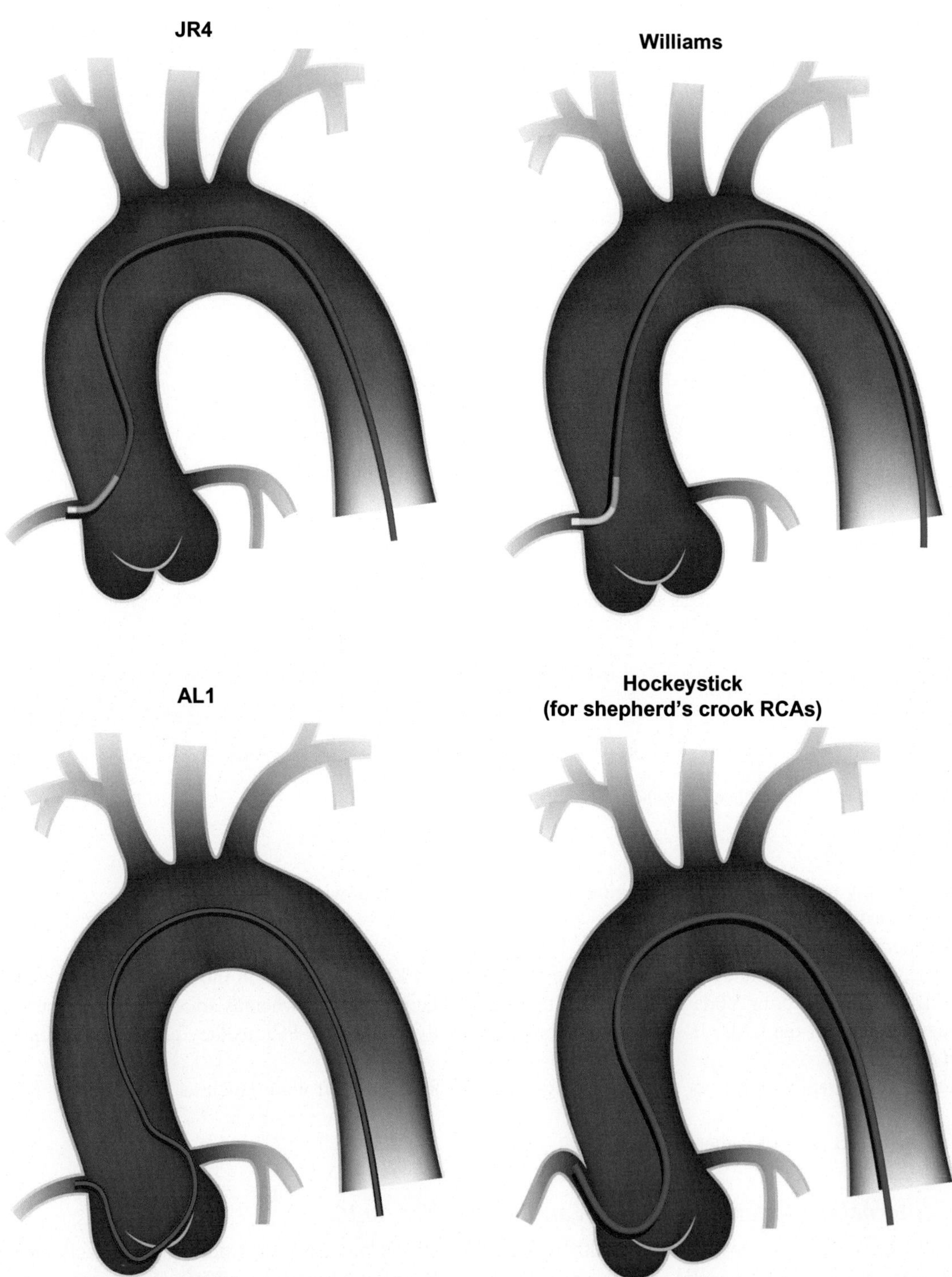

FIGURE 30.12 Catheters used to engage the right coronary artery via femoral access.

30.2.6.5 Right-sided bypass grafts (Fig. 30.18)

Those grafts usually have an inferiorly directed take-off and are usually engaged with a JR4 or a multipurpose catheter. Alternatively, there is a dedicated catheter for right coronary bypass grafts, the RCB catheter.

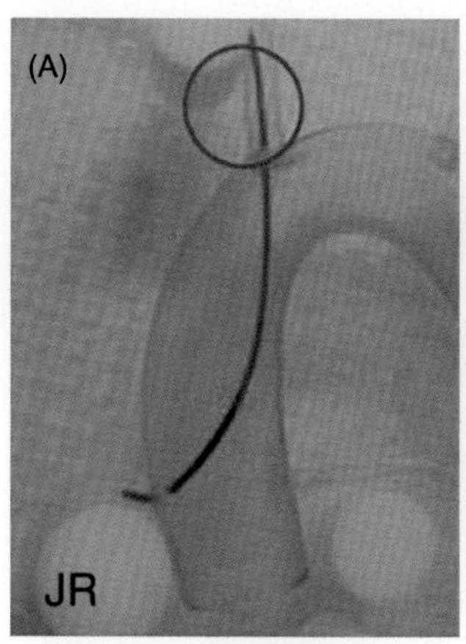

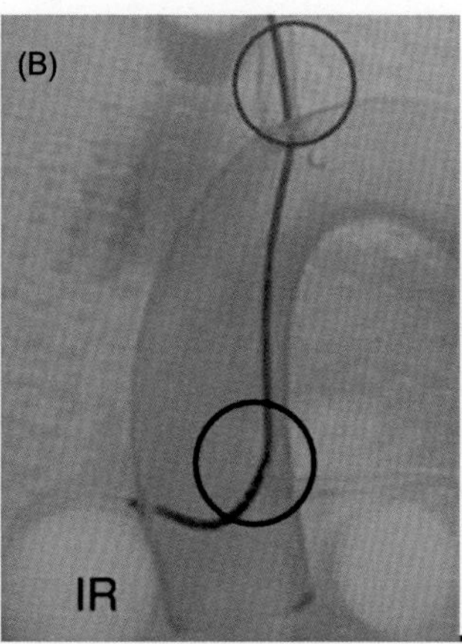

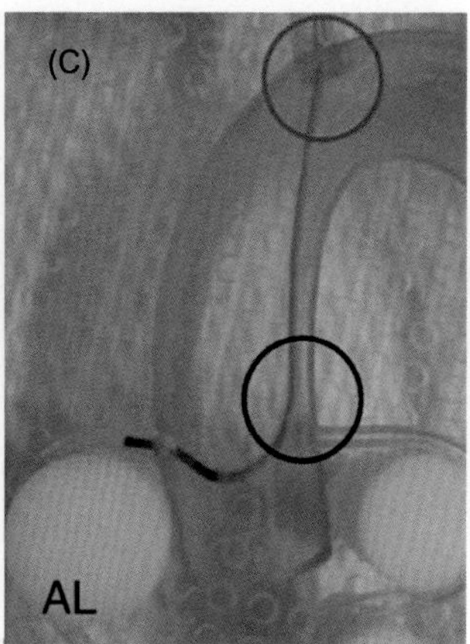

FIGURE 30.13 Catheters used to engage the right coronary artery via radial access. JR = Judkins Right; IR = Ikari Right; AL = Amplatz Left. Reprinted with permission from Ikari Y. Long-term experience using the Ikari guide catheter for radial PCI. Cath Lab Digest 2013;12:9:36-41. Copyright HMP Global.

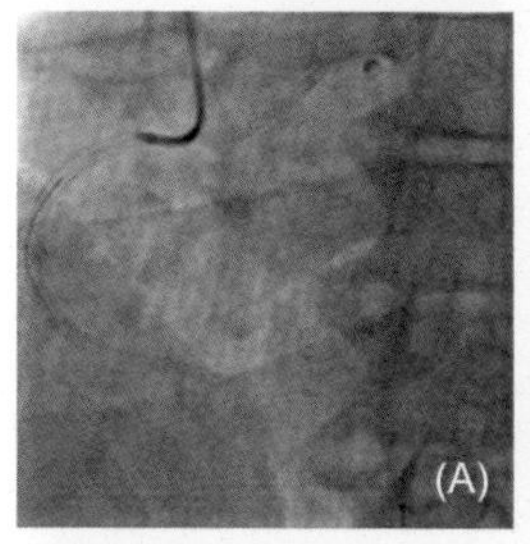

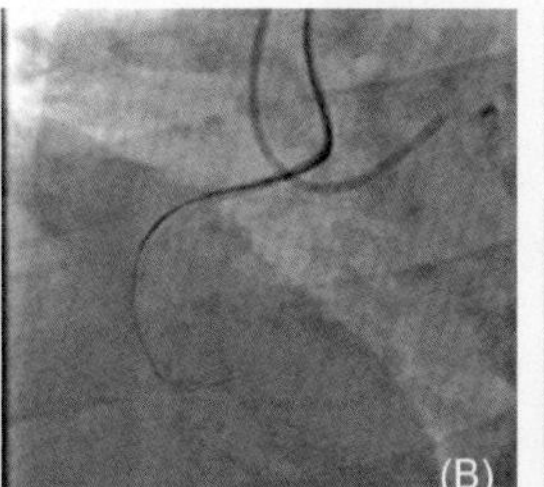

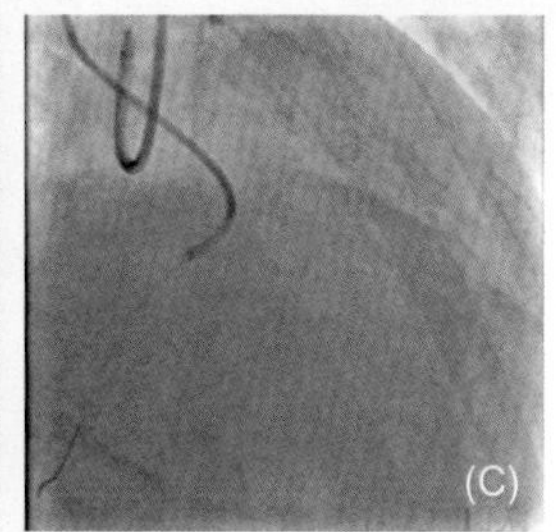

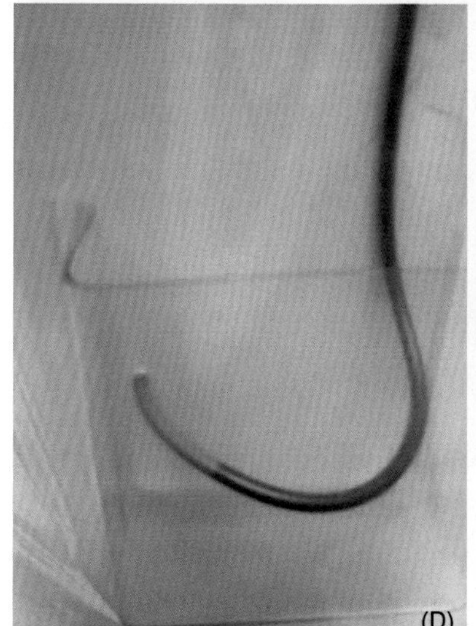

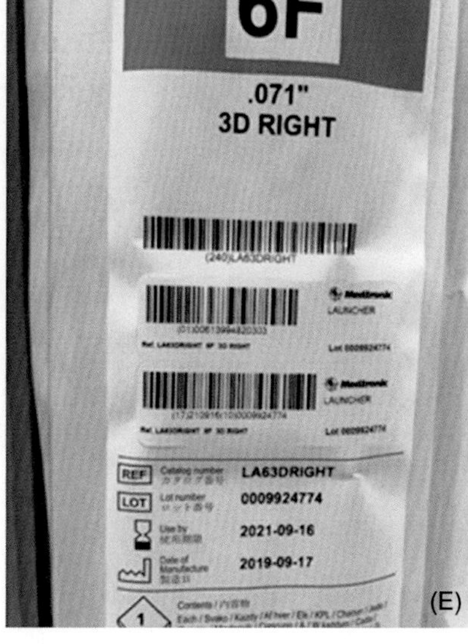

FIGURE 30.14 The 3D RIGHT guide catheter. The 3D RIGHT guide (Medtronic Launcher) is available in 6 and 7 French sizes. This catheter allows the flexibility to treat ostial as well as more distal right coronary artery disease simply by application of torque while maintaining guide stability. (A) LAO view after application of counter clock torque to treat the ostial right coronary artery. (B) LAO view after application of clockwise torque to treat more distal disease. (C) RAO view illustrating 3D RIGHT supported by the posterior aortic root. Excellent back wall support can be achieved with both radial as well as femoral access. (D) Three-dimensional construction requires special protective packaging. (E) 3D RIGHT is often confused with other right coronary catheters with similar names. *Courtesy of Dr. Jaikirshan Khatri.*

30.3 Guide catheter extensions

30.3.1 Guide catheter extension types

CTO PCI Manual Online cases: 12, 27, 36, 39, 70, 86, 87, 89, 90, 105, 106, 109, 115, 122, 123.

PCI Manual online cases: 1, 2, 4, 7, 13, 17, 18, 19, 35, 37, 42, 45, 62, 64, 79, 80.

Four guide catheter extensions are currently available in the United States (Table 30.5):

1. Guideliner V3 catheter (Teleflex, Fig. 30.19).
2. Trapliner (Teleflex, Fig. 30.20), which is a rapid exchange guide catheter extension with a guidewire trapping balloon.
3. Guidezilla II (Boston Scientific, Fig. 30.21).
4. Telescope (Medtronic, Fig. 30.22).

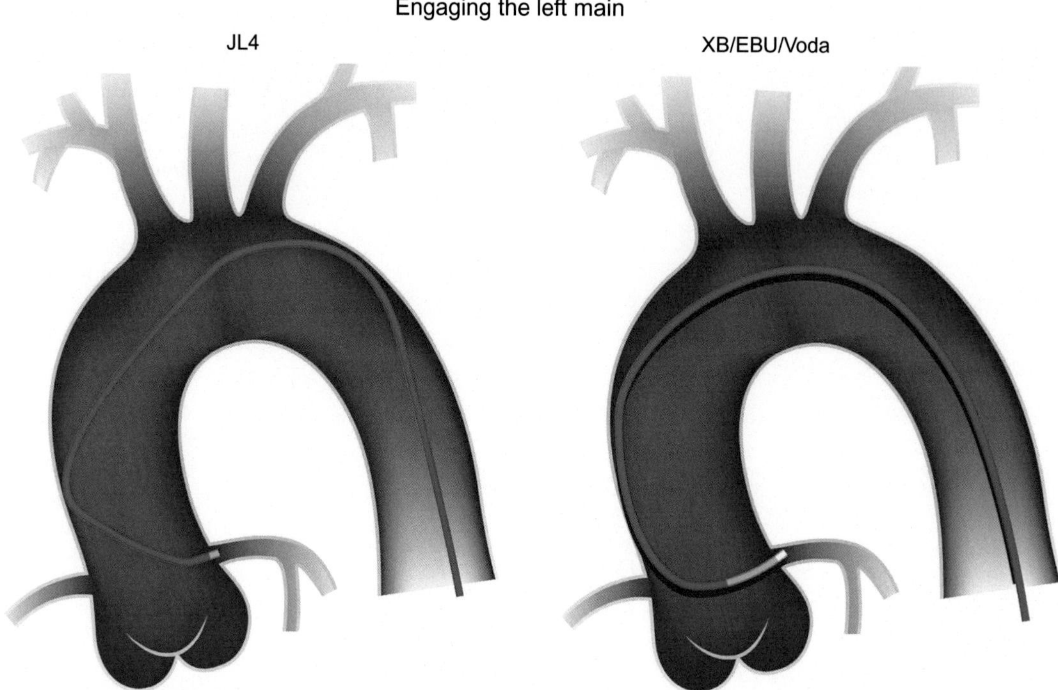

FIGURE 30.15 Catheters used to engage the left main via femoral access.

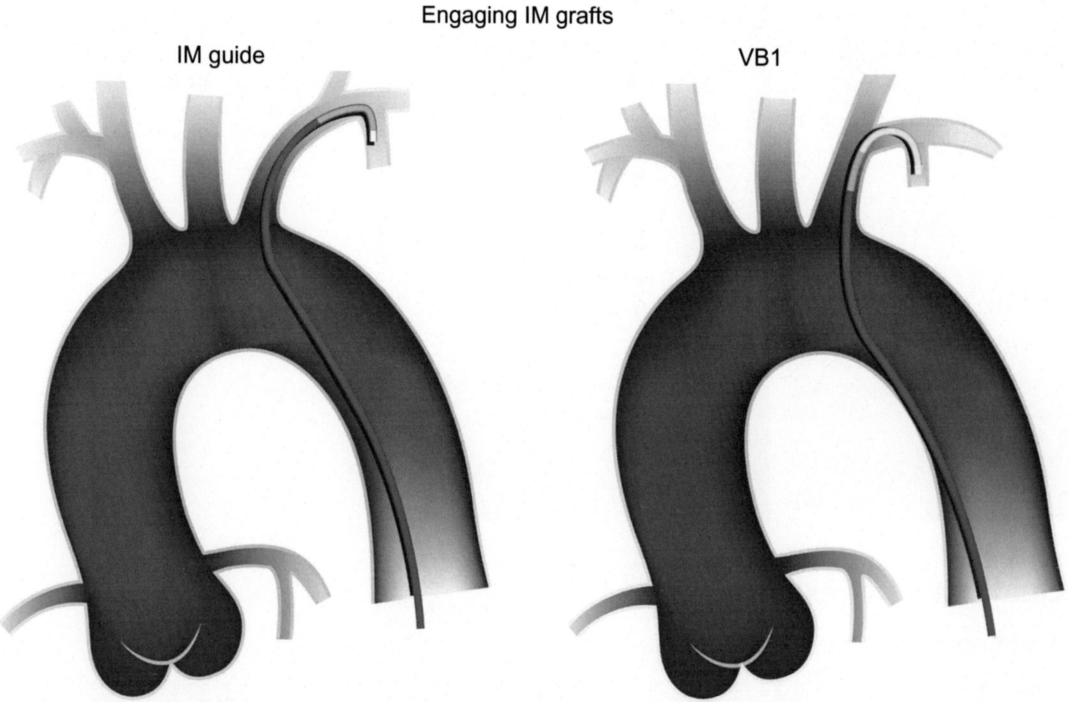

FIGURE 30.16 Catheters used to engage internal mammary grafts via femoral access.

Two other guide catheter extension, the Guidion (Interventional Medical Device Solutions, Fig. 30.23) and the Heartrail 5 in 6 system (Terumo, Fig. 30.24) [10] are not available in the United States but are available in other countries.

All guide catheter extensions (with the exception of Heartrail 5 in 6 system) consist of a push rod and a distal cylinder (25 cm long in the Guideliner V3, the Guidezilla II and the Telescope, and 13 cm in the Trapliner) that is advanced

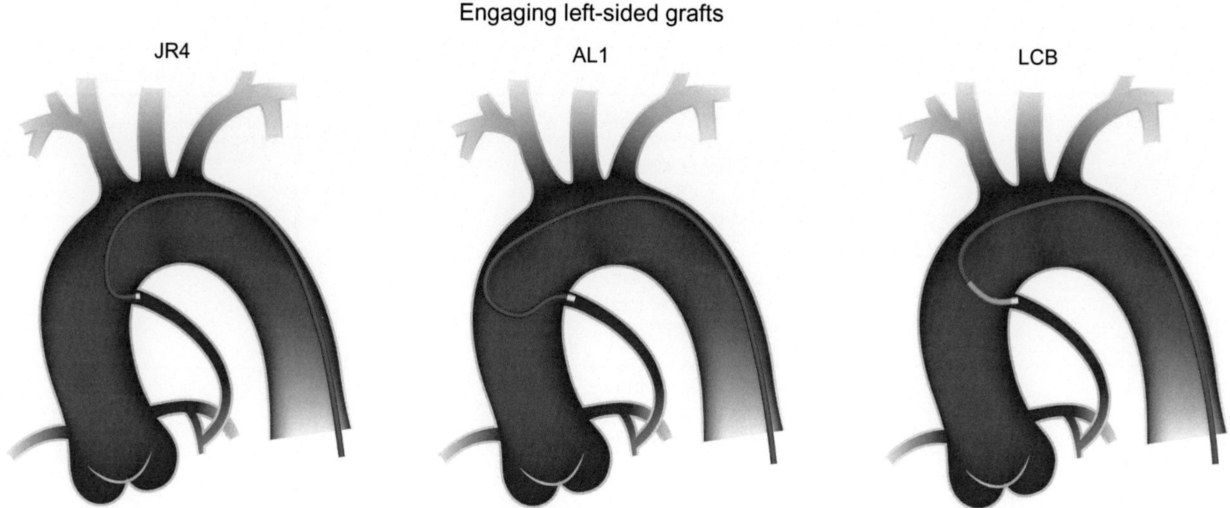

FIGURE 30.17 Catheters used to engage left-sided bypass grafts.

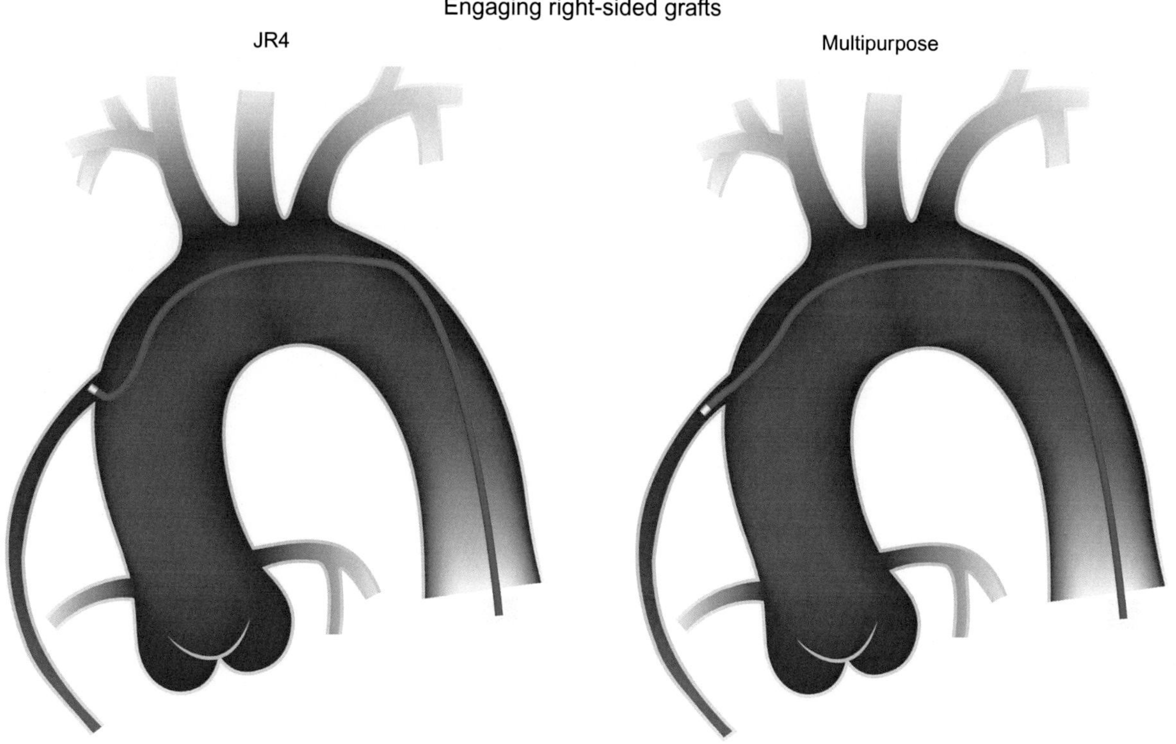

FIGURE 30.18 Catheters used to engage right-sided bypass grafts.

into the coronary artery. They are manufactured in various sizes (Table 30.5) to fit various guide catheters, resulting in an inner diameter that is approximately 2 French smaller than that of the guide catheter. In addition to the pushing rod and cylinder, the Trapliner also has a balloon proximal to the proximal collar that allows for trapping equipment (Fig. 30.20).

30.3.2 Guide catheter extensions: uses

1. Facilitate equipment delivery (balloons, stents, etc.), which is the major indication for their use [11]. Although coaxial alignment of the guide catheter is ideal, the guide catheter extension may be particularly effective in facilitating

TABLE 30.5 Overview of guide catheter extensions.

Name	Sizes	Internal diameter	Total length	Distal cylinder length
Guideliner V3	5 Fr 5.5 Fr 6 Fr 7 Fr 8 Fr	0.046 in. (1.17 mm) 0.051 in. (1.30 mm) 0.056 in. (1.42 mm) 0.062 in. (1.57 mm) 0.071 in. (1.80 mm)	150 cm	25 cm XL: 40 cm
Trapliner	6 Fr 7 Fr 8 Fr	0.056 in. (1.42 mm) 0.062 in. (1.57 mm) 0.071 in. (1.80 mm)	150 cm	13 cm
Guidezilla II	6 Fr 7 Fr 8 Fr	0.057 in. (1.45 mm) 0.063 in. (1.60 mm) 0.072 in. (1.83 mm)	145 cm	25 cm XL: 40 cm
Guidion	5 Fr 6 Fr 7 Fr 8 Fr	0.041 in. (1.04 mm) 0.056 in. (1.42 mm) 0.062 in. (1.57 mm) 0.071 in. (1.80 mm)	150 cm	25 cm
Telescope	6 Fr 7 Fr	0.056 in. (1.42 mm) 0.062 in. (1.57 mm)	150 cm	25 cm
Heartrail	5 in 6 system	0.059 in. (1.50 mm)	120 cm	No distal cylinder—Heartrail is a 120 cm long catheter inserted through a standard guide catheter

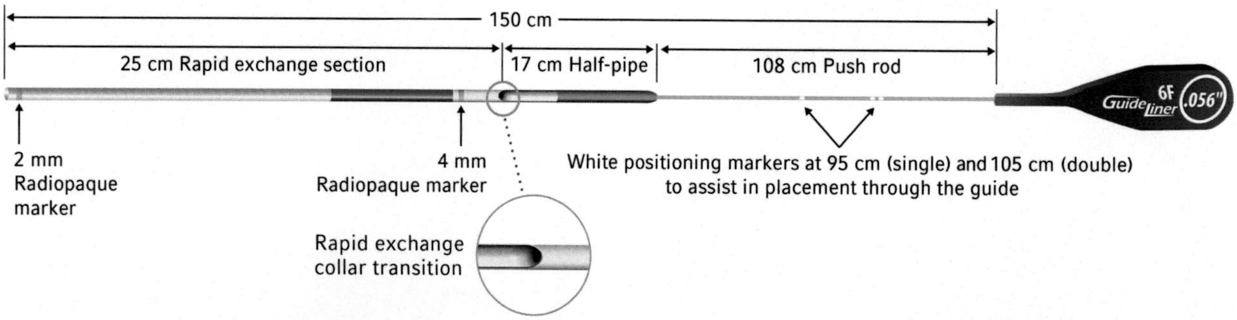

FIGURE 30.19 Illustration of the Guideliner V3 catheter.

vessel engagement and equipment delivery when guide coaxial alignment is not possible, for example, in anomalous coronary arteries (Fig. 30.25) [12] or internal mammary artery grafts [13].

2. Facilitate coronary artery engagement in challenging clinical scenarios, for example, in patients with dilated ascending aorta [14].
3. Perform thrombectomy, by advancing the guide catheter extension into the target vessel and aspirating through the guide catheter [15].
4. Facilitate retrieval of entrapped equipment, such as Rotablator burrs (see Section 19.9.5.7.3) [16].
5. Facilitate the reverse controlled antegrade and retrograde tracking (reverse CART) technique ("guide catheter extension reverse CART", Fig. 30.26) during retrograde chronic total occlusion interventions. A guide catheter extension is advanced through the antegrade guide catheter to reduce the distance that the retrograde guidewire needs to traverse [17]. Guide catheter extensions can also be used from the retrograde side to increase support for retrograde equipment delivery.

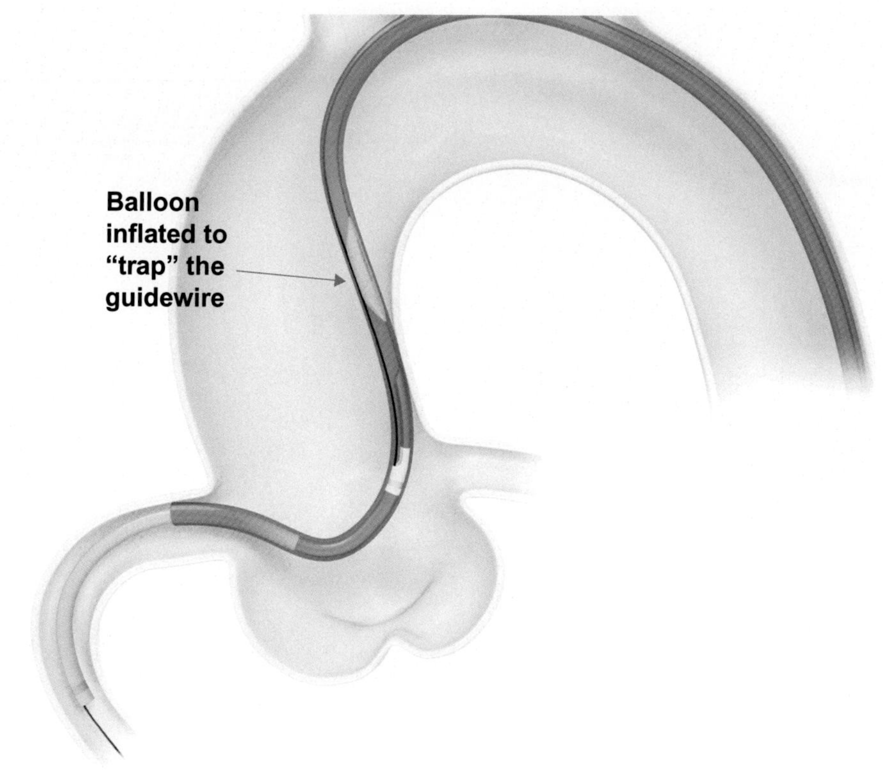

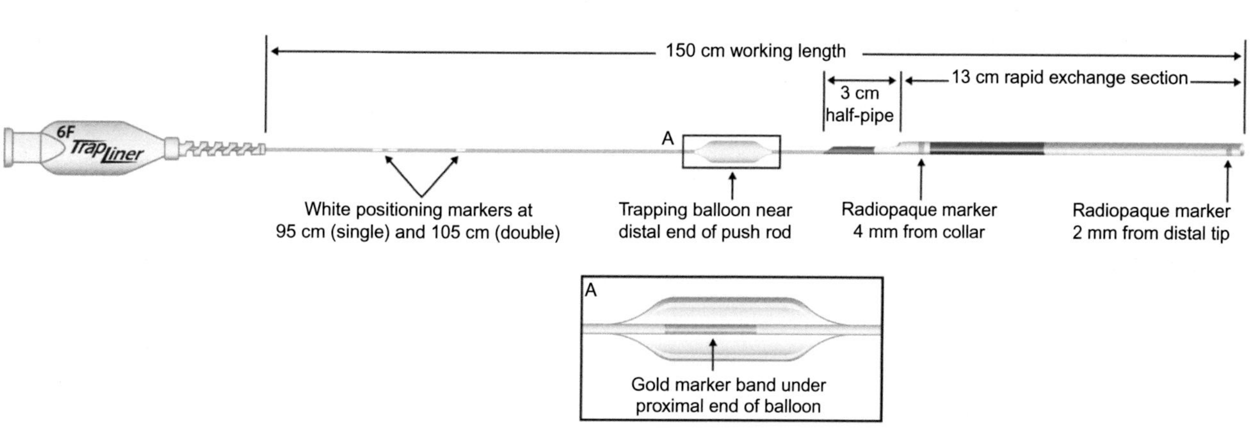

FIGURE 30.20 Illustration of the Trapliner catheter.

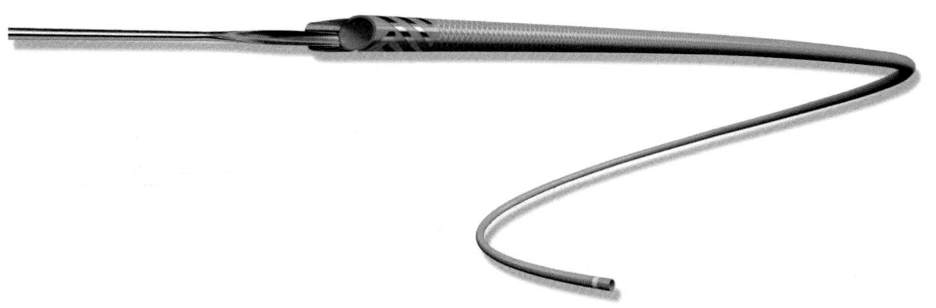

FIGURE 30.21 Illustration of the Guidezilla II catheter. *Image courtesy Boston Scientific © 2019 Boston Scientific Corporation or its affiliates. All rights reserved.*

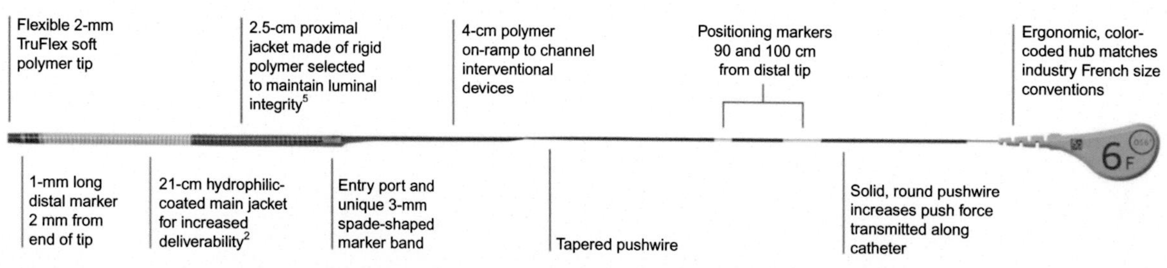

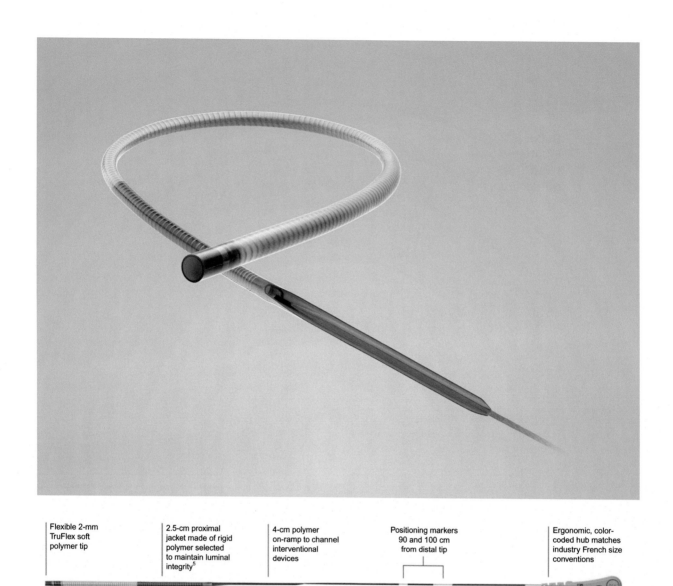

FIGURE 30.22 Illustration of the Telescope guide catheter extension. *Used with permission by Medtronic, Inc.*

Guidion

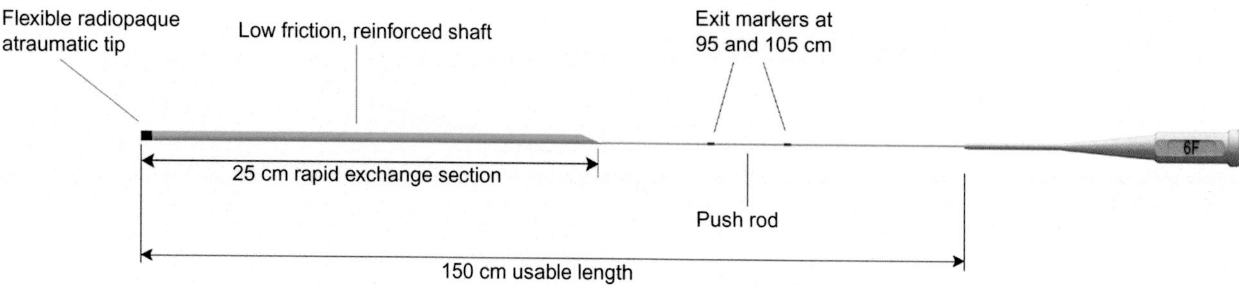

FIGURE 30.23 Illustration of the Guidion catheter. *Reproduced with permission from IMDS.*

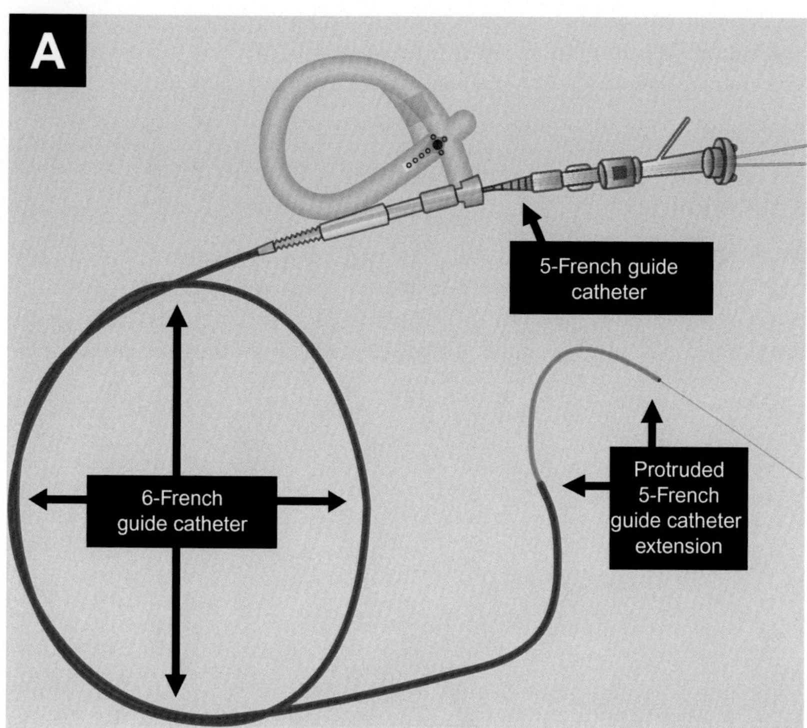

FIGURE 30.24 Illustration of the Heartrail II guide catheter extension.

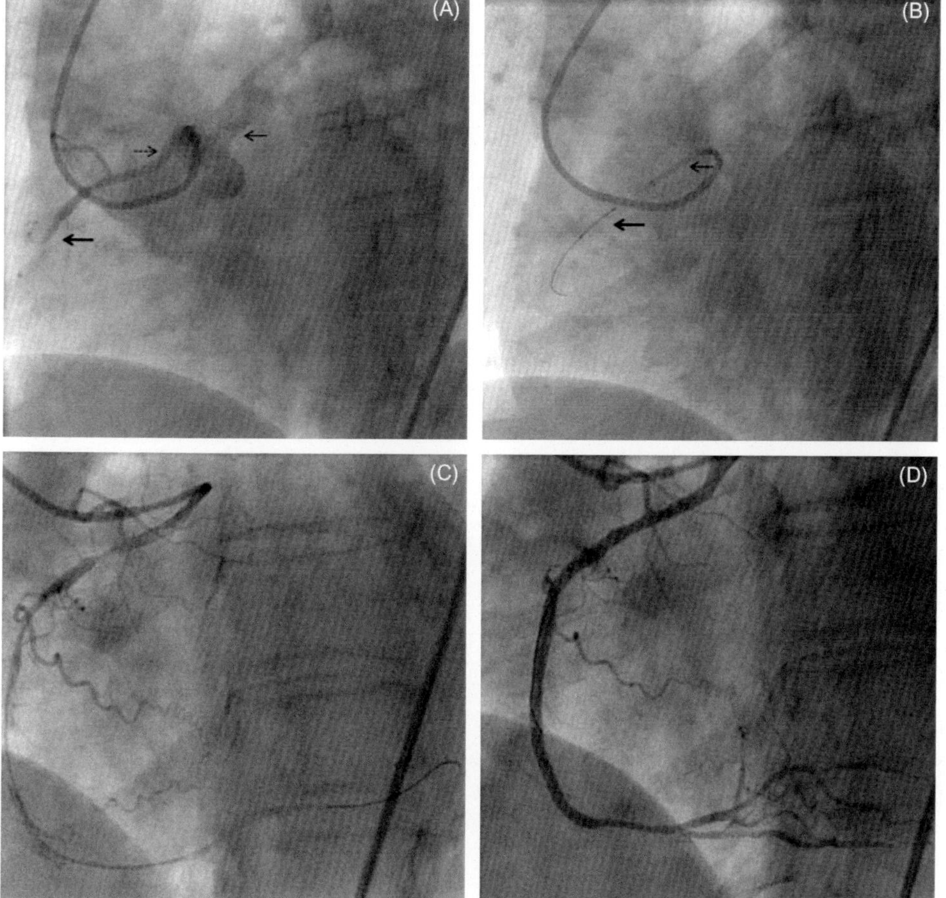

FIGURE 30.25 Use of a guide catheter extension for treating a CTO of an anomalous right coronary artery. (Panel A) Chronic total occlusion (*arrow*) of an anomalous right coronary artery arising from the left sinus of Valsalva. (Panel B) A Guideliner catheter was placed over a Fielder guidewire with the support of an uninflated balloon kept in the proximal right coronary artery for better support. (Panel C) The CTO was crossed with a Confianza Pro 9 wire with the support of the Guideliner. (Panel D) Successful recanalization of the right coronary artery with TIMI 3 flow. *Reproduced with permission from Senguttuvan NB, Sharma SK, Kini A. Percutaneous intervention of chronic total occlusion of anomalous right coronary artery originating from left sinus—use of mother and child technique using guideliner. Indian Heart J 2015;67 3: S41−2. Copyright Elsevier.*

6. Create a "homemade" snare (KAM-snare), that consists of a wire loop being trapped by an inflated balloon at the distal portion of a guide catheter extension (Section 27.1.3.5) [18].
7. Reduce hematoma formation when using antegrade dissection and reentry CTO crossing techniques by pressure dampening and reducing antegrade flow.

30.3.3 Guide catheter use extensions tips and tricks

1. To minimize the risk of the guidewire wrapping around the guide catheter extension push rod after insertion [19], the external push rod should be placed under a towel at the side of the Y-connector (Fig. 30.27).
2. Advancing the guide catheter extension may be easier to achieve by inflating a balloon half-way inside the guide catheter extension distal tip and the vessel (Fig. 30.28). The guide catheter extension is then advanced upon

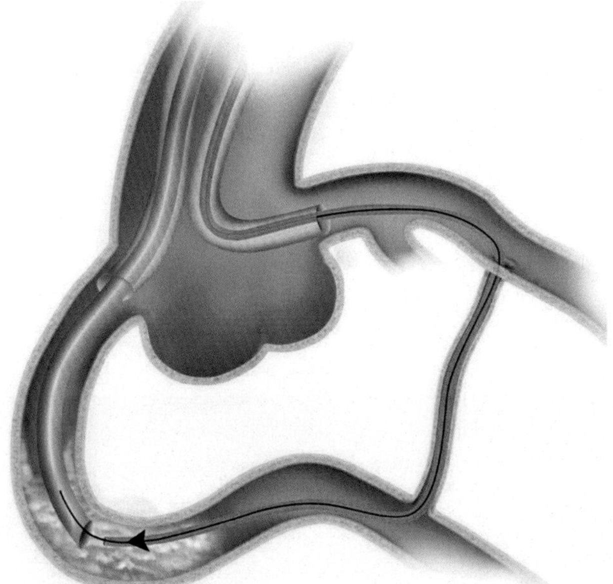

FIGURE 30.26 Illustration of the "guide extension reverse CART" for CTO PCI. *Reproduced with permission from Brilakis ES. Manual of coronary chronic total occlusion interventions. A step-by-step approach. 2nd ed. Cambridge, MA: Elsevier; 2017 (Figure 6.27).*

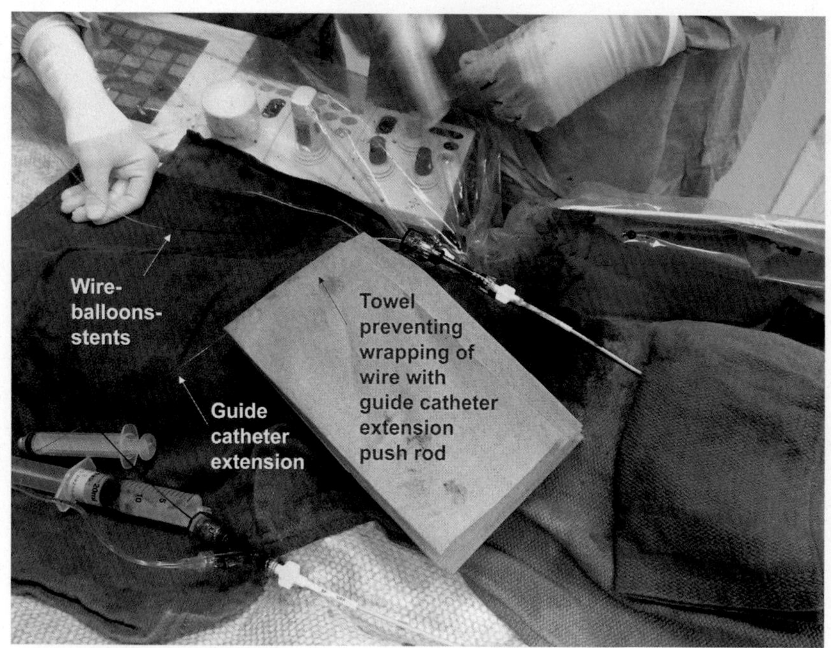

FIGURE 30.27 Guide catheter extension manipulation to minimize the risk of guidewire "wrap around" the guide catheter extension push rod. *Reproduced with permission from Brilakis ES. Manual of coronary chronic total occlusion interventions. A step-by-step approach. 2nd ed. Cambridge, MA: Elsevier; 2017 (Figure 2.9, panel A).*

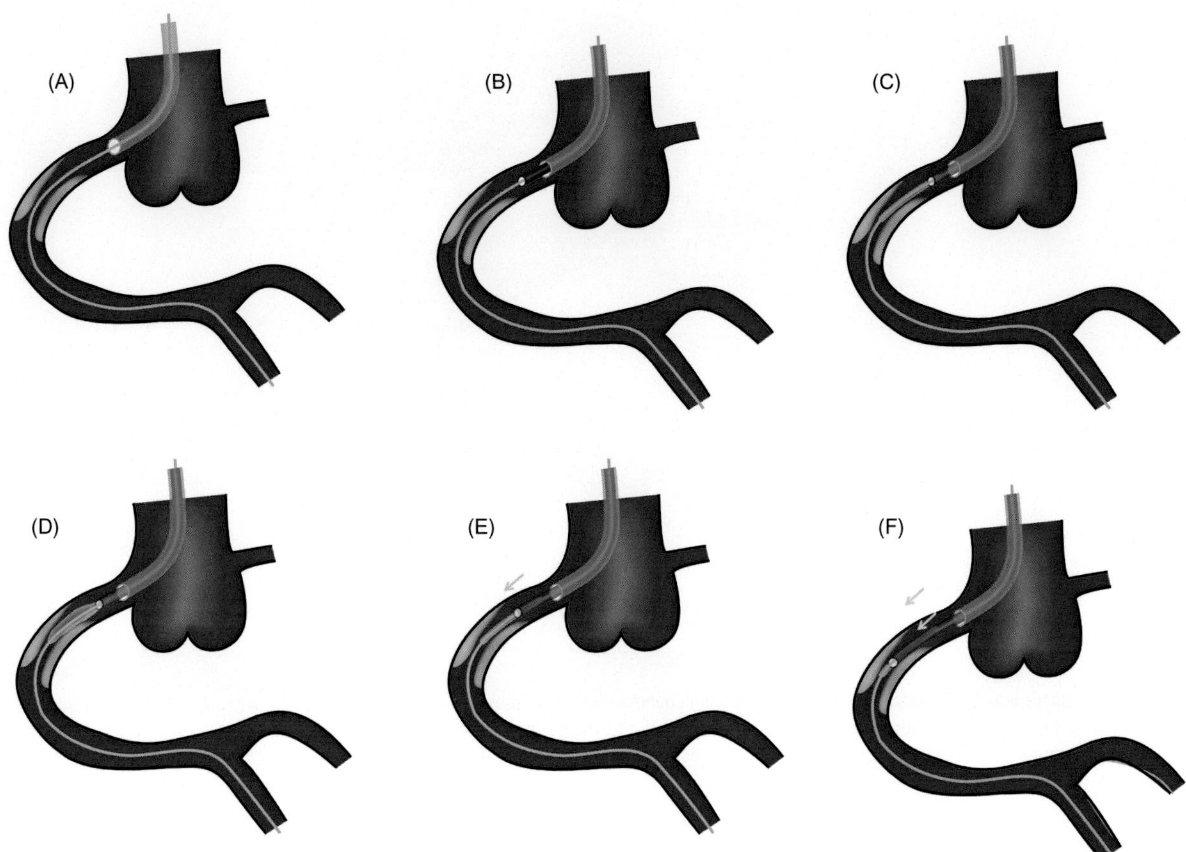

FIGURE 30.28 Delivery of a guide catheter extension to the target coronary segment using the "inchworming technique." (A) Guide catheter in position. (B) Guide catheter extension is advanced to the tip of the guide catheter. (C) A small balloon (usually 2.0 mm) is advanced halfway in and halfway out of the guide catheter extension. (D) The balloon is inflated, usually at low pressure (6–8 atm). (E) The balloon is deflated. (F) While the balloon is being deflated the guide catheter extension is advanced forward.

balloon deflation ("inchworming" technique). Advancement over a balloon catheter or microcatheter is preferred to advancement over a 0.014 in. coronary wire to minimize the risk of catching a plaque edge and causing a dissection. (*CTO PCI Manual* Online case: 44).

Online video: How to deliver a guide catheter extension using the "inchworming" technique

3. Additional strategies for delivering guide catheter extensions include:
   a. Guide extension balloon-assisted tracking (GBAT: inflation of a balloon protruding from the guide catheter extension at nominal pressure followed by advancement of both the guide catheter extension and the balloon to the desired coronary artery location [20]) (Fig. 30.29); and
   b. Distal anchor technique (Fig. 30.30) [21].

   The "inchworming" technique is generally preferred over the above techniques for delivering a guide extension, as it centers the guide catheter extension, and is hence less likely to cause vessel trauma.
4. Attempts to advance guidewires through a guide that contains a guide catheter extension smaller than the size of the guide (e.g., a 6 Fr extension within an 8 Fr guide), should be avoided, as the wire is likely to advance between the cylinder of the guide catheter extension and the guide catheter wall (Fig. 30.31).

   Furthermore, even when the guide catheter extension is of similar size to the guide catheter (e.g., a 7 Fr guide catheter extension within a 7 Fr guide catheter) there is a risk of wire damage when trying to advance it into the guide extension cylinder. Using a microcatheter to enter the guide extension cylinder with the wire withdrawn into the microcatheter can reduce this risk.
5. The proximal collar of the guide catheter extension should not be advanced outside the guide catheter.
6. Trapping of equipment using a guide catheter extension is difficult: (1) the trapping balloon needs to be placed proximal to the proximal collar; and (2) the equipment needs to be retracted to this location in order for successful trapping to occur. When the need of trapping is anticipated, use of the Trapliner is preferred.

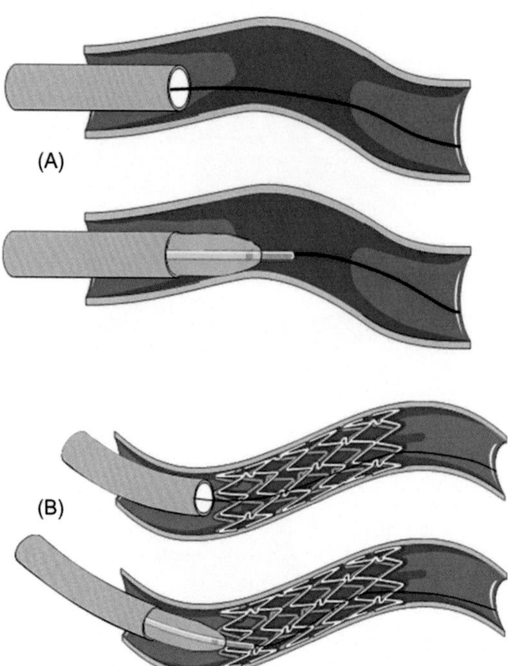

FIGURE 30.29 Illustration of guide catheter extension advancement using balloon-assisted tracking. (Panel A) Advancement of a guide catheter extension through a tortuous vessel with the compliant balloon at the tip of the guide catheter extension. (Panel B) Advancement of the guide catheter extension through previously stented segments with the leading balloon tip. *Courtesy of Dr. Basem Elbarouni.*

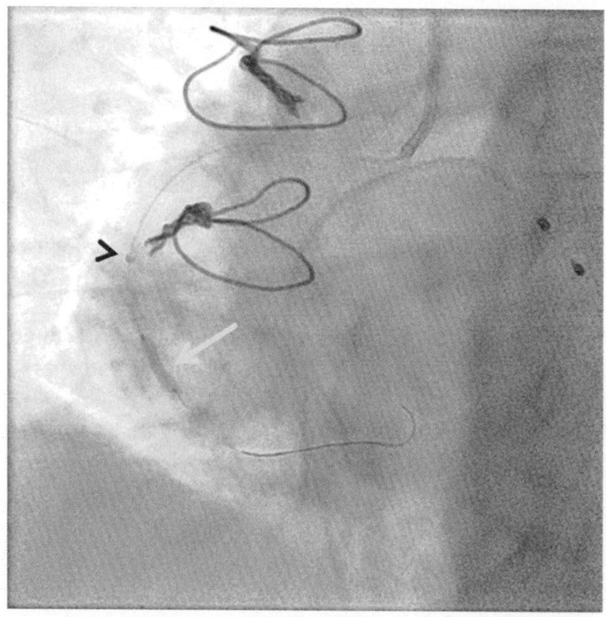

FIGURE 30.30 Illustration of the distal anchoring technique (*arrow*) to deliver a guide catheter extension (*arrowhead*). *Courtesy of Dr. Abdul Hakeem.*

7. Guide catheter extensions are very flexible and can advance even through highly tortuous lesions [22].
8. Two guide catheter extensions can be used simultaneously in a "mother–daughter–granddaughter" configuration (i.e., a 6 Fr extension through an 8 Fr extension) when multiple extreme bends need to be navigated (Fig. 30.32) [23].
9. Although distal to proximal stenting is preferred, "proximal to distal" stenting is a viable option and can be facilitated by inserting the guide catheter extension through the proximal stent [24].

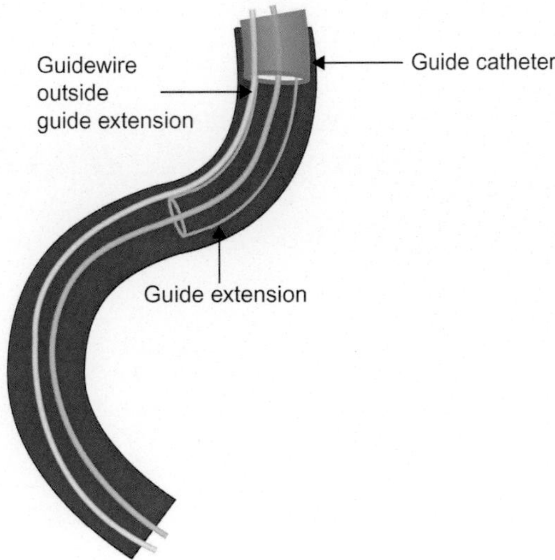

FIGURE 30.31 Guidewire inserted within the guide catheter but outside the guide catheter extension.

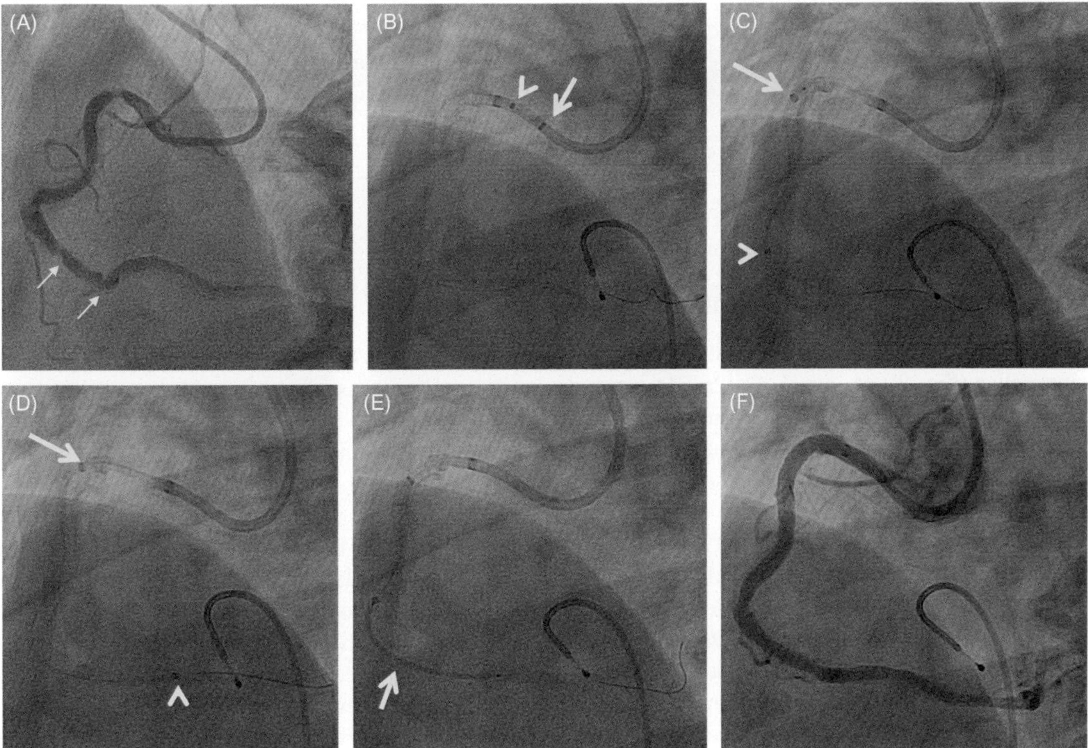

FIGURE 30.32 Illustration of the "mother–daughter–granddaughter" technique. (A) Diagnostic angiography demonstrating lesions in the distal right coronary artery (*arrows*), which was very tortuous. (B) A 6 Fr GuideLiner (*arrowhead*) is advanced inside an 8 Fr GuideLiner (*arrow*) into the guide catheter. (C) The 6 Fr GuideLiner (*arrowhead*) and the 8 Fr GuideLiner (*arrow*) are advanced into the right coronary artery. (D) The 6 Fr GuideLiner (*arrowhead*) is advanced past the distal right coronary artery lesion. (E) A stent (*arrow*) is delivered through the "mother–daughter–granddaughter" system to the distal right coronary artery. (F) Excellent final result after stent implantation. *Courtesy of Dr. William Nicholson. Reproduced with permission from Brilakis ES. Manual of coronary chronic total occlusion interventions. A step-by-step approach. 2nd ed. Cambridge, MA: Elsevier; 2017 (Figure 2.16).*

10. An alternative strategy to use of a guide catheter extension is deep intubation of the guide catheter [25]. Usually 5 or 6 French guide catheters are used for deep intubation.
11. Compatibility of the Guideliner with various equipment is shown in Fig. 30.33.

	Caravel ASAHI	Corsair ASAHI	Corsair Pro ASAHI	Minnie Teleflex	Super Cross Angled Tip Teleflex	Super Cross Straight Tip Teleflex	Turnpike Teleflex	Turnpike Gold Teleflex	Turnpike LP Teleflex	Turnpike Spiral Teleflex	TwinPass Teleflex	TwinPass Torque Teleflex	Venture RX Teleflex	Venture OTW Teleflex
Max OD -->	2.6Fr (0.85 mm, 0.034 in)	2.8Fr (0.93 mm, 0.037 in)	2.8Fr (0.93 mm, 0.037 in)	3.1Fr (1.02 mm, 0.040 in)	3.2Fr (1.07 mm, 0.042 in)	2.5Fr (0.84 mm, 0.033 in)	2.9Fr (0.97 mm, 0.038 in)	2.9Fr (0.97 mm, 0.038 in)	2.9Fr (0.97 mm, 0.038 in)	2.9Fr (0.97 mm, 0.038 in)	3.4Fr (1.14 mm, 0.045 in)	3.5Fr (1.17 mm, 0.046 in)	4.1 Fr (1.37 mm, 0.054 in)	4 Fr (1.32 mm, 0.052 in)
5F Guideliner Effective ID =0.044 in (1.12 mm)	✓	✓	✓	✓	✓	✓	✓	✓	✓	✓	No	No	No	No
5.5F Guideliner Effective ID = 0.051 in (1.30 mm)	✓	✓	✓	✓	✓	✓	✓	✓	✓	✓	No	No	No	No
6F Guideliner Effecitve ID = 0.056 in (1.42 mm)	✓	✓	✓	✓	✓	✓	✓	✓	✓	✓	✓	✓	✓	✓
7F Guideliner Effective ID = 0.062 in (1.57 mm)	✓	✓	✓	✓	✓	✓	✓	✓	✓	✓	✓	✓	✓	✓
8F Guideliner Effective ID = 0.071 in (1.80 mm)	✓	✓	✓	✓	✓	✓	✓	✓	✓	✓	✓	✓	✓	✓

	Rotablator Burr 1.25 mm	Rotablator Burr 1.5 mm	Rotablator Burr 1.75 mm	Rotablator Burr 2.0 mm	CSI OA
5F Guideliner	No	No	No	No	No
5.5F Guideliner	✓	No	No	No	✓
6F Guideliner	✓	No	No	No	✓
7F Guideliner	✓	✓	No	No	✓
8F Guideliner	✓	✓	✓	No	✓

	Stent Sizes
5F	3.0 mm
5.5F	3.5 mm
6F	4.0 mm
7F	
8F	

FIGURE 30.33 Guideliner compatibility with other equipment. *Courtesy of Dr. Bilal Murad.*

30.3.4 Guide catheter extension related complications

1. Deformation of guidewires, stents, or other equipment can occur during *advancement* through the guide catheter extension collar (Fig. 30.34, panel A) [26]. In some cases, it may be necessary to first advance the stent or other equipment into the guide catheter extension *outside* the body and then introduce everything as a single unit into the guide catheter. If any resistance is encountered during equipment advancement when using a guide catheter extension, fluoroscopic guidance should be used to visualize the stent as it enters to proximal port of the guide catheter extension.
2. Deformation of stents or other equipment can also occur while *withdrawing* the equipment back into the distal tip of the guide catheter extension after a failed attempt to advance to the target lesion (Fig. 30.34, panel B). Hence, any withdrawn stent after a failed delivery attempt should be closely inspected for damage.
3. Coronary or aorto-coronary dissection. Deep advancement of the guide catheter extension may cause coronary dissection [11], especially if contrast is injected while the pressure is dampened. Every effort should be undertaken to minimize pressure dampening, but if dampening occurs, it is important to verify that adequate antegrade flow is preserved and no vessel injury has occurred before proceeding with the intervention [11]. Injection through a guide catheter extension with dampened pressure waveform may cause dissection that can propagate either antegrade or retrograde (Fig. 30.35). In general, injections though deeply engaged guide catheter extensions should be avoided due to risk of dissection.

   In some cases, use of a guide catheter extension may lead to acute vessel closure that can be challenging or impossible to correct (Fig. 30.36) [27].
4. Because the guide catheter extension decreases the original guide size by 1-2 French (e.g., the inner diameter of a 5 French Launcher guide catheter is 0.058 in. and that of a 6 French Guideliner is 0.056 in., the inner diameter of a 6 French Launcer guide catheter is 0.071 in. and that of the 8 French Guideliner is 0.071 in.; the inner diameter of the 7 French Guideliner is 0.062 inch, which is between the inner diameter of a 5 French and a 6 French Launcher guide catheter), special attention to pressure dampening and to activated clotting time (ACT) is needed to decrease the risk of thrombus formation.

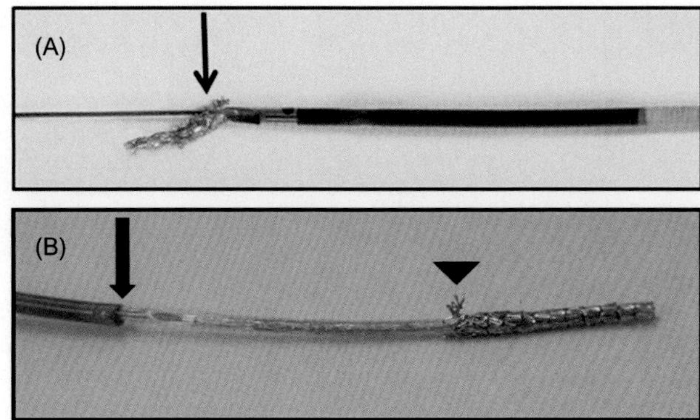

FIGURE 30.34 Complications of stent delivery through guide catheter extensions. (Panel A) Stent deformation while attempting to deliver it through a Guideliner catheter. (Panel B) Stent deformation while attempting to retrieve an undeployed stent into the distal tip of a Guidezilla catheter. The tip of the Guidezilla prolapsed on itself (*arrow*) during attempted stent retrieval, resulting in catching the proximal edge of the stent and causing deformation (*arrowhead*). *Courtesy of Dr. William Nicholson. Reproduced with permission from Brilakis ES. Manual of coronary chronic total occlusion interventions. A step-by-step approach. 2nd ed. Cambridge, MA: Elsevier; 2017 (Figure 2.13) Reproduced with permission from Papayannis AC, Michael TT, Brilakis ES. Challenges associated with use of the GuideLiner catheter in percutaneous coronary interventions. J Invasive Cardiol 2012;24:370–1.*

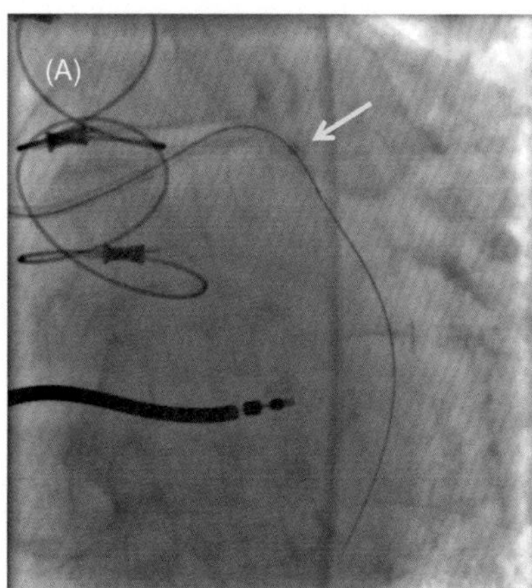

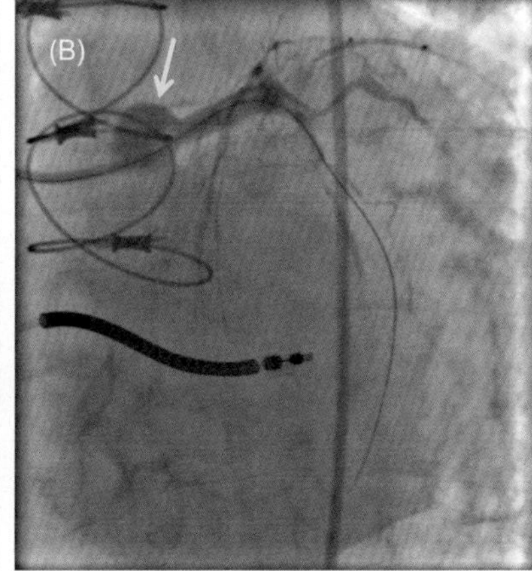

FIGURE 30.35 Illustration of aorto-coronary dissection related to use of a guide catheter extension. Retrograde dissection caused by contrast injection through a guide catheter extension with dampened waveform. (A) Guide catheter extension with its tip (*arrow*) deep-seated into the circumflex artery. (B) Retrograde dissection into aortic root (*arrow*) from contrast injection through the deep-seated guide catheter extension. *Reproduced with permission from Brilakis ES. Manual of coronary chronic total occlusion interventions. A step-by-step approach. 2nd ed. Cambridge, MA: Elsevier; 2017 (Figure 2.14).*

5. Guide catheter extension distal advancement: hold the guide catheter extension push rod during forceful contrast injection to minimize the risk of "ejecting" the guide catheter extension into the vessel [28].
6. Guide catheter extension fracture (Fig. 30.37) [29].
7. Longitudinal stent deformation upon forceful advancement of a guide catheter extension through a previously deployed stent [30,31].

30.4 Support catheters

There are several catheters that can deeply intubate the vessel (similar to a guide catheter extension), while at the same time providing additional support through contact with the vessel wall, by using a self-expanding nitinol mesh (CenterCross and MultiCross) or nitinol wires (Novacross), or by inflating a soft balloon inside the target coronary artery (Prodigy).

30.4.1 MultiCross and CenterCross

CTO PCI Manual Online case: 7. The MultiCross (Fig. 30.38) and CenterCross (Fig. 30.39) (Boston Scientific) are two support catheters that provide anchoring to the vessel wall. Both have a stabilizing self-expanding scaffold (expands up to 4.5 mm in diameter) that is deployed proximal to the target lesion. The Multicross contains three separate lumens within the scaffold, each located 120° apart [32]. The CenterCross has a single, large central lumen that can

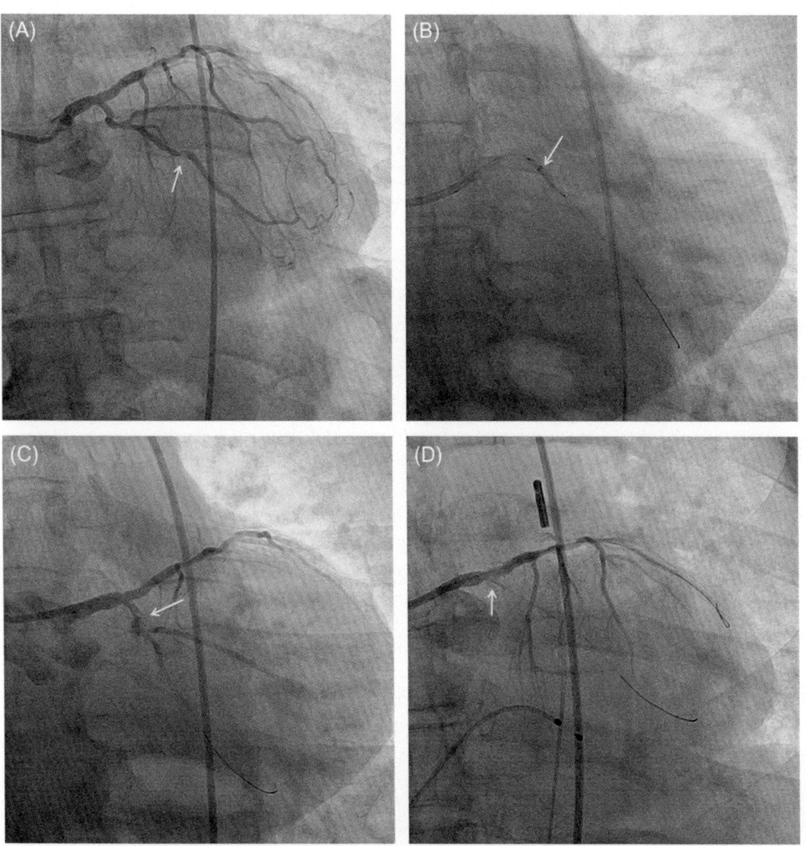

FIGURE 30.36 Coronary angiography demonstrating a lesion in the first obtuse marginal branch (*arrow*, panel A). Attempts to deliver a stent failed despite using a guide catheter extension (*arrow*, panel B), and caused proximal circumflex dissection (*arrow*, panel C) and acute vessel closure (*arrow*, panel D). Attempts to restore antegrade flow failed, and the patient underwent emergency coronary artery bypass graft surgery. *Reproduced with permission from Duong T, Christopoulos G, Luna M, et al. Frequency, indications, and outcomes of guide catheter extension use in percutaneous coronary intervention. J Invasive Cardiol 2015; 27:E211–5.*

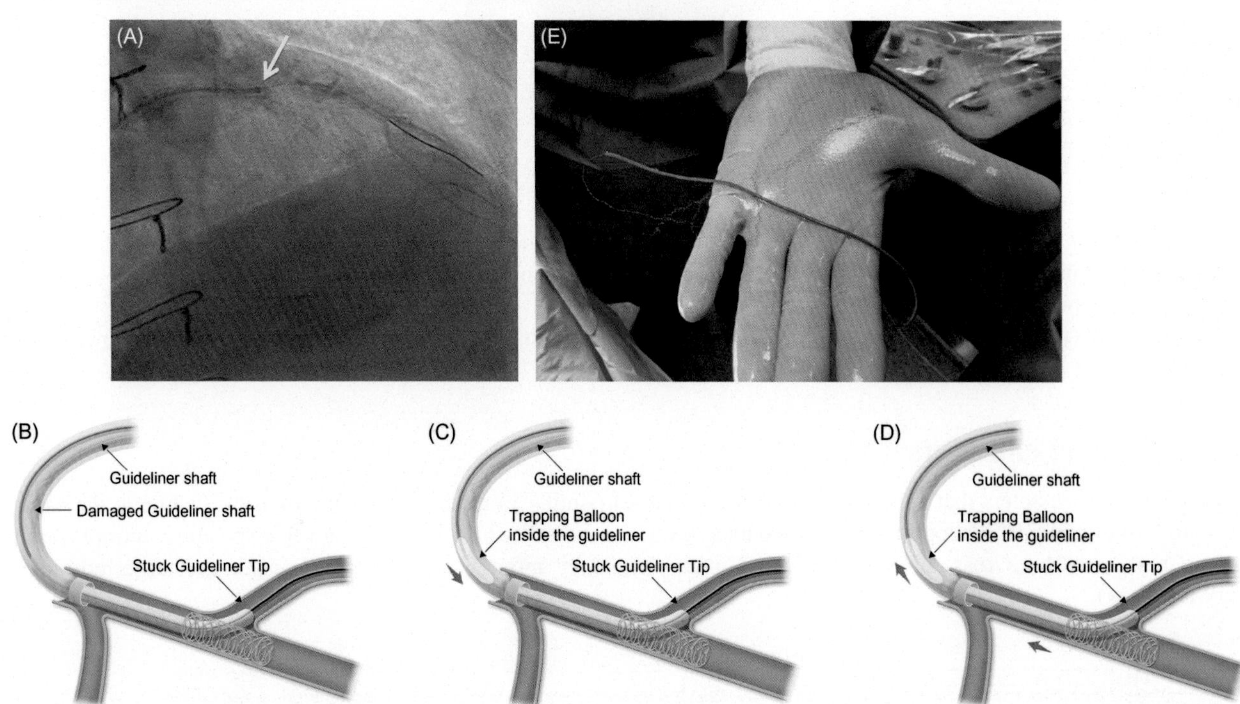

FIGURE 30.37 Retrieval of a fractured guide catheter extension. (Panel A) Entrapment of the tip of a Guideliner (*arrow*, panel A) during attempts to advance into the diagonal branch through an LAD stent. (Panel B) Fracture of the Guideliner shaft during retrieval attempts. (Panel C) Balloon advancement inside the Guideliner cylinder. (Panel D) Balloon inflation inside the cylinder, followed by withdrawal that retrieved the entrapped Guideliner. (Panel E) The retrieved fragment of the fractured Guideliner. *Courtesy of Dr. Gabriele Gasparini.*

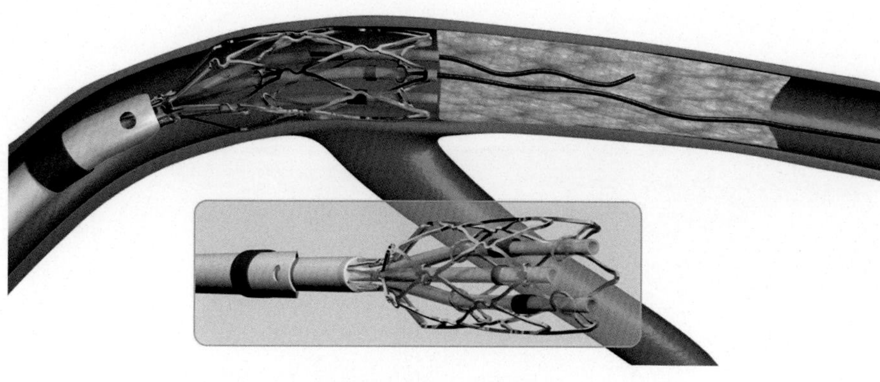

FIGURE 30.38 The MultiCross catheter. *Image courtesy Boston Scientific © 2019 Boston Scientific Corporation or its affiliates. All rights reserved.*

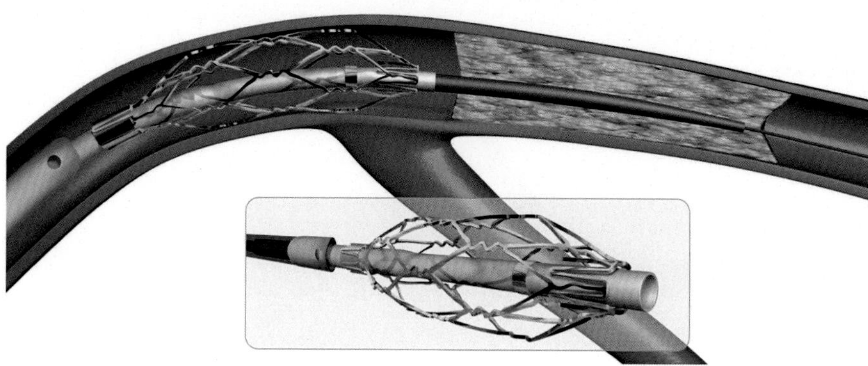

FIGURE 30.39 The CenterCross catheter. *Image courtesy Boston Scientific © 2019 Boston Scientific Corporation or its affiliates. All rights reserved.*

accommodate a microcatheter. These catheters increase the backup support and the penetration force to cross the proximal cap and traverse through the occlusion.

MultiCross and CenterCross tips and tricks

1. The CenterCross inner lumen can accommodate all $\geq$150 cm long single lumen microcatheters discussed below (Corsair, Caravel, Finecross, Turnpike, Turnpike LP, MicroCross, etc.)
2. Due to their profile, neither the CenterCross nor the MultiCross can be exchanged using the trapping technique, hence long (300 cm) guidewires should be used to remove them from the guide catheter.
3. Both catheters require at least 10 mm proximal landing zone in the target coronary vessel.

30.4.2 NovaCross

The NovaCross catheter (Nitiloop) has a 10 mm long flexible Nitinol element which upon axial compression deforms by curving outward several helical struts (Fig. 30.40) increasing support [33].

30.4.3 Prodigy

Similar to MultiCross and CenterCross, the Prodigy catheter (Fig. 30.41, Radius Medical) was designed to provide strong guidewire support [34]. It consists of a 5 French catheter with a soft, atraumatic elastomeric balloon at its distal tip that can be expanded up to 6 mm in diameter. The inflation lumen has a pressure relief valve that limits inflation pressure to 1 atm, anchoring the catheter in place while minimizing the risk of proximal vessel injury and allowing enhanced guidewire pushability.

30.5 Y-connectors with hemostatic valves

Using a Y-connector with a hemostatic valve (Fig. 30.42), such as the Co-Pilot (Abbott Vascular), Guardian (Teleflex), Watchdog (Boston Scientific), and the OKAY II (InfraRedx) can help minimize blood loss from back bleeding (which

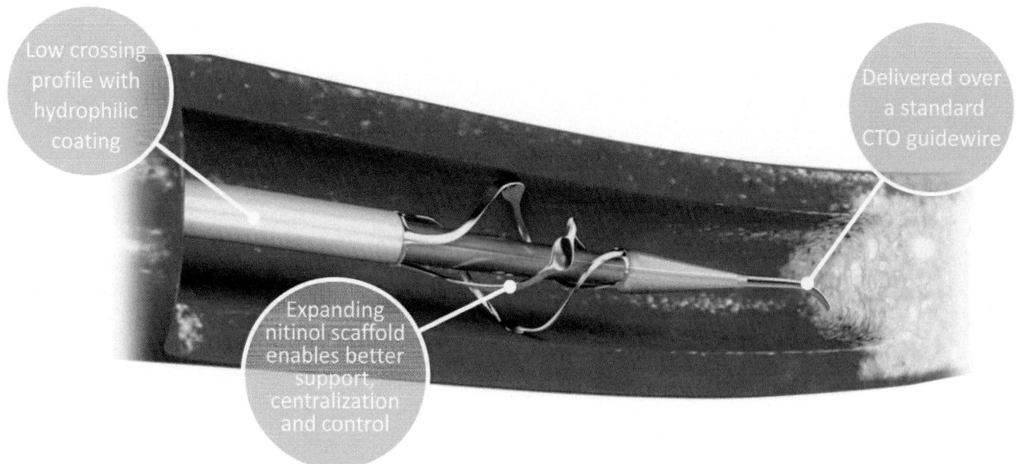

FIGURE 30.40 The NovaCross catheter. *Reproduced with permission by Nitiloop Ltd.*

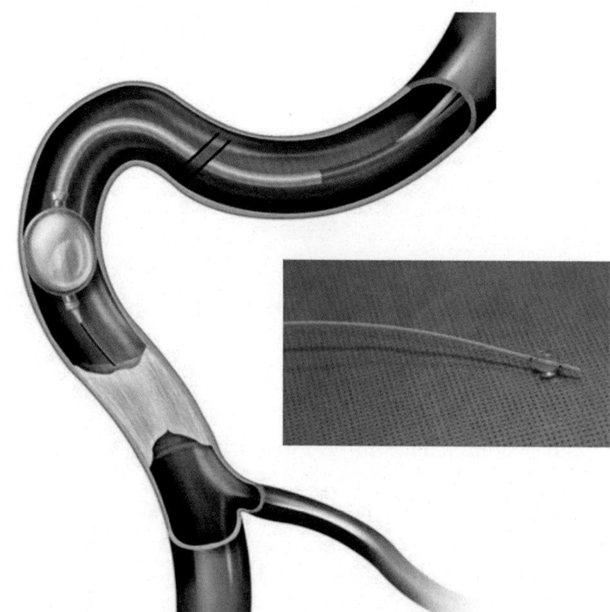

FIGURE 30.41 The Prodigy catheter. *Reproduced with permission from Radius Medical.*

is particularly important for larger guide catheters, such as 8 Fr) and is easier to use compared with standard rotating hemostatic valves.

30.6 Microcatheters

As their name implies, microcatheters (μικρός = small + καθετήρας = catheter) are small (usually 1.8 to 3.2 Fr in diameter) catheters.

Although over-the-wire balloons can be used to support guidewire advancement instead of a microcatheter, microcatheters are preferred to over-the-wire balloons because:

1. They allow better understanding of distal tip position (a marker is placed at the microcatheter tip, whereas in small balloons [≤1.5 mm diameter] the marker is located in the middle of the balloon) (Fig. 30.43) [8].
2. Are more flexible and track better than over-the-wire balloons.
3. Have less tendency to kink than over-the-wire balloons (kinking of balloon shaft prohibits future wire exchanges and often necessitates balloon catheter and wire removal and replacement with new gear, losing the crossing

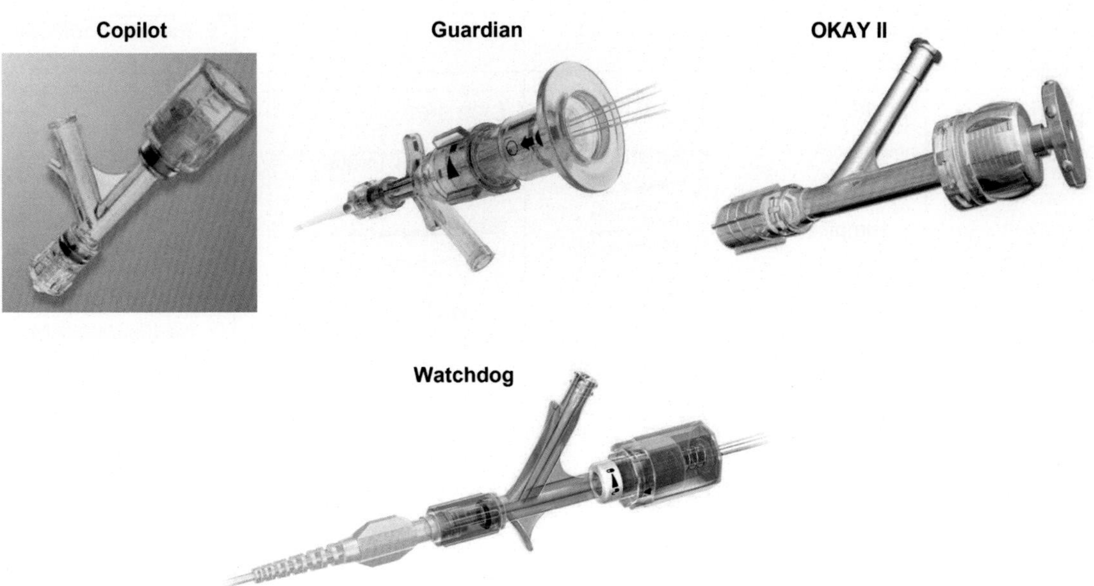

FIGURE 30.42 Types of Y-connectors with hemostatic valves available in the United States. *Image of Copilot courtesy Abbott Vascular. ©2019 Abbott. All rights reserved. Image of Watchdog courtesy Boston Scientific © 2019 Boston Scientific Corporation or its affiliates. All rights reserved. Image of OKAY II: Copyright 2019 InfraRedx, Inc. All rights reserved.*

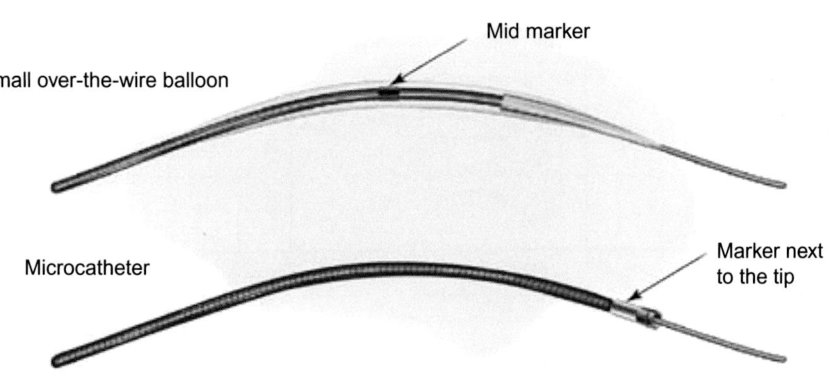

FIGURE 30.43 Comparison of over-the-wire balloons and microcatheters for facilitating wiring. *Reproduced with permission from Brilakis ES. Manual of coronary chronic total occlusion interventions. A step-by-step approach. 2nd ed. Cambridge, MA: Elsevier; 2017 (Figure 2.18).*

progress achieved). Over-the-wire balloons, however, do provide better support than many microcatheters and are significantly cheaper.

Several microcatheters are commercially available and can be classified into five categories (Fig. 30.44 and Table 30.6) [35]:

1. Large outer diameter microcatheters (such as the Corsair [Asahi Intecc] [36], Turnpike and Turnpike Spiral [Teleflex]), Teleport Control [Orbus Neich], Mamba [Boston Scientific], and M-Cath [Acrostak]).
2. Small outer diameter microcatheters, such as the Finecross (Terumo), Caravel and Corsair XS (Asahi Intecc), Turnpike LP (Teleflex), and MicroCross (Boston Scientific).
3. Angulated microcatheters, such as the Supercross (Teleflex), Venture (Teleflex) and Swift-Ninja (Merit Medical).
4. Dual lumen microcatheters, such as the TwinPass and TwinPass Torque (Teleflex), Sasuke (Asahi Intecc), Crusade (Kaneka), FineDuo (Terumo), NHancer Rx (IMDS), and ReCross (IMDS).
5. Plaque modification microcatheters, such as the Tornus (Asahi Intecc), and Turnpike Gold (Teleflex).

The major role of each microcatheter category [36,37] is shown in Fig. 30.45.

Moreover, microcatheters:

1. Provide better support and increase guidewire tip stiffness, enhancing its penetration capacity. Microcatheters should be routinely used for guidewire support during chronic total occlusion interventions [38], but they can be useful in

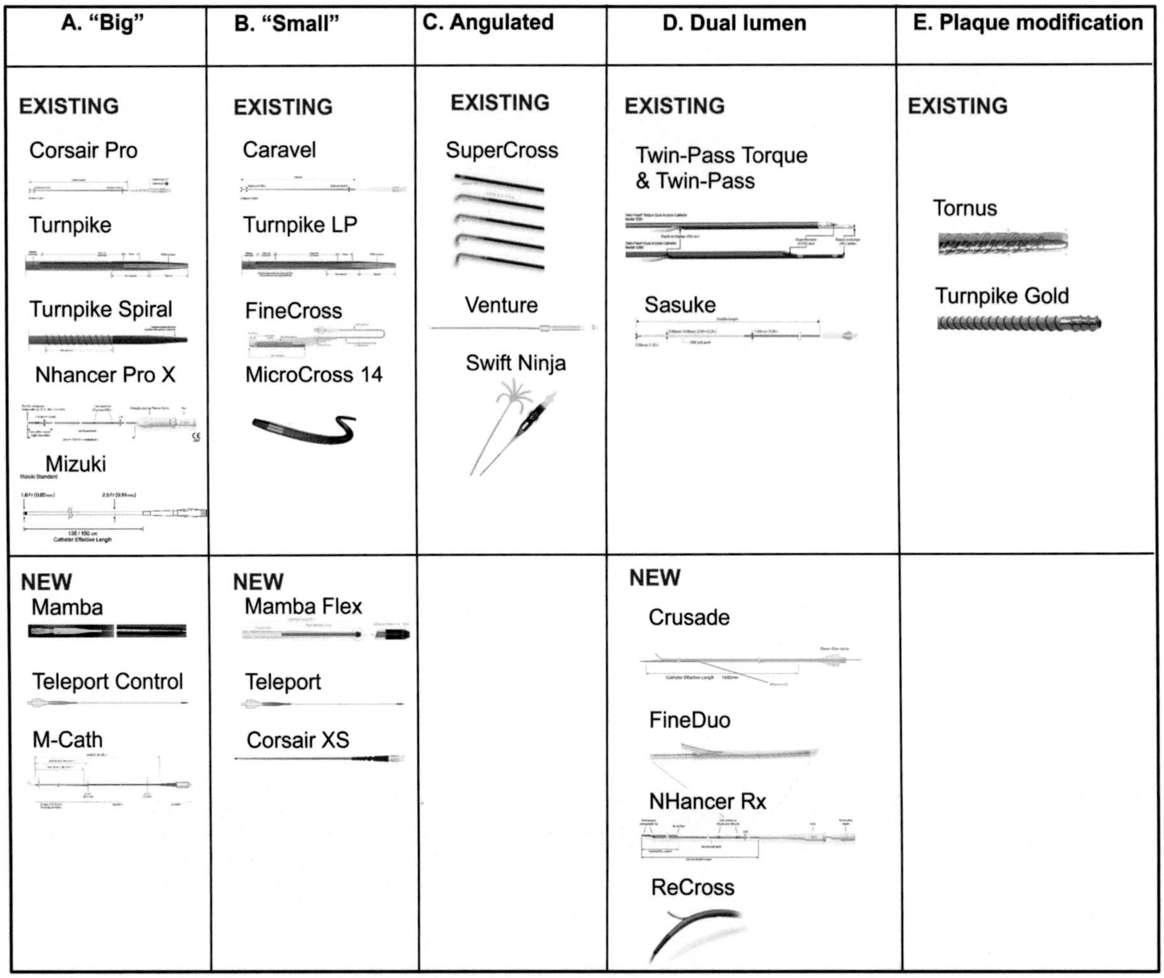

A. "Big"	B. "Small"	C. Angulated	D. Dual lumen	E. Plaque modification
EXISTING Corsair Pro Turnpike Turnpike Spiral Nhancer Pro X Mizuki	**EXISTING** Caravel Turnpike LP FineCross MicroCross 14	**EXISTING** SuperCross Venture Swift Ninja	**EXISTING** Twin-Pass Torque & Twin-Pass Sasuke	**EXISTING** Tornus Turnpike Gold
NEW Mamba Teleport Control M-Cath	**NEW** Mamba Flex Teleport Corsair XS		**NEW** Crusade FineDuo NHancer Rx ReCross	

FIGURE 30.44 Currently available microcatheter categories.

other challenging clinical cases, such as wiring through tortuosity, angulation, or areas of severe calcification. They can also be used for bifurcation stenting, after rewiring a side branch through stent struts [39].

2. Allow reshaping of the guidewire tip.
3. Facilitate guidewire exchanges.
4. Protect the proximal part of the vessel (and/or the collateral vessel in case of retrograde crossing) from guidewire-induced injuries.
5. Allow injection of medications or contrast, either in the distal true lumen, or within the intima or subintimal space during chronic total occlusion interventions (Carlino technique) [40].
6. Allow delivery of coils, fat, or thrombin in case of perforation [41]. However, 0.014 in. microcatheters (all coronary microcatheters are 0.014 in.) cannot be used for delivering 0.018 in. coils (Section 30.19.2) with the exception of the FineCross microcatheter. Larger microcatheter, such as the Progreat (2.4 Fr, Terumo), Renegade (2.5 Fr, Boston Scientific) and Transit (2.5 Fr, Cordis) or the FineCross can be used for delivery of 0.018 in. coils. This is why it is important for cardiac catheterization laboratories to have 0.014 in. microcatheter compatible coils available for treating coronary perforations (such as the Axium coils, Medtronic). Alternatively, larger microcatheters should be available.

30.6.1 Large microcatheters

Large microcatheters are usually used for guidewire support during CTO PCI, but can serve any other purpose described above. Large microcatheters can be spun to facilitate advancement, although overspinning should be avoided to prevent microcatheter structural damage (Section 27.4).

TABLE 30.6 Characteristics of currently available microcatheters.

A.Large outer diameter microcatheters	Corsair Pro	Turnpike and Turnpike Spiral	Teleport control	Mamba	M-cath	NHancer ProX NX3	Mizuki	Mizuki FX
Catheter length (cm)	135/150 cm	135/150 cm (only 135 cm for the Spiral)	135/150 cm	135 cm	135 cm	135/155 cm	135/150 cm	135/150 cm
Prox. O.D.	2.8F	2.9F	2.7F	2.9F	3.3F	3.2F	3.2F	2.5F
Distal O.D.	2.6F	2.6F	2.1F	2.4F	2.25F	2.0F	1.8F	1.8F
Tip entry O.D.	1.3F	1.6F	1.45F	1.4F	1.6F	1.5F	1.7F	1.7F
Inner lumen I.D.	0.0177 in.	0.0175 in.	0.0175 in.	0.023 in.	0.016 in.	0.020 in.with distal tip taper to 0.017 in.	0.022 in. with distal tip taper to 0.018 in.	0.022 in. with distal tip taper to 0.017 in.
Tip length	5 mm	11 mm (10 mm for LP)	6 mm	3 mm	2.5 mm	6 mm	1 mm	1 mm
Tip material	Tungsten	Mixture of tungsten and polyurethane	Mixture of Tungsten and Pebax		Pebax	Tungsten polymer blend	Pebax	Pebax
Tip construction	5 mm tungsten, 0.8 mm platinum coil	5 layer construction including dual-layer coils and a braid	Radiopaque. Dura tapered	Radiopaque, integrated tip with tapered coil		Integrated three layer hybrid tip design	Platinum ring marker 1.0 mm	Platinum ring marker 1.0 mm
Exterior coating	Hydrophilic (60 cm from distal tip)	Hydrophilic coating (distal 60 cm including tip)	Hydrophilic (60 cm of distal catheter including tip)	HydroPass Coating	Hydrophilic (43 cm from distal tip)	Hydrophilic (60 cm for 155 cm length and 25 cm for 135 cm length)	Hydrophilic (70 cm from distal tip)	Hydrophilic (70 cm from distal tip)

B.Small outer diameter microcatheters	Turnpike LP	Corsair Pro XS	Teleport	Mamba Flex	Caravel	FineCross	MicroCross 14 and 14ES
Catheter length (cm)	135/150 cm	135/150 cm	135/150 cm	135/150 cm	135/150 cm	135/150 cm	155 cm
Prox. O.D.	2.9F	2.9F	2.6F	2.9F	2.6F	2.6F	2.9F
Distal O.D.	2.2F	2.1F	2.0F	2.4F	1.9F	1.8F	1.6F
Tip entry O.D.	1.6F	1.3F	1.45F	1.4F	1.4F	1.8F	1.6F
Inner lumen I.D.	0.0175 in.	0.015 in.	0.0170 in.	0.023 in.	0.017 in. (distal) 0.022 in. (proximal)	0.0210 in.	0.0165 in.

C. Dual lumen microcatheters	TwinPass Torque	TwinPass	Sasuke	NHAncer Rx	ReCross
Catheter Length (cm)	135 cm	135 cm	145 cm	135 cm	140 cm
Prox. O.D.	3.1F	2.9F	3.2F	2.6F	3.4 x 2.6F
Dual lumen O.D.	3.5 x 3.5F	3.4 x 2.7F	3.3 x 2.5F	3.3 x 2.3F	3.3 x 2.3F
Tip entry O.D.	2.1F	2.0F	1.5F	1.5F	1.5F
Inner lumen I.D.	0.015 in. (RX) 0.0155 in. (OTW distal) 0.0165 in. (OTW proximal)	0.016 in. (RX) 0.0165 in. (OTW)	0.016 in. (tip) 0.017 in. (shaft)	0.019 in. tip and shaft lumen	0.019 in. tip and shaft lumen
Distal tip length	7 mm	20 mm	4 mm	5 mm	5 mm
Tip material	37D Pebax	47D Pebax	Tungsten	Tungsten	Tungsten
Tip construction	37D Pebax/ 40D Pebax, two eccentric Pt/Ir marker bands—one on each lumen; OTW lumen features 10° guidewire kick-out angle	47D Pebax with two Pt/Ir marker bands	5 mm tungsten, 0.8 mm platinum coil	Tungsten polymer both ports	Tungsten polymer all three ports
Exterior coating	Hydrophilic coating (distal 25 cm)	Hydrophilic coating (distal 18 cm)	Hydrophilic coating (38 mm length)	Hydrophilic (NDurance)	
Distance of OTW lumen port from tip	7 mm	20 mm	6.5 mm	6.5 mm	8 and 12 mm

30.6.1.1 Corsair Pro

The Corsair Pro microcatheter (Asahi Intecc, Fig. 30.46) was developed as a septal channel dilator to facilitate retrograde CTO PCI [42]. The Corsair "Shinka" shaft is constructed with eight thin wires wound with two larger wires, which facilitates torque transmission. The inner lumen is lined with a polymer that enables contrast injection and facilitates wire advancement. The distal 60 cm of the catheter are coated with a hydrophilic polymer to enhance crossability. The tip is tapered and soft and is loaded with tungsten powder to enhance visibility.

Two Corsair Pro lengths are currently available (135 cm long with light blue proximal hub and 150 cm long with dark blue proximal hub). The Corsair Pro catheter can be advanced by rotating in either direction, although it is braided to have better torque transmission with counterclockwise rotation.

30.6.1.2 Turnpike and Turnpike Spiral

The Turnpike (Teleflex) has a dual layer bidirectional coil (Fig. 30.47) that facilitates torque transmission, improves flexibility and prevents kinking. It also has a soft, tapered tip facilitating collateral branch crossing.

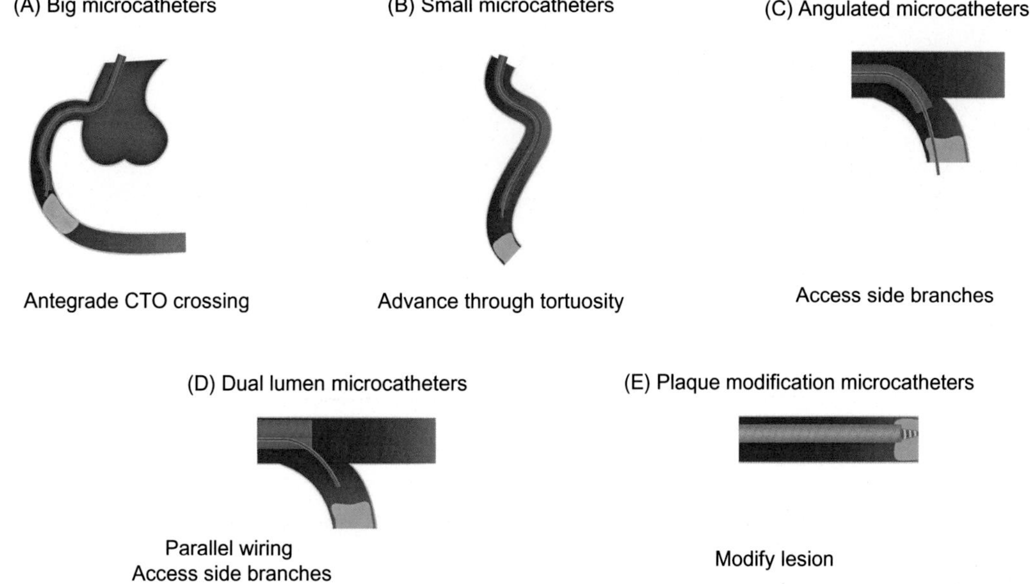

FIGURE 30.45 Main uses of coronary microcatheters in percutaneous coronary intervention. (Panel A) Big microcatheters are used for antegrade CTO crossing. (Panel B) Small microcatheters are used to cross tortuous vessels and collaterals. (Panel C) Angulated microcatheters are used for accessing side branches. (Panel D) Dual lumen microcatheters are used for parallel wiring or for accessing a side branch. (Panel E) Plaque modification microcatheters are used to modify plaque, usually in balloon uncrossable lesions.

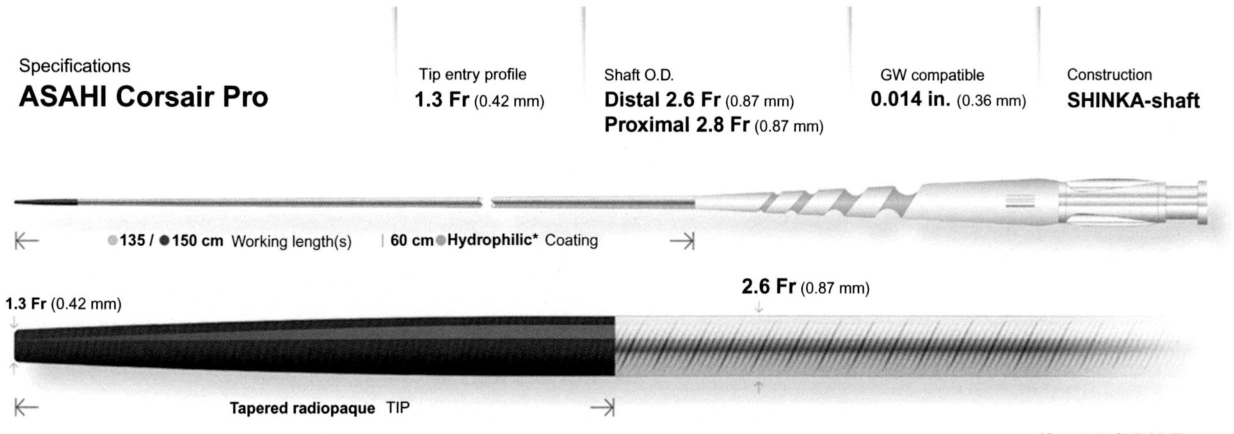

FIGURE 30.46 Illustration of the Corsair Pro microcatheter. *Reproduced with permission from Asahi Intecc.*

The Turnpike catheter is produced in four versions (Fig. 30.48): Turnpike, Turnpike LP, Turnpike Spiral and Turnpike Gold. Turnpike is the standard catheter with a 1.6 French outside diameter at the distal tip and 2.6 French outside distal shaft diameter. The Turnpike Spiral has a distal nylon coil and is preferred by many operators for antegrade CTO crossing. The Turnpike and Turnpike LP catheters can be rotated in either direction. In contrast, the Turnpike Spiral and Gold are rotated clockwise to advance and counter-clockwise for withdrawal (opposite direction compared with the Tornus catheter).

30.6.1.3 Teleport Control

The Teleport and Teleport Control microcatheters (Orbus Neich, Fig. 30.49) have a tungsten radiopaque short tip, an inner stainless-steel body of hybrid braiding and coil (Hybracoil) construction, an ultra-thin outer jacket from nylon transitioning into Pebax to facilitate greater distal flexibility, and a lubricious hydrophilic coating in the distal 60 cm.

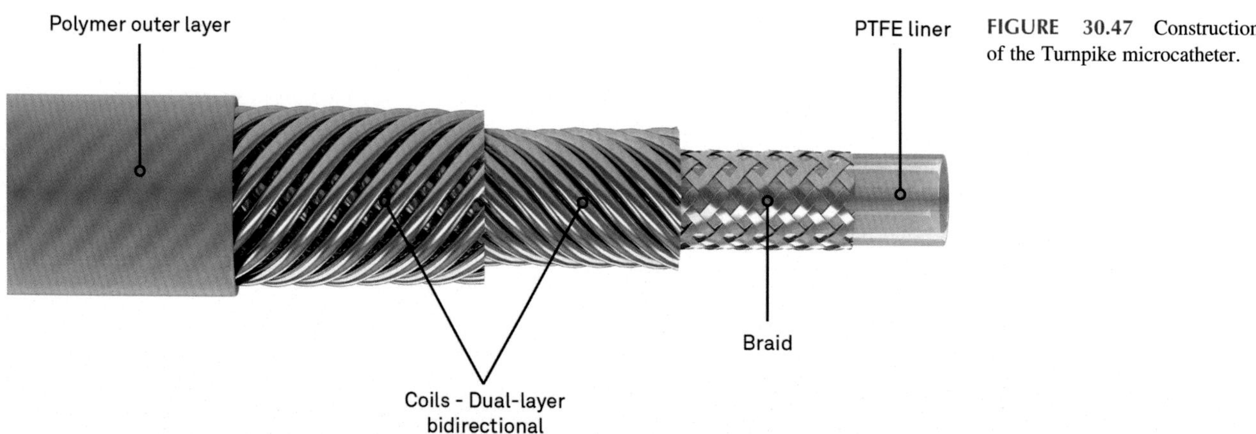

FIGURE 30.47 Construction of the Turnpike microcatheter.

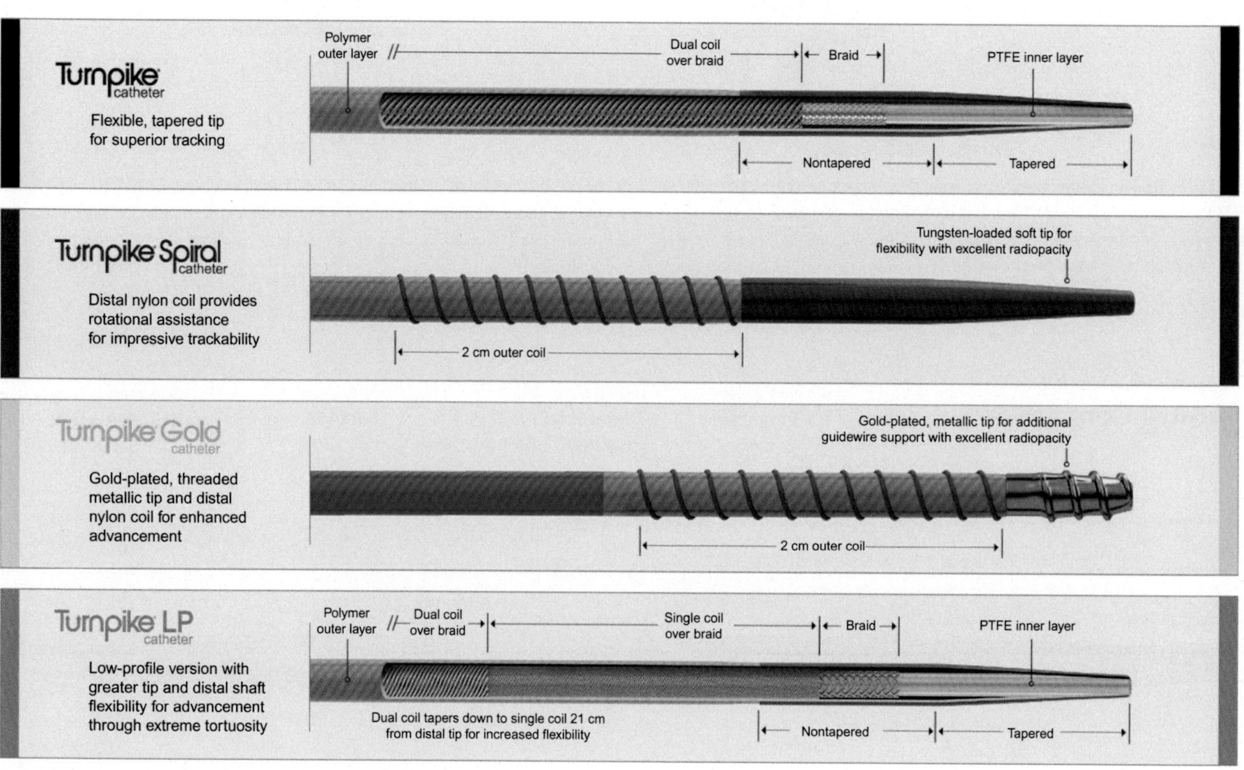

FIGURE 30.48 Illustration of the Turnpike microcatheters.

The hybrid braiding consists of 2 round and 14 flat wires outside a flat coil. The Teleport Control has a higher distal crossing profile compared with the Teleport. Both the Teleport and Teleport Control microcatheters are available in 135 and 150 cm length.

30.6.1.4 Mamba

The Mamba and Mamba Flex microcatheters (Boston Scientific, Fig. 30.50) have a flexible, tapered coil formed by 11 wires that are tightly wound on the proximal end to provide stiffness, torque and pushability and then taper to allow for a lower profile and flexibility at the distal end. They both have a durable, lubricious hydrophilic coating on the distal 60 cm. The coil support extends to 0.5 mm from the tip, improving the tip support and reducing the risk of tip separation, at the cost of lower tip flexibility.

The Mamba has a higher distal crossing profile (0.032 in., 0.81 mm) and is designed to provide strong guidewire support for antegrade crossing. It is only available in 135 cm length. The Mamba Flex has a lower distal crossing profile (0.028 in., 0.71 mm) and can be used for both retrograde and antegrade crossing, especially through areas of tortuosity. The Mamba Flex is available in both 135 and 150 cm length.

30.6.1.5 M-cath

The M-catheter (Acrostak, Fig. 30.51) is another microcatheter with low crossing profile and is currently only available in 135 cm length. Its tip is resistant to deformation, hence it is particularly suitable for heavily calcified occlusions.

30.6.1.6 Nhancer ProX

The Nhancer ProX catheter (IMDS, Fig. 30.52) is available in 135 and 155 cm lengths, and has a soft, tapered tip and tip to hub variable braid.

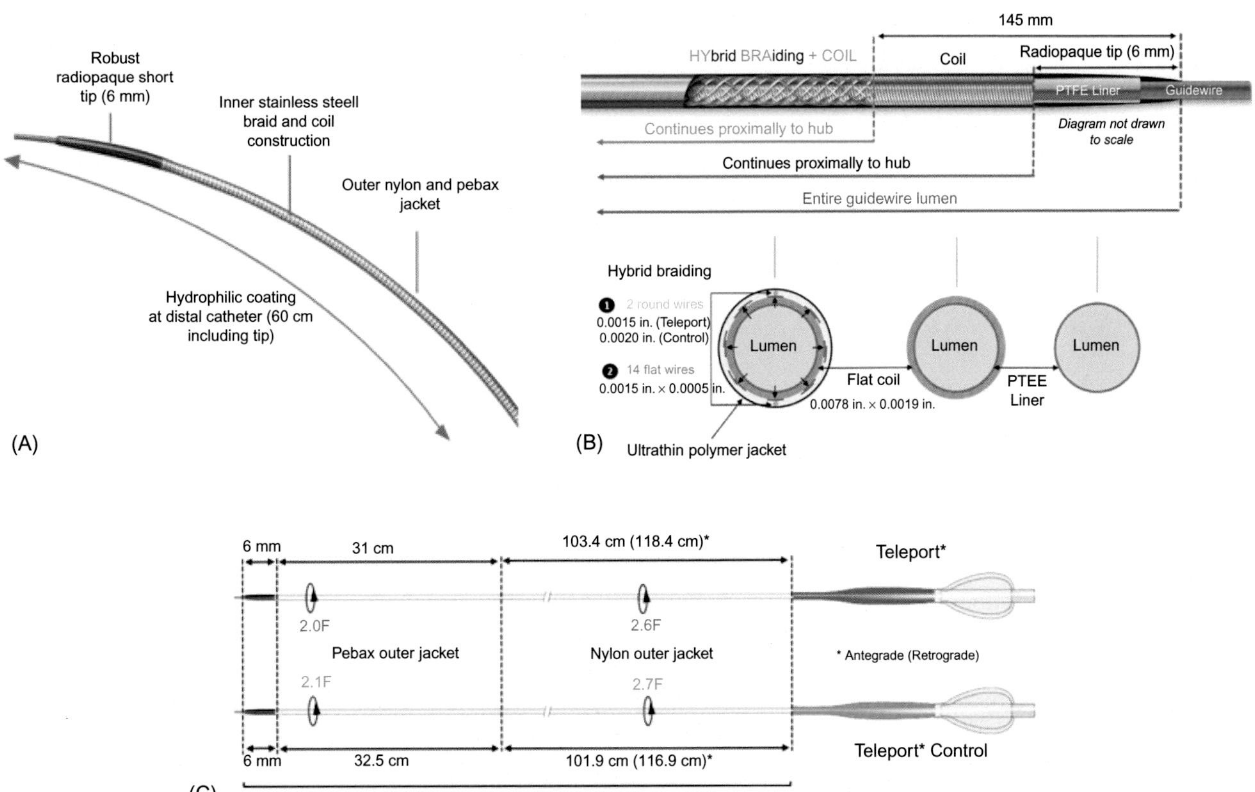

FIGURE 30.49 Illustration of the Teleport microcatheters construction. (Panel A) Main characteristics of the Teleport microcatheter. (Panel B) Construction of the Teleport microcatheter in the shaft and the tip. (Panel C) Comparison of the Teleport and the Teleport Control microcatheters. *Image used with permission of OrbusNeich Medical.*

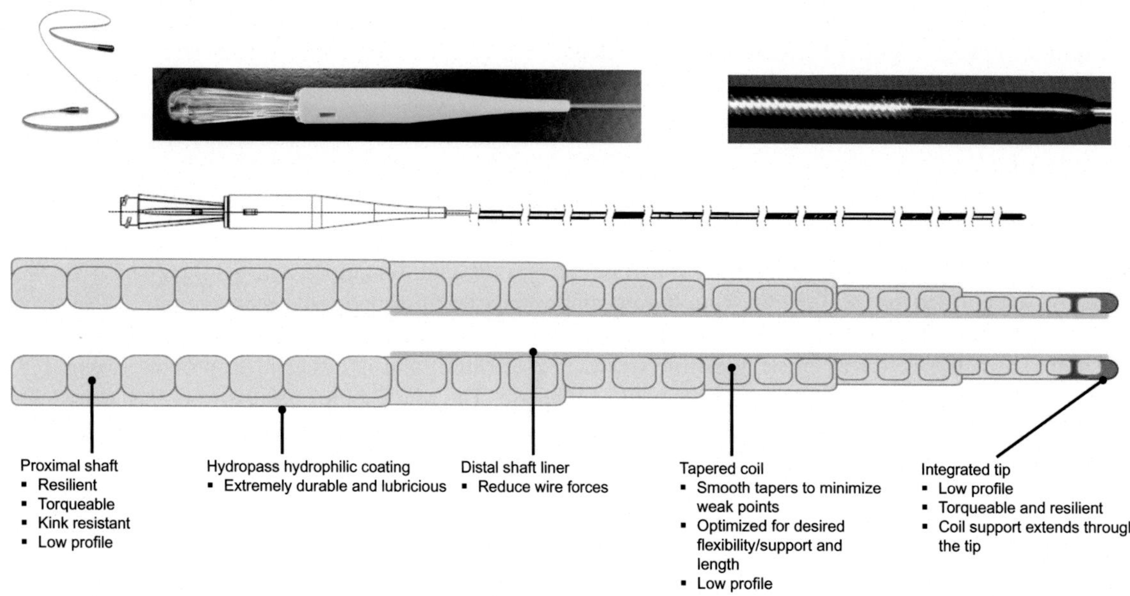

FIGURE 30.50 Illustration of the Mamba and Mamba Flex microcatheters. *Image courtesy Boston Scientific © 2019 Boston Scientific Corporation or its affiliates. All rights reserved.*

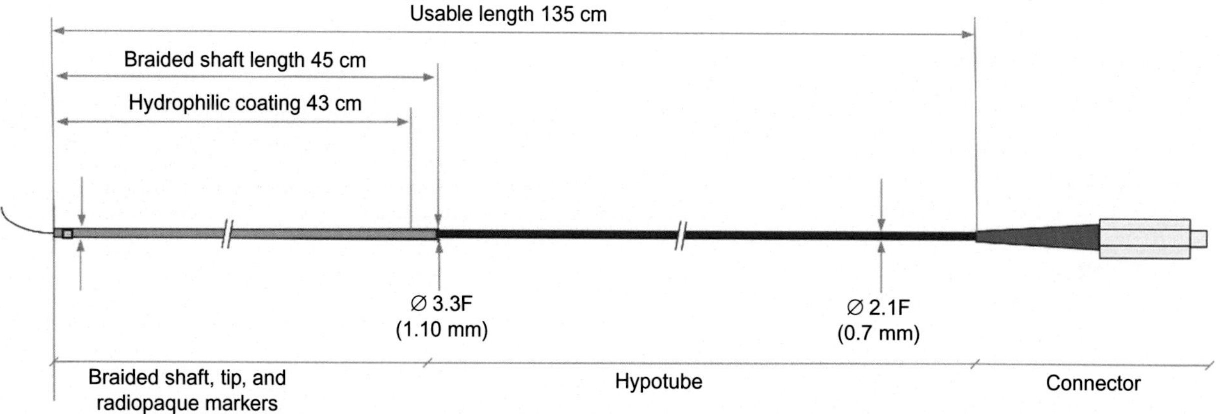

FIGURE 30.51 Illustration of the M-catheter. *Image used with permission of Acrostak.*

30.6.1.7 Mizuki and Mizuki FX

The Mizuki catheter (Kaneka, Fig. 30.53) is available in two versions, one with standard tip stiffness (Mizuki) and one with a more flexible tip (Mizuki FX). It has a hydrophilic coating and a fluoro-resin inner surface for lubricity.

30.6.2 Small outer diameter microcatheters

Small outer diameter microcatheters are easier to deliver through small, tortuous vessels and are the microcatheters of choice for retrograde chronic total occlusion interventions through epicardial collaterals.

Small outer diameter microcatheters can be subdivided into those with coils in their wall (smaller versions of the large outer diameter microcatheters Turnpike, Corsair Pro, Teleport Control, and Mamba) that can be spun for advancement and those without coils (Caravel, Finecross, MicroCross), which have smaller outer diameter but are not designed to be spun.

30.6.2.1 Turnpike LP

The Turnpike LP (low profile) (Fig. 30.48) is similar to the Turnpike catheter, but has a lower profile.

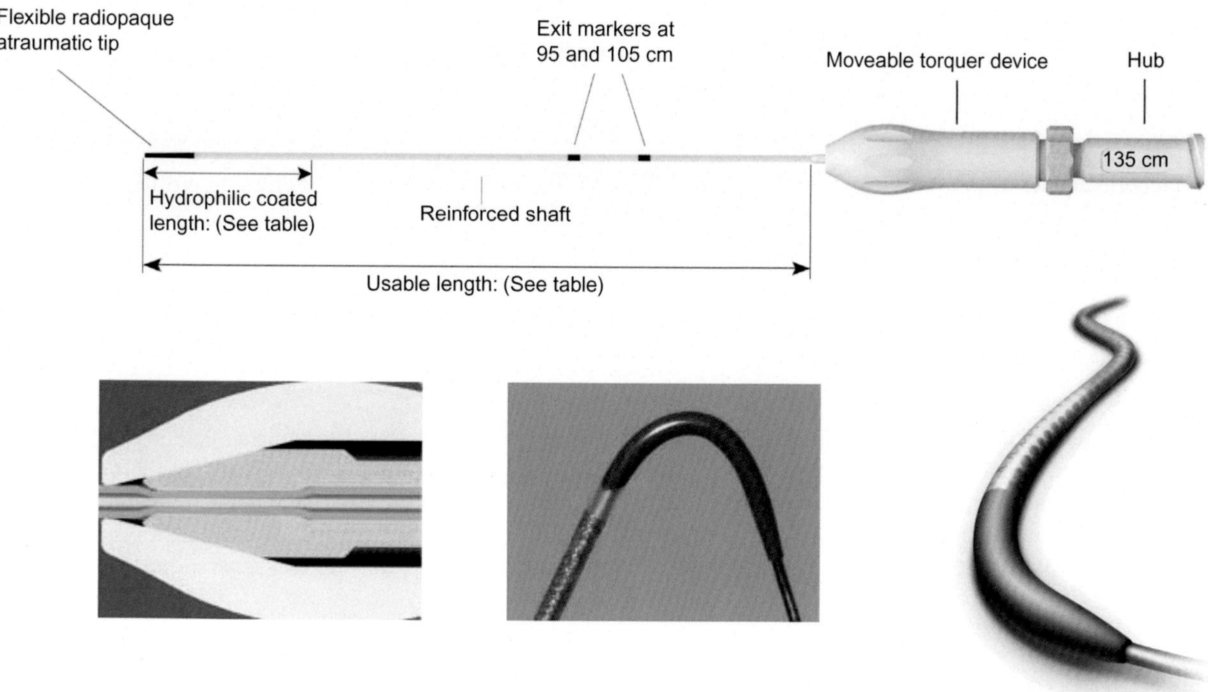

FIGURE 30.52 Illustration of the NHancer Pro X microcatheter. *Reproduced with permission from IMDS.*

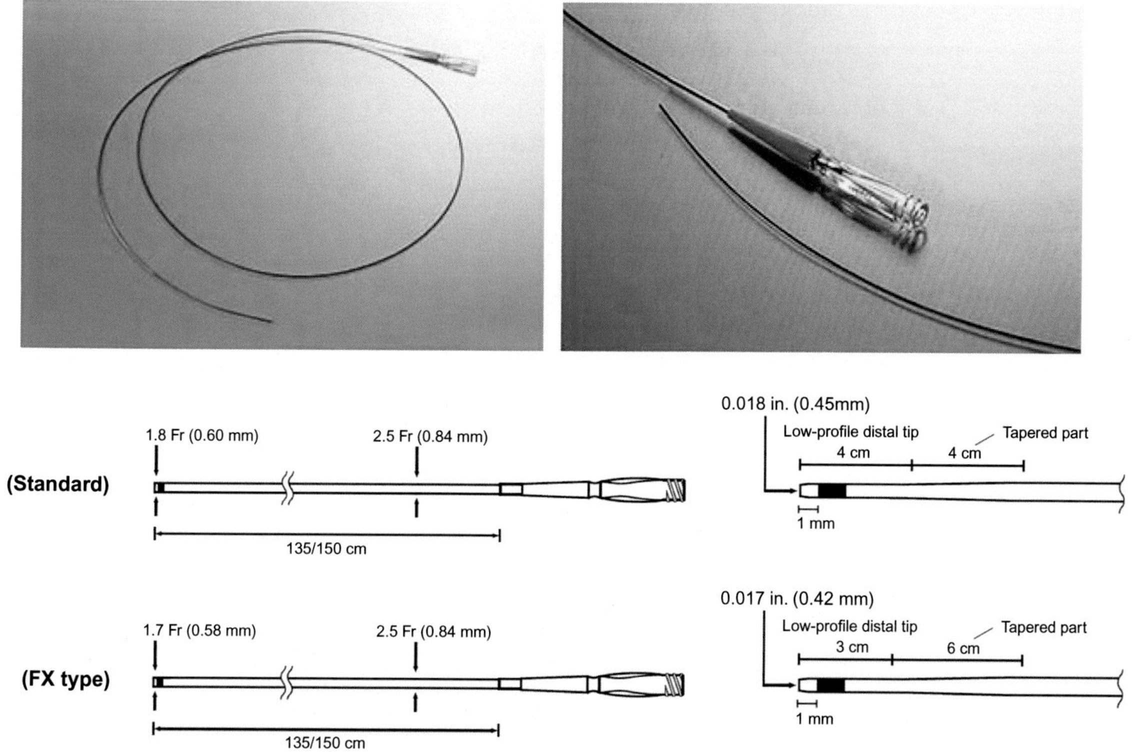

FIGURE 30.53 Illustration of the Mizuki and Mizuki FX microcatheters. *Reproduced with permission from Kaneka.*

30.6.2.2 Corsair Pro XS

The Corsair Pro XS (extra small) (Fig. 30.54) is similar to the Corsair Pro catheter, but has a lower profile.

30.6.2.3 Teleport

The Teleport catheter is similar to the Teleport Control (Fig. 30.49), but has a lower distal crossing profile.

30.6.2.4 Mamba Flex

The Mamba Flex is similar to the Mamba (Fig. 30.50), but has a lower distal crossing profile.

30.6.2.5 Caravel

The Caravel microcatheter (Fig. 30.55) was developed to advance through small and tortuous collaterals. It has a very low distal tip profile (1.4 Fr), and low distal shaft profile (1.9 Fr) with a hydrophilic coating. It also has a braided shaft. It was designed to advance with forward push but could also be gently rotated to cross challenging collaterals.

The Caravel, however, was not designed to withstand aggressive rotation and advancement. Such an approach can strain the distal tip connection to the shaft of the microcatheter and result in tip fracture and separation (Fig. 30.56) [8] (video: *CTO PCI Manual* case 87).

30.6.2.6 Finecross

The Finecross (Terumo) microcatheter is very flexible and has a low crossing profile (1.8 Fr distal tip). It has a stainless steel braid (to enhance torquability) and a distal marker located 0.7 mm from the tip (Fig. 30.57). Although the Finecross is mainly advanced using forward push, many operators are using a combination of push and rotation to facilitate advancement. Moreover, the Finecross can be used to deliver 0.018 in. coils (such as Tornado coils, Cook Medical) to treat perforations.

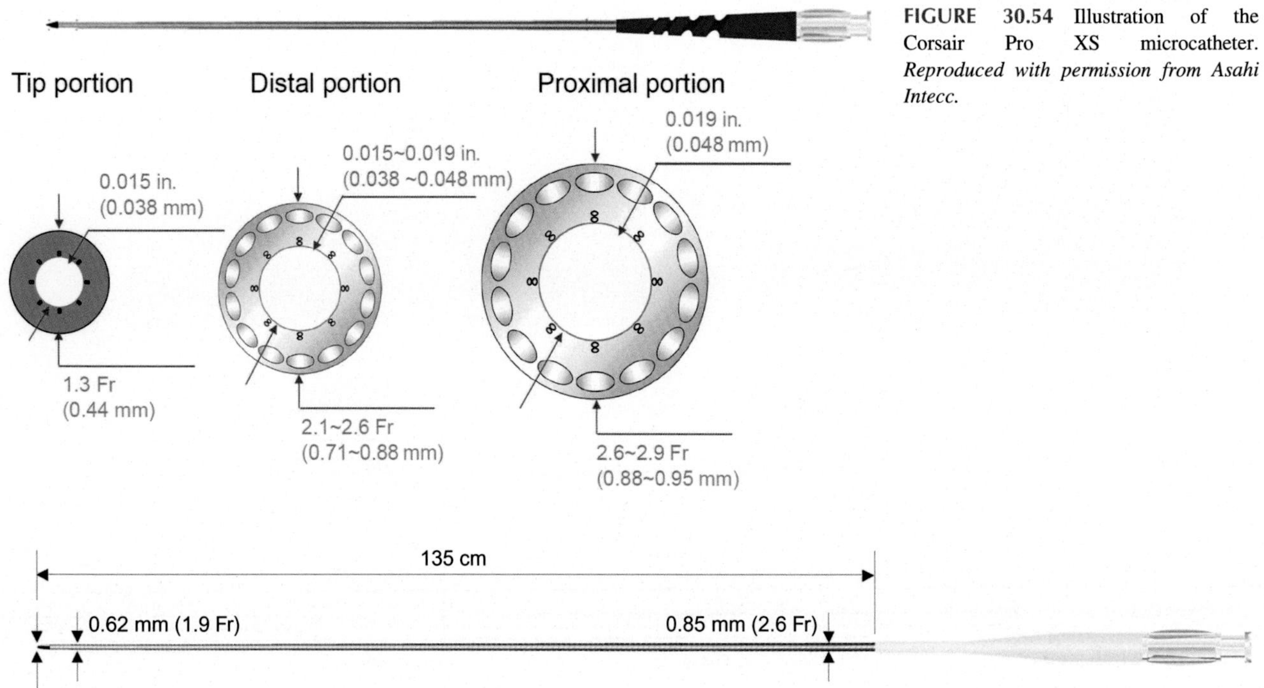

FIGURE 30.54 Illustration of the Corsair Pro XS microcatheter. *Reproduced with permission from Asahi Intecc.*

FIGURE 30.55 Illustration of the Caravel microcatheter. *Reproduced with permission from Asahi Intecc.*

30.6.2.7 Microcross

The MicroCross 14 (Boston Scientific) microcatheter is the longest (155 cm long) and has low distal crossing profile (1.6 Fr distal tip) (Fig. 30.58). It has variable pitch braid, and a hydrophilic coating. It is available in two versions, Micro 14 and Micro 14 es (extra support). Micro 14 is more flexible for advancing through tortuosity or for retrograde crossing, whereas the Micro 14 es is designed for enhanced antegrade crossing. Rotation of the catheter is possible, but the catheter is designed to advance mainly by pushing.

30.6.3 Angulated microcatheters

30.6.3.1 Venture

CTO PCI Manual Online cases: 48, 96, 97, 127.

PCI Manual Online cases: 57.

The Venture catheter (Teleflex, Fig. 30.59 has an 8 mm radiopaque torquable distal tip that has a bend radius of 2.5 mm [43–48]. The tip can be deflected up to 90° by clockwise rotation of a thumb wheel on the external handle. With rotation of the entire catheter, steering in all planes is possible. It is compatible with 6 French guiding catheters and with 0.014 in. guidewires. Both a rapid exchange and an over-the-wire catheter are available, but the over-the-wire Venture catheter should be used for CTO PCI, as it allows wire exchanges.

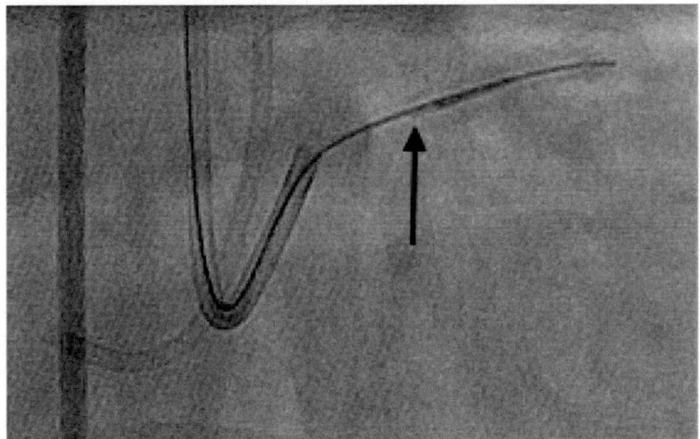

FIGURE 30.56 The tip of the Caravel microcatheter broke off from the remainder of the body of the catheter while attempting torquing through a calcified stenotic segment of the left anterior descending artery. *Courtesy of Dr. William Nicholson. Reproduced with permission from Brilakis ES. Manual of coronary chronic total occlusion interventions. A step-by-step approach. 2nd ed. Cambridge, MA: Elsevier; 2017 (Figure 2.23).*

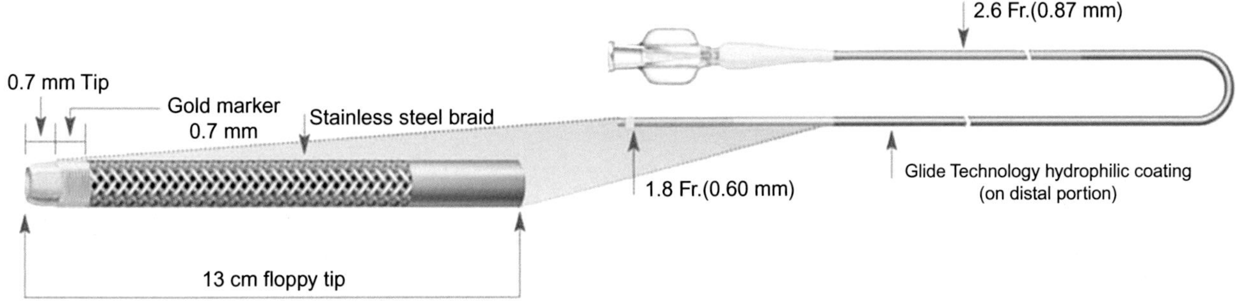

FIGURE 30.57 Illustration of the Finecross microcatheter. *©2020 Terumo Medical Corporation. All rights reserved.*

FIGURE 30.58 Illustration of the MicroCross microcatheter. *Image courtesy Boston Scientific © 2019 Boston Scientific Corporation or its affiliates. All rights reserved.*

***Venture tips and tricks*:**

1. The Venture catheter has a deflectable tip, which can be utilized to assist with accessing difficult side branch vessels. As shown in Fig. 30.60, the catheter design allows the operator to rotate the tip deflector twist knob to transmit increasing tip deflection to the distal tip of the catheter.
2. Usually the Venture catheter is delivered to the target vessel in a straight configuration over a workhorse guidewire (Fig. 30.61, panels A and B). Once it reaches the target coronary segment, the workhorse guidewire is withdrawn inside the Venture catheter (Fig. 30.61, panel C) and the tip deflector twist knob is rotated clockwise to deflect the catheter tip. The deflected catheter is rotated and withdrawn until it points to the desired direction (Fig. 30.61, panel D), followed by guidewire advancement into the lesion (Fig. 30.61, panel E). The Venture catheter can then be straightened and removed, leaving the guidewire in place (Fig. 30.61, panel F).
3. A classic example of Venture catheter use is for crossing ostial circumflex CTOs (Fig. 30.62) [45].
4. The Venture catheter can also prevent the guidewire from prolapsing into a side branch [48].
5. During retrograde CTO PCI, the Venture catheter can be used to enable wiring of collateral branches with difficult, acutely angulated takeoffs.
6. Removal of the Venture catheter using a "trapping balloon technique" requires 8 French guide catheters, given its larger profile as compared with an over-the-wire balloon or other microcatheters [48]. For the same reason an 8 French guide catheter is needed to perform the parallel-wire technique, when one of the wires is inserted through the Venture catheter.
7. The Venture catheter is stiff, which can be both an advantage and disadvantage, as it can provide extra support, but can also predispose to vessel injury. Since the bend radius is 2.5 mm, special care must be exercised when deflecting the tip in <2.5 mm diameter arteries.
8. The Venture catheter bend should be released, and the tip straightened during advancement or removal to prevent vessel damage.

30.6.3.2 SuperCross

CTO PCI Manual Online cases: 22, 83, 91, 110, 111, 134, 135.
PCI Manual Online cases: 41, 62.

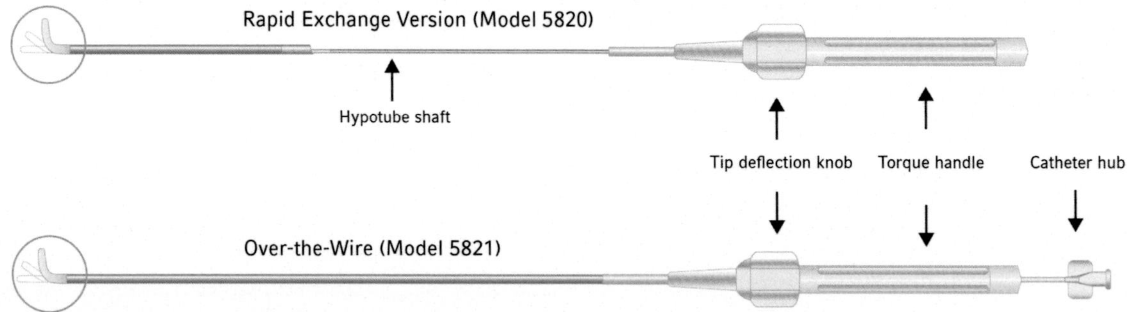

FIGURE 30.59 Illustration of the Venture microcatheter.

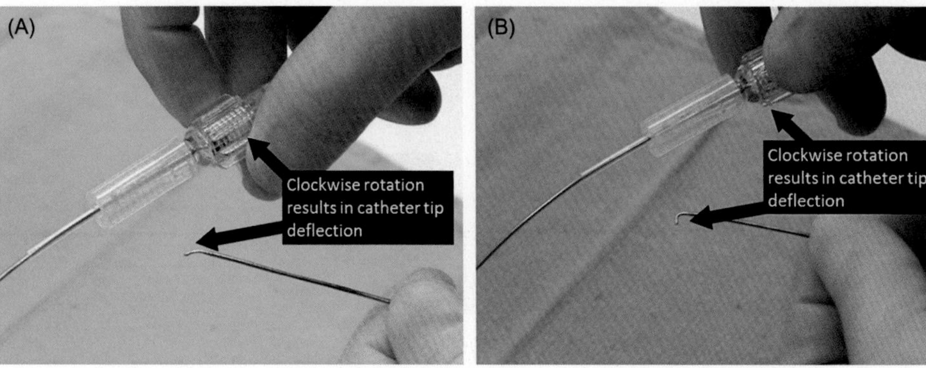

FIGURE 30.60 Illustration of Venture catheter manipulation. *Courtesy of Dr. William Nicholson. Reproduced with permission from Brilakis ES. Manual of coronary chronic total occlusion interventions. A step-by-step approach. 2nd ed. Cambridge, MA: Elsevier; 2017 (Figure 2.32).*

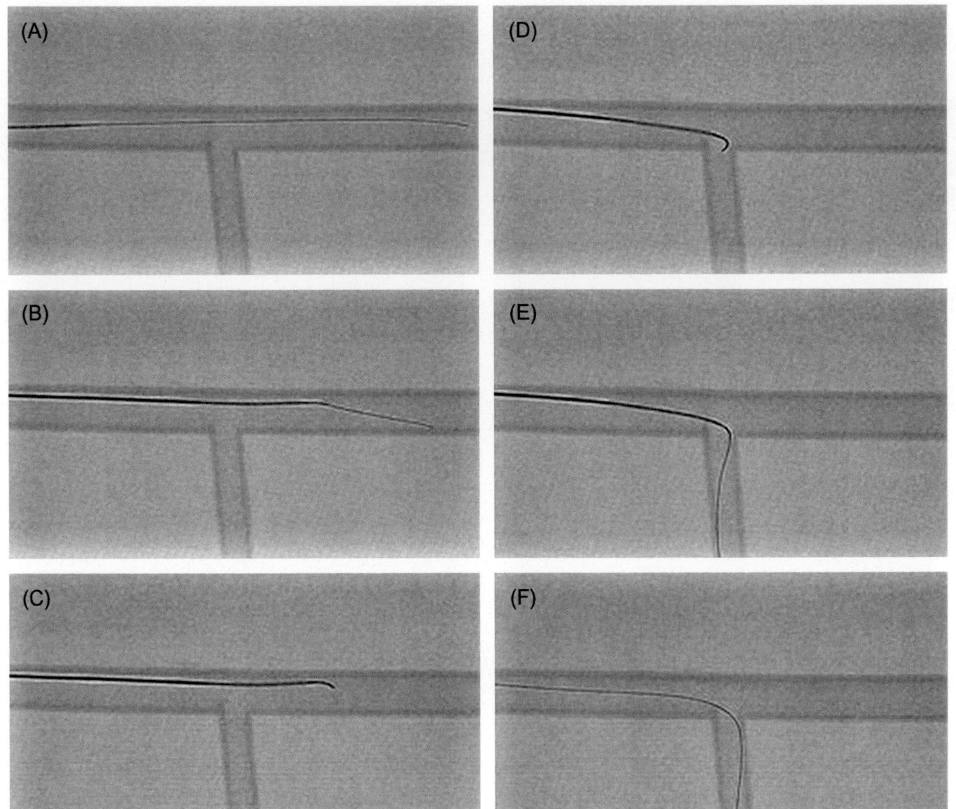

FIGURE 30.61 How to use the Venture catheter. *Courtesy of Dr. William Nicholson. Reproduced with permission from Brilakis ES. Manual of coronary chronic total occlusion interventions. A step-by-step approach. 2nd ed. Cambridge, MA: Elsevier; 2017 (Figure 2.33).*

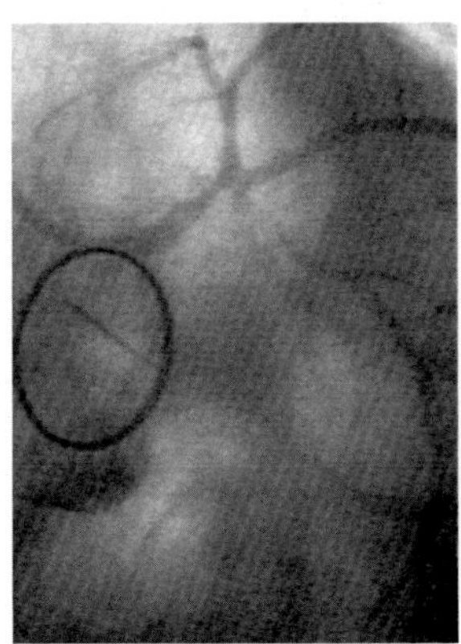
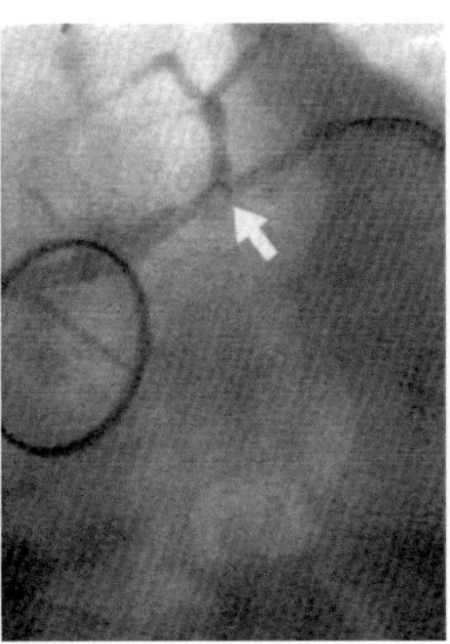
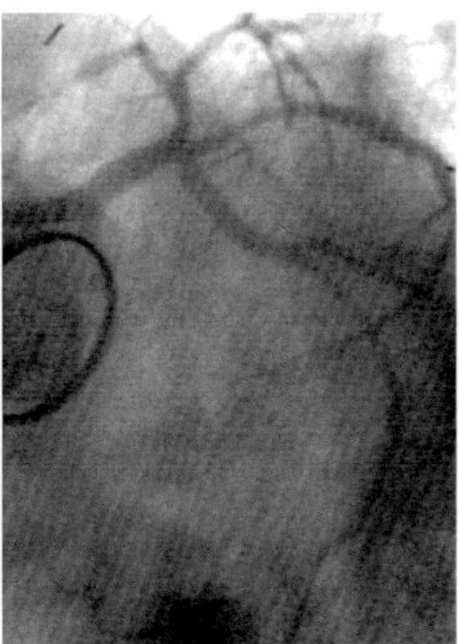

FIGURE 30.62 Example of Venture catheter use to cross an ostial circumflex CTO. *Reproduced with permission from McNulty E, Cohen J, Chou T, Shunk K. A "grapple hook" technique using a deflectable tip catheter to facilitate complex proximal circumflex interventions. Catheter Cardiovasc Interv 2006;67:46–8.*

The SuperCross (Fig. 30.63) (Teleflex) is a microcatheter that is manufactured with either a straight tip or with various distal tip angulations (45°, 90°, and 120°) and can facilitate guidewire advancement through areas of tortuosity. The SuperCross microcatheters can be helpful in scenarios similar to those in which the Venture Catheter is employed. The 120° bend can be very helpful in retrograde CTO cases with reverse angled collaterals or through saphenous vein grafts, in which the retrograde wire is biased to advance toward the distal vessel. Similarly, the SuperCross 120° bend microcatheter can be very useful when trying to treat a lesion upstream of a saphenous vein graft distal anastomosis.

30.6.3.3 Swift Ninja

The Swift Ninja (Fig. 30.64) (Merit Medical) is straight tip catheter that articulates up to 180° in opposing directions.

(A)

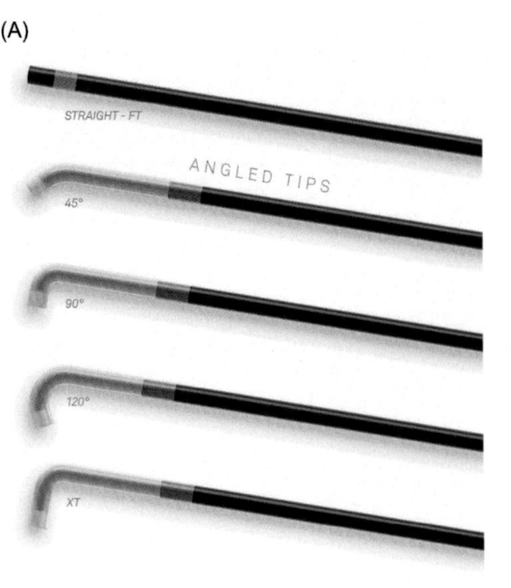

(B)

FIGURE 30.63 Illustration of the SuperCross microcatheter. (Panel A) Various types of SuperCross angulation. (Panel B) Angiographic appearance of the SuperCross microcatheter.

FIGURE 30.64 Illustration of the Swift Ninja microcatheter. *Courtesy Merit Medical.*

30.6.4 Dual lumen microcatheters

CTO PCI Manual Online cases: 11, 55, 88, 105, 106, 108, 111, 121, 122, 124, 137.
PCI Manual Online cases: 19, 41, 52.

The Twin Pass (Teleflex, Fig. 30.65), Twin Pass Torque (Teleflex, Fig. 30.65) and Sasuke (Asahi Intecc, Fig. 30.66) dual lumen microcatheters are currently available in the United States.

The Twin Pass torque has a braided shaft that facilitates torquing and positioning of the over-the-wire port, which has a 10° exit angle and is located 1.5 mm distal to the proximal marker.

The Sasuke microcatheter (Asahi Intecc, Fig. 30.66) is a dual lumen microcatheter that has the same tapered soft tip used in the Corsair and Caravel microcatheters and an oval design with a double stainless steel core in the proximal shaft to ensure kink resistance. The over-the-wire exit port is located 6.5 mm from the tip and is highly visible facilitating guidewire manipulations. It is 145 cm long with a hydrophilic coating in the most distal 38 cm.

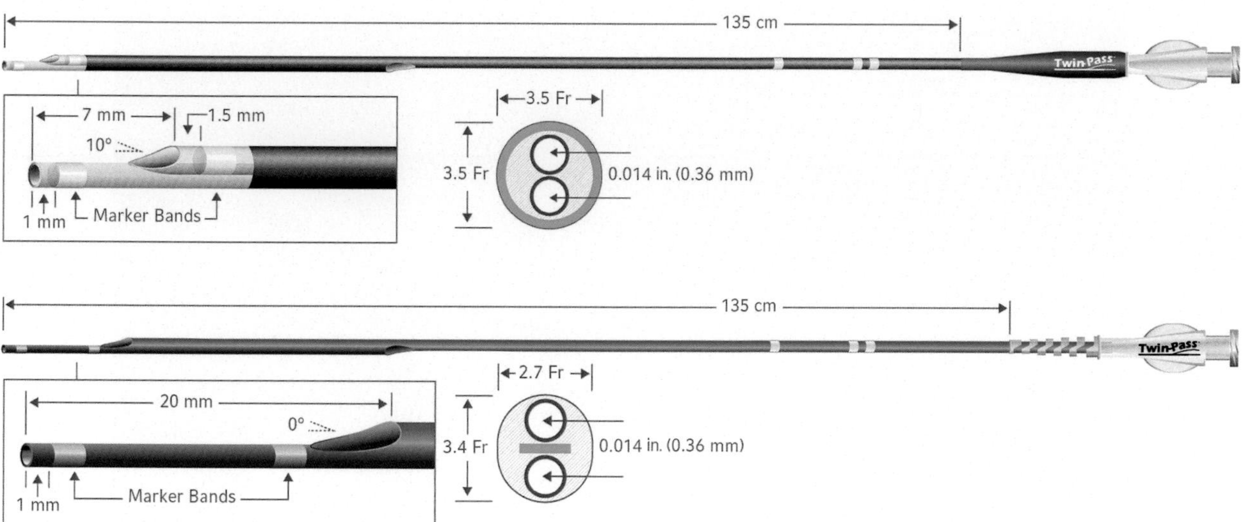

FIGURE 30.65 Illustration of the TwinPass Torque (top panel) and TwinPass (bottom panel) dual lumen microcatheters.

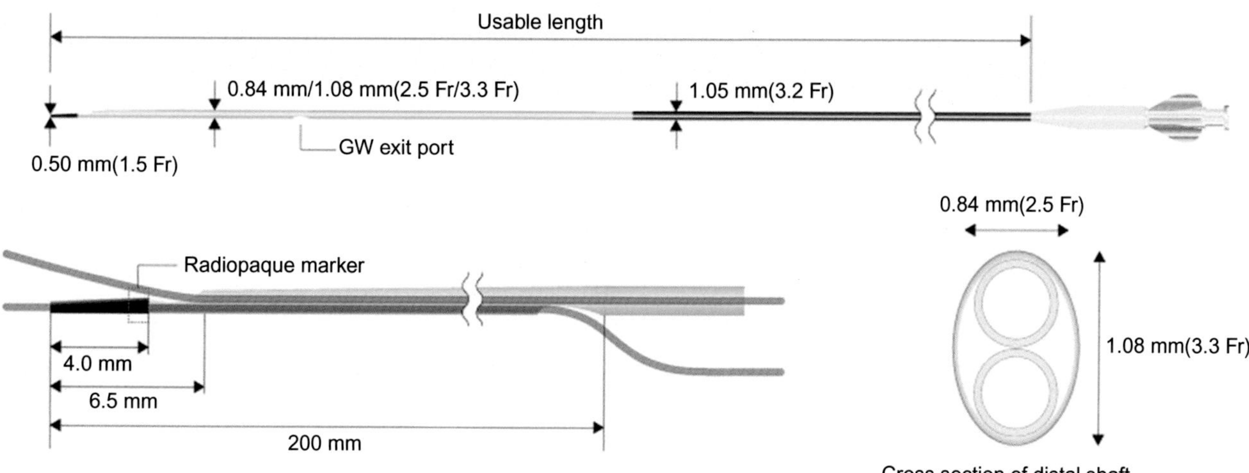

FIGURE 30.66 Illustration of the Sasuke dual lumen microcatheter. *Reproduced with permission from Asahi Intecc.*

Four other dual lumen microcatheters are available outside the United States: Crusade (Kaneka, Fig. 30.67), FineDuo (Terumo, Fig. 30.68), NHancer Rx (IMDS, Fig. 30.69), and ReCross (IMDS, Fig. 30.70).

The NHancer Rx (IMDS, Roden, The Netherlands, Fig. 30.69) is a dual lumen microcatheter with an over-the-wire lumen and a rapid exchange lumen. The NHancer Rx has a removable stylet in the over-the-wire lumen that increases pushability. The distal shaft design is oval with a small dual lumen crossing profile (2.3 Fr) that allows catheter trapping using a conventional balloon in a 6 French guiding catheter. The distal tip and over-the-wire lumen exit port are designed with soft tungsten filled material allowing visibility of the exit port on X-ray. Shaft braiding reinforcement improves the torque and lumen integrity performance.

The ReCross microcatheter (IMDS, Fig. 30.70) is the only dual lumen microcatheter with two over-the-wire lumens instead of a combination of a rapid exchange lumen and an over-the-wire lumen. As a result, guidewire exchange is feasible through both lumens. Another unique feature of the ReCross is that the over-the-wire lumen that leads to the tip-exit has a second exit port located 12 mm proximal from the tip. This exit port is oriented 180° opposite from the exit port of the second OTW-lumen that is located 8 mm from the tip. As the shape of the ReCross is oval it will tend to turn with its broader side to the coronary artery wall subintimally similar to a Stingray balloon, hence one of the opposite proximal exit ports may allow puncture for distal reentry.

All dual lumen microcatheters except the ReCross, consist of a rapid exchange delivery system in the distal segment together with an over-the-wire lumen that runs the length of the catheter. A radiopaque marker band identifies the distal

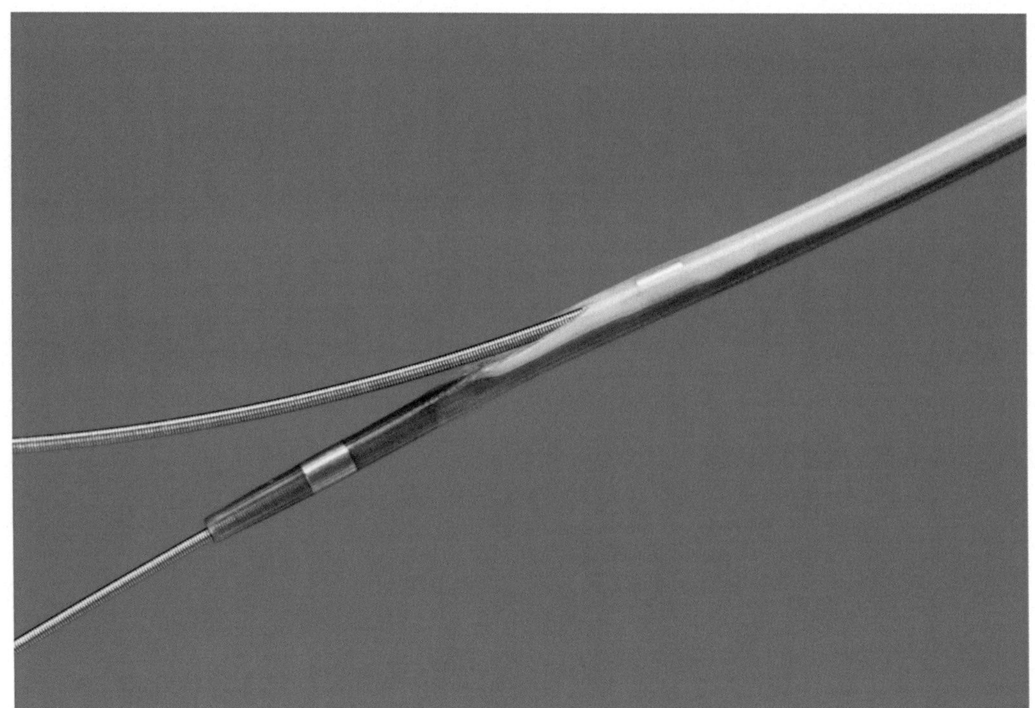

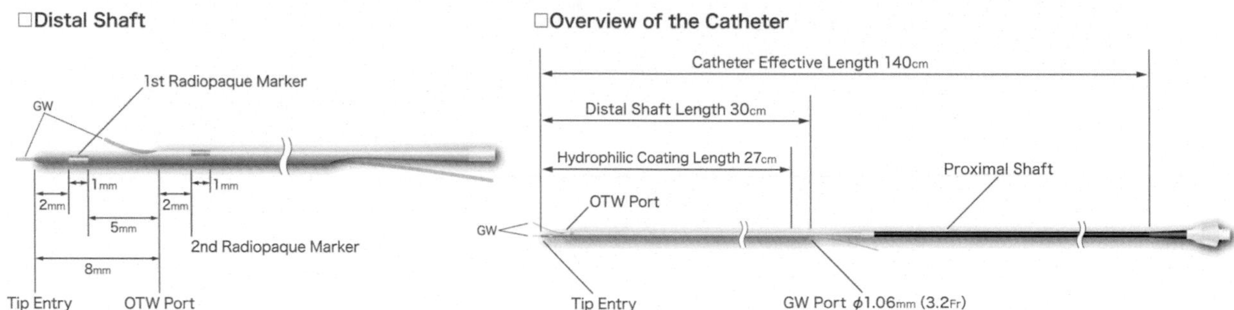

FIGURE 30.67 Illustration of the Crusade dual lumen microcatheter. *Reproduced with permission from Kaneka.*

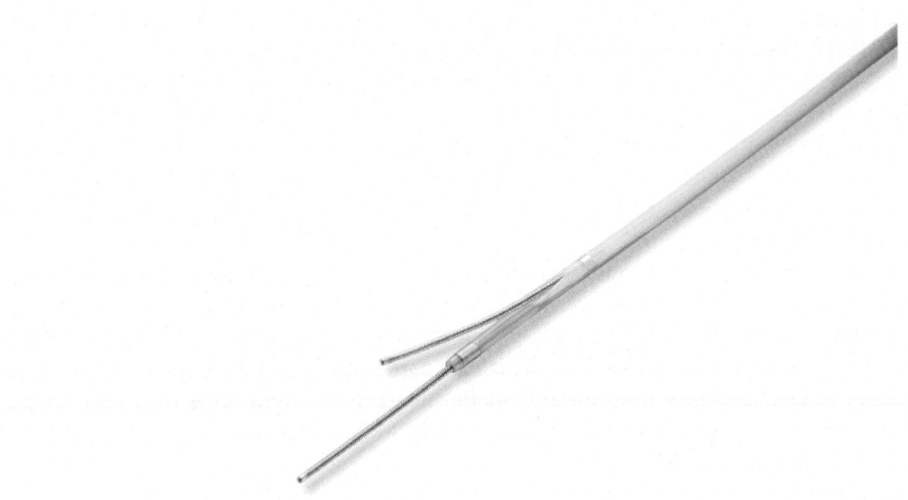

FIGURE 30.68 FineDuo® dual lumen microcatheter. *©2020 Terumo Medical Corporation. All rights reserved.*

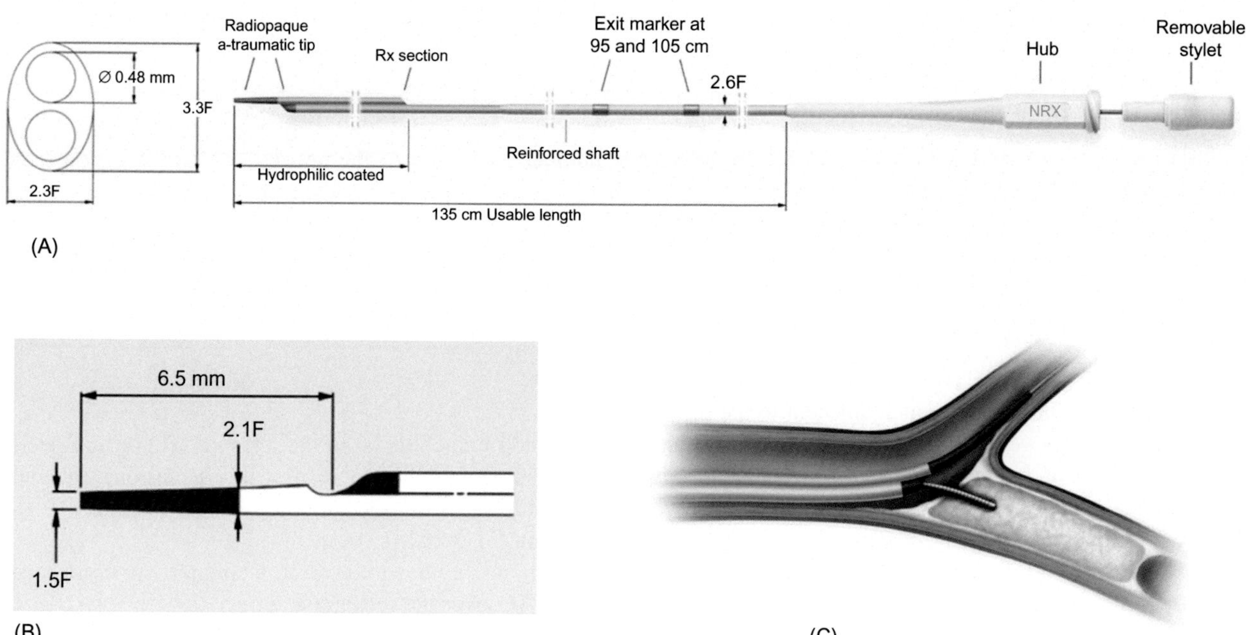

FIGURE 30.69 Illustration of the NHancer Rx dual lumen microcatheter. (Panel A) NHancer Rx microcatheter design. (Panel B) NHancer Rx microcatheter distal shaft design. (Panel C) Illustration of the NHancer Rx use in a CTO located at a bifurcation. *Reproduced with permission from IMDS.*

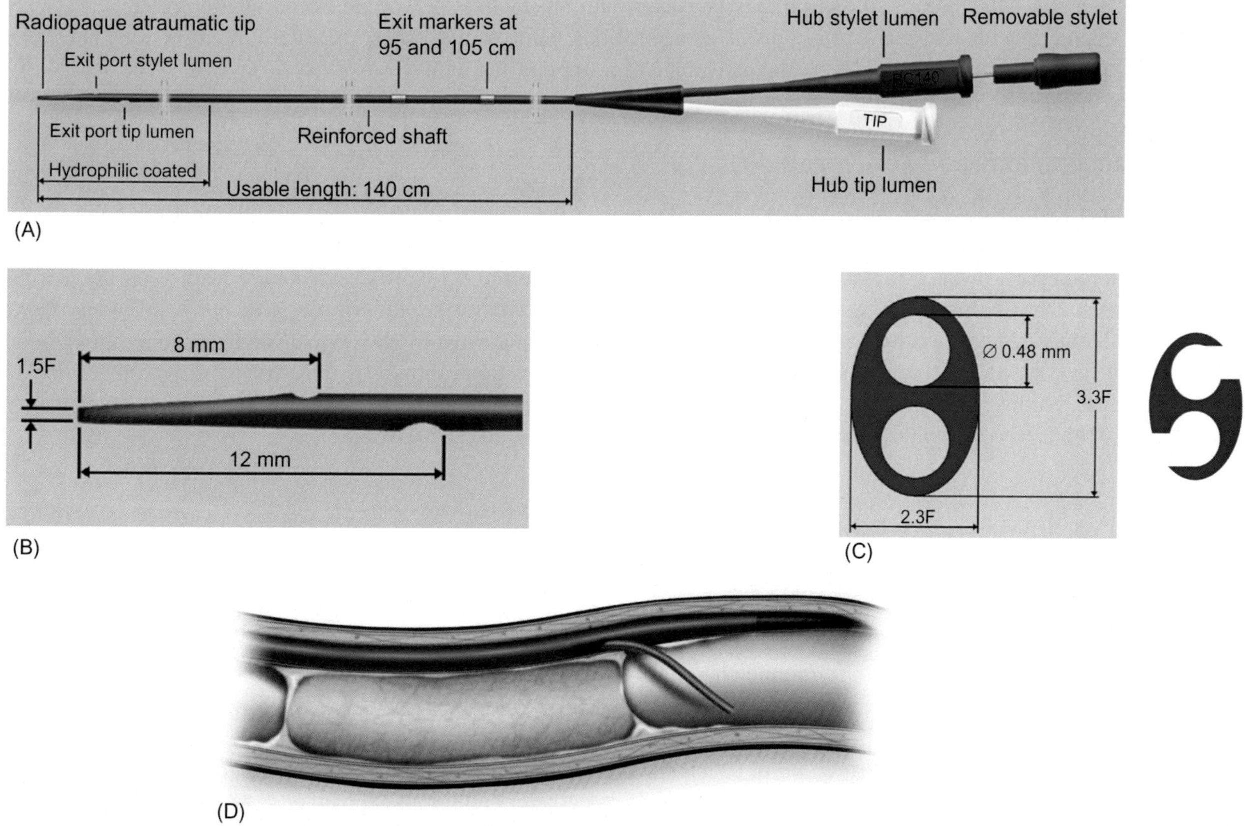

FIGURE 30.70 Illustration of the ReCross dual lumen microcatheter. (Panel A) ReCross microcatheter design. (Panel B) ReCross microcatheter distal shaft design. (Panel C) ReCross microcatheter transverse view of the shaft and the exit port positioning. (Panel D) Illustration of use of the ReCross microcatheter for reentry into the distal true lumen. *Reproduced with permission from IMDS.*

tip of each lumen; the distal band corresponds to the exit point of the rapid exchange segment and the proximal band marks the exit point of the over-the-wire lumen.

Dual lumen microcatheters have multiple uses in CTO and non-CTO PCI (Fig. 30.71):

Non-CTO PCI:

1. Inserting a second guidewire after successfully advancing another guidewire to the desired coronary location (e.g., after successfully crossing the target lesion with a standard 0.014 in. guidewire a dual lumen microcatheter can be used to insert a Rotafloppy wire across the target lesion).
2. Wiring the side branch of a bifurcation, including wiring through jailed side branches during bifurcation stenting.
3. Facilitate the reversed guidewire (otherwise called "hairpin" guidewire) technique (Section 8.6) [49–51].

CTO PCI:

1. Parallel wiring during CTO PCI.
2. Wiring CTOs with a side branch adjacent to the proximal cap.
3. Wiring the distal vessel if guidewire crossing is achieved into a side branch distal to the lesion (CTO or non-occlusive): instead of pulling back and redirecting the guidewire (risking inability to recross the occlusion), the dual lumen microcatheter enables wiring the main distal vessel without losing access to the side branch.
4. Wire various septal branches during retrograde crossing attempts in CTO PCI.
5. Antegrade wiring of the distal true lumen if the externalized retrograde guidewire crossed a collateral in close proximity to the distal cap, precluding safe antegrade dilation of the CTO over the retrograde wire.

Dual lumen microcatheters tips and tricks

1. Dual lumen microcatheters may, sometimes, be challenging to deliver. Some of them come with a stiffening mandrel (Nhancer Rx, ReCross) that, when loaded into the OTW lumen can provide support and pushability during catheter insertion.
2. Controlling the direction of the side port can be challenging and may require removing the microcatheter and reinserting it. Controlling the direction may be easier with the Twin Pass Torque microcatheter.
3. Once wiring is achieved, maintaining distal wire positions while withdrawing the dual lumen microcatheter can be difficult and is best achieved with the assistance of a trapping balloon.

30.6.5 Plaque modification microcatheters

CTO PCI Manual Online cases: 5, 73. The Tornus and Tornus 88Flex microcatheters (Fig. 30.72) consist of eight stainless steel wires stranded in a coil [52]. The Tornus is available in two sizes (2.1 and 2.6 Fr), with the latter providing more guidewire support. It has a platinum marker located 1 mm from the tip. Unlike other microcatheters, the Tornus does not have an inner polymer and as a result injection of contrast cannot be done through it, but it provides strong support. The Tornus is advanced by counterclockwise rotation and withdrawn by clockwise rotation. To avoid catheter kinking and unraveling of the stranded steel wires, no more than 20 rotations should be done in any direction.

Similar to the Tornus, the Turnpike Spiral and Turnpike Gold (Fig. 30.48) can be used to modify a resistant lesion. In contrast to Tornus, the Turnpike Spiral and Gold are advanced using clockwise rotation.

30.6.6 Microcatheter-related complications

1. Microcatheter overrotation could cause catheter deformation and entrapment, fracture proximal to the catheter tip (CTO PCI Manual online case 87), or result in the guidewire binding to the microcatheter (microcatheter "fatigue") (Fig. 30.73) [8] (Section 27.4).
2. Contrast can be injected through the microcatheter (except for the Tornus) for distal vessel visualization, but the catheter should subsequently be flushed to minimize the risk of guidewire "stickiness". Rarely the guidewire may get "stuck" requiring removal of both the microcatheter and guidewire.
3. If difficulty is encountered while attempting to advance a microcatheter or manipulate a guidewire after prolonged use, the cause may be "microcatheter fatigue" and the microcatheter should be exchanged for a new one.
4. During retrograde CTO PCI, after wire externalization and during antegrade gear delivery the tip of antegrade equipment (such as balloons and stents) should never come in contact with the tip of the retrograde microcatheter catheter over the same guidewire to avoid "interlocking" and equipment entrapment.

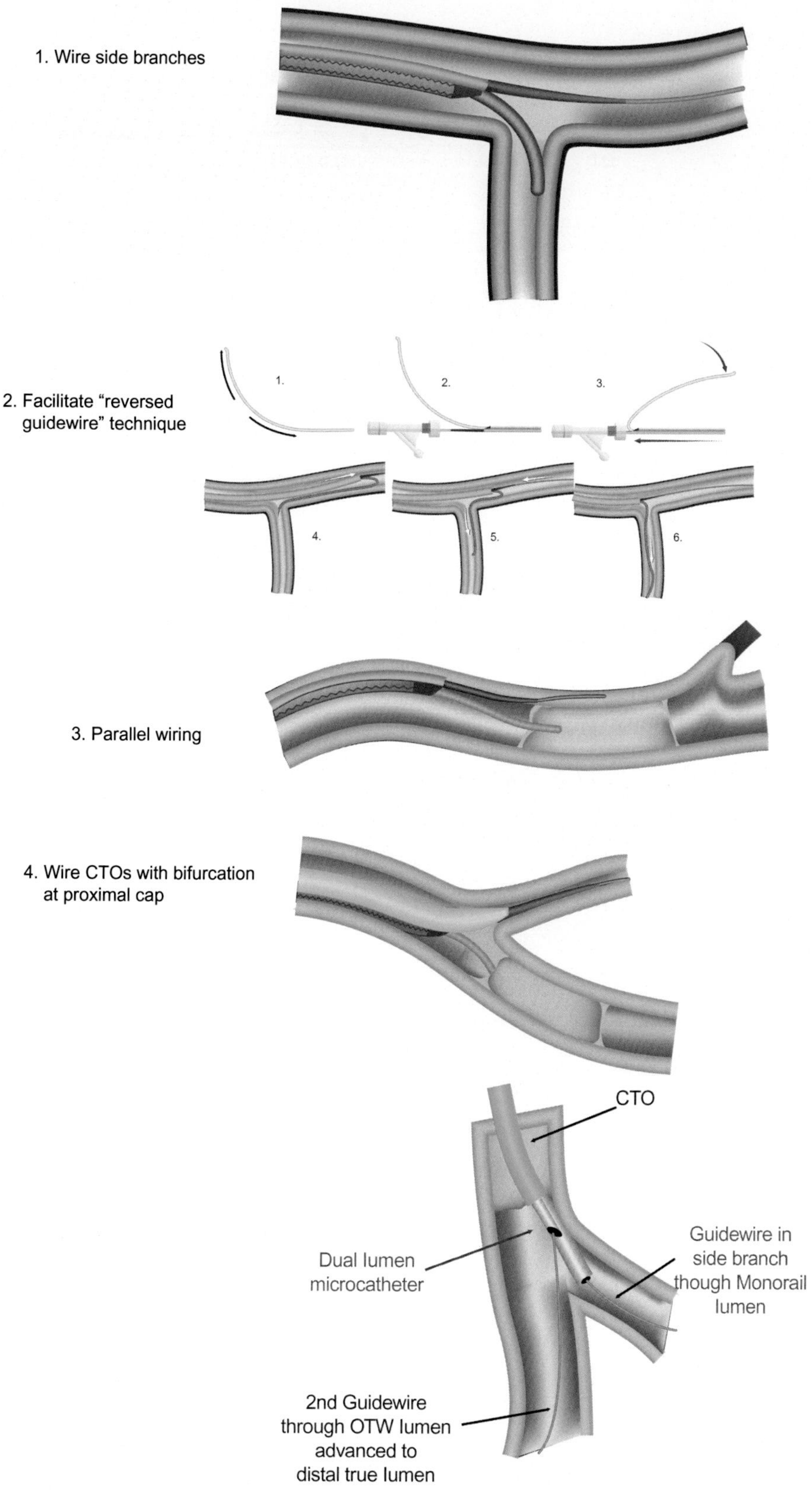

FIGURE 30.71 The various uses of the dual lumen microcatheters.

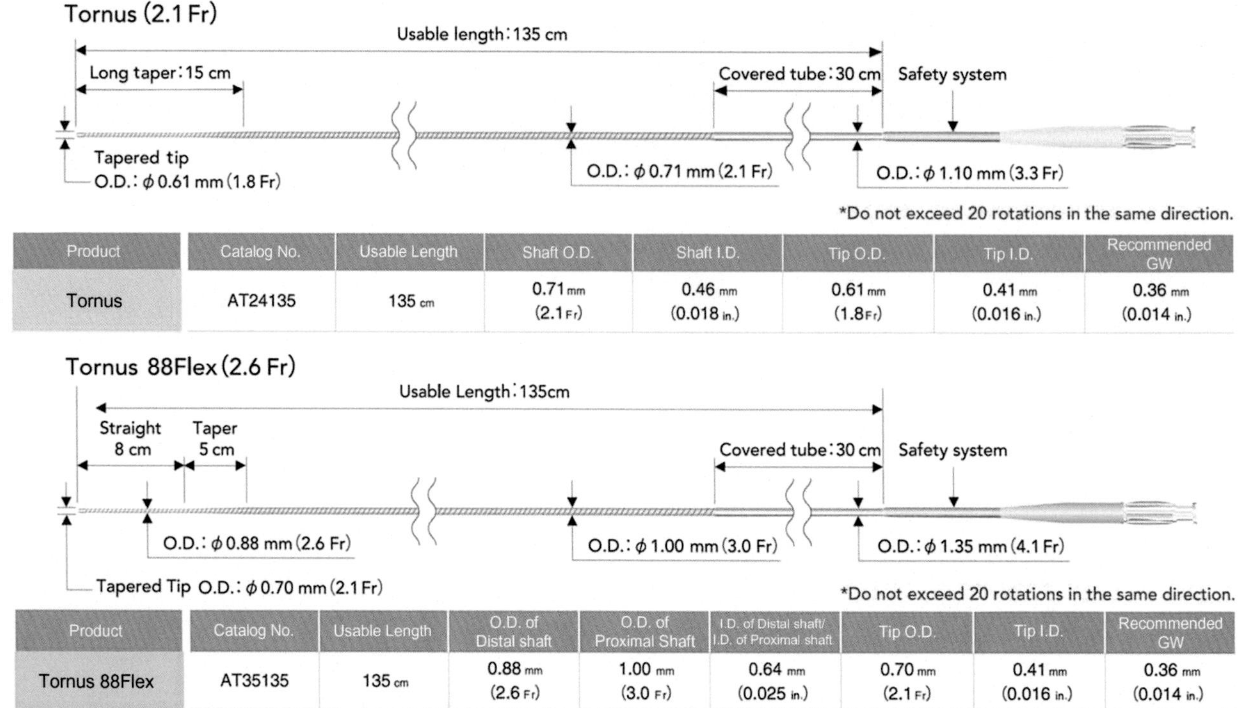

Product	Catalog No.	Usable Length	Shaft O.D.	Shaft I.D.	Tip O.D.	Tip I.D.	Recommended GW
Tornus	AT24135	135 cm	0.71 mm (2.1 Fr)	0.46 mm (0.018 in.)	0.61 mm (1.8 Fr)	0.41 mm (0.016 in.)	0.36 mm (0.014 in.)

Product	Catalog No.	Usable Length	O.D. of Distal shaft	O.D. of Proximal Shaft	I.D. of Distal shaft/ I.D. of Proximal shaft	Tip O.D.	Tip I.D.	Recommended GW
Tornus 88Flex	AT35135	135 cm	0.88 mm (2.6 Fr)	1.00 mm (3.0 Fr)	0.64 mm (0.025 in.)	0.70 mm (2.1 Fr)	0.41 mm (0.016 in.)	0.36 mm (0.014 in.)

FIGURE 30.72 Illustration of the Tornus (2.1 Fr) and Tornus 88 Flex (2.6 Fr) microcatheters. *Reproduced with permission from Asahi Intecc.*

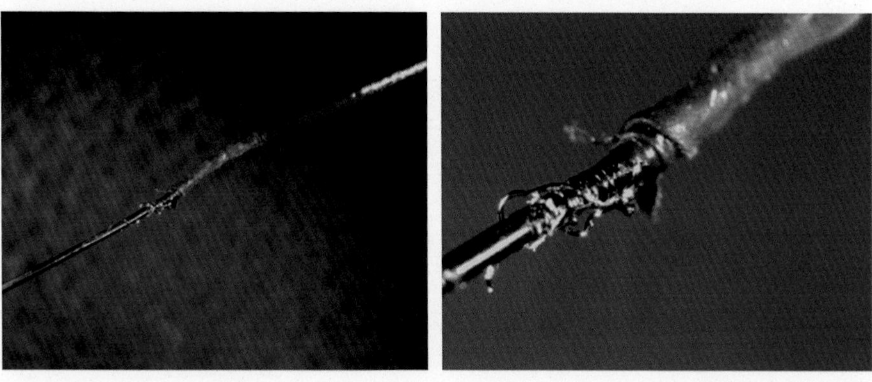

FIGURE 30.73 Images of the tip of a Corsair catheter permanently bound to a Pilot 200 guidewire with destruction of the guidewire's polymer jacket and entanglement of the guidewire coil with the tip of the *Corsair catheter. Courtesy of Dr. William Nicholson. Reproduced with permission from Brilakis ES. Manual of coronary chronic total occlusion interventions. A step-by-step approach. 2nd ed. Cambridge, MA: Elsevier; 2017 (Figure 2.21).*

30.7 Guidewires

A 0.014 in. guidewire is essential equipment for performing PCI, as discussed in Chapter 8: Wiring. A 0.035 or 0.038 in. guidewire is also essential for obtaining arterial access (Chapter 4: Access) and advancing the catheter to the coronary or bypass graft ostium (Chapter 5: Coronary and graft engagement).

There are many guidewire options and operators often have strong personal preferences. The 0.014 in. guidewires can be classified into seven categories:

1. Workhorse.
2. Polymer-jacketed.
3. Stiff.
4. Support.
5. Atherectomy.
6. Externalization.
7. Pressure guidewires.

30.7.1 Workhorse guidewires

Workhorse guidewires (Table 30.7) are used for most non-CTO PCIs. The key characteristics of workhorse guidewires are:

1. Soft tip (usually 1 g or less tip load), and
2. No polymer jacket (in contrast to polymer-jacketed guidewires, discussed in Section 30.7.2).

These characteristics of the workhorse guidewires reduce the risk of vessel injury (dissection and perforation), during wiring and equipment delivery. If more aggressive guidewires (such as polymer jacketed or stiff guidewires) are needed to wire a vessel, they should be exchanged for a workhorse guidewire (or a support guidewire, Section 30.8.4) prior to equipment delivery.

Other characteristics of workhorse guidewires include:

1. Hydrophilic coating. Several workhorse guidewires have a hydrophilic coating (which is different than a full polymer jacket). As anticipated, hydrophilic coating increases deliverability but increases the risk of vessel injury.
2. Material: most guidewires are made of stainless steel. Nitinol guidewires have better tip shape retention but can be harder to shape.
3. Construction of the tip: most guidewires have a wire coil around a central core. The central coil might extend to the tip of the guidewire (core to tip) or a shaping ribbon might be added to allow easier tip shaping, which is commonly used in nitinol guidewires. Some guidewires have multiple cores (such as the Sion, Sion blue, Suoh 03 [Asahi Intecc]) or have dual coils (such as the Samurai and Samurai RC [Boston Scientific]).

30.7.2 Polymer-jacketed guidewires

The tip of the polymer-jacketed guidewires is fully covered by a very slippery polymer. As a result, they facilitate wiring through tortuosity and through collateral vessels during CTO PCI. They, however, carry increased risk of complications, such as distal vessel perforation and/or subintimal tracking causing dissection. Polymer-jacketed guidewires should, therefore, only be used when necessary and be exchanged for a workhorse guidewire before delivering balloons and stents.

There are three major groups of polymer-jacketed guidewires (Fig. 30.74 and Table 30.8). Soft, nontapered guidewires are most commonly used during non-CTO PCI, whereas soft tapered and stiff polymer jacketed guidewires are usually used in CTO PCI.

30.7.3 Stiff-tip guidewires

Stiff-tip guidewires are designed for crossing resistant proximal caps during CTO PCI. Representative guidewires are the Gaia (Asahi Intecc) and Judo guidewires (Boston Scientific) which have stiff, tapered-tip with moderate penetrating power. The Confianza Pro 12 (Asahi Intecc), Hornet 14 (Boston Scientific), and Warrior (Teleflex) guidewires have stiff, tapered-tip with strong penetrating power. These guidewires are discussed in detail in the *Manual of CTO PCI* [8].

30.7.4 Support

Support guidewires have strong, supportive bodies and soft tips. They are designed to facilitate equipment delivery (or even guide catheter exchanges) (Table 30.9). Support guidewires include the BHW, Extra S'port, All Star, and Iron Man (Abbott Vascular), Sion Blue extra support and Grand Slam (Asahi Intecc) and Mailman (Boston Scientific).

The Wiggle guidewire (Fig. 30.75, Abbott Vascular) is a unique support guidewire with three deflections in the distal portion: the deflections start 6 cm from the tip and go over the next 6 cm giving three waves, each having an amplitude of 3 mm. These deflections reduce side-wall bias and center the delivery portion of the wire. The Wiggle guidewire can be especially useful for equipment delivery through previously placed stents and also for delivery through tortuosity and calcification. The Wiggle guidewire should not be used for primary wiring: instead another guidewire is used first, followed by exchange for the Wiggle guidewire over a microcatheter or over-the-wire balloon.

If a Wiggle guidewire is not available, bends can be placed manually into the distal portion of a workhorse guidewire to mimic the Wiggle guidewire construction.

TABLE 30.7 Overview of various workhorse guidewires.

Tip stiffness	Wire name	Manufacturer	Tip stiffness (g)	Hydrophilic tip coating	Tip technology	Tip radiopacity (cm)	Material	Comment
Ultra low tip stiffness	Suoh 03	Asahi Intecc	0.3	Yes—52 cm	ACT ONE	3	Stainless steel	Best wire for tortuous epicardial collaterals
Low tip stiffness								
1.	Balance middle weight (BMW)	Abbott	0.6	No	Shaping Ribbon	3	Nitinol	
2.	BMW universal	Abbott	0.6	Yes	Shaping Ribbon	3	Nitinol	
3.	BMW universal II	Abbott	0.7	Yes	Shaping Ribbon	3	Nitinol	
4.	BMW elite	Abbott	0.8	Yes	Core-to-tip	3	Nitinol	
5.	Turntrac	Abbott	0.8	Yes+ No (version with no coating in distal 1 cm)	Core-to-tip	3	Nitinol	
6.	Turntrac Flex	Abbott	0.6	Yes+ No (version with no coating in distal 1 cm)	Core-to-tip	3	Nitinol	
7.	Versaturn	Abbott	0.8	Yes	Core-to-tip	3	Nitinol	
8.	Hi-Torque Floppy II	Abbott	0.4	No	Shaping ribbon	3	Nitinol	
9.	Hi-Torque Powerturn	Abbott	0.9	Yes	Core-to-tip	3	Stainless steel	
10.	Hi-Torque Powerturn Flex	Abbott	0.9	Yes	Core-to-tip	3	Stainless steel	
11.	Hi-Torque Powerturn Ultraflex	Abbott	0.9	Yes	Core-to-tip	3	Stainless steel	
12.	Hi-Torque Advance	Abbott	1.0	Yes	Core-to-tip	3	Stainless steel	
13.	Light	Asahi Intecc	0.5	No	Core-to-tip	3	Stainless steel	
14.	Soft	Asahi Intecc	0.7	No	Core-to-tip	3	Stainless steel	
15.	Prowater	Asahi Intecc	0.8	No—30 mm	Core-to-tip	3	Stainless steel	
16.	Prowater Flex	Asahi Intecc	0.8	No—30 mm	Core-to-tip	3	Stainless steel	
17.	SION	Asahi Intecc	0.7	Yes—28 cm	ACT ONE	3	Stainless steel	
18.	SION blue	Asahi Intecc	0.5	No—15 mm	ACT ONE	3	Stainless steel	
19.	Samurai	Boston Scientific	0.5	No—10 mm	Inner coil	4	Stainless steel	
20.	Samurai RC	Boston Scientific	1.2	Yes—20 cm	Yes Inner coil	4	Stainless steel	
21.	Marvel	Boston Scientific	0.9	Yes—17c m	Core-to-tip	3	Stainless steel	
22.	Forte Floppy	Boston Scientific	0.6	No	Shaping ribbon	2	Stainless steel	

23.	Forte Floppy Marker	Boston Scientific	0.6	No	Shaping ribbon	2	Stainless steel	Has two 5 mm markers: the first one located 15 mm proximal to the radiopaque tip, and the second one 15 proximal to the first marker.
24.	Forte Moderate Support	Boston Scientific	0.6	No	Shaping ribbon	2	Stainless steel	
25.	Forte Moderate Support Marker	Boston Scientific	0.6	No	Shaping ribbon	2	Stainless steel	Has two 5 mm markers: the first one located 15 mm proximal to the radiopaque tip, and the second one 15 proximal to the first marker
26.	Luge	Boston Scientific	0.9	Yes-except distal 3 cm	Core-to-tip	3	Stainless steel	
27.	Runthrough NS Floppy	Terumo	1.0	Yes	Core-to-tip	3	Nitinol	
28.	Runthrough NS Extra Floppy	Terumo	0.6	Yes	Core-to-tip	3	Nitinol	
29.	Runthrough NS Hypercoat Floppy	Terumo	1.0	Yes	Core-to-tip	3	Nitinol	
30.	Spectre	Teleflex	0.9	Yes—42 cm	Core-to-tip	3	Nitinol	
31.	Cougar LS	Medtronic	0.5	Yes—42 cm	Shaping ribbon	3	Nitinol	Low support
32.	Cougar XT	Medtronic	0.5	Yes—42 cm	Shaping ribbon	3	Nitinol	Medium support
33.	Intuition	Medtronic	1.0		Core-to-tip	3	Stainless steel	
34.	Zinger Light	Medtronic	0.5	Yes		3	Stainless steel	
35.	Zinger Marker	Medtronic	0.5	Yes		3	Stainless steel	
36.	Zinger Medium	Medtronic	0.5	Yes		3	Stainless steel	
37.	ATW All Track	Cordis		No	Core-to-tip	3	Stainless steel	

(Continued)

TABLE 30.7 (Continued)

Tip stiffness	Wire name	Manufacturer	Tip stiffness (g)	Hydrophilic tip coating	Tip technology	Tip radiopacity (cm)	Material	Comment
38.	ATW Marker wire	Cordis		No	Core-to-tip	3	Stainless steel	Four radiopaque marker bands spaced 10 mm apart for lesion measurement
39.	Stabilizer Balanced Performance	Cordis		No	Core-to-tip	3	Stainless steel	
40.	Stabilizer Marker Wire	Cordis		No	Core-to-tip	3	Stainless steel	Six 1.5 mm radiopaque marker bands spaced 15 mm apart for lesion measurement

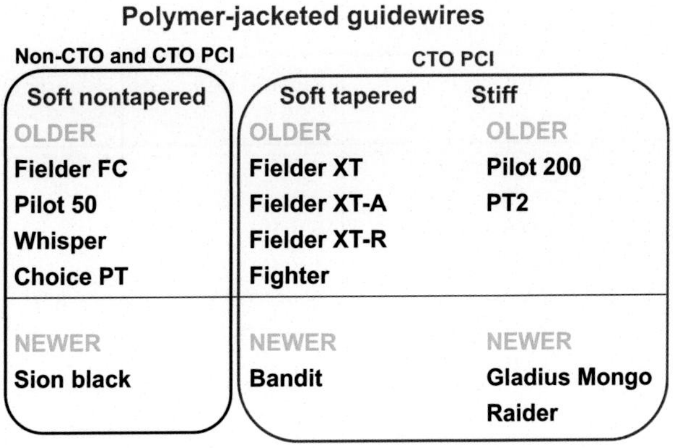

Polymer-jacketed guidewires

Non-CTO and CTO PCI	CTO PCI	
Soft nontapered	**Soft tapered**	**Stiff**
OLDER	OLDER	OLDER
Fielder FC	Fielder XT	Pilot 200
Pilot 50	Fielder XT-A	PT2
Whisper	Fielder XT-R	
Choice PT	Fighter	
NEWER	NEWER	NEWER
Sion black	Bandit	Gladius Mongo
		Raider

FIGURE 30.74 Classification of polymer-jacketed guidewires.

TABLE 30.8 Overview of polymer-jacketed guidewires.

Tip Style	Commercial name	Tip stiffness (grams)	Manufacturer	Properties
Soft, tapered	Fielder XT Fielder XT-A Fielder XT-R	0.8 1.0 0.6	Asahi Intecc	Front-line wires for antegrade CTO crossing. Can also be used for knuckle wire formation (Fielder XT) and for retrograde crossing (Fielder XT-R).
	Fighter	1.2	Boston Scientific	
	Bandit	0.8	Teleflex	
Soft, nontapered	Fielder FC	0.8	Asahi Intecc	Non-CTO PCI: wiring through tortuosity. CTO PCI: collateral vessel crossing.
	Sion Black	0.8	Asahi Intecc	
	Whisper LS, MS, ES	0.8, 1.0, 1.2	Abbott Vascular	
	Pilot 50	1.5	Abbott Vascular	
	Choice PT Floppy	2.1	Boston Scientific	
Stiff	Pilot 150 | 200	2.7 | 4.1	Abbott Vascular	Non-CTO PCI: difficult to penetrate lesions. CTO PCI: Antegrade crossing, especially when the course of the occluded vessel is unclear. Also useful for knuckle wire formation and for reentry into true lumen during antegrade dissection and reentry.
	Gladius Mongo	3	Asahi Intecc	
	PT Graphix Intermediate	1.7	Boston Scientific	
	PT2 Moderate Support	2.9	Boston Scientific	
	Raider	4.0 g	Teleflex	

30.7.5 Atherectomy guidewires

Atherectomy requires use of dedicated guidewires, that is, the Rotawire (floppy and extra support, Section 30.10.1.1.3) for rotablation and the Viper (Advance and Advance with Flex Tip, Section 30.10.2.3) for orbital atherectomy.

30.7.6 Externalization guidewires

The RG3 (330 cm long, Asahi Intecc) and R350 (350 cm long, Teleflex) wires are used for externalization in retrograde CTO PCI and are discussed in detail in the *Manual of CTO PCI* [8].

TABLE 30.9 Description of support coronary guidewires.

Wire category	Tip style	Commercial name	Tip stiffness	Manufacturer	Properties
Extra support guidewires	Soft, nontapered	BHW (nitinol) Extra S'port All Star Iron Man	0.8 gr 0.9 gr 0.8 gr 1.0 g	Abbott Vascular	190 and 300 cm long
		Grand Slam Sion blue extra support	0.7 g 0.5 g	Asahi Intecc	180 and 300 cm long
		Mailman	0.8 g	Boston Scientific	182 and 300 cm long
Wiggle guidewire	Soft, nontapered with curved distal portion	Wiggle	1.0 g	Abbott Vascular	190 and 300 cm long

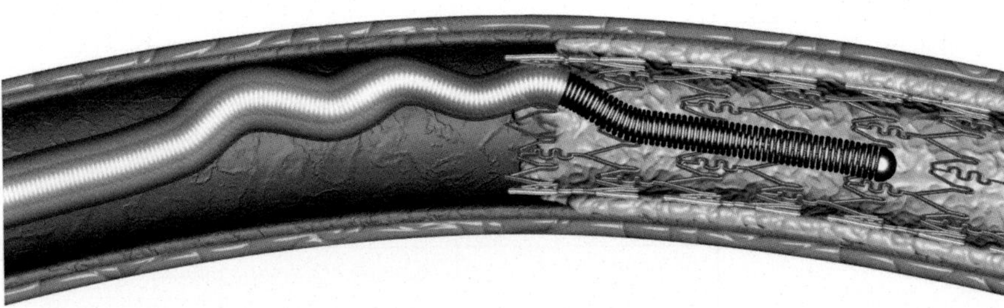

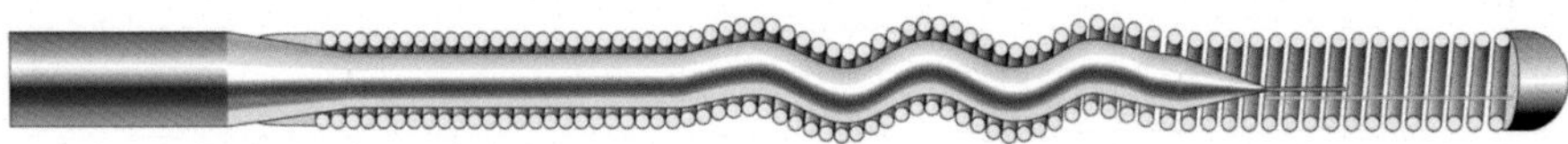

FIGURE 30.75 Illustration of the Wiggle guidewire. *Courtesy Abbott Vascular. ©2019 Abbott. All Rights Reserved.*

30.7.7 Pressure guidewires

Pressure guidewires, such as the Verrata, Verrata Plus, and Omniwire (Philips), PressureWire X (Abbott Vascular), Comet (Boston Scientific), Optowire (Opsens), and the Navvus rapid exchange FFR microcatheter can be used for physiologic assessment as discussed in Chapter 12: Coronary physiology.

30.7.8 0.035 or 0.038 in. guidewires

Using 0.035 or 0.038 in. guidewires is necessary for obtaining arterial access and for advancing catheters to the coronary ostium. There are different 0.035 in. guidewires with different tip stiffness and shape, body stiffness, and coatings (with vs without polymer jacket).

Standard 0.035 or 0.038 in. guidewires have a J-tip and are used in most procedures.

Soft tip 0.035 in. guidewires, such as the Bentson (Cook Medical) and the Wholey wire (Medtronic) are useful for crossing diseased and tortuous peripheral vessels.

Polymer-jacketed 0.035 in. guidewires also facilitate crossing diseased and tortuous vessels, but may be more likely to create a subintimal dissection compared with non-polymer jacketed guidewires. The Glidewire (Terumo) is the most commonly used such wire. The Glidewire Advantage has a polymer-jacketed tip with a stiff nonjacketed body. The Glidewire baby-J had a 1.5 mm distal J-tip radius (vs 3 mm for standard J-tip wires) and is specifically designed for use with radial access.

Stiff body 0.035 in. guidewires, such as the Amplatz SuperStiff (Boston Scientific), Supracore (Abbott Vascular), and Landerquist (Cook Medical) facilitate sheath insertion and equipment advancement through areas of calcification and tortuosity. They usually have soft tips and are inserted over a diagnostic catheter.

30.8 Embolic protection devices

Embolic protection devices are indicated for use in saphenous vein grafts to minimize the risk of distal embolization (Section 18.8.1). They can also be used for lesions with heavy thrombus burden (Section 20.8), or native coronary artery lesions at high risk of distal embolization, such as lesions with large lipid core plaque (Section 25.2.3.2.1).

Two embolic protection devices are currently available in the United States, the Filterwire (Boston Scientific) and the Spider (Medtronic). The Filterwire is delivered directly through the lesion, whereas the Spider requires crossing of the lesion with a 0.014 in. guidewire first before delivering the device.

30.8.1 Filterwire (step-by-step instructions provided in Section 18.8.1.1)

The Filterwire (Fig. 30.76) is a 0.014 in. guidewire with a filter bag attached near its distal tip, designed to contain and remove embolic material that is liberated during saphenous vein graft PCI. The guidewire is delivered through the target SVG lesion with the filter collapsed within the EZ delivery sheath. The EZ delivery sheath is then withdrawn to expand the filter during PCI. After SVG PCI is completed, the filter is collapsed again using the EZ retrieval sheath and removed along with any captured debris. There are two Filterwire sizes, the 2.25–3.5 mm device (requires reference vessel diameter at the filter loop deployment site of 2.25–3.5 mm) and the 3.5–5.5 mm device (requires reference vessel diameter at the filter loop deployment site of 3.5–5.5 mm).

30.8.2 Spider (step-by-step instructions provided in Section 18.8.1.2)

The SpiderFX Embolic Protection Device (Fig. 30.77) is a percutaneously delivered distal embolic protection device, consisting of the following components:

- One Capture Wire composed of a nitinol mesh filter with a distal floppy tip, mounted on a 190 cm or a convertible 320/190 cm PTFE-coated 0.014 in. stainless steel wire. This Capture Wire acts as the primary guidewire for other interventional devices compatible with a 0.014 in. wire. The convertible 320/190 cm Capture Wire is scored, allowing it to be snapped to a 190 cm usable length for use with rapid exchange systems, if desired. The distal portion of the Capture Wire for rapid exchange use is gold. The proximal portion for standard over-the-wire use is black. The filter features a heparin-coating, designed to maintain patency during filter deployment. The coating does not have a systemic effect on the patient anticoagulation.

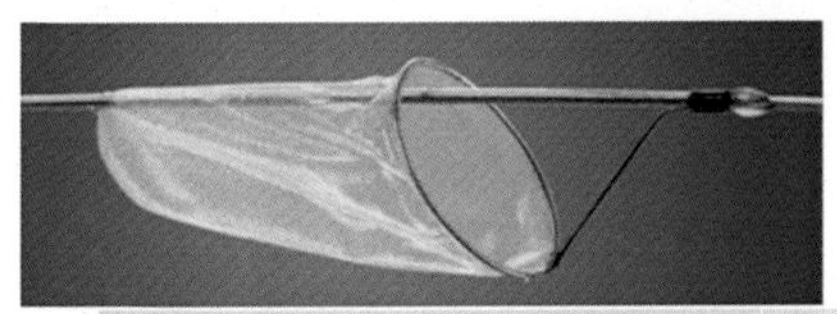

Filterwire

Design	Polyurethane filter basket
Guide catheter	6 Fr
Pore size	110 μm
Diameters	2.25–3.5 and 3.5–5.5
Length	300 cm, 190 cm
Crossing profile	3.2 Fr
Landing zone	>25 mm (2.25) or > 30 mm (3.5)

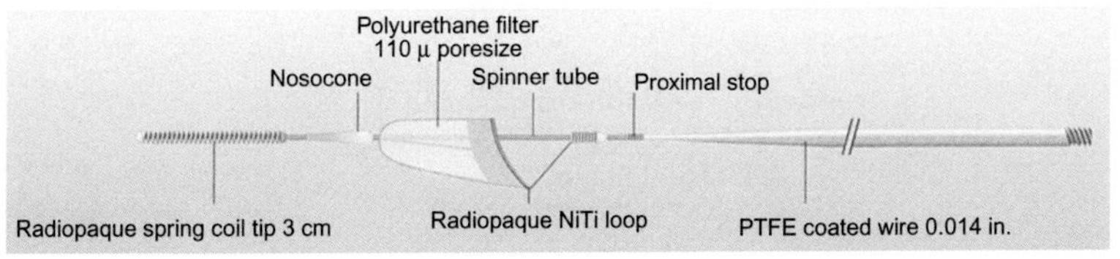

FIGURE 30.76 Illustration of the Filterwire. *Image courtesy Boston Scientific © 2019 Boston Scientific Corporation or its affiliates. All rights reserved.*

- One dual-ended SpiderFX Catheter used to exchange the primary access guidewire with the Capture Wire, deploy the Capture Wire at the desired location, and recover the Capture Wire at the end of the procedure. The green end of the SpiderFX Catheter is for delivery of the Capture Wire and the blue end of the SpiderFX Catheter is for retrieval of the Capture Wire.
- One 23 gauge blunt needle for attachment to a syringe with luer connector for priming and flushing the catheter.

The main difference of the SpiderFX device compared with the Filterwire is that it can be delivered over any 0.014 in. guidewire that is first advanced through the target SVG lesion. Moreover, the SpiderFX device has a unique 320/190 cm convertible wire: it is 320 cm long, but can be "snapped" to become 190 cm long, by firmly grasping the capture wire on each side of the score of the wire (gold and black) and snapping it.

30.9 Balloons

Balloons are the main method for plaque modification during PCI, but they serve several other roles, such as expanding the stent, delivering drugs to the vessel wall, trapping equipment into the guide catheter, and performing intravascular lithotripsy. There are eight key balloon categories (Fig. 30.78):

1. Standard balloons.
2. Small balloons.
3. Plaque modification balloons.
4. Trapping balloons.
5. Ostial Flash balloon (described in Section 30.13.2).

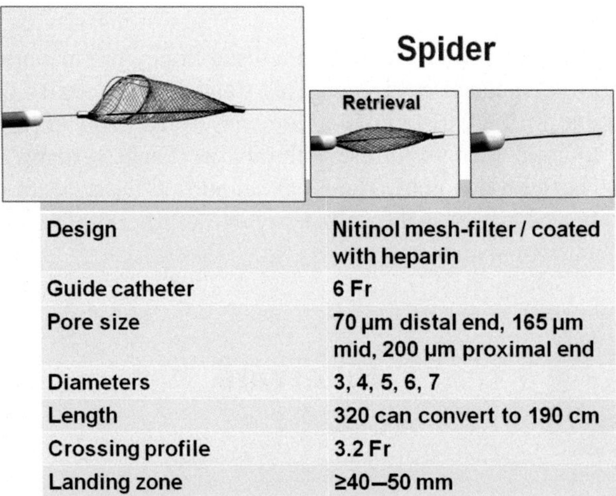

Design	Nitinol mesh-filter / coated with heparin
Guide catheter	6 Fr
Pore size	70 μm distal end, 165 μm mid, 200 μm proximal end
Diameters	3, 4, 5, 6, 7
Length	320 can convert to 190 cm
Crossing profile	3.2 Fr
Landing zone	≥40–50 mm

FIGURE 30.77 Illustration of the SpiderFX.

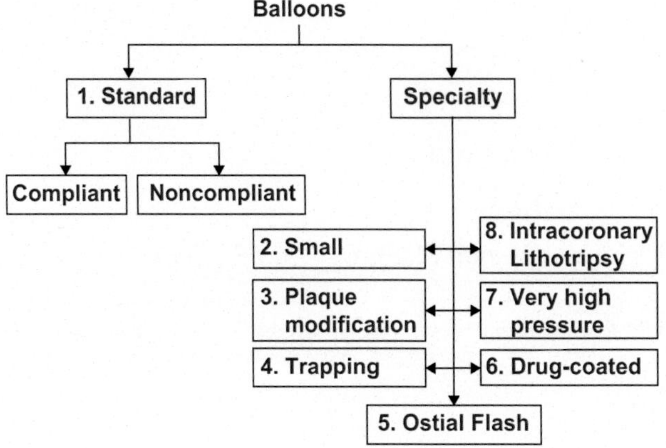

FIGURE 30.78 Balloon classification.

6. Drug-coated balloons (DCB).
7. Very-high pressure balloons.
8. Intracoronary lithotripsy balloon.

30.9.1 Standard balloons

Standard balloons are classified into semicompliant, and noncompliant. Semicompliant balloons are more flexible and easier to deliver, whereas noncompliant balloons are harder to deliver but do not expand much upon high pressure inflation and are typically used for stent postdilation and for treating balloon undilatable lesions (Section 23.1).

30.9.2 Small balloons

Small balloons are important for "challenging to cross lesions, such as balloon uncrossable lesions (Section 23.1) and jailed side branches (Chapter 16: Bifurcations).

The Sapphire Pro 1.0 mm balloon (Fig. 30.79, Orbus Neich) is the lowest profile balloon currently available in the United States and is currently the first choice for balloon uncrossable lesions.

Two other 1.0 mm balloons that are not currently available in the United States are the Ikazuchi Zero (Kaneka, Fig. 30.80) and the Ryurei 1.0 mm (Terumo).

The Nano 0.85 mm balloon (SIS) is the lowest profile balloon currently available in Europe.

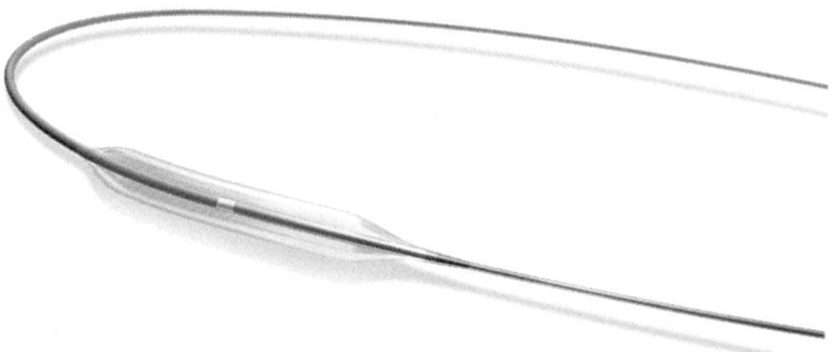

FIGURE 30.79 Illustration of the Sapphire Pro 1.0 mm balloon. *Image used with permission of OrbusNeich Medical.*

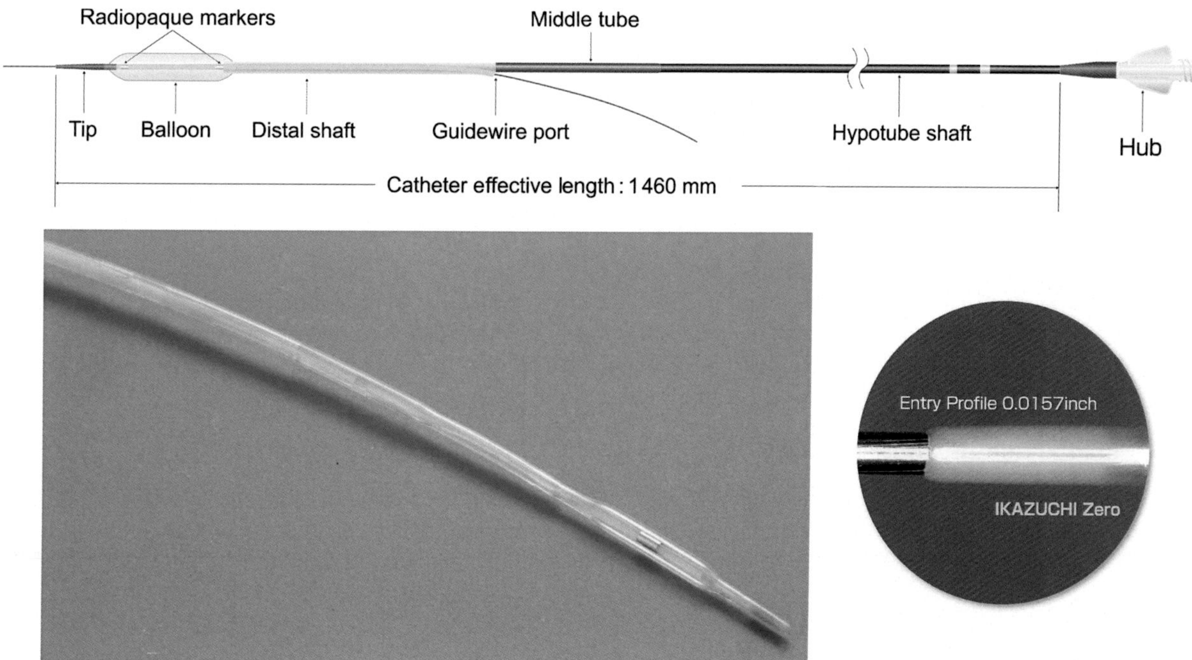

FIGURE 30.80 Illustration of the Ikazuchi Zero balloon. *Reproduced with permission from Kaneka.*

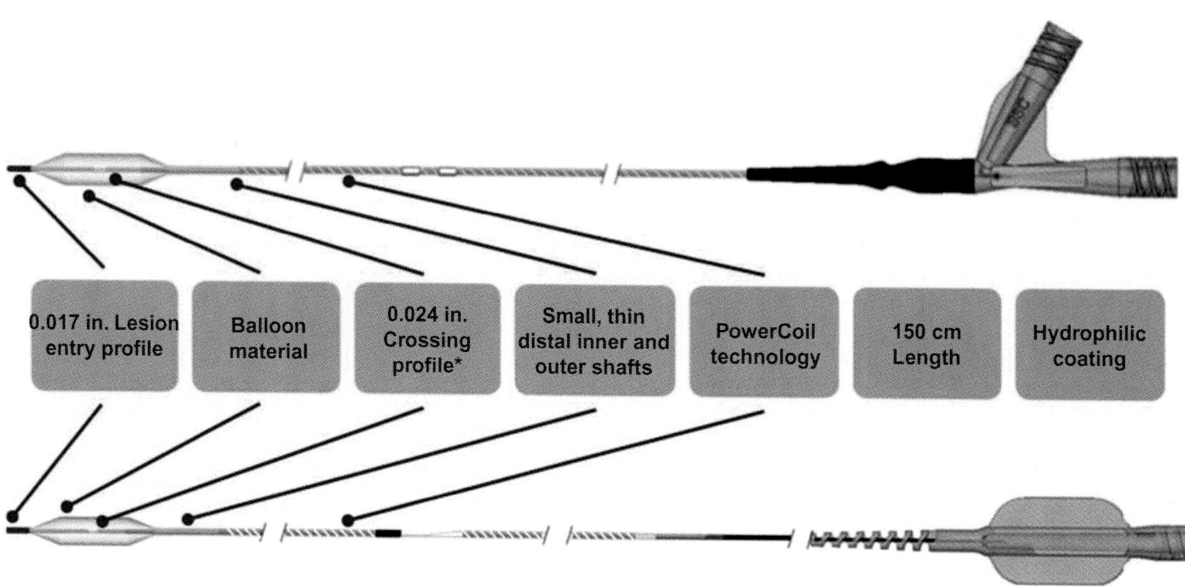

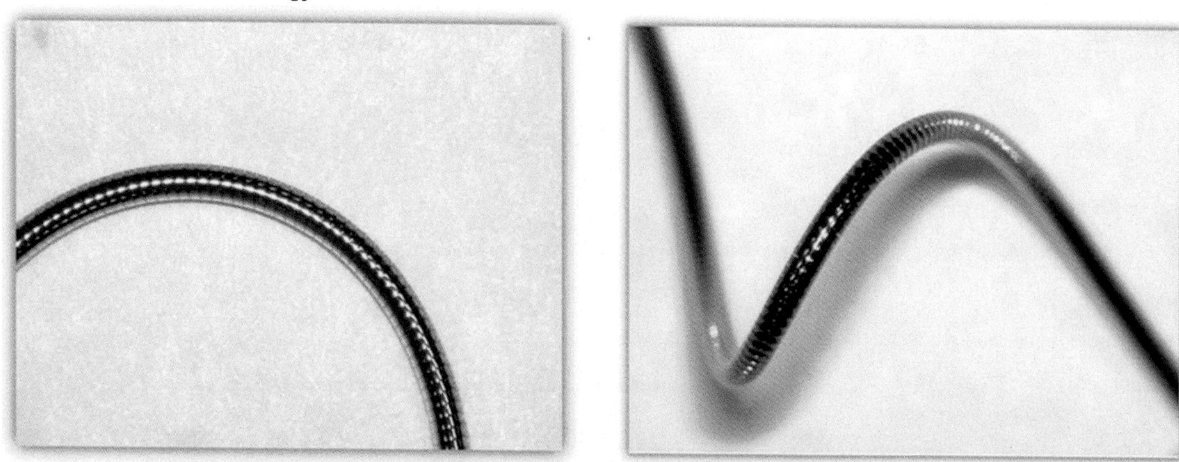

FIGURE 30.81 The Threader microdilation catheter. *Image courtesy Boston Scientific © 2019 Boston Scientific Corporation or its affiliates. All rights reserved.*

The Threader microdilation catheter (Boston Scientific, Fig. 30.81) combines a 1.20 × 12 mm balloon at its tip with a kink resistant shaft that enhances deliverability. The Threader is available both as rapid exchange and as over-the-wire system, although the rapid exchange system provides more support and is preferred by most operators.

The Glider balloon (Teleflex, Fig. 30.82) is a semicompliant balloon with skived tip and low entry profile. It is particularly well suited for crossing through stent struts, and ballooning jailed side branches.

30.9.3 Plaque modification balloons

These balloons are designed to "score" or modify the lesion to facilitate expansion and prevent "watermelon seeding". Three such balloons are currently available in the United States: Angiosculpt (Philips), Chocolate (Teleflex), and the Wolverine cutting balloon (Boston Scientific), while another one is only available in Europe: Blimp (IMDS).

30.9.3.1 Angiosculpt

The AngioSculpt scoring balloon catheter (Philips) (Fig. 30.83) is composed of a semicompliant balloon encircled by three nitinol spiral struts to score the target lesion on balloon inflation. Compared with the cutting balloon, the AngioSculpt balloon has a lower crossing profile and produces more scoring marks per millimeter of plaque.

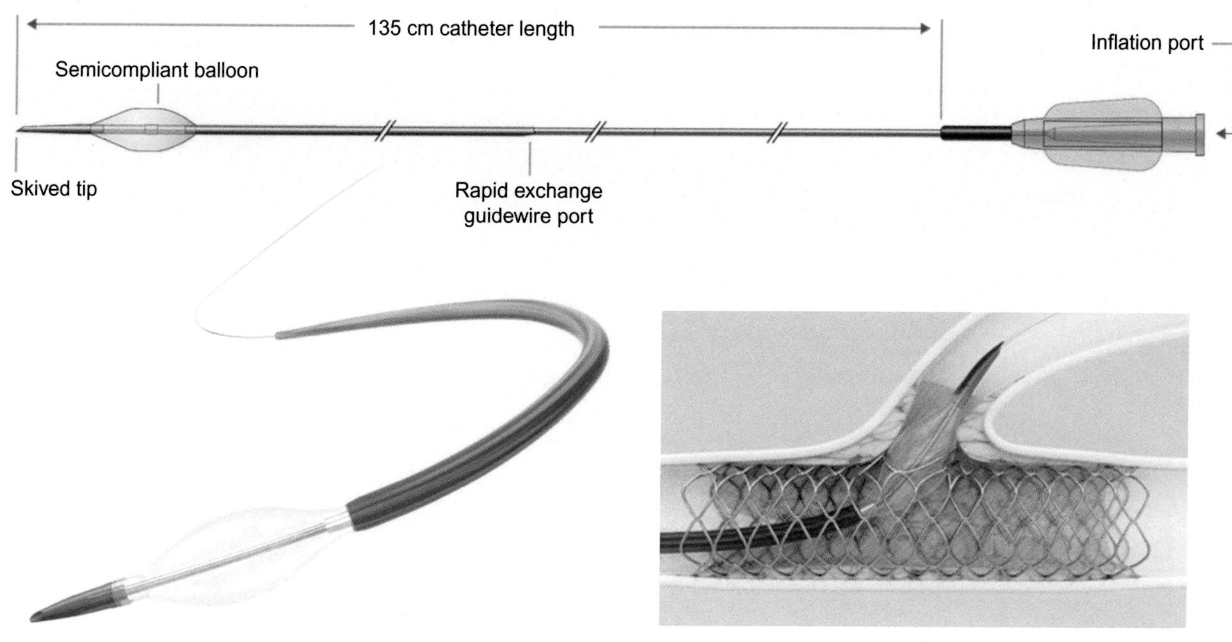

FIGURE 30.82 The Glider balloon.

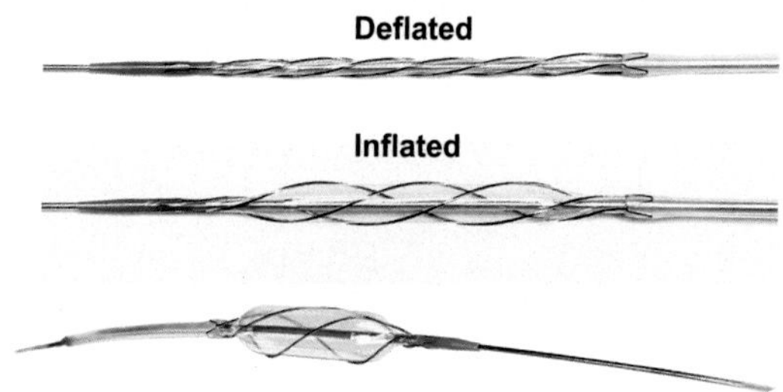

FIGURE 30.83 The Angiosculpt balloon. *Reproduced with permission from Philips.*

30.9.3.2 Chocolate

In contrast to the Angiosculpt, the Chocolate balloon (Teleflex, Fig. 30.84) is a nitinol-caged balloon. As the balloon inflates, the cage causes the balloon to form a series of segmented pillows and grooves along the entire lesion. The pillows apply force to create small dissections that facilitate effective dilatation. The grooves relieve the stress and stop dissections from propagating.

30.9.3.3 Cutting balloon

The Wolverine cutting balloon (Boston Scientific, Fig. 30.85) contains atherotomes that protrude 0.005 in. from the surface of the balloon and modify the target lesion by creating three or four endovascular radial incisions through the fibrocalcific tissue, thus allowing further expansion with conventional balloons [53]. In addition to helping expand challenging lesions the cutting balloon has been used for releasing intramural hematomas that may develop post stenting, after reentry [54], or in the setting of spontaneous coronary dissection. Cutting balloon inflation should be done very slowly (1 atm every 5 sec). Cutting balloons have a high crossing profile.

30.9.3.4 Blimp balloon

The Blimp scoring balloon catheter (IMDS, Fig. 30.86) has a short distal monorail segment with a 0.6 mm diameter, 5 mm long balloon adjacent to the coronary guidewire. The device has a high burst pressure (nominal pressure 25 atm and rated burst pressure 30 atm) to increase the ability to open a stenosis. The Blimp balloon scores the lesions by exerting force on the plaque through the guidewire, which is back-loaded from the tip of the device and exits just at the distal side of the balloon through the Rx port and is positioned along the balloon.

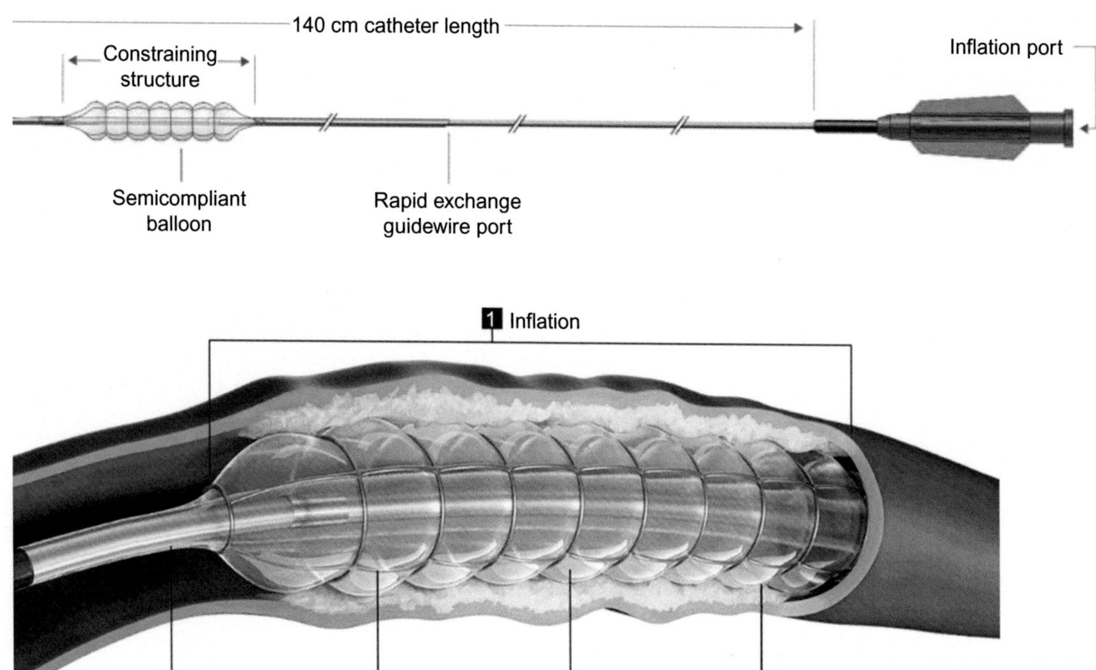

FIGURE 30.84 The Chocolate balloon.

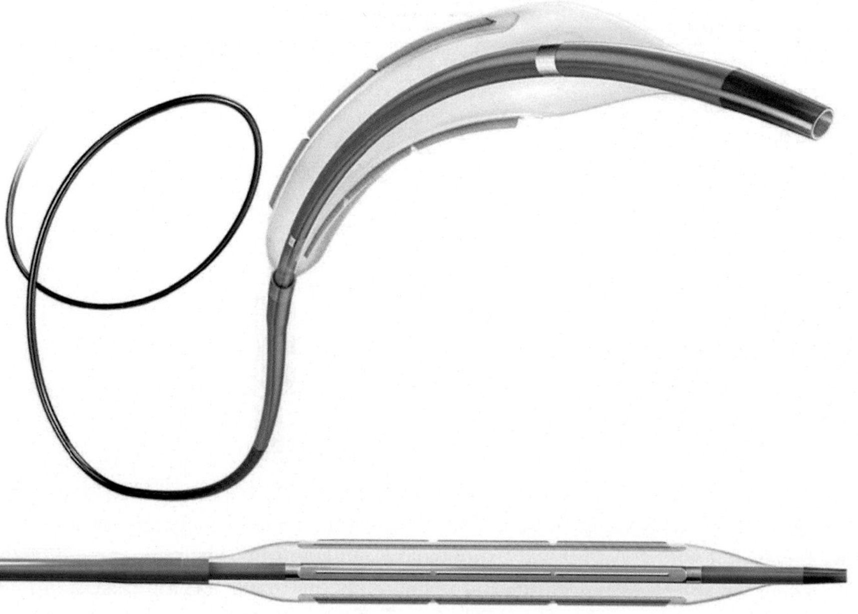

FIGURE 30.85 The Wolverine cutting balloon. *Image courtesy Boston Scientific*

30.9.4 Trapping balloons

Trapping (Section 8.9.1) can be performed with standard balloons, however dedicated trapping balloons (that do not have a wire lumen and are, therefore, resistant to kinking and deformation and have a lower profile) are also available. The only trapping balloon currently available in the United States is the Trapper (Boston Scientific, Fig. 30.87). The Trap it (IMDS, Fig. 30.88) and Kusabi (Kaneka) trapping balloons are also available in other countries. Trapping can also be performed with the Trapliner guide catheter extension (Section 30.3.1).

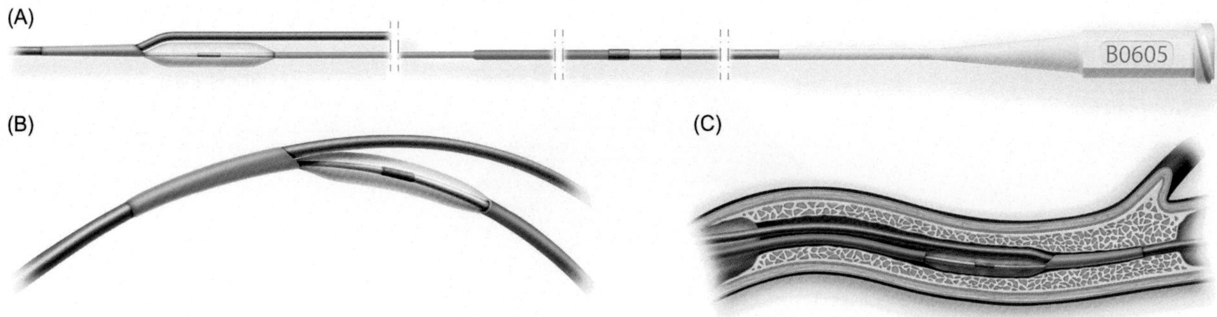

FIGURE 30.86 The Blimp balloon. *Reproduced with permission from IMDS.*

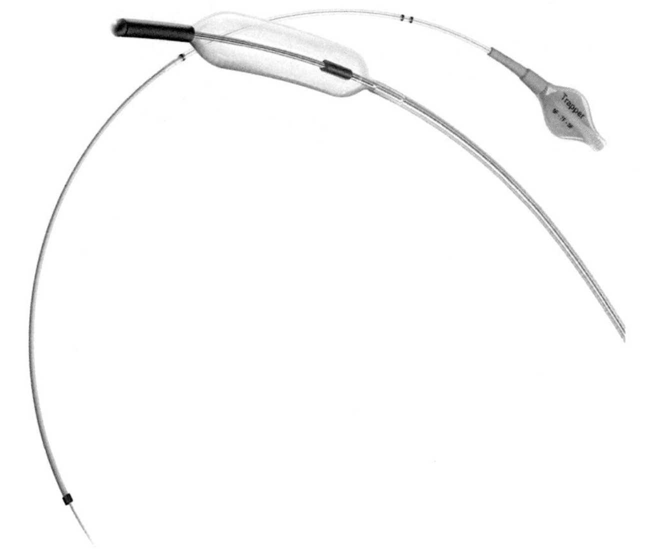

FIGURE 30.87 The Trapper balloon. *Image courtesy Boston Scientific © 2019 Boston Scientific Corporation or its affiliates. All rights reserved.*

30.9.5 Ostial Flash

The Ostial Flash balloon is designed to flare aorto-ostial stents and is discussed in Section 30.13.2.

30.9.6 Drug-coated balloons

Drug-coated balloons have a drug (paclitaxel and sirolimus are most commonly used) that is released to the vessel wall upon balloon inflation, decreasing the risk of restenosis. They are used most often for treating in-stent restenosis, but have also been used in bifurcation lesions and small vessels [55]. Coronary drug-coated balloons are not currently available in the United States, although peripheral DCBs have been used off label in some patients with large coronary vessels.

30.9.7 Very high-pressure balloon

The OPN NC balloon (SIS Medical, Switzerland) is a specialized noncompliant balloon that can deliver very high (> 40 atm) pressures, but is not currently available in the United States (Section 23.2.10) [56].

30.9.8 Intracoronary lithotripsy balloon

The intracoronary lithotripsy system (Shockwave Medical, Inc., Fig. 30.89) uses ultrasonic shockwaves to overcome extreme lesion resistance. The coronary lithotripsy balloon is not approved in the United States as of 2020, but the

FIGURE 30.88 The Trap-It balloon. *Reproduced with permission from IMDS.*

peripheral lithotripsy balloon has been used off label in large coronary arteries (smallest peripheral intravascular lithotripsy balloon is 2.5 × 40 mm). The intracoronary lithotripsy balloon can be very useful for treating balloon undilatable lesions (Section 23.2.9).

30.10 Atherectomy

30.10.1 Rotational atherectomy

CTO PCI Manual Online case: 53.
PCI Manual Online cases: 17, 23, 60. Two rotational atherectomy systems are currently available: the Rotablator classic system and the Rotapro system.

30.10.1.1 Rotablator classic system

The Rotablator classic system has seven components:

1. Burr.
2. Advancer (which can be purchased preconnected to the burr).
3. Rotawire (there are two types: Rotawire Floppy and Rotawire Extra Support).
4. WireClip torquer.
5. Console.
6. Foot pedal.
7. Rotaglide.

Specifications

Diameter (mm) - 2.5, 3.0, 3.5, 4.0

Length (mm) - 12

Guidewire compatibility (in.) - 0.014

Guide catheter compatibility - 6F

Working length (cm) - 138

Crossing profile range* (in.) - 0.044, +/-0.002

Pulses (max) - 80

*0.043 in. max. for 2.5 mm, 0.044 in. max for 3.0–3.5 mm and 0.046 in. max. for 4.0 mm

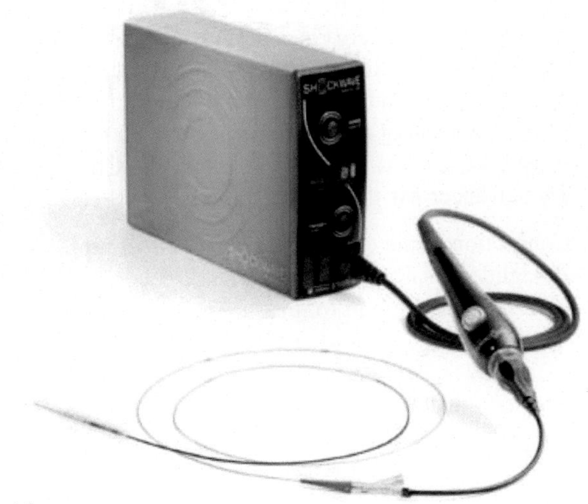

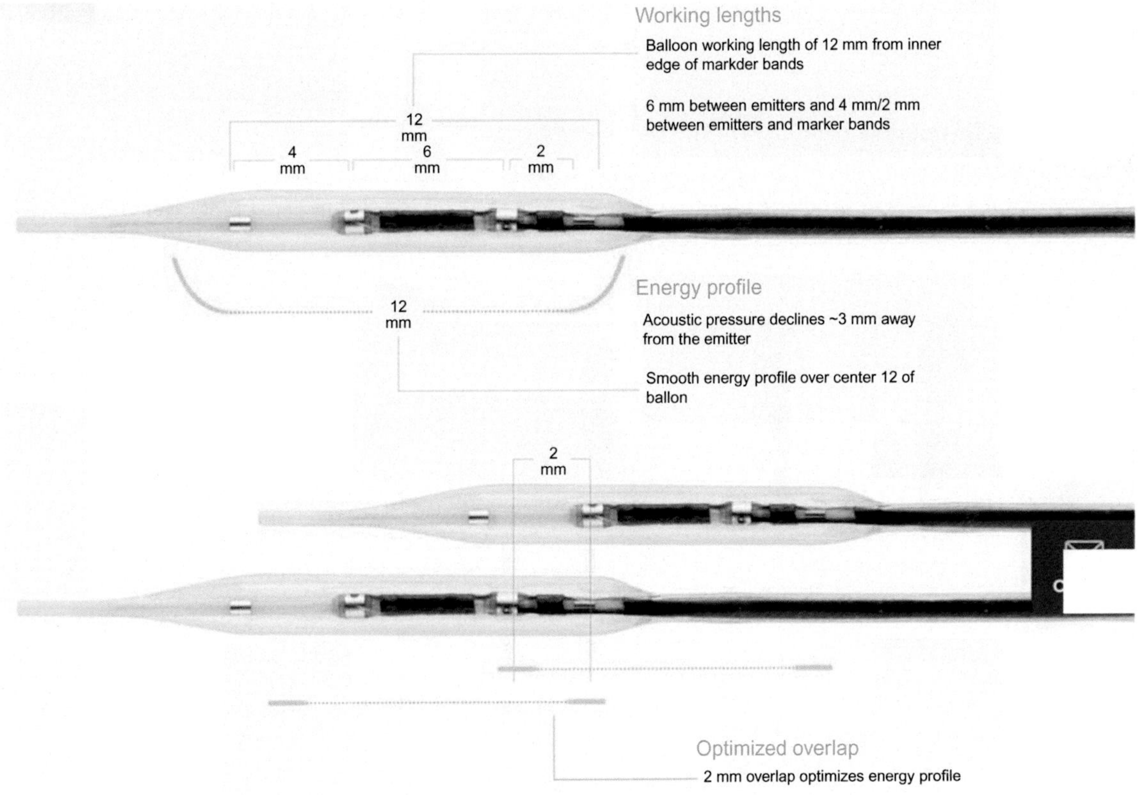

FIGURE 30.89 The intracoronary lithotripsy system. *Reproduced with permission from Shockwave Medical Inc.*

30.10.1.1.1 Burr

The Rotablator burr (Fig. 30.90) has an elliptical shape and is nickel coated. Its distal edge is coated with 2,000–3,000 microscopic diamond crystals between 20 and 30 μm in size that protrude 5 μm from the surface. The proximal edge has no coating, which explains the potential risk of device entrapment (atherectomy can only happen with the leading edge of the burr during advancement but not during withdrawal of the burr). There are eight burr sizes (1.25, 1.50, 1.75, 2.00, 2.15, 2.25, 2.38, 2.5 mm), although the 1.25 and 1.50 mm burrs are the most commonly used. The burr to artery ratio should be 0.5 or lower.

30.10.1.1.2 Advancer

The Rotablator advancer (Fig. 30.91) is used by the operator to advance and withdraw the burr during rotablation.

The driveshaft of the Rotablator burr is attached in the front of the advancer (burrs can be purchased preconnected with the advancer) and the Rotawire exits from the back of the advancer. There is a knob at the top of the advancer that is used for advancing and withdrawing the burr. There is a brake defeat button at the back to allow for the wire brake to be released and a docking port for the WireClip torquer.

The advancer has three connection ports:

1. Connection to the compressed gas cylinder port that rotates the drive shaft of the burr.
2. Rotaglide or saline infusion port.
3. Fiber-optic cable port to measure the rotating speed of the burr.

30.10.1.1.3 Rotawire

Rotational atherectomy should only be performed using a dedicated Rotawire. There are two versions of the Rotawire: the Rotawire Floppy and the Rotawire Extra Support (Fig. 30.92 and Table 30.10).

Both guidewires are 330 cm in length with a 0.009 in. shaft and 0.014 in. tip that does not allow the Rotablator burr to go past the wire tip. The Rotawire Floppy has a longer taper than the Rotawire Extra Support. The Rotawire Floppy is used in the vast majority of cases. The Rotawire Extra Support may be useful in ostial lesions to better align the guide catheter with the vessel.

30.10.1.1.4 WireClip torquer

The Rotawire can be torqued using the WireClip torquer (Fig. 30.93).

It can also prevent the guidewire from spinning during rotational atherectomy when performing rotational atherectomy while the brake is defeated, by inserting the WireClip torquer into the docking port (Fig. 30.94).

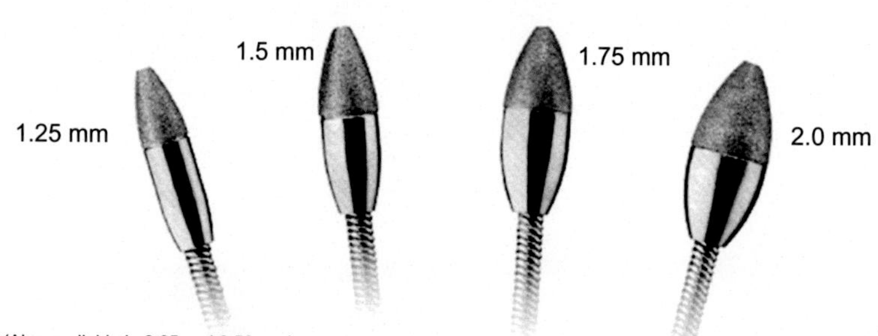

FIGURE 30.90 Rotablator burrs of different sizes. *Image courtesy Boston Scientific © 2019 Boston Scientific Corporation or its affiliates. All rights reserved.*

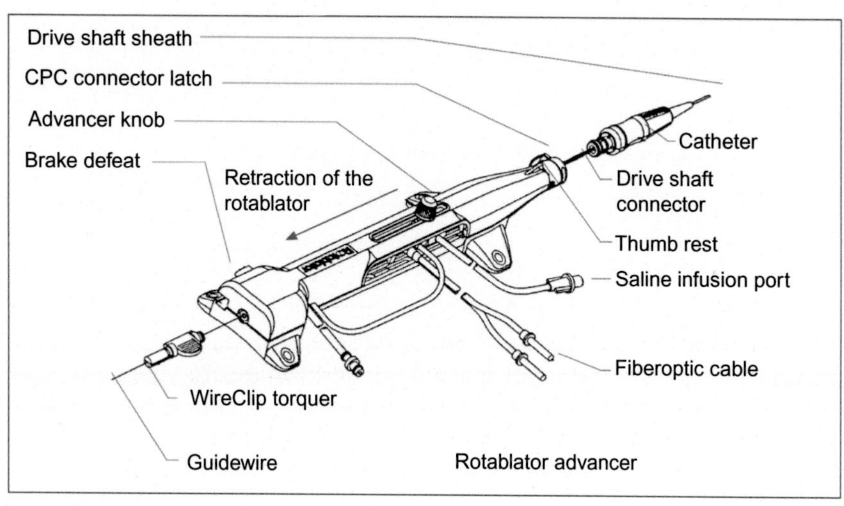

FIGURE 30.91 The Rotablator advancer. *Image courtesy Boston Scientific © 2019 Boston Scientific Corporation or its affiliates. All rights reserved.*

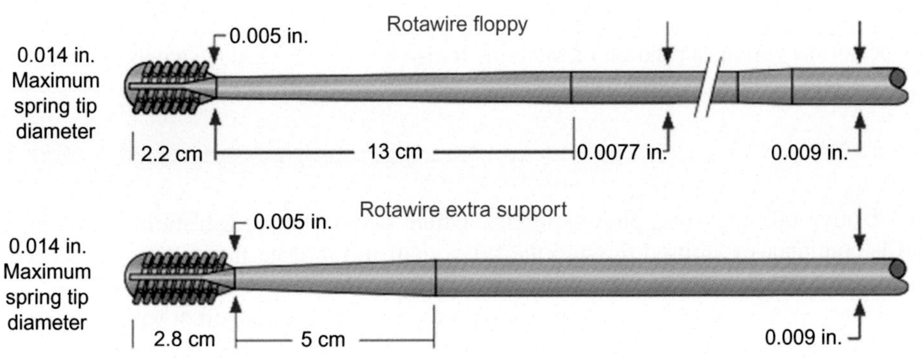

FIGURE 30.92 Illustration of the Rotawire Floppy and Rotawire Extra Support. *Image courtesy Boston Scientific © 2019 Boston Scientific Corporation or its affiliates. All rights reserved.*

TABLE 30.10 Characteristics of the Rotawire Floppy and the Rotawire extra support.

Model/description	Length	Tip length	Flexibility	Spring tip diameter	Maximum diameter
ROTAWire Extra Support Guide Wire	330 cm	2.8 cm	Stiff	0.009 in.	0.014 in.
ROTAWire Floppy Guide Wire	330 cm	2.2 cm	Flexible	0.009 in.	0.014 in.

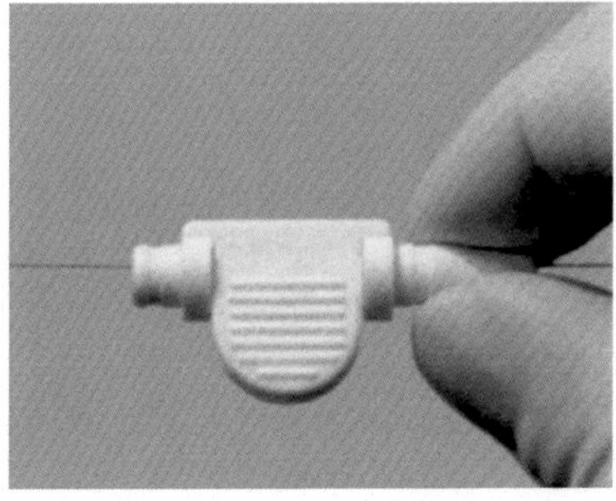

FIGURE 30.93 The WireClip torquer. *Image courtesy Boston Scientific © 2019 Boston Scientific Corporation or its affiliates. All rights reserved.*

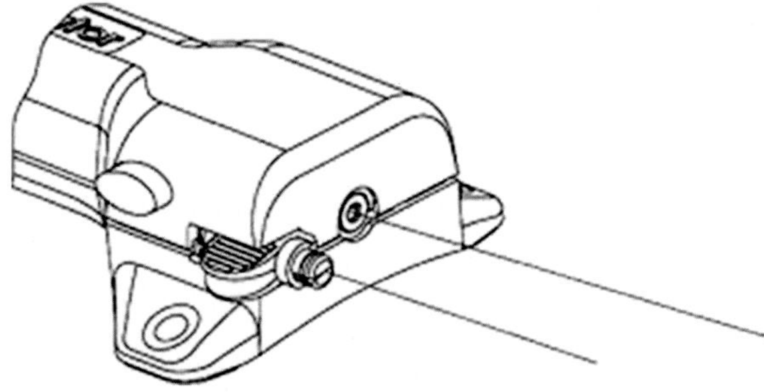

FIGURE 30.94 The WireClip torquer docking port. The brake defeat button is pressed followed by insertion of the WireClip torquer into the docking port. This allow single operator insertion and removal of Rotablator without the need for an assistant to manage the guidewire. *Image courtesy Boston Scientific © 2019 Boston Scientific Corporation or its affiliates. All rights reserved.*

30.10.1.1.5 Console

The Rotablator console (Fig. 30.95) has several indicators, the most important of which are:

1. Tachometer (speed of rotation in rpm), and
2. Dynaglide indicator.

30.10.1.1.6 Foot pedal

The foot pedal (Fig. 30.96) has the pedal that activates rotation of the burr. It also has the Dynaglide switch that allows retrieval of the burr at rotation speeds of 60,000–90,000 rpm.

30.10.1.1.7 Rotaglide

The Rotaglide is a solution made of olive oil, egg yolk, phospholipids, sodium deoxycholate, L-histidine, disodium EDTA, and sodium hydroxide and is continuously infused through the drive shaft to lubricate the system and reduce friction and heat generation.

It is contraindicated in patients who are allergic to eggs, in whom normal saline can be infused, often with verapamil and nitroglycerin added to the infused saline solution. Normal saline is used instead of Rotaglide in countries where Rotaglide is not available.

30.10.1.2 Rotapro

The Rotapro system has six components:

1. Burr.
2. Advancer (which can be purchased preconnected to the burr).
3. Rotawire (there are two types: Rotawire Floppy and Rotawire Extra Support).
4. WireClip torquer.
5. Console.
6. Rotaglide.

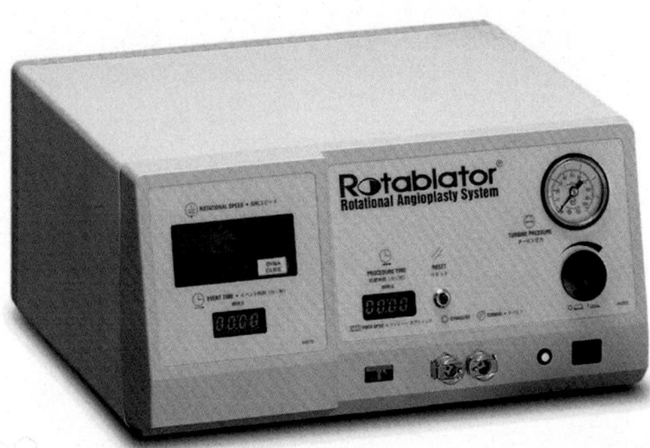

FIGURE 30.95 The Rotablator console. *Image courtesy Boston Scientific © 2019 Boston Scientific Corporation or its affiliates. All rights reserved.*

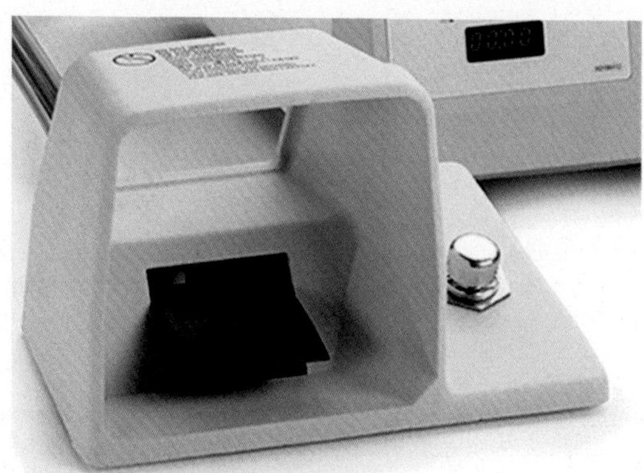

FIGURE 30.96 The Rotablator foot pedal. *Image courtesy Boston Scientific © 2019 Boston Scientific Corporation or its affiliates. All rights reserved.*

The difference of the Rotapro system compared with the Rotablator classic system is in the advancer and the console. The other components (burr, Rotawire, WireClip torquer, and Rotaglide) are identical to the Rotablator classic system. The Rotapro system does not have a foot pedal.

1. *Burr*: burrs are the same as for Rotablator classic.
2. *Advancer (which can be purchased preconnected to the burr)*
   Pushing the knob on the Rotapro advancer (Fig. 30.97) can activate and stop rotation, obviating the need for a foot pedal.
3. *Rotawire*: is the same as for Rotablator classic.
4. *The WireClip torquer*: is the same as for Rotablator classic.
5. *The console*.
   The Rotapro console (Fig. 30.98) is smaller and provides enhanced feedback of decelerations.
6. *Rotaglide*: is the same as for Rotablator classic.

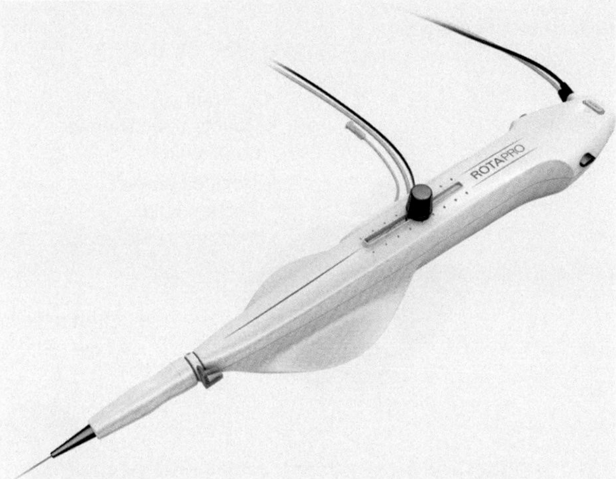

FIGURE 30.97 The Rotapro advancer. *Image courtesy Boston Scientific © 2019 Boston Scientific Corporation or its affiliates. All rights reserved.*

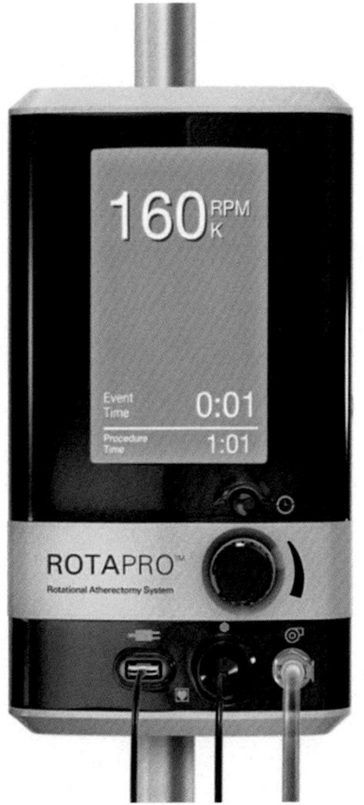

FIGURE 30.98 The Rotapro concole. *Image courtesy Boston Scientific © 2019 Boston Scientific Corporation or its affiliates. All rights reserved.*

30.10.2 Orbital atherectomy

CTO PCI Manual Online case: 39, 126, 136, 150.

PCI Manual Online cases: 3, 6, 20, 21, 37, 46, 59, 64, 99. The orbital atherectomy system has four components (Fig. 30.99).

30.10.2.1 Diamondback 360 coronary orbital atherectomy device

The orbital atherectomy device (OAD) is a hand-held, over-the-wire device that includes a sheath-covered drive shaft and a diamond-coated crown. The diamond coating on the crown provides an abrasive surface that reduces coronary plaque within coronary arteries. The GlideAssist feature facilitates advancing and retracting the OAD crown over the guide wire.

The OAD is electrically powered and consists of the following:

- Crown.
- Crown advancer knob.
- Drive shaft.
- Sheath (covering the drive shaft up to the crown).

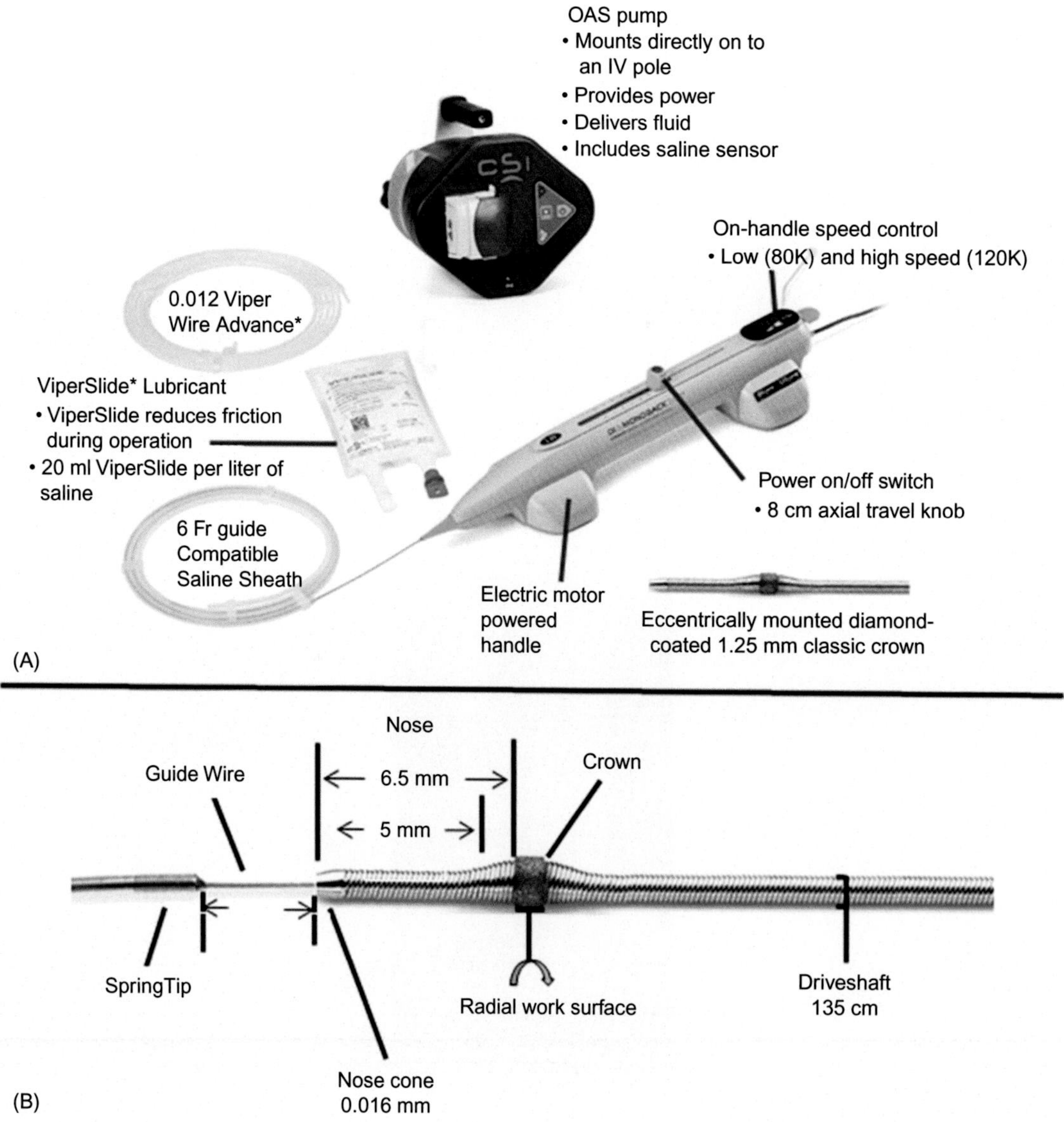

FIGURE 30.99 Illustration of the orbital atherectomy system components. *©2019 Cardiovascular Systems, Inc. CSI, Diamondback 360, GlideAssist, ViperWire Advance and ViperSlide are registered trademarks of Cardiovascular Systems, Inc., and used with permission.*

- Electrical power cord.
- Saline tubing for connecting with the orbital atherectomy system pump.
- On/off button at the top of the advancer knob.
- Two speed control buttons (80,000 or 120,000 rpm).
- The speed control buttons can also be used to activate the GlideAssist function (rotation of the crown at 5,000 rpm).
- Crown advancement measurement indicators.
- Manual guidewire brake.

30.10.2.2 Pump

The OAS pump (Fig. 30.100) provides the saline pumping mechanism and power to the OAD. It is reusable, and portable and attaches to a standard five-wheel rolling intravenous pole or table mount pole. The OAS pump includes a built-in, audible 25 seconds spin time notification, OAS power and priming buttons, and status indicators.

30.10.2.3 ViperWire Advance guidewire

The OAD is designed to track and spin only over a ViperWire Advance guidewire. There are two ViperWire Advance guidewires:

1. ViperWire Advance, which is a moderately stiff, stainless steel guidewire, with a silicone coating, epoxy proximal tip bond, and a radiopaque distal spring tip. Its diameter is 0.012 in. in the shaft and 0.014 in. at the tip.
2. ViperWire Advance with Flex Tip, which is a nitinol guidewire with 1 g tip load. Its body is more flexible than the ViperWire Advance reducing wire bias and it has very good torque response.

30.10.2.4 ViperSlide

The ViperSlide contains 10% soybean oil, 1.2% egg yolk phospholipid, 2.25% glycerin, sodium hydroxide and water for injection. Use 20 mL ViperSlide per liter of saline.

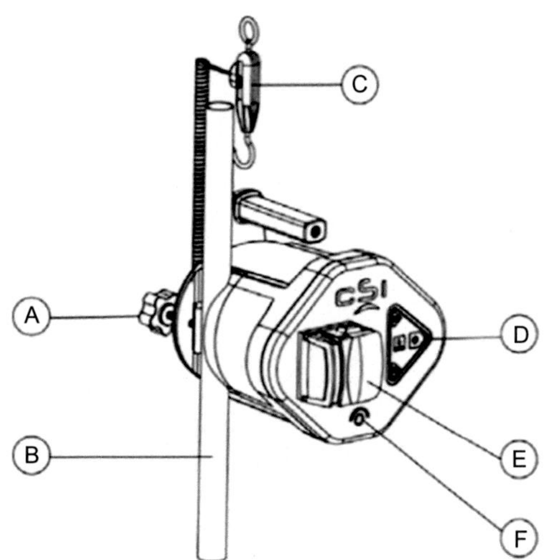

A. IV pole screw clamp
B. IV pole or table mount (not included)
C. Low saline level sensor and connector cord
D. Control panel
E. OAS pump door
F. OAD connection

FIGURE 30.100 Illustration of the orbital atherectomy pump. *©2019 Cardiovascular Systems, Inc. CSI, Diamondback 360, GlideAssist, ViperWire Advance and ViperSlide are registered trademarks of Cardiovascular Systems, Inc., and used with permission.*

30.11 Laser

Online video "Impact of contrast on laser activation"

CTO PCI Manual online cases: 1, 5, 18, 27, 47, 52, 73, 86.

PCI Manual online cases: 23, 35, 55, 64, 67, 88.

Excimer laser coronary atherectomy (ELCA, Philips, Fig. 30.101) can be used to treat balloon uncrossable (Section 23.1) or undilatable (Section 23.2) lesions (including in-stent undilatable lesions) and occasionally intracoronary thrombus (Section 20.9.6).

ELCA uses ultraviolet energy (wavelength: 308 nm) delivered by a xenon-chlorine pulsed laser catheter with pulse frequency of 25–80 Hz and fluence of 30–80 mJ/mm^2. ELCA uses energy for disruption and disintegration of the molecular bonds within the atherosclerotic plaque in a highly controlled manner through ablation rather than burning. Since laser catheters can be advanced over any 0.014 in. guidewire, ELCA is highly valuable for the treatment of balloon uncrossable (and also balloon undilatable lesions) [57]. In some instances, the laser catheter can facilitate lesion crossing by modifying the proximal cap, without actually crossing the lesion.

The laser catheters currently available for coronary use range in diameter from 0.9 to 2.0 mm and are available as rapid exchange (0.9, 1.4, 1.7, and 2.0 mm) or over-the-wire (0.9 mm). The catheter used in coronary lesions is nearly always the 0.9 mm X-80 catheter that is compatible with 6 French guide catheters. Since the coronary laser catheters allow lasing for only 10 seconds before the laser is switched off for 5 seconds, many operators prefer to use the 0.9 mm Turbo-Elite catheter (Spectranetics) off-label, which does not have lasing duration limits and allows repetition up to 80 Hz. Laser is usually performed with saline infusion, although laser has been used with simultaneous contrast injection to expand under-expanded stents (Section 23.2.6) [58].

30.12 Thrombectomy devices

Thrombectomy is discussed in detail in Section 20.9.6.

Two types of thrombectomy systems are currently used: standard aspiration thrombectomy systems and the Penumbra system. Laser (Section 30.11) can also be used to "vaporize" thrombus. Rheolytic thrombectomy was associated with higher mortality in the AiMI trial and is currently rarely utilized [59].

30.12.1 Aspiration thrombectomy

CTO PCI Manual Online cases: 6, 39, 103.

PCI Manual Online cases: 11, 14, 21, 35, 43, 54, 55, 61, 65, 78, 84, 94, 95. All aspiration thrombectomy catheters have similar construction with a Monorail guidewire lumen and an over-the-wire thrombectomy lumen. Several types are currently available (Fig. 30.102, Table 30.11).

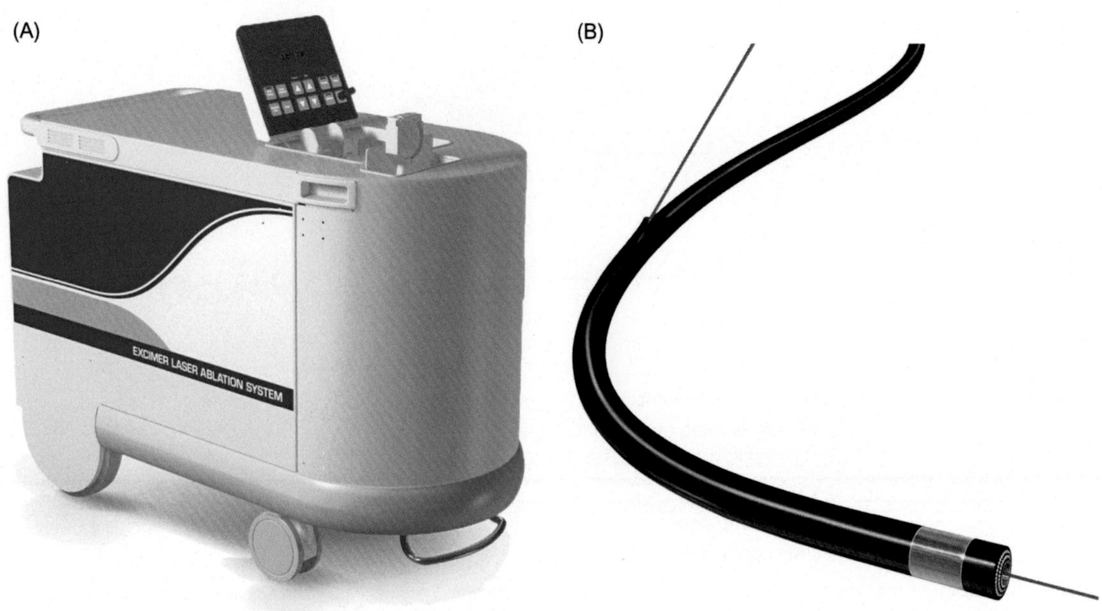

FIGURE 30.101 The laser console (panel A) and catheter (panel B). *Reproduced with permission from Philips.*

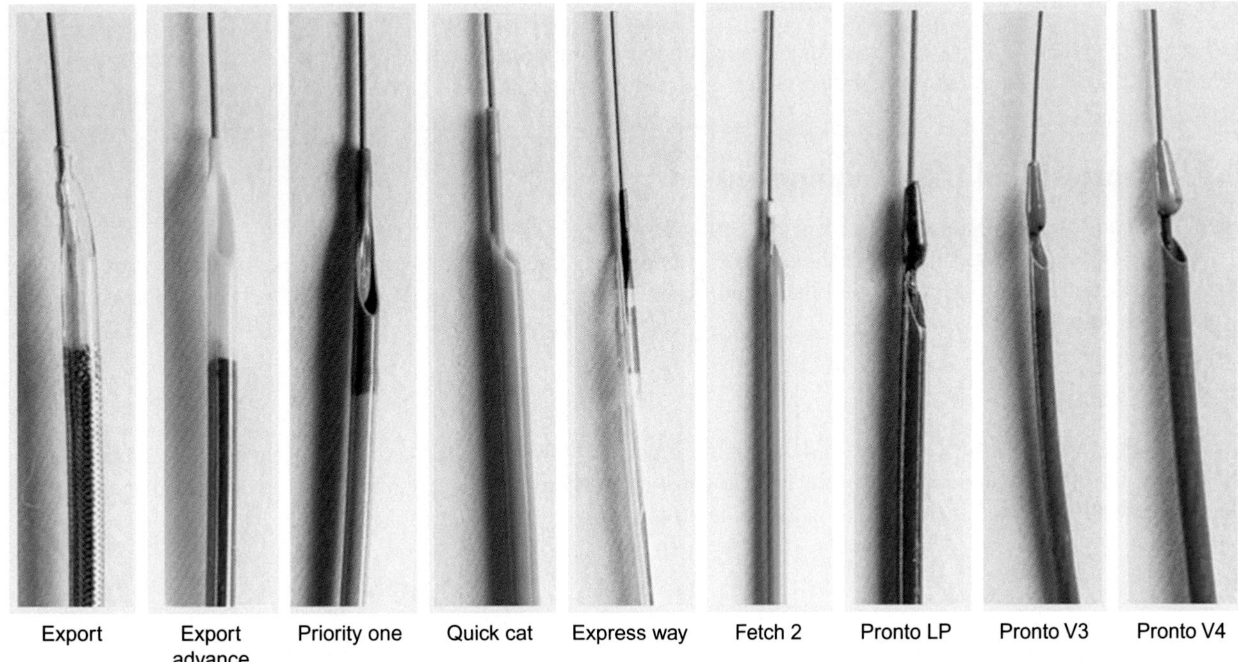

FIGURE 30.102 Commercially available aspiration thrombectomy catheters. *Reproduced with permission from* Cardiovascular Intervention: A Companion to Braunwald's Heart Disease *(Figure 14.1). Copyright Elsevier.*

TABLE 30.11 Aspiration thrombectomy catheters.

Device	Manufacturer	Guide needed	Distal extraction area	Proximal extraction area	Stylet
ASAP	Merit Medical	6 Fr			No
ASAP LP	Merit Medical	6 Fr			No
Export advance	Medtronic	6 Fr	0.93 mm^2	0.98 mm^2	Yes
Export AP	Medtronic	6 Fr	0.87 mm^2	0.85 mm^2	No
Export 7 Fr	Medtronic	7 Fr			No
Xpress-way	Atrium	6 Fr			Yes
Fetch 2	Boston Scientific	6 Fr			No
Priority One 6 Fr	Terumo	6 Fr			No
Priority One 7 Fr	Terumo	7 Fr			No
Pronto LP	Teleflex	5 Fr			Yes
Pronto V3	Teleflex	6 Fr	0.90 mm^2	0.90 mm^2	No
Pronto V4 5.5 Fr	Teleflex	6 Fr		1.04 mm^2	No
Pronto V4 6 Fr	Teleflex	6 Fr		1.25 mm^2	No
Pronto V4 7 Fr	Teleflex	7 Fr		1.45 mm^2	No
Pronto V4 8 Fr	Teleflex	8 Fr		2.17 mm^2	No
QuickCat	Philips	6 Fr			No

30.12.2 Penumbra

The Penumbra system (Fig. 30.103) consists of an aspiration thrombectomy catheter (Cat Rx) similar to those described in Section 20.12.1) and a pump that creates suction and retrieves thrombus and blood into a canister.

30.13 Aorto-ostial lesion equipment

Treatment of aorto-ostial lesions is described in Section 15.1.

There are two devices specifically designed for facilitating treatment of aorto-ostial lesions, the Ostial Pro (designed to facilitate accurate stent placement) and the Ostial Flash (designed to achieve flaring of the portion of the stent hanging out in the aorta).

30.13.1 Ostial Pro

The Ostial Pro (Merit Medical, Fig. 30.104) is a nitinol device with distal, self-expanding legs that are advanced just distal to the tip of the guide catheter to prevent the entry of the guiding catheter into the target vessel, mark the plane of the aortic wall, and align the tip of the guide catheter with the aorto-ostial plane [60,61].

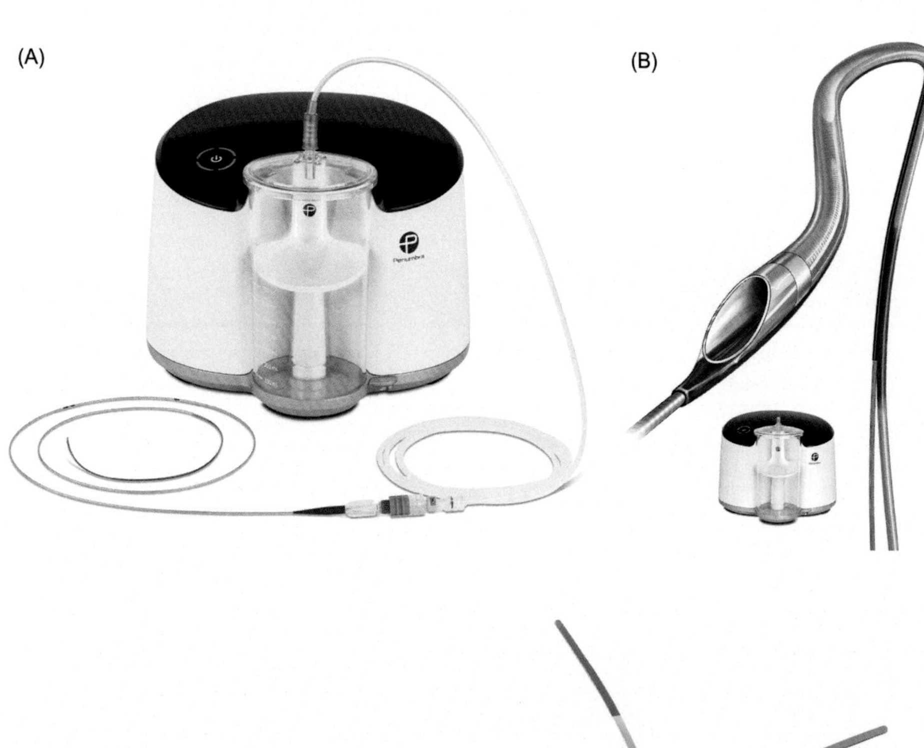

FIGURE 30.103 The Penumbra thrombus aspiration system. *Reproduced with permission from Penumbra, Inc.*

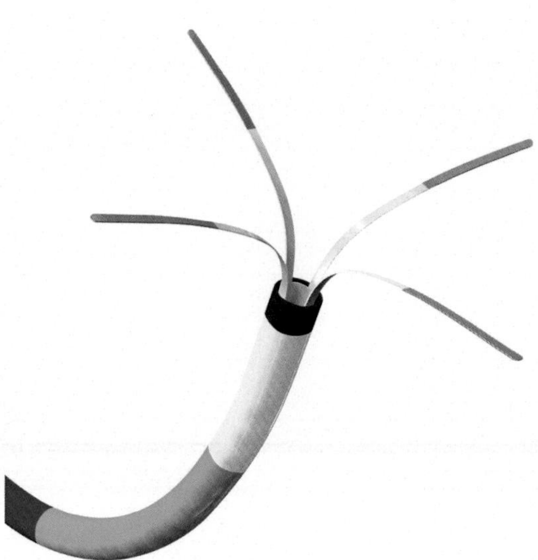

FIGURE 30.104 The Ostial Pro system. *Merit Medical, Reprinted by Permission.*

30.13.2 Ostial Flash

CTO PCI Manual Online cases: 37, 118.
PCI Manual Online cases: 28, 70, 83, 85. The Ostial Flash balloon flares the proximal stent struts against the aortic wall and facilitates expansion and reengagement with a catheter (Section 15.1.10.5) (Fig. 15.15). Ostial Flash is a dual balloon angioplasty catheter with a larger proximal low pressure balloon and a higher pressure distal balloon [62,63]. There are three markers which can be visualized under fluoroscopy: a proximal marker that marks the proximal end of the anchoring balloon and should be located in the aorta, a mid marker that marks the proximal end of the angioplasty balloon and should be placed at the vessel ostium, and a distal marker that marks the distal end of the angioplasty balloon. The distal balloon length is 8 mm and is available in diameters from 3.0 to 4.5 mm. The proximal anchoring balloon can expand up to 14 mm. The distal balloon is inflated with an inflating device whereas the proximal balloon is inflated with a 1 cc syringe.

30.14 Stents

Latest generation drug-eluting stents (DES) should be used in nearly all PCIs. The only lesion subgroup in which DES have shown similar outcomes with bare metal stents are saphenous vein grafts (Section 18.10.2) [64,65]. Covered stents are discussed in section 30.18: Complication management.

Step-by-step description of stent delivery and deployment is provided in Chapter 10: Stenting.

30.15 Vascular closure devices

Use of vascular closure devices can be used to achieve hemostasis at the end of the procedure. Step-by-step description of vascular closure is provided in Chapter 11: Access closure.

30.15.1 Femoral access

Use of closure devices allows faster patient ambulation and improves patient comfort, and may reduce the risk for complications, although the latter is debated [66,67]. The Angioseal and Perclose are most commonly used. The MANTA device (Teleflex) is also available for large bore access closure.

30.15.1.1 Angioseal

The Angioseal VIP device (Fig. 30.105) achieves hemostasis by "sandwiching" the arteriotomy between a polymer anchor (i.e., inserted inside the artery) and a collagen sponge (located outside the artery) that are connected with a suture.

The Angioseal VIP system has the following components:

1. Guidewire.
2. Insertion sheath.
3. Dilator (also called arteriotomy locator, as it allows estimation of the depth of insertion into the femoral artery). The indicator has a side hole that aligns with a hole in the insertion sheath 1.5 cm from the tip of the sheath.
4. Angioseal device (that contains the polymer anchor and the hemostatic collagen sponge connected with a suture) (Fig. 30.106).

All components of the Angioseal VIP device (anchor, suture, collagen) are absorbed within 60–90 days from deployment.

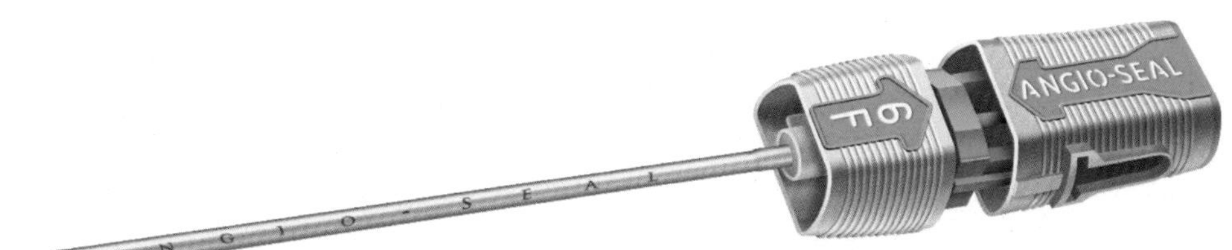

FIGURE 30.105 The Angioseal VIP system. *©2020 Terumo Medical Corporation. All rights reserved.*

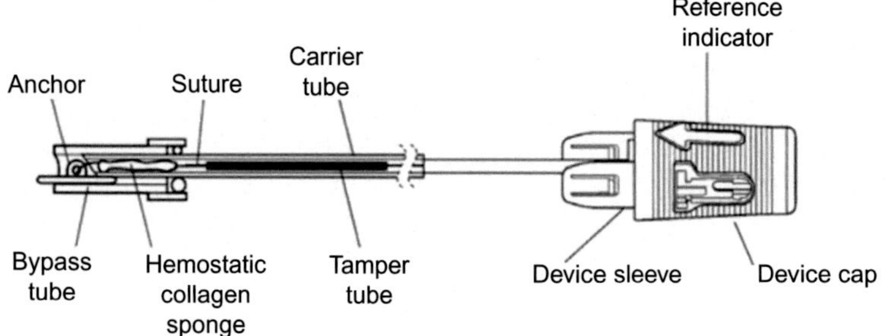

FIGURE 30.106 Illustration of the Angioseal device. *©2020 Terumo Medical Corporation. All rights reserved.*

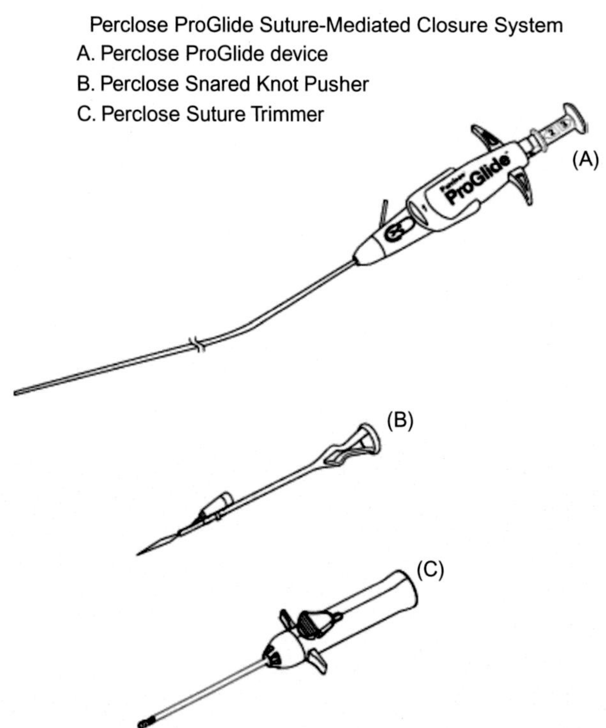

FIGURE 30.107 Illustration of the Perclose device. *Courtesy Abbott Vascular. ©2019 Abbott. All Rights Reserved.*

There are two Angioseal VIP devices, one 6 French (that can also be used to seal arteriotomies caused by 7 Fr sheaths) and one 8 French. The 6 French is compatible with 0.035 in. (but not 0.038 in. guidewires), whereas the 8 French sheath is compatible with 0.038 in. guidewires. The Angioseal VIP is packaged with the appropriate sized wires.

Step-by-step instructions for use of the Angioseal VIP system are provided in Section 11.1.6.

30.15.1.2 Perclose

The Perclose system (Fig. 30.107) is designed to deliver a single monofilament polypropylene suture across the vessel wall. It can be used to close both arterial (5–21 Fr) and large bore venous (5–24 Fr) access sites. It has three components:

1. Perclose ProGlide device
2. Perclose Snared Knot Pusher
3. Perclose Suture Trimmer

Step-by-step instructions for use of the Perclose system are provided in Section 11.1.7.

30.15.2 Radial access

Several radial compression devices are available for achieving radial hemostasis, such as the TR band (Terumo), Prelude Sync (Merit Medical) and Tracelet (Medtronic), as described in Section 11.2. Many devices have an air chamber that is inflated applying pressure to the radial access site.

30.16 CTO PCI dissection/reentry equipment

The CrossBoss catheter and Stingray balloon and guidewire (Boston Scientific) can be used for antegrade dissection and reentry during CTO PCI. For additional information on how to use these devices please refer to the *Manual of CTO PCI* [8].

30.16.1 CrossBoss

CTO PCI Manual Online cases: 8, 9, 16, 17, 19, 25, 32, 40, 50. The CrossBoss catheter Fig. 30.108) is a stiff, metallic, over-the-wire catheter with a 1 mm blunt, rounded, hydrophilic-coated distal tip that can advance through the occlusion when the catheter is rotated rapidly using a proximal torquer device ("fast spin" technique). If the catheter enters the subintimal space, it creates a limited dissection plane making reentry into the distal true lumen easier. The risk of perforation is low provided that the CrossBoss catheter is not advanced into side branches. If the CTO is crossed subintimally, the Stingray LP balloon and guidewire can be used to assist with reentry into the distal true lumen, as described below [68–70].

30.16.2 Stingray

CTO PCI Manual Online cases: 9, 16, 17, 21, 25, 27, 32, 34, 35, 43, 45, 48, 72, 79, 80, 82, 83, 92, 93, 97, 98, 100, 105, 120, 129, 131, 135, 142, 144, 146, 148, 149.
PCI Manual Online case: 40.

The Stingray LP (low-profile) balloon is 2.5 mm wide when inflated and 10 mm in length and has a flat shape with two side exit ports. Upon low-pressure (4 atm) inflation it orients one exit port automatically toward the true lumen, especially when the space created by dissection is small (Fig. 30.109) [71]. The Stingray guidewire is a stiff guidewire with a 20 cm distal radiopaque segment and a 0.009 in. tapered tip with a 0.0035 in. distal prong. The Stingray guidewire can be directed toward one of the two side ports of the Stingray LP balloon under fluoroscopic guidance to reenter into the distal true lumen [68–70]. Use of these devices is discussed in detail in the *Manual of CTO PCI* [8].

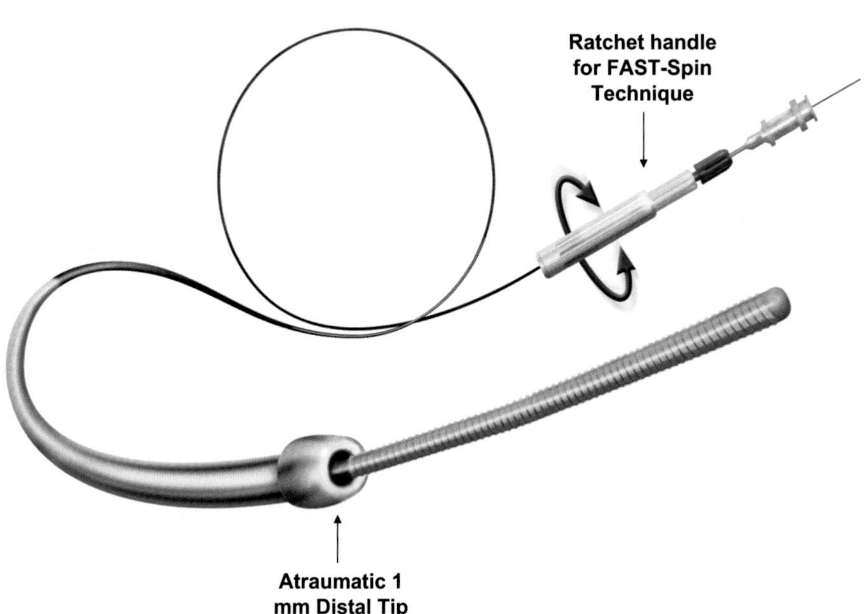

FIGURE 30.108 Illustration of the CrossBoss catheter. *Image courtesy Boston Scientific © 2019 Boston Scientific Corporation or its affiliates. All rights reserved.*

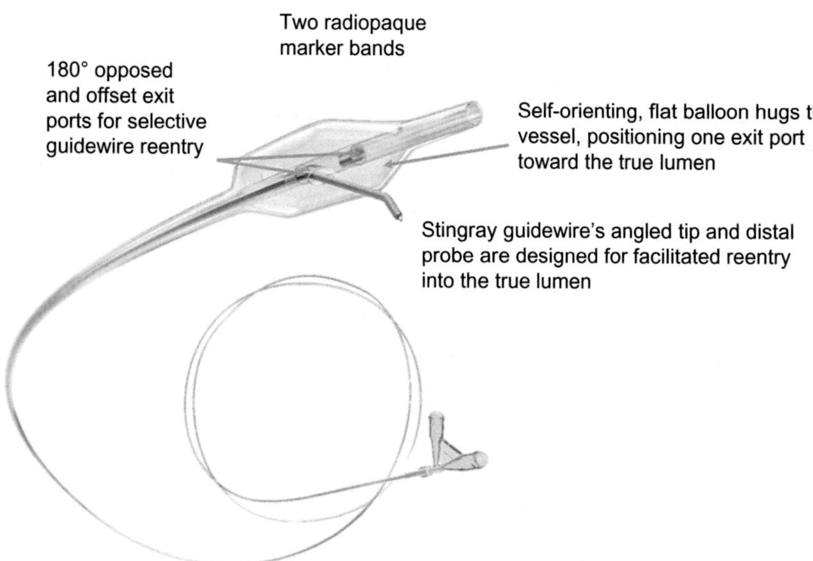
Two radiopaque marker bands
180° opposed and offset exit ports for selective guidewire reentry
Self-orienting, flat balloon hugs the vessel, positioning one exit port toward the true lumen
Stingray guidewire's angled tip and distal probe are designed for facilitated reentry into the true lumen

FIGURE 30.109 Illustration of the Stingray LP catheter and the Stingray guidewire. *Image courtesy Boston Scientific © 2019 Boston Scientific Corporation or its affiliates. All rights reserved.*

30.17 Intravascular imaging

Intravascular ultrasound (IVUS) and optical coherence tomography (OCT) are the currently available intravascular imaging modalities to guide PCI. Chapter 13: Coronary intravascular imaging provides detailed information on how to use intravascular imaging, including a comparison of IVUS and OCT (Section 13.2).

30.17.1 Intravascular ultrasound (Section 13.4)

There are three types of IVUS systems:

1. Solid-state, phased array system (Eagle Eye Platinum, Philips). This catheter is also available with a short tip (2.5 mm tip-to-imaging distance).
2. Rotational systems, such as the Refinity (45 MHz, Philips), OptiCross HD (60 MHz, Boston Scientific), and the Kodama HD IVUS (40–60 MHz, ACIST).
3. Dual imaging systems:
   - DualPro (InfraRedx), that combines IVUS (35–65 MHz) with NIRS (near-infrared spectroscopy), providing simultaneous IVUS and NIRS assessment. Large lipid core plaques as identified by NIRS have been associated with increased risk of distal embolization and no-reflow (Section 25.2.3.2).
   - Novasight hybrid system that provides simultaneous IVUS and OCT imaging (Conavi Medical).

The solid state IVUS system does not require preparation and is hence faster to use but is more difficult to deliver compared with the rotational systems, has lower resolution, and does not allow length measurements, uncles coregistration is performed (Section 13.4.1).

30.17.2 Optical coherence tomography (Section 13.3)

Optical coherence tomography (OCT) provides high-resolution images which greatly facilitates interpretation but requires contrast injection during acquisition to clear the blood from the coronary artery. One OCT imaging system is available in the United States (Dragonfly Optis Imaging catheter, Abbott Vascular), as well as one combined IVUS-OCT system (Novasight, Conavi Medical).

30.18 Complication management

Covered stents, coils, Amplatzer vascular plugs, pericardiocentesis tray, and snares are needed to treat perforations and equipment loss or entrapment. Although this equipment is rarely needed, it must be available in all cardiac catheterization laboratories.

30.18.1 Covered stents

CTO PCI Manual Online cases: 26, 39, 40, 42, 63, 89, 90, 131, 134.
PCI Manual Online cases: 45, 55, 80.

As of 2020, two coronary covered stents are commercially available in the United States, the Graftmaster Rx (Abbott Vascular) and PK Papyrus (Biotronik). They are both approved through a humanitarian device exemption [72] for use in large vessel perforations. Another covered stent, the BeGraft coronary stent (Bentley InnoMed GmbH, Hechingen, Germany), is available in Europe.

30.18.1.1 Graftmaster

1. The Graftmaster Rx (Fig. 30.110) consists of two stainless steel stents with a middle layer of ePTFE.
2. It is bulky and difficult to deliver; hence excellent guide catheter support is important.
3. It is available in diameters of 2.8–4.8 mm and lengths between 16 and 26 mm.
4. Requires a 6 French guide catheter for the 2.8–4.0 mm stents and a 7 French guide catheter for the 4.5 and 4.8 mm stents.
5. The Graftmaster Rx may be difficult to advance through previously deployed stents, necessitating techniques such as distal anchor and use of a guide catheter extension.
6. Minimum inflation pressure is 15 atm, but even higher pressures for up to 60 seconds (and use of intravascular ultrasound) are preferred to ensure adequate stent expansion.
7. After expansion the stent may shorten up to 1.6 mm on each side (for a total of 3.2 mm at nominal pressure, which is 15 atm). Hence, adequate overlap of stents is important to cover long areas of perforation.
8. Use of a *dual catheter ("ping-pong guide") technique* (Section 26.3) is often required to minimize bleeding into the pericardium while preparing for covered stent delivery and deployment [73], although delivery of both a balloon and a covered stent is feasible through 8 French guide catheters.
9. Postdilation of the shoulders of the stent may be necessary to fully appose the stent to the vessel wall if extravasation persists behind the stent despite covering the perforation.

30.18.1.2 PK Papyrus

1. The PK Papyrus (Fig. 30.111) covered coronary stent system is a balloon-expandable covered stent mounted on a rapid-exchange delivery catheter [74]. The stent is manufactured by covering the PRO-Kinetic Energy ultrathin strut, amorphous silicon carbide coated cobalt chromium stent with an electrospun polymeric matrix composed of siloxane-based polyurethane. More specifically, the electropsun cover consists of individual nonwoven fibers approximately 2 μm in thickness. When the stent is fully expanded, the cover is approximately 90 μm in thickness [74].

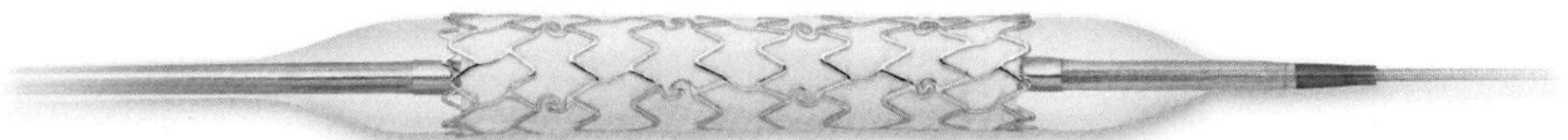

FIGURE 30.110 Illustration of the Graftmaster Rx covered stent. *Courtesy Abbott Vascular. ©2019 Abbott. All Rights Reserved.*

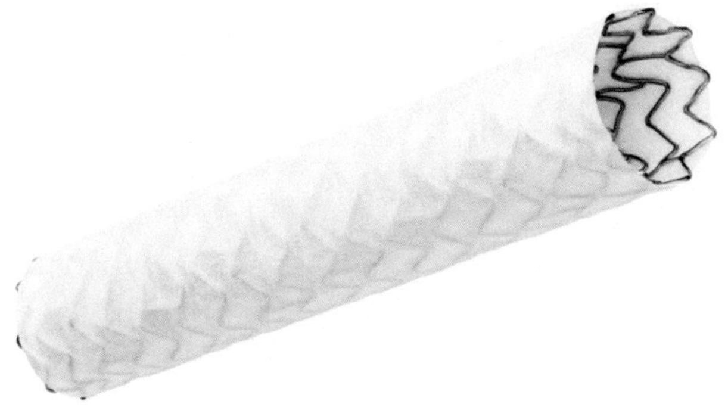

FIGURE 30.111 Illustration of the PK Papyrus covered stent. *Image used with permission from BIOTRONIK, Inc.*

2. The device size matrix ranges from 2.5 to 5.0 mm stent diameters and lengths of 15, 20, and 26 mm. The PK Papyrus stent may be postdilated to a maximum stent expansion diameter of 3.50 mm for the 2.5 and 3.0 mm stents; 4.65 mm for the 3.5 and 4.0 mm stents; and 5.63 mm for the 4.5 and 5.0 mm stents. PK Papyrus is compatible with 5 French guiding catheters for diameters 2.5−4.0 mm and 6 French for diameters 4.5 and 5.0 mm [74].
3. PK Papyrus is easier to deliver compared with the Graftmaster but can occasionally be dislodged from the balloon during attempts to deliver it to the perforation site.

30.18.1.3 BeGraft

The BeGraft coronary stent (Bentley InnoMed GmbH, Germany) is a cobalt chrome (L-605), open-cell platform covered with a single-layer of a PTFE membrane (thickness of 89 ± 25 μm), which is clamped at the proximal and distal stent ends [75]. The single layer design results in a crossing profile between 1.1−1.4 mm with a guide catheter compatibility of 5 French for all sizes.

30.18.2 Coils

CTO PCI Manual Online cases: 41, 43, 61, 105, 139, 140.
PCI Manual Online case: 39. Online video: "How to deliver and deploy an Axium coil".

Coils should be available for use in case of distal (Section 26.4) or collateral vessel perforation. They can also be used to stop a large vessel perforation (Section 26.3) by occluding the vessel. Coils are permanent embolic agents that can be deployed either through 0.014 in. compatible microcatheters (neurovascular coils [Table 26.3]), such as the Axium (Medtronic, Fig. 30.112) or the Smart coil (Penumbra) or through the Finecross or larger 0.018 in. microcatheters (standard coils, such as Interlock, Boston Scientific, Fig. 30.113) [76]. Coils are usually made of stainless steel or platinum alloys and some of them have polymers or synthetic wool or dacron fibers attached along the length of the wire to increase thrombogenicity. Once advanced into the target vessel, the coils assume a preformed shape, sealing the perforation. Particular attention needs to be made when coiling branches to prevent the coil from prolapsing into the main vessel.

Coils tips and tricks [77]:

1. Since coils are used very infrequently in cardiac catheterization laboratories, it is important for each operator to be familiar with the principles underlying their use and with one to two specific coil types, so that coils can be delivered rapidly in case of perforation.
2. There are two broad categories of coils according to *mechanism of release*: pushable and detachable. *Pushable* coils are inserted into a microcatheter and pushed with a coil pusher or the front end of a guidewire until they exit into the vessel, hence deployment can be unpredictable and is irreversible. *Detachable* coils are released using a dedicated release device once their position into the target vessel is confirmed; conversely if their position is not satisfactory, they can be retrieved. Detachable coils are preferred for treating perforations, as they allow optimal and predictable positioning.

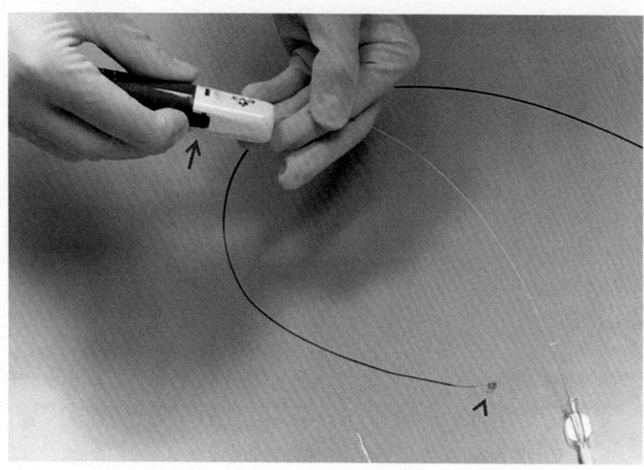

FIGURE 30.112 Illustration of the Axium detachable coil system (Medtronic), that is compatible with 0.014 in. microcatheters. The coil (*arrowhead*) is released pulling the lever (*arrow*) on the delivery device.

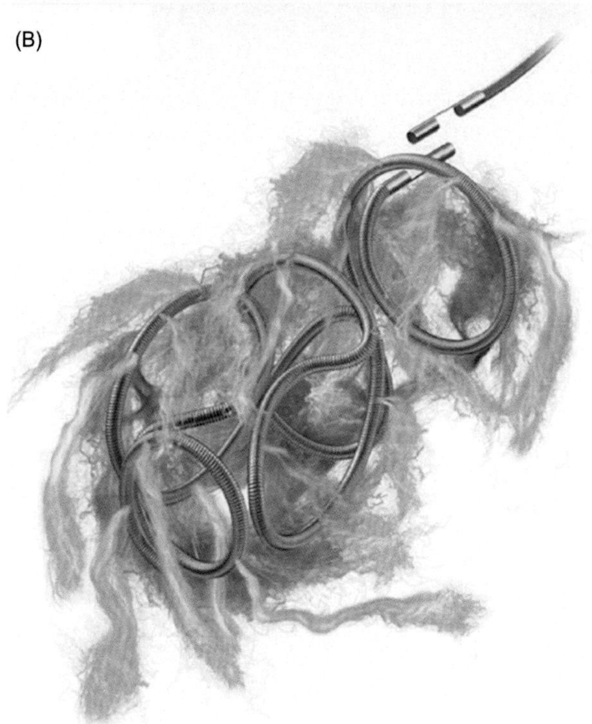

FIGURE 30.113 Example of a detachable coil that can be used for embolization in case of distal coronary perforation (Interlock, *Boston Scientific*). Panel A demonstrates deployment of the coil whereas panel B illustrates the coil configuration after delivery. *Image courtesy Boston Scientific* © *2019 Boston Scientific Corporation or its affiliates. All rights reserved.*

3. There are also two broad categories of coils according to the *size of the delivery microcatheter. Coils compatible with 0.014 in. microcatheters (such as Axium, Medtronic) are preferred*, as they can be delivered through the standard microcatheters used for CTO PCI (such as the Corsair, Caravel, Turnpike) without requiring change to a larger microcatheter. Coils compatible with 0.018 in. microcatheters (such as the Interlock [Boston Scientific], Azur [Terumo], and Micronester [Cook]), cannot be delivered through the standard microcatheters used during CTO PCI with the exception of the Finecross, and require change to a larger microcatheter, such as the Progreat (Terumo), Renegade (Boston Scientific), or Transit (Cordis).
4. Simultaneous balloon inflation (to stop bleeding into the pericardium) and coil delivery (through a microcatheter) can be achieved through a single guide catheter ("block and deliver" technique, Section 26.4) [78,79].

30.18.3 Amplatzer vascular plugs

CTO PCI Manual online case: 127.

Another method for occluding a vessel (either emergently in case of perforation or electively, for example, in a saphenous vein graft that has had multiple failures after recanalization of the native coronary artery) is using the Ampltatzer vascular plugs. The Amplatzer vascular plugs are disks made of a mesh of braided nitinol that are attached to a PTFE-coated delivery wire with a stainless-steel micro screw, which allows the operator to release the plug into the final position by rotating the cable in a counter clockwise fashion using a supplied torque device. The plug can be retrieved and readjusted as needed before final release [80]. There are four types of Amplatzer vascular plugs (Table 30.12). The Amplatzer Vascular Plug II is used in most cases when saphenous vein grafts need to be occluded (CTO PCI Manual online case 127).

Size selection: the diameter of the Amplatzer Vascular Plug is selected to be 30%–50% larger than the vessel diameter at the occlusion site. Moreover, the operator should ensure that the occlusion site has sufficient length to accommodate the deployed device length without obstructing other vessels or anatomical structures.

30.18.4 Pericardiocentesis tray

Pericardiocentesis can be performed using a standard 18 gauge needle, a J-tip 0.035 in. guidewire and a standard pigtail catheter, however having all equipment assembled in a premanufactured pericardiocentesis kit can facilitate and speed up the procedure. Use of a micropuncture needle kit and echocardiographic guidance (if time allows) can increase the safety of the procedure.

30.18.5 Snares

Snares are often needed to retrieve entrapped or lost equipment (CTO Manual online case: 74, 104, 122; PCI Manual online case 77, 93). They are also used for capturing the retrograde guidewire during retrograde CTO PCI (CTO Manual online cases 18, 77, 101, 116).

30.18.5.1 Snare types

There is a wide variety of commercially available snares for retrieving equipment from the coronary and peripheral circulation, such as the Amplatz Goose Neck snare (Medtronic), the Microsnare Elite (Teleflex) and the En Snare (Merit Medical) (Fig. 30.114) [81].

These snares consist of a wire loop typically made of nitinol, that is advanced through a microcatheter (or alternatively through a diagnostic or guide catheter), positioned around the lost device, and then pulled back, trapping the device against the catheter. The catheter/loop assembly is subsequently removed from the body, along with the lost device. Three-loop (tulip) snares are preferred, as they have three overlapping loops (instead of one in the Goose Neck snare), hence increasing the likelihood of retrieving the lost device. If a commercially manufactured snare is not available, a loop snare can be created in the catheterization laboratory using an exchange-length coronary guidewire and a multipurpose catheter through the distal tip of which the guidewire tip is reinserted (Fig. 30.115).

Another way to create a "homemade" snare ('KAM-snare') is by using a wire, a balloon and a guide catheter extension: the snare consists of a wire loop being trapped by an inflated balloon (at 8–12 atm) at the distal site of a guide catheter extension (Fig. 30.116) [18].

Online video: "How to make a homemade snare"

30.18.5.2 Snaring technique

(Fig. 30.117)

The snare is withdrawn and collapsed into the introducer tool (Fig. 30.117, panels A, B, and C). It is then introduced into a diagnostic or guide catheter (panels D and E) and advanced until it exits from the distal guide tip (panel F). The snare is then positioned close to the lost device and the snare is withdrawn into the diagnostic or guide catheter until the device is captured between the snare and the catheter tip. The device is then withdrawn along with the catheter and the snare.

30.18.5.3 Snaring tips and tricks

1. The size of the snare depends on the size of the vascular space that contains the lost or entrapped device. For example, small (2–4 mm) snares are used within the coronary arteries and large snares (27–45 mm or 18–30 mm) are used in the aorta.

TABLE 30.12 Overview of the Amplatzer vascular plugs.

	AVP I	AVP II	AVP III	AVP IV
	A A B C	A B C C		B C D C A E
Construction	Single-layered cylindrical disk	Densely braided multilayer nitinol mesh with three components generating six barrier planes for acceleration of vascular occlusion	Oblong cross-sectional shape, multiple Nitinol mesh layers and extended rims	Two tapering lobe design
Diameter	4–16 mm	3–22 mm	4–14 mm	4–8 mm
Maximum delivery sheath length	100 cm	100 cm		155 cm
Minimum delivery catheter or delivery sheath ID (inch)	4–8 mm plug: 0.056 10–12 mm plug: 0.067 14–16 mm plug: 0.088	3–8 mm plug: 0.056 10–12 mm plug: 0.070 14–16 mm plug: 0.086 18–22 mm plug: 0.098		0.038 in. diagnostic catheter
Comment	Excellent for short landing zones	Significantly reduces time to occlusion	Provides the fastest 7 occlusion of all AMPLATZER vascular plugs and is ideal for high-flow situations	The flexible mesh of the AMPLATZER Vascular Plug 4 and the floppy distal section of the delivery wire enable the device to travel through tortuous anatomy.

Images of the Amplatzer vascular plugs courtesy Abbott Vascular.

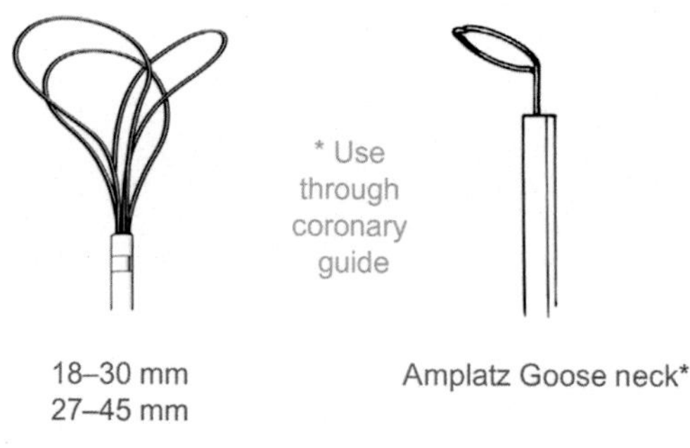

FIGURE 30.114 Illustration of three-loop and single-loop snares. *Reproduced with permission from Brilakis ES. Manual of coronary chronic total occlusion interventions. A step-by-step approach. 2nd ed. Cambridge, MA: Elsevier; 2017 (Figure 2.62).*

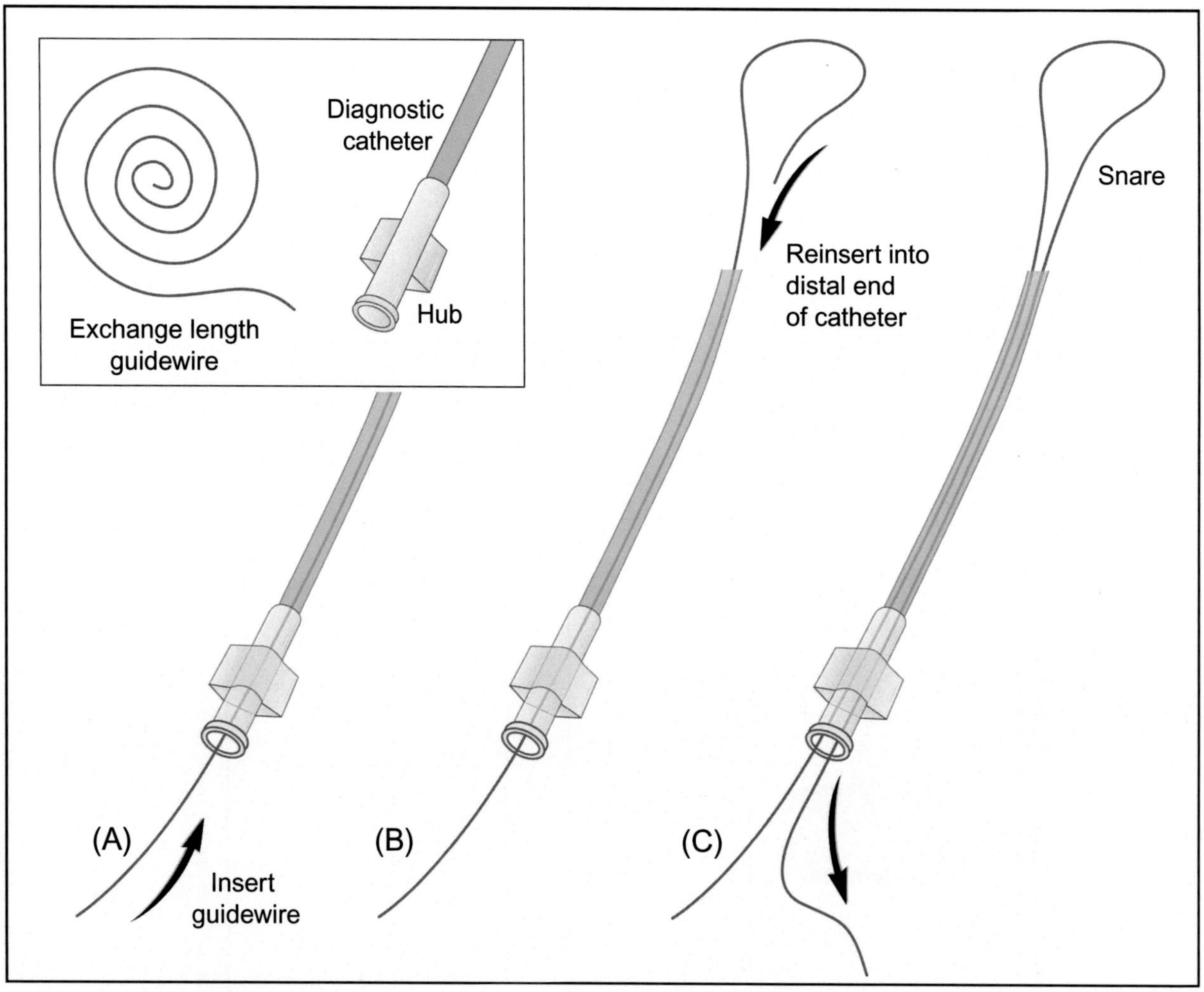

FIGURE 30.115 How to create a loop snare using a catheter and a long 0.014 in. coronary guidewire.

2. Each snare comes with a delivery sheath, which is usually discarded. The snare is advanced through a diagnostic or guide catheter and withdrawn, trapping the lost device between the loops of the snare and the tip of the catheter.
3. Do not discard the snare collapsing tool, as it is necessary for reintroducing the snare into the guide catheter, if needed.

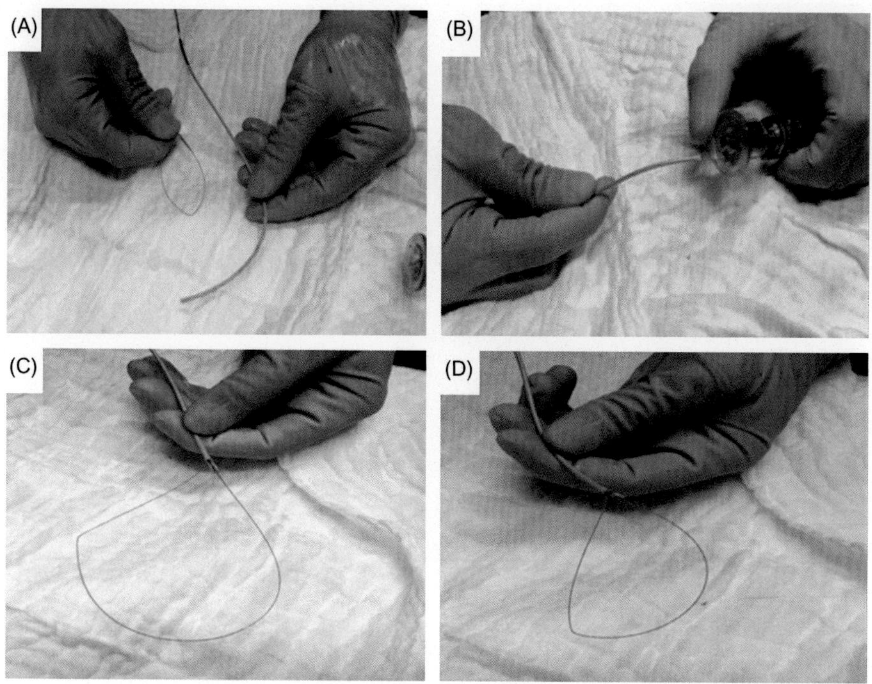

FIGURE 30.116 How to make a "KAM-snare." The "KAM-snare" consists of a guidewire being inserted within the monorail lumen of a conventional angioplasty balloon. The distal end of the guidewire is then shaped as a loop (panel A). The looped guidewire with the railed balloon are introduced into a guide catheter extension at the proximal entry site. Hence, the balloon entraps the distal returning end of the guidewire loop within the guide catheter extension (panel B). The "KAM-snare" is inserted into the proximal site of the Y-connector (panel B). By either pulling or pushing the exterior proximal end of the guidewire the diameter of the "KAM-snare" either increases or decreases (panels C and D). *Courtesy of Dr. Kambis Mashayekhi.*

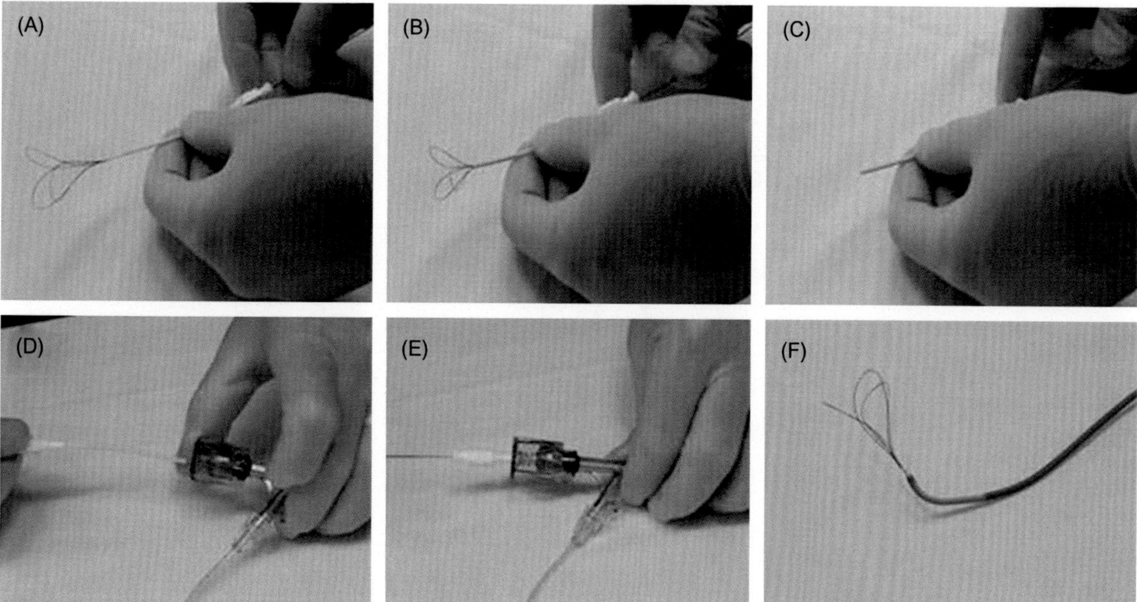

FIGURE 30.117 How to insert a three-loop snare into a guide catheter. *Courtesy of Dr. William Nicholson. Reproduced with permission from Brilakis ES. Manual of coronary chronic total occlusion interventions. A step-by-step approach. 2nd ed. Cambridge, MA: Elsevier; 2017 (Figure 2.63).*

30.19 Radiation protection

Every effort should be undertaken to minimize patient and operator dose, as described in detail in Section 28.2. In addition to fixed shields, there are several radiation protection pads that decrease scatter radiation from the patient, such as the RadPad (Worldwide Innovations & Technologies, Inc.), and the Zero Gravity ceiling suspended radiation protection system (Biotronik) (Fig. 30.118).

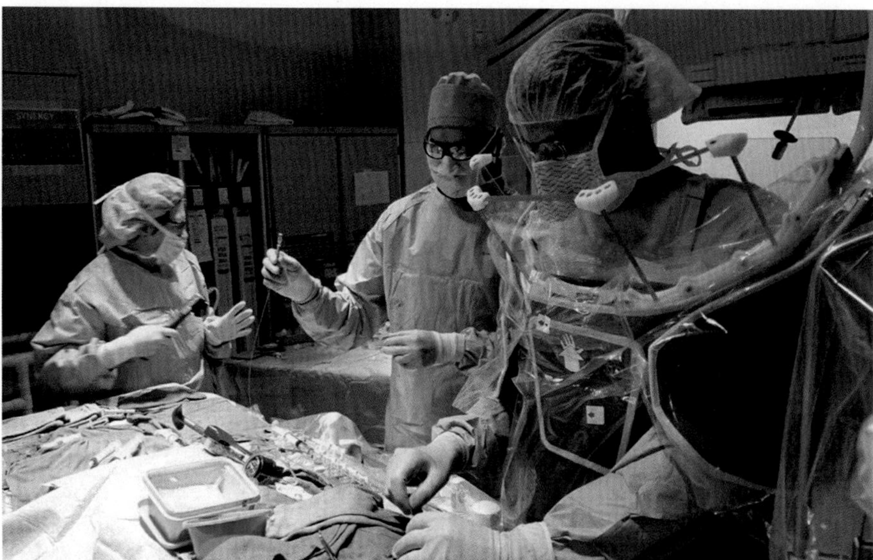

FIGURE 30.118 Illustration of the Zero Gravity system.

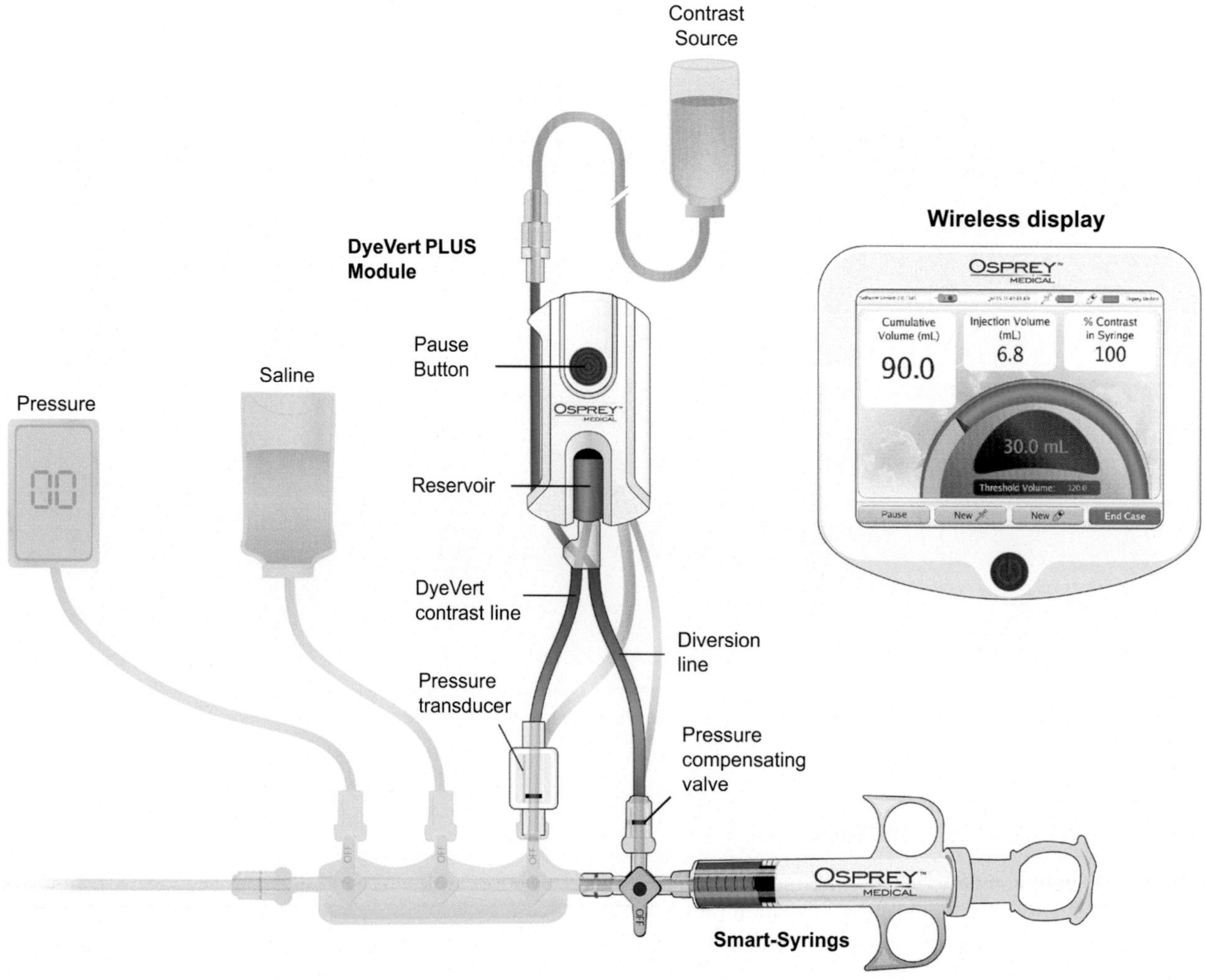

Contrast Source
DyeVert PLUS Module
Pause Button
Reservoir
OSPREY MEDICAL
Pressure
Saline
DyeVert contrast line
Diversion line
Pressure transducer
Pressure compensating valve
Smart-Syrings
Wireless display
OSPREY MEDICAL
Cumulative Volume (mL)
90.0
Injection Volume (mL)
6.8
% Contrast in Syringe
100
30.0 mL
Threshold Volume:
Pause
New
New
End Case

FIGURE 30.119 Illustration of the Dyevert Plus system. *Reproduced with permission from Osprey Medical.*

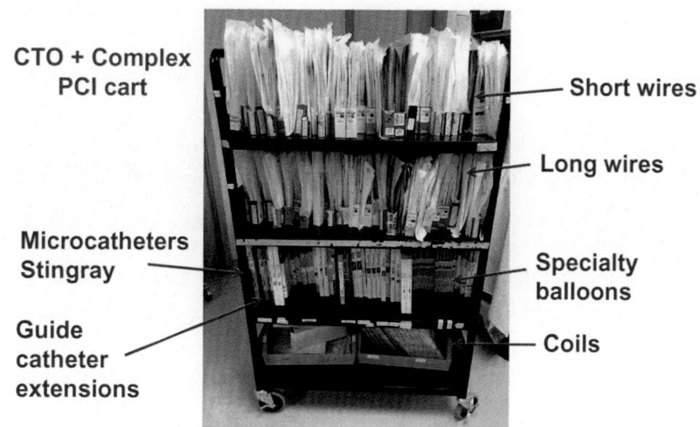

FIGURE 30.120 Example of a "CTO and complex PCI cart."

30.20 Hemodynamic support devices

PCI Manual Online cases: 45 (IABP), 31, 36, 41, 59, 63, 73, 91, 92 (Impella), 27, 40, 51, 72 (VA-ECMO)
CTO PCI Manual Online cases: 11, 14, 20, 31, 46, 126, 136 (Impella), 29, 67, 138, 146 (VA-ECMO)

Hemodynamic support can be used prophylactically, after occurrence of a complication during PCI, or in patients with cardiogenic shock. Given the potential for serious, life-threatening, complications during complex PCI, availability of hemodynamic support devices is important as part of comprehensive high risk and complex PCI programs [82]. A description of currently available hemodynamic support device and step-by-step instructions are provided in Chapter 14: Hemodynamic support.

30.21 Contrast management

Prevention of acute kidney injury after cardiac catheterization is discussed in detail in Section 28.3. Preprocedural hydration and reducing contrast volume are the hallmarks of prevention. One way to reduce contrast volume is by using the Dyevert Plus system (Osprey Medical, Minneapolis, Minnesota, Fig. 30.119). The Dyevert Plus is a device that is connected between the injection syringe and the manifold via a four-way stopcock, allowing diversion of excess contrast during manual injection. The fraction of contrast that would not contribute to coronary opacification, but rather would reflux into the aortic root, is diverted into a reservoir chamber. The system also provides real time measurement of the contrast volume administered to the patient. Initial studies suggest 20%–40% reduction in contrast volume with use of the device [83,84].

30.22 Brachytherapy

Coronary brachytherapy is useful for treatment of recurrent in-stent restenosis [85,86]. Beta radiation with a strontium-90/yttrium source is the only currently available system, but is only available at a limited number of centers. Patients who undergo brachytherapy should receive indefinite dual antiplatelet therapy to minimize the risk of stent thrombosis.

30.23 The "CTO–Complex PCI cart"

Having a dedicated "CTO-Complex PCI cart" (Fig. 30.120) with all commonly used CTO and PCI equipment (including equipment for managing complications, such as covered stents and coils) can facilitate and expedite performance of these procedures. The CTO–Complex PCI cart should be readily available during the procedure, restocked regularly, and known to the operators and the catheterization laboratory staff.

References

[1] Brilakis ES. The essential equipment for CTO interventions. Cardiol Today Interv 2013;2013 May–June.

[2] Joyal D, Thompson CA, Grantham JA, Buller CEH, Rinfret S. The retrograde technique for recanalization of chronic total occlusions: a step-by-step approach. JACC Cardiovasc Interv 2012;5:1–11.

[3] Brilakis ES, Grantham JA, Thompson CA, et al. The retrograde approach to coronary artery chronic total occlusions: a practical approach. Catheter Cardiovasc Interv 2012;79:3–19.

[4] Brilakis ES, Grantham JA, Rinfret S, et al. A percutaneous treatment algorithm for crossing coronary chronic total occlusions. JACC Cardiovasc Interv 2012;5:367–79.

[5] Agelaki M, Koutouzis M. Balloon-assisted tracking for challenging transradial percutaneous coronary intervention. Anatol J Cardiol 2017;17:E1.

[6] Ryan TJ. The coronary angiogram and its seminal contributions to cardiovascular medicine over five decades. Circulation 2002;106:752–6.

[7] Tajti P, Sandoval Y, Brilakis ES. "Around the world"—How to reach native coronary artery lesions through long and tortuous aortocoronary bypass grafts. Hellenic J Cardiol 2018;59:354–7.

[8] Brilakis ES. Manual of coronary chronic total occlusion interventions. A step-by-step approach. 2nd ed. Cambridge, MA: Elsevier; 2017.

[9] Wu EB, Chan WW, Yu CM. Retrograde chronic total occlusion intervention: tips and tricks. Catheter Cardiovasc Interv 2008;72:806–14.

[10] Takahashi S, Saito S, Tanaka S, et al. New method to increase a backup support of a 6 French guiding coronary catheter. Catheter Cardiovasc Interv 2004;63:452–6.

[11] Luna M, Papayannis A, Holper EM, Banerjee S, Brilakis ES. Transfemoral use of the GuideLiner catheter in complex coronary and bypass graft interventions. Catheter Cardiovasc Interv 2012;80:437–46.

[12] Senguttuvan NB, Sharma SK, Kini A. Percutaneous intervention of chronic total occlusion of anomalous right coronary artery originating from left sinus—use of mother and child technique using guideliner. Indian Heart J 2015;67:S41–2.

[13] Vishnevsky A, Savage MP, Fischman DL. GuideLiner as guide catheter extension for the unreachable mammary bypass graft. Catheter Cardiovasc Interv 2018;92:1138–40.

[14] Williams R, Shouls G, Firoozi S. Mother-and-child telescopic guide-catheter extension to identify severe left main stem disease in a patient with a severely dilated aortic root. J Invasive Cardiol 2018;30:E71–2.

[15] Stys AT, Stys TP, Rajpurohit N, Khan MA. A novel application of GuideLiner catheter for thrombectomy in acute myocardial infarction: a case series. J Invasive Cardiol 2013;25:620–4.

[16] Sakakura K, Taniguchi Y, Tsukui T, Yamamoto K, Momomura SI, Fujita H. Successful removal of an entrapped rotational atherectomy burr using a soft guide extension catheter. JACC Cardiovasc Interv 2017;10:e227–9.

[17] Xenogiannis I, Karmpaliotis D, Alaswad K, et al. Comparison between traditional and guide-catheter extension reverse controlled antegrade dissection and retrograde tracking: insights from the PROGRESS-CTO registry. J Invasive Cardiol 2019;31:27–34.

[18] Yokoi K, Sumitsuji S, Kaneda H, et al. A novel homemade snare, safe, economical and size-adjustable. EuroIntervention 2015;10:1307–10.

[19] Hashimoto S, Takahashi A, Yamada T, et al. Spontaneous rotation of the monorail-type guide extension support catheter during advancement of a curved guiding catheter: the potential hazard of twisting with the coronary guidewire. Cardiovasc Interv Ther 2018;33:379–83.

[20] Elbarouni B, Moussa M, Kass M, Toleva O, Vo M, Ravandi A. GuideLiner balloon assisted tracking (GBAT): a new addition to the interventional toolbox. Case Rep Cardiol 2016;2016:6715630.

[21] Andreou C, Karalis I, Maniotis C, Jukema JW, Koutouzis M. Guide extension catheter stepwise advancement facilitated by repeated distal balloon anchoring. Cardiovasc Revasc Med 2017;18:66–9.

[22] Repanas TI, Christopoulos G, Brilakis ES. "Candy cane" guide catheter extension for stent delivery. J Invasive Cardiol 2015;27:E169–70.

[23] Finn MT, Green P, Nicholson W, et al. Mother-daughter-granddaughter double guideliner technique for delivering stents past multiple extreme angulations. Circ Cardiovasc Interv 2016;9.

[24] Mamas MA, Fath-Ordoubadi F, Fraser DG. Distal stent delivery with Guideliner catheter: first in man experience. Catheter Cardiovasc Interv 2010;76:102–11.

[25] Aznaouridis K, Bonou M, Masoura K, Vaina S, Vlachopoulos C, Tousoulis D. Successful stent delivery through a slaloming coronary path. J Invasive Cardiol 2019;31:E43.

[26] Papayannis AC, Michael TT, Brilakis ES. Challenges associated with use of the GuideLiner catheter in percutaneous coronary interventions. J Invasive Cardiol 2012;24:370–1.

[27] Duong T, Christopoulos G, Luna M, et al. Frequency, indications, and outcomes of guide catheter extension use in percutaneous coronary intervention. J Invasive Cardiol 2015;27:E211–15.

[28] Chang YC, Fang HY, Chen TH, Wu CJ. Left main coronary artery bidirectional dissection caused by ejection of guideliner catheter from the guiding catheter. Catheter Cardiovasc Interv 2013;82:E215–20.

[29] Chen Y, Shah AA, Shlofmitz E, et al. Adverse events associated with the use of guide extension catheters during percutaneous coronary intervention: reports from the Manufacturer and User Facility Device Experience (MAUDE) database. Cardiovasc Revasc Med 2019;20:409–12.

[30] Alkhalil M, Smyth A, Walsh SJ, et al. Did the use of the Guideliner V2(TM) guide catheter extension increase complications? A review of the incidence of complications related to the use of the V2 catheter, the influence of right brachiocephalic arterial anatomy and the redesign of the V3(TM) Guideliner and clinical outcomes. Open Heart 2016;3:e000331.

[31] Waterbury TM, Sorajja P, Bell MR, et al. Experience and complications associated with use of guide extension catheters in percutaneous coronary intervention. Catheter Cardiovasc Interv 2016;88:1057–65.

[32] Mitsutake Y, Ebner A, Yeung AC, Taber MD, Davidson CJ, Ikeno F. Efficacy and safety of novel multi-lumen catheter for chronic total occlusions: from preclinical study to first-in-man experience. Catheter Cardiovasc Interv 2015;85:E70–5.

[33] Walsh S, Dudek D, Bryniarski L, et al. Efficacy and safety of novel NovaCross microcatheter for chronic total occlusions: first-in-human study. J Invasive Cardiol 2016;28:88–91.

[34] Moualla SK, Khan S, Heuser RR. Anchoring improved: introduction of a new over-the-wire support balloon. J Invasive Cardiol 2014;26:E130–2.

[35] Vemmou E, Nikolakopoulos I, Xenogiannis I, et al. Recent advances in microcatheter technology for the treatment of chronic total occlusions. Expert Rev Med Devices 2019;16:267–73.

[36] Kandzari DE, Grantham JA, Karmpaliotis D, et al. Safety and efficacy of dedicated guidewire and microcatheter technology for chronic total coronary occlusion revascularization: principal results of the Asahi Intecc Chronic Total Occlusion Study. Coron Artery Dis 2018;29:618–23.

[37] Mohandes M, Rojas S, Guarinos J, et al. Efficacy and safety of Tornus catheter in percutaneous coronary intervention of hard or balloon-uncrossable chronic total occlusion. ARYA Atheroscler 2016;12:206–11.

[38] Brilakis ES, Mashayekhi K, Tsuchikane E, et al. Guiding principles for chronic total occlusion percutaneous coronary intervention. Circulation 2019;140:420–33.

[39] Fujimoto Y, Iwata Y, Yamamoto M, Kobayashi Y. Usefulness of Corsair microcatheter to cross stent struts in bifurcation lesions. Cardiovasc Interv Ther 2014;29:47–51.

[40] Amsavelu S, Carlino M, Brilakis ES. Carlino to the rescue: use of intralesion contrast injection for bailout antegrade and retrograde crossing of complex chronic total occlusions. Catheter Cardiovasc Interv 2016;87:1118–23.

[41] Fischell TA, Moualla SK, Mannem SR. Intracoronary thrombin injection using a microcatheter to treat guidewire-induced coronary artery perforation. Cardiovasc Revasc Med 2011;12:329–33.

[42] Tsuchikane E, Katoh O, Kimura M, Nasu K, Kinoshita Y, Suzuki T. The first clinical experience with a novel catheter for collateral channel tracking in retrograde approach for chronic coronary total occlusions. JACC Cardiovasc Interv 2010;3:165–71.

[43] McClure SJ, Wahr DW, Webb JG. Venture wire control catheter. Catheter Cardiovasc Interv 2005;66:346–50.

[44] Naidu SS, Wong SC. Novel intracoronary steerable support catheter for complex coronary intervention. J Invasive Cardiol 2006;18:80–1.

[45] McNulty E, Cohen J, Chou T, Shunk K. A "grapple hook" technique using a deflectable tip catheter to facilitate complex proximal circumflex interventions. Catheter Cardiovasc Interv 2006;67:46–8.

[46] Aranzulla TC, Colombo A, Sangiorgi GM. Successful endovascular renal artery aneurysm exclusion using the Venture catheter and covered stent implantation: a case report and review of the literature. J Invasive Cardiol 2007;19:E246–53.

[47] Aranzulla TC, Sangiorgi GM, Bartorelli A, et al. Use of the Venture wire control catheter to access complex coronary lesions: how to turn procedural failure into success. EuroIntervention 2008;4:277–84.

[48] Iturbe JM, Abdel-Karim AR, Raja VN, Rangan BV, Banerjee S, Brilakis ES. Use of the venture wire control catheter for the treatment of coronary artery chronic total occlusions. Catheter Cardiovasc Interv 2010;76:936–41.

[49] Ide S, Sumitsuji S, Kaneda H, Kassaian SE, Ostovan MA, Nanto S. A case of successful percutaneous coronary intervention for chronic total occlusion using the reversed guidewire technique. Cardiovasc Interv Ther 2013;28:282–6.

[50] Kawasaki T, Koga H, Serikawa T. New bifurcation guidewire technique: a reversed guidewire technique for extremely angulated bifurcation—a case report. Catheter Cardiovasc Interv 2008;71:73–6.

[51] Suzuki G, Nozaki Y, Sakurai M. A novel guidewire approach for handling acute-angle bifurcations: reversed guidewire technique with adjunctive use of a double-lumen microcatheter. J Invasive Cardiol 2013;25:48–54.

[52] Fang HY, Lee CH, Fang CY, et al. Application of penetration device (Tornus) for percutaneous coronary intervention in balloon uncrossable chronic total occlusion-procedure outcomes, complications, and predictors of device success. Catheter Cardiovasc Interv 2011;78:356–62.

[53] Barbato E, Shlofmitz E, Milkas A, Shlofmitz R, Azzalini L, Colombo A. State of the art: evolving concepts in the treatment of heavily calcified and undilatable coronary stenoses - from debulking to plaque modification, a 40-year-long journey. EuroIntervention 2017;13:696–705.

[54] Vo MN, Brilakis ES, Grantham JA. Novel use of cutting balloon to treat subintimal hematomas during chronic total occlusion interventions. Catheter Cardiovasc Interv 2018;91:53–6.

[55] Megaly M, Rofael M, Saad M, et al. Outcomes with drug-coated balloons in small-vessel coronary artery disease. Catheter Cardiovasc Interv 2019;93:E277–86.

[56] Raja Y, Routledge HC, Doshi SN. A noncompliant, high pressure balloon to manage undilatable coronary lesions. Catheter Cardiovasc Interv 2010;75:1067–73.

[57] Karacsonyi J, Karatasakis A, Danek BA, Banerjee S, Brilakis ES. Laser applications in the coronaries. Textbook of atherectomy. Shammas NW, editor. HMP Communications; 2016.

[58] Karacsonyi J, Danek BA, Karatasakis A, Ungi I, Banerjee S, Brilakis ES. Laser coronary atherectomy during contrast injection for treating an underexpanded stent. JACC Cardiovasc Interv 2016;9:e147–8.

[59] Ali A, Cox D, Dib N, et al. Rheolytic thrombectomy with percutaneous coronary intervention for infarct size reduction in acute myocardial infarction: 30-day results from a multicenter randomized study. J Am Coll Cardiol 2006;48:244–52.

[60] Fischell TA, Malhotra S, Khan S. A new ostial stent positioning system (Ostial Pro) for the accurate placement of stents to treat aorto-ostial lesions. Catheter Cardiovasc Interv 2008;71:353–7.

[61] Fischell TA, Saltiel FS, Foster MT, Wong SC, Dishman DA, Moses J. Initial clinical experience using an ostial stent positioning system (Ostial Pro) for the accurate placement of stents in the treatment of coronary aorto-ostial lesions. J Invasive Cardiol 2009;21:53–9.

[62] Nguyen-Trong PJ, Martinez Parachini JR, Resendes E, et al. Procedural outcomes with use of the flash ostial system in aorto-coronary ostial lesions. Catheter Cardiovasc Interv 2016;88:1067–74.

[63] Desai R, Kumar G. Flash ostial balloon in right internal mammary artery percutaneous coronary intervention: a novel approach. Cureus 2017;9:e1537.

[64] Brilakis ES, Edson R, Bhatt DL, et al. Drug-eluting stents versus bare-metal stents in saphenous vein grafts: a double-blind, randomised trial. Lancet 2018;391:1997–2007.

[65] Colleran R, Kufner S, Mehilli J, et al. Efficacy over time with drug-eluting stents in saphenous vein graft lesions. J Am Coll Cardiol 2018;71:1973–82.

[66] Robertson L, Andras A, Colgan F, Jackson R. Vascular closure devices for femoral arterial puncture site haemostasis. Cochrane Database Syst Rev 2016;3:CD009541.

[67] Jiang J, Zou J, Ma H, et al. Network meta-analysis of randomized trials on the safety of vascular closure devices for femoral arterial puncture site haemostasis. Sci Rep 2015;5:13761.

[68] Werner GS. The BridgePoint devices to facilitate recanalization of chronic total coronary occlusions through controlled subintimal reentry. Expert Rev Med Devices 2011;8:23–9.

[69] Brilakis ES, Lombardi WB, Banerjee S. Use of the Stingray guidewire and the Venture catheter for crossing flush coronary chronic total occlusions due to in-stent restenosis. Catheter Cardiovasc Interv 2010;76:391–4.

[70] Brilakis ES, Badhey N, Banerjee S. Bilateral knuckle" technique and Stingray re-entry system for retrograde chronic total occlusion intervention. J Invasive Cardiol 2011;23:E37–9.

[71] Michael TT, Papayannis AC, Banerjee S, Brilakis ES. Subintimal dissection/reentry strategies in coronary chronic total occlusion interventions. Circ Cardiovasc Interv 2012;5:729–38.

[72] Romaguera R, Waksman R. Covered stents for coronary perforations: is there enough evidence? Catheter Cardiovasc Interv 2011;78:246–53.

[73] Ben-Gal Y, Weisz G, Collins MB, et al. Dual catheter technique for the treatment of severe coronary artery perforations. Catheter Cardiovasc Interv 2010;75:708–12.

[74] Kandzari DE, Birkemeyer RPK. Papyrus covered stent: device description and early experience for the treatment of coronary artery perforations. Catheter Cardiovasc Interv 2019;94:564–8.

[75] Kufner S, Schacher N, Ferenc M, et al. Outcome after new generation single-layer polytetrafluoroethylene-covered stent implantation for the treatment of coronary artery perforation. Catheter Cardiovasc Interv 2019;93:912–20.

[76] Pershad A, Yarkoni A, Biglari D. Management of distal coronary perforations. J Invasive Cardiol 2008;20:E187–91.

[77] Brilakis ES, Karmpaliotis D, Patel V, Banerjee S. Complications of chronic total occlusion angioplasty. Interv Cardiol Clin 2012;1:373–89.

[78] Tarar MN, Christakopoulos GE, Brilakis ES. Successful management of a distal vessel perforation through a single 8-French guide catheter: combining balloon inflation for bleeding control with coil embolization. Catheter Cardiovasc Interv 2015;86:412–16.

[79] Garbo R, Oreglia JA, Gasparini GL. The Balloon-Microcatheter technique for treatment of coronary artery perforations. Catheter Cardiovasc Interv 2017;89:E75–83.

[80] Lopera JE. The amplatzer vascular plug: review of evolution and current applications. Semin Intervent Radiol 2015;32:356–69.

[81] Malik SA, Brilakis ES, Pompili V, Chatzizisis YS. Lost and found: coronary stent retrieval and review of literature. Catheter Cardiovasc Interv 2018;92:50–3.

[82] Kirtane AJ, Doshi D, Leon MB, et al. Treatment of higher-risk patients with an indication for revascularization: evolution within the field of contemporary percutaneous coronary intervention. Circulation 2016;134:422–31.

[83] Gurm HS, Mavromatis K, Bertolet B, et al. Minimizing radiographic contrast administration during coronary angiography using a novel contrast reduction system: a multicenter observational study of the DyeVert plus contrast reduction system. Catheter Cardiovasc Interv 2019;93:1228–35.

[84] Tajti P, Xenogiannis I, Hall A, et al. Use of the DyeVert System in chronic total occlusion percutaneous coronary intervention. J Invasive Cardiol 2019;31:253–9.

[85] Negi SI, Torguson R, Gai J, et al. Intracoronary brachytherapy for recurrent drug-eluting stent failure. JACC Cardiovasc Interv 2016;9:1259–65.

[86] Megaly M, Glogoza M, Xenogiannis I, et al. Outcomes of intravascular brachytherapy for recurrent drug-eluting in-stent restenosis. Catheter Cardiovasc Interv 2020.

Index

Note: Page numbers followed by "*f*" and "*t*" refer to figures and tables, respectively.

A

B

D

E

F

G

H

I

J

K

L

M

N

O

P

Printed in the United States
By Bookmasters